PIPELINE DESIGN & CONSTRUCTION: A Practical Approach

Third Edition

By

M. Mohitpour
H. Golshan
A. Murray

ASME PRESS

Library of Congress Cataloging-in-Publication Data

Mohitpour, Mo
Pipeline design & construction : a practical approach / by M. Mohitpour, H. Golshan, A. Murray. –3rd ed.
p. cm.
ISBN 0-7918-0257-4
1. Pipelines–Design and construction. I. Golshan, H. (Hossein) II. Murray, A. (Matthew Alan) III. Title

TJ930 .M57 2007

621.8'672–dc21   2007057760

TABLE OF CONTENTS

ACKNOWLEDGMENTS

There were many contributors who helped with the preparation of this book, and the authors would very much like to acknowledge and thank all those listed below. First are the sponsors and the original contributors without whom this book would not have been possible. The backbone of this book is the material initiated and prepared by the primary author (M. Mohitpour) for the course "Innovation in Pipeline Design and Construction Course" at the Faculty of Continuing Education, University of Calgary, Alberta, Canada. This course was initiated in October 1988 by NOVA Gas International Ltd., now TransCanada Corporation (due to a merger of NOVA and TransCanada in July 1998), and has been since offered annually. The funds thus generated by the author from the course were allocated to a scholarship fund established under the auspices of the American Society of Mechanical Engineers (ASME), OMAE Calgary Chapter set up at the University of Calgary, Alberta, Canada. Also authors proceeds from the sale of the 1st edition of the book was directed to ASME Pipeline Scholarship Fund that as well was set up at the University of Calgary.

PRIMARY SPONSOR (1ST EDITION)

The authors are grateful to TransCanada for sponsorship of the entire project, specifically for services received, financial sponsorship and above all permission granted for use of internally developed materials for preparation of the first edition of this book. Special thanks is due to the leadership of TransCanada (present & former), specifically Ron Turner, Ms. Shelagh Ricketts, Messrs. Ardean Braun, Andrew Jenkins, David Montemurro, Dave Cornies, at TransCanada for their continual support of the project.

PROJECT CONTRIBUTORS

Thanks are due to the authors' colleagues at TransCanada and those of the former NOVA, who in many ways contributed to the preparation and delivery of the material in this book. Acknowledgment is due to Marezban Canteenwalla, Dr. Iain Colquhoun, Dave Detchka, Bob De Wolff, John Kazakoff, Michael McManus, Cliff Mitchel, Mark Wright, Neal Russell, Tom Slimmon, Keith Coulson, Rick Spittal, William Thompson, Bill Trefanenko, Trent Van Egmond, Doug Waslen, Robert Worthingham, and Chuck Middleton. Without their initial contributions, the original lecture series at the University of Calgary, which led to the eventual preparation of this book, would not have occurred.

GRAPHICS DESIGNER AND TECHNICAL WRITER/EDITOR

A project such as this, of course, owes its completeness to the technical writer/editor and reviewers, who kept a quality check on the timeliness of content and accuracy of the information included.

Therefore, it is with deep gratitude that the authors acknowledge ASME publications, specifically Mr. Philip DiVietro, Ms. Mary Grace Stefanchik and Ms. Tara Smith for their tremendous encouragement and commitment toward completion of this work.

We are also grateful to Ms. Daphne McIntyre and Ms. Karla Ferbey for their diligence and technical know-how in preparing the text. We would like to thank Ms. Camilla Williams (Robinson) of TransCanada for her detailed review of the book's content and Joel Brimacombe (University of Saskatchewan) for creating many of the figures appearing in this book.

ERRATA

Acknowledgment is due to many colleagues and associates whom have reported typographical errors and omissions noted in the 1st edition of the book. We are grateful to Messrs E.J. Seiders, Doug James, Bill Leighty, Chia Hong Kiat, Marina Marchenkova, Keith Coulson, and Bill Tyson.

M. Mohitpour, Ph.D., P.E., F.I.Mech.Eng, Fellow.EIC, FASME
mo.mohitpour@shaw.ca
H. Golshan, Ph.D, P.Eng.
hossein_golshan@transcanada.com
A.Murray, Ph.D, P.Eng.
ma-murray@shaw.ca

PERMISSIONS ACKNOWLEDGMENT

The authors wish to thank all of the organizations that kindly granted their permission to reprint their figures and tables in this book. Details regarding these items and their formal acknowledgment appear below:

Chapters 3, 4 and 8   **The McGraw-Hill Companies**
Figures 3–3, 4–14 and 4–26: From Handbook of Natural Gas Engineering by D. Katz, et al. copyright 1990. Figure 8–14 From book "Profitable Procurement Strategies", 1998. These figure are "reproduced with the permission of The McGraw-Hill Companies".

Chapter 5   **Marcel Dekker, Inc.**
Figure 5–7: Reprinted from Chemical Engineering Fluid Mechanics, p. 239, by R. Darby, courtesy of Marcel Dekker, Inc.

Chapter 7 and 11   **CSA International**
Chapter 7 tables (7–1, 7–2, 7–3, 7–4) and Figures 7–11, 7–18 as well as Chapter 11, figures 11–6 and 11–7. With the permission of CSA International, material is reproduced from CSA Standard CAN/CSA-Z662, Oil and Gas Pipeline Systems, which is copyrighted by CSA International, 178 Rexdale Blvd., Etobicoke, Ontario, M9W 1R3.

Chapter 12   **Elsevier Science**
Figures 12–34 a and b: Reprinted from PHYSICA, Vol. 25, Michels et al, "Compressibility Isotherms of hydrogren...", T-S Diagrams, pp. 25, Copyright 1959, with permission from Elsevier Science.

The authors also gratefully acknowledge the following organizations for permission granted to reproduce various items that appear in the text:

American Gas Association—PRCI
American Petroleum Institute
CRC Press LLC
Crane Company
Daniel Industries Canada
David Brown Union Pumps (Canada) Limited Entec Inc.
Gas Processors and Suppliers Association (GPSA)

Hydraulic Institute
John M. Campbell & Co.
Institute of Gas Technology
Institute of Materials
KTAB-TV
National Fire Protection Association
Pipeline and Gas Journal, Petroleum Engineer Publishing Company
Standby Systems
Welding Technology Institute of Australia

FOREWORD

"Pipeline Design & Construction - A Practical Approach" was first published in October 2000 at the time of ASME International Pipeline Conference in Calgary, Alberta, CANADA. The second edition of this book was published in October 2003.

This publication has been a resounding success due to its practical approach in the development of pipeline systems from inception through to design, construction, operations and maintenance. The authors have substantially upgraded the book for the 3rd edition.

"Pipeline Design & Construction - A Practical Approach" evolved from training courses initiated by M. Mohitpour in 1988 in response to the pipeline community's need to educate those in the industry's hierarchy, and the training and development needs of those entering the profession. These courses have been offered worldwide since that time by the authors.

The content of the book, generated by the considerable knowledge and experience of the authors, is augmented by current industry practices, some of which has been made available by TransCanada Corporation, one of the world's largest pipeline companies. I am very pleased that this experience and knowledge continues to be available to the pipeline industry through the publication of this third edition.

Shelagh Ricketts
Vice President: Systems Design & Operations
TransCanada Corporation

September 2006, Calgary, CANADA

DEDICATION

This book is dedicated to all those pipeliners who baffle a novice on "pigging a pipeline." These pipeliners are the real contributors to our technological advancements because without them progress in the pipelining industry would have been very limited. Where there is the largest advanced network of pipelines, there is also the most progress in technological development. It is dedicated to an industry whose breadth of expertise has been a principal party to these advancements. It is dedicated to the future of the pipeline industry.

PREFACE

"Pipeline Design and Construction: A Practical Approach" is designed to assist the education and learning of those interested in designing, building and managing pipelines. The book provides a practical way to learn about the elements that make up a single-phase liquid and gas pipeline system, as well as a rational way to design, construct, commission, and assess pipelines and related facilities. It is a reference material for those involved in the industry and a tool for training new entrants or for refreshing the knowledge of professionals. Materials for compilation of the book have been gathered from the authors' collective experience totaling more than 65 years of service in the industry, covering all aspects of gas and liquid transmission; compression, pumping, protection and integrity; procurement services; construction, commissioning and operation; as well as management of pipeline projects. It also draws upon materials researched by the authors from outside sources and materials developed by the authors' employer, TransCanada.

The layout of the book generally presents, in a logical manner, the sequential steps in the design, construction and integrity maintenance of the pipeline.

Where possible, mathematical models are presented from basic principles developed by the authors or obtained from other sources. Examples and case studies are described in some detail for illustrative purposes. In some chapters, application - oriented examples, with sketches and descriptions of systems, are presented and discussed. References and bibliographical guides are presented to the reader for additional information.

In this book, a mix of imperial and metric units is utilized; however, corresponding metric conversions are provided for imperial units. The use of both systems is justified because the industry uses them interchangeably.

While every care has been exercised by the authors to contact copyright holders and obtain permissions and reference materials, avoid errors and omissions, and provide information adequately, it is not intended that specific examples, or applications be copied for turnkey use. Readers are encouraged to check formulations and other details prior to use. Notifications of corrections, omissions and attributions are welcomed by the authors.

METRIC CONVERSION OF SOME COMMON UNITS

TO CONVERT FROM CUSTOMARY UNIT	TO DEFINE UNIT 1	SYMBOL	MULTIPLY BY
barrel per hour	liters per second	L/s	0.044 163
barrel per day	cubic meters per day	m^3/d	0.158 987
MMBOD	cubic meters per day	m^3/d	$0.158\ 987 \times 10^6$
Btu/second	kilowatt	kW	1.055 056
Btu/hour	watt	W	0.293 071
Btu/lb_m	kilojoule per kilogram	kJ/kg	2.326
Btu/lb_m-°F R	kilojoule per kilogram-kelvin	kJ / (kg.K)	4.1868
Btu/lb_m-mole-°R	joule per mole-kelvin	J / (mol.K)	4.1868
Btu/°R	kilojoule per kelvin	kJ/K	1.8991
Btu/ft^2-hr.	joule per sq. meter-second	J / (m^2.s)	3.154 591
Btu/ft-hr-°F	joule per meter-second-kelvin	J / (m/s.K)	1.730 735
$\dfrac{Btu}{ft^2\text{-hr-}°F}$	joule per square meter-second kelvin	J / (m^2.s.k)	5.678 263
foot-pound force (ft. lb_f)	joule	J	1.355 818
$foot^2$	square meter	m^2	0.092 903
$foot^3$	cubic meter	m^3	0.028 316 85
$foot^3$/minute	liter per second	L/s	0.471 947
$foot^3$/hour	cubic meter per day	m^3/d	0.679 604
MMSCFD	cubic meter per second	m^3/s	0.327 774
gallon/minute (GPM)	liter per second	L/s	0.063 090
$inch^2$	square centimeter	cm^2	6.451 600
$inch^3$	cubic centimeter	cm^3	16.387 064
kilowatt-hour (kWh)	megajoule	MJ	3.6
mile per hour	kilometer per hour	km/h	1.609 344
pound	kilogram	kg	0.453 592 37
pound force/$foot^2$ (psf)	pascal	Pa	47.880 258
pound mass/$foot^3$ (lb_m/ft^3)	kilogram per cubic meter	kg/m^3	16.018 463
pound mass/gallon	kilogram per liter	kg/L	0.119 826
pound mass/hour	kilogram per hour	kg/h	0.453 592
psi	kilopascal	kPa	6.894 757

TO CONVERT FROM CUSTOMARY UNIT	TO DEFINE UNIT 1	SYMBOL	MULTIPLY BY
psi/foot	kilopascal per meter	kPa/m	22.620 59
psi/mile	pascal per meter	Pa/m	4.284 203
Watt-hour	kilojoule	kJ	3.6
yard^2	square meter	m^2	0.836 127
yard^3	cubic meter	m^3	0.764 555
acre	square meter	m^2	4,046.856
atmosphere (std)	kilopascal	kPa	101.325
barrel (42 gal)	cubic meter	m^3	0.158 987
Btu (International Table)	kilojoule	kJ	1.055 056
calorie (Thermochemical)	joule	J	4.184
degree F	degree Celsius	°C	5/9 (°F-32)
degree R	degree kelvin	K	5/9
foot	meter	m	0.3048
gallon (U.S. liquid)	liter	L	3.785 412
horsepower (U.S.)	kilowatt	kW	0.7457
inch (U.S.)	millimeter	mm	25.4
inch of mecury (60°F)	kilopascal	kPa	3.376 85
inch of water (60°F)	kilopascal	kPa	0.248 843
mil	micrometer	μm	25.4
mile (U.S. statute)	kilometer	km	1.609 344
ounce (U.S. fluid)	milliliter	mL	29.573 53
poise	pascal-second	Pa.s	0.1
stokes	square centimeter per second	cm^2/s	1
ton, long (2,240 lb$_m$)	ton	t	1.016 047
ton, short (2,000 lb$_m$)	ton	t	0.907 184 74
ton of refrigeration	kilowatt	kW	3.516 853
yard (U.S.)	meter	m	0.9144

NOTE: Multiply factors for compounds units. For example:

1. To convert lb/ft^3 to kg/m^3, multiply

$$\frac{1 \text{ lb}}{\text{ft}^3} \times \frac{0.45536 \text{ kg}}{\text{lb}} \times \frac{\text{ft}^3}{(0.3048)^3 \text{ m}^3}$$

2. To convert a viscosity at 25°C of 0.548 centistoke to viscosity in centipoise, obtain

$$\frac{0.548 \text{ centistoke}}{1} \times \frac{1 \text{ cm}^2/\text{s}}{1 \text{ centistoke}} \times \frac{\text{mm}^2/\text{s}}{100 \text{ cm}^2/\text{s}} = 0.00548 \text{ mm}^2/\text{s}$$

Now multiply by the flow density $\rho \frac{\text{kg}}{\text{mm}^3}$ to determine the viscosity in centipoise:

$$\frac{0.00548 \text{ mm}^2}{\text{s}} \times \rho \frac{\text{kg}}{\text{mm}^3} = 0.00548\rho \text{ centipoise}$$

Chapter 1

ELEMENTS OF PIPELINE DESIGN

INTRODUCTION

This chapter provides an overview of elements that systematically influence pipeline design through to construction, operation, and maintenance. Subsequent chapters provide detailed information on these topics.

Pipelines affect daily lives in most parts of the world. Modern people's lives are based on an environment in which energy plays a (predominant) role. Oil and gas are major participants in the supply of energy, and pipelines are the primary means by which they are transported. It is no coincidence that an extensive pipeline network goes hand-in-hand with a high standard of living and technological progress.

Among other uses, oil and gas are utilized to generate electrical power. Using electricity/oil and gas directly, houses are heated, meals are cooked, and a comfortable living environment is created. Petrochemical processes also use oil and gas to make useful products.

To fulfil the oil and gas demand for power generation, recovery processes, and other uses, pipelines are utilized to transport the supply from their source. These pipelines are mostly buried and operate without disturbing normal pursuits. They carry large volumes of natural gas, crude oil, and other products in continuous streams.

Construction procedures for most pipeline systems can be adapted to consider specific environmental conditions and are tailored to cause minimal impact on the environment.

Unattended pumping stations move large volumes of oil and petroleum products under high pressure. Similarly, natural gas transmission systems, supported by compressor stations, deliver large volumes of gas to various consumers.

Many factors have to be considered in the engineering and design of long-distance pipelines, including the nature and volume of fluid to be transported, the length of pipeline, the types of terrain traversed, and the environmental constraints.

To obtain optimum results for a pipeline transmission system, complex economic and engineering studies are necessary to decide on the pipeline diameter, material, compression/pumping power requirements, and location of the pipeline route.

Major factors influencing pipeline system design are:

- Fluid properties
- Design conditions
- Supply and demand magnitude/locations
- Codes and standards
- Route, topography, and access
- Environmental impact
- Economics
- Hydrological impact
- Seismic and volcanic impacts

- Material
- Construction
- Operation
- Protection
- Long-term Integrity

FLUID PROPERTIES

The properties of fluid to be transported have significant impact on pipeline system design. Fluid properties are either given for the system design or have to be determined by the design engineer. The following properties have to be calculated for gas at a specific pressure and temperature:

- Specific volumes
- Super compressibility factor
- Specific heat
- Joule-Thompson coefficient
- Isentropic temperature change exponent
- Enthalpy
- Entropy
- Viscosity

For liquid (such as oil or water):

- Viscosity
- Density
- Specific heat

ENVIRONMENT

The environment affects both below- and above-ground pipeline design. For below-ground pipelines, the following properties have to be determined during system design:

- Ground temperature
- Soil conductivity
- Soil density
- Soil specific heat
- Depth of burial

In most cases, only air temperature and air velocity have a significant impact on the design of above-ground facilities.

For both below- and above-ground pipelines, ground stability influences pipeline design/the pipeline support system. Significant variations in ground elevation particularly affect liquid pipeline design.

EFFECTS OF PRESSURE AND TEMPERATURE

General

Temperature and pressure influence all fluid properties. A temperature rise is generally beneficial in liquid pipelines as it lowers the viscosity and density, thereby lowering the pressure drop. A temperature rise lowers the transmissibility of gas pipelines due to an increase in pressure drop. This results in a net increase in compressor power requirements for a given flow rate. The value of absolute (dynamic) viscosity for gas increases with an increase in pressure and temperature. Such an increase will result in an increase in frictional loss along the length of the pipeline.

Pipeline temperature also impacts the environment. For example, heated liquid lines that are not insulated can cause crop damage in farmland during summer seasons, as shown in Figure 1-1 (Mohitpour 1991). In winter, the cold soil temperature can affect the pipe and the fluid being transported. Cooling of noninsulated liquid pipelines by frozen ground increases liquid viscosity and density, thereby requiring greater pumping power. Figure 1-2 shows the temperature pattern around a buried pipeline during the cold season.

Liquids that have a constant shear rate with respect to shear stress at any given temperature are termed Newtonian fluids (e.g., water, crude oil), and the viscosity is a function of temperature only, increasing with decreasing temperatures. Non-Newtonian, fluids such as bitumen have viscosities that are not only a function of temperature, but also of shear rate (Figure 1-3) and, in some cases, time (i.e., shrinkage) (Kung and Mohitpour 1987; Withers and Mowll 1982).

There are a number of different fluids that can exhibit Non-Newtonian behaviour. These include dilutants (e.g., starch, water, quicksand), pseudoplastic fluids (e.g., lime solution) and Bingham plastics (Lestor 1958).

For Non-Newtonian fluids, the viscosity has to be carefully considered. Since the shear rate changes with different fluid velocities, the viscosity curve of a specific fluid can be determined at a known fluid velocity along the fluid temperature profile of a pipeline.

Typical Flow Equations

The interrelationship of pressure, temperature, and other parameters for liquid and gas pipeline design can be summarized by the review of typical relevant flow equations. The following equations explain the relationships of pressure and temperature, pipe characteristics such as diameter and pipe roughness, flow rate, pipeline length and elevation profiles, and the properties of the fluid to be transported. They will be described in further detail later on in this book.

Liquid Pipelines

Equation for steady state isothermal liquid flow in a pipeline (ASCE 1975):

$$Q_1 = C_1 \times 10^6 \sqrt{1/f}\, d^{2.5} \frac{\left[P_i - P_d - C_2 G(h_i - h_d)\right]^{0.5}}{\Delta LG} \qquad (1-1)$$

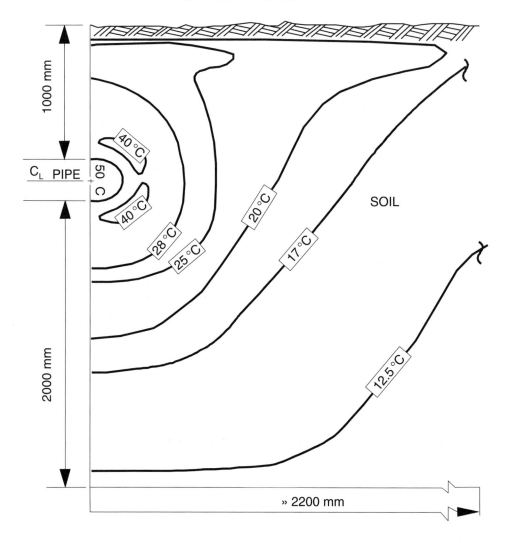

AIR TEMPERATURE = 25 °C

GROUND SURFACE » 18 °C

Figure 1-1. Temperature isotherms around a buried pipeline after 30 days (summer season)

Friction factor f for a fully turbulent flow is given by the Colebrook-White equation:

$$\sqrt{1/f} = 4 \log\left[\left(\frac{k}{3.7\mathrm{d}}\right) + \frac{1.25}{\mathrm{Re}} \times \sqrt{1/f}\right] \qquad (1-2)$$

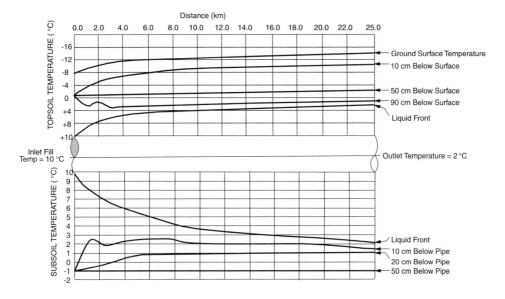

Figure 1-2. Temperature along a buried pipeline (winter season)

Gas Pipelines

The equation for steady state isothermal flow of a compressible fluid in a pipeline is written as (ASCE 1975, Katz et al., 1959):

$$Q_1 = C_3 \left(\frac{T_b}{P_b}\right) d^{2.5} \sqrt{1/f} \left[\frac{P_i^2 - e^s P_d^2}{G \, T_{fa} \, \Delta L_e \, Z_a}\right]^{0.5} \qquad (1-3)$$

For fully turbulent flow, Nikuradse rough pipe flow law for the determination of friction factor is applicable:

$$\sqrt{1/f} = 4 \log\left(\frac{3.7d}{k}\right) \qquad (1-4)$$

where
Q_1 = flow rate (turbulent region)
f = friction factor
k = pipe roughness
d = pipe internal diameter
P_d = downstream or outlet pressure
P_i = inlet or upstream pressure
G = specific gravity of fluid
h_d = downstream elevation
h_i = upstream elevation
ΔL = pipe segment length
$\Delta L_e = \Delta L(e^s - 1)/s$
Re = Reynolds Number

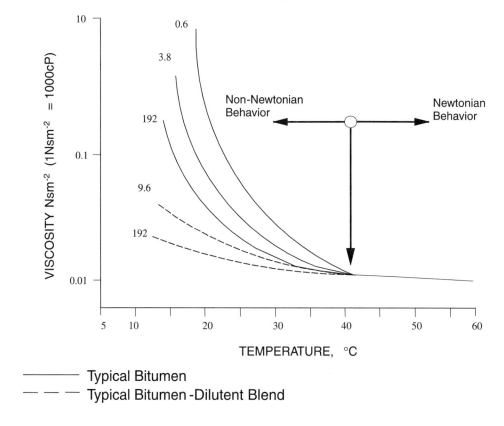

Figure 1-3. Viscosity characteristics for typical non-newtonian fluids

$e = 2.718 \ldots$, natural logarithm base

$s = $ gas density factor (to allow for elevation change) $= C_4 \left(\dfrac{h_i}{T_{fa} Z_a} \right) G$

T_b, P_b = base temperature and pressure
T_{fa} = flow temperature (average for segment)
Z_a = compressibility factor (average for segment)
C_1, C_2, C_3, C_4 = constants

The expression $\sqrt{1/f}$ is also sometimes referred to as the transmission factor, F.

From the analysis of Equation (1-1), applicable to liquid lines, it can be inferred that for constant friction and elevation, pressure loss (P_i - P_d) is directly proportional to flow. Such a pressure loss (converted to head) when plotted on a distance elevation scale will represent a straight line called the hydraulic gradient. In pipeline design, the hydraulic gradient must never cross the pipeline elevation profile, else the liquid will not be able to clear the elevation high point.

The analysis of gas pipeline Equation (1-3) indicates that for constant friction factor the pressure loss and flow do not have a linear relationship.

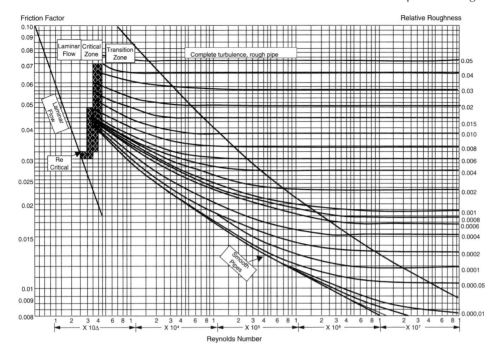

Figure 1-4. Moody diagram friction factor for flow of fluids in pipelines [Moody 1944]

Reviewing Equations 1-1 through 1-4, it is also significant to note the interrelationship between friction factor and pipe roughness, diameter, and Reynolds Number (Re), which is related to fluid Viscosity (μ), density (ρ) and flowing velocity (V). Such a relationship is well depicted by the Moody diagram (Moody 1944) shown in Figure 1-4.

Changes in elevation alter the flow velocity. Also, changes in temperature along the pipeline alter viscosity and density. The result in each case is that the friction changes. It is important not to confuse friction with pipe roughness. Roughness is the physical characteristic of a pipe and is generally constant at any given time and location, but it is friction that changes within the pipeline.

SUPPLY/DEMAND SCENARIO, ROUTE SELECTION

Supply and delivery points, as well as demand buildup, affect the overall pipeline system design. The locations of supply and delivery points determine the pipeline route and the locations of facilities and control points (e.g., river crossings, energy corridors, mountain passes, heavily populated areas). The demand buildup determines the optimum pipeline facilities size, location, and timing requirements.

Following the identification of supply and delivery points, and as a prelude to pipeline design, a preliminary route selection is undertaken. Such a preliminary route selection is generally undertaken as follows:

1. Identification of supply and delivery points (1:50000 map).
2. Identification of control points on the map.

3. Plot of shortest route considering areas of major concern (high peaks, waterlogged terrain, lakes, etc.).
4. Plot of the selected route on aerial photographs and analysis of the selected route using a stereoscope to ascertain vegetation, relative wetness, suitability of terrain, construction access, and terrain slopes, etc.
5. Refinement of the selected route to accommodate better terrain, easier crossings, etc.

This preliminary route selection is often examined by aerial reconnaissance and on-site visits to ensure the pipeline route and potential facility locations are feasible prior to detailed survey. Determining the pipeline route will influence design and construction in that it affects requirements for line size (length and diameter), as well as compressor or pumping facilities and their location. Hydraulics, operational aspects, and the requirement for special studies in areas where the pipeline traverses unstable ground in highly wet and corrosive soil are generally established at this stage.

The economically optimum sizes of facilities required for the entire range of possible flow rates are determined by the supply and demand buildup data. This data also influences the timing and location of the facilities and whether additional metering stations, compression/pumping facilities, or looping will be required.

CODES AND STANDARDS

Pipelines and related facilities expose the operators, and potentially the general public, to the inherent risk of high-pressure fluid transmission. As a result, national and international codes and standards have been developed to limit the risk to a reasonable minimum. Such standards are mere guidelines for design and construction of pipeline systems. They are not intended to be substitutes for good engineering practices for safe designs.

Major codes affecting pipeline design are listed in Table 1-1. Some Federal and other governmental authorities have the right to issue regulations defining minimum requirement for the pipeline and related facilities. These regulations are legally binding for the design, construction, and operation of pipeline system facilities, which are under the jurisdiction of the relevant authority.

There are also a number of other authorities (e.g., utility boards) who have jurisdiction over specific concerns with regard to pipeline design and construction. These authorities have the right to enforce their own regulations, setting minimum requirements for pipeline facilities within their jurisdiction.

ENVIRONMENTAL AND HYDROLOGICAL CONSIDERATIONS

Environmental

The environmental evaluation of a pipeline route is an integral component of its design and construction. It requires special planning to ensure effective and successful protection and reclamation procedures. Initially, resources in the immediate area of the pipeline route are identified and assessed to determine potential impacts. Although site-specific, resources that are usually considered in an evaluation are wildlife, fisheries, water crossings, forest

TABLE 1-1. List of major organization codes and standards affecting pipeline design, construction and operation

Acronym	Organization/topic
ACI	American Concrete Institute
AGA	American Gas Association
ANSI	American National Standard Institute
API	American Petroleum Institute
ASME	American Society of Mechanical Engineers
ASTM	American Society for Testing Materials
CSA	Canadian Standards Association
DEP	Design Engineering Practice
IEEE	Institute of Electronic and Electrical Engineers
IP	Institute of Petroleum
ISA	Instrument Society of America
ISO	International Standards Organization
MSS	Manufacturers Standardization Society
NACE	National Association of Corrosion Engineers
NAG	Normas Argentinas de Gas
NEMA	National Electrical Manufacturing Association
NFPA	National Fire Protection Association
SIS	Standards Institute of Sweden
SSPC	Steel Structures Painting Council
ANSI B16.5	Pipe Flanges and Flanged Fittings
ASTM A 350	Pipe Flanges and Flanged Fittings Material
MSS SP-25	Standard Marking System for Valves, Fittings, Flanges, and Unions
MSS SP-44	Steel Pipe Line Flanges
API 5L	API Specifications for Line Pipe
API 6D	Specifications for Pipeline Valves, End Closures, Connectors, and Swivels
API 1104 (NAG 100)	Welding of Pipeline and Related Facilities
ASTM A 333 or ASTM A 106	Materials for Surface Installations Piping
ANSI B16.9	Butt Welding Elbows/Tees
ASTM A 234 or ASTM A 420	Butt Welding Elbows/Tees Materials
ASTM A 350 or ASTM A 105	Forged Fittings<NPS 2 Material
ANSI B16.11	Forged Fittings<NPS 2
ANSI/ASME B31.4	Pipeline Transportation Systems for Liquid Hydrocarbons and other Liquids
ANSI/ASME B31.8	Gas Transmission and Distribution Piping Systems
ANSI B16.9	Factory-Made Wrought Steel Buttwelding Fittings
ANSI B16.10	Factory-Made and End-to-End Dimensions of Ferrous Valves
DEP 31.38.01.10	Piping Classes and Basis of Design
ANSI B16.34	Steel Valves, Flanges and Buttwelding End
ANSI B1.1	Unified Screwed Threads
ANSI B95.1	Terminology for Pressure Relief Devices
Division I, Rules for Construction of Pressure Vessels	ASME Boiler and Pressure Vessel Code Section VIII
Qualification Standards for Welding and Brazing Procedures, Welders, Brazers, Welding, and Brazing Operators	ASME Boiler and Pressure Vessel Code Section IX
MSS SP-53	Quality Standard for Steel Castings and Forgings for Valves, Flanges and Fittings, and other Piping Components, Magnetic Particle Examination Method
MSS SP-54	Quality Standard for Steel Castings and Forgings for Valves, Flanges and Fittings, and other Piping Components, Radiographic Examination Method
MSS SP-55	Quality Standard for Steel Castings for Valves, Flanges and Fittings, and other Piping Components (Visual Method)
MSS SP-75	Specification for High Test Wrought Buttwelding Fittings
API RF 6F	Recommended Practice for Fire Test for Valves
API 601	Dimensions for Spiral Wound Gaskets
API RP 520	Recommended Practice for the Design and Installation of Pressure Relieving Systems in Refineries
API 526	Flanged Steel Safety Relief Valves
API 527	Commercial Seat Tightness of Safety Relief Valves with Metal-to-Metal Seats
ASTM A106	Seamless Carbon Steel Pipe for High Temperature Service

TABLE 1-1. (Continued)

Acronym	Organization/topic
ASTM A234	Pipe Fittings of Wrought Carbon Steel and Alloy Steels for Moderate and Elevated Temperatures
ASTM A694	Forging, Carbon and Alloy Steel, for Pipe Flanges, Fittings and Valves, and Parts for High-Pressure Transmission
ASTM A193	Alloy Steel and Stainless Steel Bolting Materials for High-Temperature Service
ASTM A194	Carbon and Alloy Steel Nuts and Bolts for High-Pressure and High-Temperature Service
ASTM A370	Methods and Definitions for Mechanical Testing of Steel Products
ASTM E384	Test Method for Microhardness of Materials
CAN/CSA 2662	Oil and Gas Pipeline Systems
CAN/CSA Z245.20-M92	External Fusion Bond Epoxy Coated Steel Pipe
CAN/CSA Z245.21-M92 and DIN 30670/DIN 30671	External Polyethylene and Thermoplastic Coating for Line Pipes
API RP 5L2	Recommended Practice for Internal Coating of Line Pipe for Gas Transmission Service
SSPC-SP1	Surface Preparation - Solvent Cleaning
SSPC-SP10	Surface Preparation Specification No. 10, Near-White Blastcleaning
SSPC-VIS-1	Pictorial Surface Preparation for Painted Surfaces
SIS 05-5800-1967	Pictorial Surface Preparation Standard for Painting Steel Surfaces
SPSC-SP3	Power Tool Cleaning as per Steel Structure Painting Council
ASTM G8-72	Standard Test Methods for Cathodic Disbonding of Pipeline Coatings
NACE RP-01-92	Control of External Corrosion on Underground or Submerged Piping System
NACE RP-05-72	Design, Installation, Operation, and Maintenance of Impressed Current Deep Groundbed
NACE RP-01-77	Mitigation of Alternating Current and Lightning Effects on Metallic Structures and Corrosion Control Systems
NACE RP-02-86	Mitigation of Alternating Current and Lightning Effects on Metallic Structures and Corrosion Control Systems
NACE RP-02-74	High Voltage Electrical Inspection of Pipeline Coatings Prior to Installation
IP Part 1	Model Code of Safe Practice, Electrical Safety Code
IP Part 15	Area Classification Code for Petroleum Installation
DEP 33.64.10.10	Electrical Engineering Guideline
API 500C	Hazardous Area Classification
AGA 3	Measurement of Gas by Orifice Meters
AGA 8	Determination of Supercompressibility Factors for Natural Gas
AGA 9	Measurement of Gas by Turbine Meters
ANSI B40.1	Gauges and Pressure Indicating Dial Type, Elastic Element
IP Part 1	Model Code of Safe Practice, Electrical Ch. 3 Instrumentation
ISA	Standards and Practices for Instrumentation
ASTM-A36	Structural Steel Manual of Steel Construction
ASTM-A82	Specification for Cold-Drawn Steel Wire for Concrete Reinforcement
ASTM-A184	Specifications for Fabricated Deformed Steel Bar Mats for Concrete Reinforcement
ASTM-A185	Specification for Welded Steel Wire Fabric for Concrete Reinforcement
ASTM-A196	Specification for Steel Wire, Deformed, for Concrete Reinforcement
ASTM-A615	Specification for Deformed and Plain Billet-Steel Bars for Concrete Reinforcement
ASTM-C33	Specification for Concrete Aggregates
ASTM-C39	Test for Compressive Strength of Cylindrical Specimens
ASTM-C94	Specification for Ready-Mixed Concrete
ASTM-C136	Test for Sieve or Screen Analysis of Fine and Coarse Aggregates
ASTM-C150	Specification for Portland Cement
ASTM-C172	Method of Sampling Freshly Mixed Concrete
ASTM-D422	Particle Size Analysis of Soil
ASTM-D698	Moisture Density Relations of Soil and Soil Aggregate Mixtures Using 5.5 lb. Rammer
ASTM-D1557	Test for Moisture-Density Relations of Soils and Soil Aggregate Mixtures Using 10 lb. Rammer and 18 in Drop
ASTM-D2049	Standard Method of Test for Relative Density of Cohesion Soils
ASTM-D4318	Test for Liquid Limit, Plastic Limit and Plasticity Index of Soil
ISO 9001	Quality Systems for Design/Development Production, Installation, and Servicing

cover, and archaeological and palaeontological resources. A soil and vegetation evaluation is also conducted to determine soil handling and reclamation procedures.

Land use in the immediate area of the pipeline route is also identified and evaluated to ensure that conflicts do not arise with other companies or individuals. Protection procedures are based upon these resource assessments and are then integrated into the design parameters of the pipeline construction and specifications.

Timing for pipeline construction is also taken into consideration in the evaluation of resources, as seasons can effect the selection of the most appropriate mitigative procedures. Construction techniques that are effective during the summer may not be appropriate or effective in the winter.

Alternatively, dealing with site-specific resource impacts may create conflicting timing constraints with those of proposed construction schedules.

Specialized techniques to reclaim the construction right-of-way are implemented following mainline construction. Every attempt is made to match existing species for revegetation purposes to ensure successful reclamation of noncultivated lands and to return cultivated lands to their previous agricultural productivity.

During construction of the pipeline, environmental inspection is ongoing to ensure compliance with environmental design and protection procedures, and to maintain consistency with various regulatory approvals. Problems that are identified during the design and construction phase of the project are reviewed internally and may initiate environmental research, which in turn may modify future design criteria.

In selecting a pipeline right-of-way (ROW) that is economical and complies with environmental regulations, an environmental impact assessment is usually undertaken for the purpose of determining/developing environmental quality management guidelines for pipeline construction and operation. These guidelines usually include the following:

1. Compliance
   - Legislation compliance
   - Environmental guideline compliance
   - Environmental coordination, audit, and training
   - Recommendation by volcanologist/geotechnical/seismic consultants
2. Guidelines for environmental protection
3. Guidelines for soil erosion protection
   - Erosion process and types
   - Erosion risk assessment
   - Erosion protection
4. Guidelines for water quality protection
   - Baseline water quality
   - Site selection
   - Water analysis/quality index
   - Impact and mitigation measures
5. Guidelines for archeological heritage protection
   - Historical resources
   - Archeological studies
   - Regulations
6. Environmental protection resources and methods
   - ROW preparation and pipe installation
   - ROW width during construction
   - Grading
   - Rip rap

- Sand plugs
- Temporary cover
- Sediment/silt traps
- Canals/disturbances
- Volcanic areas (if applicable)
- Permanent physical erosion control method
- Soft plugs
- Revegetation/regrading
- Drainage system
- Berm/gabions
- Ditch foam plugs
- Agronomic erosion control
- Revegetation measures schedule
- Procedures and methods
- Revegetation plan

Hydrological

A pipeline may be subject to buoyant forces due to water migration and flooding or water crossings. The design must consider the potential for damage to the pipeline due to flood plains and the need for dredging of water crossings, scouring, or channel shifting in order to determine the correct pipeline design and installation technique. The design may need to consider the determination of pipe depth, buoyancy control, pipeline installation, and construction methodology in areas of hydrological concern. Usually facilities are designed such that they will not sustain any unanticipated damage from a 1:100 year flood, and will fully and continually operate under 1:50 year flood conditions. Hydrological conditions and scouring evaluations should be undertaken for major river crossings with established 1:50 years and 1:100 years flood plains.

ECONOMICS

General

The economics of transporting fluids by pipelines affect almost all design and construction parameters. For any pipeline project, the objective of economic analysis is to determine which of the alternative design and construction solutions offers the best economic advantage (Mohitpour 1977). Economic analysis is carried out to determine the optimum choice between size of pipelines (diameter, wall thickness, and material) and compression/pumping power requirements. An economic analysis is also needed for utility purposes, to define tariffs that have to be charged for the transmission of fluid to achieve a stipulated economic performance of a pipeline system investment (Mohitpour and McManus 1995).

The economic feasibility of a pipeline project is usually established before any optimization takes place. One criterion that is often used for acceptance or rejection of a project is the expected rate of return on the invested capital. Once the feasibility is proven, optimum choices between line size and pumping/compression requirements are determined. Timing considerations may be optimized and tariffs may be determined (mostly for utility companies).

Estimates have to be made for the overall investment requirement and associated operating costs of the proposed pipeline system. These are the two principal components of owning and operating a pipeline system. The relevant costs for a pipeline project have to be carefully considered, covering all phases from planning to operating and maintaining the pipeline over the course of its life. Elements that influence the cost estimate and therefore the economic analysis are outlined in the following subsections.

Direct Costs

Direct costs cover expenditure directly related to the design, construction, and operation of a pipeline system, including the following:

- Line pipe
- Compression/pumping facilities
- Meter stations
- Valve and fittings
- Protection facilities (coating plus cathodic protection)
- Scraping/cleaning facilities if any
- Pressure reduction facilities
- Power generation (if applicable)
- Construction costs
- Engineering costs
- Survey costs
- Cost of ancillary facilities
- Legal and land costs (ROW acquisition)
- Docks, wharves
- Leak detection system
- Logistic costs (those associated with material and equipment transportation)
- Operating and maintenance costs (those associated with local taxes, fuel, energy, material, and labor costs)
- Other costs (line fill, working capital required to operate the pipeline, etc.)

Indirect Costs

Indirect costs affect the financing of a pipeline project, and include costs associated with the acquisition of necessary funds to cover the purchase of materials and construction. Indirect costs also cover the interest on money borrowed to finance the pipeline project.

Economic Analysis

Oil and gas producing companies may use an economic analysis to justify a pipeline project over to alternative solutions. Oil can be transported by road or rail. Gas may be fed into an existing pipeline network owned by a utility company. In the latter case, a tariff for self-ownership is determined and compared to the tariff charged by the utility company for transportation of gas to the delivery point. The system design is influenced by the economic analysis since a number of technical solutions are investigated by varying pipeline size, pumping or compression facilities, and the timing of the facility expansion or addition.

Utility companies are interested in establishing the appropriate tariff to charge a customer for transporting the customer's fluid in the company's lines. For a selected pipeline size, given that the volume of fluid to be transported can have an upward trend (as a result of demand buildup), the tariff generally decreases as the throughput increases (Figure 1-5). If the pipeline design or operating conditions are changed, the tariff will change. As a result, it is at the design stage that a pipeline economic study is undertaken to determine the viability of the pipeline system configuration for all modes of volumetric flows.

When the economic viability of a pipeline project is established, the analysis is usually rerun using the optimal technical configuration and the most refined cost information available. The following additional cost parameters need to be considered:

- Equity investment as a portion of total investment (loan-equity ratio)
- Interest on borrowed capital
- Duration of borrowed capital (i.e., repayment schedule)
- Depreciation rate of facilities investment for book purposes
- Depreciation for tax purposes
- Escalation rate of operation and maintenance cost
- Required rate of return on investment
- Escalation rate on tariff
- Expected life of the pipeline until abandonment

After evaluation of all the above parameters, the factor by which the economic viability of a pipeline system is measured is the rate of return. The higher the rate of return, the more attractive the project is from an investment point of view. Calculation of investment rate of

TABLE 1-2. Relationship between cost parameter and cost component for cost of service

Item	Description
Gross Facilities Cost	Capital Cost + AFUDC + Working Capital
Allowance for Fund During Construction (AFUDC)	RR x (month (12) x Construction Cost
Rate of Return (RR)	(Debt%) x Debt Interest + (Equity%) x Equity Interest
Depreciation	(Original Cost - Salvage Cost)/Project Life
Net Base Rate	Gross Facilities Cost - Accumulated Depreciation
Income Tax	[Tax Rate/(1 - Tax Rate)] x Taxable Income
Taxable Income	Return + Depreciation - Interest Expenses - Capital Cost Allowance (CCA)
Interest Expenses (service cost)	(Embedded Cost of Debt) x Debt% x Rate Base
Capital Cost Allowance (CCA)	(Gross Facilities Cost - Accumulated Depreciation - AFUDC) x CCA rate
Operation and Maintenance (O&M) Cost	Administration and General + Actual Operation Costs
Annual Cost of Service	Depreciation + O&M + Taxes + Return
Accumulated Cost of Service (ACOS)	Cumulative Cost of Service/Project Life
	$\sum_{j=1}^{n} \frac{ACOS_j}{(1+i)^{j-1}}$
Present Value of Cost of Service (PVCOS) or Average Discounted Unit Cost of Service (ADUCS) based on volume	$PVCOS \left[\sum_{j=1}^{n} \frac{AVOL_j}{(1+i)^{j+1}} \right]$
	where i = Discount Rate and AVOL = Accumulated Volume n = Life of Project (years)

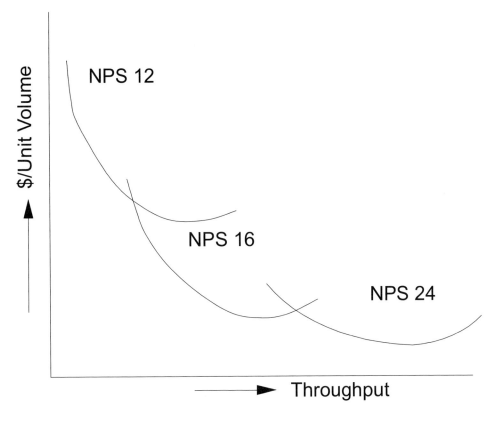

Figure 1-5. Tariff vs. throughput

return is based on computing relevant cash streams for each calendar year and discounting accordingly. This type of computation is best suited for computer applications due to the repetitive nature of the calculations.

Table 1-2 depicts some of the major cost parameters and their inter relationships with cost components (Asante et al. 1993). Typical results of such an analysis for a liquid pipeline system are shown in Figure 1-6.

MATERIALS/CONSTRUCTION

For long-distance pipeline systems, the significant cost in terms of capital investment is the cost of the pipe material and installation. Pipeline pressure, grade, installation location, and technique affect the cost and design (Mohitpour and McManus 1995).

Pipe material/grade affect the wall thickness and determine the choice of and limit on the welding/installation technique. For a given design pressure and pipe diameter, the wall thickness decreases with a higher grade material. However, higher grades of steel are usually accompanied by cost premiums and more stringent construction techniques, which translate into higher costs (Figure 1-7).

The location of the pipeline or surrounding environment determines allowable material (e.g., Category I versus Category II) and labor/equipment, including construction materials requirements.

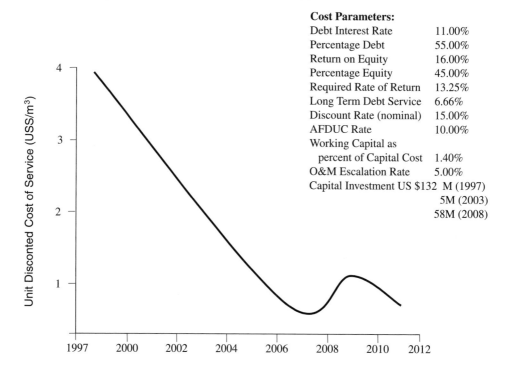

Cost Parameters:

Debt Interest Rate	11.00%
Percentage Debt	55.00%
Return on Equity	16.00%
Percentage Equity	45.00%
Required Rate of Return	13.25%
Long Term Debt Service	6.66%
Discount Rate (nominal)	15.00%
AFDUC Rate	10.00%
Working Capital as percent of Capital Cost	1.40%
O&M Escalation Rate	5.00%
Capital Investment US $132 M (1997)	
	5M (2003)
	58M (2008)

Figure 1-6. Example of cost of service curve

Depending on the application (Mohitpour 1987), the material requirements, as indicated by the notch toughness, are divided into three categories:

Category I   Requirement: No notch toughness requirement.

Typical application: Low vapor pressure fluids (water, crude oil, etc.).

Category II   Requirement: Notch toughness in the form of energy absorption and fracture appearance.

Typical application: Buried and above-ground pipelines (-5°C to -45°C). high vapor pressure (HVP).

Category III   Requirement: Proven notch toughness in the form of energy absorption only.

Typical application: High vapor pressure liquids.

OPERATION

The conditions under which a pipeline operates are set at the design stage (Stuchly 1977; Mohitpour and Kung 1986). The design stage should also determine the most stringent conditions the pipeline would operate under and provide for facilities to prevent failure, including line rupture. An example of the latter is sudden valve closure in liquid pipelines where valves are located downstream in a downhill terrain. Such an example is shown in the paper by Kung and Mohitpour (1986) entitled "Non-Newtonian Liquid Pipeline Hydraulics Design and Simulation Using Micro computers." Sudden valve closure at such a location

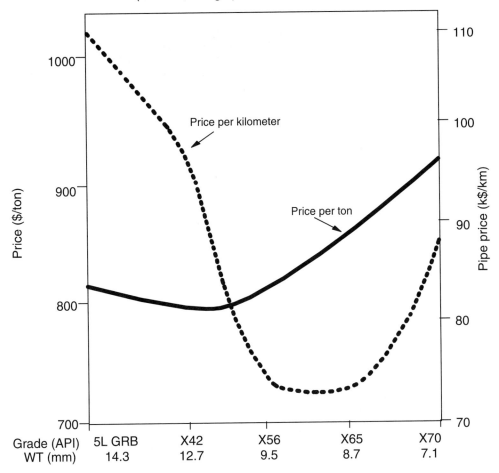

(NPS 16, design pressure 10200 kPa)

Figure 1-7. Typical pipeline grade cost

could cause pipeline pressure to exceed the maximum design pressure set by the pipe strength and wall thickness. Without surge-mitigating facilities the pipeline could rupture, resulting in environmental and maintenance implications.

Sudden valve closure in liquid pipelines (e.g., crude oil) may create a low- pressure situation that in extreme cases can lead to vapor pockets in the line. Since the collapse of vapor pockets can damage the pipeline, this condition has to be avoided.

PIPELINE PROTECTION

External Protection

Buried pipelines are subject to external corrosion caused by the action and composition of the soils surrounding them. During the design stage, the available types of external coating material and cathodic systems required to protect the pipeline from external corrosion are evaluated. The coating and cathodic protection are chosen according to economics and ability to protect the pipeline.

External coating is usually a plastic material that is wrapped or extruded onto the pipe or fusion-bonded to the surface. External coatings have to be designed to serve as a corrosion barrier and to resist damage during transportation, handling, and backfilling. Therefore, in some cases corrosion protection coatings are combined with other external coatings, such as insulation, rockshield, or concrete.

Internal Protection

Fluids containing corrosive components such as salt water, hydrogen sulphide (H_2S), or carbon dioxide/monoxide can cause internal corrosion. Many of the internal corrosion problems can be corrected in the design stage. This is done by proper design and selection of materials appropriate for the fluid to be transported. An example is the pipeline transportation of sour gas. The types of corrosion that can occur in sour gas pipelines are:

- Hydrogen-induced corrosion
- Hydrogen-induced cracking
- Sulphide stress cracking (hydrogen embrittlement)
- Pitting corrosion
- General corrosion
- Erosion corrosion

Hydrogen-induced cracking such as blistering has been observed in both low- and high-yield-strength steels under both stressed and nonstressed conditions. Hydrogen blistering and cracking results from the diffusion of atomic hydrogen, produced by the corrosive elements in a wet H2S environment, into the steel, where it is absorbed in laminations or inclusions in the pipe wall.

The atomic hydrogen changes to nondiffusible molecular hydrogen, building up high localized pressures that cause blisters or cracks in the pipe walls. The design will set the stage for the protection of the pipe against such a failure by specifying the limits on the following:

- Quantities of cerium or other rare earth metals to spheroidize manganese sulphides
- Level of sulphur content
- Level of copper content (up to 0.3%) to reduce the hydrogen absorption properties of the steel

The exact mechanism of sulphide stress cracking or hydrogen embrittlement is not clearly understood; however, it is generally agreed that it is influenced by three factors—environmental, metallurgical, and stress-related. Environmental factors include pH, H2S concentration and temperature. Metallurgical variables include strength or hardness, ductility, composition, heat treatment, and microstructure. The susceptibility of a steel to sulphide stress cracking increases with increasing hardness and stress and also with decreasing pH level of liquids. Sulphide stress cracking is evidenced as a reduction in the normal ductility and the embrittlement of steel.

A specification for the line pipe material developed at the design stage ensures that the pipe produced is suitable for the operating temperature that will be encountered and is not susceptible to hydrogen-induced corrosion. Pitting corrosion results from chemical attack at low points where fluids settle and accumulate in the piping system. Sulphate-reducing bacteria may also cause pitting corrosion.

General corrosion results from chemical attack and usually occurs on the upper half of the pipe wall adjacent to low areas where the pipe wall is alternately wet and dry. The use of corrosion inhibitors has proven to be the only effective method to mitigate internal corrosion in wet sour gas pipelines. On-stream pigging facilities are generally incorporated into the design of the system to permit the removal of liquid accumulation on a schedule. On-stream pigging also improves the distribution of the corrosion inhibitors and is a valuable aid in the mitigation of internal corrosion.

Erosion corrosion results from impingement of fluids/chlorides on the pipe surface at high flowing velocities. Piping is generally sized to limit flowing velocities below the critical velocity at which corrosion erosion will begin to occur. Critical velocity is defined as the point at which velocity is a significant factor in the removal of inhibitor films or corrosion products.

PIPELINE INTEGRITY MONITORING

No matter how well pipelines are designed and protected, once in place they are subjected to environmental abuse, external damage, coating disbondment, soil movement/ instability (Wong et al. 1988) and third-party damage. Figure 1-7 illustrates an example of a type of damage recorded on transmission lines in the United States (Crouch et al. 1994).

The goal of any pipeline integrity program is to prevent structural integrity problems from having a significant effect on public safety, the environment, or business operations

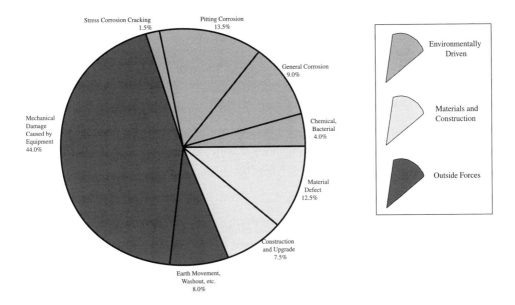

Figure 1-8. Pipeline incidents related to line pipe

by identifying and performing the most effective inspection, monitoring, and repair activities.

Integrity Assessment Methods

There are several techniques available to assess the integrity of the pipeline once it's in place. These are summarized as follows:

- Visual inspection
- Depth of cover survey
- External nondestructive testing (NDT)
  - Radiography
  - Magnetic particle testing
  - Dye penetrant inspection
  - Ultrasonic inspection
- Cathodic protection monitoring
- Coating disbondment and damage survey
- Hydrostatic testing
- Geometry in-line inspection (ILI) tools
  - Caliper pig
  - x, y, z geometry (inertial guidance) tool
- Ultrasonic in-line inspection tools
  - Conventional magnetic flux
  - High-resolution magnetic flux (3D)

Utilization of high-resolution tools facilitates an accurate prediction of external and internal corrosion areas (Grimes 1992).

Risk Assessments

Risk assessment is an integrity management tool and its purpose is to identify and quantify the risks associated with pipeline operation, such that remedial action can be performed in a timely manner. This is achieved through the ranking of potential risk to safety, environment, and operations.

Several risk assessment methods are used by the industry. The most common are failure probability methods and ranking systems. The most appropriate method depends upon several factors, including system complexity, availability of historical data, and rigor required by the analysis (Trefanenko et al. 1992).

Pipeline integrity and management decisions are made much easier by the risk assessment and prioritization process, which establishes a firm, documented basis for determining expenditures and schedules (Urednicek et al. 1991).

Pipeline Repairs

Once an integrity assessment method establishes a requirement for pipeline repair, there are several methods that are commonly used by the industry to restore pipeline integrity:

- Local coating repairs
- Coating rehabilitation
- Sleeve repair
- Cutout repairs

Use of each repair system depends on the extent of damage or corrosion problem, but repairs are carried out to restore the integrity of the pipeline to assure its intended operational capacity.

REFERENCES

Asante, B., Luk, W., and Lixing, M., 1993, "Pipeline System Optimization: A Case Study of Quinghai Gas Pipeline," *Proc. 12th International ASME OMAE Conference,* Glasgow, Scotland, Volume V, Pipeline Technology, pp. 385–393.

ASCE, 1975, Committee on Pipeline Planning, "Pipeline Design for Hydrocarbon Gases and Liquids: Report of the Task Committee on Engineering Practice in the Design of Pipelines," New York, NY.

Crouch, A. E., Bubenik, T. A., and Barnes, B., 1994, "Outlook on In-Line Inspection for Stress Corrosion Cracking," *Pipeline Pigging & Integrity Monitoring Conference,* Houston, TX.

Grimes, K., 1992, "Inspection Technologies for a Wide Range of Pipeline Defects," *Pipeline Pigging & Inspection Technology Conference,* Houston, TX.

Katz, D.L., et al. 1959, *Handbook of Gas Engineering,* McGraw Hill Book Co., New York, NY.

Kung, P., and Mohitpour M., 1987, "Non-Newtonian Liquid Pipeline Design and Simulation Using Microcomputers," *Proc. ETCE Conference,* New Orleans, LA., PD-Vol 3, pp. 73–78.

Lester, C.B., 1958, *Hydraulics for Pipelines,* Oilden Publishing Co., Houston, TX.

Mohitpour, M., 1977, "Some Technical and Economic Aspects of High Pressure, Long Distance, Large Diameter Gas Transmission Pipelines," *Proc. 16th AIRAPT Conference,* University of Colorado, Boulder, Colorado.

_____., 1987, "A Guideline Manual for Design of Hydrogen Pipeline Systems," *NOVA Internal Report,* Calgary, Alberta, Canada.

_____, 1991, "Temperature Computation in Fluid Transmission Pipelines," *Proc. ASME ETCE Conference,* Volume 34, Pipeline Engineering, pp. 78–84.

Mohitpour, M., and Kung P., 1986, "Gas Pipeline Hydraulics Design and Simulation: A Microcomputer Application," *Proc. Conference Society for Computer Simulation (SCI),* San Diego, CA.

Mohitpour, M., and McManus, M., 1995, "Pipeline System Design, Construction and Operation Rationalization," *Proc. ASME 14th OMAE Conference,* Copenhagen, Denmark, Vol. V, Pipeline Technology, pp. 459–467.

Moody, L. F., 1944, "Friction Factors for Pipe Flow," Transaction ASME, 66, p. 671.

Stuchly, J. M., 1977, Elements of Pipeline Engineering, Canuck Engineering Ltd., unpublished Training Course Material.

Trefenanko, B., Coutts, R., Ronsky, D., and McManus, M., 1992, "Risk Assessment an Integrity Management Tool," *Pipeline Risk Assessment, Rehabilitation and Repair Conference,* Houston, TX.

Urednicek M., Coote, R. I., Coutts, R., 1991, "Risk Assessment and Inspection for Structural Integrity Management," *Pipeline Pigging & Inspection Technology Conference,* Houston, TX.

Withers, V. R., and Mowll, R. T. L., 1982, "How to Predict Flow of Viscous Crude," Pipeline Industry.

Wong, F., Mohitpour, M., St. J. Price, J., Porter J., and Teskey, W. F., 1988, "Pipeline Integrity Analysis and Monitoring Systems Show Deformation Behaviour," *Proceedings of the 7th Annual ASME-OMAE Conference,* Volume V, Houston, TX, pp. 153–158.

Chapter 2

PIPELINE ROUTE SELECTION, SURVEY, AND GEOTECHNICAL GUIDELINES

INTRODUCTION

A pipeline system survey is sanctioned prior to construction to obtain a complete and comprehensive record of all physical aspects associated with the system. To design and construct a pipeline project the following survey activities are required:

- Preliminary route selection
- Engineering survey
- Legal survey
- Construction survey
- As-built survey

PRELIMINARY ROUTE SELECTION

Route selection is a process of identifying constraints, avoiding undesirable areas, and maintaining the economic feasibility of the pipeline. Having to divert the pipeline around obstacles can be very costly. For example, an NPS 42 (Nominal Pipe Size 42 inches diameter) pipeline costs approximately $1,000 per meter to build. As illustrated in Figure 2-1 all the components/constraints interact throughout the phases of route selection.

The ideal route, of course, would be a straight line from the origin to the terminal point. However, physiographic, environmental, design, and construction constraints usually alter the route. The following factors must be considered prior to selecting the optimal route for the pipeline:

- Cost efficiency
- Pipeline integrity
- Environmental impacts
- Public safety
- Land-use constraints
- Restricted proximity to existing facilities

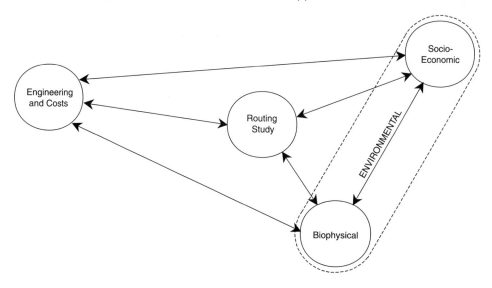

Figure 2-1. Major considerations during all phases of route selection [Passey and Wooley 1980]

The preliminary route selection involves planning the route on available maps or photo mosaics in the office. The office work is followed by a field reconnaissance to verify the acceptability of the route.

KEY FACTORS FOR ROUTE SELECTION

The initially selected routes should address the intrinsic and extrinsic constraints inherent to pipeline construction and operation by avoiding or minimizing the various geographic and regulatory restrictions. The following elements headline restrictions, construction challenges, factors that affect routes, and factors that can actually help routing.

Related to rivers, creeks, lakes, and swamps:

- Unnecessary crossings
- Braided channels
- Areas of erosion potential
- Bedrock
- Natural meander progression

Related to physiography:

- Excessively steep slopes
- Side slopes
- Rocky slopes
- Erosive soils

- Rocky soils
- Sandy soils
- Earthquake intensities/locations
- Fault locations/types/movements

Related to the environment:

- Fish spawning areas
- Endangered species habitats
- Historical and archaeological sites
- Merchantable timberlands

Other factors that affect route selection and are considered during the selection process are:

- Existing corridors
- Road and railway crossings
- Areas of population concentration
- Restricted areas such as national parks
- Native reservations
- Forest regeneration sites
- Temporary and permanent access
- Camp locations (if applicable)
- Construction schedules

Considering all of the above factors, possible routes are first selected in the office; this involves a number of steps.

Map Selection

1. Obtain the best available topographic maps or photo mosaics for the area. Often, better maps and photos enable a better route to be selected in the office. This will save field survey costs.
2. Use a scale of 1:50,000 if possible.
3. Large-scale topographic maps (1:50,000 or 1:250,000) may be obtained from the local map retail outlets (in North America and Europe).
4. Well site and pipeline plans are obtained from the appropriate Regulatory Department (e.g., in Alberta it is the Alberta Energy and Utilities Board).
5. The Department of Energy and Natural Resources, or its equivalent, can provide the locations of any licenses of occupation (LOCs) and other government interests (e.g., cutting blocks in Forest Management Areas).
6. The Land Titles office is contacted to obtain existing plans, to find out if other surveys have been performed in the area, and to get the names of other companies or people with an interest in the area.
7. Existing alignments and specific drawings are obtained for pipeline loops when trying to determine tie-in points to existing facilities.
8. General area photo-mosaics are obtained from the Government Bureau of Surveying and Mapping, or, if not available, by having an aerial survey performed.

Plotting Restrictions

Plot and highlight all of the following on the photo-mosaics or large-scale topographical maps:

- Townships, ranges, sections
- Major highways, roads
- Railways
- Rivers, canals, creeks, lakes
- Well sites, access roads
- Existing pipelines
- Utility easements
- Wildlife and environmental reserves
- Parks (recreational areas)
- Licenses of occupation
- Historical reserves
- Forest reserve sites

Establishing the Route

1. Identify areas to avoid and areas that are advantageous between the two points A and B, the terminii of the line.
2. Draw a line from point A to B, while adhering to the previously identified restrictions.
3. Draw alternate routes.
4. Draft a preliminary route sketch (examples of preliminary route sketches are given in Figures 2-2 and 2-3).

Evaluating the Preliminary Route(s)

1. Note the river or major creek crossings.
2. Check the proximity to population concentrations and restricted areas (national parks, military reserves, etc.).
3. Take initial environmental considerations into account.
4. Complete an initial terrain analysis.
5. Complete a "rule of thumb" cost estimate.
6. Establish the preliminary length of the route.

Refining the Preliminary Route(s)

1. Evaluate the various considerations/restrictions for each alternate route or option.
2. Look for opportunities to combine the best parts of various alternatives.
3. Redraw or refine the route.
4. Re-evaluate each route and choose the most optimum solution from a cost perspective.

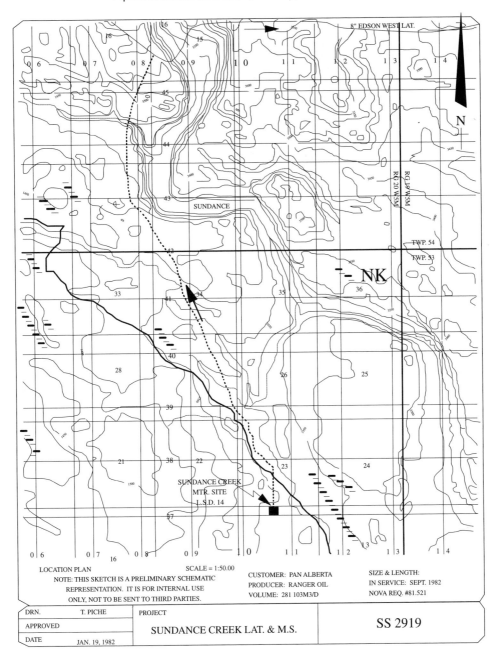

Figure 2-2. Preliminary route sketch (topographic map) [Passey and Wooley 1980]

Obtain Existing Aerial Photographs

1. Obtain suitable photographs from government photo libraries.
2. Evaluate river crossings and steep terrain.
3. When existing government photos are outdated, the proposed route may be reflown and rephotographed, if the project is large enough to warrant it.

Figure 2-3. Preliminary route sketch (photo-mosaic)

Flying the Preliminary Route

1. Confirm or reject speculations.
2. Reconsider any alternate routes.
3. Adjust the line as necessary and add any pertinent notes (i.e., abandoned farms, new well sites, etc.).

Finalizing the Office Route Selection

1. Follow the procedure used in "Refining the Preliminary Routes".

Field Reconnaissance

An aerial survey/ground inspection is made to identify problem areas along each route and to evaluate key factors affecting it. The factors to be considered during a field reconnaissance are as follows:

- General topography (land form)
- Relief conditions and drainage patterns
- Geotechnical implications (slope stability)

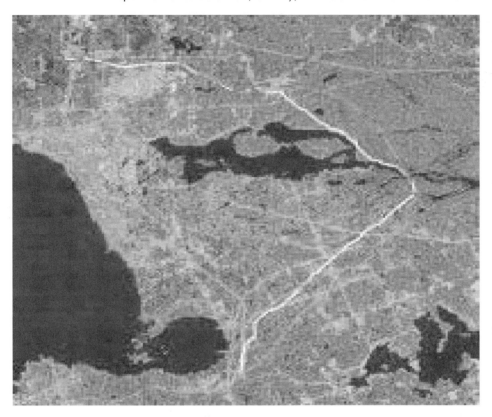

Figure 2-4. Preliminary route sketch (satellite image)

- Surface conditions (topsoil, soil type, location of bedrock)
- Water courses
- Vegetation (forests, muskeg)
- Presence and type of agriculture
- Environmentally sensitive areas
- Historical and archaeological sites
- Man-made obstacles
- Existing facilities
- Availability of existing corridors
- Access during construction
- Winter or summer construction considerations
- Crossing angles:
  - 90° at roads
  - 90° at railroads
  - 90° at rivers, creeks, and canals
  - 70° to 90° at foreign pipelines
- Proximity restrictions:
  - distance between a road boundary and a pipe to be no less than 30 meters
  - minimum distance from a road boundary before deflecting the pipeline to be no less than 30 meters
- Pipe bending requirements

After the field reconnaissance, detailed studies of particular problem areas, such as geotechnical assessments of river crossings and environmental studies, may be required. At this time initial contact with landowners can commence. The route selected may then be refined to accommodate requests from the landowners. Once all of the assessments and studies have been completed, the route can be refined in preparation for the engineering survey.

Geotechnical Guidelines Affecting Pipeline Route Selection

The following basic geotechnical guidelines should be taken into consideration during the selection of a pipeline route:

1. Wherever possible, pipelines must be located so as to avoid side or cross slopes. If a cross slope is unavoidable, the fall line of the slope must be at 90° to the centerline of the pipe. However, pipeline construction procedures require that the working side of the right-of-way must be essentially level in order to install and backfill the pipeline. In areas with excessive cross slope, considerable grading is required to construct the pipeline. The centerline of the pipeline should be parallel to the fall line of the slope in order to reduce the chance of pipe rupture due to slope movement. The alignment of the pipe should also minimize grading and slope disturbance.

2. Wherever possible, pipelines should avoid unstable slopes. This does not necessarily include old, inactive slides, as many slopes will have experienced some form of instability. It is more important to avoid slopes displaying signs of recent movement. Some of the more obvious signs of possible recent slope movement are cracks, scarps, curved trees, evidence of toe erosion, and the exit of groundwater onto the slope (in the form of a spring).

3. If there is evidence of recent instability, the nature of the mass movement must be assessed. Deep-seated movements often require relocation of the proposed pipeline route, particularly since the cost of slope stabilization can be excessive and may not be reliable. In some cases, shallow movements are not a major stabilization problem and can be accommodated in the design.

4. The most cost-effective approach must be taken. The cost of stabilization measures such as grading and drainage control (surface and subsurface) has to be compared with the cost of rerouting the pipeline to a more stable area.

5. A detailed subsurface investigation and stability analysis is required for any slope suspected of being potentially unstable along the preferred pipeline route.

6. One of the most important steps in the selection of a pipeline route is the determination of river crossing locations. River crossings can have a significant effect on both the cost of a pipeline and the total length of the line. Some of the basic rules of route selection at river crossings are:
   - It is essential to find the most suitable riverbed. Bedrock requires expensive blasting while very silty river beds can require large excavations.
   - Pipelines should cross rivers at right angles in order to minimize the width of the crossing and to avoid side slopes on the approaches.
   - Active bank erosion can lead to exposing and damaging the pipeline as well as impacting the environment.

- Fast-flowing sections of the river should be avoided as they can make construction difficult.
- The crossing should be located in a straight section of the river to minimize active bank erosion.

ENGINEERING SURVEY

Prior to detailed engineering and preparation of alignment sheets and crossing drawings, an engineering survey is necessary to acquire and document the following information:

- A chainage (a measurement along the length of the pipeline) of the proposed ditchline, taking note of all physical features, boundaries, road, railway, utility and stream crossings, and areas requiring buoyancy control
- An elevation profile for the entire route
- Detailed profiles at crossings
- Site information at all proposed facilities

The engineering survey can be conducted congruent with the legal survey. The following steps are required in an engineering survey:

1. Establishing the route
2. Running the survey line
3. Contour chainage
4. Line profile
5. Individual profiles

Establishing the Route

Preliminary Reconnaissance Route
When establishing the route and possible alternatives, it is important to consider the following: Adherence to the preliminary route previously established, as closely as possible (note all deviations); utilization of the photo-mosaics or topographic maps compiled during the preliminary route selection; and use of any related legal plans for paralleling existing right-of-ways.

Terrain Selection
Avoid the restrictions and areas of concern detailed in the preliminary route selection as often as possible. Figure 2-5 illustrates the routing alternatives required to address these concerns.

Method

Flare sightings, flagged sights, or calculated angles should be used to delineate the route.

Running the Survey Line

a). Procedure

The following procedures should be followed when performing a line survey:

1. Run the survey line to the centerline of the ditch.
2. Show all side bends as deflections (right or left) rather than internal or external angles. State the amount of deflection and direction for the chaining crew.
3. Limit all deflection angles to the maximum pipe bend, which varies with the pipe diameter.
4. Break up large deflection angles to accommodate the limits dictated by individual pipe bends.

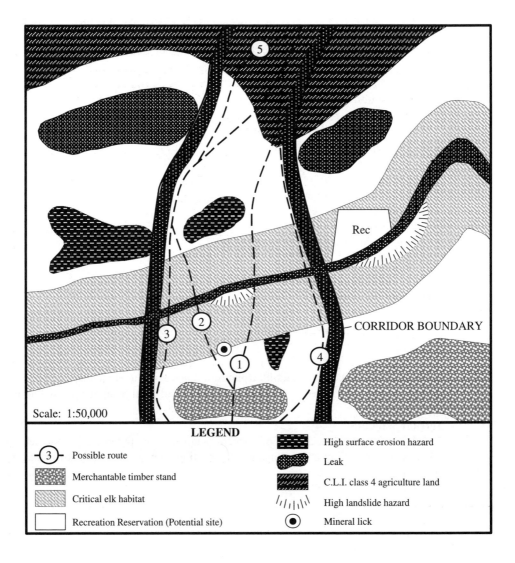

Figure 2-5. Route alternatives

Table 2-1 can be used as a guideline for the maximum pipe bends and minimum chord lengths for specific pipe diameters.

b). Crossing Angles

When practical, use perpendicular crossings at all major water courses, roads, canals, railways, highways, foreign pipelines, and other utilities.

Crossing angles should be no less than a 70° to the centerline of the watercourse, road, etc., or as governed by the authorities having jurisdiction.

Crossing angles at seismic faults should be designed and implemented to reduce stresses in the pipeline upon any fault movement.

Contour Chaining

The contour chain should be developed at no less than 100-meter intervals and the following items recorded at the appropriate chainage location:

- Topographical features
- Description of forestation and vegetation
- Centerline of roads, trails, railroads, etc.
- Cultivated and pasture land, muskeg, swamp, bogs
- Fences, power lines, cables
- Pipelines and other buried utilities
- Seismic lines
- Creeks, rivers, gullies, crests, steep slopes, toes, faults, water's edge, banks, canals
- Points of tangents (POTs) points of inflection (PIs)
- Right-angle location to all buildings within the boundary of the proposed pipeline

Ensure appropriate stationing at POTs, PIs, hills, and gullies for the line profile.

Notes should be made regarding the need for any, or all of the following:

- Swamp weight requirements
- Extra cover locations
- Heavy walled pipe locations (creeks, gullies, etc.)

TABLE 2-1. Maximum pipe bends and minimum chord lengths

Pipe Diameter	Minimum Chord Length	Maximum Pipe Bends
NPS 4	18 m	40°
NPS 6	18 m	35°
NPS 8	18 m	35°
NPS 10	18 m	30°
NPS 12	18 m	30°
NPS 16	23 m	30°
NPS 18	23 m	25°
NPS 20	23 m	25°
NPS 24	23 m	20°
NPS 30	23 m	20°
NPS 36	23 m	15°
NPS 42	23 m	12°

Note: NPS = Nominal Pipe Size; for example NPS 24 is a pipe with a diameter of 24 inches.

Line Profile

In order to develop line profiles it is essential to utilize available geodetic benchmarks (railways, highways, etc.). The profile can then be developed by:

- Obtaining grade elevations at every full station
- Obtaining grade elevations at:
  - Centerline of railroads, roads, trails
  - Hilltops, bottom of valleys, gullies
  - Water's edge, banks, pipelines, and other utilities
  - Crests and toes of steep slopes

Elevations should be taken at all benchmarks locations, left at all roads, creeks and pipeline crossings for future reference.

Individual Profiles

Individual profiles should be correlated with adjacent line profile elevations. Individual profiles are required for:

- Railroads
- Highways, secondary roads
- Pipeline crossings
- Canals, irrigation ditches
- Rivers, named creeks, sizable unnamed creeks
- Coulees, gullies
- Meter stations

Railroads

- Profile the ditchline about 40 meters beyond the railroad boundaries
- Show the crossing angle
- Show horizontal chainages and elevations of all track centerlines, railway beds, shoulders, toes of shoulders, ditches, and boundaries
- Indicate the legal description where possible
- Leave top of rails elevated

Highways, Secondary Roads

- Profile the ditchline about 50 meters beyond the boundaries
- Indicate the crossing angle
- Show elevations and horizontal chainages for all highway centerlines, shoulders, ditches, boundaries, and medians
- Classify the road type as asphalt, oiled, or gravel
- Profile the roads 500 meters along the centerline of the highway and on each side of the centerline of the proposed pipeline crossing

Foreign Pipelines

- Profile the ditchline about 40 meters beyond the right-of-way boundaries
- Indicate the crossing angle
- Expose the foreign pipeline by hand to establish an accurate cover and elevation

- Indicate the pipe size and owner
- Include the legal description

Canals, Irrigation Ditches

- Profile the ditchline about 40 meters beyond the boundaries
- Indicate the crossing angle
- Show elevations and horizontal chainages to the bottom, edge of the water, banks, and toes of banks
- Show the direction of water flow

Rivers and Creeks

- Establish a benchmark near the edge of the water
- Indicate all banks, toes, side slopes, cross slopes, changes in slope, drainage patterns, breaks, and springs
- Indicate the edge of the water, high water marks, beaches, and bottom elevations (as many as possible)
- Define the main channel
- Tie-in all iron pins
- Employ horizontal chainages for accuracy
- Profile rivers and creeks approximately 300 meters upstream and downstream from the crossing

Various methods of obtaining river bottom elevations can be employed, including:

1. Echo soundings with a fathometer
2. Soundings with a wire and weight
3. Rod readings (see Figure 2-6)

1. Echo Soundings

- Using a fathometer, triangulate positions of fixation
- Establish the water-level elevation
- Read bottom elevations off the fathometer graph

2. Rod Readings

- Use direct measurement in shallow rivers
- Bore holes through any ice cover

The horizontal distance on soundings can be controlled by the following methods:

- Triangulation
- Direct chainage
- Stadia

Figure 2-6. Taking a rod reading at a creek crossing

LEGAL SURVEY

The land to be purchased or leased by the pipeline company must be surveyed. Survey plans must be prepared for filing at a land titles office and to form part of the purchase/lease agreement. Permanent survey monuments are required at existing property boundaries, points of intersection (changes in direction), and at any intermediate points necessary to clearly define a boundary. The work can legitimately be completed only by a qualified legal surveyor.

CONSTRUCTION/AS-BUILT SURVEY

Prior to construction, the permanent right-of-way and temporary lands available during construction must be clearly marked. Underground utilities and other structures requiring protection from equipment must be identified.

After clearing and grading the right-of-way, the ditchline is staked. Each stake will be marked with the chainage at regular intervals (50 meters or less is common), consistent with those on the alignment sheets. The stakes should be offset from the work's side boundary. The stakes must be maintained throughout the construction period and utilized for reporting progress and directing construction activities. In addition, items such as pipe wall thicknesses/grade changes, buoyancy control measures, and extra depth requirements must be clearly staked. In general, the work is a transfer of information from the construction drawings to the construction right-of-way. Additional information, such as cut-and-fill requirements, is marked on the stakes.

As-building is the process of transferring the actual "as-constructed" information on to drawings for future reference. The objective is to provide an accurate record of the completed work. The process involves surveying and documenting the entire system, including, but not limited to, the following:

- Changes in pipe specifications
- Actual pipe profiles at crossings
- The location, spacing and number of buoyancy control devices
- Each mainline weld. The pipe joint and heat numbers are recorded at and for each weld. Documentation of weld locations enable future reconnaissance of individual welds or pipe joints, such as during in line inspections. Weld numbers are also recorded on each radiograph.

In general, all details and dimensions given on the construction drawings are either confirmed or revised to form a permanent record of the facility as it was constructed.

The following is a step-by-step outline of the construction/as-built survey procedure:

1. Lay out the boundaries
2. Mark the centerline
3. Consolidate the information

Lay Out the Boundaries

Staking the right-of-way boundaries involves delineating and flagging the boundaries and locating foreign lines as follows:

(a) Boundaries

- Stake all cultivated areas every 100 meters (or closer if necessary)
- Flag forested areas every 50 to 100 meters
- Mark deflections
- Flag all restricted right-of-ways

(b) Foreign Line Crossings

- Locate and flag all foreign pipeline crossings
- Flag foreign lines across the full extent of the right-of-way
- Ensure that crossings are highlighted by keeping in close contact with the appropriate inspector
- Locate all buried cables (telephone, electrical, etc.)

Survey Posts (Iron Pins) and Markers

Locate and mark all iron pins (these are usually countersunk in cultivated areas) in a manner so as to avoid damaging the pin.

Field Changes

Note all discrepancies between the engineering survey and the construction survey (e.g., new pipeline crossings, new access roads, new building sites, etc.) and notify the chief inspector and the project engineer.

Grading Stakes

Stake the cut and fill requirements for grading operations at valve sites and meter stations.

Ditchline Stakes

- Stake the centerline ahead of the ditching operation.
- Utilize legal survey pins.
- Mark deflections on the stake.
- Divide large deflection angles into smaller increments to accommodate maximum pipe bends.
- Locate cut and fill stakes at all side valve locations to ensure a horizontal alignment at the tie-in points.
- Ensure cut and fill stakes are provided where the line enters and exits a facility (e.g., meter stations, block valve sites, etc.) to guarantee proper alignment of the pipe.

Centerline Chainages

The centerline chainage is completed in two stages: (1) the topographic/construction layout; and (2) the as-built chainages. The topographic/construction layout procedure is required to establish the location of the line as engineered. The centerline chainage is considered as-built when the pipe is in the ground and all the measurements pertaining to the pipe are correlated to the centerline chainage and recorded. The completion of each stage results in an as-built linear record of all details directly related to the pipeline and to the features encountered along the pipeline route.

Topographic Chainage

- Use contour plus stationing in 100-meter intervals, and indicate all full even stations. This chainage is known as the "contour station" chainage.
- Chain in all topographic features, such as:

  - Right-angle ties and deflections
  - Highways, secondary roads, access roads, trails, road allowances, cutlines
  - Canals, water courses, drainage ditches
  - Rivers, creeks, streams (banks and water's edge)
  - Vegetation (bush, cultivation, and pasture)
  - Muskeg, sloughs
  - Fences, property lines, municipal leases, cutlines
  - Railroads
  - All foreign pipelines
  - All utility lines (e.g., electrical, telephone, seismic lines)

Construction Layout

The construction layout involves staking all items pertaining to the pipe, referenced on the working side. Details for the construction layout are found on alignment sheets, detailed drawings, and legal plans.

- Line pipe

  - Stake out where the wall thickness changes
  - Reference the start and end stations
  - Stake the start and end of all sags and side bends

- Location of heavy wall pipe

  - Stake the locations of the heavy wall pipe, which often include:

    - Test sections
    - Road crossings (including proposed roads or highways)
    - River and creek crossings
    - Air relief valves
    - Railroad crossings
    - Valve locations

  - Stake the location of transition pieces

- Extra depth locations

  - Stake the start and end points and the minimum cover required at the centerline for the following anomalies:

    - Rivers, creeks, streams
    - Trails, cutlines, seismic lines
    - Drainage ditches, water courses
    - Foreign pipelines
    - Valve and meter station sites
    - Landowner's requests (e.g., at proposed drains or deep ploughing)

- River and swamp weights

  - Stake out the start and end stations
  - Information on the lath (survey stake) must contain the number of weights, center-to-center distance between weights, and the type of weights

  - Test leads

    - Stake the location and type, whether the leads are above or below ground, and the number of wires to be installed

  - Casing

    - Casing is usually used at railroad and highway crossings

- Stake the start and end points and the depth

- Stubs, sales taps, farm taps

  - Stake the location, size, and direction
  - Rockshield/padding may have to be staked

As-Built Chainage

- Chain each and every item noted above that can be accurately surveyed prior to being installed in the line
- In addition to the items noted above, accurately chain in the following:

  - A record of the manufacturer and differences in the longitudinal seam, such as the spiral or long seam
  - All changes in the pipe heat numbers; a joint-to-joint chainage may be required
  - All welds; incorporate the X-ray tag numbers

Figure 2-7 illustrates an example of as-built chainage notes.

1. Pipe Chainages

- The procedure for chaining specific sections of pipe is typically as follows:

  - Utilize benchmarks established during the location survey
  - Chain several lengths of pipe after the welding but ahead of any wrap and placement into the ditch
  - Chain along the top dead center of the pipe
  - Set Station $0 + 000$ for each section at the first weld past the start of the section; end the stations at the loose end
  - Back-chain the first joint of each section from Station $0 - 000$
  - Reference Station $0 + 000$ to the contour station chainage (see earlier section) once the pipe has been welded into place
  - Measure all lengths of pipe (pups); the lengths are used during consolidation of data and for pipe reconciliation
  - Take elevations from the top of the pipe and the adjacent grade along the entire system

- Record the following:

  - Changes in wall thickness and seam type
  - Heavy wall sections
  - Transition pieces
  - Drag sections
  - Sags, overbends, side bends
  - All x-ray weld numbers

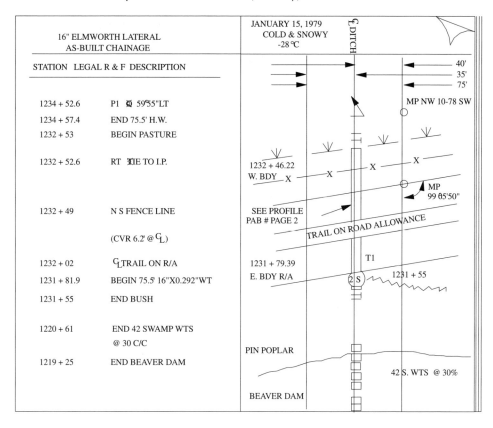

Figure 2-7. As-built chainage notes

- Manufacturer's pipe number
- Contractor's joint number

2. Consolidation

- Consolidate the pipe chainages into contour station chainages
- If discrepancies occur, adjust the pipe chainage to correspond with the contour station chainage at a tie-in point. Discrepancies should not exceed one meter
- Keep field notes in separate pipe data book
- Cross-reference the field notes to the appropriate profile notes and drawings

Individual Crossing Profiles

- As-built crossing profiles are required at all crossings
- Cross-reference individual crossing profiles with the mainline profile
- Verify or revise all information given on the alignment sheets and design drawings

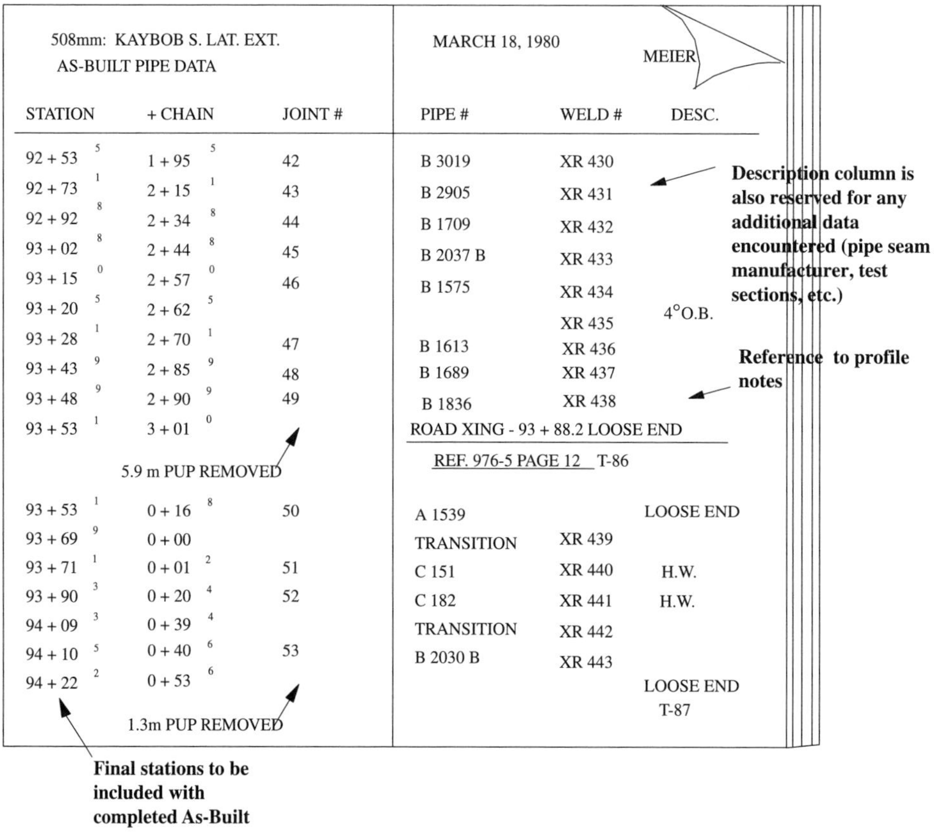

508mm: KAYBOB S. LAT. EXT. AS-BUILT PIPE DATA			MARCH 18, 1980		MEIER	
STATION	+ CHAIN	JOINT #	PIPE #	WELD #	DESC.	
92 + 53 ^{5}	1 + 95 ^{5}	42	B 3019	XR 430		
92 + 73 ^{1}	2 + 15 ^{1}	43	B 2905	XR 431		Description column is also reserved for any additional data encountered (pipe seam manufacturer, test sections, etc.)
92 + 92 ^{8}	2 + 34 ^{8}	44	B 1709	XR 432		
93 + 02 ^{8}	2 + 44 ^{8}	45	B 2037 B	XR 433		
93 + 15 ^{0}	2 + 57 ^{0}	46	B 1575	XR 434		
93 + 20 ^{5}	2 + 62 ^{5}			XR 435	4°O.B.	
93 + 28 ^{1}	2 + 70 ^{1}	47	B 1613	XR 436		Reference to profile notes
93 + 43 ^{9}	2 + 85 ^{9}	48	B 1689	XR 437		
93 + 48 ^{9}	2 + 90 ^{9}	49	B 1836	XR 438		
93 + 53 ^{1}	3 + 01 ^{0}		ROAD XING - 93 + 88.2 LOOSE END			
	5.9 m PUP REMOVED		REF. 976-5 PAGE 12 T-86			
93 + 53 ^{1}	0 + 16 ^{8}	50	A 1539		LOOSE END	
93 + 69 ^{9}	0 + 00		TRANSITION	XR 439		
93 + 71 ^{1}	0 + 01 ^{2}	51	C 151	XR 440	H.W.	
93 + 90 ^{3}	0 + 20 ^{4}	52	C 182	XR 441	H.W.	
94 + 09 ^{3}	0 + 39 ^{4}		TRANSITION	XR 442		
94 + 10 ^{5}	0 + 40 ^{6}	53	B 2030 B	XR 443		
94 + 22 ^{2}	0 + 53 ^{6}				LOOSE END T-87	
	1.3m PUP REMOVED					

Final stations to be
included with
completed As-Built

Figure 2-8. Individual crossing profile notes

Consolidation of Information
Pipe Tally Sheet

- Compile tally sheets from field as-built notes
- Include all valve assemblies, note wall thicknesses, casings, transition pieces, etc.
- Check tally sheet pipe totals against alignment sheet totals

Drawing Update

- Mark up the construction alignment sheets from the as-built chainage and as-built profile notes
- Mark up individual crossing drawings and permit maps
- Revise profile sheets and change to as-built status
- Check the alignment sheet material list, pipe totals, and ROW details
- Update all existing facility drawings with any new facilities

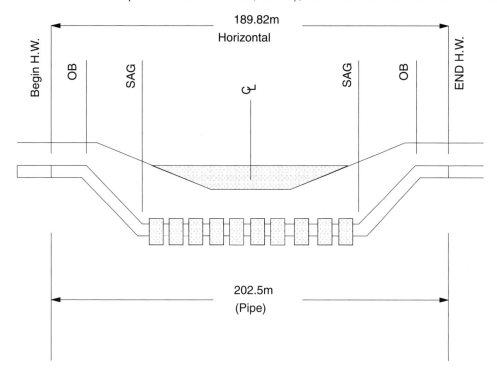

Figure 2-9. Profile of a river or creek crossing

- All as-built drawings must be checked by a senior survey technician
- Ensure that all drawings are referenced with the respective profile book and page numbers

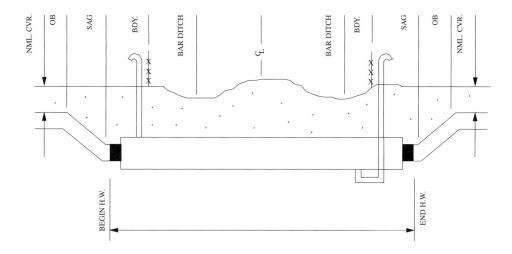

Figure 2-10. Profile of a cased crossing

NAME: 16" ELMWORTH LATERAL A.F.E. No 8170	PIPE RECONCILIATION AS-BUILT & PIPE PURCHASED			DATE: MARCH 6, 1980 SHEET 1 OF 2	
AS-BUILT STATION TO STATION	LINE PIPE 16" X .219" wt	LINE PIPE	HEAVY WALL 16" X .292" wt	CASING PIPE 20" X .375" wt	PIPE TRANS PIECE
0 + 00 TO 0 + 86	Side Valve Assembly		64.2	not field in	
0 + 86 TO 0 + 90			(approx.)		4'
0 + 90 TO 49 + 20	4829.9				
49 + 20 TO 49 + 24					4'
49 + 24 TO 50 + 03			79.2		
50 + 03 TO 50 + 07					4'
50 + 07 TO 69 + 16	1908.77				
69 + 16 TO 69 + 20					4'
69 + 20 TO 71 + 56			236.23		
TO 114	4327.4				4'
704 + 50 TO 759 + 56					
759 + 56 TO 759 + 60					
759 + 60 TO 953 + 11	19351.5				
953 + 11 TO 953 + 15					4'
953 + 15 TO 956 + 73			357.7		
956 + 73 TO 956 + 77					4'
956 + 77 TO 1016 + 07	5330.3				
1016 + 07 TO 1010 + 11					4'
	96524.04		2391.56		96.0

NOTE: 119.3' HW + 2-4' TRANS TO BE INSTALLED AT LATER DATE AT APPROX. STA. 465 + 70 - 466 + 90. IT WILL REPLACE LINE PIPE.

Figure 2-11.   Pipe tally sheet

GEOTECHNICAL DESIGN

Slope Stability

As mentioned earlier, during the pipeline route selection process it is important to avoid unstable or potentially unstable slopes wherever possible (as shown in Figure 2-12). The distribution and characteristics of naturally-occurring landslides can be many and varied, and can depend on local soil and groundwater conditions and landforms. During the design phase of the project, very close field inspection is required to detect signs of slope movement that could threaten pipeline integrity. Slopes that show no signs of recent instability may become potentially unstable as a result of pipeline construction.

The worst type of slope failure affecting a buried pipeline is a deep-seated failure where the failure surface passes well beneath the pipe (see Figure 2-12). When choosing a pipeline route it is vital to avoid areas where deep-seated slope instability is encountered. Pipelines, however, can be designed and constructed to traverse potentially unstable slopes, including old inactive landslides, without initiating renewed movement.

Slope stabilization procedures commonly used to improve the stability of a pipeline right-of-way include careful selection of spoil disposal sites and slope grading.

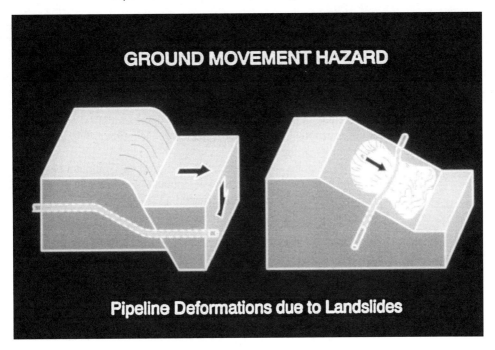

Figure 2-12. Ground movement hazard

Certain pipeline construction practices can contribute to right-of-way instability, such as disposal of spoil on right-of-way slopes as well as the improper restoration of the original contours on steep slopes.

There is considerable evidence that many right-of-way slope failures are the result of improper placement of excess spoil generated by grading. Excess soil graded from the top of a slope is sometimes pushed downhill and placed on the flatter areas of the slope. These flatter areas, in many instances, form the tops of slump blocks of old landslides. The additional weight of the graded material can cause renewed movement along pre-established failure planes and possibly result in a pipeline failure due to stress on the pipe.

It is important to select suitable disposal areas for spoil material to ensure it will not adversely affect the stability of the slope. The most cost-effective alternative is usually to locate spoil areas at the base of the slope since the material can simply be pushed down the slope. However, where space is unavailable, it is sometimes necessary to haul all spoil up the slope for disposal within the right-of-way well at the back of the crest of the hill.

A common pipeline practice in the past was to attempt to restore slopes to their original contours after a pipeline had been installed, thus minimizing the visual environmental impact of construction. This procedure, in fact, caused a great many slope failures and associated environmental disturbances.

Pipeline contractors tend to grade slopes flatter during winter construction to make access, pipe bending, and ditching easier. The frozen soils are then replaced on the slope after pipe installation. Upon thawing, the loose, usually saturated, soil mass overlying the pipe is highly susceptible to failure. In many cases, the natural subsurface groundwater exit on the slope is blocked by the replaced soil mass, further contributing to instability.

Since most pipeline specifications require that the pipeline be buried in undisturbed soils, failure of the overlying loose fill does not necessarily involve the pipe. However, from the surface, it is very difficult to determine whether the pipe is involved in the instability. As a result, it is very often assumed that the pipe may be involved in the slide, and remedial measures may be initiated.

Slope grading associated with pipeline construction is undertaken for a variety of reasons. These include improving access along the pipeline right-of-way where slopes are too steep for construction equipment and vehicles, reducing the bend angles required, and improving the stability of the slope.

In the majority of cases, spoil generated by this grading should not be replaced on the slope, particularly during winter construction. Suitable spoil areas that do not adversely affect the stability of the slope should be selected.

Drainage and Erosion Control

The control of surface and subsurface drainage within a pipeline right-of-way is an important aspect of pipeline design. Severe right-of-way erosion, pipeline exposure, and slope instability can, in many cases, be avoided with the incorporation of suitable drainage and erosion-control measures. Diversion berms, gabions, ditch plugs/subdrains are generally installed for this purpose.

Diversion Berms

The use of diversion berms to control surface water within a pipeline right-of-way has been standard practice in the pipeline industry for many years. Diversion berms consist of shallow earth-filled dikes, which are placed at intervals on a slope to collect and direct surface flow off the right-of-way and away from the pipeline.

Some typical problems associated with diversion berms being improperly designed or constructed are as follows:

- A berm of insufficient height will permit surface flow to breach and flow over the ditch.
- In cold climates, during winter construction, organics and snow in the berm fill can result in settlement upon thawing. Maximum settlement usually occurs at the pipe ditch, thereby collecting and redirecting surface flow down the ditchline.
- Berms constructed with an excessive downhill gradient can result in erosion of the uphill side of the berm.

Some general construction guidelines that should help improve the long-term performance of diversion berms are:

- Berms should be constructed with mineral soils, with every effort made to minimize organics and snow being mixed in.
- The top width and height of berms should be approximately 0.75 m for summer construction and 1 meter for winter construction to compensate for settlement during thawing.
- Nominal compaction of berm material in lifts will increase its resistance to erosion.

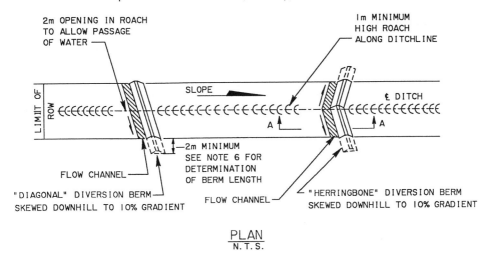

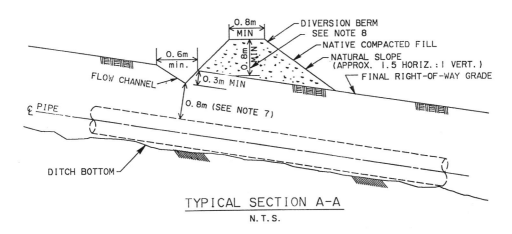

Figure 2-13. Typical plan and profile of a diversion berm

- The downslope gradient of the berm should be approximately 5% to limit erosion from surface runoffs.
- The berms should extend across the full width of the right-of-way to prevent the flow of water back onto the right-of-way.
- To a large extent, the spacing of the berms depends on the local topography and drainage. Berm spacing should be reduced as the slope increases. For instance, a slope in excess of 30% may require 10 to 20 meter spacing, while a 15% slope may require only a 60 meter spacing (see Table 2-2).

Diversion berms may be constructed in a herringbone or diagonal pattern (see Figures 2-14 and 2-15). Diagonal diversion berms are used in situations where the existing topography and slope drainage suggest a preferred direction of runoff. Berms constructed in a herringbone pattern are used in situations where there is no preferred direction of runoff, or where the berms are located across a slope with side cuts on both ends of the right-of-way. This pattern is preferred in these cases since it diverts the water away from the ditch and does not concentrate on just one side of the right-of-way.

TABLE 2-2. Typical diversion berm spacing

Slope	Soil Erosion Potential		
	High (Fine sands & silts)	Moderate (Clays & coarse sands)	Low (Gravel & exposed bedrock)
Gentle (<5%)	45 m	50 m	Not necessary
Moderate (5–10%)	30 m	45 m	50 m
Steep (>10%)	305/% grade = m	305×1.5/% grade = m	305×2/% grade = m

Gabions

Areas subjected to severe erosion by stream flow or concentrated surface drainage may require a more robust type of erosion protection. One method often used consists of gabion baskets (fabricated from wire mesh), which are filled with stone and placed along the uphill side of a diversion berm where concentrated surface flows would severely erode the berm.

Gabions are also used along banks of streams and rivers to prevent toe erosion. Generally, gabions are used to adequately protect the bank or diversion berm when available stones are too small for use as riprap. A gravel blanket or filter cloth is typically laid under the gabions to create a flat surface for tying the baskets together and as a filter medium to prevent loss of underlying fines.

Figure 2-14. Herringbone berm pattern

Figure 2-15. Diagonal berm pattern

Ditch Plugs

When designing a pipeline in sloping terrain, one must also recognize the potential for surface and subsurface seepage to collect and flow within the loose pipe backfill. If this seepage is not controlled, it can lead to backfill erosion and pipe exposure. The installation of ditch plugs or impervious seepage barriers at intervals along the pipe ditch will effectively block subsurface seepage within the pipe backfill and force it to the surface, where it can be diverted off the right-of-way by a diversion berm.

It has been a common practice in the industry to use sack breakers to control seepage within the pipe ditch. These are of sand-filled jute sacks keyed into the pipe ditch walls and hand-placed in the form of a pyramid.

There are a number of problems associated with this type of plug, including:

- Filling the sacks with soil and placing them in the ditch by hand is labor-intensive, time-consuming, and costly.
- To ensure that the sack breaker effectively impedes the flow of water down the ditchline, it is essential that the sacks be carefully keyed into the ditch walls and ditch bottom, with the bags placed to interlock with one another. This is necessary to ensure that there are no short-circuit seepage paths through the sack breaker. In the past, observations made in the field suggest that close inspection is required to ensure proper sack breaker installation.
- During winter construction, significant problems have been experienced with excavating an adequate key into the frozen ditch walls.

Figure 2-16. Gabion baskets installed on the Gas Andes pipeline right-of-way

In recent years, this type of design has become rare because of its relative ineffectiveness. An improved design for a ditch plug has been developed to replace the conventional sack breaker. The design substantially reduces the amount of hand labor required and does not require a high degree of inspection to ensure its effectiveness.

This design utilizes a dry mixture of bentonite clay with fine gravel or concrete sand. The mixture is placed to form a plug within the pipe ditch. When the bentonite comes in contact with water, it will become saturated and swell to form an impervious barrier. The abundance of bentonite clays in Alberta (Canada) make this a

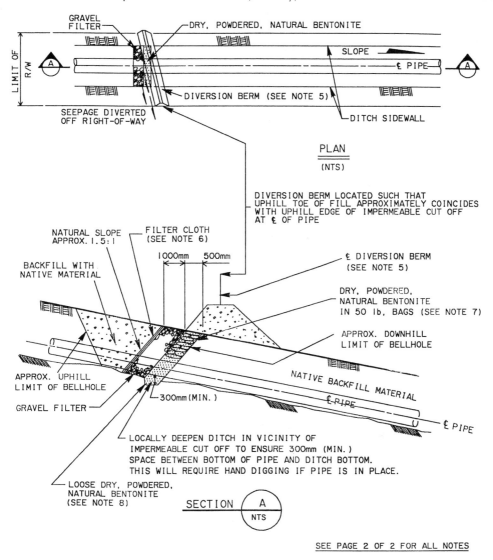

Figure 2-17. Typical bentonite ditch plug

particularly effective design material for this region. In many cases, it is cheaper to eliminate the sand and gravel and use only pure bentonite, thereby eliminating the need for mixing equipment and making the installation easier.

The following is a construction sequence illustrating the methods employed to install a bentonite ditch plug:

- To deepen the ditch beneath the pipe in the vicinity of where the bentonite plug is to be placed, the area is dug by hand.
- The sides of the ditch wall are marked at one meter intervals, parallel to the back slope, in order to establish material quantities.

- A filter cloth is placed in the open hole. The cloth is tacked or weighted from the surface in order to ensure that there is sufficient enclosure of the granular zone by the cloth.
- A 0.5 – 1 m layer of gravel is placed on top of the filter cloth in order to contain the bentonite, as dry bentonite has a tendency to flow. This gravel is also placed upstream of the bentonite to permit the controlled exit of seepage to the surface where it can be (directed) off the right-of-way.
- Pure bentonite is used as a plug.
- The end of the filter cloth is pulled back on top of the gravel and bagged bentonite is then poured up to the level of gravel.
- The preceding three steps are repeated until the level of gravel is up to within 150 mm of the surface.
- The gravel is then totally enclosed by the filter cloth to prevent infiltration of fines into the free-draining gravel.
- A 150 mm layer of gravel is then placed on top of the filter cloth and, finally, the remaining open portion of the bell hole is backfilled with on-site material.

Ditch plugs constructed in this manner are not required to be keyed into the ditch walls and ditch bottom. The swelling of the mixture of bentonite and sand, or pure bentonite, ensures a good contact with the ditch walls and ditch bottom and results in a complete impervious barrier.

The following points should be made about ditch plugs:

- First, wherever a ditch plug is proposed, a diversion berm is recommended immediately downslope to direct seepage that has been forced to the surface off the right-of-way.
- Athough most ditch plug locations should be shown on the drawings, particularly at major slopes, changes can be expected in the field. The best time to finalize ditch plug locations is immediately after ditch excavation, when springs, weak soils conditions, etc., are most obvious.
- If a ditch plug is required after the ditch has been backfilled, it is much more difficult and costly to install; the ditch width will typically be much wider and will require more materials to construct.

Subdrains

There are instances where it is necessary to lower groundwater levels within a pipeline right-of-way slope to improve stability/control surface erosion. One method that has proven effective in controlling shallow groundwater flows within the pipeline right-of-way is the installation of subdrains.

The subdrain consists of a perforated, galvanized, corrugated metal pipe placed in a trench excavated across the right-of-way and backfilled with clean granular fill. The upper portion of the subdrain trench is backfilled with local fine-grained soils to inhibit the infiltration of surface water.

The subdrain is located across the pipeline right-of-way, skewed downhill so that it has a downslope gradient of approximately 5% exiting off the right-of-way. At the point where it crosses the pipeline ditch, a ditch plug is installed directly downslope to prevent water from being directed down the pipe ditch.

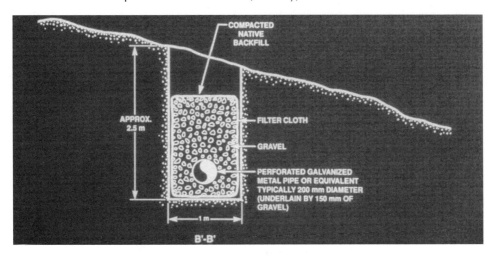

Figure 2-18. Section through a subdrain

The following illustrate the typical procedure involved in constructing a subdrain:

- An open ditch is excavated on a downhill gradient of approximately 5% across the right-of-way. Local deepening of the ditch beneath the pipe is done by hand. In some cases, it is advantageous to step the side walls of the ditch in order to keep the filter cloth from sliding down the sides.
- Filter cloth for the subdrains is laid in the open subdrain ditch and approximately 150 mm of clean gravel is placed at the ditch bottom.
- The perforated pipe, generally 219.1 mm OD, is then placed in the ditch, with the perforations on the underside at the four and eight o'clock positions.
- The remainder of the trench is then backfilled with gravel to within 0.5 m of grade.
- The filter cloth is then overlapped to totally enclose the gravel.
- The backfilling is then completed by placing local fine-grained fill, nominally compacted, to original grade. The end of the drain pipe that protrudes above ground is cut off just above grade and capped.

The depth of the other end of the subdrain may vary depending on field conditions; however, it is typically 3 m deep, with the ditch being one meter wide.

Final subdrain locations should be selected once the pipe ditch is excavated and significant seepage zones identified. They can also be installed within areas of shallow instability on an existing pipeline right-of-way without major disruption of pipeline service.

Construction

In areas such as major river crossings and steep slopes that have particular geotechnical concerns, detailed field inspection during construction is required in order to:

- Ensure sufficient grading on steep slopes for improved slope stability,

- Determine the required locations of ditch plugs, subdrains, and diversion berms after the ditch has been opened and conditions evaluated,
- Assess the need for erosion protection features such as riprap and gabions,
- Guarantee that all of the above measures are properly undertaken or installed.

During construction, one must also ensure that all cut and fill operations are performed to the recommended design. Side slopes and cut slopes should consist of undisturbed soils and not loose fill, especially during winter construction, to prevent downslope movement upon thawing.

Operation and Maintenance

Periodical surveillance of a pipeline right-of-way to identify potential geotechnical problem areas can play an important role in the continued safe operation and maintenance of a pipeline. These inspections are especially important after spring breakup or after a period of heavy rainfall, in order to identify areas of slope instability or erosion. Any problem area that has developed must be investigated and immediate remedial action should be taken at sites judged to be critical; where the problem is not considered critical, the site is closely monitored. This inspection should also include a close look at the effectiveness of any geotechnical measures that were implemented.

Hydrologic Design Considerations

Pipeline river crossings can have a major influence on route selection and may significantly affect the cost of the pipeline. The following is a typical sequence of activities associated with the design of the hydrologic aspects at a major pipeline river crossing:

- Aerial and ground reconnaissance is performed to determine the suitability of the crossing location and gather data on bed and bank material and conditions affecting bank migration.
- Design flows are computed (1:50 year return period for gas pipelines; and 1:100 year return period for oil pipelines).
- River crossing designs are prepared; these include:

  - Computing the 1:50 (or 1:100 in the case of an oil pipeline) year water level. Using the 1:50 year water level, the general and local scouring that theoretically would occur during flood conditions are calculated. The scour at bankful flow is also calculated, as it may be more severe than at the design flood stage. Potential cutoff channels are determined and their effect on scour and pipeline design assessed.
  - Based on the results of preceding studies, the required burial depth is determined to ensure that the pipe is below the calculated scour depth.
  - From a comparison of photographs taken at different times and from site reconnaissance, the sag-bend locations are determined and the need for bank toe protection measures is assessed.

Minor stream crossings do not normally require such an extensive design, and standard designs can be adopted for them.

REFERENCES

Champlin, J. B. F., 1973, "Ecological Studies along Transmission Lines in South Western United States," Journal of Environmental Sciences, 16, pp. 11–18.

Environment Canada, 1978, "Ecological Land Survey Guidelines for Environmental Impact Assessment," Environmental Management Service and Federal Environmental Assessment Review Office, Ottawa.

Passey, G. H., and Wooley, D. R., 1980, "The Route Selection Process - A Biophysical Perspective," Alberta Energy and Natural Resources, Edmonton.

Chapter *3*

NATURAL GAS TRANSMISSION

INTRODUCTION

In this chapter, the general flow equation for compressible flow in a pipeline will be derived from basic principles. Having obtained the general flow equation, the way in which flow efficiency is affected by varying different gas and pipeline parameters will be examined. Different flow regimes in gas transmission systems (i.e., partially turbulent and fully turbulent flow) will be presented. Some of the widely used transmission equations and their applications, advantages, and limitations will be outlined.

This will be followed by a discussion of pipes in series, pipeline looping, gas velocity, line packing, pipeline maximum operating pressure, and some pipeline codes. The impact of gas temperature on the flow efficiency and gas temperature profile (i.e., heat transfer from a buried pipeline and Joule-Thompson effects) will be discussed.

Finally, some major economic aspects and considerations in the design of gas pipeline systems will be presented.

GENERAL FLOW EQUATION — STEADY STATE

In this section, the general flow equation for compressible fluids in a pipeline at steady-state condition is derived. First, the general Bernoulli equation will be obtained using a force balance on a segment of the pipeline. The Bernoulli equation is then used to derive the general flow equation for compressible fluids (natural gas) in a pipeline.

Consider a pipeline that transports a compressible fluid (natural gas) between points 1 and 2 at steady-state condition, as shown in Figure 3-1

where ρ = gas density
          P = gas pressure
          A = pipeline cross-sectional area
          u = gas velocity

at steady-state condition

$$\frac{dm}{dt} = 0 \qquad (3-1)$$

where m is the mass of gas flowing in the pipeline and t is time. The mass flow rate of gas at point 1 can be defined as

$$\dot{m} = \rho_1 \cdot A_1 \cdot u_1 \qquad (3-2)$$

57

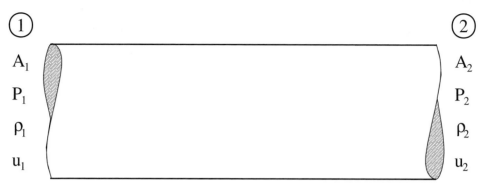

Figure 3-1. Steady state flow of a compressible fluid in a pipeline

Likewise, the mass flow rate at point 2 is

$$\dot{m} = \rho_2 \cdot A_2 \cdot u_2$$

It then follows that

$$\rho_1 \cdot A_1 \cdot u_1 = \rho_2 \cdot A_2 \cdot u_2$$

If the pipe has a constant diameter, then

$$\rho_1 \cdot u_1 = \rho_2 \cdot u_2$$

or, in general

$$\dot{m} = \rho \cdot A \cdot u$$

or

$$\frac{\dot{m}}{A} = \rho \cdot u$$

$$\rho \cdot u = C$$

where C is a constant.
It is also known that

$$\rho = \frac{1}{v}$$

where v = the gas specific volume
so

$$\frac{u}{v} = C \qquad (3-3)$$

From Newton's Law of Motion for a particle of gas moving in a pipeline [see Figure 3-2a]:

$$dF = a \cdot dm$$

where $a = du/dt$ is the acceleration:

$$dF = \frac{du}{dt} \cdot dm = \frac{du}{dt} \cdot \rho \cdot A \cdot dy = \rho \cdot A \cdot du \cdot \frac{dy}{dt}$$

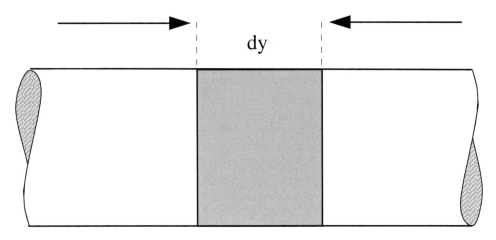

$$dm = \rho.A.dy$$

Figure 3-2a. Movement of a gas particle in a pipeline

and

$$\frac{dy}{dt} = u$$

therefore

$$dF = \rho \cdot A \cdot u \cdot du$$

In U.S. units, using the proportionality constant g_c, the above equation could be written as follows:

$$dF = \frac{A}{g_c} \cdot u \cdot \rho \cdot du = \frac{A}{g_c} \cdot \frac{u}{\upsilon} \cdot du \qquad (3-4)$$

The impact of all existing forces (i.e., pressure, weight, friction, etc.) exerted on a particle of gas in a nonhorizontal pipeline [Figure 3-2b] can be considered as follows:

The forces F_1 and F_2 acting on the gas particle due to the gas pressure P_1 and P_2 can be defined as:

$$dF_1 = A \, dP_1$$

and

$$dF_2 = A \, dP_2$$

The force F_3 exerted on the gas due to the weight W of the gas particle is

$$F_3 = W \cdot \sin \alpha$$

in differential form

$$dF_3 = dW \cdot \sin \alpha$$

where the weight of the gas is

$$dW = \frac{g_L}{g_C} \cdot A \cdot dy \cdot \rho \qquad (3-5)$$

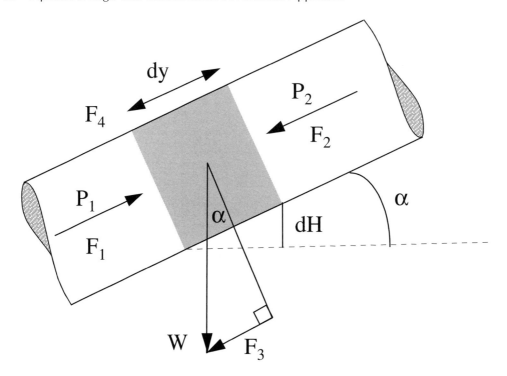

Figure 3-2b. Demonstration of all forces acting on a gas particle moving in a nonhorizontal pipeline

(g_L is local acceleration of gravity)
furthermore

$$\sin \alpha = \frac{dH}{dy}$$

where dH is the change in elevation. Upon substitution for both dW and $\sin \alpha$

$$dF_3 = \frac{g_L}{g_C} \cdot A \cdot \rho \cdot dH \qquad (3-6a)$$

or

$$dF_3 = \frac{g_L}{g_C} \cdot \frac{A}{v} \cdot dH \qquad (3-6b)$$

Finally, the friction force is defined as:

$$dF_4 = \pi \cdot D \cdot dy \cdot \tau \qquad (3-7)$$

where $\pi D dy$ is the surface area and τ is the shear stress.

The summation of all the forces acting on the element of the gas should be equal to zero, therefore:

$$\frac{A}{g_c} \cdot \frac{u}{v} \cdot du + A dP + \frac{g_L}{g_c} \cdot \frac{A}{v} dH + \pi D dy \cdot \tau = 0 \qquad (3-8)$$

This is the general form of the Bernoulli equation. In most cases, it is assumed that the numerical values of g_L and g_C are equal. Then

$$\frac{A}{g_c} \cdot \frac{u}{v} \cdot du + AdP + \frac{A}{v}dH + \pi Ddy \cdot \tau = 0 \qquad (3-9)$$

Multiply both sides by v/A:

$$\frac{1}{g_c} \cdot udu + vdP + dH + \frac{\pi Ddyv}{A} \cdot \tau = 0 \qquad (3-10)$$

where udu = kinetic energy; vdP = pressure energy; dH = potential energy; $(\pi Ddy \times v/A)\tau$ = friction or losses

The friction term or losses created by moving a fluid in a pipeline is defined by the Fanning equation as follows:

$$dF_{\text{Fanning}} = \frac{2fu^2}{g_c \cdot D} \cdot dL \qquad (3-11)$$

where u = average gas velocity
$\quad f$ = friction factor
$\quad D$ = pipeline diameter
$\quad L$ = pipeline length

Substituting the Fanning equation for losses in the general energy equation will result in

$$\frac{1}{g_c} \cdot udu + vdP + dH + \frac{2fu^2}{g_c \cdot D} \cdot dL = 0 \qquad (3-12)$$

Dividing both sides of the equation by v^2:

$$\frac{1}{g_c} \cdot \frac{u}{v^2} \cdot du + \frac{dP}{v} + \frac{dH}{v^2} + \frac{2f}{g_c \cdot D} \cdot \frac{u^2}{v^2} \cdot dL = 0 \qquad (3-13)$$

The final form of the equation can be obtained by integrating each term, assuming $u/v = \dot{m}/A = C$ = constant.

Kinetic Energy Term

$$\int_1^2 \frac{1}{g_c} \cdot \frac{u}{v^2} \cdot du = \int_1^2 \frac{1}{g_c} \cdot \frac{u}{v} \cdot \frac{du}{v}$$

since

$$\frac{u}{v} = C$$

$$\int_1^2 \frac{C}{g_c} \cdot \frac{du}{v} = \frac{C}{g_c} \int_1^2 \frac{du}{v}$$

since

$$v = \frac{u}{C}$$

$$\frac{C}{g_c} \int_1^2 \frac{du}{u/C} = \frac{C^2}{g_c} \int_1^2 \frac{du}{u}$$

then

$$Kinetic\ energy = \frac{C^2}{g_c} \ln \frac{u_2}{u_1} \qquad (3-14)$$

Pressure Energy Term

$$\int_1^2 \frac{dP}{v} = \int_1^2 \rho \cdot dP$$

From the real gas law

$$PV = nZRT$$

where Z is the compressibility factor of the gas and R is the gas constant for

$$n = \frac{m}{M}$$

and

$$\rho = \frac{m}{V}$$

the equation for the density of a gas is:

$$\rho = \frac{P \cdot M}{Z \cdot R \cdot T} \qquad (3-15)$$

where M is the average molecular weight of the gas.
After substitution into $\int_1^2 \rho dP$

$$\int_1^2 \frac{P \cdot M}{Z \cdot R \cdot T} \cdot dP = \frac{M}{Z_{ave} \cdot R \cdot T_{ave}} \int_1^2 PdP$$

$$= \frac{M}{Z_{ave} \cdot R \cdot T_{ave}} \cdot \frac{P_2^2 - P_1^2}{2} \qquad (3-16)$$

where T_{ave}, is defined as follows:

$$T_{ave} = \frac{T_1 + T_2}{2} \qquad (3-17)$$

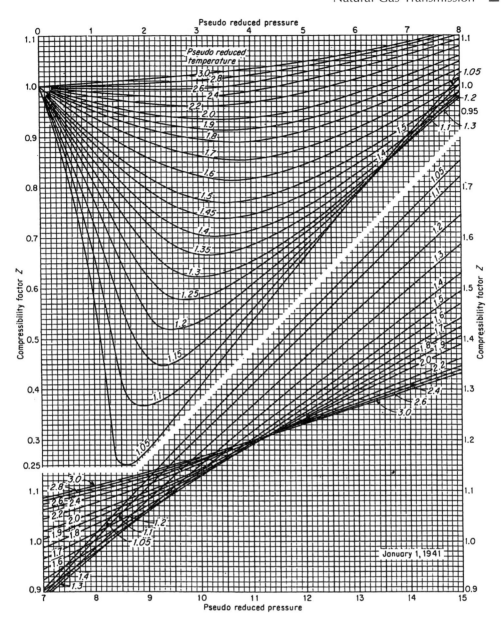

Figure 3-3. Compressibility factor for natural gases [Katz et al., 1959 Handbook of Gas Engineering, reproduced with permission from McGraw-Hill Co.]

(T_1 and T_2 are the upstream and downstream gas temperatures) and P_{ave} is obtained based on the relation $\int_1^2 PdP$:

$$P_{\mathrm{ave}} = \frac{\int_1^2 P \cdot P \cdot dP}{\int_1^2 PdP} = \frac{\int_1^2 P^2 \cdot dP}{\int_1^2 PdP}$$

or

$$P_{ave} = \frac{2}{3}\left[P_1 + P_2 - \frac{P_1 \cdot P_2}{P_1 + P_2}\right] \qquad (3-18)$$

(P_1 and P_2 are the upstream and downstream gas pressures).

Having obtained T_{ave} and P_{ave} for the gas, the average compressibility factor, or Z_{ave}, can be obtained for lean sweet natural gases with an excellent accuracy using Kay's rule and compressibility factor charts. The Z factor can also be calculated with one of the widely used equations of state, such as *AGA-8, BWRS, RK, SRK,* or any other such formula.

To calculate Z_{ave} for a natural gas using Kay's rule, T_{ave} and P_{ave} of the gas are needed, and also pseudocritical pressure and temperature of the natural gas. Pseudocritical values can be obtained with Kay's rule as follows:

$$T_C' = T_{CA} \cdot y_A + T_{CB} \cdot y_B + T_{CC} \cdot y_C + \dots \qquad (3-19)$$

$$P_C' = P_{CA} \cdot y_A + P_{CB} \cdot y_B + P_{CC} \cdot y_C + \dots \qquad (3-20)$$

where
T_C' = average pseudocritical temperature of the gas
P_C' = average pseudocritical pressure of the gas
$T_{CA}, T_{CB}, T_{CC},$ = critical temperature of each component
$P_{CA}, P_{CB}, P_{CC},$ = critical pressure of each component
$y_A, y_B, y_C,$ = mole fraction of each component

Finally, pseudoreduced pressure and temperature can be obtained as follows:

$$P_r' = \frac{P_{ave}}{P_C'} \qquad (3-21)$$

and

$$T_r' = \frac{T_{ave}}{T_C'} \qquad (3-22)$$

The values of P_r' and T_r' can be used in compressibility factor charts to calculate Z_{ave} (as shown in Figure 3-3).

Example 3.1
What is the compressibility factor for a natural gas with the following analysis at 1,000 psia and 100°F?

Gas Composition	
COMPONENT	Mole %
C_1	85
C_2	10
C_3	5

Using Kay's rule to calculate pseudocritical properties (see Table 3-1):

$$P_C' = P_{CA} \cdot Y_A + P_{CB} \cdot Y_B + P_{CC} \cdot Y_C +$$

$$T_C' = T_{CA} \cdot Y_A + T_{CB} \cdot Y_B + T_{CC} \cdot Y_C +$$

$P_C' = 666 \times 0.85 + 707 \times 0.10 + 617.4 \times 0.05 = 667.67$ psia, pseudocritical pressure

$$T_C' = 343.3 \times 0.85 + 549.8 \times 0.10 + 666.0 \times 0.05$$

$$= 380.085°R \text{ pseudocritical temperature}$$

$$P_r' = \frac{P_{ave}}{P_C'} = \frac{1,000}{667.67} = 1.498$$

$$T_C' = \frac{T_{ave}}{T_C'} = \frac{460 + 100}{380.085} = 1.474$$

using the appropriate chart (see Figure 3-3), which covers both the P_r' and T_r' range value, $Z_{ave} = 0.85$ is obtained.

The model mentioned above for the calculation of the compressibility factor is a quick and accurate model for dry and sweet natural gases, and is most suitable for hand calculations. In large gas transmission networks with hundreds of pipe segments, while dividing each pipe into smaller segments to consider temperature changes in the pipeline, this model becomes a cumbersome and time-consuming procedure to calculate the compressibility factor. For these networks, where all simulations are automated, apply accurate equations of state to calculate the compressibility factor. There are a large number of equations of state that are suitable for a limited range (depending on pressure, temperature, and gas composition), but could deviate to produce inaccurate results if used without these considerations. One such equation, which is commonly used in gas transmission systems and has a proven accuracy of better than 0.3% for compressibility factor, is an AGA equation given by Starling and Savage (1994). This equation covers most of the existing pressure and temperature ranges used in gas transmission lines. For further information on different equations of state (e.g., for gases containing H_2S and CO_2) refer to (Campbell et al., 1994).

TABLE 3-1. Critical properties of constituents of natural gas (Courtesy Campbell Petroleum Series)

Compound	Molecular Weight	Critical Temperature		Critical Pressure	
		°R	°K	psia	MPa
C_1	16.043	343	191	666	4.60
C_2	30.070	550	305	707	4.88
C_3	44.097	666	370	617	4.25
iC_4	58.124	734	408	528	3.65
nC_4	58.124	765	425	551	3.80
iC_5	72.151	829	460	491	3.39
nC_5	72.151	845	470	489	3.37
nC_6	86.178	913	507	437	3.01
nC_7	100.205	972	540	397	2.74
nC_8	114.232	1,024	569	361	2.49
nC_9	128.259	1,070	595	332	2.29
nC_{10}	142.286	1,112	618	305	2.10
nC_{11}	156.302	1,150	639	285	1.97
nC_{12}	170.338	1,185	658	264	1.82
N_2	28.016	227	126	493	3.40
CO_2	44.010	548	304	1,071	7.38
H_2S	34.076	672	373	1,300	8.96
O_2	32.000	278	155	731	5.04
H_2	2.016	60	33	188	1.30
H_2O	18.015	1,165	647	3,199	22.06
Air	28.960	238	132	547	3.77
He	4.000	9	5	33	0.23

NATURAL GAS HIGHER AND LOWER HEATING VALUES

Both the higher and the lower heating values of the natural gas could be calculated using direct summation method (weighted average based on the mole fraction of each component) with a reasonable accuracy as follows:

$$HHV = \sum_{i=1}^{n}(HHV)_i \cdot x_i$$

$$LHV = \sum_{i=1}^{n}(LHV)_i \cdot x_i$$

If further accuracy is required for the calculation of the heating values, we can use the AGA – 8D equation of state where it does not assume $Z = 1.0$ at standard condition, (it calculates the actual Z). For further information see the AGA Report No-10, 2003.

The difference between the higher and the lower heating values is the latent heat of the condensation of water which exists in the exhaust gases. If we extract this much of heat, the summation is the higher heating value, but if the exhaust gases leave the system as vapour then the lower heating value is obtained. For pipeline natural gas the ratio of LHV / HHV is normally within the 90% range. We should emphasize that we use the HHV of the natural gas for the billing purposes while the LHV is the basis for the calculation of the fuel at the compressor stations

Table 3-1a is extracted from GPSA which gives the values for both HHV and LHV in SI and Imperial units for the components of the natural gas.

TABLE 3-1a. Higher and Lower Heating Values of Natural Gas Constituents, SI & Imperial Units (Courtesy of GPSA)

Gas Component	HHV in kJ/m^3	HHV in BTU/ft^3	LHV in kJ/m^3	LHV in BTU/ft^3
C_1	37694	1010.0	33936	909.4
C_2	66032	1769.6	60395	1618.7
C_3	93972	2516.1	86456	2314.9
nC_4	121779	3262.3	112384	3010.8
iC_4	121426	3251.9	112031	3000.4
nC_5	149654	4008.9	138380	3706.9
iC_5	149319	4000.9	138044	3699.0
nC_6	177556	4755.9	164402	4403.8
nC_7	205431	5502.5	190398	5100.0
nC_8	233286	6248.9	216374	5796.1
nC_9	261189	6996.5	242398	6493.2
nC_{10}	289066	7742.9	268396	7189.6
N_2	0	0.0	0	0.0
CO_2	0	0.0	0	0.0
CO	11959	320.5	11959	320.5
He	0	0.0	0	0.0
H_2	12091	324.2	10230	273.8
H_2S	23791	637.1	21912	586.8
H_2O	0	0.0	0	0.0
O_2	0	0.0	0	0.0

Potential Energy Term

Integration of the potential energy term of Equation (3-13) will result in:

$$\int_1^2 \frac{dH}{v_2} = \int_1^2 \rho^2 \, dH = \int_1^2 \left(\frac{P \cdot M}{ZRT}\right)^2 dH = \frac{P_{\text{ave}}^2 \cdot M^2}{Z_{\text{ave}}^2 \cdot R^2 \cdot T_{\text{ave}}^2} \Delta H \qquad (3-23)$$

where $\Delta H = H_2 - H_1$.

There is no simple mathematical relationship between elevation change, gas pressure and gas temperature, so the relationship $(P^2 \cdot M^2 / Z^2 \cdot R^2 \cdot T^2)$ can be taken out of the integral in the form of average values while maintaining reasonable accuracy.

Friction Loss Term

The integral of the energy losses can be evaluated as follows:

$$\frac{2fC^2}{g_c D} \int_1^2 dL = \frac{2fC^2}{g_c D} L \qquad (3-24)$$

where L is the pipeline length.

The general form of the flow equation is obtained by adding all the terms together and setting them equal to zero

$$\frac{C^2}{g_c} \ln \frac{u_2}{u_1} + \frac{M}{2RZ_{\text{ave}} T_{\text{ave}}} \left(P_2^2 - P_1^2\right) + \frac{M^2 \cdot P_{\text{ave}}^2}{R^2 \cdot T_{\text{ave}}^2 \cdot Z_{\text{ave}}^2} \Delta H + \frac{2fC^2}{g_c \cdot D} L = 0 \qquad (3-25)$$

The above equation can be further simplified if the kinetic energy term is neglected (for almost all high-pressure gas transmission lines, the contribution of the kinetic energy term compared to the other terms is insignificant). Therefore

$$\frac{M}{2RZ_{\text{ave}} T_{\text{ave}}} \left(P_2^2 - P_1^2\right) + \frac{M^2 \cdot P_{\text{ave}}^2}{R^2 \cdot T_{\text{ave}}^2 \cdot Z_{\text{ave}}^2} \Delta H + \frac{2fC^2}{g_c \cdot D} L = 0 \qquad (3-26)$$

The above equation can be even further simplified upon the following substitutions:

$$C = \frac{\dot{m}}{A}, \quad C^2 = \left(\frac{\dot{m}}{A}\right)^2, \quad A = \frac{\pi D^2}{4}, \quad \text{for a pipe}$$

Moreover, the gas relationship at a base or standard condition is

$$P_b \cdot Q_b = \dot{n}_b \cdot Z_b \cdot R \cdot T_b$$

where Q_b is the volumetric gas flow.

If

$$\dot{n}_b = \frac{\dot{m}}{M}$$

and

$$C = \frac{\dot{m}}{A}, C^2 = \frac{\dot{m}^2}{(\pi D^2/4)^2}$$

then

$$C^2 = \frac{16 Q_b^2 \cdot M^2 \cdot P_b^2}{\pi^2 \cdot R^2 \cdot T_b^2 \cdot Z_b^2 \cdot D^4}$$

Gas gravity is defined as

$$G = \frac{M_{gas}}{M_{air}}$$

where $M_{air} \approx 29$.

Upon substitution and rearrangement to solve for Q_b, Equation (3-26) would be

$$Q_b^2 = \frac{\pi^2 \cdot R \cdot g_c}{32} \cdot \frac{Z_b^2 \cdot T_b^2}{P_b^2} \left\{ \frac{P_1^2 - P_2^2 - \dfrac{58G \cdot \Delta H \cdot P_{ave}^2}{R \cdot T_{ave} \cdot Z_{ave}}}{58 Z_{ave} \cdot T_{ave} \cdot G \cdot L} \right\} \cdot \frac{D^5}{f} \qquad (3-27)$$

By taking the square root of Q_b, the general flow equation of natural gas in a pipeline is

$$Q_b = \pi \sqrt{\frac{g_c \cdot R}{1,856}} \cdot \frac{Z_b \cdot T_b}{P_b} \sqrt{\frac{P_1^2 - P_2^2 - \dfrac{58G \cdot \Delta H \cdot P_{ave}^2}{R \cdot T_{ave} \cdot Z_{ave}}}{Z_{ave} \cdot T_{ave} \cdot G \cdot L}} \cdot \sqrt{\frac{1}{f}} \cdot D^{2.5} \qquad (3-28)$$

The above equation can be used in Imperial or S.I. units; for any size or length of pipe; for laminar, partially turbulent or fully turbulent flow; and for low, medium, or high-pressure systems.

Definition of Parameters (Imperial Units):

Q_b = gas flow rate at base conditions, MMSCFD or MCF/HR
g_c = proportionality constant, 32.2 ($lb_m \times ft/lb_f \times sec^2$)
$Z_b \approx$ compressibility factor at base condition $Z_b \approx 1$
T_b = temperature at base condition, 520°R
P_b = pressure at base condition, 14.7 psia
P_1 = gas inlet pressure to the pipeline, psia
P_2 = gas exit pressure, psia
G = gas gravity, dimensionless
ΔH = elevation change, ft
P_{ave} = average pressure, psia
R = gas constant, 10.73, (psia $\times ft^3$/lb moles $\times °R$)
T_{ave} = average temperature, °R
Z_{ave} = compressibility factor at P_{ave}, T_{ave}, dimensionless
L = pipeline length, ft or miles
f = friction coefficient, dimensionless
$\sqrt{\dfrac{1}{f}}$ = transmission factor, dimensionless
D = inside diameter of the pipeline, inch

Equation (3-28) can be written in the following form, taking all constants as C', therefore

$$Q_b = C' \cdot \sqrt{\dfrac{P_1^2 - P_2^2 - \dfrac{58G \cdot \Delta H \cdot P_{ave}^2}{R \cdot T_{ave} \cdot Z_{ave}}}{Z_{ave} \cdot T_{ave} \cdot G \cdot L}} \cdot \sqrt{\dfrac{1}{f}} \cdot D^{2.5} \qquad (3-29)$$

In transmission lines, if the pipeline is horizontal or ΔH is insignificant compared to the value of $P_1^2 - P_2^2$, or

$$P_1^2 - P_2^2 \rangle\rangle \dfrac{58G \cdot \Delta H \cdot P_{ave}^2}{R \cdot T_{ave} \cdot Z_{ave}},$$

then the elevation term can be omitted and Equation (3-29) becomes:

$$Q_b = C' \cdot \sqrt{\dfrac{P_1^2 - P_2^2}{Z_{ave} \cdot T_{ave} \cdot G \cdot L}} \cdot \sqrt{\dfrac{1}{f}} \cdot D^{2.5} \qquad (3-30)$$

The above equation shows the effect of $\sqrt{\dfrac{1}{f}}$ and D on the flow of gas in a pipeline. The expression $\sqrt{\dfrac{1}{f}}$ is the transmission factor and is an important parameter that represents the transmissivity of gas in a pipeline. Diameter is another major factor in pipeline design; it can be seen that if the diameter of the pipeline is doubled, the gas flow rate will be increased by $(2)^{2.5} = 5.66$ times.

This demonstrates the importance of considering possible future expansions when selecting pipeline diameter. For example, if a 20 inches gas pipeline is changed to a 30 inches pipeline, gas flow rate is increased by almost 2.756 times, assuming the remaining parameters are constant.

Unlike liquids, Equation (3-30) shows that a system operating at a lower temperature results in higher flows or lower pressure drops. In contrast, higher temperatures will increase the gas viscosity, which will reduce the flow capacity of the pipeline. The impact of other parameters, such as G and Z, will be discussed later.

Considering the previous equation (3-30), Q_b^2 can be calculated as

$$Q_b^2 = C'^2 \cdot \dfrac{P_1^2 - P_2^2}{Z_{ave} \cdot T_{ave} \cdot G \cdot L} \dfrac{1}{f} \cdot D^5$$

Rearranging the equation:

or in general

$$P_1^2 - P_2^2 = K Q_b^n \qquad (3-31)$$

where P_1 = pipeline inlet pressure
P_2 = pipeline exit pressure
K = pipeline total resistance. $K = R \times L$, where R is resistance per foot of pipeline and L is the length of pipeline in feet
Q_b = gas flow rate at base condition
n = gas flow exponent (having values between 1.74 and 2)

TABLE 3-2. Formulas and transmission factors for commonly used flow equations (Courtesy IGT)

Equation	Formula[a]	Transmission Factor
Fritzsche[b]	$Q_b = 1.720\left(\frac{T_b}{P_b}\right)\left[\frac{(P_1^2 - P_2^2)D^{5.7}}{T_f L}\right]^{0.538}\left(\frac{1}{G}\right)^{0.462}$	$5.145\,(R_e D)^{0.071}$
Fully Turbulent	$Q_b = 0.4696\left(\frac{T_b}{P_b}\right)\left(\frac{P_1^2 - P_2^2}{GT_f Z_{ave} L}\right)^{0.500} \cdot \log\left(3.7\frac{D}{K_e}\right) \cdot D^{2.5}$	$4\log\,(3.7D/k_e)$
Panhandle B	$Q_b = 2.431\left(\frac{T_b}{P_b}\right)^{1.02}\left(\frac{P_1^2 - P_2^2}{G^{0.961} LT_f Z_{avg}}\right)^{0.510} \cdot D^{2.53}$	$16.49(R_e)^{0.01961}$
Colebrook-White	$Q_b = 0.4696\left(\frac{T_b}{P_b}\right)\left(\frac{P_1^2 - P_2^2}{GT_f Z_{avg} L}\right)^{0.500} \cdot \left[-4\log\left(\frac{K_e}{3.7D} + \frac{1.4126\sqrt{\frac{1}{f}}}{R_e}\right)\right] \cdot D^{2.5}$	$\sqrt{\frac{1}{f}} = -4\log\left(\frac{K_e}{3.7D} + \frac{1.4126\sqrt{\frac{1}{f}}}{R_e}\right)$
IGT Distribution	$Q_b = 0.6643\left(\frac{T_b}{P_b}\right)\left(\frac{P_1^2 - P_2^2}{T_f L}\right)^{5/9}\left(\frac{D^{8/3}}{G^{4/9}\mu^{1/9}}\right)$	$4.619\,(R_e)^{0.100}$
Mueller	$Q_b = 0.4937\left(\frac{T_b}{P_b}\right)\left(\frac{P_1^2 - P_2^2}{T_f L}\right)^{0.575}\left(\frac{D^{2.725}}{G^{0.425}\mu^{0.150}}\right)$	$3.35\,(R_e)^{0.130}$

Panhandle A[b]

$$Q_b = 2.450\left(\frac{T_b}{P_b}\right)^{1.0788}\left(\frac{P_1^2-P_2^2}{T_f L Z_{avg}}\right)^{0.539}\left(\frac{D^{2.618}}{G^{0.4601}}\right)$$

$$6.872\,(R_e)^{0.0730}$$

Pole

$$Q_b = C\left(\frac{h_w D^5}{GL}\right)^{0.500}$$

Pipe Diameter, (in.)	C[c]	Pipe Diameter, (in.)	$\sqrt{\frac{1}{f}}$
3/4 to 1	1.732	3/4 to 1	9.56
1 1/4 to 1 1/2	1.905	1 1/4 to 1 1/2	10.51
2	2.078	2	11.47
3	2.252	3	12.43
4	2.338	4	12.90

Spitzglass (High Pressure)[d]

$$Q_b = 3.415\left[\frac{(P_1^2-P_2^2)D^5}{GL(1+3.6/D+0.03D)}\right]^{0.500}$$

$$\left(\frac{354}{1+3.6/D+0.03D}\right)^{0.500}$$

Spitzglass (Low Pressure)[d]

$$Q_b = 3.550\left[\frac{h_w D^5}{GL(1+3.6/D+0.03D)}\right]^{0.500}$$

$$\left(\frac{354}{1+3.6/D+0.03D}\right)^{0.500}$$

Weymouth

$$Q_b = 1.3124\left(\frac{T_b}{P_b}\right)^{0.500}\left[\frac{(P_1^2-P_2^2)D^{16/3}}{GT_f L}\right]^{0.500}$$

$$11.19\,D^{1/6}$$

a The units of the quantities in all of these equations are: D = in.; h_w = in. wc; L = ft; P_1, P_2, P_b = psia; Q_b = MCF/hr; μ = lb$_m$/ft sec; and T_f, T_b = °R.

b The constants 1.720 and 2.450 include: $\mu = 7.0 \times 10^{-6}$ lb$_m$/ft. sec.

c Includes: P_b = 14.73 psia; T_b = 492°R; and T_f = 500°R.

d The constants 3.415 and 3.550 include: P_b = 14.7 psia; T_b = 520°R; and T_f = 520°R.

TABLE 3-3. Common flow equation resistance factors *(Courtesy IGT)*

Equation	Resistance Factor (per foot)[a,b,c]	Flow-Rate Exponent for Simplified Flow Equation
Fritzsche	$4.82 \times 10^{-4} \, BG^{0.859}$	1.86
Fully Turbulent (AGA)	$3.639 \times 10^{-3} \, BGZ_{avg} / (\log 3.7D/K_e)^2$	2.00
Panhandle B	$1.405 \times 10^{-4} \, T_f Z_{avg} G^{0.961} / D^{4.961}$	1.961
Colebrook-White	$3.639 \times 10^{-3} \, BGZ_{avg} / (\log(K_e/3.7D + 1.4126F/R_e))^2$	2.00
IGT Distribution	$3.418 \times 10^{-3} \, T_f G^{0.80} \, \mu^{0.20} / D^{4.80}$	1.80
Mueller	$6.922 \times 10^{-3} \, T_f G^{0.739} \mu^{0.261} / D^{4.739}$	1.74
Panhandle A[d]	$2.552 \times 10^{-4} \, T_f Z_{avg} G^{0.855} / D^{4.856}$	1.855
Pole[e]	$G / C^2 D^5$	2.00
Spitzglass High-Pressure[f]	$8.575 \times 10^{-2} \, (1 + 3.6 / D + 0.03D) \, G/D^5$	2.00
Spitzglass Low-Pressure[f]	$7.935 \times 10^{-2} \, (1 + 3.6 / D + 0.03D) \, G/D^5$	2.00
Weymouth	$4.659 \times 10^{-4} \, T_f \, G/D^{16/3}$	2.00

[a] The following standard conditions are included in the constant unless otherwise specified: $T_b = 520 \,°R$; and $P_b = 14.73$ psia.
[b] $B = T_f/D^5$.
[c] $F = (1/f)^{0.5}$.
[d] Constant includes: $7.0 \times 10^{-6} \, lb_m/ft$ sec.
[e] Values of C can be obtained from Table 3-2.
[f] Constant includes: $P_b = 14.7$ psia; $T_b = 520 \,°R$; and $T_f = 520 \,°R$.

In the above equation, K and n have different values depending on the type of equation.

A list of different pipeline equations, together with the pipeline resistance and flow exponent, is given in IGT's "Gas Distribution Home Studies Course" (Wilson et. al. 1991).

IMPACT OF GAS MOLECULAR WEIGHT AND COMPRESSIBILITY FACTOR ON FLOW CAPACITY

The general flow equation for a horizontal pipe (Equation 3-30) indicates that the pipeline flow capacity is proportional to $\sqrt{1/GZ}$. This relationship shows that lighter natural gas with a higher percentage of methane, and hence lower gas gravity, exhibits a higher flow capacity. In this situation, the compressibility factor Z will also be high (close to unity). However, when heavier hydrocarbons ($C2+$) are introduced, the gas gravity begins to increase, which decreases the value of $\sqrt{1/G}$. Yet the presence of heavier hydrocarbons also reduces the value of Z, hence increasing $\sqrt{1/Z}$.

The overall impact on $\sqrt{1/GZ}$ is determined by the rate at which Z is reduced when $C2+$ are added. This rate is not only affected by gas composition, but also by pressure and temperature conditions. Figure 3-4 shows the change in Z over a range of different temperatures and pressures (base condition is 100% methane). At pressures in the range of 1,350–2,149 psia, the decrease in Z balances the increase

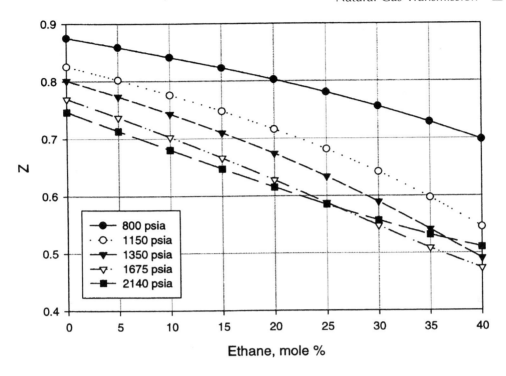

Figure 3-4. Compressibility factor for methane-ethane mixtures at 35°F [Weiss 1998]

in G, which means that the overall value of $\sqrt{1/GZ}$ is changed very little. However, at pressures exceeding 2,140 psia, the higher rate of Z reduction results in an overall increase of $\sqrt{1/GZ}$. Likewise, at pressures below 1,350 psia, the decrease in Z is relatively insignificant, so $\sqrt{1/GZ}$ tends to decrease. For operating pressures between 800 and 2,140 psia, it is recommended that careful analysis be performed to conclude how the flow capacity will be affected by the addition of ethane or heavier hydrocarbons.

Figure 3-5 shows the way in which flow capacity, $\sqrt{1/GZ}$ or $\sqrt{1/MZ}$, responds as a function of mole percent of ethane added to methane for a range of pressures on a standard volume basis. As expected, the flow decreases when only a small percentage of ethane is present, and when operating pressures are low. For example, a gas mixture of 25% ethane and 75% methane at 800 psia exhibits a reduction in standard volume flow rate of 4%. However, at 1,150 psia there is no change, and at 1,350 psia, the standard volume flow rate is increased by nearly 2%.

It is very important to remember that the basis for almost all pipeline transportation comparisons is mass or energy (i.e., heating value of the natural gas). Therefore, volume flow capacity is not usually the best means to compare two different conditions and could sometimes be misleading.

Figure 3-6 plots mass flow capacity versus ethane percentages (or heavier hydrocarbons) over a range of operating pressures. This is based on a mass flow capacity of $Q \times \rho$ where standard density is only a function of molecular weight (M), so if

$$\text{flow capacity} \, \alpha \sqrt{\frac{1}{MZ}}$$

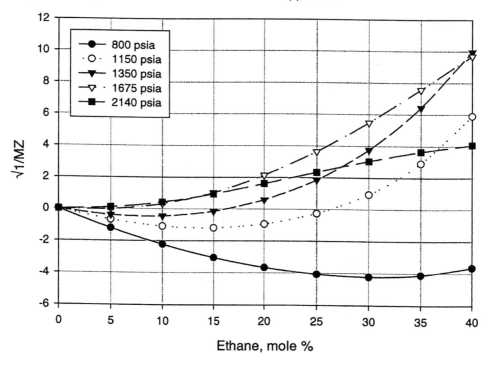

Figure 3-5. Standard volume flow capacity for methane-ethane mixture at 35°F [Weiss 1998]

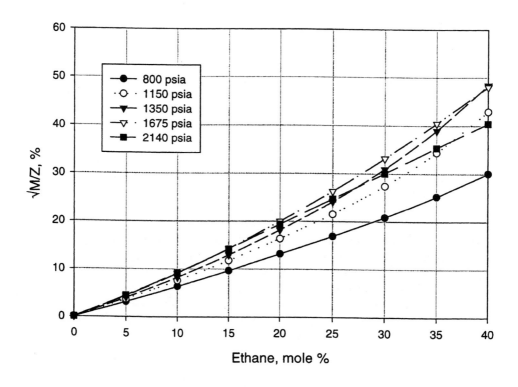

Figure 3-6. Mass flow capacity for methane-ethane mixtures at 35°F [Weiss 1998]

then

$$\text{mass flow capacity} \, \alpha \sqrt{\frac{M}{Z}}$$

On a total energy basis the approximate amount of energy that can be transported with a methane-ethane mixture (or other heavier hydrocarbons) increases at the same rate as the mass flow capacity increases. Now, if the same case as previously stated for 25% ethane at 800 psia is considered, it can be concluded that the heating value has increased by almost 18%. Therefore, the net effect is a 14% increase in energy flow (i.e., when a 4% decrease in standard volume flow rate is subtracted) or heating value (Weiss 1998).

FLOW REGIMES

In high-pressure gas transmission lines with moderate to high flow rates, two types of flow regimes are normally observed:

Fully Turbulent Flow (Rough Pipe Flow)
Partially Turbulent Flow (Smooth Pipe Flow)

The regime of flow is defined by the Reynolds number, which is a dimensionless expression:

$$\text{Re} = \frac{\rho \cdot D \cdot u}{\mu} \qquad (3-32)$$

where ρ = fluid density, $\text{lb}_\text{m}/\text{ft}^3$
D = pipeline internal diameter, ft
u = fluid average velocity, ft/sec.
μ = fluid viscosity, $\text{lb}_\text{m}/\text{ft.sec}$

For Reynolds numbers less than 2,000 the flow is normally laminar, or stable. When the Reynolds number exceeds 2,000, the flow is turbulent, or unstable. In high-pressure gas transmission lines, only two types of flow regimes are observed: fully turbulent flow and partially turbulent flow.

Partially Turbulent Flow Regime

Partially Turbulent Flow is defined by the Prandtl - Von Karman equation as follows:

$$\sqrt{\frac{1}{f}} = 4 \log_{10} \frac{\text{Re}}{\sqrt{\frac{1}{f}}} - 0.6 \qquad (3-33)$$

where f = friction factor, dimensionless; and Re = Reynolds number, dimensionless.

This equation is obtained based on theory and experiments for the case in which the flow is fully turbulent in the central region of the pipe, with a laminar sublayer covering the interior surface of the pipe.

Equation (3-33) is plotted on a semi-log graph, where the straight line shows the maximum limit of partially turbulent flow (see Figure 3-7). All points to the right-hand side

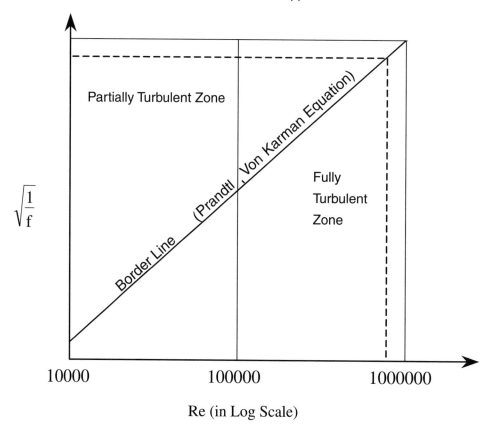

$$\sqrt{\frac{1}{f}}$$

Re (in Log Scale)

Figure 3-7. Representation of fully turbulent/partially turbulent zones by the Prandtl-Von Karman Equation

of the line exhibit fully turbulent flow, and those to the left side remain partially turbulent. Points located on the line are in the transition zone.

Example 3.2

What is the maximum Re number for which a flow regime remains partially turbulent, given a transmission factor of $\sqrt{1/f} = 18$.
Using the Prandtl - Von Karman Equation:

$$\sqrt{\frac{1}{f}} = 4 \log_{10} \left(\frac{Re}{\sqrt{\frac{1}{f}}} \right) - 0.6$$

$$\sqrt{\frac{1}{f}} = 18,$$

then

$$\log_{10}(Re) = 5.9053$$

$$Re = 804,081$$

If the calculated Re number for an actual pipeline with transmission factor 18 exceeds this value, the flow regime is fully turbulent.

Fully Turbulent Flow Regime

The transmission factor for fully turbulent flow is given by the Nikuradse equation as follows:

$$\sqrt{\frac{1}{f}} = 4\log_{10}\left[3.7\frac{D}{K_e}\right] \qquad (3-34)$$

where $\sqrt{1/f}$ = transmission factor, dimensionless
D = pipeline inside diameter, inch
K_e = effective roughness, inch
K_e/D = relative roughness, dimensionless

The effective roughness term K_e is comprised of the following terms:

$$K_e = K_s + K_i + K_d \qquad (3-35)$$

where K_s = surface roughness
K_i = interfacial roughness
K_d = roughness due to bends, welds, fittings, etc.

Generally, in high-pressure gas transmission lines with high flow rates, where the regime of flow is fully turbulent and the natural gas is almost dry, the values of K_i and K_d are negligible compared to K_s. Therefore, the effective roughness of the pipeline is almost equal to the internal surface roughness of the pipe. The value of K_s or K_e is important in fully turbulent flow because without the laminar sublayer, the surface roughness of the pipe plays an important role in determining the flow and pressure drop in the pipe.

The Nikuradse equation shows that if the effective roughness of the pipeline is increased, the transmission factor decreases and results in higher pressure drops. However, by decreasing the value of K_e, higher transmission factors or lower pressure drops are obtained. For internally uncoated commercial pipes, when a number for K_e is unavailable, a value of 700 μ inches (micro inches = 10^{-6} inches) may be assumed.

The effective roughness values that are normally measured and used for uncoated commercial pipes are within the range of 650–750 μ inches. Different studies (Golshan and Narsing 1994) have proven that these values could be increased between a range of 30–50 μ inches per year due to erosion, corrosion, contamination, and other associated problems, which finally result in higher fuel consumption and compression power requirements to overcome the higher pressure drops. Another way to reduce the effective or surface roughness of a pipeline is to internally coat the pipes. Materials such as epoxy/polyamide coatings reduce the surface roughness to within a range of 200 – 300 micro inches. It has been further proven (Golshan and Narsing 1994) that the rate of deterioration for internally coated pipes is much slower than the uncoated pipes (i.e., within the range of 50 – 75 micro-inches for every five years).

Some of the benefits of internal coating have been confirmed by experimental studies conducted by various pipeline companies. The amount of reduction in surface roughness could substantially increase the pipeline capacity (a comparison of the effect of the change of K_e on different transmission equations is presented later in this text). Another benefit associated with internal coating is the protection against corrosion, which is caused by atmospheric oxidation during storage or the presence of corrosive components in the transported material during service.

Due to the high cost of internally coating pipelines, the final decision about whether or not to coat is essentially an economic one. It requires a detailed evaluation of the costs and benefits of internal coating over the projected life of the pipeline (Asante 1994).

Simplified Equation for the Calculation of Reynolds Number in Gas Transmission Systems

As discussed earlier, Reynolds number is defined as:

$$\mathrm{Re} = \frac{\rho \cdot D \cdot u}{\mu}$$

where

$$u = \frac{Q}{\pi D^2 / 4}$$

therefore

$$\mathrm{Re} = \frac{\rho \cdot D \cdot Q}{\mu \cdot \pi D^2 / 4}$$

where $\rho Q = \rho_b Q_b$ at steady-state conditions

$$\mathrm{Re} = \frac{4 Q_b \cdot \rho_b}{\mu \cdot \pi \cdot D} \qquad (3-36)$$

and if

$$\rho_b = \frac{P_b \cdot M}{Z_b \cdot R \cdot T_b}$$

where the value of $Z_b \approx$ one, and $M = 29G$
then

$$\mathrm{Re} = \frac{4 Q_b \cdot 29G \cdot P_b}{\mu \cdot \pi \cdot D \cdot R \cdot T_b}$$

substituting for π, $R = 10.73 \dfrac{\mathrm{psia \cdot} ft^3}{\mathrm{lbmoles \cdot ^\circ R}}$, $T_b = 520$ °R, $P_b = 14.7$ psia, and $\mu = 7.23 \times 10^{-6} \dfrac{\mathrm{lb}\ \mathrm{m}}{\mathrm{ft\ .sec}}$ (viscosity normally assumed for natural gases);

$$\mathrm{Re} = 45 \cdot \frac{Q_b \cdot G}{D} \qquad (3-37)$$

where Q_b = gas flow rate, ft³/hr (standard conditions)
           G = gas gravity, dimensionless
           D = inside diameter of the pipe, inches

This is a simplified equation that gives the Re number in terms of pipeline parameters with reasonable accuracy. The Re number can be used to check the flow regime of a gas transmission line.

Example 3.3

What would be the regime of flow in a 56 inches gas transmission line (ID = 54 inches), $G = 0.64$, when the gas flow rate is $Q_b = 1,500,000$ m^3/hr?

$$1m^3 = 35.31 \text{ ft}^3$$

$$Re = \frac{45 \cdot 1,500,000 \cdot 35.31 \cdot 0.64}{54}$$

$$Re = 28,248,000$$

assume commercial pipe with $K_e = 700 \ \mu$ inches. Use the Nikuradse relationship to calculate $\sqrt{1/f}$:

$$\sqrt{\frac{1}{f}} = \log\left[3.7\frac{54}{0.0007}\right]$$

$$\sqrt{\frac{1}{f}} = 21.82$$

Prandtl - Von Karman equation could now be used to find the Re number at the transition zone, which is:

$$\sqrt{\frac{1}{f}} = 4\log\frac{Re}{\sqrt{\frac{1}{f}}} - 0.6$$

$$21.82 = 4\log\frac{Re}{21.82} - 0.6$$
$$\log \ Re = 6.94385$$
$$Re = 8,787,291$$

The actual Re obtained based on pipeline properties is much larger than 8,787,291, so the flow regime is fully turbulent.

WIDELY USED STEADY-STATE FLOW EQUATIONS

A more simplified form of the general flow equation (3-28) in Imperial Units can be written as follows:

$$Q_b = 38.774\frac{T_b}{P_b} \cdot \sqrt{\frac{1}{f}} \cdot \left[\frac{P_1^2 - P_2^2 - 0.0375G \cdot \Delta H \cdot \frac{P_{ave}^2}{T_{ave} \cdot Z_{ave}}}{Z_{ave} \cdot T_{ave} \cdot G \cdot L}\right]^{\frac{1}{2}} \cdot D^{2.5} \qquad (3-38)$$

assuming that the potential energy term is

$$E = 0.0375G \cdot \Delta H \cdot \frac{P_{ave}^2}{T_{ave} \cdot Z_{ave}} \qquad (3-39)$$

then

$$Q_b = 38.774 \frac{T_b}{P_b} \cdot \sqrt{\frac{1}{f}} \cdot \left[\frac{P_1^2 - P_2^2 - E}{Z_{ave} \cdot T_{ave} \cdot G \cdot L} \right]^{\frac{1}{2}} \cdot D^{2.5} \qquad (3-40)$$

where

Q_b = gas flow rate at base conditions, SCF/D
T_b = temperature at base condition, $520\ °R$
P_b = pressure at base condition, 14.7 psia
$\sqrt{\frac{1}{f}}$ = transmission factor, dimensionless
P_1 = gas inlet pressure, psia
P_2 = gas exit pressure, psia
G = gas gravity, dimensionless
ΔH = elevation change, ft.
P_{ave} = average pressure, psia
T_{ave} = average temperature, $°R$
Z_{ave} = average compressibility factor, dimensionless
L = pipeline length, miles
D = pipeline inside diameter, inch

The following are some of the most common and widely used flow equations that are suitable for the design of large-diameter, high-pressure gas transmission lines. For further information, see IGT "Home Study Course" (Wilson et al. 1991) or IGT "Technical Reports on Steady-Flow in Gas Pipelines."

Partially Turbulent Equations
Panhandle A
The Panhandle A equation is normally appropriate for medium to relatively large diameter pipelines with moderate gas flow rate, operating under medium to high pressure.
It is defined in Imperial Units as

$$Q_b = 435.83 \left(\frac{T_b}{P_b} \right)^{1.0788} \left[\frac{P_1^2 - P_2^2 - E}{G^{0.8539} \cdot L \cdot T_{ave} \cdot Z_{ave}} \right]^{0.5394} \cdot D^{2.6182} \qquad (3-41)$$

where transmission factor is defined as

$$\sqrt{\frac{1}{f}} = 6.872\ Re^{0.07305} \qquad (3-42)$$

or

$$\sqrt{\frac{1}{f}} = 7.211 \left(\frac{Q_b \cdot G}{D} \right)^{0.07305} \qquad (3-43)$$

where Q_b is in SCF/D. All parameters are the same as in Equation 3-38.

AGA Partially Turbulent

The AGA partially turbulent equation is highly dependent on the Reynolds number. It is used for medium-diameter, medium-flow and high-pressure systems

In Imperial Units it is defined as follows:

$$Q_b = 38.774 \frac{T_b}{P_b} \cdot \left[\frac{P_1^2 - P_2^2 - E}{Z_{ave} \cdot T_{ave} \cdot G \cdot L} \right]^{0.5} \cdot 4D_f \log \frac{Re}{1.4126\sqrt{\frac{1}{f}}} \cdot D^{2.5} \qquad (3-44)$$

where the transmission factor is:

$$\sqrt{\frac{1}{f}} = 4 \cdot D_f \cdot \log \frac{Re}{1.4126\sqrt{\frac{1}{f}}} \qquad (3-45)$$

D_f is the drag factor that normally appears in partially turbulent flow equations and compensates for the inefficiencies due to the bends, welds, fittings, etc., and has a numerical value in the range of 0.92 to 0.97. Q_b is obtained in *SCF/D*; all other parameters are the same as in Equation 3-38.

Fully Turbulent Equations

Panhandle B

The Panhandle B equation is normally suitable for high-flow-rate, large-diameter (i.e., pipes larger than NPS 24), and high-pressure systems. The degree of accuracy depends on how precisely the pipeline efficiency is measured.

The equation has the following form in Imperial Units:

$$Q_b = 737.02 \left(\frac{T_b}{P_b}\right)^{1.02} \left[\frac{P_1^2 - P_2^2 - E}{G^{0.961} \cdot L \cdot T_{ave} \cdot Z_{ave}} \right]^{0.510} \cdot D^{2.53} \qquad (3-46)$$

where the transmission factor is:

$$\sqrt{\frac{1}{f}} = 16.49(Re)^{0.01961} \qquad (3-47)$$

or

$$\sqrt{\frac{1}{f}} = 16.70 \left(\frac{Q_b \cdot G}{D}\right)^{0.01961} \qquad (3-48)$$

The efficiency in Panhandle B equations is defined as:

$$\eta = \frac{Q_{actual}}{Q_{theoretical}} \qquad (3-49)$$

where η could be multiplied in the equation to calculate more accurate values for Q_b. All other parameters are the same as in Equation 3-38 and Q_b is in *SCF/D*.

Weymouth

The Weymouth equation is normally used for high-flow-rate, large-diameter, and high-pressure systems. This equation tends to overestimate the pressure drop predictions, and contains a lower degree of accuracy relative to the other equations. Weymouth is commonly used in distribution networks for the sake of safety in predicting pressure drop.

The Weymouth equation has the following form in Imperial Units:

$$Q_b = 432.7 \frac{T_b}{P_b} \cdot \left[\frac{P_1^2 - P_2^2 - E}{G \cdot L \cdot T_{ave} \cdot Z_{ave}} \right]^{1/2} \cdot D^{2.667} \qquad (3-50)$$

where the transmission factor is defined as

$$\sqrt{\frac{1}{f}} = 11.19 D^{1/6} \qquad (3-51)$$

Q_b is in SCF/D, and all other parameters have the same units as in Equation 3-38.

AGA Fully Turbulent

The AGA fully turbulent is the most frequently recommended and widely used equation in high-pressure, high-flow-rate systems for medium- to large-diameter pipelines. It predicts both flow and pressure drop with a high degree of accuracy, especially if the effective roughness values used in the equation have been measured accurately.

The AGA fully turbulent equation has the following form in Imperial Units:

$$Q_b = 38.774 \frac{T_b}{P_b} \cdot \left[\frac{P_1^2 - P_2^2 - E}{G \cdot L \cdot T_{ave} \cdot Z_{ave}} \right]^{0.5} \left[4 \log \frac{3.7D}{K_e} \right] \cdot D^{2.5} \qquad (3-52)$$

where the transmission factor is defined using the Nikuradse equation:

$$\sqrt{\frac{1}{f}} = 4 \log \frac{3.7D}{K_e}$$

Q_b is obtained in SCF/D, and all other parameters are the same as Equation 3-38.

Colebrook-White

This equation combines both partially turbulent and fully turbulent flow regimes and is most suitable for cases where the pipeline is operating in the transition zone. This equation is again used for large-diameter, high-pressure, and medium- to high-flow-rate systems. It predicts a higher pressure drop or lower flow rates than the AGA fully turbulent equation.

This equation has the following form in Imperial Units:

$$Q_b = 38.774 \frac{T_b}{P_b} \cdot \left[\frac{P_1^2 - P_2^2 - E}{Z_{ave} \cdot T_{ave} \cdot G \cdot L} \right]^{0.5} \left[-4 \log \left(\frac{K_e}{3.7D} + \frac{1.4126 \sqrt{\frac{1}{f}}}{Re} \right) \right] \cdot D^{2.5}$$

$$(3-53)$$

where the transmission factor is defined as:

$$\sqrt{\frac{1}{f}} = -4 \log \left(\frac{K_e}{3.7D} + \frac{1.4126 \sqrt{\frac{1}{f}}}{Re} \right) \qquad (3-54)$$

Q_b is obtained in SCF/D, and all other parameters are the same as in Equation 3-38.

Example 3.4

A gas transmission line is to be constructed to transport 1,500,000 m³/hr of natural gas from a gas refinery to the first compressor station located 100 km away. The route is almost horizontal with no considerable elevation changes. Determine the size of the pipeline required to transport the gas if the pipeline inlet pressure is 1,140 psia, and a 300 psia pressure drop is allowable. Use Weymouth, Panhandle B, and AGA fully turbulent equations to compare the diameters predicated by each flow equation. Assume an effective roughness value of $K_e = 700$ micro inches for the line.

Additional data:

$$T_{ave} = 522.6\ ^\circ R$$
$$G = 0.64$$
$$T_b = 520\ ^\circ R$$
$$P_b = 14.7\ \text{psia}$$
$$Z_{ave} = 1.0$$

Solution: Using equations in Imperial Units:

$$L = 100\ \text{km} = 62.1504\ \text{miles}$$
$$P_1 = 1,140\ \text{psia}$$
$$P_2 = 1,140 - 300 = 840\ \text{psia}$$
$$D = ?\ \text{(inches, inside diameter)}$$
$$Q = 1,500,000 \times 35.31 = 52,965,000\ SCF/HR = 1,271,160,000\ SCF/D$$
$$E = \text{zero}$$

A. Using Weymouth Equation:

$$Q_b = 432.7 \frac{T_b}{P_b} \cdot \left[\frac{P_1^2 - P_2^2 - E}{Z_{ave} \cdot T_{ave} \cdot G \cdot L} \right]^{1/2} \cdot D^{2.667}$$

upon substitution

$$ID = 37.287\ \text{inches, NPS 40 or 42}$$

B. Using Panhandle B Equation:

$$Q_b = 737.02 \left(\frac{T_b}{P_b}\right)^{1.02} \left[\frac{P_1^2 - P_2^2 - E}{G^{0.961} \cdot L \cdot T_{ave} \cdot Z_{ave}} \right]^{0.510} \cdot D^{2.53}$$

and upon substitution of data

$$ID = 35.380\ \text{inches, } NPS\ 36$$

using 100% efficiency for Panhandle B.

C. AGA fully turbulent equation:

$$Q_b = 38.774 \frac{T_b}{P_b} \cdot \left[\frac{P_1^2 - P_2^2 - E}{Z_{ave} \cdot T_{ave} \cdot G \cdot L} \right]^{0.5} \left[4 \log \frac{3.7D}{K_e} \right] \cdot D^{2.5}$$

and upon substitution of data:

$$ID = 36.765\ \text{inches, } NPS\ 40\ \text{or } 42$$

It can be concluded that the Weymouth equation is normally the one that is the most conservative type of equation. An efficiency factor is always needed for the Panhandle B to compare it with the AGA equation. In this case the efficiency has been assumed to be 1.0

for an uncoated pipe with 700 micro inches of surface roughness, which is not a practical assumption. The efficiency could be less than 95%.

SUMMARY OF THE IMPACT OF DIFFERENT GAS AND PIPELINE PARAMETERS ON THE GAS FLOW EFFICIENCY

The percentage impact of different parameters on the flow capacity of a pipeline is listed in Table 3-4 (Asante 1996). A further numerical analysis on the performance of the pipeline, or flow capacity, using three different major gas flow equations (AGA fully turbulent, Panhandle B, and Colebrook-White), together with the impact of different gas/pipeline parameters are given in Appendix B.

PRESSURE DROP CALCULATION FOR PIPELINES IN SERIES AND PARALLEL

Pipelines in Series

For pipelines in series with different diameters and lengths (see Figure 3-8), pressure drops are calculated as follows.

Using the simplified form of the general flow equation (3-31):

$$P_1^2 - P_2^2 = K_1 \, Q_b^n$$
$$P_2^2 - P_3^2 = K_2 \, Q_b^n$$
$$P_3^2 - P_4^2 = K_3 \, Q_b^n$$

Where K_1, K_2, and K_3 are pipeline resistance at each segment and n is the flow exponent depending on the type of equation.

If the three equations are added together, then

$$P_1^2 - P_4^2 = (K_1 + K_2 + K_3) Q_b^n \qquad (3-55)$$

Let

$$K_T = K_1 + K_2 + K_3 \qquad (3-56)$$

then

$$P_1^2 - P_4^2 = K_T \cdot Q_b^n \qquad (3-57)$$

Pipelines in Parallel (Looping)

Consider two different pipe segments connected in parallel, as shown in Figure 3-9.

The governing equation to calculate pressure drop for each segment would be:

$$P_1^2 - P_2^2 = K_1 \, Q_{b1}^n$$
$$P_1^2 - P_2^2 = K_2 \, Q_{b2}^n$$

where

$$Q_{b1} + Q_{b2} = Q_b$$

TABLE 3-4. Percent impact of different parameters on pipeline flow capacity (Asante 1996)

Pipe Parameters	Unit Variation of Parameter	Percent Change in Flow	Remarks	Applicable Analytical Equation
Inside Diameter	Consecutive standard pipe size	40 – 70%	40% reflects consecutive pipe sizes, such as 48/42; 70% reflects nonconsecutive sizes such as 30/24	$\frac{Q_2}{Q_1}=\frac{\log\frac{3.7D_2}{K_e}D_2^{2.5}}{\log\frac{3.7D_1}{K_e}D_1^{2.5}}\approx 1.02\left(\frac{D_2}{D_1}\right)^2$
Wall Thickness (w.t.)	10% change	<0.5%	Impact is minimal for variations usually encountered in planning. R is the ratio.	$\frac{Q_2}{Q_1}=\left[1-\left(\frac{R-1}{\frac{D_o}{2wt_1}-1}\right)\right]^{2.5}$
Pipe Grade	Consecutive standard grades	<1%		$\frac{Q_2}{Q_1}=\frac{wt_2}{wt_1}$ or $\frac{Grade\,1}{Grade\,2}$
Roughness	100 micro inch change	1.25 – 1.5%	Impact on flow is also dependent on roughness ratio	$\frac{Q_2}{Q_1}=\frac{\log\frac{3.7D}{K_{e2}}}{\log\frac{3.7D}{K_{e1}}}$
Drag Factor	1% change	1%	Drag factor varies from 0.92 – 0.98 for typical pipelines operating in the partially turbulent flow regime.	$\frac{Q_2}{Q_1}=\frac{Df_2}{Df_1}$
Gas Parameters Specific Gravity	0.01 change	0.8%	Minimal impact	$\frac{Q_2}{Q_1}=\sqrt{\frac{G_1}{G_2}}$
Compressibility	0.05 change	1.5%	For $0.825 \leq Z \leq 0.925$	$\frac{Q_2}{Q_1}=\sqrt{\frac{Z_1}{Z_2}}$
Viscosity	10% change	<1% in partially turbulent	Insignificant, minimal impact in partially turbulent flow. No effect in fully turbulent region.	—
Heat Transfer Parameters Soil Temperature	1°C change	3/D%	Change in flow is minimal	—
Soil Thermal Conductivity	50% change	<1%	Impact is minimal	—
Burial Depth	50% change	<1%	Impact is minimal	—

System Operating Parameters	Typical Unit Variation of Parameter	Percent Change in Flow	Remarks	Applicable Analytical Equation
Operating Pressure	100 kPa change (14.5 psia)	2.0 – 2.47%	For MOP* range of 5,000–10,000 kPa. The higher the MOP, the lower the impact. Optimum pressure drop assumed is 15–30 kPa/km.	$\frac{Z_2 Q_2}{Z_1 Q_1}=\frac{P_2^2-P_d^2}{P_1^2-P_d^2}$
Pressure Buffers Delivery Pressure	50 kPa change (7.25 psia) 100 kPa change (14.5 psia)	≈1% ≈1%	(Same conditions as stated above apply)	$\frac{Z_2 Q_2}{Z_1 Q_1}=\frac{P_s^2-P_2^2}{P_s^2-P_1^2}$
Operating Temperature (Avg.)	10°C change	3%	Also slightly dependent on average compressibility. May change slightly with change in gas composition.	—
Elevation Changes	100 m change (328 ft)	1.5%	—	—

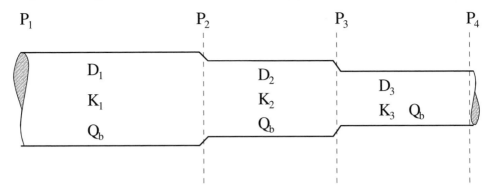

Figure 3-8. System of pipeline with different lengths and diameters connected in series

In general

$$P_1^2 - P_2^2 = K \, Q_b^n$$

where K is the total resistance of a pipe substituted for the loop. Upon rearrangement of the equations:

$$Q_{b1} = \frac{\sqrt[n]{P_1^2 - P_2^2}}{\sqrt[n]{K_1}} \qquad (3-58)$$

$$Q_{b2} = \frac{\sqrt[n]{P_1^2 - P_2^2}}{\sqrt[n]{K_2}} \qquad (3-59)$$

$$Q_b = \frac{\sqrt[n]{P_1^2 - P_2^2}}{\sqrt[n]{K}} \qquad (3-60)$$

now substitute the values for Q_{b1}, Q_{b2}, and Q_b:

$$\frac{\sqrt[n]{P_1^2 - P_2^2}}{\sqrt[n]{K}} = \frac{\sqrt[n]{P_1^2 - P_2^2}}{\sqrt[n]{K_1}} + \frac{\sqrt[n]{P_1^2 - P_2^2}}{\sqrt[n]{K_2}}$$

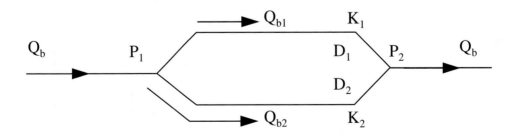

Figure 3-9. System of pipelines connected in parallel (looped)

$$\frac{1}{\sqrt[n]{K}} = \frac{\sqrt[n]{K_1} + \sqrt[n]{K_2}}{\sqrt[n]{K_1} \cdot \sqrt[n]{K_2}}$$

or

$$K = \frac{K_1 \cdot K_2}{\left(K_1^{1/n} + K_2^{1/n}\right)^n} \qquad (3-61)$$

For $n = 2$, the following equation giving total resistance of two pipelines in parallel is obtained:

$$K = \frac{K_1 \cdot K_2}{\left(\sqrt{K_1} + \sqrt{K_2}\right)^2} \qquad (3-62)$$

In the above equation, K is the total resistance of the two pipelines looped together. If the two pipes in parallel have equal diameters, then $K = 1/4\ K_1$, which means that the resistance of the looped system against flow is equal to 1/4 of a single line.

Example 3.5

What would be the downstream pressure of a gas pipeline transporting 200,000 m^3/hr of natural gas ($G = 0.65$)? Inlet pressure to the pipeline is $P_1 = 1{,}000$ psia, $L = 200$ miles, $D_1 = 20$ inches (19.5 inches inside diameter), and average flowing gas temperature is assumed to be $T_f = 520\ °R$ and $Z_{avg} = 1.0$. What would be the downstream pressure if the existing system is looped with a 16-inch pipeline (inside diameter 15.5 inches)? Use the Panhandle A flow equation.

Data given:
$P_1 = 1{,}000$ psia
$G = 0.65$
$D_1 = 20$ inch (inside diameter = 19.5 inches)
$L = 200$ miles
$Q_b = 200{,}000$ m^3/hr
$D_2 = 16$ inch (inside diameter = 15.5 inches)
$T_f = 520\ °R$
$P_2 = ?$ (psia)
The simplified form of the general flow equation will be used:

$$P_1^2 - P_2^2 = K_1\ Q_b^n$$

with the Panhandle A flow equation, pipeline resistance and flow exponent can be obtained from Tables 3-2 and 3-3:

$$R = 2.552 \times 10^{-4} \cdot T_f \cdot Z_{avg} \cdot \frac{G^{0.855}}{D^{4.856}}$$

$$n = 1.855$$

where R = pipeline resistance per foot
T_f = average flowing gas temperature in $°R$
G = gas gravity
D = pipe inside diameter in inches
n = flow exponent

P_1 = inlet pressure, psia
P_2 = exit pressure, psia
K = total pipeline resistance, $K = RL$ (L is in ft)
Q_b = gas flow rate at base condition and it should be in MCF/hr.

Therefore

$$K_1 = RL = 2.552 \times 10^{-4} \cdot 520 \frac{(0.65)^{0.855}}{(19.5)^{4.856}} \cdot 200 \cdot 5280$$

$$K_1 = 0.052745$$
$$Q_b = 200,000 \cdot 35.31 = 7,062 \text{ MCF/ hr}$$

$$P_1^2 - P_2^2 = K_1 \, Q_b^n$$
$$(1,000)^2 - P_2^2 = 0.052745(7062)^{1.855}$$
$$P_2 = 522 \text{ psia}$$

Now if the system is looped:
Total resistance of the second pipe is:

$$K_2 = R_2 \cdot L = 2.55 \times 10^{-4} \cdot 520 \frac{(0.65)^{0.855}}{(15.5)^{4.856}} \cdot 200 \cdot 5280$$

$$K_2 = 0.160728$$

For a looped system, the total resistance is:

$$K_T = \frac{K_1 \cdot K_2}{\left(K_1^{1/n} + K_2^{1/n}\right)^n}$$

where n = 1.855.

$$K_T = \frac{(0.052745)(0.160728)}{\left(0.052745^{1/1.855} + 0.160728^{1/1.855}\right)^{1.855}}$$

$$K_T = 0.023430$$

$$P_1^2 - P_2^2 = K_T \, Q_b^{1.855}$$
$$(1,000)^2 - P_2^2 = 0.023430(7062)^{1.855}$$
$$P_2 = 823 \text{ psia (downstream pressure after looping)}$$

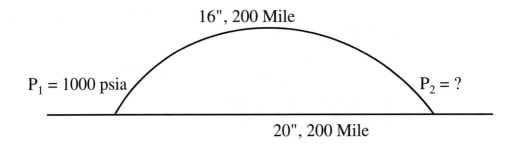

Figure 3-10. Pipeline system after looping

Pipeline Segmental Looping

In many cases, it may not be necessary to loop the entire pipeline to obtain the desired flow or downstream pressure. Therefore, only a segment of the pipeline is looped to meet requirements.

Assume that the existing line has length L, diameter D_1, and a total resistance of $K_1 + K_1'$, where inlet and exit pressures are P_1 and P_2, respectively. It is intended to increase the existing gas flow rate from Q_1 to Q_2 (i.e., Q_{b1} to Q_{b2}) without any changes in downstream pressure. A pipeline loop with diameter D_2 and length X will be added to the existing pipeline in order to increase Q_1 to Q_2 without any changes to the downstream pressure. The value of X, the length of the pipeline to be looped to the existing system, must be determined. Note that using, larger diameter pipes will reduce the required length of the segment to be looped.

To obtain the total pipeline resistance, start with one of the major transmission equations and continue to develop the equation to calculate the length of the loop. For this example, the Weymouth equation will be used.

The Weymouth equation in Imperial Units, is written as:

$$P_1^2 - P_2^2 = \frac{0.000466 \, G \cdot T_f \cdot L}{D_1^{16/3}} \cdot Q_1^2$$

or

$$P_1^2 - P_2^2 = K \cdot Q_1^2$$

and $K = K_1 + K_1'$ (total resistance of the single line)

In the total resistance formula of the Weymouth equation, the value of 0.000466. GT_f is a constant, which could be assumed as:

$$C = 0.000466 \cdot G T_f$$

then

$$K_1' = \frac{C \cdot X}{D_1^{16/3}}, \; K_2' = \frac{C \cdot X}{D_2^{16/3}}, \; K_1 = \frac{C(L - X)}{D_1^{16/3}}, \; \text{and} \; K = \frac{C \cdot L}{D_1^{16/3}}$$

The equivalent resistance for the looped segment is:

$$K_e = \frac{\dfrac{C \cdot X}{D_1^{16/3}} \cdot \dfrac{C \cdot X}{D_2^{16/3}}}{\left(\sqrt{\dfrac{C \cdot X}{D_1^{16/3}}} + \sqrt{\dfrac{C \cdot X}{D_2^{16/3}}} \right)^2} \qquad (3-63)$$

after simplification K_e would be:

$$K_e = \frac{C \cdot X}{\left(D_1^{8/3} + D_2^{8/3} \right)^2} \qquad (3-64)$$

and

$$K_{E \; Total} = K_e + K_1 \; (\text{i.e., pipes in series})$$

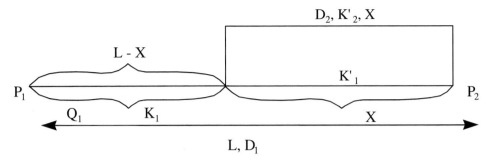

Figure 3-11. Pipeline segmental looping

then

$$K_E = \frac{C \cdot X}{\left(D_1^{8/3} + D_2^{8/3}\right)^2} + \frac{C(L - X)}{D_1^{16/3}} \qquad (3 - 65)$$

and

$$P_1^2 - P_2^2 = K_E \cdot Q_2^2$$

Dividing the flow equations for the existing pipeline and after segmental looping:

$$\frac{P_1^2 - P_2^2}{P_1^2 - P_2^2} = \frac{\dfrac{CL}{D_1^{16/3}}}{K_E} \cdot \left(\frac{Q_1}{Q_2}\right)^2$$

or

$$1 = \frac{\dfrac{CL}{D_1^{16/3}}}{\dfrac{C \cdot X}{\left(D_1^{8/3} + D_2^{8/3}\right)^2} + \dfrac{C(L-X)}{D_1^{16/3}}} \cdot \left(\frac{Q_1}{Q_2}\right)^2$$

Finally, the equation could be written as follows:

$$X = L \cdot \frac{\left(Q_1/Q_2\right)^2 - 1}{\left[\dfrac{1}{1+\left(\dfrac{D_2}{D_1}\right)^{8/3}}\right]^2 - 1} \qquad (3 - 66)$$

X = length of the pipeline to be looped, miles
L = length of the existing pipeline, miles
Q_1 = initial gas flow rate, MMSCFD
Q_2 = final gas flow rate, MMSCFD
D_1 = existing pipeline inside diameter, inches
D_2 = looped segment inside diameter, inches

For AGA fully turbulent equation, a value of 2.5 instead of 8/3 is used as the exponent in the denominator of Equation (3-66).

Example 3.6

A gas transmission line of OD = 16 inches (ID = 15.5 inches), Q_1 = 70,000 m³/hr, and L = 300 km is to be used without any pressure changes in the delivery of 120,000 m³/hr of natural gas. What would be the length of pipeline with identical diameter that should be looped to the existing pipeline to satisfy the increased capacity (according to the Weymouth equation).

Solution:

Using Equation (3-66).

Q_1 = 70,000 m³/hr
Q_2 = 120,000 m³/hr
D_1 = 15.5 inches
D_2 = 15.5 inches
L = 300 km
X = ? km

$$X = 300 \frac{\left(\frac{70.000}{120.000}\right)^2 - 1}{\left(\frac{1}{1+\left(\frac{15.5}{15.5}\right)^{8/3}}\right)^2 - 1}$$

$$X = 300 \frac{(7/12)^2 - 1}{\left(\frac{1}{2}\right)^2 - 1} = 263.88 \text{ km, segment to be looped}$$

Some Important Considerations Regarding Pipeline Looping

Equation (3-66) demonstrates that looping will increase pipeline flow capacity without any changes to the upstream and downstream pressures. Likewise, if the flow is kept constant, adding a loop results in less of a pressure drop along the pipeline. However, Equation (3-66) also implies that the impact of looping on the flow capacity (or pressure drop along the pipeline) is independent of the location of the loop. In practical pipeline operations this is not the case, and the placement of the looping can have a significant impact on the response of the system. The behavior of the system is greatly affected by changes in temperature and the compressibility factor of the gas along the pipeline.

There are two important parameters to consider when choosing the location for a pipeline loop: temperature and pressure. Considering pipeline pressure, the magnitude of the pressure drop is higher at the downstream section of the pipeline because the gas has expanded. Hence, considering pressure alone, looping at the downstream portion of the pipeline is more efficient.

However, temperature must also be considered. At the upstream part of the pipeline, particularly downstream of a compressor station, the gas temperature is typically much higher than in other places along the line. Adding a loop in areas where the gas temperature is hotter increases the heat transfer from the pipeline to the immediate environment. This is especially true if the ground temperature is significantly less than the gas temperature. When higher rates of cooling occur, the pressure drop along the pipeline is considerably less. Therefore, it is most often recommended that under steady-state conditions, pipeline looping be in an upstream region, such as immediately downstream of a compressor station, especially if the gas is hot. It should be noted that a comprehensive simulation involving a gas temperature profile giving consideration to elevation changes is always necessary to determine the exact location of looping in steady-state operations. Typically, in situations where the gas temperature is very close to the ground temperature, or the difference is less than 5 – 10°C, temperature is no longer an important consideration when choosing a location for a loop.

In the following example, the results of hydraulic simulations for three different cases are presented to further clarify the effects of different loop locations for a pipeline at steady state. In all simulations, Hydraulic Analysis and Resources Tool (HART) simulation software was used. It is TransCanada Pipeline's steady-state simulation software developed in-house for use in designing the company's pipeline network.

Parameters for Case Study:

- Pipeline length = 100 km (62.15 miles)
- Gas flow rate is constant = 289.542 MMSCFD
- Gas inlet pressure = 1,200 psia
- Gas inlet temperature = 45°C or 113°F
- Pipeline *OD* = 20 inches
- Pipeline *ID* = 19.44 inches
- Pipeline roughness = 750 μ inches
- Soil temperature = 10°C (50°F)

Case I: No Loop

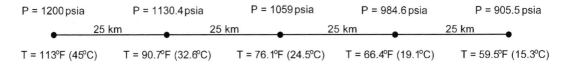

Case II: Looping First 25 Km of the Pipeline (Upstream)

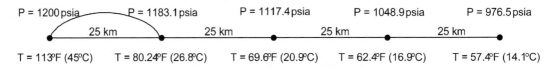

Case III: Looping Last 25 Km of the Pipeline (Downstream)

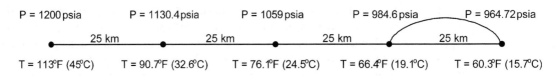

For these three cases, it can be concluded that for the same gas flow rate, looping upstream of the pipeline gives the highest delivery pressure or the least pressure drop.

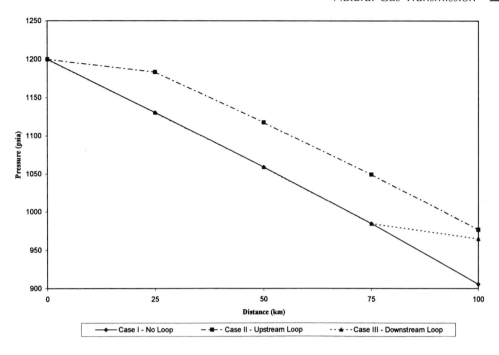

Figure 3-12. Comparison of pressure drops for cases I, II, and III

The collective results of the above cases are compared in Figure 3-12.

PIPELINE GAS VELOCITY

The equation to determine the gas velocity in a pipeline is obtained as follows:

$$u_s = Q_S/A \qquad (3-67)$$

where u_s = gas velocity at any section
$\quad\quad\quad Q_s$ = gas flow rate at any section
$\quad\quad\quad A$ = cross sectional area.
at steady state

$$Q_b \cdot \rho_b = Q_s \cdot \rho_s$$

and also

$$\rho_s = \frac{P_s \cdot M}{R \cdot T_S}; \; \rho_b = \frac{P_b \cdot M}{R \cdot T_b}$$

or

$$\frac{\rho_b}{\rho_S} = \frac{P_b \cdot T_S}{P_S \cdot T_b}$$

(consider the compressibility factor as 1).

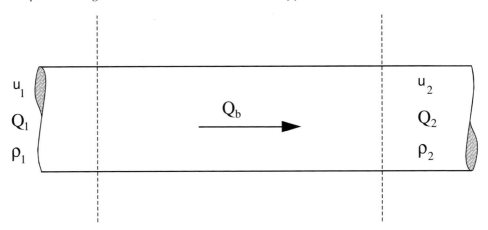

Figure 3-13. Change of gas velocity in a gas transmission line

Combining these expressions, where $A = \frac{\pi D^2}{4}$

$$u_s = \frac{Q_b}{\frac{\pi D^2}{4}} \cdot \frac{P_b \cdot T_S}{T_b \cdot P_S}$$

Substitute $P_b = 14.7$ psia, $T_b = 520\ °R$, and assume a flowing gas temperature of $T_s = 520\ °R$:

$$u_s = 0.75 \frac{Q_b}{P \cdot D^2} \qquad (3-68)$$

where u_s = gas velocity at any segment, ft/sec
       Q_b = gas flow rate at base condition, ft³/hr
       P = pressure at any section, psia
       D = pipeline inside diameter, inches.

Equation (3-68) gives a good estimate of the gas velocity in the pipeline at any segment. If the flowing gas temperature is different from T_b, then:

$$u_s = 0.75 \frac{T_f}{T_b} \cdot \frac{Q_b}{P \cdot D^2} \qquad (3-69)$$

where T_f = flowing gas temperature, $°R$
       T_b = base temperature, 520 °R.

If the effect of the compressibility factor is also considered at any segment, the gas velocity equation could be written as follows:

$$u_s = 0.75 \frac{T_f}{T_b} \cdot \frac{Q_b \cdot Z}{P \cdot D^2} \qquad (3-70)$$

or

$$u_s = 1.44 \times 10^{-3} \frac{Q_b \cdot Z \cdot T_f}{P \cdot D^2} \qquad (3-71)$$

where Z is the compressibility factor at any section.

Example 3.7

What would be the maximum and minimum velocity of gas in a pipeline

where $Q_b = 1,500,000$ m³/hr
$D = 56$ inches
$ID = 55$ inches
$P_1 = 1,140$ psia
$P_2 = 840$ psia
$T_f = T_b = 520$ °R

$$u_s = 0.75 \left(Q_b / P \cdot D^2 \right).$$

Minimum velocity occurs at the beginning of the pipe, where the pressure is higher:

$$u_s = 0.75 \frac{1,500,000 \cdot 35.31}{1,140 \cdot (55)^2} = 11.52 \, \text{ft/sec}$$

and maximum velocity occurs at the end of the pipe, where the pressure is less:

$$u_s = 0.75 \frac{1,500,000 \cdot 35.31}{840 \cdot (55)^2} = 15.63 \, \text{ft/sec}$$

It is sometimes necessary to calculate gas velocity in miles/hr while using gas flow rate in MMSCFD. The following equation can be used with these units:

$$u_s = 40.909 \frac{Q_b \cdot Z \cdot T_f}{P \cdot D^2} \qquad (3-72)$$

where u_s = gas velocity at any segment, miles/hr
Q_b = gas flow rate at standard condition, MMSCFD
Z = compressibility factor, dimensionless
T_f = flowing gas temperature, °R
P = pipeline pressure at any segment, psia
D = pipe inside diameter, inches.

EROSIONAL VELOCITY

When a fluid passes through a pipeline with a high velocity it can cause both vibration and erosion in the pipeline, which will erode the pipe wall over time. If the gas velocity exceeds the erosional velocity calculated for the pipeline, the erosion of the pipe wall is increased to rates that can significantly reduce the life of the pipeline. Therefore, it is always necessary to control gas velocity in gas transmission lines to prevent it from rising above this limit.

The erosional velocity for compressible fluids is expressed as:

$$u_e = \frac{C}{\rho^{0.5}} \qquad (3-73)$$

where, in Imperial Units,
u_e = erosional velocity, ft/sec
ρ = gas density, lb$_m$/ft³.

and C is a constant defined as $75 < C < 150$. The recommended value for C in gas transmission pipelines is $C = 100$ (see Beggs 1991).

Gas density is obtained by:

$$\rho = \frac{P \cdot M}{Z \cdot R \cdot T}$$

Substituting for $C = 100$ and $M = 29\,G$:

$$u_e = \frac{100}{\sqrt{\dfrac{29G \cdot P}{Z \cdot R \cdot T}}} \qquad (3-74)$$

In the above equation:

u_e = erosional velocity, ft/sec
G = gas gravity, dimensionless
P = minimum pipeline pressure, psia
Z = compressibility factor at the specified pressure and temperature, dimensionless
T = flowing gas temperature, °R
R = 10.73 (ft^3 × psia/lb moles ×°R).

The recommended value for the gas velocity in gas transmission mainlines is normally 40% to 50% of the erosional velocity (i.e., a value of 10–13 m/s or 33–43 ft/sec is an acceptable value for design purposes). This value could be increased to 15–17 m/s for nonmajor mainlines or laterals.

Example 3.8

What would be the erosional, maximum, and minimum velocities in a gas pipeline if the gas inlet pressure is 1,000 psia, gas exit pressure is 700 psia, gas gravity is 0.65, compressibility factor is 0.9, isothermal flowing gas temperature is 520 °R, pipe $OD = 56$ inches ($ID = 55$ inches), and gas flow rate is assumed to be $Q_b = 1,500,000$ m^3/hr.

Solution

Using the equation for erosional velocity:

$$u_e = \frac{100}{\sqrt{\dfrac{29GP}{Z \cdot R \cdot T}}}$$

In this equation

$G = 0.65$
$P = 700$ psia (minimum pipeline pressure)
$Z = 0.9$
$R = 10.73$
$T = 520.$

Then erosional velocity is:

$$u_e = \frac{100}{\sqrt{\dfrac{29 \cdot 0.65 \cdot 700}{0.9 \cdot 10.73 \cdot 520}}} = \frac{100}{1.621} = 61.7\,\text{ft/s (erosional velocity)}$$

maximum velocity in the pipeline is:

$$u_{max} = \frac{0.75\,Q_b Z}{P \cdot D^2} = \frac{0.75 \cdot 1{,}500{,}000 \cdot 35.31 \cdot 0.9}{700 \cdot (55)^2}$$

$$u_{max} = 16.9\,\text{ft/s}$$

minimum velocity is:

$$u_{min} = \frac{0.75\,Q_b Z}{P D^2} = \frac{0.75 \cdot 1{,}500{,}000 \cdot 35.31 \cdot 0.9}{1{,}000 \cdot (55)^2}$$

$$u_{min} = 11.8\,\text{ft/s}$$

Since the maximum velocity is considerably lower than the erosional velocity, the system is in the safe velocity region.

Erosional Gas Flow Rate

The erosional gas flow rate is defined based on the pipeline erosional velocity as:

$$Q_e = u_e \cdot A \qquad\qquad (3-75)$$

where Q_e = erosional gas flow rate, ft³/sec
$\quad u_e$ = erosional velocity, ft/sec
$\quad A$ = pipeline cross-sectional area, ft².

OPTIMUM PRESSURE DROP FOR DESIGN PURPOSES

The optimum pressure drop per unit length of pipe is an important factor used to design the most cost-effective system. Maintaining the optimal pressure drop along each section of the pipeline system is necessary to minimize the required facilities and operating expenses (including pipeline, compressor, and fuel-consumption costs).

Studies performed by the Pipeline System Design department of TransCanada Pipelines have proven that a pressure drop of 15 – 25 kPa/km (3.5 – 5.85 psi/mile) is optimal. This means that, when the final system design is complete, the pressure drop in all sections of the pipeline system should be within this range. Pressure drops in excess of 25 kPa/km will cause the downstream compressors to work at a greater load factor, which will result in higher fuel costs. Excessive pressure drops also introduce a greater potential for operating problems. Pressure drops below 15 kPa/km are an indication that too many facilities have been installed. For further information see (Hughes 1993).

PIPELINE PACKING

A gas pipeline, which transports gas from point 1 to point 2 with pressure P_1 and P_2, respectively, will have some natural gas "packed" inside, at an average pressure of P_{ave}. The volume of the gas packed inside the pipe can be determined using the following equation:

$$P_{ave} \cdot V = n_T \cdot Z_{ave} \cdot R \cdot T_{ave} \qquad\qquad (3-76)$$

where

$$P_{\text{ave}} = \frac{2}{3}\left[P_1 + P_2 - \frac{P_1 \cdot P_2}{P_1 + P_2}\right]$$
$$T_{\text{ave}} = \frac{T_1 + T_2}{2}$$

$$V = \frac{\pi D^2}{4} \cdot L$$

(Z_{ave} can be obtained from Kay's rule)

where P_{ave} = average pressure, psia
        D = pipe inside diameter, ft
        L = pipeline length, ft.
        n_T = total number of moles of gas, lb. moles
        Z_{ave} = average compressibility factor, dimensionless
        R = gas constant, 10.73 (ft$^3 \times$ psia/lb moles$\times$°R)
        T_{ave} = average gas temperature, °R.

Then the total number of moles n_T packed between points 1 and 2 at average pipeline conditions is:

$$n_T = \frac{\pi D^2 \cdot P_{\text{ave}} \cdot L}{4 \cdot Z_{\text{ave}} R \cdot T_{\text{ave}}} \qquad (3-77)$$

This value can then be used to determine the gas volume V_b, existing in the pipeline at base condition (i.e., $P = 14.7$ psia and $T = 520$ °R):

$$V_b = \frac{n_T R T_b}{P_b}$$

$$V_b = \frac{n_T \cdot 10.73 \cdot 520}{14.7}$$

For a fairly accurate calculation of the storage capacity of pipelines at packed and unpacked conditions when gas is flowing, the Clinedinst equation can be used, which considers the variation in gas compressibility. For further information see (Katz 1959).

DETERMINING GAS LEAKAGE USING PRESSURE DROP METHOD

The Pressure Drop Method can be used to determine the volume of gas that will escape from a pipeline due to leakage. Consider the pipeline in Figure 3-14 at two different time intervals.
The number of moles of gas in the pipeline at time t = zero is:

$$P_1 \cdot V_1 = n_1 \cdot R \cdot T_1$$

where P_1 = initial pressure, psia
        V_1 = initial gas volume, ft^3
        n_1 = initial number of moles, lb.moles
        R = gas constant, 10.73 (psia$\times$ft^3/lb moles$\times$°R)
        T_1 = initial gas temperature, °R.

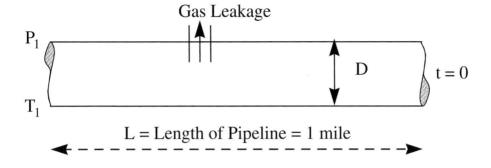

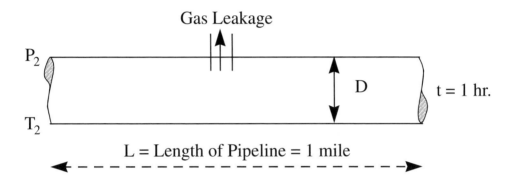

Figure 3-14. Leakage in a gas pipeline

Assuming that the pipeline inside diameter is D (ft), and the pipeline length is L (ft), then

$$P_1 \cdot \frac{\pi D^2}{4} \cdot L = n_1 \cdot R \cdot T_1$$

where

$$\frac{\pi D^2}{4} \cdot L = V_1 \ (\text{initial gas volume in the pipeline})$$

therefore, the number of moles of gas initially in the line is

$$n_1 = \frac{\pi D^2 \cdot P_1 \cdot L}{4 \cdot R \cdot T_1} \qquad (3-78)$$

After one hour, due to leakage the volume of gas in the pipeline is reduced to

$$P_2 \cdot V_2 = n_2 \cdot R \cdot T_2$$

and

$$P_2 \cdot \frac{\pi D^2}{4} \cdot L = n_2 \cdot R \cdot T_2$$

or

$$n_2 = \frac{\pi D^2 \cdot P_2 \cdot L}{4 \cdot R \cdot T_2} \qquad (3-79)$$

The amount of gas that has escaped from the pipeline due to leakage is n and it is equal to:

$$n = n_1 - n_2$$

substitute for n_1 and n_2:

$$n = \frac{\pi D^2 \cdot L}{4 \cdot R} \left(\frac{P_1}{T_1} - \frac{P_2}{T_2} \right) \qquad (3-80)$$

The volume of gas that has escaped to the atmosphere at standard condition (SC) would be:

$$V_b = \frac{n R T_b}{P_b}$$

where P_b = gas pressure at SC, 14.7 psia
V_b = leaked gas volume at SC, ft^3
n = number of moles of leaked gas, lb moles
R = gas constant, 10.73 (psia×ft^3/lb moles×°R)
T_b = gas temperature at SC, 520 °R.

therefore

$$\frac{P_b \cdot V_b}{R \cdot T_b} = \frac{\pi D^2}{4} \cdot L \cdot \frac{1}{R} \cdot \left(\frac{P_1}{T_1} - \frac{P_2}{T_2} \right) \qquad (3-81)$$

cancel R and substitute the constants:

$$\frac{14.7 \cdot V_b}{520} = \frac{\pi D^2}{4} \cdot 5,280 \left(\frac{P_1}{T_1} - \frac{P_2}{T_2} \right)$$

If the length of the pipeline is $L = 1$ mile (5,280 ft) and one year is assumed to be 8,640 hrs, then

$$V_b = \frac{\pi D^2}{4} \cdot \frac{5,280 \cdot 520 \cdot 8,640}{144 \cdot 1,000,000 \cdot 14.7} \left(\frac{P_1}{T_1} - \frac{P_2}{T_2} \right)$$

Then the gas volume leaked in one year is

$$V_b = 8.797 D^2 \left(\frac{P_1}{T_1} - \frac{P_2}{T_2} \right) \qquad (3-82)$$

where V_b = gas volume leaked, MMSCF/year
D = pipeline ID, inches
P_1 = pipeline initial pressure, psia
P_2 = pipeline final pressure, psia
T_1 = pipeline initial temperature, °R
T_2 = pipeline final temperature, °R.

The above equation gives the volume of gas escaped due to leakage from one mile of pipeline over a period of one year.

For a more accurate analysis, the compressibility of gas at both initial and final states should be considered, which results in the following equation:

$$V_b = 8.797D^2 \left(\frac{P_1}{Z_1 T_1} - \frac{P_2}{Z_2 T_2} \right) \qquad (3-83)$$

where Z_1 = compressibility factor at initial state, dimensionless
Z_2 = compressibility factor at final state, dimensionless.

And if the change in gas pressure and temperature before and after leakage is small, then $Z_1 = Z_2 = Z$, therefore

$$V_b = \frac{8.797D^2}{Z} \left(\frac{P_1}{T_1} - \frac{P_2}{T_2} \right) \qquad (3-84)$$

where Z is the average compressibility factor at , $T_{ave} = (T_1 + T_2)/2$, and $P_{ave} = (P_1 + P_2)/2$.

WALL THICKNESS/PIPE GRADE

In general, system hydraulic designers are concerned with optimizing facilities sizing based on leased capital and operating costs. Hence they need to understand the interrelationships between pipe diameter, pipe grade/wall thickness, pressure, horsepower, flow, and economy factors.

Formulations for pipe wall thickness and grade selections based on pressure and other loads, detailed in Chapter 7 ("Pipeline Mechanical Design"), are repeated below to some extent to give system designers a better understanding of their impact on system optimization.

Both U.S. and Canadian Codes are referred to so as to provide an understanding of code differences to hydraulic system designers.

The wall thickness of gas transmission pipelines varies with the pipe grade, location, and design pressure. The design pressure, which is specified by the system designers for any specific location, should not be less than the maximum operating pressure (MOP) of the pipeline at the location where all the forces are considered. The pipe wall thickness and material selected by the designer should provide adequate strength to prevent deformation and collapse by handling stresses, external reactions, and thermal expansions and contractions.

The designer should also consider the low temperature properties of materials (refer to Chapter 8, "Materials Selection and Quality Management"), since they are likely to be exposed to low-temperature media during installation, pressure testing, start-up, and operation. According to the pipeline codes (ANSI - ASME B31.8 or CSA Z662-96), the stress design requirements to be considered are limited to normal design conditions for operating pressure, thermal expansion and contraction ranges, temperature differential, and other forces acting on the pipeline. Additional loadings could include:

1. Occasional extreme loads (earthquake)
2. Slope movements
3. Fault movements
4. Seismic-related earth movements
5. Thaw settlements

6. Frost heave
7. Loss of support
8. Cyclical traffic loads
9. Construction and maintenance deformations
10. Mechanical vibrations
11. Hydraulic shock
12. Vortex shedding

These loadings require supplemental design criteria (such as heavier wall thickness or stronger material) to ensure a safe and operational pipeline.

Relationship between Design Pressure and Wall Thickness

The equation that relates design pressure to wall thickness can be derived by performing a force balance on a pipe segment under a specified design pressure, as shown in Figure 3-15.

Force F_1 that is exerted on the pipe wall due to the design pressure is:

$$F_1 = \pi \cdot D_o \cdot L \cdot P_{\text{design}}$$

Force F_2 is the pipeline specified minimum yield strength over the specified thickness

$$F_2 = S[\pi D_o L - \pi D_i L]$$

and

$$F_2 = S[\pi(D_i + 2t)L - \pi D_i L]$$

after simplifications

$$F_2 = 2\pi L S t$$

to balance the forces $F_1 = F_2$, or

$$\pi D_o L P_{\text{design}} = 2\pi L S t$$

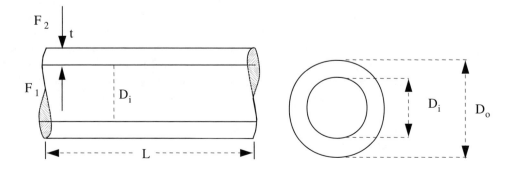

Figure 3-15. Representation of forces acting on a segment of pipeline

which gives P_{design} as:

$$P_{design} = \frac{2St}{D_o} \qquad (3-85)$$

in this equation:

P_{design} = pipeline design pressure, psia
S = specified minimum yield strength of the pipe, psia
t = pipeline thickness, inches
D_o = pipeline outside diameter, inches.

The design pressure for a given wall thickness, considering all these safety factors, will then be determined by

$$P_{design} = \frac{2St}{D_o} \cdot F \cdot L \cdot J \cdot T \qquad (3-86)$$

where
F = design factor
L = location factor
J = joint factor
T = temperature correction factor or temperature derating factor.

The Canadian Standards Association (CSA) recommends the following values:

- The design factor (F) = 0.80:
- The location factor (L) depends on both population and other factors, such as roads, railways, and stations:

Class 1	Deserted	$L = 1.00$
Class 2	Village	$L = 0.90$
Class 3	City	$L = 0.70$
Class 4	City (densely populated)	$L = 0.55$

The values of location factors (L) between ASME and CSA Codes are compared below, see also Chapter 7, Table 7-2:

	CSA Recommended Values	ASME Recommended Values
Class 1	0.80	0.72
Class 2	0.72	0.60
Class 3	0.56	0.50
Class 4	0.44	0.40

- The joint or welding factor is given as follows:

Pipe Type	Joint Factor
Seamless	1.00
Electric welded	1.00
Submerged arc welded	1.00
Furnace butt welded	0.60

TABLE 3-5. Temperature correction factor for gas transmission lines [ASME, B31.8]

Temperature (°F)	Temperature Correction Factor
Up to 250	1.00
300	0.97
350	0.93
400	0.91
450	0.87

- Temperature correction (or derating) factor for different temperatures are given in Table 3-5. In gas transmission lines or distribution networks where the temperature is normally kept below 250°F (120°C), the temperature correction factor is taken as unity.

The minimum wall thickness for steel pipes is given in Table 3-6 for both mainline pipe and compressor station piping (CSA Z245.1-M90, standards for steel pipe).

* The minimum wall thickness is consistent with CAN/CSA-Z245.1-M90 standards for steel pipe, and are based on the ability of the pipe to resist handling damage, buckling, and wrinkling when subjected to field bending and welding.

TABLE 3-6. Minimum wall thickness for steel pipe [courtesy of CSA, Z662-96]

Nominal Pipe Size (NPS)	Pipe O.D. (inch)	Minimum Wall Thickness (inches*)	
		Line Pipe	Compressor Station Pipe
3	3.5	0.126	0.217
4	4.5	0.126	0.236
6	6.626	0.126	0.280
8	8.626	0.157	0.323
10	10.752	0.189	0.366
12	12.752	0.189	0.406
14	14	0.209	0.437
16	16	0.220	0.500
18	18	0.220	0.500
20	20	0.220	0.500
22	22	0.236	0.500
24	24	0.252	0.500
26	26	0.264	0.500
30	30	0.287	0.500
34	34	0.311	0.500
36	36	0.323	0.500
42	42	0.354	0.500
48	48	0.402	0.500

TABLE 3-7. API standard for 5LX type pipe

Specification	Minimum Specified Yield Strength (psi)
API 5LX Grade X42	42,000
API 5LX Grade X46	46,000
API 5LX Grade X52	52,000
API 5LX Grade X56	56,000
API 5LX Grade X60	60,000
API 5LX Grade X65	65,000
API 5LX Grade X70	70,000
API 5LX Grade X80	80,000
API 5LX Grade X90	90,000

Example 3.9

What is the design pressure for a NPS 42 pipeline with a wall thickness of 0.354 inches and a location factor of 0.90. The pipeline is of the type 5LX-X80, and the flowing gas temperature is less than 250°F. Assume a design factor of $F = 0.80$.

Solution

The 5LX type of pipes have a welding or joint factor of 1.0 and the X80 grade has a minimum specified yield strength of 80,000 psi:

The following equation is applicable for pipe grade 5LX-X80, joint factor $J = 1.0$, and the value of $S = 80,000$ psia.

$$P_{\text{design}} = \frac{2\,St}{D_o}\, F \cdot L \cdot J \cdot T$$

where $P_{\text{design}} = ?$, psia
$S = 80,000$ psia
$t = 0.354$ inches
$D_o = 42$ inches
$F = 0.80$
$L = 0.90$
$J = 1.0$
$T = 1.0$

$$P_{\text{design}} = (2 \cdot 80,000 \cdot 0.354/42) \cdot 0.80 \cdot 0.90 \cdot 1.0 \cdot 1.0 = 971 \text{ psi}.$$

The equivalent value of ASME B.31.8 would be:

$$P_{\text{design}} = (2 \cdot 80,000 \cdot 0.354/42) \cdot 0.60 \cdot 1.0 \cdot 1.0 = 809 \text{ psi}$$

Weight of Steel

In gas transmission lines 50–55 % of the cost of the project is the weight of steel used in the construction of the pipeline. Therefore, it is important to have a good estimate of the amount of steel that will be required. From the following equations, the weight and cost of steel for a pipeline can be calculated.

Consider a pipeline where $OD = D_1$, $ID = D_2$, pipeline length $= L$, and density of steel $= \rho$, then the volume V_s and weight W_s of the steel are:

$$V_s = \frac{\pi}{4} \cdot L \left(D_1^2 - D_2^2 \right) \qquad (3-87)$$

and

$$W_s = \frac{\pi}{4} \cdot L \left(D_1^2 - D_2^2 \right) \rho \qquad (3-88)$$

or

$$W_s = \frac{\pi}{4} \cdot L \cdot \rho \cdot (D_1 - D_2)(D_1 + D_2)$$

since $D_1 - D_2 = 2t$ (t is the pipe thickness)

$$W_s = \pi L \rho t (D_2 + t). \qquad (3-89)$$

From this equation, it is obvious that pipeline thickness has a significant impact on the weight of steel. Consequently, for high-pressure gas transmission lines, designers may spend more money to use a high grade of pipe. The higher grade of pipe permits a smaller wall thickness, and the savings on the weight of steel greatly outweigh the money spent for an advanced grade of pipe. For example, when there is a shift from pipe grade X70 to X80, a material cost increase of 2–3% is observed, while at the same time a savings of 12–13% can be obtained due to the reduction of the thickness of the pipe. (Later in the text, when discussing construction techniques, additional factors such as handling thin wall pipe and advanced joining/welding procedures are considered).

Example 3.10

What would be the difference in the cost of steel for a NPS 42, $L = 900$ km gas transmission line using X70 or X80 pipe grades. Assume that the maximum operating pressure of the line is 1,440 psia, L Factor $= 0.90$, and D Factor $= 0.80$. The density of both X70 and X80 steel is $\rho \approx 7.801$ ton/m^3.

Additional data:

Cost of pipe in place: X70 @ $1,000/ton; X80 @ $1,030/ton

Solution
Case I, X70

Pipe thickness for X70 grade is:

$$t = \frac{D \cdot P}{2 \cdot S \cdot F \cdot L} = \frac{42 \times 1,440}{2 \times 70,000 \times 0.72} = 0.60 \text{ inch}$$

Pipe inside diameter $(D_2) = 42 - 2 \cdot 0.60 = 40.8$ inches

$$W_s = \pi \cdot L \cdot \rho \cdot t \cdot (D_2 + t)$$

$$W_s = \pi \cdot 900,000 \cdot 7.801 \cdot \frac{0.60 \cdot 2.54}{100} \cdot (40.8 + 0.60) \cdot \frac{2.54}{100}$$

$$W_s = 353,298.23 \text{ ton}$$

$$\text{Cost of X70 steel} = 353,298.23 \cdot \frac{\$1,000}{\text{ton}} = 353.3 \text{ MM\$}$$

Case II, X80 Steel

Pipe thickness for X80 grade is:

$$t = \frac{D \cdot P}{2 \cdot S \cdot F \cdot L} = \frac{42 \cdot 1440}{2 \cdot 80000 \cdot 0.72} = 0.525 \text{ inch}$$

Pipe inside diameter $(D_2) = 42 - 2 \times 0.525 = 40.95$ inches

$$W_s = \pi \cdot L \cdot \rho \cdot t \cdot (D_2 + t)$$

$$W_s = \pi \cdot 900,000 \cdot 7.801 \cdot \frac{0.525 \cdot 2.54}{100} \cdot (40.95 + 0.525) \cdot \frac{2.54}{100}$$

$$W_s = 309,695.98 \text{ ton}$$

$$\text{Cost of X80 steel} = 309,695.98 \text{ ton} \cdot \frac{\$1,030}{\text{ton}} = 318.99 \text{ MM\$}$$

(Note: The cost of steel would be higher if the equivalent 0.60 location factor recommended by ASME B.31.8 were used instead of the location factor 0.72 recommended by CSA.)

TEMPERATURE PROFILE

Temperature has considerable influence on the economic and technical evaluations involved in the design of pipelines and related facilities. Computing the flow temperature for a fluid pipeline at steady state is adequate for most design purposes. This method fails when soil-pipe-environment interaction information is required for time- and temperature-dependent parameters. The following set of equations provides a comprehensive formulation for computing a steady-state temperature profile along a buried pipeline.

Consider a segment of transmission line between points 1 and 2 (Figure 3-16) and assume that the ground or soil temperature is T_g, then:
the energy equation is

$$dq = -\dot{m} \cdot C_p \cdot dT \qquad (3-90)$$

and the heat transfer equation is

$$dq = U \cdot dA \cdot (T - T_g) \qquad (3-91)$$

Combining Equations (3-90) and (3-91):

$$-\dot{m} \cdot C_p \cdot dT = U \cdot dA \cdot (T - T_g) \qquad (3-92)$$

where
q = heat transfer rate, BTU/hr
m = gas mass flow rate, lb_m/hr
C_p = gas average heat capacity, BTU/$\text{lb}_m \times °F$
U = overall heat-transfer coefficient, BTU/hr. $\text{ft}^2 \times °F$
T = gas temperature at any segment, $°F$

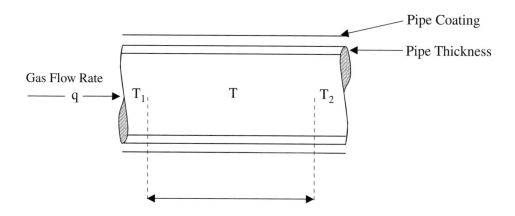

T_g (Soil Temperature)

Gas Flow Rate

Pipe Coating

Pipe Thickness

T_g (Ground or Soil Temperature Assumed to be Constant Over a Reasonable Length)

Figure 3-16. Heat transfer from a buried gas pipeline

T_g = ground temperature (constant throughout), °F
A = surface area of the pipe, ft².

After rearranging:

$$\frac{dT}{T - T_g} = \frac{-U}{\dot{m} \cdot C_p} \cdot dA$$

integrating from T_1 to T_2:

$$\int_{T_1}^{T_2} \frac{dT}{T - T_g} = \frac{-U}{\dot{m} \cdot C_p} \cdot A$$

gives

$$T_2 - T_g = (T_1 - T_g)e^{-UA/\dot{m}C_p} \qquad (3-93)$$

where $A = \pi dL$ (d = pipe outside diameter, ft; and L = pipe length, ft).

Notice that, as pipe length increases, $e^{-UA/\dot{m}C_p}$ approaches zero, so $T_2 \rightarrow T_g$. This means that for longer pipes, the gas temperature cools closer to ground temperature over the length of the pipeline (see Figure 3-17).

For further information, see Holman (1997) or the paper included in Appendix C by (Mohitpour 1991).

Equation (3-93) was derived without considering gas expansion (Joule-Thompson effect). In the following section, the effect of gas expansion will be considered in heat transfer for a more accurate analysis of the flowing gas temperature.

Joule-Thompson Effect

The Joule-Thompson effect describes the temperature loss due to the pressure drop that occurs when gas expands in a pipeline. The Joule-Thompson factor is defined as:

$$j = \frac{\Delta T_j}{\Delta P} \quad \text{or} \quad \frac{\Delta T_j}{\Delta L} \tag{3 – 94}$$

Applying this factor to Equation (3-92) over the length dL:

$$-\dot{m} \cdot C_p \cdot dT = U \cdot dA \cdot (T - T_g) + \dot{m} \cdot C_p \cdot j \cdot dL \tag{3 – 95}$$

Substituting $dA = \pi \times d \times dL$:

$$+\dot{m} \cdot C_p \cdot dT = -U \cdot \pi \cdot d \cdot dL(T - T_g) - \dot{m} \cdot C_p \cdot j \cdot dL$$

dividing by $\dot{m} \cdot C_p$:

$$dT = \frac{-\pi \cdot d \cdot U}{\dot{m} \cdot C_p} \cdot (T - T_g)dL - j \cdot dL$$

Assuming U and C_p are constant, let:

$$a = \frac{\pi \cdot d \cdot U}{\dot{m} \cdot C_p}$$

$$dT = -a \, dL[(T - T_g) + j/a] \tag{3 – 96}$$

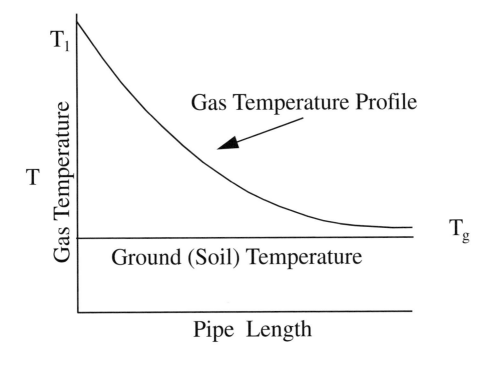

Figure 3-17. Pipeline temperature profiles

rearranging and taking the integral from points 1 to 2:

$$\int_{T_1}^{T_2} \frac{dT}{[(T - T_g) + j/a]} = \int_{L_1}^{L_2} -a\, dL$$

$$T_2 = \frac{T_1 - T_g + j/a}{e^{aL}} + T_g - j/a \qquad (3 - 97)$$

where T_1 = inlet gas temperature, °F
T_2 = exit gas temperature, °F
T_g = ground temperature, °F
j = Joule-Thompson coefficient, °F/ft (sometimes °F/psi)
a = a constant value for constant U and C_p , ft^{-1}
L = pipeline length, ft.

To determine the rate at which heat passes from the pipe coating to the soil in the case of buried pipelines:

$$q = K \cdot S \cdot (T - T_g) \qquad (3 - 98)$$

where q = heat transfer rate
K = soil thermal conductivity
S = conduction shape factor for buried pipes
T = pipe wall temperature
T_g = soil (ground) temperature.
The conduction shape factor S for buried pipelines is given by

$$S = \frac{2\pi L}{\cosh^{-1}(h/r)} \qquad (3 - 99)$$

where L = pipeline length
h = distance from centre of the pipe to the ground surface
r = pipe radius.

For further details on conduction shape factors, see (Holman 1997).

It should be noted that the flowing gas temperature in transmission lines does not remain constant over the length of the pipeline. When performing simulations for very long pipelines, and for downstream portions of the line where temperature changes are not very high, it is normally sufficient to use a constant average temperature to calculate the pressure drop. However, for transmission networks where pipes are interconnected and lengths are small (30 – 40 km or less), it is generally recommended that the system be simulated, using a temperature profile model instead of assuming a constant flowing gas temperature. This is also true for upstream portions of the pipeline, where considerable temperature changes often occur. Creating such a profile involves dividing the pipeline into smaller segments to more accurately calculate the pressure drop.

Variations in ground temperature and its thermal conductivity are other important factors to be considered and carefully measured. Ground temperature can vary both with location and season. Sometimes there are variations of 3–5°C in ground temperature between locations as close as 50-km apart. In most areas the ground temperature also varies from summer to winter by as much as 10°C. These are important factors to consider when performing pipeline simulations, as gas temperature plays an important role in the calculation of both pressure drop for transmission lines and power requirement for gas compressors.

The procedure outlined above provides an accurate prediction of fluid temperature under steady-state conditions for buried pipelines when the ground temperature and thermal conductivity remain almost constant over the length of the pipeline. However, this approach cannot adequately provide an analytical understanding of soil/pipe or environment/pipe interactions due to environmental temperature changes and effects. Complete temperature distribution patterns for various segments (fluid/pipe/soil/ environment) along pipeline lengths are required to adequately design the pipeline for operation under varying environmental conditions. For example, monthly temperature variations can affect pipeline operation as well as the determination of accurate line capacity for a gas pipeline system.

To predict time-dependant temperature patterns in a pipeline segment, a finite-element technique may be employed as an alternative method for solution of transient heat-transfer problems. This involves replacing the pipe/soil area with a continuum of finite pipe/soil elements, with finer elements placed near discontinuities/boundaries. The methodology of time-dependant heat flux in an axisymmetric body is then solved using a numerical recursive procedure. This method is fully described by (Mohitpour 1991). As the procedure is essential to some pipeline temperature simulation, the entire paper is included in Appendix C.

OPTIMIZATION PROCESS

Preliminary Hydraulic Simulations

The pipeline optimization process begins with a preliminary analysis in which rough hydraulic simulations are performed to examine the pipe size, station spacing, and maximum operating pressure for the pipeline. The volume, compression ratio, and maximum operating pressure are held constant while the capital cost of each system is estimated as a function of pipe size. Then, for each pipe size, the systems are assessed at several volumes above and below the design volume.

Preliminary Economic Evaluation

The preliminary economic evaluation determines the first-year unit cost of service, based on the following assumptions:

- There is no volume buildup in the pipe,
- The facilities are designed to meet the design volume,
- The time value of money is neglected.

Operating and maintenance costs, depreciation, fuel costs, taxes, and return on investment are assumed to be constant during the preliminary evaluation.

J-Curve Analysis

The results of the preliminary simulations and economic evaluation are often presented in the form of J-curves. J-curves graph the first-year unit volume cost of service versus the flow for different sizes of pipe (or compressor). The pipe (or compressor) size and maximum operating pressure are kept constant for each curve. J-curves allow for

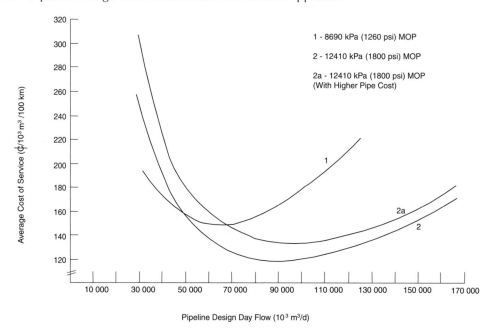

Figure 3-18a. Cost of service vs. design day flow for 1,219 mm (NPS 48) diameter pipe

comparisons of different pipe and compressor sizes to determine the most cost-effective solution.

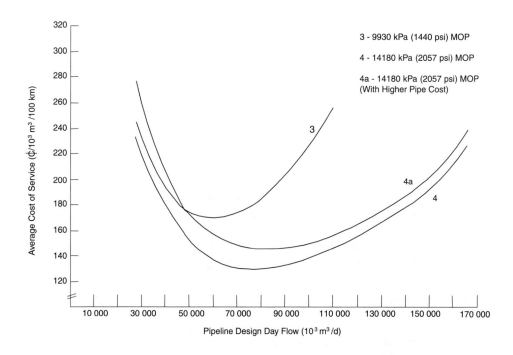

Figure 3-18b. Cost of service vs. design day flow for 1,067 mm (NPS 42) diameter pipe

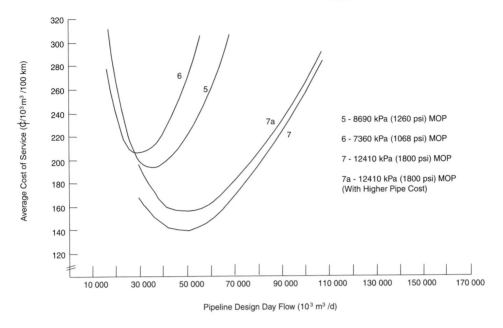

Figure 3-18c. Cost of service vs. design day flow for 914 mm (NPS 36) diameter pipe

Figures 3-18a, 3-18b, and 3-18c show J-curves for various pipe sizes at different operating pressures. Each curve was generated by first fixing the pipe size and then the maximum operating pressure (MOP) and compression ratio. Once the parameters were set, a station spacing simulation was run for a range of volumes. The cost of service was calculated for each volume using the attendant capital costs and fuel rates. Depreciation was held constant over the entire project life. Inflation was ignored. Municipal and Federal taxes were included at current rates. The unit length of the pipeline was taken as 100 km. The cost of service was calculated for one year to yield the cost per unit volume per unit length. This was plotted against the volume. The process was then repeated for other pipe sizes.

Figure 3-18a shows the J-curves for NPS 48 pipe, Figure 3-18b shows NPS 42 pipe, and Figure 3-18c shows NPS 36 pipe.

The most efficient operating range for each curve, defined by a relatively flat level, are each side of the maximum. If one figure overlays the next, it is evident that each system has a different efficient operating range. For a given flow, the system (pipe size and MOP) with the lowest cost becomes the starting point for further analysis.

J-curves are a simplistic approach to arrive at a preliminary system. They do not account for volume variations, load factors, the time value of money, or many other parameters. Therefore, for a more detailed evaluation, the remaining parameters must be evaluated in the next step of the optimization process.

Optimization Parameters

At this point in the analysis, a number of alternatives have been eliminated and some parameters might have been established. The remaining alternatives need to be evaluated for the maximum and minimum flow conditions. In general, a system should be flexible enough to operate under all expected flow conditions, but should operate most efficiently under the daily average volume. Loss of efficiency under maximum or minimum volumes can be tolerated provided the periods are short lived.

In order to evaluate the remaining alternatives, more parameters need to be identified.

Routing

For new pipeline systems, the routing of the system is an important consideration, although the possibility of routing a commodity through an existing system should also be explored.

Factors to be considered when selecting an appropriate pipeline route include terrain, topography, environmental constraints, population centers, and the existence of corridors (see Chapter 2). The importance of each factor can vary from location to location, but costs, schedules, and a reasonable effort to minimize environmental impact ultimately determine the route.

For an addition to an existing system, the routing considerations may be as simple as paralleling the existing system. However, situations may arise where a new route offers significant savings or other benefits over following the same route.

Design Year

Generally, there is a projection of throughput requirements over a number of years. The designer is faced with the task of deciding which year to use in the forecast as the design year. For a new pipeline system, the capacity incorporated in the initial design may or may not be considered as one of the optimization parameters.

In the case of an existing system, the concern is whether to provide additional capacity for only one year or for several years.

Maximum Operating Pressure (MOP)

For a long pipeline with few delivery points, the higher the MOP the more efficient the system. However, the length of the line, the number of delivery points, required delivery pressure, available supply pressure, local code restraints and cost premiums can all influence the MOP. Although the MOP is a function of many variables, it can be considered an optimization parameter.

When looping a system, the MOP of the existing pipeline should not be the only choice for the new loop. If a total loop-out is likely within a few years, one option is a higher pressure line eventually isolated from the existing system.

It is also possible to consider more than one MOP along the pipeline system. If long sections have no delivery points, a high MOP may be appropriate. At major delivery points, low MOPs may be more economic, since there is no benefit to adding energy only to waste it through a pressure-reduction device at the delivery location.

Load Factor

Pipeline volume information is compiled to provide a forecast by year for the design day maxima and minima, as well as for the average daily volumes. The load factor is defined as the average daily volume divided by the peak daily volume. If a pipeline system does not have storage facilities to accommodate peak demand, other pipeline facilities must be sized to transport this gas. However, as the load factor approaches 1.0 (always meeting demand), special allowances for peak demand are no longer necessary.

Storage facilities are often installed to reduce the number of other facilities required. Pipeline systems that have a higher storage capacity typically need fewer additional facilities. Most often the combined capital and operating costs of storage facilities are less than the costs of other pipeline facilities. However, there is rarely a linear relationship between storage costs and other facility costs, so it is necessary to determine an optimum combination of the two.

Depending on the fluid, various storage options are available. Options for gas include:

- Salt caverns
- Underground caverns
- Abandoned reservoirs
- Liquefied natural gas pipelines
- Compressed natural gas pipelines
- Pipeline bottles

Options for liquid include:

- Salt caverns
- Underground caverns
- Abandoned reservoirs
- Tanks
- Open reservoirs

If the load factor is low, the range over which compressor or pump stations must operate efficiently is very wide and multiunit stations are common. The number of operating units is varied to allow the stations to meet changing demand. Systems with high load factors often have single unit stations. However, standby units are included to increase the reliability of the system. Generally the unit volume cost of transportation decreases as the load factor increases.

Pipe Size

For short systems, the optimum pipe size is the minimum size required to meet the maximum input pressure, the minimum output pressure, and the demand. Of course, the size is limited by standard diameters manufactured by pipe mills.

For long systems, the ideal line will include a combination of pipe and compressor or pumping stations. The possible combinations are almost limitless. Designers have developed some simple rules of thumb to provide the initial direction.

Optimum pipe diameter is affected by the desired pressure drop along the pipeline. For an efficient pipeline system, experience has shown that the pressure drop along a gas pipeline should range between 15 and 25 kPa/km (3.50 – 5.85 psi/mile). Below this range, the pipe is underutilized and the capital expenditure is excessive. Above this range, friction losses are high and the fuel consumption is excessive. This range will vary depending on the specific costs of materials and services for different projects.

For liquid lines, the spacing between pump stations should be no less than 40 kilometers (25 miles) and no greater than 160 kilometers (100 miles).

When looping an existing system, the pressure drop can be used as a guide to selecting the length and diameter of the loop. Another rule is to limit the minimum loop diameter to at least the diameter of the largest existing line being looped. This is especially important if the throughput is increasing annually.

Another factor to consider when looping is whether to add enough loop to meet the design requirements (a neat loop) or to loop between existing valve sites (a practical loop).

Compressor/Pump Station Spacing

For gas pipelines where intermediate compression is required, there is considerable flexibility in the location of the compressor stations. Shifting the location changes the pressure at the inlet to the station and hence the compression ratio. If the stations are limited by the compression ratio, then the compression ratio is included as an optimization parameter.

If the flow is relatively low, either centrifugal or reciprocating compressors can be used. For the higher flows typically found in mainline pipeline systems, centrifugal compressors are generally used. Reciprocating compressors should be used for power requirements less than 5,500 kilowatts (7,375.6 HP). Compression ratios for centrifugal compressors are generally less than 1.5. Staging is required for larger ratios. The most efficient centrifugal compression ratios tend to be in the 1.25 to 1.35 range. Reciprocating compression ratios can well exceed 1.5 but they tend to be limited by the flow. Capital costs and fuel costs are the key factors in determining the optimum spacing between compressor stations.

For liquid pipelines, pump location is somewhat limited by requirements for minimum inlet and maximum outlet pressures. However, it is possible to select pipe sizes that allow pump stations to operate below the maximum pressure differential and continue to maintain an economic viability. Generally, pump stations should be located at least 40 kilometers (25 miles) apart.

Environmental/Socioeconomic Factors

The optimum spacing of compressor/pump stations is primarily based on system optimization and hydraulic requirements. However, sometimes it is necessary to adjust the station locations in response to environmental or socioeconomic factors. Such changes in location could affect the overall design efficiency, and it is possible that the compressor/pump units and cooling/heating requirements may be different. This usually increases the capital costs.

Compressor/Pump Drivers

The decision to use natural gas or electricity-powered drivers is dictated by availability, current and projected prices, and capital cost.

Heating and Cooling

High gas temperatures in pipelines result in high pressure losses. Significant savings can be achieved by cooling the gas downstream of the compressor. Although money must be invested to install and operate coolers, usually the resulting reduction in pipe size or compression requirements justifies the cost.

The absence of cooling may have many undesirable effects. For instance, the inlet and outlet temperatures of the pipeline can progressively rise at successive compressor stations along the line and result in increasing pressure losses. This effect is termed "cascading." Consistently high discharge temperatures can also cause metallurgical problems. Pipeline coating may also be damaged, especially at high discharge temperatures of 65°C (149°F) or more.

The opposite is generally true for liquid pipelines. In this case, high temperatures reduce liquid viscosity and thereby initiate savings by lowering the friction losses.

For small-diameter liquid pipelines, the conductive heat losses to the soil around the pipe generally exceed the heat generated by friction. As the diameter increases, the frictional heating increases at a faster rate than the conductive heat losses (assuming the lines are operating near capacity). As a rough guideline, diameters of 400 mm (16 inch) or less have conductive losses in excess of the frictional gains. Lines with a diameter of 500 mm (20 inch) or more will generate more frictional heating than the conductive heat losses.

Comparing Alternatives

Developing Design Alternatives

Considering all of the possible combinations of parameters, it can be seen that many different alternatives exist for the pipeline design. The objective of the optimization process is to methodically screen the available options to eventually arrive at the best combination of parameters. This involves deciding which parameters are important to the project and evaluating their potential impact to the overall design.

The first step is to establish the routing. Make an educated guess at the contributing parameters and try to determine two or three parameters that will be most affected by a route change. Identify possible routes based on geotechnical, environmental, and landowner concerns. Establish a rough facility list for each route based on peak daily flows only for the design year. For each possibility, calculate the rough capital costs and approximate the fuel and other operating and maintenance costs. Comparisons of the alternatives based on costs and major parameters should indicate the optimum route. However, it is prudent to identify at least one alternative. The bulk of the work continues with respect to the preferred route, but prior to finalizing the design, it may be necessary to review the alternative.

The second step is to determine the rough economics of storage versus compression/pumping facilities. Using an educated guess for the design year, MOP, and station spacing, run hydraulic simulations for each year up to the design year. Use the results of the simulations to establish the various facility and storage requirements for each load factor. Use the storage and facility requirements to estimate the costs of operating of each alternative system, including operating and maintenance costs, taxes, depreciation, and return on investment. The estimated costs should account for inflation. To facilitate a reasonable comparison between the alternatives, convert the annual costs to present value and add them to achieve the total present-value cost for each alternative. The alternative with the lowest total present-value cost should generally be selected.

The third step in the optimization process is to estimate a compression ratio for various station spacings. Prior to final design, any alternative can be re-evaluated for sensitivity to the spacing.

The last step is to reevaluate the alternatives using different design years, MOPs and pipe sizes, and, consequently, compression requirements. As the design work progresses, new alternatives may be investigated to further optimize the parameters.

The optimization process can often be simplified by analyzing some of the parameters at a later date in a "suboptimization" study. For example:

1. It would be wise to select one type of driver for compressors or pumps and treat the evaluation of alternative energy sources as a separate study,
2. The decision to heat or cool the fluid might be deferred to a later stage.

Analysis of the Alternatives

The previous section briefly touched on the analysis of alternatives. This section will elaborate on the process.

First, a separate analysis of each alternative is required for the design year. Hydraulic simulations at peak flow conditions must be performed for each alternative in the design year. The data are then used to size and locate the compressor/pump stations. Additional simulations for the preceding years will be required to determine the station installation (and power upgrade) schedule.

Given the installation schedule, transient simulations are used to check the stations' ability to meet the seasonal average flow conditions on an annual basis. Once the transient conditions are met, the fuel requirements can be determined.

At this stage, the facility installation schedule is used to estimate the capital costs, the fuel costs, and other operating and maintenance costs for each alternative. The estimates are projected to the year of installation in terms of as-spent costs. The higher cost alternatives are eliminated. The lower cost alternatives undergo further evaluation.

For each of the remaining alternatives, calculate the total present-value cost. Include depreciation, financing, return on equity, and taxes, if applicable. The discount rate used to calculate the present value of the annual costs is usually the company's current cost of money (i.e., a combination of debt financing at the rates available plus equity financing at the minimum acceptable rate of return).

When comparing alternatives, ideally all cases should have identical throughput. Realistically, each throughput varies slightly from the other. Thus, the alternatives with a lower capacity must be expanded to match the highest capacity to ensure a fair comparison. The additional capital, fuel, and operating and maintenance costs for each alternative must be estimated and incorporated into the total. The alternatives can then be reevaluated using the new cumulative present-value costs. Choose the alternative with the lowest total present-value cost.

Finally, vary two or three parameters to test the sensitivity of the first choice against the next best two or three alternatives. That is, reevaluate the first choice to ensure it remains the best alternative after some of the parameters are altered.

Selection of the Design Alternative

The optimum solution is normally the alternative with the lowest cumulative present value of the total annual costs. However, the total present-value costs of the two or more alternatives are frequently within a few percent (say 5%). For practical purposes, these costs are equal and the designers will extend the selection process by investigating other factors to arrive at the best alternative for the project.

One such factor to consider might be a comparison of the lowest initial capital investment. The alternatives with the lowest initial investment may be attractive if there is a risk that the project will not develop as forecasted. Similarly, if there is a high probability that future demands will exceed those forecasted, the alternative with the largest-diameter pipe may be selected for its greater expansion potential. Ease of operation and reliability are other considerations.

Sensitivity Analyses

Consider the optimization process up to this point. Some of the parameters were kept constant while others were varied in an effort to select the optimum design alternative. Further assessment of the last group of alternatives leads to the question: "What if the parameters previously kept constant are varied?"

- What if the throughput requirements build up faster or slower?
- What if the capital costs are higher or lower?
- What if the fuel or operating and maintenance costs are higher or lower?
- What if inflation factors change?
- What if financing costs change?
- What if the discount factor changes?

Rarely will the chosen alternative remain the best choice under the scrutiny of each and every question. A sensitivity analysis must be performed to discover how the top few chosen alternatives respond to changes in parameters that are not so easily predicted. The result may be that one alternative is less sensitive to fluctuations in various conditions, so it becomes the best choice.

The optimization process requires careful consideration of the major parameters influencing the system to obtain a low-cost transportation system. The ultimate effort expended in the process clearly depends on the number and severity of the parameters that must be addressed.

Timing Consideration

Timing can be important to the selection of the "optimum" design. The effects of the "time value of money" have a considerable effect on the cash flow. Delays in construction can significantly increase the capital cost of a project.

Regulatory Authorities

Political considerations and regulatory authorities often limit the selection of alternatives. For example, if a regulatory authority decides that only one route will be acceptable regardless of price, environmental considerations, etc., then this route is selected. The alternative is to attempt to change the regulatory decision. There can be significant cost to this process.

Capability Simulations and Outages

The system flow requirements are usually contractually determined. Detailed hydraulic simulations will determine the system's ability to meet the required flows with all facilities in operation. However, this is not accurate, since it is rarely true that all facilities will be operating at any given time.

To more precisely predict the effective capability of a system, the designer requires a historical assessment of the reliability of the various components. Then the capability of the system must be calculated for the outage of either a single unit, station, or loop, or any combination of components. Probability theories are used to assess the likelihood of each event, and thereby the effective capability, C, over a period of time. The effective capability is determined by:

$$C = \sum \frac{n!}{r!(n-r)!} p^r q^{(n-r)} \qquad (3-100)$$

where
n = total number of units on the pipeline
r = number of nonfunctional units, $0 \leq r \leq n$
q = typical unit availability (total hours available per month/744 hrs per month)
$p = 1.0 - q$
C = capability with r units down.

If the effective capability C is below the average daily flow, standby units at the compressor or pump stations will be necessary to meet the required annual flow. If C is greater than the average daily flow and less than the design daily maximum, an acceptable level of outages should be defined, and standby units added only as required.

GAS TRANSMISSION SOLVED PROBLEMS

Problem 1

Natural gas is pumped from a gas refinery toward the city gate of a town through a 16 inches (ID = 15.5 inches) gas pipeline (Figure 3-19). The distance between the refinery and city gate station (CGS) is 304 km. The inlet pressure to the CGS is 250 psig. Due to the expansion of the distribution network and petrochemical plant, a 50% increase on gas flow rate is required. The so-called expansion and increase on gas flow rate is supplied by connecting a 56 inches (ID = 55 inches) gas line to the 16 inches pipeline at point 2, diverting gas from Compressor Station No. 1 to this point (e.g., the construction of 56 inches pipeline is complete but it is not transporting gas before the diversion of gas from C/S No. 1 to it). Determine the inlet gas pressure to CGS and the inlet gas temperature to C/S No. 1. Further data follows:

1. Panhandle A equation could be used for all of the calculations.
2. $T_f = 100°F$.
3. Exit gas temperature at gas refinery is 150°F.

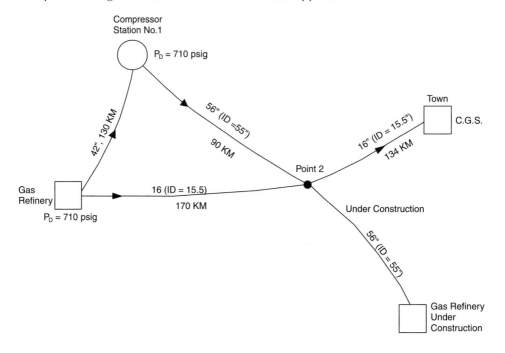

Figure 3-19. Pipeline schematic

4. $G = 0.65$.
5. Exit pressure at Compressor Station No. 1 = 710 psig.
6. Minimum required pressure at town CGS is between 200 and 400 psig.
7. Overall heat-transfer coefficient $U = 1.0$ BTU/hr·ft²·°F.
8. Gas heat capacity $C_p = 0.30$ BTU/lbm·°F
9. Ground temperature $T_g = 40$°F.
10. $Z = 1$.
11. Q_b gas flow rate (total) through NPS 42 = 1,000,000 m³/hr at SC.

Solution

Resistance for Panhandle A equation is given by (Wilson et al. 1991, Chapter 5):

$$R = 2.552 \times 10^{-4}(T_f \cdot G^{0.855} / D^{4.856}) \quad (\text{assuming the } Z_{ave} = 1.0)$$
$$T_f = 100 + 460 = 560\,^{\circ}R,\ G = 0.65,\ D = 15.5\,\text{inches}$$
$$R = 2.552 \times 10^{-4} \cdot 560\,[(0.65)^{0.855}\,T_f/(15.5)^{4.856}] = 1.64 \times 10^{-7}$$
$$K = R \cdot L = 1.64 \times 10^{-7}\frac{304}{1.609} \cdot 5{,}280\,(1\,\text{mile} = 5{,}280\,\text{ft})$$
$$K = \left(1.64 \times 10^{-7}\right) \cdot (304)[(5280/1.609)] = 0.164$$

using Equation (3-31) for Panhandle:

$$P_1^2 - P_2^2 = KQ^n$$
$$P_1 = 710 + 14.7 = 724.7\ \text{psia}$$
$$P_2 = 250 + 14.7 = 264.7\ \text{psia}$$
$$n = 1.855, K = 0.164$$

$$(724.7)^2 - (264.7)^2 = 0.164 (Q_b)^{1.855}$$

$$Q = 2,978.53 \, \text{MCF} / \text{hr} = 84,354 \, \text{m}^3 / \text{hr}.$$

Now from the same Panhandle A equation, K can be determined for 90 km of 56 inches pipeline:

$$K = \frac{2.552}{10^4} \cdot 560 \cdot \frac{(0.65)^{0.855}}{(55)^{4.856}} \cdot \frac{90}{1.609} \cdot 5,280$$

$$K = 1.033 \times 10^{-4}$$

pressure P_1 at Compressor Station No. 1 is 710 psig, so

$$P_1 = 724.7 \, \text{psia}, P_2 = ?$$
$$K = 1.033 \times 10^{-4}$$

for a 50% flow increase:

$$Q_b = 2,978.53(1.5) = 4,468 \, \text{MCF/hr}$$

$$P_1^2 - P_2^2 = K Q^n \quad \text{or}$$

so

$$(724.7)^2 - P_2^2 = 1.0333 \times 10^{-4} \cdot (4,468)^{1.855}$$

$$P_2 = 724.3 \, \text{psia}$$

This means that, by diverting gas through the 56 inches gas pipeline, only a 0.4-psi pressure drop is observed.

Then, for the 134 km, 16 inches pipeline:

$$P_1^2 - P_2^2 = K Q^n \quad \text{or}$$

$$(724.3)^2 - P_2^2 = 0.164 \cdot \frac{134}{304} \cdot (4,468)^{1.855}$$

$P_2 = 314.8$ psia at the town CGS, which is within the acceptable range.

On the other side, for the 42 inches gas pipeline:

$$Q_b = 1,000,000 \times 35.331$$
$$Q_b = 35,310,000 \, \text{ft}^3 / \text{hr}$$

and at the base condition:

$$P Q_b = n \, RT$$

$$14.7 \cdot 35,310,000 = n \cdot 10.73 \cdot 520$$

$$n = 93,027.6 \, \text{moles/hr}$$

If for a long transmission line with moderate pressure drop the Joule-Thompson effect is neglected, then the heat transfer from the pipeline to the surrounding will be calculated using Equation (3-93):

$$T_2 - T_g = (T_1 - T_g) e^{-UA / \dot{m} C_p}$$

This is the general equation that defines the gas temperature profile in a long gas transmission line.

For the existing pipeline with

$L = 130$ km

$U = 1.0$ BTU/hr $\cdot$ ft^2 $\cdot$ °F

$C_p = 0.30$ BTU/lb$_m$ $\cdot$ °F

$n = 93,027.6$ lb moles/hr

$\dot{m} = n \times M = 93,027.6 \times 18.85 = 1,753,570$ lb$_m$/hr

$M = 29G = 29 \times 0.65 = 18.85$

$A = \pi DL = \pi 42/12 \cdot 130/1.609 \cdot 5,280 = 4,688,338$ ft^2

$-UA/\dot{m} \times C_p = -1.0 \times 4,688,338/1,753,570 \times 0.3 = -8.9$

$e^{-8.911} \approx 0.$

therefore

$$T_2 - T_g = 0$$

and

$$T_2 - T_g = 40 \overset{\circ}{} F$$

Therefore, the gas will reach ground temperature.

Problem 2

A system of gas transmission lines is used for the gas injection to an oil reservoir, transporting gas from a gas reservoir 350 miles away (see Figure 3-20). Based on previous calculations, an injection rate of 1,000,000 m^3/hr of gas at SC is required, and the pressure at the injection point should be at least 1,000 psia. Before the gas injection operation the gas well, which is located at the depth of 6,000 ft, is closed; the gas reservoir pressure is 2,650 psia, $G = 0.60$, $T_{ave} = 120$ °F, and $Z_{ave} = 0.822$. Use AGA fully turbulent flow equation with a pipeline roughness of 750 microinches.

1. Calculate the Closing Well Head Pressure (CWHP) of the gas well?

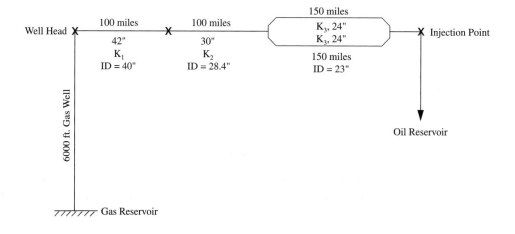

Figure 3-20. Pipeline configuration

2. If the gas well is opened and a Flowing Well Head Pressure (FWHP) of 2,000 psia is obtained for the injection of 1,000,000 m³/hr gas into the oil reservoir, what would be the pressure at the injection point?

3. If the present transmission system is replaced by a single gas line, what would be the equivalent diameter of the new pipeline and which system is cheaper based on $25,000/inch km of the pipeline cost?

Solution

1. In order to find the CWHP of the gas well, Bernoulli's equation could be written for no flow condition:

$$\frac{1}{g_c} \cdot u \cdot du + vdP + \frac{g}{g_c} dh = 0$$

where $u = 0$ for no flow, so:

$$vdP + \frac{g}{g_c} dh = 0$$

or $vdP + dh = 0$ for a closed well. Integrating from the top of the well (1) to the bottom (2):

$$\int_1^2 \frac{dP}{\rho} + \Delta h = 0$$

since $\rho = \frac{P \cdot M}{Z \cdot R \cdot T}$, or $\rho = \frac{29 \cdot G \cdot P}{Z \cdot R \cdot T}$,

$$\int_1^2 \frac{ZRT}{29G} \frac{dP}{P} = -\Delta h$$

assuming that variations of T and Z are insignificant, and replacing them with T_{ave} and Z_{ave}

$$\frac{R \cdot T_{ave} \cdot Z_{ave}}{29G} \ln P_2/P_1 = -\Delta h$$

$$\ln \frac{P_2}{P_1} = \frac{-29G\Delta h}{RT_{ave}Z_{ave}}$$

then

$$P_2 = P_1 e^{-29G \cdot \Delta h / RT_{ave} \cdot Z_{ave}}$$

where P_2 = pressure at the bottom hole, psia
P_1 = pressure at the well head, psia
G = gas gravity, dimensionless
Δh = well depth, ft. ($h_2 - h_1$ or $-6,000$ ft.)
T_{ave} = average gas temperature, °R
Z_{ave} = average gas compressibility factor, dimensionless

R = gas constant, 10.73 psia $\cdot$ ft³ / lb mole $\cdot$ ° R

$$P_2 = P_1 e^{-29G \cdot \Delta h / RT_{ave} \cdot Z_{ave}}$$

$$2,650 = P_1 e^{\frac{29 \times 0.6 \times 6,000}{10.73 \times 144 \times 580 \times 0.822}}$$

$$2,650 = P_1 \times 1.15,226$$

$$P_1 = 2,300 \text{ psia, CWHP}$$

2. Now the well is opened and the flow of gas has started, using AGA flow equation (assuming the same T_{ave} and Z_{ave} as in 1 above):

$$K = 3.639 \times 10^{-3} \frac{T_{ave} \cdot G \cdot Z_{ave}}{D^5} \times \frac{1}{\left(\log \frac{3.7D}{K_e}\right)^2} \times L$$

L = ft, T_{ave} = 580 R°, D = inch, Q = MCF/hr

$$K = 3.639 \times 10^{-3} \frac{580 \cdot 0.60 \cdot 0.822}{D^5} \times \frac{1}{\left(\log \frac{3.7D}{K_e}\right)^2} \times L$$

For 100 miles, a 42 inches pipe (ID = 40″):

$$K_1 = 3.639 \times 10^{-3} \frac{580 \cdot 0.60 \cdot 0.822}{(40)^5} \times \frac{1}{\left(\log \frac{3.7 \times 40}{750 \times 10^{-6}}\right)^2} \times 5280 \times 100$$

$$K_1 = 0.00019143$$

For 100 miles, a 30 inches pipe (ID = 28.4″):

$$K_2 = 0.00112320$$

For 150 miles, a 24 inches pipe (ID = 23″):

$$K_3 = 0.00334201$$

K_3 is looped, therefore:

$$K_e = \frac{K_3 \cdot K_3}{\left(\sqrt{K_3} + \sqrt{K_3}\right)^2}$$

$$K_e = \frac{K_3}{4} = 0.00083550$$

The total resistance of the system would be:

$$K_E = 0.00019143 + 0.00112320 + 0.00083550$$

$$K_E = 0.00215013$$

Flow should be in MCF/hr, therefore

$$Q = 1,000,000 \ \mathrm{m}^3/\mathrm{hr} = \frac{1,000,000 \times 35.31}{1000}$$

$$= 35,310 \ \mathrm{MCF/hr}$$

and $P_1 = 2000$ psia, $P_2 = ?$

$$P_1^2 - P_2^2 = K \cdot Q_b^n$$

or for AGA fully turbulent equation

$$P_1^2 - P_2^2 = K \cdot Q_b^2$$

$$(2000)^2 - P_2^2 = 0.00215013(35310)^2$$

$P_2 = 1149$ psia, at the injection point

3. To find the diameter of the single pipeline that has the same resistance, the following equation could be used (AGA fully turbulent equation):

$$K = 3.639 \times 10^{-3} \times \frac{T_{\mathrm{ave}} \cdot G \cdot Z_{\mathrm{ave}}}{D^5} \times \frac{1}{\left(\log \frac{3.7D}{K_e}\right)^2} \times L$$

For a single line of $L = 350$ miles and the same value for $K = 0.00215013$, D could be obtained

$$ID = 31.9 \ \mathrm{inches}$$

This will need a 34″ line.
Cost of initial pipe is:

$$\mathrm{Cost \ in \ \$} = (100 \times 1.609 \times 42 + 100 \times 1.609 \times 30 + 150 \times 2 \times 1.609 \times 24)$$

$$\times 25000 \frac{\$}{\mathrm{inch} \cdot \mathrm{km}}$$

$$= \$579,000,000$$

Cost of single 34″ line is:

$$\mathrm{Cost \ in \ \$} = 350 \times 1.609 \times 34 \times 25000$$

$$= \$478,677,500$$

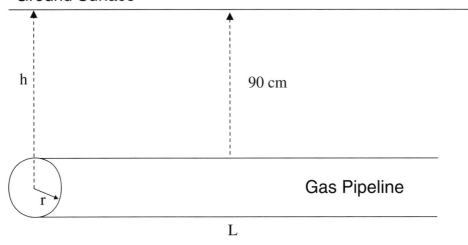

Figure 3-21. Schematic of the buried pipeline

The corresponding injection pressure for the single line would be:

$$P_1^2 - P_2^2 = K \cdot Q_b^2$$

$$(2000)^2 - P_2^2 = 0.00215620 \times (35,310)^2$$

$$P_2 = 1145 \text{ psia}$$

Problem 3

For a 36 inches gas pipeline, the overall heat-transfer coefficient is known and it is equal to $U_o = 1.66 \ W/m^2 \cdot K$. The burial depth from the ground surface to the upper side of the pipeline is 90 cm, and the heat-transfer resistance of the gas side, pipeline wall, and coating are neglected. Calculate the soil thermal conductivity in the area of operation (see Figure 3-21).

Solution

The conduction shape factor is given by Equation (3-99) as:

$$S = \frac{2\pi L}{\cosh^{-1}\left(\frac{h}{r}\right)}$$

where S = conduction shape factor
   L = pipeline length
   h = distance from center of the pipe to the ground surface
   r = pipe radius.

The equation for heat transfer from the pipe wall to the ground is given by Equation (3-98) as

$$q = K \cdot S \cdot (T - T_g)$$

where K is soil thermal conductivity. Furthermore, the general heat-transfer equation from the gas side to the soil [Equation (3-91)] is:

$$q = U_o \cdot A_o \cdot \Delta T_m$$

where, upon assumption of no resistance within the gas, pipe, and pipe coating:

$$K \cdot S = U_o \cdot A_o$$

From the above equation, U_o and A_o are known, S could be calculated as follows:

$$h = 90 \text{ cm} + 18 \text{ inches} \cdot 2.54 = 135.72 \text{ cm} = 1.3572 \text{ m}$$
$$r = 18 \text{ inches} = 45.72 \text{ cm} = 0.4572 \text{ m}$$

then

$$\cosh^{-1}(h/r) = \cosh^{-1}\left(\frac{1.3572}{0.4572}\right) = \cosh^{-1}(2.9685)$$

furthermore, $\cosh^{-1}(x)$ could be calculated as follows:

$$\cosh^{-1}(x) = \ln\left(x \pm \sqrt{x^2 - 1}\right)$$

in this case, the only acceptable solution is

$$\cosh^{-1}(2.9685) = \ln\left[2.9685 + \sqrt{(2.9685)^2 - 1}\right] = 1.7515$$

then

$$S = \frac{2\pi L}{\cosh^{-1}\left(\frac{h}{r}\right)}$$

or for $L = 1$ meter of pipeline

$$S = \frac{2\pi \cdot 1}{1.7515} = 3.58723 \text{ shape factor per meter of pipeline}$$

and

$$A_o = \pi D_o L$$

or

$$A_o = \pi \cdot \frac{36 \cdot 2.54}{100} \cdot 1 = 2.87267 \text{ m}^2$$

and finally

$$K = \frac{U_o A_o}{S}$$

upon substitution, the value of K or soil thermal conductivity will be determined.

$$K = \frac{1.66 \cdot 2.87267}{3.58723}$$
$$K = 1.33 \text{ W/m}^2 \cdot \text{K}$$

REFERENCES

American Gas Association, AGA Report No. 10, Speed of Sound in Natural Gas, 2003.

Asante, B., 1994, "Justification for Internal Coating of Natural Gas Pipelines," Design Method and Technology/Facilities Planning Department, *Nova Gas Transmission Limited (NGTL)*, Internal Reports, Calgary, Alberta, Canada.

_____, 1996, "Design Critieria and Parameters Impact Study," System Design Department, *Nova Gas Transmission Limited (NGTL)*, Internal Reports, Calgary, Alberta, Canada.

Beggs, H. D., 1991, *Gas Production Operations*, OGCI Publications, Consultants International Inc., Tulsa, OK.

Campbell, J. M., R. A. Hubbard, and R. N. Maddox, 1994, *Gas Conditioning and Processing, Volume 1*, Campbell Petroleum Series, Norman, OK.

Canadian Standard Association (CSA), 1996, *Z662-94 Oil and Gas Pipeline Systems* (Oil and Gas Industry Systems), Rexdale, Ontario, Canada.

Golshan, H., and M. Narsing, 1994(a), "Study of Pipeline Deterioration due to Age (Phase 1)," Design Methods & Technology/Facilities Planning, *Nova Gas Transmission Limited* Internal Reports, Calgary, Alberta, Canada.

_____, 1994(b), "Impact of Different Gas & Pipeline Parameters on Three Major Gas Flow Equations," Facilities Planning, *Nova Gas Transmission Limited, (NGTL)*, Calgary, Alberta, Canada.

Holman, J. P., 1997, *Heat Transfer*, 8th Edition, McGraw-Hill, New York, NY.

Hydraulic Analysis and Resources Tool (HART) Simulation Software, 1995, "Users Guide Manual," System Design Department, *Nova Gas Transmission Limited*, Calgary, Alberta, Canada.

Hughes, T., 1993, "Optimum Pressure Drop Project," Facilities Planning Department Internal Reports, *Nova Gas Transmission Limited (NGTL)*, Calgary, Alberta, Canada.

Katz, D. L., et al., 1959, *Handbook of Natural Gas Engineering*, McGraw-Hill Book Company, New York, NY.

Kern, D. Q., 1997, *Process Heat Transfer*, McGraw-Hill Pub., New York, NY.

Mohitpour, M., 1991, "Temperature Computations in Fluid Transmission Pipelines," ASME.

Starling, K. E., and J. L. Savage, 1994, *Compressibility Factors of Natural Gas and Other Related Hydrocarbon Gases*, AGA Transmission Measurement Committee Report No. 8.

Weiss, M., 1998, "Mixtures for Storage and Pipeline Transport of Gases," *NOVA Research & Technology Corporation (NRTC)*, Internal Report, NOVA Chemicals, Calgary, Alberta, Canada.

Wilson, G. G., R. T. Ellington, and J. Farwalther, 1991, *Institute of Gas Technology Education Program*, Gas Distribution Home Study Course.

Chapter *4*

GAS COMPRESSION AND COOLERS

INTRODUCTION

Compression is required in gas pipelines to overcome pressure losses that occur over the length of the pipeline. Gas is generally received from receipt points along the pipeline and delivered to sales stations at specified flows and pressures. Between these points, a pressure drop occurs due to gas expansion, friction loss, a change in elevation, or a change in temperature. Altering the flow will change the pressure in the pipeline. The following methods can be applied to maintain the required pressure at an existing delivery point when there is an increase in flow rate beyond the design point:

- Loop the pipeline
- Add a compressor station
- Utilize both a loop and compression

The evaluation of the method that will be economically more feasible depends on many factors, including:

- Capital expenditures
- Fuel cost
- Emissions
- Maintenance
- Future expansions

TYPES OF COMPRESSORS

Compressors can be divided into three major categories:

1. Positive displacement
2. Dynamic
3. Injectors

Positive displacement or intermittent flow compressors trap a quantity of gas within an enclosed volume. By reducing the volume, they increase the pressure of the trapped gas; compressed gas is then delivered to a compressor discharge point.

Positive displacement or intermittent flow compressors are divided into two distinct types: Reciprocating compressors and Rotary compressors.

In reciprocating compressors, a piston reduces the volume of gas within a cylinder. Valves are required in cylinders to direct the flow of gas and to prevent backflow.

In rotary compressors, rotors are equipped with either vanes or lobes. They trap gas in a fixed or variable volume between themselves and an outside casing. The gas moves from inlet to outlet as the rotors turn. In this type of compressor, valves are not required, and it is typically used for air compression in plants.

Dynamic or continuous flow compressors (also turbocompressors) increase the pressure of the gas via inertia forces (i.e., increasing gas velocity and changing the energy into pressure).

Dynamic compressors are also divided into two major types: Centrifugal (radial) compressors and Axial compressors.

In centrifugal compressors, velocity is added to the gas by the blades of a rotating impeller. As they spin, centrifugal forces push the gas molecules outward, which increases the radius of rotation and hence increases the tangential speed of the gas molecules. An increase in speed produces an acceleration, which activates inertial forces that act on gas molecules, and compresses them. Pressure is recovered partly in the impeller and partly in the radial diffuser surrounding the impeller, or in the discharge volute diffuser at the outlet end of the compressor.

In axial compressors, a spinning rotor transfers its energy into the gas stream during compression. In this type of compressor, gas flow is parallel to the shaft.

Injectors use the kinetic energy of one stream of fluid to compress another fluid. These types of compressors are not used in natural gas transmission systems. Therefore, positive displacement and dynamic compressors are discussed in this chapter.

Figure 4-1 summarizes the different types of compressors available.

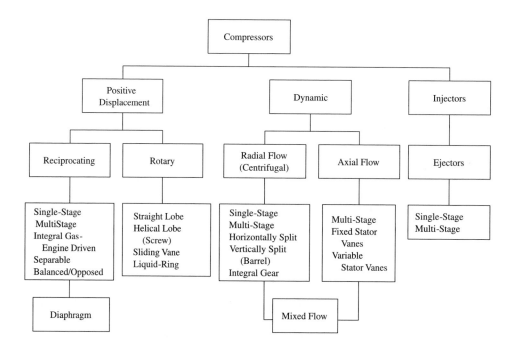

Figure 4-1. Types of compressors [Courtesy GPSA, Engineering Data Book, 1994]

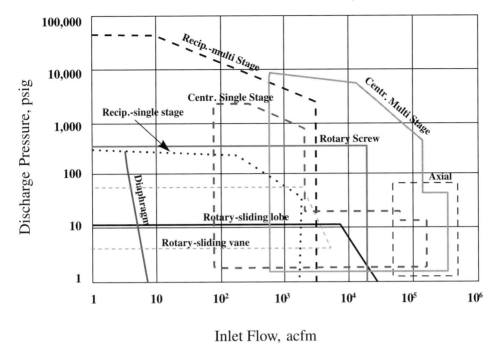

Figure 4-2. Compressor coverage chart [Courtesy of GPSA, Engineering Data Book, 1994]

In today's expanding industry, compressors can be used for a variety of purposes. The parameters typically used to choose a special type of compressor include discharge pressure (pressure ratio), compressor head, inlet flow, and operations reliability. Figure 4-2 illustrates the operating range of different types of compressors in terms of inlet flow and compressor discharge pressure.

COMPRESSOR DRIVERS

Compressors are generally coupled to another machine that drives the shaft. The following are the most common types of compressor driving machines:

- Gas turbine
- Electric motors
- Steam turbines
- Turbo expander

Gas combustion turbines (gas turbines) are the most common types of drivers used in remote areas, especially on gas transmission systems. These turbines are generally the most appropriate drivers for centrifugal gas compressors.

In gas turbines, a power turbine delivers shaft power to the pipeline compressor. Gas turbines are relatively compact, having a high power to weight ratio, and they are well-suited for the high speeds required by the centrifugal compressors. They operate throughout a broad range, matching the operating range of the compressors. The range is 60–105% of the compressor rated speed.

The gas turbines used for mechanical drive applications consist of two main components: a gas generator and a free power turbine. The gas generator, the "jet engine" of the gas turbine, produces hot exhaust gases. The power turbine is aerodynamically coupled to the gas generator, and, through an expansion process, uses these hot gases to produce shaft power. In the power generation process, air is drawn into the gas turbine through the intake by a multistage axial flow compressor. As the air passes through the compressor section, the pressure and temperature increase. The compressed air mixed with fuel is injected into the combustion section and ignited. The resulting high-pressure/high-temperature combustion gases pass through a high-pressure turbine, which rotates due to the energy extracted from the combustion gases. The high-pressure turbine is directly connected to and drives the axial flow compressor (i.e., the air compressor). The remaining

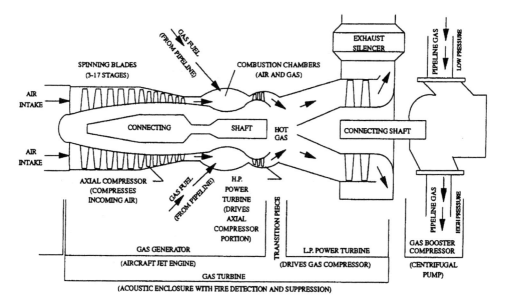

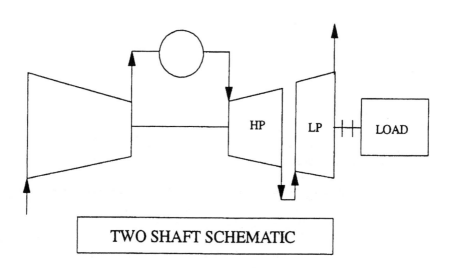

Figure 4-3. Gas turbine schematic [Courtesy Practical Pipeline System Design and Innovations, Department of Mechanical Engineering, University of Calgary]

energy in the exhaust gases rotates the free power turbine, which produces shaft power to drive the pipeline compressor. From the total energy produced by the gas generator, normally the air compressor consumes two-third and only one-third is delivered to the pipeline compressor.

Gas generators are generally classified as industrial or aeroderivative. The current aeroderivative types of gas turbines are basically modified aircraft jet engines. They are lightweight, compact, and have high thermal efficiencies. Aircraft engines are often adapted for industrial use.

The industrial type of gas turbines are heavier engines with lower thermal efficiency but longer life (see NOVA Gas Transmission Limited, 1995).

Figure 4-3 shows a schematic of a two-shaft gas turbine driving a compressor.

The following sections describe the gas turbine components and auxiliary systems starting with the air intake system. They are listed in the order that the airflow path follows through the unit (for further details, see NOVA Gas Transmission Limited, 1995).

- Air intake system
- Compressor section
- Diffuser
- Combustion section
- Turbine section
- Free power turbine
- Instrumentation and controls
- Auxiliary systems

Air Intake System

The air intake system consists of an inlet filter and ducting, inlet bellmouth, and inlet guide vanes. The filter system is usually of the pulse-air type and is self-cleaning. Secondary removable filter elements are often located in series with the main filter system. Downstream of the filters on larger units are usually a kind of baffles used to reduce air turbulence and noise. The straightened airflow enters the turbine through an inlet bellmouth where the air is accelerated. The air then passes over a set of inlet guide vanes that direct the air to the axial flow compressor at an optimum angle. Variable inlet guide vanes are sometimes used to improve performance for off-design airflow. The inlet system also includes a water wash system to wash and remove any buildup on the axial compressor blades. In some cases, units are equipped with an anti-ice system in the intake. Heat exchangers or hot air nozzles located in the inlet air stream prevent ice from developing during very severe conditions.

Compressor Section

Compression is the first step in the gas turbine operating cycle. Compression occurs in the axial flow compressor, which is located immediately downstream of the inlet guide vanes. It consists of multiple stages of airfoil-shaped blades that are attached to rotating disks. Between each stage of rotating disks is a stator assembly. The stator consists of stationary vanes that direct the airflow at an optimum angle toward the next stage. Some gas turbines contain a series of variable stator vanes that optimize flow (axial velocity) at various load conditions to prevent blade stalling and compressor surge. The number of stages can vary depending on the gas turbine type and power.

The compressor blades impart energy to the air to create a pressure increase. To perform this function the blades must rotate at very high speeds. Therefore, the blades must be lightweight to minimize centrifugal pull on the compressor disk. High fatigue strength and erosion and corrosion resistance are also important blade material properties. The stationary blades that diffuse the airflow, resulting in a pressure rise, are called a stator assembly. These blades also direct the flow to the next set of rotating blades at an optimum angle.

Diffuser

The diffuser is located downstream from the compressor section. As the air flows through the diffuser, it expands, which reduces its velocity. The air then flows to the combustor.

Combustion Section

The combustor is directly downstream of the axial compressor section. The combustion process adds heat energy to the gas turbine air. This heat must be added to the gas turbine cycle to provide energy to drive the free power turbine and, therefore, the pipeline compressor. Heat is generated through the combustion of a natural gas/air mixture, which is ignited in the combustor.

Primary combustion uses about 20% of the airflow from the axial compressor. Thirty percent of the air from the compressor is introduced into the combustion chambers for secondary combustion. Optimum fuel-air ratios are maintained for adequate combustion. The remaining air is mixed with the combustion products to cool them. Enough cooling must occur to prevent the turbine nozzles and blades from overheating.

Combustion materials must withstand extremely high temperatures. The optimum temperature should be determined to control and reduce the production of pollutants such as nitrogen oxides (NO_x) and carbon monoxide (CO). A high combustion temperature results in an excess NO_x pollutant, while a low temperature results in larger amounts of CO and unburned hydrocarbon. The optimum temperature is normally achieved with premix and lean fuel-to-air mixtures.

Turbine Section

The turbine extracts energy from the hot and high-pressure combustion gases. The turbine that drives the axial flow compressor uses almost two third of this energy. The remaining energy rotates the free power turbine that drives the pipeline compressor. The turbine section has stages of rotating blades and stationary blades of increasing diameter. This allows the hot gases to expand across each stage, providing energy to the turbine rotors.

Free Power Turbine

Gas turbines designed for mechanical drive applications include a free power turbine, which is independent of the gas generator. The free power turbine is connected to the gas generator while the rotor is aerodynamically coupled. The free power turbine rotor is directly coupled to the driven equipment, such as a pipeline compressor. Hot exhaust gases from the gas generator are ducted into the free power turbine inlet.

The free power turbine consists of one to three stages of stationary nozzles and rotors. As the gas passes through each stage it undergoes an expansion. The energy extracted from the gas during this expansion process provides the torque required to turn the power turbine and the pipeline compressor.

Instrumentation and Controls

The gas turbine control system includes a fuel-metering system and a computing section. The computer analyses data from the sensors and performs all the computations required for operating the gas turbine at the desired set points. The position of a fuel-metering valve controls the power output of the gas turbine. When more power is required, a signal is sent to the valve, which continues to open until a desired power output or power turbine speed set point is achieved. At this point a signal is relayed to the valve to maintain its position.

Various sensors on the gas turbine monitor critical operating parameters to ensure that limits are not exceeded. Such parameters include rotor speeds, exhaust gas temperature, vibration, bearing temperatures, and lube oil temperatures, and pressures.

Auxiliary Systems

The main auxiliary systems of a gas turbine include the start system, the fuel system, and the hydraulic and lubrication system.

COMPRESSOR STATION CONFIGURATION

Compressors for gas transmission service are installed in various configurations. However, station layout is generally standardized to the maximum extent possible, incorporating all requirements of CSA Z-184 and other applicable codes. The layout of a compressor station is given in Figure 4-4, and a brief description of the configuration follows.

Description of Station Layout

Station layout, equipment, and piping common to both reciprocating and centrifugal stations will be described from the suction side valve to the discharge side valve, with reference to Figure 4-4.

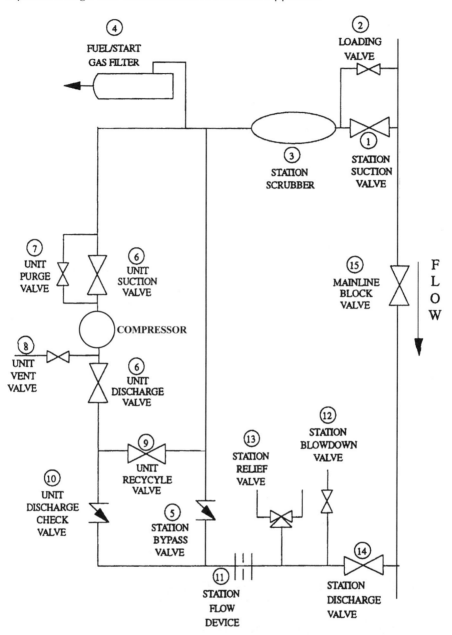

Figure 4-4. Station piping schematic [Courtesy of the Mechanical Engineering Department, University of Calgary, Practical Pipeline System Design and Innovations Program]

1. A station suction side valve is installed to isolate the station from the pipeline.
2. A station-loading valve is used to purge and pressurize the compressor station piping.
3. A station scrubber is installed to clean the gas, knocking out liquids and solid contaminants to protect the compressors.

4. A fuel/start gas filter further cleans the gas to the quality required for the compressor driver fuel/start gas systems.
5. A bypass line with check valve is installed between the suction and discharge header to allow bypassing of compressors when they are not operating, and to allow gas on the suction side to be vented in the event of a blowdown.
6. Unit suction and discharge valves are used for isolating the compressor units while the yard piping remains pressurized.
7. A unit purge valve is used to purge and pressurize the compressor in preparation for operation.
8. A unit vent valve is used during purging of the compressor, and to depressurize the compressor after shutdown.
9. A unit recycle valve is used to unload the compressor while it is starting or stopping, and occasionally during operation, to keep the dynamic compressor from surging.
10. A unit discharge check valve is installed to prevent backflow through the compressor while it is starting or stopping.
11. A flow measurement device, typically an orifice meter, is installed to measure the flow of gas going through the compressor station.
12. A blowdown valve is installed to vent all high pressure piping in the yard. Valves 1, 2, 6, 7, 8, 12, 14, and 15 are connected to the station Emergency Shutdown System (ESD), and are sequenced on gas or fire detection, power failure or manual operation of the ESD.
13. A station relief valve is installed to protect the station piping/equipment from overpressure in the event of failure of the protective and pressure control systems.
14. A station discharge side valve is installed to isolate the station from the pipeline.
15. A mainline block valve is installed in the pipeline and is closed to divert gas through the station when compression or free flowing through the station is desired, and opened when the flow is to bypass the station.

THERMODYNAMICS OF ISOTHERMAL AND ADIABATIC GAS COMPRESSION

Theoretically, there are three forms of natural gas compression:

1. Isothermal
2. Adiabatic reversible (isentropic)
3. Polytropic

Isothermal and adiabatic gas compression will be discussed in this section. The polytropic process will be covered separately.

Isothermal Gas Compression

In isothermal gas compression, natural gas is compressed while heat is removed from the system to keep the gas temperature constant. As will be proved later, in this type of compression the least mechanical power is required to compress a gas from P_1 to P_2 compared to other types of compression. In reality, keeping the gas temperature constant during the entire compression process is almost impossible. Although theoretical equations can be derived for isothermal gas compression, it is not the type of compression that occurs in a real gas compressor.

Isentropic and Polytropic Gas Compression

In adiabatic gas compression, natural gas is compressed while there is no heat transferred between the system and the surroundings, or $dq = 0$. Compressor manufacturers often use the terms "adiabatic" and "isentropic" interchangeably.

Polytropic process is the general case for which no specific conditions other than mechanical reversibility are imposed.

Theoretically, to calculate the mechanical power required to compress a gas from P_1 to P_2, the adiabatic process is applied. Ultimately some mechanical losses will need to be considered based on the type of compressor.

Using an adiabatic/polytropic gas compression process while considering mechanical losses given by the manufacturer, theoretical equations that best represent the actual power requirements of the system can be obtained.

In adiabatic gas compression, the relationship between the pressure and volume of the gas is given as follows:

$$P \cdot V^K = C \tag{4-1}$$

where P = gas pressure
V = gas volume
K = adiabatic gas exponent
C = constant

Using thermodynamic relationships, the above equation can be derived for a natural gas compressed adiabatically. Applying the first law of thermodynamics:

$$dU = dq - dW \tag{4-2}$$

where dU = change in internal energy
dq = heat transfer between system and surrounding
dW = work done by the system or on the system.

For an adiabatic process, $dq = 0$

$$dU = -dW \tag{4-3}$$

Internal energy for a gas is given by

$$dU = C_V \cdot dT \tag{4-4}$$

where C_V = specific heat at constant volume
dT = temperature change

and work is expressed by the following relationship:

$$dW = P \cdot dV \tag{4-5}$$

then

$$C_V \cdot dT = (-PdV)$$

Using ideal gas relationships, for one mole of gas

$$P \cdot V = R \cdot T$$

$$P = \frac{RT}{V}$$

therefore

$$C_V \cdot dT = -R \cdot T \frac{dV}{V}$$

and

$$\frac{dT}{T} = -\frac{R}{C_V} \cdot \frac{dV}{V}$$

According to the definition

$$\frac{C_P}{C_V} = K \qquad (4-6)$$

where C_P and C_V are heat capacities at constant pressure and volume, respectively, and K is the adiabatic gas exponent. The adiabatic gas exponent can be easily obtained from Equation 4-6 or from existing tables for C_p. Graphs are also available to obtain K if molecular weight and temperature are known. The procedures to calculate the adiabatic gas exponent are given in subsequent sections. From thermodynamic relations, the enthalpy of a gas is defined as:

$$H = U + P \cdot V \qquad (4-7)$$

where H = enthalpy of gas
U = internal energy of gas
P = gas pressure
V = gas volume

taking the derivative of Equation (4-7)

$$dH = dU + d(P \cdot V)$$

and substituting for an ideal gas, knowing that $dH = C_P \, dT$

$$C_P \cdot dT = C_V \cdot dT + RdT$$

then

$$C_P - C_V = R \qquad (4-8)$$

Dividing both sides by C_V

$$\frac{C_P}{C_V} - \frac{C_V}{C_V} = \frac{R}{C_V}$$

$$K - 1 = \frac{R}{C_V} \qquad (4-9)$$

and substituting for $\frac{-R}{C_v}$ in equation $\frac{dT}{T} = \frac{-R}{C_v} \cdot \frac{dV}{V}$,

$$\frac{dT}{T} = -(K-1)\frac{dV}{V}$$

upon integration

$$\ln\frac{T_2}{T_1} = -(K-1)\ln\frac{V_2}{V_1}$$

Therefore

$$\frac{T_2}{T_1} = \left(\frac{V_1}{V_2}\right)^{K-1} \tag{4-10}$$

Using the ideal gas law

$$\frac{P_2 \cdot V_2}{P_1 \cdot V_1} = \frac{T_2}{T_1}$$

and substituting for T_2/T_1 from Equation (4-10):

$$\frac{P_2 \cdot V_2}{P_1 \cdot V_1} = \left(\frac{V_1}{V_2}\right)^{K-1}$$

or

$$P_2 \cdot V_2{}^K = P_1 \cdot V_1{}^K \tag{4-11}$$

or, in general for adiabatic gas compression, the following relationship could be obtained:

$$P \cdot V^K = C$$

(Smith and Van Ness 1959).

The above relationship between the pressure and volume of a natural gas defined for an adiabatic process can be used for all gas compression calculations to obtain head or power requirements and other unknowns with some modifications.

The pressure/volume relationship in a polytropic process is defined as:

$$P \cdot V^n = C \tag{4-12}$$

where P = gas pressure
V = gas volume
n = polytropic gas exponent
C = constant

All equations derived for adiabatic processes could be easily converted to polytropic processes using the relationship between K and n, which will be defined later.

Work-Isothermal Gas Compression

As previously discussed, it is practically impossible to compress a gas isothermally, but the theoretical equations for this type of compression can be derived. It can be shown that the work required to compress a gas isothermally is less than for other types of compression.

This is because gas is continuously cooled during an isothermal compression process to keep its temperature constant. To derive the work required to compress a gas isothermally, the general energy equation will be used:

$$\frac{1}{g_c} u \cdot du + V \cdot dP + dP_E + df + dW = 0 \qquad (4-13)$$

where

$$\frac{1}{g_c} u \cdot du = \text{ kinetic energy term}$$

$V\,dP$ = pressure energy term
dP_E = potential energy term
df = summation of all losses
dW = work done on the system.

In Equation (4-13), some of the energy terms are negligible and can be either neglected or added to the power requirements.

The change in kinetic energy can be neglected since the inlet and outlet nozzles of the compressor are designed in such a way as to minimize velocity changes. The change in potential energy is practically zero as the gas inlet and outlet nozzles are at the same level. Finally, all the friction losses can be added to the final power requirements, depending on the type of the compressor and its efficiency. After these simplifications, Equation (4-13) would change to:

$$-dW = V \cdot dP \qquad (4-14)$$

Depending on the type of thermodynamic process, the work required to compress a gas could be calculated from Equation (4-14). In an isothermal process for an ideal gas, $P \cdot V = n \cdot R \cdot T$ or $V = (n \cdot R \cdot T / P)$, which, if substituted into Equation (4-14), gives

$$-dW = n \cdot R \cdot T \cdot \frac{dP}{P} \qquad (4-15)$$

or

$$-W = n \cdot R \cdot T \int_1^2 \frac{dP}{P}$$

and

$$-W = n \cdot R \cdot T \cdot \ln \frac{P_2}{P_1} \qquad (4-16)$$

Furthermore, the number of moles of gas is $n = (m/M)$, where for 1 lb_m of gas $n = (1/M)$ or $n = (1/29G)$. Therefore, in imperial units

$$-W = \frac{1}{29G} \cdot 10.73 \cdot 144 \cdot T \cdot \ln \frac{P_2}{P_1}$$

or

$$-W = \frac{53.28}{G} \cdot T \cdot \ln \frac{P_2}{P_1} \qquad (4-17)$$

where $-W$ = work done (head) on the compressor to compress 1 lb_m of gas from P_1 to P_2 at constant temperature, $ft\cdot lb_{f/lb_m}$

G = gas gravity, dimensionless
T = gas temperature (constant), $^\circ R$
P_1 = suction pressure, psia
P_2 = discharge pressure, psia.

It should be noticed that work is normally stated as $-W$, as it is work being done on the system.

Work-Adiabatic Gas Compression

As previously discussed, the pressure-volume relationship in an adiabatic process is defined as $P\cdot V^K = C$, which if substituted into Equation (4-14), gives the work required to compress a gas adiabatically:

$$-W = \int_1^2 V \cdot dP \qquad (4-18a)$$

substituting for V

$$-W = \int_1^2 C^{1/K} \cdot P^{-1/K} \cdot dP$$

and after integration

$$-W = C^{\frac{1}{K}} \cdot \frac{K}{K-1} \left[P_2^{\frac{K-1}{K}} - P_1^{\frac{K-1}{K}} \right]$$

after further simplification

$$-W = C^{\frac{1}{K}} \cdot \frac{K}{K-1} \cdot P_1^{\frac{K-1}{K}} \left[\left(\frac{P_2}{P_1}\right)^{\frac{K-1}{K}} - 1 \right]$$

From equation $PV^K = C$, the volume can be expressed as $V = C^{1/K} \cdot P^{-1/K}$, giving

$$-W = P_1 \cdot V_1 \cdot \frac{K}{K-1} \cdot \left[\left(\frac{P_2}{P_1}\right)^{\frac{K-1}{K}} - 1 \right] \qquad (4-18b)$$

and after substitution for $P_1 \cdot V_1 = n \cdot R \cdot T_1$

$$-W = n \cdot R \cdot T_1 \cdot \frac{K}{K-1} \cdot \left[\left(\frac{P_2}{P_1}\right)^{\frac{K-1}{K}} - 1 \right] \qquad (4-18c)$$

Now, if Equation (4-18c) is further simplified knowing that $n = m/M$, and $m = 1$ lb$_m$ of gas, then $n = 1/M$ or $n = 1/29G$, so

$$-W = \frac{1}{29G} \cdot 10.73 \cdot 144 \cdot T_1 \cdot \frac{K}{K-1} \cdot \left[\left(\frac{P_2}{P_1}\right)^{\frac{K-1}{K}} - 1 \right]$$

then in Imperial Units

$$-W = \frac{53.28}{G} \cdot T_1 \cdot \frac{K}{K-1} \cdot \left[\left(\frac{P_2}{P_1}\right)^{\frac{K-1}{K}} - 1 \right] \qquad (4-19)$$

where $-W$ = work (head) to be done on the compressor to adiabatically compress gas from P_1 to P_2, ft·lb$_f$/lb$_m$

G = gas gravity, dimensionless

T_1 = suction temperature, °R

K = adiabatic gas exponent, dimensionless

P_1 = suction pressure, psia

P_2 = discharge pressure, psia

P_2/P_1 = compression ratio (CR), dimensionless

Example 4.1

Find the work required to compress a gas from a suction pressure of $P_1 = 200$ psia to a discharge pressure of $P_2 = 800$ psia, if gas gravity is $G = 0.60$, $K = 1.30$, and suction temperature is 520 °R, for: (a) isothermal compression; and (b) adiabatic compression.

(a) Isothermal Compression:

$$-W = \frac{53.28}{G} \cdot T_1 \cdot \ln\frac{P_2}{P_1}$$

$$-W = \frac{53.28}{0.60} \cdot 520 \cdot \ln\frac{800}{200} = 6,4015 \quad \text{ft} \cdot \text{lb}_f/\text{lb}_m$$

(b) Adiabatic Compression:

$$-W = \frac{53.28}{G} \cdot T_1 \cdot \frac{K}{K-1} \cdot \left[\left(\frac{P_2}{P_1}\right)^{\frac{K-1}{K}} - 1 \right]$$

$$-W = \frac{53.28}{0.60} \cdot 520 \cdot \frac{1.3}{1.3-1} \cdot \left[\left(\frac{800}{200}\right)^{\frac{1.3-1}{1.3}} - 1 \right]$$

$$-W = 75,440 \quad \text{ft} \cdot \text{lb}_f/\text{lb}_m$$

From comparing the results for the two types of compression, it can be seen that more work (higher head) is required for adiabatic compression than for isothermal compression to obtain the same final pressure.

TEMPERATURE CHANGE IN ADIABATIC GAS COMPRESSION

METHOD 1:

In adiabatic gas compression, gas temperature increases according to the equations described below.

Using Equation (4-11) for both suction and discharge conditions:

$$P_1 \cdot V_1{}^K = P_2 \cdot V_2{}^K$$

and from the real gas laws the following equation can be written:

$$\frac{P_1 \cdot V_1}{P_2 \cdot V_2} = \frac{Z_1 \cdot T_1}{Z_2 \cdot T_2} \qquad (4-20)$$

and

$$\frac{V_1}{V_2} = \frac{P_2}{P_1} \cdot \frac{Z_1 \cdot T_1}{Z_2 \cdot T_2}$$

if this is substituted into Equation (4-11), then

$$\left(\frac{P_2}{P_1} \cdot \frac{Z_1 \cdot T_1}{Z_2 \cdot T_2}\right)^K = \frac{P_2}{P_1}$$

Rearranging:

$$\frac{Z_2 \cdot T_2}{Z_1 \cdot T_1} = \left(\frac{P_2}{P_1}\right)^{\frac{K-1}{K}} \qquad (4-21)$$

where T_1 = gas suction temperature, °R
T_2 = gas discharge temperature, °R
P_1 = gas suction pressure, psia
P_2 = gas discharge pressure, psia
Z_1 = gas compressibility factor at suction condition, dimensionless
Z_2 = gas compressibility factor at discharge condition, dimensionless
K = adiabatic gas exponent, dimensionless

Equation (4-21) can be used to calculate compressor discharge temperature (acceptable accuracy when the compressor unit adiabatic efficiency is not given), if the compressor suction pressure, temperature, and ultimately compressibility factor at suction conditions are known. Determining the compressor discharge temperature from this equation requires a trial-and-error solution, since the values of Z_2 and T_2 are both unknown. Equation (4-21) can be solved by first assuming a value of $Z_2 = 1.0$ to start the iteration. Different values of Z_2 should then be applied using a trial-and-error approach until a satisfactory discharge temperature is obtained (i.e., within the convergence limit).

If the variations in temperature and pressure are not significant (i.e., $Z_1 \approx Z_2$), then Z_1 and Z_2 could be canceled out of the equation. Gas discharge temperature can then be calculated using the following equation:

$$\frac{T_2}{T_1} = \left(\frac{P_2}{P_1}\right)^{\frac{K-1}{K}} \qquad (4-22)$$

where the parameters are the same as before (discharge temperature less accurate than Eq. (4–21)).

Example 4.2

What would be the gas discharge temperature in Example 4.1, assuming insignificant changes in compressibility factor?

Solution

Use the simplified Equation (4-22)

$$\frac{T_2}{T_1} = \left(\frac{P_2}{P_1}\right)^{\frac{K-1}{K}}$$

and

$$\frac{T_2}{520} = \left(\frac{800}{200}\right)^{\frac{1.3-1}{1.3}} = 1.377$$

$$T_2 = 716\,^{\circ}R = 256\,^{\circ}F$$

Method 2: Use of Isentropic (Adiabatic) Efficiency and Ideal Gas Law

This method is a reasonably accurate method (especially at lower Compression Ratios, less than 1.5) to calculate the gas discharge temperature from a compressor station and it is used in most of the commercial tools simulating gas transmission pipelines.

The basic steps to derive the equation are as follows:

According to the definition of the Adiabatic or Isentropic Efficiency:

$$\eta_a = \frac{Adiabatic\ Head}{Actual\ Head} \tag{4-23}$$

And

$$(Head)_{Adiabatic} = P_1 \cdot V_1 \cdot \frac{K}{K-1} \cdot \left[(CR)^{\frac{K-1}{K}} - 1\right] \tag{4-24}$$

Where CR is the Compression Ratio

$$(Head)_{Actual} = \dot{n} \cdot Cp \cdot \Delta T = \dot{n} \cdot Cp \cdot (T_d - T_s) \tag{4-25}$$

Where T_s and T_d are suction and discharge temperatures. Using the ideal gas law

$$P_1 \cdot V_1 = \dot{n} \cdot R \cdot T_1 = \dot{n} \cdot R \cdot T_s \tag{4-26}$$

Upon substitution in Eq (4-23):

$$\eta_a = \frac{\dot{n} \cdot R \cdot Ts \cdot \frac{K}{K-1} \cdot \left[(CR)^{\frac{K-1}{K}} - 1\right]}{\dot{n} \cdot Cp \cdot (T_d - T_s)} \tag{4-27}$$

Or

$$\eta_a = \frac{R \cdot T_s \frac{K}{K-1} \cdot \left[(CR)^{\frac{K-1}{K}} - 1 \right]}{Cp \cdot (T_d - T_s)} \qquad (4-28)$$

For an ideal gas system, we obtained the following relationships:

$$C_p - C_v = R, K = \frac{C_p}{C_v} \text{ and } R\left(\frac{K}{K-1}\right) = C_p \qquad (4-29)$$

Then

$$\eta_a = \frac{C_p \cdot T_s \cdot \left[(CR)^{\frac{K-1}{K}} - 1 \right]}{Cp \cdot (T_d - T_s)} \qquad (4-30)$$

Which will result in?

$$T_d - T_s = \frac{T_s \cdot \left[(CR)^{\frac{K-1}{K}} - 1 \right]}{\eta_a} \qquad (4-31)$$

From equation (4-31), we can obtain the gas discharge temperature if the gas is assumed to be ideal.

Equation (4-31) could be further improved for a higher accuracy if we multiply the average compressibility factor at both suction and discharge of the compressor into the RHS of the equation as follows:

$$T_d - T_s = \frac{T_s \cdot \left[\frac{Z_s + Z_d}{2} \right] \cdot \left[(CR)^{\frac{K-1}{K}} - 1 \right]}{\eta_a} \qquad (4-32)$$

The results obtained from equation (4-32) are quite accurate and very close to the results obtained from the full enthalpy change method.

Method 3: Full Enthalpy Change Method

This method which assumes real gas relationship, calculates the gas discharge temperature in a gas compressor based on the enthalpy change of the gas from the suction to the discharge of the compressor.

As there are no simplifications in this method, the gas discharge temperature is very accurate, but of course it is a more time consuming method.

The method is based on the following two equations to iteratively calculate the gas discharge temperature

$$\Delta H = H_d - H_s \qquad (4-33)$$

And

$$H = H^O + RT \left[(Z - 1) - \int_0^\rho \frac{T}{\rho} \left(\frac{\partial Z}{\partial T} \right) \partial \rho \right] \qquad (4-34)$$

Where H is the enthalpy of the gas.

(For further information see AGA Report No - 10, 2003)

The value of H_s (Enthalpy) at the suction condition (at known P_s and T_s) is determined, the value of H_d (Enthalpy) at the discharge condition (P_d known, T_d unknown) should be iterated for T_d to converge (the adiabatic efficiency of the compressor is known).

For lower compression ratios (CR < 1.6), the results obtained from this method are very close to the results obtained from Method 2 with average compressibility factor correction (Equation 4-32).

The following table (Table 4.1) summarizes the results obtained for the gas discharge temperature for the three common methods (i.e. Method 2 with and without compressibility factor correction and Method 3, full Enthalpy Change) at three different compression ratios.

TABLE 4.1: Comparison of Gas Discharge Temperature for Method 2 and Method 3

P_s kPa · g	P_d kPa · g	CR	T_s deg, c°	T_d without Z correction, c°	T_d with Z correction, c°	T_d Enthalpy Change Method, c°
6000	8273.7	1.37	21.1	52.7	49.4	49.4
5000	8273.7	1.64	21.1	71.1	66.6	66.7
2050	8273.7	3.89	21.1	159.3	153.2	151.8

Compressor Head and Horsepower

Head is the amount of work or energy injected into the gas to raise its pressure from P_1 to P_2. It has the units of kJ/kg in SI or ft · lb_f/lb_m in Imperial Units (see equations derived and presented for different types of compression earlier in this chapter). On the other hand, horsepower (HP) is defined as:

$$HP = \frac{mass \ flow \cdot head}{thermal \ efficiency \ of \ compression} \qquad (4-35)$$

The common unit for the calculation of the power requirements of a compressor in Imperial Units is HP/1 MMSCFD of gas. To calculate the power requirement of a compressor, the same equation derived for the calculation of head can be used and then multiplied by the gas mass flow, as shown below.

The number of lb moles in 1,000,000 cubic feet of gas at standard condition is

$$n = \frac{PV}{RT} = \frac{(14.7)(1,000,000)}{(10.73)(520)} = 2,634.6 \ \text{lb moles}$$

The mass of 1 lb mole of gas is $n = m/M$ or $m = 29G$. It is further known that 1 HP·HR = 1.98×10^6 ft·lb_f. Therefore, substituting the values in Equation (4-19) derived for adiabatic work (head) in ft·lb_f/lb_m, the resulting equation has the unit of power (HP/1 MMSCFD) as follows:

$$-W = \frac{53.28}{G} \cdot T_1 \cdot \frac{K}{K-1} \cdot \left[\left(\frac{P_2}{P_1} \right)^{\frac{K-1}{K}} - 1 \right] \text{ ft} \cdot \text{lb}_f / \text{lb}_m$$

$$HP = \frac{K}{K-1} \cdot T_1 \cdot \frac{53.28}{G} \cdot \frac{29G \cdot 2,634.6}{24 \times 1.98 \cdot 10^6} \left[\left(\frac{P_2}{P_1} \right)^{\frac{K-1}{K}} - 1 \right] \text{ HP} / 1\text{MMSCFD}$$

which will give

$$HP = 0.0857 \cdot \frac{K}{K-1} \cdot T_1 \cdot \left[\left(\frac{P_2}{P_1} \right)^{\frac{K-1}{K}} - 1 \right] \qquad (4-36)$$

where HP = adiabatic power requirement, HP/1 MMSCFD
K = adiabatic gas exponent
T_1 = suction temperature, °R
P_1 = suction pressure, psia
P_2 = discharge pressure, psia
P_2/P_1 = compression ratio (CR), dimensionless

This of course assumes a thermal efficiency equal to unity. In Equation (4-36), the impact of the compressibility factor has not been considered; in most cases, it is significant. To consider the Z factor in the equation, the real gas relationship of $P_1 V_1 = nRZ_1 T_1$ should have been used in the initial derivation. Therefore

$$HP = 0.0857 \cdot \frac{K}{K-1} \cdot Z_1 \cdot T_1 \cdot \left[\left(\frac{P_2}{P_1} \right)^{\frac{K-1}{K}} - 1 \right] \qquad (4-37)$$

where all the parameters are the same as in Equation (4-36), and Z_1 is the compressibility of gas at suction condition. It is sometimes recommended that an average for compressibility factor be taken at suction and discharge conditions. Equation (4-37) then becomes

$$HP = 0.0857 \cdot \frac{K}{K-1} \cdot T_1 \cdot \frac{Z_1 + Z_2}{2} \left[\left(\frac{P_2}{P_1} \right)^{\frac{K-1}{K}} - 1 \right] \qquad (4-38)$$

where Z_1 and Z_2 are compressibility factors at suction and discharge conditions, respectively.
Equation (4-38) can be further improved if the adiabatic efficiency of the compression is considered as follows:

$$HP = 0.0857 \cdot \frac{K}{K-1} \cdot T_1 \cdot \frac{Z_1 + Z_2}{2} \cdot \frac{1}{\eta_a} \cdot \left[\left(\frac{P_2}{P_1} \right)^{\frac{K-1}{K}} - 1 \right] \qquad (4-39)$$

where η_a is defined as adiabatic (isentropic) efficiency, which is typically in the range of 0.75 to 0.80.

Adiabatic (isentropic) efficiency η_a is defined as the ratio between adiabatic (isentropic) head and the actual head as follows:

$$\eta_a = \frac{(\text{HEAD})_{\text{ADIABATIC}}}{(\text{HEAD})_{\text{ACTUAL}}}$$

For example, if in a situation a gas compressor theoretically needs 30,000 ft·lb$_f$/lb$_m$ of head to produce a CR $= P_2/P_1$, but in an operational case the head is 40,000 ft·lb$_f$/lb$_m$, then the adiabatic efficiency of this system is $\eta_a = 30,000/40,000 = 0.75$. This means that, this compressor will only use 75% of the head produced to convert to pressure, the other 25% is being wasted.

Adiabatic (isentropic) efficiency of a gas compressor could also be represented by the following equation:

$$\eta_a = \frac{T_1 \left[\left(\dfrac{P_2}{P_1} \right)^{\frac{K-1}{K}} - 1 \right]}{T_2 - T_1} \qquad (4 - 39\text{a})$$

where
η_a = adiabatic (isentropic) efficiency, dimensionless
T_1 = suction temperature, R°
T_2 = discharge temperature, R°
P_1 = suction pressure, psia
P_2 = discharge pressure, psia
K = adiabatic gas exponent, dimensionless

THERMODYNAMICS OF POLYTROPIC GAS COMPRESSION

A polytropic process is the general case in which no specific conditions for compression, except mechanical reversibility, are imposed. For the adiabatic process, the relationship between P and V was defined as $P \cdot V^K = C$, where there is no heat transfer between the system and its surroundings ($dq = 0$). There is no such condition for a polytropic process—heat can be transferred between the system and its surroundings.

In a polytropic process, the relationship between P and V is defined by Equation (4-12), which is

$$P \cdot V^n = C$$

Polytropic Gas Exponent and Polytropic Efficiency

As previously discussed, the pressure/volume relationship in adiabatic and polytropic processes are defined as $P \cdot V^K = C$ and $P \cdot V^n = C$, respectively. It was further described that the adiabatic gas exponent $K = C_P/C_V$ can be readily obtained from the gas composition and its temperature. There are also charts available that give adiabatic or polytropic efficiency if either one of them is given (e.g., Figure 4-5).

Approximate Heat-Capacity Ratios of Hydrocarbon Gases

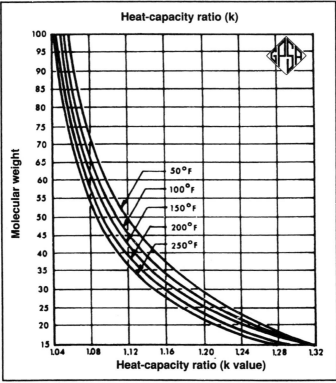

Figure 4-5. Approximate heat-capacity ratios of hydrocarbon gases [Courtesy GPSA, Engineering Data Book, 1994]

For adiabatic compression the change in temperature was expressed as:

$$\left(\frac{T_2}{T_1}\right)_a = (CR)^{\frac{K-1}{K}}$$

whereas in a polytropic process the relationship is:

$$\left(\frac{T_2}{T_1}\right)_P = (CR)^{\frac{n-1}{n}}$$

For two compressors with the same compression ratios and the same suction temperature, the gas discharge temperature will not be the same if one compressor follows an adiabatic process and the other a polytropic process. In order to derive the relationship between K and n, an identical discharge temperature is assumed for both processes, while multiplying the exponent of (CR) in the polytropic process by a constant, which is called η_P or polytropic efficiency, as follows:

$$\eta_P \cdot \frac{n-1}{n} = \frac{K-1}{K}$$

TABLE 4-2. Molar heat capacity MC$_P$ (Ideal-Gas State), Btu/(lb mol· °R) [Courtesy of GPSA, Engineering Data Book, 1994]

Gas	Chemical Formula	Mol wt	0 °F	50 °F	60 °F	100 °F	150 °F	200 °F	250 °F	300 °F
Methane	CH$_4$	16.043	8.23	8.42	8.46	8.65	8.95	9.28	9.64	10.01
Ethyne (Acetylene)	C$_2$H$_2$	26.038	9.68	10.22	10.33	10.71	11.15	11.55	11.90	12.22
Ethene (Ethylene)	C$_2$H$_4$	28.054	9.33	10.02	10.16	10.72	11.41	12.09	12.76	13.41
Ethane	C$_2$H$_6$	30.070	11.44	12.17	12.32	12.95	13.78	14.63	15.49	16.34
Propene (Propylene)	C$_3$H$_6$	42.081	13.63	14.69	14.90	15.75	16.80	17.85	18.88	19.89
Propane	C$_3$H$_8$	44.097	15.65	16.88	17.13	18.17	19.52	20.89	22.25	23.56
1-Butene (Butylene)	C$_4$H$_8$	56.108	17.96	19.59	19.91	21.18	22.74	24.26	25.73	27.16
cis-2-Butene	C$_4$H$_8$	56.108	16.54	18.04	18.34	19.54	21.04	22.53	24.01	25.47
Trans-2Butene	C$_4$H$_8$	56.108	18.84	20.23	20.50	21.61	23.00	24.37	25.73	27.07
iso-Butane	C$_4$H$_{10}$	58.123	20.40	22.15	22.51	23.95	25.77	27.59	29.39	31.11
n-Butane	C$_4$H$_{10}$	58.123	20.80	22.38	22.72	24.08	25.81	27.55	29.23	30.90
iso-Pentane	C$_5$H$_{12}$	72.150	24.94	27.17	27.61	29.42	31.66	33.87	36.03	38.14
n-Pentane	C$_5$H$_{12}$	72.150	25.64	27.61	28.02	29.71	31.86	33.99	36.08	38.13
Benzene	C$_6$H$_6$	78.114	16.41	18.41	18.78	20.46	22.45	24.46	26.34	28.15
n-Hexane	C$_6$H$_{14}$	86.177	30.17	32.78	33.30	35.37	37.93	40.45	42.94	45.36
n-Heptane	C$_7$H$_{16}$	100.204	34.96	38.00	38.61	41.01	44.00	46.94	49.81	52.61
Ammonia	NH$_3$	17.0305	8.52	8.52	8.52	8.52	8.52	8.53	8.53	8.53
Air		28.9625	6.94	6.95	6.95	6.96	6.97	6.99	7.01	7.03
Water	H$_2$O	18.0153	7.98	8.00	8.01	8.03	8.07	8.12	8.17	8.23
Oxygen	O$_2$	31.9988	6.97	6.99	7.00	7.03	7.07	7.12	7.17	7.23
Nitrogen	N$_2$	28.0134	6.95	6.95	6.95	6.96	6.96	6.97	6.98	7.00
Hydrogen	H$_2$	2.0159	6.78	6.86	6.87	6.91	6.94	6.95	6.97	6.98
Hydrogen sulfide	H$_2$S	34.08	8.00	8.09	8.11	8.18	8.27	8.36	8.46	8.55
Carbon monoxide	CO	28.010	6.95	6.96	6.96	6.96	6.97	6.99	7.01	7.03
Carbon dioxide	CO$_2$	44.010	8.38	8.70	8.76	9.00	9.29	9.56	9.81	10.05

(1) Exceptions: Air, Keehan and Keyes, Thermodynamic Properties of Air, Wiley, 3rd Printing 1947, Ammonia, Edw. R. Grabl. Thermodynamic Properties of Ammonia at High Temperatures and Pressures, Petr. Processing, April 1953. Hydrogen Sulfide-J.R. West, Chem. Eng. Progress, 44, 287, 1948.
(2) Data source: Selected Values of Properties of Hydrocarbons, API Research Project 44.

or

$$\eta_P = \frac{K-1}{K} \cdot \frac{n}{n-1} \qquad (4-40)$$

where K = adiabatic gas exponent, dimensionless
n = polytropic gas exponent, dimensionless
η_P = polytropic efficiency, dimensionless

In Equation (4-40), the value of $K = C_P/C_V$ can be obtained from Table 4-2. For ideal gases, Equation (4-8) or $C_P - C_V = R$, where $R = 1.986$ BTU/lb mole · °R can be used.

Figure 4-5 can also be used to directly obtain an approximate value of K if the gas molecular weight and temperature are known.

Power in a Polytropic Process

The same procedure as that used earlier for an adiabatic process, can be used to derive the power requirements for a polytropic process. The equations to calculate *HP* requirements are as follows:

$$HP = 0.0857 \cdot \frac{n}{n-1} \cdot T_1 \cdot Z_1 \cdot \frac{1}{\eta_P} \left[\left(\frac{P_2}{P_1} \right)^{\frac{n-1}{n}} - 1 \right] \qquad (4-41)$$

or

$$HP = 0.0857 \cdot \frac{n}{n-1} \cdot T_1 \cdot \frac{Z_1 + Z_2}{2} \cdot \frac{1}{\eta_P} \left[\left(\frac{P_2}{P_1} \right)^{\frac{n-1}{n}} - 1 \right] \qquad (4-42)$$

where HP = polytropic power required to compress a gas from P_1 to P_2, HP/1 MMSCFD
n = polytropic gas exponent, dimensionless
T_1 = suction temperature, °R
Z_1 = compressibility factor at suction condition, dimensionless
Z_2 = compressibility factor at discharge condition, dimensionless
η_P = polytropic efficiency, dimensionless
P_1 = suction pressure, psia
P_2 = discharge pressure, psia

Equation (4-42) may be rearranged using the relationship between n and K, defined in Equation (4-40), and substituting for η_P:

$$HP = 0.0857 \cdot T_1 \cdot \frac{Z_1 + Z_2}{2} \cdot \frac{K}{K-1} \left[\left(\frac{P_2}{P_1} \right)^{\frac{n-1}{n}} - 1 \right] \qquad (4-43)$$

The polytropic efficiency is established by the manufacturer, and is generally a function of the capacity at the inlet of the compressor. Figure 4-6 gives the relationship between η_a, η_P, CR, and adiabatic (isentropic) gas exponent (K).

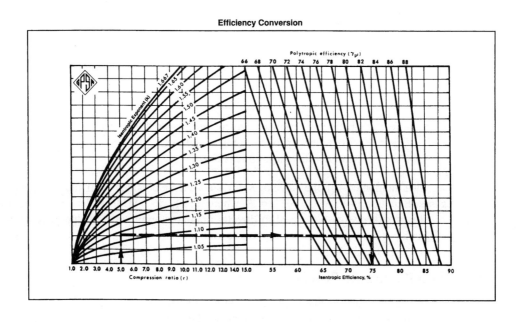

Figure 4-6. Efficiency conversion [Courtesy GPSA, Engineering Data Book, 1994]

GAS COMPRESSORS IN SERIES

Limitations on the compression ratio and temperature of a gas compressor sometimes require designers to use a number of compressors in series to obtain the required discharge pressure:

1. Compression ratio limitations: For safety reasons, manufacturers generally recommend a compression ratio of less than six for gas compressors. At high compression ratios a significant amount of force is exerted on the shaft and other mechanical components of the compressor, which makes the design of compressors complicated and expensive, and sometimes compressor operation is unsafe. Specially, at high pressures and high flow rates along large gas transmission lines, the value of the compression ratio is normally limited to between 1.2 and 2.0.
2. Temperature limitations: Compressor manufacturers normally recommend a maximum discharge temperature of 300 °F (350 °F for oxygen-free gases). This value should be lowered down to a limit of 250–275°F for gases with a small amount of oxygen.

Extra compressors must be added in series if either the compression ratio ($CR \leq 6$) or temperature ($T_d \leq 300$) rules are violated.

Theory of Equal Compression Ratios for Gas Compressors in Series

When a number of compressors are operating in series to bring the gas to a required discharge pressure, there are usually many possible configurations. For example, consider a case where natural gas needs to be compressed from 100 psia to 1,600 psia using two compressors in series. Assuming the limiting conditions $CR \leq 6$ and $T_d \leq 300°F$ are maintained, there are an infinite number of potential combinations. Some of these possibilities are shown in Figure 4-7.

Among the many possibilities for raising the pressure from $P_S = 100$ psia to $P_d = 1,600$ psia using two compressors in series, the combination that minimizes the HP requirements for the gas compression should be chosen. This combination is the one with equal compression ratios, which gives $CR_1 = CR_2 = 4$. The mathematical proof for this operation is shown below.

Consider the setup as shown in Figure 4-8 for two compressors in series with an intercooler that cools the gas to its initial temperature (T_S). Assume the gas cooler exerts no pressure drop and the HP requirements for each compressor are $(HP)_1$ and $(HP)_2$, respectively.

The power required to compress gas in the first compressor is $(HP)_1$ and it is calculated using Equation (4-36):

$$(HP)_1 = 0.0857 \cdot \frac{K}{K-1} \cdot T_S \cdot \left[\left(\frac{P_i}{P_S} \right)^{\frac{K-1}{K}} - 1 \right]$$

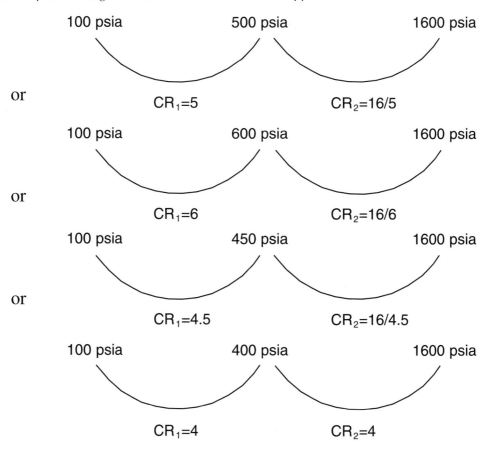

Figure 4-7. Some possible compressor ratio combinations for two compressors in series to compress gas from 100 psia to 1,600 psia

Likewise, the power requirement for the second compressor is given by:

$$(HP)_2 = 0.0857 \cdot \frac{K}{K-1} \cdot T_S \cdot \left[\left(\frac{P_d}{P_i} \right)^{\frac{K-1}{K}} - 1 \right]$$

where the total work is:

$$HP = 0.0857 \cdot \frac{K}{K-1} \cdot T_S \cdot \left[\left(\frac{P_d}{P_S} \right)^{\frac{K-1}{K}} - 1 \right]$$

The value of the term "$0.0857 \cdot K/(K-1) \cdot T_S$" is constant, and if $P_i/P_S = x$ and $P_d/P_i = y$, then $(P_i/P_S) \cdot (P_d/P_i) = x \cdot y$. If it is assumed that $P_d/P_S = C_1$ is constant, then

$$x \cdot y = C_1 \text{ (constant)}$$

from the above equations:

$$HP = (HP)_1 + (HP)_2 \tag{4 – 44}$$

and if the values for $(HP)_1$ and $(HP)_2$ are substituted, then:

$$HP = C\left[x^{\frac{K-1}{K}} + y^{\frac{K-1}{K}} - 2\right] \qquad (4-45)$$

now substitute for $y = \dfrac{C_1}{x}$

$$HP = C\left[x^{\frac{K-1}{K}} + \left(\frac{C_1}{x}\right)^{\frac{K-1}{K}} - 2\right] \qquad (4-46)$$

To determine the minimum power, $\dfrac{d(HP)}{dx} = 0$, so

$$\left(x^{\frac{K-1}{K}}\right)^2 = C_1^{\frac{K-1}{K}}$$

or

$$x = \sqrt{C_1}$$

and from equation $x \cdot y = C_1$, $y = \sqrt{C_1}$, so

$$x = y$$

or

$$\frac{P_i}{P_S} = \frac{P_d}{P_i} \qquad (4-47)$$

This proves that $CR_1 = CR_2$ yields minimum power requirements. It can be further proved that for more than two compressors in series, minimum power requirements are achieved when the relationship $CR_1 = CR_2 = CR_3$ is true. The general form of Equation (4-47) is then

$$\left(\frac{P_d}{P_S}\right) = (CR)^n \qquad (4-48)$$

or

$$\left(\frac{P_d}{P_S}\right)^{\frac{1}{n}} = CR \qquad (4-49)$$

where P_d = gas discharge pressure, psia
          P_S = gas suction pressure, psia
          n = number of compressors in series, dimensionless
      CR = compression ratio for each compressor (all equal), dimensionless

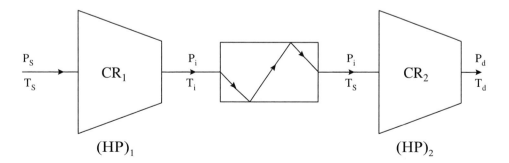

Figure 4-8. Compressors in series with no pressure drop in the intercooler

Example 4.3

The required discharge pressure in a gas compressor is P_d = 1,600 psia, and K = 1.4; the suction pressure is P_S = 100 psia; and suction temperature is T_S = 40°F. The gas is cooled down to its initial temperature at each stage. For a maximum allowable discharge temperature of 300°F, how many compressors in series will be required?

Solution

Check the conditions $CR \leq 6$ and $T_d \leq 300°F$:

Using Equation 4-49, CR can be determined:

$$CR = \left(\frac{P_d}{P_S}\right)^{\frac{1}{n}}$$

If the number of compressors n = 1, then:

$$CR = \left(\frac{1,600}{100}\right)^{\frac{1}{1}} = 16$$

This is too high. Next, try n = 2 for two compressors in series, which gives:

$$CR = \left(\frac{1,600}{100}\right)^{\frac{1}{2}} = 4$$

Therefore, CR = 4, which is acceptable. Now the discharge temperature should be checked using Equation (4-22):

$$\frac{T_2}{T_1} = \left(\frac{P_2}{P_1}\right)^{\frac{K-1}{K}}$$

$$\frac{T_i}{460 + 40} = (4)^{\frac{1.4-1}{1.4}} = 1.486$$

$$T_i = 743°R = 283°F$$

T_i = 283°F < 300°F, so a combination of two compressors in series is acceptable.

Example 4.4

What would be the setup of the compressors in Example 4.3, if suction temperature is increased to 70°F?

Solution

The CR is still acceptable, but a new discharge temperature should be checked:

$$\frac{T_2}{T_1} = \left(\frac{P_2}{P_1}\right)^{\frac{K-1}{K}}$$

or

$$\frac{T_i}{460 + 70} = (4)^{\frac{1.4-1}{1.4}} = 1.486$$

$$T_i = 787.6°R = 327.6°F$$

which is unacceptable, based on the temperature condition of $T_d \leq 300°F$. For this case, there are two solutions:

1. The addition of a precooler to the system to reduce the suction temperature and permit operation of only two compressors with $T_d \leq 300°F$. This temperature can easily be determined by choosing $T_d = 300°F$ and calculating T_S as follows:

$$\frac{T_2}{T_1} = \left(\frac{P_2}{P_1}\right)^{\frac{K-1}{K}}$$

$$\frac{300 + 460}{T_S} = (4)^{\frac{1.4-1}{1.4}} = 1.486$$

$$T_S = 511.5°R = 51.5°F$$

$T_S = 51.5°F$ is the maximum suction temperature allowable.

2. Without using a precooler, the number of compressors must be increased. Try three compressors instead of two, so that $n = 3$:

$$CR = \left(\frac{P_d}{P_S}\right)^{\frac{1}{n}}$$

$$CR = \left(\frac{1,600}{100}\right)^{\frac{1}{3}} = 2.52$$

therefore $CR_1 = CR_2 = CR_3 = 2.52$. Now check that the discharge temperature is less than 300°F:

$$\frac{T_2}{T_1} = (CR)^{\frac{K-1}{K}}$$

or

$$\frac{T_i}{460 + 70} = (2.52)^{\frac{1.4-1}{1.4}} = 1.3022$$

or

$$T_i = 690.2°R = 230°F$$

which is acceptable. It should be noted that intercooling between stages is still required.

Effect of Intercooler Pressure Drop on the Compression Ratio

For the previous case, it was assumed that no pressure drop occurs across the intermediate gas cooler. Therefore, the inlet and exit pressures of the intercooler were equal, and all the calculations were based on this assumption. However, if there is a pressure drop equal to ΔP across the gas cooler, the impact of this pressure drop on the entire system needs to be accounted for (see Figure 4-9).

For minimum power requirements the equal compression ratio rule is valid, therefore:

$$CR_1 = \frac{P_i}{P_S}$$

$$CR_2 = \frac{P_d}{P_i - \Delta P}$$

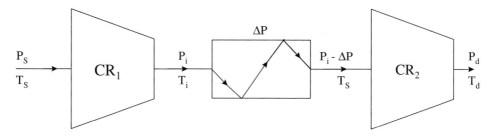

Figure 4-9. Compressors in series with pressure drop across the intercooler

$CR_1 = CR_2$, so

$$\frac{P_i}{P_S} = \frac{P_d}{P_i - \Delta P} \qquad (4-50)$$

or

$$P_i^2 - \Delta P \cdot P_i - P_S \cdot P_d = 0 \qquad (4-51)$$

If Equation (4-51) is solved for P_i, then:

$$P_i = \frac{\Delta P \pm \sqrt{(\Delta P)^2 + 4 \cdot P_S \cdot P_d}}{2} \qquad (4-52)$$

The only valid solution is

$$P_i = \frac{\Delta P + \sqrt{(\Delta P)^2 + 4 \cdot P_S \cdot P_d}}{2} \qquad (4-53)$$

In Equation (4-53), $(\Delta P)^2 < 4 \cdot P_S \cdot P_d$, so:

$$P_i = \frac{\Delta P}{2} + \sqrt{P_S \cdot P_d} \qquad (4-54)$$

Therefore, it could be concluded that in systems with a gas cooler, if the cooler produces a pressure drop equal to ΔP, compression ratios can be equalized if a value equal to $(\Delta P/2)$ is added to the exit pressure from the first compressor, this will then adjust $CR_1 = CR_2 = \ldots = CR_n$.

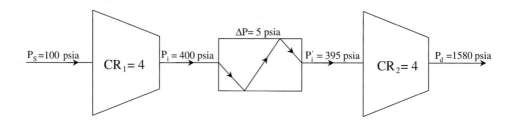

Figure 4-10.   Example 4.5 — Initial compressor configuration for $CR = 4$

Example 4.5

For the following two compressors in series, $P_S = 100$ psia, $P_d = 1,600$ psia, and the intercooler imposes $\Delta P = 5$ psia pressure drop. Determine the best operating CR for this system.

Solution

When the cooler imposes a 5-psi pressure drop, the discharge pressure will drop by 20 psi from 1,600 psia to 1,580 psia. To maintain $P_d = 1,600$ psia, it is necessary to increase $CR_2 = 1,600/395 = 4.05$, which upsets the balance of equal CRs and minimum power consumption.

Therefore, the procedure mentioned above will be used to adjust the compression ratios to equal values, and also to maintain the discharge pressure at $P_d = 1,600$ psia. From Equation (4-54), P_i should be

$$P_i = \frac{\Delta P}{2} + \sqrt{P_S \cdot P_d}$$

$$P_i = \frac{5}{2} + \sqrt{100 \times 1,600}$$

$$P_i = 402.5 \text{ psia}$$

so

$$CR_1 = \frac{402.5}{100} = 4.025$$

and

$$P_i = 402.5 - 5 = 397.5 \text{ psia}$$

or

$$CR_2 = \frac{1,600}{397.5} = 4.025$$

Notice $CR_1 = CR_2 = 4.025$. The system is shown in Figure 4-11.

Equal Compression Ratio Rule for More than Two Compressors in Series

As mentioned earlier, if more than two compressors are set up in series, the least amount of power is required when the compression ratios are equal. However, as the number of

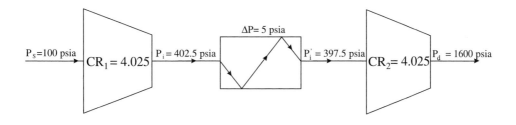

Figure 4-11. Example 4.5 – Compressor configuration correcting for 5-psi pressure drop across the intercooler, $CR = 4.025$

compressors increases, the equations become more complicated. Consider the system in Figure 4-12, with three compressors and two gas coolers in series.

The corresponding compression ratios are:

$$CR_1 = \frac{P_i'}{P_S}$$

$$CR_2 = \frac{P_i''}{P_i' - \Delta P}$$

$$CR_3 = \frac{P_d}{P_i'' - \Delta P}$$

for minimum power consumption $CR_1 = CR_2 = CR_3$, so

$$\frac{P_i'}{P_S} = \frac{P_i''}{P_i' - \Delta P} = \frac{P_d}{P_i'' - \Delta P} \qquad (4-55)$$

In Equation (4-55), the values of P_S, P_d and ΔP are known and P_i' and P_i'' need to be determined. Factoring and simplification give

$$P_i'^3 - \Delta P \cdot P_i'^2 - P_S \cdot \Delta P \cdot P_i' - P_S^2 \cdot P_d = 0 \qquad (4-56)$$

Equation (4-56) can be solved using one of the methods for the solution of cubic equations (there is no exact solution for this equation, but iterative methods can be used to solve it). When P_i' is obtained, P_i'' can be calculated from Equation (4-55). All the CRs can then be determined ($CR_1 = CR_2 = CR_3$).

It should be noted that for compressors in series, gas coolers with low-pressure drops should be chosen (within the range of 1–5 psi) to minimize the total pressure drop within the system, as higher-pressure drops are multiplied in each stage and will finally result in a significant drop in pressure.

Impact of Compressor Station Yard Losses on the Compression Ratio of Centrifugal and Reciprocating Compressors

When compressor systems are compared based on overall station horsepower requirements, the losses in station piping, valves, or headers must be considered along with compressor horsepower requirements.

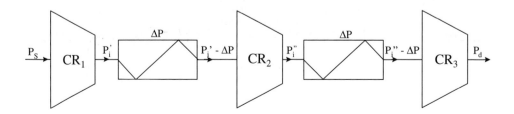

Figure 4-12. Three compressors in series

The compression ratio (*CR*) for centrifugal units in series can be obtained from the following equation:

$$CR = \left[C_o \times (1 + \text{ fraction of loss between each unit})^{n-1} \right]^{\frac{1}{n}} \qquad (4-57)$$

where   *CR* = actual compression ratio for each unit, dimensionless
         n = number of compressors in series
and C_o is defined by the following equation:

$$C_o = \frac{\text{station discharge pressure} + \text{discharge losses}}{\text{station suction pressure} - \text{suction losses}} \qquad (4-58)$$

For reciprocating compressors, the *CR* for each reciprocating unit is defined by the equation:

$$CR = \frac{\text{station discharge pressure} + \text{discharge losses}}{\text{station suction pressure} - \text{suction losses}} \qquad (4-59)$$

which is just like saying $CR = C_o$ for reciprocating compressors.

Example 4.6

Consider a typical station, which has three centrifugal units arranged in series. If the following losses are experienced in the operation, determine the *CR* for the centrifugal units.

- Suction pressure to station = 600 psia
- Discharge pressure from station = 900 psia
- Scrubber and suction piping loss = 3 psi
- Suction valve loss = 5 psi
- Discharge valve loss = 5 psi
- Discharge piping loss = 2 psi
- Loss between each series unit = 0.5 %

Solution

For centrifugal compressors, *CR* would be obtained as follows:

$$C_o = \frac{\text{station discharge pressure} + \text{discharge losses}}{\text{station suction pressure} - \text{suction losses}}$$

$$C_o = \frac{900 + 2 + 5}{600 - 5 - 3} = 1.532$$

and

$$(1 + \text{ fraction of loss between each unit})^{n-1} = (1.005)^{3-1} = 1.01$$

then

$$CR = \left[C_o \times (1.01) \right]^{\frac{1}{n}} = \left[1.532 \times 1.01 \right]^{\frac{1}{3}} = 1.157$$

If losses within the station are not considered, the calculated CR would be lower and will be obtained as:

$$CR = \left(\frac{P_d}{P_S}\right)^{\frac{1}{n}}$$

or

$$CR = \left(\frac{900}{600}\right)^{\frac{1}{3}} = 1.145$$

This compression ratio would not exactly provide the necessary discharge pressure of 900 psia (Segeler 1965).

CENTRIFUGAL COMPRESSOR HORSEPOWER

The three major parameters most often used in centrifugal compressor design for a given discharge pressure and capacity are: brake (shaft) horsepower, discharge temperature, and operating speed.

Brake Horsepower

The horsepower that was calculated through the adiabatic and polytropic equations is the compressor's gas horsepower, or GHP. The compressor brake horsepower (BHP) equals GHP divided by mechanical efficiency, as shown in Equation (4-60):

$$BHP = \frac{GHP}{\eta_M} \qquad (4-60)$$

or

$$BHP = GHP + Losses \qquad (4-61)$$

where η_M is the compressors mechanical efficiency. For most centrifugal compressor applications, mechanical losses are relatively small, and an average mechanical efficiency of 99% may be assumed for estimation purposes.

Compressor Head and GHP

The equations for both GHP and compressor head (ft·lb$_f$/lb$_m$) were presented earlier in this chapter. More of the relationships between GHP and compressor head will be discussed here.

The head, in feet (ft·lb$_f$/lb$_m$), that a centrifugal gas compressor stage consisting of an impeller and diffuser will develop can be related to the impeller peripheral velocity by

$$Head = \mu' \cdot \frac{u^2}{g_c} \qquad (4-62)$$

where   Head = compressor head, ft·lb$_f$/lb$_m$ (ft)
μ' = pressure coefficient obtained from compressor charts, dimensionless

u = peripheral velocity of impeller, $\pi{\cdot}D{\cdot}N/720$, feet per second, where D = impeller diameter in inches, and N = rotational speed in RPM

g_c = proportionality constant, 32.2 $lb_m{\cdot}ft/lb_f{\cdot}sec^2$

The value of the pressure coefficient μ' is a characteristic of the stage design (Segeler 1965).

The GHP can be defined either by a relationship in terms of enthalpy change, or in terms of compressor head as follows:

$$GHP = \frac{m \cdot \Delta H}{33,000} \qquad (4-63)$$

where   GHP = gas horsepower, HP

m = gas flow rate, lb_m/min

ΔH = enthalpy change, $ft{\cdot}lb_f/lb_m$

or

$$GHP = \frac{m \cdot \text{Head}}{33,000 \times \eta_a} \qquad (4-64)$$

where   Head = compressor adiabatic head, $ft{\cdot}lb_f/lb_m$

η_a = adiabatic efficiency, dimensionless.

Compressor Power De-Ration for Elevation and Temperature

The available power for gas compressors will change with both ambient temperature and the site elevation. Gas compressors are generally rated at the sea level (zero elevation) and a temperature of 15.0 C°. Variations in both temperature and altitude will impact the air density, hence the available power of the gas compressor. As the impact of these two parameters are significant, they have to be considered in the available (site) power calculations.

The following equations are recommended to be used to correct for both elevation and temperature changes.

To consider the impact of elevation on the available power, we have

$$(POWER)_a = (POWER)_r \; x \; 10^{-\frac{h}{62,900}} \qquad (4-65a)$$

Where:

$(POWER)a$ = Engine power available, HP
$(POWER)r$ = Engine rated power (sea level), HP
h = Elevation, ft

Or

$$(POWER)_a = (POWER)_r \; x \; 10^{-\frac{h}{19171.92}} \qquad (4-65b)$$

Where:

$(POWER)a$ = Engine power available, kW
$(POWER)r$ = Engine rated power (sea level), kW
h = Elevation, m

There is another equation available for power correction due to elevation, and is defined as follows:

$$(POWER)_a = (POWER)_r \; x \; \left(1 - \frac{F_h \cdot h}{100,000}\right) \qquad (4-65c)$$

Where

$(POWER)a$ = Engine power available, HP
$(POWER)r$ = Engine rated power (sea level), HP
h = Elevation, ft
F_h = Elevation de-rating factor, the theoretical value is 3.6% per 1000 ft (for engines not specified)

Example 4.7

What is the available power at an elevation of 5000 ft for an engine with 3.6% de-rating factor. Check both equations (4.65a, and 4.65c).
From equation (4.65a) we have:

$$\frac{(POWER)a}{(POWER)r} = 10^{\frac{-5000}{62,900}} = 0.833$$

From equation (4.65c) we have:

$$\frac{(POWER)a}{(POWER)r} = 1 - \frac{3.6 \; x \; 5000}{100,000} = 0.82$$

For the temperature correction we can use the following relationship

$$\frac{(POWER)a}{(POWER)r} = 1 - \frac{Ft(T - Tr)}{1000} \qquad (4-66)$$

Where

$(POWER)a$ = Engine power available, HP
$(POWER)r$ = Engine rated power, HP
F_t = Temperature de-rating factor % per 10 F°
T = Ambient temperature, F°
T_r = Engine rated temperature, 60 F°

The common F_t values used in compressor industry are generally between 1 - 2% per 10 F° or 1.8 - 3.6 per 10 C°, unless otherwise a value is recommended by the compressor manufacturer. The theoretical value which is commonly used (if no information is provided) is 2% divided by 10 F°.

If the ambient temperature is specified below the rated temperature (T_r), compressor available power is increased and if the ambient temperature is higher than the rated temperature (T_r), the compressor available power will decrease.

For example, suppose the ambient temperature specified is 50 C° or 122 F°, the rated temperature (T_r), is 60 F°, and recommended value for F_t is 1.5. The overall de-ration would be:

$$\frac{(POWER)_a}{(POWER)_r} = 1 - \frac{Ft \cdot (T - T_r)}{1000}$$
$$= 1 - \frac{1.5(122 - 60)}{1000}$$
$$= 0.907$$

Which means that the unit should be almost de-rated by 10% at this ambient temperature.

Rotational Speed in Centrifugal Gas Compressors

The peripheral velocity and the diameter of the impellers determine the RPM of a centrifugal compressor. Generally, the peripheral velocity is related to the head to be developed and the impeller diameter is determined by the capacity to be handled, as measured at inlet conditions; thus

$$N = \frac{1,300}{D} \cdot \sqrt{\frac{\text{Head per stage}}{\mu'}} \qquad (4-67)$$

where N = speed of the rotation of impellers, RPM
$\qquad\qquad$ D = impeller diameter, inch
Head per stage = head per stage, ft·lb$_f$/lb$_m$
$\qquad\qquad$ μ' = pressure coefficient, which will be given for each compressor, dimensionless.

Relationships between Flow, Head, Power, and Rotational Speed in Centrifugal Gas Compressors

For centrifugal gas compressors, the following relationships are extensively used to estimate the compressor performance. These relationships are defined as follows:

1. The gas volumetric flow is directly proportional to the impeller rotational speed, that is:

$$\frac{Q_1}{Q_2} = \frac{N_1}{N_2} \qquad (4-68)$$

2. The head is proportional to the square of the impeller rotational speed:

$$\frac{(\text{Head})_1}{(\text{Head})_2} = \left(\frac{N_1}{N_2}\right)^2 \qquad (4-69)$$

3. The shaft horsepower required for compression is proportional to the impeller rotational speed cubed, or

$$\frac{(\text{HP})_1}{(\text{HP})_2} = \left(\frac{N_1}{N_2}\right)^3 \qquad (4-70)$$

where Q_1 = initial flow, MMSCFD
$\qquad$ Q_2 = final flow, MMSCFD
$\qquad$ (Head)$_1$ = initial compressor head, ft·lb$_f$/lb$_m$
$\qquad$ (Head)$_2$ = final compressor head, ft·lb$_f$/lb$_m$
$\qquad$ (HP)$_1$ = initial compressor shaft power, HP
$\qquad$ (HP)$_2$ = final compressor shaft power, HP
$\qquad$ N_1 = initial impeller speed, RPM
$\qquad$ N_2 = final impeller speed, RPM

ENTHALPY/ENTROPY CHARTS (MOLLIER DIAGRAM)

A quick way to solve a compressor problem is by using a Mollier diagram or enthalpy/ entropy charts. These charts can be used to calculate compressor head, horsepower, discharge temperature, and gas cooling requirements. If kinetic and potential energy terms in the general energy equation are neglected, the following is obtained:

$$\Delta H = q - W \qquad (4-71)$$

where ΔH = change in the enthalpy of the system, BTU/lb$_m$
     q = heat transfer rate between system and surrounding, BTU/lb$_m$
     W = work done on the system, BTU/lb$_m$.

For an isothermal process, the heat transfer rate is defined as

$$q = T \cdot \Delta S \qquad (4-72)$$

where ΔS is the entropy change of the system in BTU/lb$_m\cdot$°R, and T is the absolute temperature in °R. Substituting Equation (4-72) into (4-71)

$$\Delta H = T \cdot \Delta S - W \qquad (4-73)$$

Equation (4-73) can be used to calculate head or work done on a compressor for an isothermal process. On the other hand, for an adiabatic process, $q = 0$; thus

$$\Delta H = -W \qquad (4-74)$$

The following example shows how a Mollier diagram can be used to solve a compressor problem.

Example 4.8

Determine the isothermal and adiabatic work required to compress 1 MMSCFD of a natural gas from P_S = 100 psia to a discharge pressure of P_d = 1,600 psia. Assume a suction temperature of 80°F. Use a Mollier diagram for G = 0.60 natural gas, and assume that there is an intercooler, which cools gas to its initial temperature of 80°F also calculate the duty of the intercooler, see figure 4.13. Neglect the effect of Z (Katz 1959).

Solution

The number of compressors needed to be in series as well as the intermediate gas temperature (i.e., gas exit temperature from the first compressor) should first be checked.

Compression could not be accomplished in one stage as CR = 16 is too high. Trying compressors in series:

$$CR = \left(\frac{P_d}{P_S}\right)^{\frac{1}{n}}$$

for two compressors:

$$CR = \left(\frac{1,600}{100}\right)^{\frac{1}{2}} = 4 (\le 6), \quad \text{which is acceptable.}$$

To check the intermediate temperature or exit temperature from the first compressor, the adiabatic gas exponent, K, is required. The composition of the gas is unknown, so calculation of C_P and C_V is not possible. Using Figure 4-5, knowing the gas molecular weight is $M = 0.60 \times 29 = 17.4$, the value of K can be obtained. For an average molecular weight of 17.4, and a suction temperature of 80°F, $K = 1.28$, then

$$\frac{T_2}{T_1} = (CR)^{\frac{K-1}{K}}$$

$$\frac{T_2}{540} = (4)^{\frac{1.28-1}{1.28}}$$

$$T_2 = 731\,^{\circ}R = 271\,^{\circ}F$$

which is an acceptable temperature ($\leq 300°F$). The compressor arrangement for an adiabatic process is shown in Figure 4-13.

Now, the enthalpy/entropy diagram for an adiabatic process can be used (see Figure 4-14). The following steps are used in the solution of the problem:

Step 1:

Knowing both suction pressure ($P_S = 100$ psia) and suction temperature ($T_S = 80°F$) to the compressor, locate Point 1, which has an enthalpy of $H_1 = 380$ BTU/lb·mole.

Step 2:

When gas enters the first compressor, it goes through an adiabatic gas compression cycle in which the entropy remains constant as:

$$\Delta S = \int_{T_1}^{T_2} \frac{dq}{T}$$

where, for an adiabatic process $\Delta S = 0$, so $S_1 = S_2$ (constant entropy process). Now, move from Point 1 ($P_S = 100$ psia, $T_S = 80°F$), to Point 2 ($P_i = 400$ psia) on a constant entropy line (vertical line). Point 2 has a discharge pressure of $P_i = 400$ psia and a temperature of 260°F from the chart, and is the exit point from compressor number 1. Point 2 also has an enthalpy of $H_i = 1,990$ BTU/lb·mole.

The enthalpy change within the first compressor is therefore:

$$\Delta H = H_i - H_1 = 1,990 - 380 = 1,610 \text{ BTU/lb} \cdot \text{mole}$$

Step 3

In this step, gas should be cooled down to its original temperature while the cooler has a zero pressure drop $\Delta P = 0$. Now, move on the constant pressure line of $P_i' = 400$ psia, while reducing the temperature from 260°F to 80°F. At 80°F, Point 3 is obtained: $P_i' = 400$ psia, $T_i' = T_S = 80°F$, and $H_i' = 220$ BTU/lb$_m$·mole.

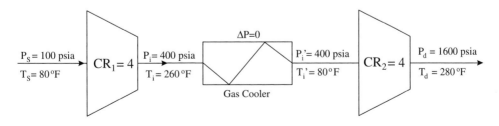

Figure 4-13. Example 4-8 — Compressor arrangement for an adiabatic process

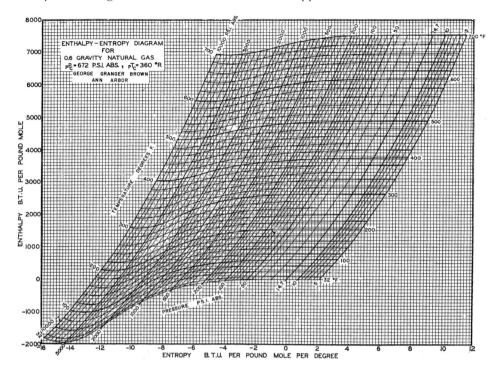

Figure 4-14. Enthalpy-entropy diagram for 0.6-gravity natural gas [Brown, 4.16. Courtesy AIME, Handbook of Natural Gas Engineering, D.L. Katz et al., 1959, McGraw-Hill Co.]

At this point, the amount of heat removed from the gas could be calculated (duty of the intercooler) as follows:

$$\Delta H = H_i' - H_i = 220 - 1,990 = -1,770 \text{ BTU/lb} \cdot \text{mole (heat removed)}$$

Step 4

The gas will be compressed again when it goes through the second compressor. Gas will enter the second compressor from Point 3 and will be adiabatically (i.e., constant entropy line) compressed to $P_d = 1,600$ psia. At Point 4: $P_d = 1,600$ psia, $T_d = 280°F$, and $H_d = 1,920$ BTU/lb·mole. Now

$$\Delta H = H_d - H_i' = 1,920 - 220 = 1,700 \text{ BTU/lb} \cdot \text{mole}$$

To calculate total head (work done on the compressor) and HP requirements;

$$-W = \Delta H_T = \left(H_i - H_1\right) + \left(H_d - H_i'\right)$$

or

$$\Delta H_T = 1,610 + 1,700 = 3,310 \text{ BTU/lb} \cdot \text{mole}$$

Converting the above head into horsepower:

$$HP = 0.0432 \cdot \Delta H_T$$

where   HP = horsepower

         ΔH_T = total enthalpy change, BTU/lb · mole

$$HP = 0.0432 \times 3,310 = 143 \frac{HP}{1 \; MMSCFD}$$

The duty of the gas cooler was calculated to be $-1,770$ BTU/lb·mole

Therefore, it is necessary to remove 1,770 BTU/lb·mole for each 1 MMSCFD of gas to drop its temperature from 260°F to 80°F. If the total gas flow rate is known, then using the heat transfer and energy relationship, the size of the intercooler can be determined.

To determine the minimum head or power requirements, as in an isothermal process, Figure 4-14 can be used again.

For an isothermal process Point 1, which is ($P_S = 100$ psia, $T_S = 80$°F), can be specified. At this point $H_1 = 380$ BTU/lb·mole and $S_1 = -3.2$ BTU/lb·mole·°R.

To find Point 2, move on the constant temperature line of 80°F, to hit $P_d = 1,600$ psia. At this point $H_2 = -480$ BTU/lb·mole, and $S_2 = -9.75$ BTU/lb·mole·°R.

For an isothermal process the following relationship is known:

$$\Delta S = \int_1^2 \frac{dq}{T}$$

$$q = T \cdot \Delta S$$

and from the general energy equation:

$$\Delta H = q - W$$

then

$$-W = \Delta H - T \cdot \Delta S$$

Therefore, compressor HP is

$$HP = 0.0432(\Delta H - T \cdot \Delta S)$$

If the values for ΔH, ΔS, and T are substituted

$$\Delta H = H_2 - H_1 = -480 - 380 = -860 \frac{BTU}{lb \cdot mole}$$

$$\Delta S = -9.75 - (-3.20) = -6.55 \frac{BTU}{lb \cdot mole \cdot °R}$$

$$HP = 0.0432[-860 - 540(-6.55)] = 115.7 \frac{HP}{1 \; MMSCFD}$$

For an isothermal process the system is under permanent cooling to maintain a constant temperature.

CENTRIFUGAL COMPRESSOR PERFORMANCE CURVE

Compression Stages

Dynamic compressors may consist of one or more compression stages. To achieve the desired discharge pressure, an actual compression is performed within those stages. A stage consists of a stage inlet with or without guide vanes, an impeller, a vaned or vaneless diffuser, and an outlet from the compression stage. The outlet may be a return channel to the inlet of the next compression stage, or a volute redirecting the compressed fluid into the compressor discharge nozzle. The actual energy that is injected into the gas stream for the purpose of compression comes only from the impeller.

In the impeller, gas is accelerated and compressed at the same time. The gas contains the highest energy level at the outlet of the impeller. This energy level is a combination of increased pressure and kinetic energy. The function of a diffuser located downstream from the impeller is to convert kinetic energy into pressure energy. It is only a conversion of the existing energy, not an injection of a new energy into the gas stream. Approximately two third of the pressure energy is generated in the impeller. The remaining third of the pressure energy is recovered in the diffuser by slowly reducing the gas velocity (see Figure 4-15).

Characteristic Curves for Centrifugal Gas Compressors

A centrifugal compressor takes atmospheric air, compresses it, and subsequently discharges it to a receiver. The compressor and receiver are initially at the atmospheric pressure. When the compressor starts operating at a constant RPM, the inlet and discharge pressures are equal. The only resistance to flow is the friction in the airflow path. Consequently, the only head produced by the compressor is that which is needed to overcome friction. If friction losses are small, the produced head will be close to zero. This head is plotted as Point 1 in Figure 4-16.

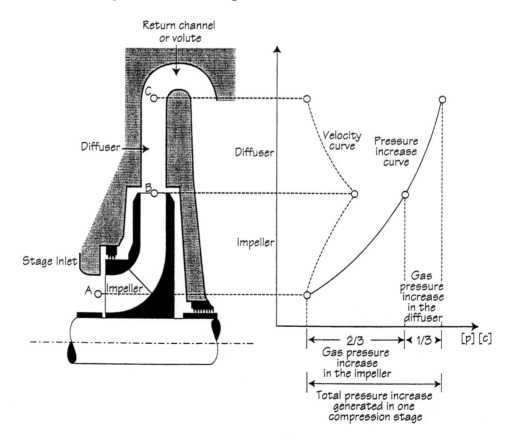

Figure 4-15. Centrifugal compressor stage diagram [Courtesy of NGTL, Technical Training Department, Airdrie Service Centre, Alberta, Canada, 1996]

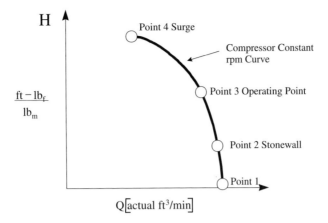

As the compressor delivers air to the receiver, the pressure in the receiver begins to rise, gradually increasing resistance. At first, due to this resistance, the flow drops very slightly (Point 2, Figure 4-16). This point is called the Stonewall Point. The curve between Points 1 and 2 is essentially a straight, vertical line. As the amount of air in the receiver increases, the pressure also increases, providing greater pressure differential from inlet to discharge (Point 3, Figure 4-16).

As the amount of air in the receiver increases, a level of discharge pressure is reached, beyond which the compressor operation becomes unstable. This is Point 4, Figure 4-16, and is called the surge point.

When these points are plotted, remembering that the compressor RPM is constant, they can be connected with a single curve as shown in Figure 4-16. The stonewall (Point 2) is the maximum stable compressor flow point. The stability of the compressor operation can still be maintained at this minimum head point, but the compressor is choking. A flow increase beyond the stonewall is practically impossible, operation beyond it is unpredictable and inefficient. The vertical slope of the head/flow curve results in a lack of stability in this zone. A minimal variation in pumped volume corresponds with large variations in pressure.

Manufacturers usually stop their curves before they reach the stonewall zone. The curve normally covers the area where performance can be reasonably predicted.

The surge point (Point 4) is the minimum stable compressor flow point. The stability of the compressor operation can still be maintained at this highest head point. As the pressure in the receiver increases, the volume flow to the receiver decreases, while a differential pressure across the compressor increases. Eventually, the differential pressure reaches a level too high for the compressor to handle. The compressor no longer performs with stability and compression ceases momentarily. The generated pressure at the impeller discharge drops slightly and a partial flow reversal occurs. This pressure pulsation associated with a back-and-forth flow motion is called surging and can damage the compressor. Operation within the surge zone must be prevented.

If the compressor speed (RPM) is now increased to another level, keep it constant and perform the same plottings to generate another performance curve. The family of these curves is shown in Figure 4-17. A detailed performance curve (wheel map) together with the efficiency lines and surge limit is shown in Figure 4-18.

It should be noted that in dynamic gas compressors the maximum and minimum speeds have certain limits. The maximum speed is limited by the maximum permissible stress level

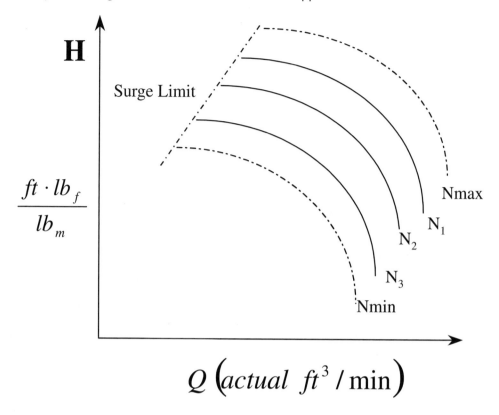

Figure 4-17. Centrifugal compressor performance map [Courtesy of NGTL, Technical Training Department, Airdrie, Alberta, Canada]

caused by centripetal forces in the rotors. The minimum operational speed is defined by a safety margin from critical speeds if the first bending mode is below the operating speed (Nova Gas Transmission Limited 1995). The centrifugal compressor can be adapted to varying conditions by changing speed or by using inlet guide vanes.

A smaller horsepower variation occurs for a given change in compression ratio in a reciprocating machine, as compared to in a centrifugal one. However, since centrifugal compressors can operate in series, adding or subtracting units can produce greater variations in the ratios. The "stability limit" in a centrifugal gas compressor represents the maximum pressure ratio at a given speed, beyond which the compressor becomes unstable and surging may result. When designing and selecting equipment, placing the stability limit beyond the operating range will avoid instability and surging during normal operation.

Where constant speed drivers are used with centrifugal compressors, an additional range of operation over the design head and volume curve (i.e., the characteristic curve) can be obtained by employing "inlet guide vanes." These are adjustable stationary blades located on the suction side of the impeller, which impart a swirl to the entering gas either in the direction of the impellers rotation or against it. The range of the ratios made available by this means is adequate for most pipeline operations; however, using guide vanes does sacrifice efficiency.

Centrifugal units and reciprocating compressors alike must be selected with operating characteristics that allow for future expansion. The initial range can usually be selected to include some expansion, allowing for a change of impellers for further expansion. In extreme cases, a completely new unit may be installed on the existing prime mover, possibly with a change in rated compressor speed.

Impellers with backward-bending vanes have a pressure-volume characteristic that, at constant speed, causes the discharge pressure to decrease gradually with increasing volume. Thus, at the design inlet temperature and pressure, it is not possible to overload a properly selected prime mover since both the discharge pressure and brake horsepower will decrease appreciably as the suction volume increases above 115–120% of its normal level.

Discharge pressure in centrifugal gas compressors differs greatly with small changes in speed. The prime movers of these compressors are generally designed for operation between 95% and 105% of normal speed.

Surge in Dynamic Gas Compressors

Dynamic compressors can operate only over a definite stable range, limited by an instability or "surge" point at the high-head, low-flow end of the characteristic curve, as well as by a "stonewall" or "choke" at the low-head, high-flow end of the curve. At stonewall condition, the performance curve of the compressor turns almost vertically downward, as a result of the sonic velocity that is reached inside the machine and prevents any capacity increase.

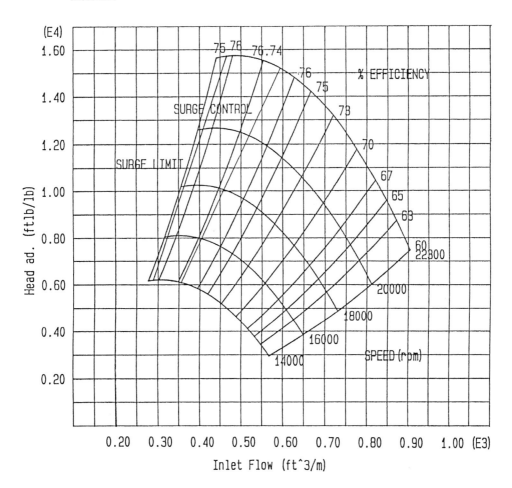

Figure 4-18. Centrifugal compressor detailed performance map (wheel map) [Courtesy of NGTL, Technical Training Department, Airdrie, Alberta, Canada]

The surge point at the opposite end of the performance curve is a complicated phenomenon, characterized by alternating flow reversals between compressors and the discharge system. This phenomenon is sometimes created by the lack of flow to a compressor that has a much higher capacity, causing a kind of vacuum in the system. This can usually be quite violent, especially on axial and multistage centrifugal gas compressors, and could lead to severe damage in the compressor.

One of the most important control aspects of dynamic compressors is detection and prevention of surge, usually by opening a recycle (bypass) valve to recirculate additional gas from discharge to suction. This will move the operating point further to the right, away from the surge line. Sometimes, recycling will cause overheating in the system, which may end up with a cooling system to reduce the gas temperature. Surge of centrifugal or axial compressors used for air compression (e.g., gas turbine compressors) may be prevented by blowing air to the atmosphere, which will eliminate overheating problems that often cause problems in a closed recycle system.

INFLUENCE OF PIPELINE RESISTANCE ON CENTRIFUGAL COMPRESSOR PERFORMANCE

Compressors are used in natural gas transportation systems to increase gas pressure to overcome friction losses. Natural gas in a transmission system travels through pipes, elbows, tees, aerial coolers, and scrubbers. While gas travels along the pipeline, a pressure drop is observed. The drop is a direct result of friction losses incurred by gas movement. A relation between gas pressure drop and the actual volume being pumped through the system is known as a system resistance curve. The curve varies with the square of the actual volume flow, and it is a parabola (see Figure 4-19).

The System Resistance Curve can be plotted by using one of the major flow equations for a transmission line (such as AGA) while plotting $\Delta P = P_1 - P_2$ or P_1/P_2 on the y-axis versus volumetric flow rate Q on the x-axis. The pressure drops in areas such as bends, beads, valves, coolers, and scrubbers should be added to the total pipeline pressure drop. In some cases, it is possible to plot enthalpy change (head) versus volumetric flow to attain the System Resistance Curve.

Figure 4-19 shows four compressor curves for various RPMs, with the System Resistance Curve superimposed on the plot. The only steady points available for the compressor for operation are points where the compressor performance curves intersect the System Resistance Curve. Operation under steady-state condition at points other than these is not possible, since only these points satisfy both the compressor and the system.

This limitation is of course not true for transient operations. If only intersect points between the System Resistance Curve and the compressor performance curve were possible for operation, then the compressor operational envelope may look too wide. The System Resistance Curve could also change with the changes in the compressor station's performance. But at any given time, the System Resistance Curve is well defined, and does not change momentarily. In Figure 4-19, assume that point A is the compressor design point. Point B is a lower flow point that satisfies both the compressor performance and the current System Resistance Curve. To achieve a decrease in a volumetric flow through the compressor, the speed must be decreased. In the dynamic compressors, head and volume are interrelated. Any change in volumetric flow is inevitably associated with head changes.

Head is somehow "built in" to positive displacement compressors. Volume can be controlled by controlling RPM without affecting the compression ratio. This is why

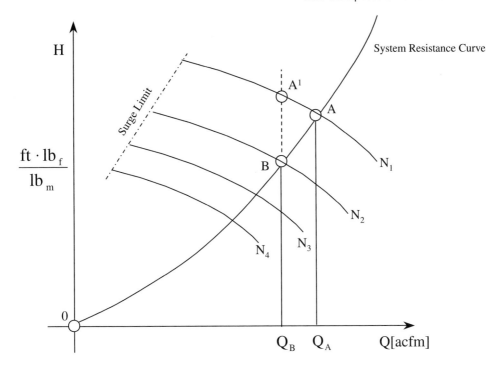

Figure 4-19. System resistance curve superimposed on compressor curves for four different rpms [Courtesy of NGTL, Technical Training Department, Airdrie, Alberta, Canada]

dynamic compressors did not readily replace positive displacement machines. Positive displacement compressors are much simpler to control (Nova Gas Transmission 1995).

Impact of Flow Variations on Multistage Compressors

If volumetric flow to a centrifugal compressor is reduced by x%, a certain reduction in volume flow to the second stage occurs. Due to the x% flow reduction in the first stage, the discharge pressure from the first stage increases. The smaller the compressed volume, the higher the generated head. Consequently, the volume flow to the second stage is further reduced by an additional increase in discharge pressure from the first stage. Therefore, the volume flow to the second stage is further reduced (i.e., more than x%). The third stage experiences a similar effect.

When two or more stages are combined for a compression process, the operational envelope of the compressor is narrower than the operational envelope of the individual stages (Nova Gas Transmission 1995).

Effect of Variations of Gas Parameters on Compressor Performance

Gas parameters such as molecular weight, inlet gas temperature, gas compressibility factor (Z), and isentropic exponent all have an impact on the performance of the compressor. Molecular weight of the gas is a factor that has a significant impact on the compressor's performance. Consider a compressor designed to compress low molecular weight gas. If it is then used in a new application to compress a higher molecular weight gas at identical inlet volume conditions, the head/volume ratio will have no effect on the first stage performance. The generated head will be the same as that of the gas of the design molecular weight. However, the discharge pressure from the first stage will be higher. In a single stage unit, variations of gas molecular weight do not affect the head/volume ratio, although speeds may have to be corrected for proper performance.

The higher discharge pressure from the first stage causes the volume flow to the second stage to decrease. The first stage observes just one change: higher molecular weight. The second stage observes three changes: higher molecular weight, reduced inlet volumetric flow, and higher inlet pressure. These factors contribute to an increase in discharge pressure and a decrease in the volumetric flow to the third stage. This effect amplifies from stage to stage. In extreme cases the last stage may be pushed into surge. Multistage gas compressors with higher than rated gas molecular weight have a narrower operational envelope.

Lowering the gas molecular weight from rated also decreases the range of the multistage compressor performance envelope. In terms of the head/volume ratio, the first stage is not affected by the lower gas molecular weight. However, a lower discharge pressure from the first stage and a higher volumetric flow to the second stage will be observed. This effect amplifies again from stage to stage. In extreme cases, the last stage may choke in the stonewall zone. Multistage gas compressors with lower than rated gas molecular weight have a narrower operational envelope.

Generally, it can be concluded that multistage gas compressors, when compressing gas with a molecular weight greater or smaller than the rated or designed molecular weight, have an operational envelope narrower than the design envelope. In extreme cases, the impact can be so severe that when the first stage is at stonewall, the last stage is at surge, or vice versa. In such a case, the width of the operational envelope is practically zero, and a compressor rerating may be necessary.

The above statements can be best understood if they are described using the adiabatic head and power equation. The energy equation for adiabatic compression was given by Equation (4-18a):

$$-W = \int_1^2 V \cdot dP$$

upon substitution, the following equation was obtained (Equation 4-18b):

$$-W = P_1 \cdot V_1 \cdot \frac{K}{K-1} \cdot \left[\left(\frac{P_2}{P_1} \right)^{\frac{K-1}{K}} - 1 \right]$$

Substituting for $P_1 \cdot V_1 = n \cdot R \cdot Z_1 \cdot T_1$, and $n = m/M$:

$$-W = \frac{m \cdot R \cdot Z_1 \cdot T_1}{M} \cdot \frac{K}{K-1} \cdot \left[\left(\frac{P_2}{P_1} \right)^{\frac{K-1}{K}} - 1 \right]$$

In the above equation [this is Equation (4-18c) with the introduction of compressibility factor], if there is an increase in the molecular weight of the gas while gas discharge pressure P_2 is kept constant, then there will be a decrease of the compressor head. Remember that for hydrocarbons, when the heavier ends are increased, there is a decrease in compressibility factor, which has a multiple effect resulting in further reduction of head when M is increased. There are of course some insignificant changes to the adiabatic gas exponent that will result in an increase of volumetric flow, or, in extreme cases, "choke." On the other hand, if gas molecular weight is decreased (corresponding to an increased compressibility factor) for a constant discharge pressure, compressor head will be increased, leading to lower volumetric flow, or, in extreme cases, "surge."

Consider a compressor working with a gas of rated (design) molecular weight, generating a fixed discharge pressure of P_2. Now assume that the gas composition has changed, resulting in a molecular weight increase. If the discharge pressure cannot change quickly, the compressor will choke. Since the discharge pressure remains virtually unchanged, the compressor needs to generate less head to maintain the same discharge pressure. For a constant RPM curve, less head means more volume.

Now consider a second case where the compressor is again working with a gas of the rated (design) molecular weight, generating a constant discharge pressure of P_2. This time, assume that the molecular weight of the gas has suddenly decreased. If the discharge pressure cannot change quickly, the compressor must generate more head to maintain the same discharge pressure. For a constant RPM curve, more head means less volume. If the required head increase corresponds to a significant volume change, the compressor will surge.

From the same equation described before:

$$-W = \frac{m \cdot R \cdot Z_1 \cdot T_1}{M} \cdot \frac{K}{K-1} \left[\left(\frac{P_2}{P_1} \right)^{\frac{K-1}{K}} - 1 \right]$$

it can be concluded that for a decrease in molecular weight, a higher compression power is required (when M is reduced, the compressibility factor normally increases, resulting in more power) for the same discharge pressure. This dependence is shown in Figure 4-20; for further details, see (Weiss 1998).

Variations of other inlet parameters, such as gas temperature, compressibility factor, or isentropic exponent, also reduce the operational envelope. However, the impact of these variations on the compressor performance is not as significant as the molecular weight change. Variations in inlet pressure will not impact the flow stability, although some speed correction may be required.

In conclusion, it can be said that a certain range of off-design conditions can be successfully handled by multistage gas compressors. Compressors do have some flexibility. However, the more operating conditions differ from rated values, the less flexibility will be available. If the operational conditions are significantly different from the design values, a compressor rerate should be considered.

There are further considerations to be taken into account regarding the specifications of compressors arranged in a "parallel" or "series" mode of operation. Units arranged in a parallel configuration need compressor wheels capable of generating similar heads. If not, either the second unit will not be able to get online before going into surge (first unit generates too high a head), or the second unit, while going online, will push the first into surge (it generates too high a head in comparison with the first unit).

Compressor stability in parallel arrangement translates into a compatibility of generated heads, and there are no restrictions on actual volumetric flows. Compressor stability in series arrangement translates into a compatibility of actual volumetric flows, and there are no restrictions on generated heads (Nova Gas Transmission 1995).

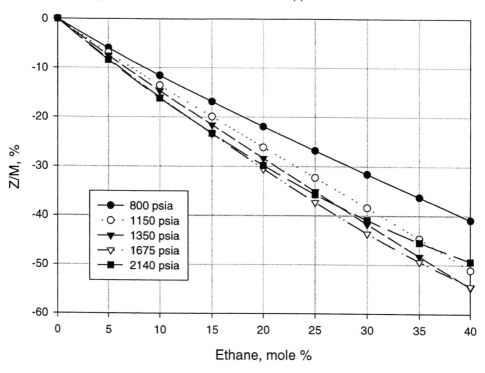

Figure 4-20. Required compression power for methane-ethane mixtures at 35°F [Courtesy of M. Weiss, NRTC Reports, NOVA Research & Technology Corporation, 1998]

Calculating Fuel for Centrifugal Compressor Units Based on Power and Speed

Currently the majority of hydraulic simulators calculate fuel as a function of the power only, using the straight line equation:

$$f = A + B \cdot (POWER) \qquad (4-75)$$

Where

f = Fuel Consumption, kJ/h
A = Intercept, kJ/h
B = Slope, kJ/kWh
POWER = Compressor unit power, kW

If the lower heating value kJ/m^3 of the natural gas is known (i.e. for fuel calculations we always use the LHV of the natural gas), then the fuel consumption could be calculated in volumetric units of MMSCMD or MMSCFD.

A more accurate approach to relate fuel and power is generally represented as a polynomial relationship.

$$f = A + B \cdot (POWER) + C \cdot (POWER)^2 \qquad (4-76)$$

Where A, B, and C are coefficients.

Another measure of fuel consumption is heat rate which is given by:

$$h = \frac{f}{(POWER)}$$

Where:

h = Heat rate, kJ/kWh

f = Fuel consumption, kJ/h

POWER = Compressor power, kW

However, for Gas-Turbine driven centrifugal compressors, heat rate is a function of both power and speed (RPM), as shown in Figure 4.21. Compressor unit vendors typically supply this type of performance curves. These curves are normally supplied at different ambient temperatures. Therefore, we can conclude that the fuel consumption in a compressor unit is a function of power, speed and ambient temperature. The heat rate curves could be used to determine coefficients A and B in equation (4-75) for the fuel calculations. Field data could be also used to determine the coefficients in equations (4-75) and (4-76) to relate fuel consumption to the power. However, it is worth mentioning that equations which only relate fuel consumption to power, somehow over-simplify the fuel calculations. The most accurate approach is to use the vendor's heat rate curves at different ambient temperatures.

Figure 4.21 represents the heat rates graph for a compressor unit at -40°C ambient temperature (i.e. a unit on TransCanada Pipeline System). This graph could be used to calculate compressor unit fuel consumption at a given ambient temperature, compressor speed, and power.

Example 4.9:
Calculate the fuel (MMSCFD) consumed by the compressor unit where the heat rate curve is given in Figure 4.21 at the speed of 5500 RPM and 10,000 kW power consumption. The lower heating value (LHV) of the natural gas used as fuel is 33500 kJ/m³.

The heat rate value obtained from Figure 4.21 at the speed of 5500 and 10,000 kW power is 10,000 kJ/kWh.

If we multiply power by heat rate:

$$10,000 \; kW \; x \; 10,000^{kJ}\!/_{kWh} = 100,000,000^{kJ}\!/_{h}$$
$$\text{Fuel Consumption} = \text{Power } x \text{ Heat Rate}$$

To change the fuel consumption from energy to volumetric flow, we need the LHV, which is given 33500 kJ/m³.

Therefore,

$$\begin{array}{c}\text{Fuel Consumption}\\\text{(Volumetric)}\end{array} = \frac{kJ/h}{kJ/m^3} = {m^3}\!/_{h}$$

Or

$$\begin{array}{c}\text{Fuel Consumption}\\\text{(Volumetric)}\end{array} = \frac{100,000,000}{33500} = 2985.07^{m^3}\!/_{h}$$

Which is 71641.8 m^3/day or 2.53 MMSCFD

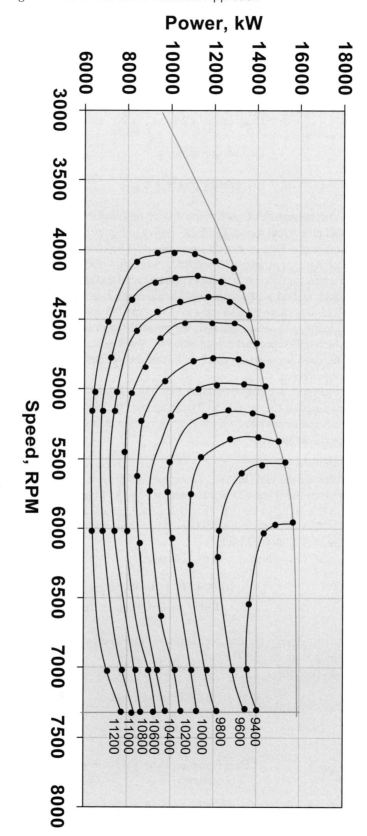

Figure 4-21. Heat Rates [kJ/kWh] at -40 C

The heat rates graph given in Figure 4.21 could be plotted in a different way as heat rate vs. power at constant speed. A family of these graphs could be generated for different speeds for a given ambient temperature (a joint research work TransCanada/NRTC). The results for a single speed of 5500 RPM is given in Figure 4.22. The same procedure could be used to generate the same family of graphs for heat rate vs. speed for a constant power at a given ambient temperature as in Figure 4.23.

Compressor Station Optimization

In pipeline operations it is possible to adjust compressor unit set points to determine the optimum compressor operating point to minimize system-wide fuel consumption or maximize the throughput. In a small gas transmission pipeline when we are dealing with one or two stations with a few units in each station, it is sometimes possible to manually adjust the operating points for the best operation mode.

In large pipeline systems with many compressor stations and several units in each station, the manual optimization is almost impossible. For this task we need optimization algorithms to select the best operating mode for the pipeline, depending on the objective requested (Fuel Minimization/Throughput Maximization). There are many optimization modules (commercial software programs) available in industry for piping system simulations and optimization.

To briefly demonstrate the process for the optimization of a compressor station containing a number of units, we have considered a simple example of a station having two compressor units operating in parallel mode (i.e. a real system) on a NPS 42 pipeline.

For this system, the current operating point (Point 1) is given on both units, while creating a fixed amount of head and pumping a fixed amount of gas through the station. The objective is to search for a better operating point (Point 2) on both units to minimize the power required (compared to Case 1) which basically minimizes the fuel consumption while maintaining the same amount of head and gas throughput.

Example 4.10:
The following compressor station which consists of two centrifugal units (Unit A, Figure 4.25 and Unit B, Figure 4.26, see the wheel map for each unit) operating in parallel mode on a NPS 42 pipeline, receives gas at a suction pressure of 4480 kP_a (650 psia) and creates a head of 47.5 kJ/kg (see the wheel maps).

The total flow through the station is 12.07 $a \cdot m^3/s$ at the suction condition. The current operating condition which is Point 1 for each unit is shown on the wheel map.

Determine the current power usage by this system, and find out the possibility of saving some power by operating in a better operational mode (Point 2) while maintaining the same head and throughput.

Additional data is given in Figure 4.24

From the wheel maps and operating points (1), we can find the $a \cdot m^3/s$ flow through each unit, which is

$$
\begin{aligned}
\text{Flow, Unit A} &= 8.65a \cdot m^3/s \\
\text{Flow, Unit B} &= 3.42a \cdot m^3/s \\
\text{Total Flow} &= 8.65 + 3.42 = 12.07a \cdot m^3/s
\end{aligned}
$$

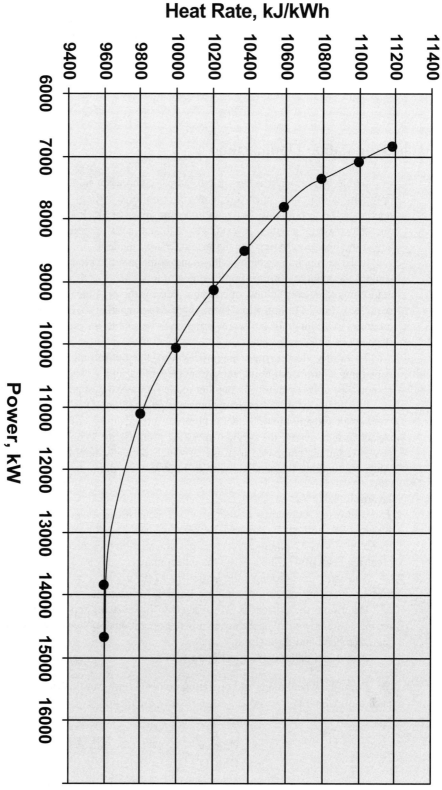

Figure 4-22.   Heat Rate vs Power for Speed = 5500

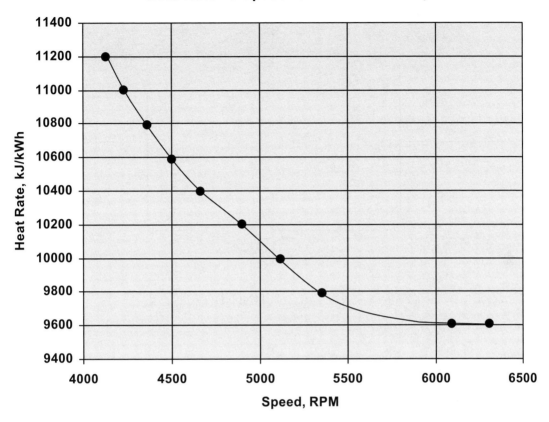

Figure 4-23. Heat Rate vs Speed for Power = 12000, kW

Total flow at standard condition could be obtained from the following relationship

$$\frac{Ps \cdot Qs}{P_{ST} \cdot Q_{ST}} = \frac{Zs \cdot Ts}{Z_{ST} \cdot T_{ST}}$$

$$\frac{4480 \; x \; 12.07}{101.325 \; x \; QST} = \frac{0.89 \; x \; 293.15}{1.0 \; x \; 288.15}$$

$$Q_{ST} \; = \; 589.4 \; \text{Standard} \; {}^{m^3}\!/s \; = \; 50924 \; e^3 m^3/d \; = \; 1798.12 \; MMSCFD$$

For this operating point (Point 1), the power consumption by each unit could be obtained as follows:

$$POWER = \frac{HEAD \; x \; \dot{m}}{\eta a} = \frac{\frac{kJ}{kg} \cdot \frac{kg}{s}}{\eta a} = kW$$

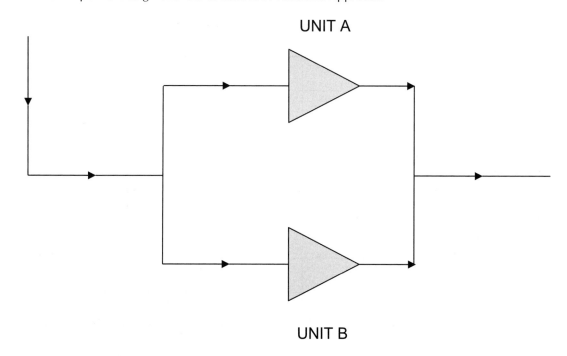

UNIT A

UNIT B

ADDITIONAL DATA

$$T_s \quad = \quad 20\ C° \ (68\ F°)$$

$$P_s \quad = \quad 4480\ Kpa.a \ (650\ psia)$$

$$Z_s \quad = \quad 0.90$$

$$Z_d \quad = \quad 0.88$$

$$Z_{ave} \quad = \quad 0.89$$

$$G \quad = \quad 0.60$$

$$HEAD \quad = \quad 47.50 \ kJ/kg$$

Figure 4-24. Station layout

Case 1, Operating Point 1.

Flow split is as follows:

$$Unit\ A: \quad \frac{8.65}{12.07} = 71.67\%$$

$$Unit\ B: \quad \frac{3.42}{12.07} = 28.33\%$$

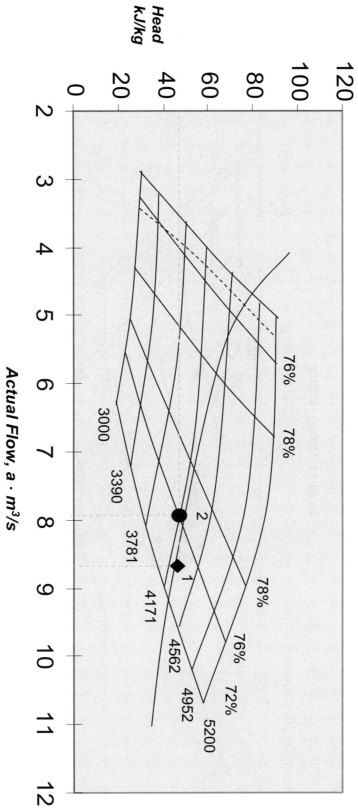

Figure 4-25.   Unit A, Wheel Map

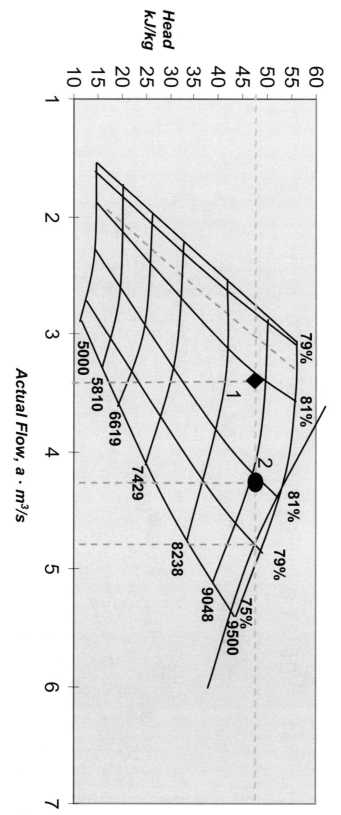

UNIT B

Head kJ/kg

Actual Flow, a · m³/s

Figure 4-26. Unit B, Wheel Map

To find the total mass flow through both units;

$$P_{ST} \cdot Q_{ST} = \frac{\dot{m}}{Mw} \cdot R \cdot T$$

$$14.696 \; x \; 1798.12 \; x \; 10^6 = \frac{\dot{m}}{28.96 \; x \; 0.60} \; x \; 10.73 \; x \; 520$$

$$\dot{m} = 82,293315 \; lbm/day$$

$$= 37,361,165 \; kg/day$$

$$= 432.421 \; {}^{kg}\!/_{s}$$

Based on the percent of flow split between two units.
Mass flow through Unit A = 309.916 kg/s
Mass flow through Unit B = 122.505 kg/s
From the wheel maps for both units A and B, knowing the operating point (Point 1, Case 1), we can find head, efficiency, and speed of each unit.

$$Unit \; A \left\{ \begin{array}{l} HEAD = 47.5kJ/kg \\ \eta_a = 0.752 \end{array} \right\}$$

$$Unit \; B \left\{ \begin{array}{l} HEAD = 47.5kJ/kg \\ \eta_a = 0.808 \end{array} \right\}$$

We can now calculate the power that each unit consumes:

$$POWER = \frac{HEAD \; x \; \dot{m}}{\eta_a}$$

Or

$$Unit \; A, \; POWER = \frac{47.5 \; x \; 309.916}{0.752} = 19576 \; kW$$

$$Unit \; B, \; POWER = \frac{47.5 \; x \; 122.505}{0.808} = 7202 \; kW$$

$$TOTAL \; POWER = 26778 \; kW \; or \; 35909 \; HP$$

This is the total power consumed by both units A and B at this operating point (Point 1).

To further optimize the operating condition without any changes in the total head of 47.5 kJ/kg, we have to change the distribution of the flow between the two units and also change the adiabatic efficiencies and operating speeds for each unit. The objective is to find another operating point (Point 2), to produce the same head and the same throughput while consuming less power (less fuel). We can manually set the operating point to Point 2, and check the total required power.

For gas compressor units in parallel, we have to be careful to have exactly the same head for each unit when we try to find an optimum operating point. This means that for parallel units with common suction (same suction pressure); the discharge pressure should be exactly the same in order to be able to operate all the units. This would result in the same head for all units. The flow could be distributed as required to force the optimum operation.

Case II

In this case, we move the operating point (Point 1) to a new location on the same head line. This will create the operating point (Point 2) with the following characteristics:

$$Unit\ A \begin{cases} HEAD = 47.5kJ/kg \\ FLOW = 7.82a \cdot^3/s \\ \eta_a = 0.762 \end{cases}$$

$$Unit\ B \begin{cases} HEAD = 47.5kJ/kg \\ FLOW = 4.25a \cdot m^3/s \\ \eta_a = 0.810 \end{cases}$$

The flow percentage per unit is as follows:

$$Unit\ A = 64.79\% \ or \ 280.166 \ kg/s$$
$$Unit\ B = 35.21\% \ or \ 152.255 \ kg/s$$

Therefore the power consumption for each unit could be calculated:

$$Unit\ A,\ POWER = \frac{47.5 \ x \ 280.66}{0.762} = 17464 \ kW = 23420 \ HP$$

$$Unit\ B,\ POWER = \frac{47.5 \ x \ 152.255}{0.810} = 8929 \ kW = 11973 \ HP$$

This means that, if we operate the station (units A and B) at the new operating point (Point 2), the total power consumption for creating the same head and flowing the same amount of gas through the station compared to Case I is 385 kW (516 HP) lower.
This would results in lower fuel consumption too.

RECIPROCATING COMPRESSORS

Reciprocating compressors trap gas in a cylinder and compress it with a piston to a desired pressure, at which point it is released through discharge valves. In practice, the volume handled by the cylinder is less than the piston displacement volume, which is due to "clearance," "space between valves and gas passages," and "preheating." Clearance is the space left between the piston and the cylinder to prevent collision. Percentage clearance, "Cl," is normally defined as:

$$Cl = \frac{clearance\ volume\ inch^3}{piston\ displacement\ volume\ inch^3} \times 100$$

Preheating of the gas occurs when it enters the cylinder, which has already been warmed by the previous compression cycle. This will increase the volume of the gas, which will ultimately cause a reduction in the actual flow capacity.

A new term can be defined to measure the impact of all the parameters that will cause a change in the actual volume of a reciprocating compressor. This term is called volumetric efficiency (VE), and it is given by the following equation:

$$VE = 100 - A - Lu - CR - Cl \left[\frac{Z_S}{Z_d} \cdot (CR)^{\frac{1}{K}} - 1 \right] \qquad (4-77)$$

where VE = volumetric efficiency of the compressor, percentage
  A = the effect of leakage, losses, etc., percentage
  Lu = the effect due to the lack of lubrication, percentage
  Cl = percentage of clearance, percentage
  Z_S = compressibility factor at suction condition, dimensionless
  Z_d = compressibility factor at discharge condition, dimensionless
  CR = compression ratio, dimensionless
  K = adiabatic gas exponent, dimensionless.

In practice and for normal operations, it is quite common to assume a value of 96% for the entire term 100 - A - Lu, which changes the equation to:

$$VE = 96 - CR - Cl \left[\frac{Z_S}{Z_d} \cdot (CR)^{\frac{1}{K}} - 1 \right] \qquad (4-78)$$

The ideal capacity of a reciprocating compressor can be obtained from the following equation:

$$PD = \frac{stroke \times RPM \times \pi \cdot D^2}{4 \times 1,728} \qquad (4-79)$$

where PD = piston displacement volume, ft^3/min
  stroke = travel length of piston, inches
  RPM = number or revolutions of piston per minute
  D = cylinder inside diameter, inches.

If the diameter of the piston rod that supports the movement of the piston is considered, the following equation can be used to calculate the displacement volume:

$$PD = \frac{stroke \times RPM \times \left(D^2 - d^2 \right) \cdot \pi}{4 \times 1,728} \qquad (4-80)$$

where d = rod diameter, inches.

For double-acting reciprocating compressors, the displacement volume will be doubled as follows:

$$PD = \frac{2 \times stroke \times RPM \times \left(D^2 - d^2 \right) \cdot \pi}{4 \times 1,728} \qquad (4-81)$$

When both the ideal displacement volume and the compressor VE are known, the actual cylinder capacity or the actual gas throughput can be calculated:

$$ACFM = PD \times VE \qquad (4-82)$$

where ACFM = actual cubic feet per minute of gas through compressor, ft³/min
PD = piston displacement volume (ideal volume), ft³/min
VE = volumetric efficiency of the compressor, percent.

It should be noted that to obtain higher accuracy in the calculations, should be taken into account the following corrections for a reciprocating gas compressor
(a) Preheating correction: The following relationship can be used to account for the warming of the incoming gas to a reciprocal compressor:

$$T_C = 0.20\left[\frac{T_S + T_d}{2} - T_S\right] \qquad (4-83)$$

where T_C = corrected temperature, °F
T_S = suction temperature, °F
T_d = discharge temperature, °F.

The value of T_C should then be added to the suction temperature, while using equations to calculate gas discharge temperature and compressor HP. It should be further noted that this is a trial-and-error calculation in which the uncorrected suction temperature is used to calculate discharge temperature for the first trial.
(b) Pressure correction: As mentioned previously for all types of compressors, the pressure drops in the compressor suction and discharge valves should be considered as:

$$P_S = P_1 - \Delta P_1$$

and

$$P_d = P_2 + \Delta P_2$$

where P_S = actual suction pressure, psia
P_1 = suction pressure before adjustment, psia
ΔP_1 = pressure drop in suction valve, psi
P_d = actual discharge pressure, psia
P_2 = discharge pressure before adjustment, psia
ΔP_2 = pressure drop in discharge valve, psi.

After necessary corrections, the actual compression ratio (CR) of the compressors that should be used for power calculations is based on the corrected values of P_S and P_d, or

$$CR = \frac{(P_d = P_2 + \Delta P_2)}{(P_S = P_1 - \Delta P_1)} \qquad (4-84)$$

which will result in a slightly higher compression ratio. It is quite common to add another two to three psi to P_d and subtract this amount from P_S to consider the pressure drop in piping and fittings in the outlet and inlet of the compressor.

GAS COMPRESSION SOLVED PROBLEMS

Problem 1

Calculate both GHP and BHP and also discharge temperature for a gas compressor handling a mixture of gases, which has the following gas analysis:
The process operating conditions are:

Gas Analysis	Mole %
C_2	5
C_3	89
nC_4	6

$T_S = 41°F$
$P_S = 20.3$ psia
$P_d = 101.5$ psia
$\eta_P = 0.77$

Molar gas flow rate is 2,400 moles/hour. Specifications of the compressor are given in Figures 4-27, 4-28 and Table 4-3.
Solution

Gas Analysis	M	Mole %	P_C, psia	T_C, °R	C_P, BTU/ lb·mole × °R at T_S
C_2	30.07	5	707	550	11.98
C_3	44.09	89	617	666	16.58
nC_4	58.12	6	551	765	22.33

$$M = (30.07)(0.05) + (44.09)(0.89) + (58.12)(0.06) = 44.23$$

The compressibility factor at suction condition can be calculated using Kay's rule and compressibility charts:

$$P'_C = P_{CA} \cdot Y_A + P_{CB} \cdot Y_B + P_{CC} \cdot Y_C + \ldots$$

$$P'_C = 707 \times 0.05 + 617 \times 0.89 + 551 \times 0.06 = 617.6 \text{ psia}$$

and

$$T'_C = T_{CA} \cdot Y_A + T_{CB} \cdot Y_B + T_{CC} \cdot Y_C + \ldots$$

$$T'_C = 550 \times 0.05 + 666.0 \times 0.89 + 765 \times 0.06 = 666.2°R$$

$$P'_r = \frac{P}{P'_c} = \frac{20.3}{617.6} = 0.0328$$

$$T'_r = \frac{T}{T'_c} = \frac{41 + 460}{666.2} = 0.752$$

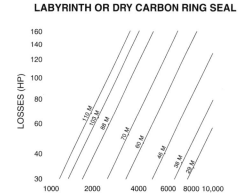

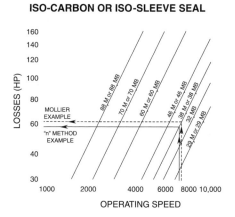

Figure 4-27. Compressor losses

from compressibility factor charts (Figure 3-3) $Z_S = 0.97$. Gas inlet to the compressor is $2,400/60 = 40$ moles/min, or

$$P_1 \cdot Q_1 = n_1 \cdot R \cdot Z_1 \cdot T_1$$

$$20.3 \times Q_1 = 40 \times 10.73 \times 0.97 \times 501$$

$$Q_1 = 10,275 \ \text{ft}^3/\text{min} = 14.796 \ \text{MMCF/D at suction condition}$$

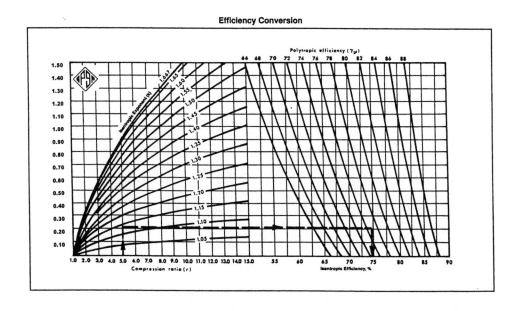

Figure 4-28. Temperature rise factor [Courtesy GPSA, Engineering Data Book, 1994]

or at standard condition:

$$\frac{P_1 \cdot Q_1}{P_b \cdot Q_b} = \frac{n_1 \cdot R \cdot Z_1 \cdot T_1}{n_b \cdot R \cdot Z_b \cdot T_b}$$

or

$$\frac{20.3 \times 14.796}{14.7 \times Q_b} = \frac{0.97 \times 501}{1 \times 520}$$

$$Q_b = 21.863 \ \text{MMSCF/D}$$

To calculate the adiabatic gas exponent (K):

$$C_{P\text{ave}} = 11.98 \times 0.05 + 16.58 \times 0.89 + 22.33 \times 0.06 = 16.71 \ \text{BTU/lb} \cdot \text{mole} \cdot {}^{\circ}\text{F}$$

$$C_{V\text{ave}} = C_{P\text{ave}} - R = 16.71 - 1.986 = 14.72 \ \text{BTU/lb} \cdot \text{mole} \cdot {}^{\circ}\text{F}$$

$$K = \frac{C_{P\text{ave}}}{C_{V\text{ave}}} = \frac{16.71}{14.72} = 1.135$$

and for the calculation of GHP:

$$\text{CR} = (P_d/P_S)^{1/n}$$

for $n = 1$ (i.e., one compressor with CR = 5)

$$\text{CR} = \frac{101.5}{20.3} = 5$$

compressor discharge temperature should be checked for CR = 5:

$$\frac{Z_2 \cdot T_2}{Z_1 \cdot T_1} = \left(\frac{P_2}{P_1}\right)^{\frac{K-1}{K}}$$

To calculate T_2, the above equation can be solved through iterations or by using Figure 4-28 (i.e., temperature rise factor, Figure 4-28, is exactly a reproduction of Figure 4-6 or efficiency conversion, with the only difference that the values of temperature rise factor are added on the Y-axis), which will give a value for T_2 with reasonable accuracy. Using Figure 4-28 and having $\eta_P = 0.77$, $K = 1.135$, and CR = 5, temperature rise factor $X = 0.21$, and $\eta_a = 0.75$, an approximated T_2 can be obtained from the following equation:

$$T_2 = \frac{X}{\eta_a} \cdot T_1 + T_1$$

therefore

$$T_2 = \frac{0.21}{0.75} \cdot 501 + 501 = 641.3 \, {}^{\circ}\text{R} = 181.3 \, {}^{\circ}\text{F}$$

This discharge temperature is within the acceptable limit ($T_d < 300°F$). This value of T_2 can now be used to determine the compressibility factor at discharge condition:

$$ptT_r' = \frac{T}{T_C'} = \frac{641.3}{666.2} = 0.963$$

$$P_r' = \frac{P}{P_C'} = \frac{101.5}{617.6} = 0.164$$

then, $Z_2 = Z_d = 0.93$

$$Z_{ave} = \frac{Z_1 + Z_2}{2} = \frac{0.97 + 0.93}{2} = 0.95$$

Equations for polytropic efficiency are given as:

$$\eta_P = \frac{\dfrac{n}{n-1}}{\dfrac{K}{K-1}}$$

in which $K = 1.135$, $\eta_P = 0.77$, so $n = 1.1826$.
The compressor GHP can be obtained from a form of Equation (4-36):

$$GHP = 0.0857 \times \frac{n}{n-1} \times \frac{1}{\eta_P} \times \frac{Z_1 + Z_2}{2} \times T_1 \cdot \left[\left(\frac{P_2}{P_1} \right)^{\frac{n-1}{n}} - 1 \right]$$

which, upon substitution of parameters, gives

$$GHP = \frac{0.0857 \times 6.47}{0.77} \times 501 \times 0.95 \times \left[(5)^{0.1544} - 1 \right] = 96.74 \text{ HP}/1 \text{ MMSCFD}$$

and total GHP $= 96.74 \times 21.863 = 2,116$ HP.

Using Table 4-3, which is the specification table for this compressor, with an inlet flow of 10,275 ft³/min, $\eta_P = 0.77$, the frame is 38 M with RPM = 8,100 (9 stages), which can produce 10,000–12,000 ft head.

Friction losses for this type of compressor could be found using Figure (4-27). If Iso-Carbon Seal is used, for a 38 M frame, 8,100 RPM compressor, losses would be close to 70 HP, therefore:

$$BHP = GHP + \text{ losses}$$
$$BHP = 2,116 + 70 = 2,186 \text{ HP}$$

TABLE 4-3. Compressor specifications

Frame	Normal Inlet Flow Range (ft³/min)	Nominal Polytropic Head per Stage (H_p)	Nominal Polytropic Efficiency (η_p)	Nominal Maximum No. of Stages	Speed of Nominal Polytropic Head/Stage
29 M	500 – 8,000	10,000	0.76	10	11,500
38 M	6,000 – 23,000	10,000/12,000	0.77	9	8,100
46 M	20,000 – 35,000	10,000/12,000	0.77	9	6,400
60 M	30,000 – 58,000	10,000/12,000	0.77	8	5,000
70 M	50,000 – 85,000	10,000/12,000	0.78	8	4,100
88 M	75,000 – 130,000	10,000/12,000	0.78	8	3,300
103 M	110,000 – 160,000	10,000	0.78	7	2,800
110 M	140,000 – 190,000	10,000	0.78	7	2,600
25 MB (H)(HH)	500 – 5,000	12,000	0.76	12	11,500
32 MB (H)(HH)	5,000 – 10,000	12,000	0.78	10	10,200
38 MB (H)	8,000 – 23,000	10,000/12,000	0.78	9	8,100
46 MB	20,000 – 35,000	10,000/12,000	0.78	9	6,400
60 MB	30,000 – 58,000	10,000/12,000	0.78	8	5,000
70 MB	50,000 – 85,000	10,000/12,000	0.78	8	4,100
88 MB	75,000 – 130,000	10,000/12,000	0.78	8	3,300

Problem 2

A gas compressor uses a lean natural gas for gas injections to an oil reservoir. Calculate gas discharge temperature, adiabatic head, and compressor horsepower if:

- Suction pressure = 1,327 psia
- Suction temperature = 98.6°F
- Gas molecular weight = 19.64
- Discharge pressure = 2,408 psia
- Capacity = 347.5 MMSCFD
- Polytropic efficiency = 0.766

and gas composition is:

Gas Analysis	M	Mole Fraction	P_C, psia	T_C, °R
C_1	16.043	0.8588	666	343
C_2	30.070	0.0605	707	550
C_3	44.097	0.030	617	666
iC_4	58.124	0.0052	528	734
nC_4	58.124	0.0100	551	765
iC_5	72.151	0.0029	491	829
nC_5	72.151	0.0028	489	845
C_6	86.178	0.0016	437	913
C_7^+	100.205	0.0012	397	972
N_2	28.013	0.0008	493	227
CO_2	44.010	0.0255	1,071	548
H_2S	34.076	0.0007	1,300	672

Solution

Using Kay's rule and compressibility charts to calculate Z at suction condition:

$$P'_C = P_{CA} \cdot Y_A + P_{CB} \cdot Y_B + P_{CC} \cdot Y_C + \ldots$$

$$T'_C = T_{CA} \cdot Y_A + T_{CB} \cdot Y_B + T_{CC} \cdot Y_C + \ldots$$

upon substitution, $P'_C = 680$ psia, and $T'_C = 381$ °R, then:

$$P'_r = \frac{P}{P_c} = \frac{1,327}{680} = 1.95$$

$$T'_r = \frac{T}{T'_c} = \frac{98.6 + 460}{381} = 1.47$$

where $Z_1 = Z_S = 0.80$.

To find the adiabatic gas exponent K, Figure 4-5 can be used, since the gas molecular weight and its suction temperature are known. This gives $K = 1.27$. It would have been more accurate to determine K if an average temperature between suction and discharge conditions were taken, but in this case, K at suction will be used.
To find compressor horsepower:

$$CR = \left(\frac{P_d}{P_S}\right)^{1/n}$$

for $n = 1$

$$CR = \left(\frac{2,408}{1,327}\right)^{1} = 1.815$$

Discharge temperature can now be checked using the same method of temperature rise factor as for problem 1, then

$$T_2 = \frac{X}{\eta_a} \cdot T_1 + T_1$$

Knowing that CR = 1.815, $K = 1.27$, $\eta_P = 0.766$, then $X = 0.14$, and $\eta_a = 0.75$, (from Figure 4-28), therefore:

$$T_2 = \frac{0.14}{0.75}(460 + 98.6) + (460 + 98.6)$$

$$T_2 = 662.9\,°R = 202.9\,°F$$

The temperature is within the acceptable limit. The compressibility factor at discharge condition could be calculated as follows:

$$T_r' = \frac{T}{T_C'} = \frac{662.9}{381} = 1.74$$

$$P_r = \frac{P}{P_C'} = \frac{2{,}408}{680} = 3.54$$

where $Z_2 = Z_d = 0.86$ from compressibility charts. Now

$$Z_{\text{ave}} = \frac{Z_d + Z_S}{2} = \frac{0.80 + 0.86}{2} = 0.83$$

To calculate compressor adiabatic head, use Equation (4-36), and include compressibility factor at suction condition:

$$-W = \frac{53.28}{G} \cdot Z_1 \cdot T_1 \cdot \frac{K}{K-1} \cdot \left[\left(\frac{P_2}{P_1} \right)^{\frac{K-1}{K}} - 1 \right]$$

where Z_1 and T_1 are the compressibility factor and temperature, respectively, at suction conditions, then:

$$-W = \frac{53.28}{19.64/29} \cdot 0.80 \cdot (98.6 + 460) \cdot \frac{1.27}{1.27 - 1} \cdot \left[\left(\frac{2{,}408}{1{,}327} \right)^{\frac{1.27-1}{1.27}} - 1 \right]$$

$$-W = 22{,}335 \text{ ft} \cdot \text{lb}_f/\text{lb}_m \text{ (Head)}$$

To calculate compressor gross horsepower (GHP), the polytropic gas exponent will be calculated as follows:

$$\eta_P = \frac{\frac{n}{n-1}}{\frac{K}{K-1}}$$

$$0.766 = \frac{\frac{n}{n-1}}{\frac{1.27}{1.27-1}}$$

where $n = 1.384$, the polytropic gas exponent.

$$\text{GHP} = 0.0857 \cdot T_S \cdot Z_{\text{ave}} \cdot \frac{n}{n-1} \cdot \frac{1}{\eta_P} \left[(\text{CR})^{\frac{n-1}{n}} - 1 \right]$$

$$GHP = 0.0857 \times 558.6 \times 0.83 \times \frac{1.384}{1.384 - 1} \times \frac{1}{0.766} \left[(1.815)^{\frac{1.384-1}{1.384}} - 1 \right]$$

$$GHP = 33.62 \ HP/1 \ MMSCFD$$

or total GHP = $33.62 \times 347.5 = 11,683 \ HP$

To choose a suitable compressor, the value in actual ft^3/min (ACFM) is needed:

$$Q_b = 347.5 \ MMSCFD$$

$$Q_b = \frac{347.5 \times 10^6}{24 \times 60} = 241,319.5 \ ft^3/min \text{ at standard condition}$$

and

$$\frac{P_b \cdot Q_b}{P_1 \cdot Q_1} = \frac{T_b \cdot Z_b}{T_1 \cdot Z_1}$$

$$\frac{14.7 \times 241,319.5}{1,327 \times Q_1} = \frac{520 \times 1}{558.6 \times 0.80}$$

$$Q_1 = 2,297.35 \ ACFM.$$

Problem 3

A compressor station with centrifugal compressors has a performance curve as shown in Figure 4-29. The station is discharging natural gas for injection purposes via a 30 km, 6 inches pipeline with injection pressure characteristic as shown in Figure 4-30. What would be the approximate compressor speed (RPM) for compression and injection of 32 MMSCFD of gas into the reservoir, and what is the minimum possible injection rate without recycling? Also, find the compressor horsepower for the injection of 32 MMSCFD of gas into the reservoir.
Additional data:

- Gas molecular weight = 20.32
- Suction temperature = 100°F
- Suction pressure = 180 psia
- Average flowing gas temperature in the pipeline = 120°F
- Pipeline efficiency = 90%
- Pipe wall thickness = 0.250 inch
- Pipe outside diameter = 6.625 inches
- Compressors adiabatic efficiency = 0.77

Use the Panhandle B flow equation for the calculations.

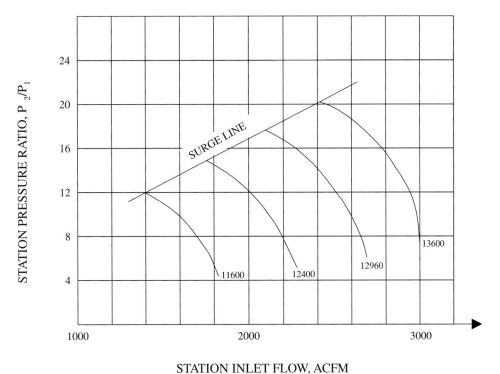

Figure 4-29. Compressor performance curve

Solution

Using Figure 4-30, P_3 (injection pressure) can be obtained based on 32 MMSCFD gas injection. P_3 will be 2,300 psia. If Panhandle B equation is used with $P_3 = 2,300$ psia, P_2 or gas discharge pressure from the compressor can be calculated:

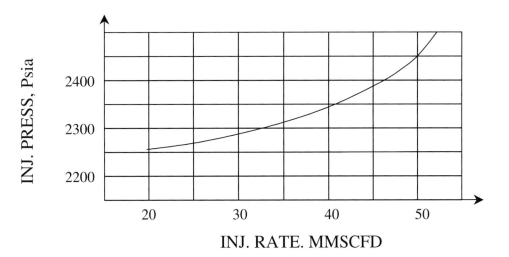

Figure 4-30. Compressor injection pressure characteristics

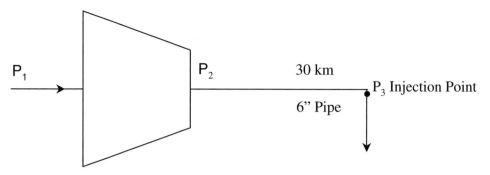

Figure 4-31. Compressor and pipeline configuration

$$Q_b = 737.02 \left(\frac{T_b}{P_b}\right)^{1.02} \cdot \left[\frac{P_2^2 - P_3^2 - E}{Z_{\text{ave}} \cdot T_{\text{ave}} \cdot G^{0.961} \cdot L}\right]^{0.510} \cdot \frac{1}{\eta} \cdot D^{2.53}$$

Now if the parameters are substituted in Panhandle B equation, assuming no elevation changes;

$$32 \times 10^6 = 737.02 \left(\frac{520}{14.7}\right)^{1.02} \cdot \left[\frac{P_2^2 - (2,300)^2}{0.77 \times 580 \times (0.70)^{0.961} \times \frac{30}{1.609}}\right]^{0.510} \times \frac{1}{0.90}$$

$$\times (6.125)^{2.53}$$

then $P_2 = 2{,}426$ psia.

The compressibility factor was obtained using Figures 4-32 for a gas with $G = 0.70$ and the compressibility chart given in Chapter 3, Figure 3-3. To avoid iteration, a pressure drop of almost 5 psi/km was assumed, giving an average flowing gas pressure of 2,375 psia. The average flowing gas temperature of 120°F was used. Using a gas gravity of 0.7, $T_C' = 390°R$, and $P_C' = 665$ psia, and having obtained T_C' and P_C', the reduced temperature and pressure were calculated:

$$T_r' = \frac{T_{\text{ave}}}{T_C'} = \frac{460 + 120}{390} = 1.49$$

$$P_r' = \frac{P_{\text{ave}}}{P_C'} = \frac{2,375}{665} = 3.57$$

Then, Figure 3-3 was used to determine $Z_{\text{ave}} = 0.77$.

To calculate the ACFM gas entering the compressor:

$$32 \text{ MMSCFD} = 22{,}222 \text{ SCF/min}$$

this volume at suction condition is:

$$\frac{P_1 \cdot Q_1}{P_b \cdot Q_b} = \frac{Z_1 \cdot T_1}{Z_b \cdot T_b}$$

$$\frac{180 \times Q_1}{14.7 \times 22,222} = \frac{0.98 \times 560}{1 \times 520}$$

$$Q_1 = 1,915.31 \text{ ACFM}$$

($Z_1 = 0.98$ is calculated at a suction pressure of 180 psia and suction temperature of 100°F for a gas with $G = 0.70$, using Figures 4-32 and 3-3).

Using the station inlet flow, 1,915 ACFM, and the compression ratio, which is $P_2/P_1 = 2,426/180 = 13.48$, Figure 4-29 can be used to determine the compressor's RPM. The value of the RPM is 12,400.

Now, move along the 12,400 RPM curve to intersect the surge line. The flow value is 1,730 ACFM, which represents the minimum flow inlet to the station without recycling. At flows lower than this value, the system will surge.

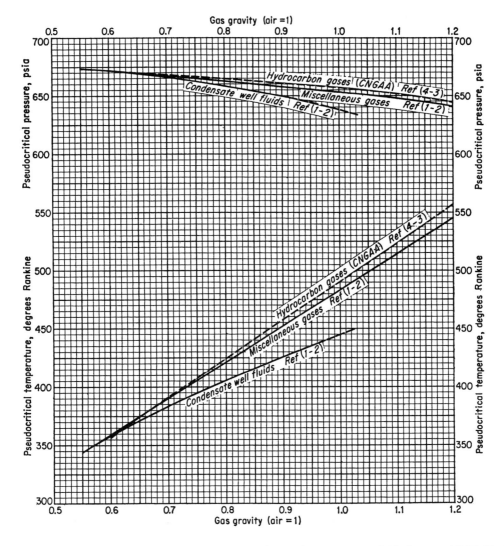

Figure 4-32. Pseudocritical properties of natural gases [Brown, 4.16, Courtesy AIME, Handbook of Natural Gas Engineering, D.L. Katz et al., 1959, McGraw-Hill Co.]

Compressor horsepower can be calculated from the following equation:

$$HP = 0.0857 \times \frac{K}{K-1} \cdot \frac{Z_1 + Z_2}{2} \cdot \frac{1}{\eta_a} \cdot T_S \cdot \left[(CR)^{\frac{K-1}{K}} - 1 \right]$$

To calculate horsepower from the above equation, the values of K and Z_2 need to be calculated, and the rest of the parameters are known.

For this case, at least two compressors in series are required (since CR = 13.48 is too high) with intercooling, which gives a CR of:

$$CR = \left(\frac{P_2}{P_1} \right)^{\frac{1}{n}}$$

for $n = 2$:

$$CR = (13.48)^{\frac{1}{2}} = 3.672$$

which is acceptable.

To find the discharge temperature, the temperature rise factor method is used as in previous problems:

$$T_2 = \frac{X}{\eta_a} \cdot T_1 + T_1$$

in which $T_1 = 560°R$, $\eta_a = 0.77$, and $K = 1.27$ (i.e., for natural gas with $M = 20.32$, and suction temperature = 100°F). The value of X from the temperature rise factor chart (Figure 4-28) is 0.31, therefore:

$$T_2 = \frac{0.31}{0.77} \times 560 + 560 = 785°R = 325°F$$

This can be assumed to be an acceptable intermediate temperature since no specifications for maximum discharge temperature were given. Using $T_2 = 785°R$ and $P_2 = 660.96$ psia ($180 \times 3.672 = 660.96$ psia), the compressibility factor at discharge condition is $Z_2 = 0.97$, therefore:

$$HP = 0.0857 \times \frac{1.27}{1.27 - 1} \cdot \frac{0.98 + 0.97}{2} \times \frac{1}{0.77} \times 560 \times \left[(3.672)^{\frac{1.27-1}{1.27}} - 1 \right]$$

$$HP = 91.06 \ HP/MMSCFD$$

and the total power requirement = $91.06 \times 2 \times 32 = 5,828$ HP.

*Note: There will be some differences between the powers of the first and second compressors due to some pressure changes, which will cause a change in the compressibility factor.

Problem 4

Methane is to be compressed from atmospheric pressure (14.7 psia) to 4,354 psia in a multistage gas compressor. Calculate the optimum number of stages and the ideal intermediate pressures, as well as the work required per lb_m of gas. Assume compression to be isentropic and that gas behaves as an ideal gas. Indicate on a T/S diagram the effect of imperfect intercooling on the work done at each stage. Gas suction temperature is 60°F.

Solution

Use the following equation to find CR:

$$CR = \left(\frac{P_d}{P_S}\right)^{\frac{1}{n}}$$

$$CR = \left(\frac{4,354}{14.7}\right)^{\frac{1}{n}}$$

for $n = 1$, 2 and 3 high compression ratios are obtained, these exceed the limit CR ≤ 6. For $n = 4$, CR = 4.15, which is acceptable if the gas discharge temperature is also within the acceptable range. The process is shown in Figure 4-33.

For methane at 60°F, the adiabatic gas exponent is

$$K = \frac{C_P}{C_V} = \frac{C_P}{C_P - R} = \frac{8.42}{8.42 - 1.986} = 1.31$$

and gas discharge temperature (assuming $Z_1 = Z_2$) is

$$\frac{T_2}{T_1} = (CR)^{\frac{K-1}{K}}$$

$$\frac{T_2}{520} = (4.15)^{\frac{1.31-1}{1.31}} = 728\,^\circ R = 268\,^\circ F$$

which is acceptable. The work required for each lb_m of gas is:

$$-W = \frac{53.28}{G} \cdot T_S \cdot \frac{K}{K-1} \cdot \left[(CR)^{\frac{K-1}{K}} - 1\right]$$

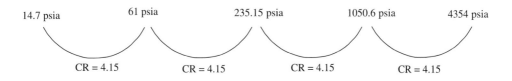

14.7 psia   61 psia   235.15 psia   1050.6 psia   4354 psia

CR = 4.15   CR = 4.15   CR = 4.15   CR = 4.15

Figure 4-33. Problem 4 — Multistage compressor configuration for CR=4.15 and n=4

$$-W = \frac{53.28}{16/29} \times 520 \times \frac{1.31}{1.31-1} \left[(4.15)^{\frac{1.31-1}{1.31}} - 1 \right]$$

$$-W = 84{,}969 \text{ ft} \cdot \text{lb}_f/\text{lb}_m \text{ (head)}$$

and total work required is

$$-W = 84{,}969 \times 4 = 339{,}877 \text{ ft} \cdot \text{lb}_f/\text{lb}_m \text{ (total head for all 4 stages)}$$

From H/S or T/S diagrams given earlier in this chapter, it is obvious that the area under the curve, which shows energy requirements of the compressor, will be increased due to imperfect cooling.

Problem 5

A natural gas compressor that is located at an elevation of 3,000 feet above sea level (barometric pressure at this level is 13.14 psi) increases gas pressure from 0 psig to 140 psig. Total gas flow through the compressor is 40 MMSCFD, and an intercooler reduces gas temperature to 95°F while imposing a 5-psi pressure drop. If the gas suction temperature is 70°F, and the adiabatic gas exponent is $K = 1.26$, determine the total power requirements of this compressor.

Solution

$$P_S = 0 + 13.14 = 13.14 \text{ psia}$$

$$P_d = 140 + 13.14 = 153.14 \text{ psia}$$

$$\text{CR} = \left(\frac{P_d}{P_S} \right)^{\frac{1}{n}} = \left(\frac{153.14}{13.14} \right)^{\frac{1}{n}}$$

For $n = 1$, CR = 11.66, which is too high. For $n = 2$, CR = 3.41, which is acceptable. The system configuration is as shown in Figure 4-34:

In this system with a 5-psi pressure drop in the intercooler and equal $CR_1 = CR_2 = 3.41$, the required discharge pressure of 153.14 psia will not be obtained. Therefore, CR_2 should be increased to a higher value of 153.14/39.86=3.82 in order to meet the requirements.

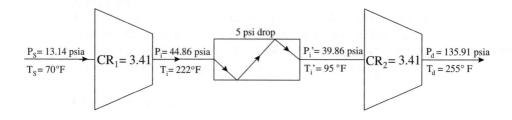

Figure 4-34. System layout (Case 1)

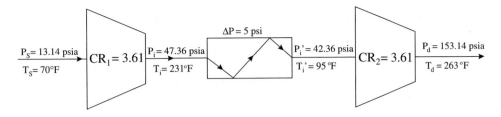

Figure 4-35. System layout (Case 2)

This change will offset the principle of minimum power requirement (or equal CR), and in the long-term, more power will be consumed. The solution is to divide the cooler pressure drops in half, and increase both CRs equally, as follows (this was formerly proved):

$$\frac{\Delta P}{2} = \frac{5}{2} = 2.5 \text{ psi}$$

$$44.86 + 2.5 = 47.36 \text{ psia}$$

The new compression ratio is CR $= \frac{47.36}{13.14} = 3.61$
With this new CR $= 3.61$, the requirements will be met (see Figure 4-35).
This new combination requires less power than any other configuration.

To check the discharge temperature from each compressor, assuming no changes in compressibility factor at this range,

$$\frac{T_i}{T_S} = (CR_1)^{\frac{K-1}{K}}$$

$$\frac{T_i}{460 + 70} = (3.61)^{\frac{1.26-1}{1.26}} = 691\,^\circ R = 231\,^\circ F$$

and for the second compressor:

$$\frac{T_d}{T_i'} = (CR_2)^{\frac{K-1}{K}}$$

$$\frac{T_d}{95 + 460} = (3.61)^{\frac{1.26-1}{1.26}} = 723\,^\circ R = 263\,^\circ F$$

and the power requirements:

$$(HP)_1 = 0.0857 \cdot \frac{K}{K-1} \cdot T_{S1} \cdot \left[(CR_1)^{\frac{K-1}{K}} - 1 \right]$$

$$(HP)_1 = 0.0857 \times \frac{1.26}{1.26 - 1} \times 530 \left[(3.61)^{\frac{1.26-1}{1.26}} - 1 \right]$$

$$(HP)_1 = 67 \text{ HP/1 MMSCFD}$$

or total power for the first compressor is:

$$(HP)_{1T} = 67 \times 40 = 2,680 \text{ HP}$$

and for the second compressor:

$$(HP)_2 = 0.0857 \cdot \frac{K}{K - 1} \cdot T_{S2} \left[(CR_2)^{\frac{K-1}{K}} - 1 \right]$$

$$(HP)_2 = 0.0857 \times \frac{1.26}{1.26 - 1} \times 555 \left[(3.61)^{\frac{1.26-1}{1.26}} - 1 \right]$$

$$(HP)_2 = 70.16 \text{ HP/1 MMSCFD}$$

Total horsepower for the second compressor is:

$$(HP)_{2T} = 70.16 \times 40 = 2,806 \text{ HP}$$

(higher power is due to the higher gas inlet temperature) and finally total power requirements for the system are:

$$(HP)_T = (HP)_{1T} + (HP)_{2T} = 2,680 + 2,806 = 5,486 \text{ HP}$$

It should be noted that, in the power calculation of the compressor, adiabatic efficiency and compressibility factor are both assumed to be equal to unity. This will of course (especially for adiabatic efficiency case) have introduced errors into the calculations. To consider the impact of elevation on the power, Equation (4-65a) could be used to adjust power requirements as follows:

$$(HP)_{\text{at elevation } h} = (HP)_{\text{at sea level}} \times 10^{\frac{-h}{62900}}$$

Or there should be a higher power requirement (GHP) at this elevation to do the same task. At 3,000 ft, the correction factor would be:

$$10^{\frac{-3,000}{62,900}} = 0.896$$

Therefore, the power requirement at this level, is $\frac{5486}{0.896} = 6,123$ HP

Problem 6

In a liquid propane gas plant, 28 MMSCFD of propane (C_3H_8) at 20 psia, and $-20°F$ is compressed by a gas compressor called C-101, as shown in Figure 4-36. The discharge of compressor C-101 enters condenser E-101 and leaves this condenser as a saturated liquid at 140°F. Finally, saturated liquid propane at 140°F enters vessel V-101. Determine:

1. Enthalpy difference (ΔH) through C-101 (assuming isentropic compression) and E-101, the compressor discharge temperature, and vessel pressure in V-101. Use the pressure/enthalpy diagram for propane (Figure 4-37).
2. Calculate the required BHP at compressor shaft ($\eta_P = 0.78$), assuming $Z \approx 1.0$. Fifty horsepower (50 hp) losses for seals and bearings should be considered.

Solution

Using the pressure/enthalpy diagram for propane (Figure 4-37) and a saturated liquid propane temperature of 140°F in V-101, the pressure of the propane is 310 psia.

Assuming there is no pressure drop in the piping, the outlet of the condenser E-101 should have the same pressure of $P = 310$ psia, and temperature of 140°F. There is no pressure drop in the condenser and its piping, so the inlet pressure of propane gas to the condenser should also be 310 psia. The inlet temperature to E-101, which is the same as the discharge temperature of C-101, is to be determined. There are two ways to determine the C-101 discharge temperature:

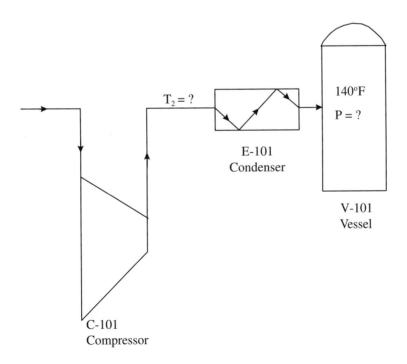

Figure 4-36. System layout

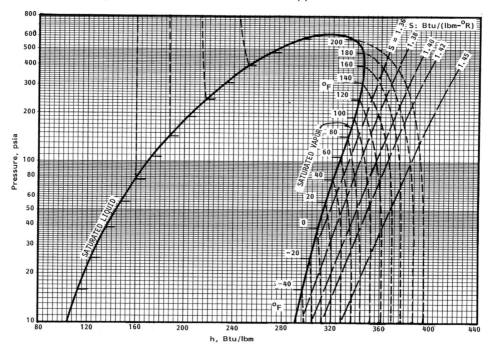

Figure 4-37. Pressure/enthalpy diagram for propane [Courtesy John M. Campbell]

1. Iterative procedure using the following equation:

$$\frac{T_2 \cdot Z_2}{T_1 \cdot Z_1} = (\mathrm{CR})^{\frac{K-1}{K}}$$

2. Pressure/enthalpy diagram for propane (i.e., Figure 4-37).

Method 2, the pressure/enthalpy diagram, will be used here to determine the propane gas discharge temperature. First, the suction point of the compressor is located on the chart with its pressure and temperature (i.e., $P_S = 20$ psia, $T_S = -20°$F). This point corresponds to an enthalpy of $H = 304$ BTU/lb$_m$ and an entropy of $S = 1.37$ BTU/lb$_m$·°R (in the total vapor area). Compression is isentropic (adiabatic), therefore move along the constant entropy line of $S = 1.37$ until it intersects with a compressor discharge pressure of $P_d = 310$ psia. This point has a temperature of $T_d = 170°$F, $H = 364$ BTU/lb$_m$, and $S = 1.37$ BTU/lb$_m$·°R. The results can be summarized as follows:

At suction condition:

$P_S = 20$ psia
$T_S = -20°$F
$H_S = 304$ BTU/lb$_m$
$S_S = 1.37$ BTU/lb$_m$·°R

At discharge condition:

$P_d = 310$ psia
$T_d = 170°$F
$H_d = 364$ BTU/lb$_m$
$S_d = 1.37$ BTU/lb$_m$·°R

The enthalpy change is therefore equal to $H = 364 - 304 = 60$ BTU/lb$_m$ in the compressor.

The gas flow rate through the system is 28 MMSCFD, and:

$$P_b \cdot Q_b = n_b \cdot R \cdot T_b$$

$$n_b = \frac{14.7 \times 28 \times 10^6}{10.73 \times 520} = 73,768.73 \text{ lb} \cdot \text{moles/day}$$

$$\dot{m} = \frac{73,768.73 \times 44}{24} = 135,243 \text{ lb}_m/\text{hr}$$

Total enthalpy change within the compressor could be obtained as follows:

$$\Delta H_T = 60 \times 135,243 = 8,114,560 \text{ BTU/hr.}$$

To find the duty of the condenser, move from the vapor point at $P = 310$ psia, and $T = 170°$F (i.e., inlet gas to the condenser E-101) on the constant pressure line of 310 psia horizontally. The intersection with the saturated liquid line on the left side of the curve (because gas leaves the condenser as a saturated liquid) will give the exit point from the condenser, which has:

$P_3 = 310$ psia

$T_3 = 170°$F

$H_3 = 232$ BTU/lb$_m$

The total heat to be removed from the propane inside E-101, to change it from vapor to saturated vapor, and finally to saturated liquid, can now be determined:

$$\Delta H = H_d - H_3 = 364 - 232 = 132 \text{ BTU/lb}_m$$

$$\Delta H_T = 132 \times 135,243 = 17,852,076 \text{ BTU/hr}$$

To calculate total compressor horsepower:

$$CR = \left(\frac{P_d}{P_S}\right)^{\frac{1}{n}}$$

$$CR = \left(\frac{310}{20}\right)^{\frac{1}{n}}$$

for $n = 1$, CR $= 15.5$ is too high, but for $n = 2$, CR $= 3.94$, and the exit temperature from this system is 170°F.

The adiabatic gas exponent for propane can be calculated as follows:

$$T_{ave} = \frac{T_S + T_d}{2} = \frac{-20 + 170}{2} = 75°\text{F}$$

This temperature is used to read $C_p = 15.65$ for propane:

$$K = \frac{C_P}{C_P - R} = \frac{15.65}{15.65 - 1.986} = 1.145$$

$$\eta_P = 0.78 = \frac{\frac{n}{n-1}}{\frac{K}{K-1}}$$

or

$$\frac{n}{n-1} = 6.159, \quad \text{and} \quad \frac{n-1}{n} = 0.162$$

Using the equation for power calculation, for the first compressor:

$$(HP)_1 = 0.0857 \cdot \frac{n}{n-1} \cdot \frac{1}{\eta_P} \cdot T_S \cdot \left[(CR)^{\frac{n-1}{n}} - 1 \right]$$

and upon substitution:

$$(HP)_1 = 0.0857 \times \frac{6.159}{0.78} \times 440 \left[(3.94)^{0.162} - 1 \right]$$

$$(HP)_1 = 74 \text{ HP/1 MMSCFD}$$

Power for the second compressor is:

$$(HP)_2 = 0.0857 \times \frac{6.159}{0.78} \times 530 \left[(3.94)^{0.162} - 1 \right]$$

$$(HP)_2 = 89.14 \text{ HP/1 MMSCFD}$$

and finally, total power requirement is:

$$(HP)_T = (74 + 89.14) \times 28 = 4{,}567.82 \text{ total gas horsepower}$$

$$BHP = GHP + \text{Losses}$$

therefore,

$$BHP = 4{,}567.82 + 50 = 4{,}617.82 \text{ BHP}$$

Problem 7

A single-acting reciprocal air compressor supplies 0.1 m³/s of air at standard condition compressed to 380 kN/m² from 101.3 kN/m² pressure. If the suction temperature is 289°K, the stroke is 250 mm, and the speed is four cycles per second, find the cylinder diameter. Assume the cylinder clearance is 4% and compression is isentropic ($K = 1.4$). What is the total theoretical head required for the compression?

Conditions

- Maximum allowable temperature = 300°F
- Take $Z_S = Z_d = 1.0$

Solution

The compression ratio is:

$$CR = \frac{380}{101.3} = 3.75$$

$$T_2 = T_1(CR)^{\frac{K-1}{K}} = 289(3.75)^{\frac{1.4-1}{1.4}} = 421°K = 298°F$$

and

$$\frac{P_b \cdot Q_b}{P_S \cdot Q_S} = \frac{T_b}{T_S}$$

where P_S, Q_S, and T_S are at suction condition:

$$P_b = P_S = 101.3 \ \text{kN/m}^2$$

Therefore

$$\frac{Q_b}{Q_S} = \frac{T_b}{T_S} \quad \text{or} \quad Q_S = \frac{Q_b \cdot T_S}{T_b}$$

$$Q_S = \frac{0.1}{4} \cdot \frac{289}{273} = 0.0265 \ \text{m}^3, \quad \text{volume per stroke}$$

The sweep volume of the compressor is:

$$V_S = \frac{\text{actual volume per stroke}}{\text{designed sweep volume}}$$

or

$$V_E = 96 - CR - Cl\left[\frac{Z_S}{Z_d} \cdot (CR)^{\frac{1}{K}} - 1\right]$$

$Z_S = Z_d = 1.0$
$CR = 3.75$
$K = 1.4$
$Cl = 4\%$

then

$$V_E = 96 - 3.75 - 4\left[\frac{1}{1}(3.75)^{\frac{1}{1.4}} - 1\right]$$

$$V_E = 86\%$$

Therefore, designed sweep volume = 0.0265/0.86 = 0.0308 m³ and the cross-sectional area of cylinder is

$$\frac{\pi \cdot d^2}{4} \cdot \text{ stroke } = \text{ designed sweep volume}$$

$$\frac{\pi \cdot d^2}{4} \cdot \frac{250}{1,000} = 0.0308$$

so the cylinder diameter is

$$d = 0.4 \text{ m} = 40 \text{ cm}$$

Compressor head per cycle is given by Equation (4-36):

$$-W = P_1 \cdot V_1 \cdot \frac{K}{K-1} \cdot \left[(CR)^{\frac{K-1}{K}} - 1\right]$$

$$-W = 101,300 \times 0.0265 \times \frac{1.4}{1.4 - 1} \times \left[(3.75)^{\frac{1.4-1}{1.4}} - 1\right]$$

$$-W = 4,311 \text{ joule/cycle}$$

and total theoretical head is 4×4,311 = 17,244 J = 17.24 kJ.

Problem 8

A two-stage vapor compression cycle with pure propane refrigerant is used for cooling a natural gas stream to −30°C to meet the required dew point (see Figure 4-38). Use the pressure/enthalpy diagram for propane (Figure 4-37) to locate the values of pressure, temperature, and enthalpy of all streams marked on the flow diagram, considering the following assumptions:

1. The only available cooling media for the condenser is dry air with a maximum temperature of 50°C.
2. Approaches of 10°C for the condenser and 5°C for the chiller between the cold side inlet and hot side outlet are used
3. Compression occurs isentropically.
4. Temperature drop of stream 5 due to interstage mixing is negligible.
5. Pressure drop in piping and exchangers may be neglected.

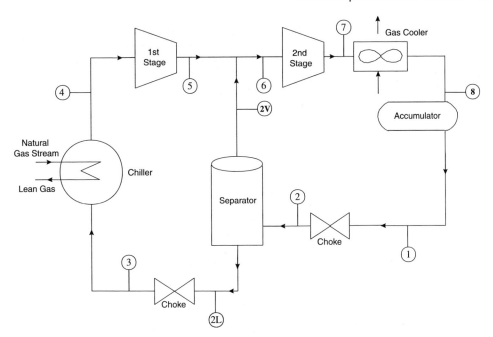

Figure 4-38. Problem 8 — System layout

Solution

Step 1 Propane leaving the accumulator at Point 1 is saturated liquid at $T = 60°C$ (i.e., $T = 50$ maximum air temperature + 10) or 140°F. This point can be located on the saturated liquid curve and has a pressure of 310 psia and enthalpy of 232 BTU/lb$_m$ (see Figure 4-37).

Step 2 Start with Point 4, which is the discharge fluid from the chiller or inlet to the gas compressor. The system is designed to have saturated vapor leaving the chiller at $-35°C$ (i.e., based on 5°C approach) or $-31°F$. For saturated propane vapor at $-31°F$, the pressure from the chart is 20 psia and enthalpy is $H = 300$ BTU/lb$_m$.

Step 3 At this step, because both compressor discharge pressure and suction pressure are known, the number of compressors can be determined:

$P_d = 310$ psia
$P_S = 20$ psia

$$CR = \left(\frac{310}{20}\right)^{\frac{1}{n}}$$

for $n = 1$, CR is too high and is unacceptable. For $n = 2$, CR = 3.94, which is acceptable. Two compressors should therefore be in series with CR = 3.94. The exit pressure from the first compressor is 20 × 3.94 = 78.8 psia.

Step 4 At this step, the values of P, T, and H for Point 5, or discharge from the first compressor, can be determined. The pressure of this point is known to be 78.8 psia. The compression is adiabatic inside the compressor, so move from Point 4, or suction of the compressor (i.e., $P = 20$ psia, $T = -31°F$, a point on saturated vapor line), on the constant entropy line of $S = 1.37$ BTU/$lb_m·°R$, until the intersection with the pressure line of $P = 78.8$ is found. Point 5 is determined as $T = 70°F$, $P = 78.8$ psia, and $H = 332$ BTU/lb_m.

Step 5 Now, consider Point 1, where saturated liquid of propane, at $P = 310$ psia and $T = 140°F$, enters from the accumulator into the adiabatic choke. The main feature of the adiabatic choke is that it suddenly drops the pressure of the stream with no enthalpy change, or $\Delta H = 0$. This means that stream 2, which leaves adiabatic choke, has a lower pressure and temperature and has been divided into two phases of liquid and vapor. The pressure is normally dropped to the extent that stream 2V has the same pressure as the discharge of the first compressor. Therefore, move from Point 1 vertically and downward (i.e., constant enthalpy line) to intersect with the pressure line of 78.8 psia in the two-phase region. This point is Point 2 after the adiabatic choke. Move horizontally on the pressure line of 78.8 psia to intersect both legs of the curve. On the saturated liquid leg, Point 2L is located with $P = 78.8$ psia, $T = 41°F$, and $H = 160$ BTU/lb_m. On the saturated vapor leg, Point 2V is located with $P = 78.8$ psia, $T = 41°F$, and $H = 322$ BTU/lb_m.

Step 6 Now move to Point 6. It was assumed that the temperature of stream 5 will not significantly change due to mixing, therefore, Points 5 and 6 have the same pressure, temperature, and enthalpy of $P = 78.8$ psia, $T = 70°F$, and $H = 332$ BTU/lb_m.

Step 7 Having obtained 6, the discharge temperature, pressure, and enthalpy of the second compressor can be determined. Move on the constant entropy line of $S = 1.37$ BTU/$lb_m·°R$ from Point 6 ($P = 78.8$ psia and $T = 70°F$) to intersect the pressure line of $P = 310$ psia. The temperature at this point is $T = 170°F$ and $H = 356$ BTU/lb_m.

Step 8 Point 3 can now be determined. Adiabatic choking from Point 2L (i.e., $\Delta H = 0$), moving vertically downward until the line intersects with the pressure line of $P = 20$ psia. This is Point 3, which is in the two-phase region and has $P = 20$ psia, $T = -31°F$, and $H = 160$ BTU/lb_m. Moving horizontally on this line to intersect the saturated vapor leg, Point 4 or gas inlet to the first compressor is located. A summary of the results is given in Table 4-4.

TABLE 4-4. Summary of results

Stream Number	P, psia	T, °F	H, BTU/lb_m
1	310	140	232
2	78.8	41	232
2V	78.8	41	322
2L	78.8	41	160
3	20	−31	160
4	20	−31	300
5	78.8	70	332
6	78.8	70	332
7	310	170	356
8	310	140	232

GAS COOLERS

INTRODUCTION

Gas coolers are widely used in the gas transportation industry. They can be used as precoolers (at the suction of a compressor station) or as intercoolers (between compressors in series) to protect the system from overheating. They may also be used as after-coolers (at the discharge of a compressor station) to protect the pipeline external coating from damage at high temperatures (exceeding 65°C or 149°F). Gas cooling at the discharge of compressor stations will also help reduce the pressure drop along the pipeline because the gas will be flowing at a colder temperature. After-coolers will also reduce the power requirements at the downstream compressor station when it receives gas at a lower suction temperature.

There are two types of gas coolers: air-cooled heat exchangers and water-cooled heat exchangers. Depending on the climate and geographical conditions, either type of heat exchanger (or a combination of both) can be used to providethe cooling requirements. It is not intended to compare the economics of the two types of exchangers in this chapter. However, it should be said that, as a rule, the operating costs for water-cooled systems are much higher than those for air-cooled exchangers.

If the ambient temperature permits, especially in remote areas, the exchanger of choice in a gas transmission line is usually an air-cooled heat exchanger. The following sections discuss the air-cooled heat exchangers (aerial-coolers) in more detail. For further information on air-cooled heat exchangers, refer to (GPSA 1994); for both air-cooled and water-cooled exchangers, see (Kern 1997) or (Mukherjee 1997).

AIR COOLED HEAT EXCHANGERS

Horizontal air-cooled heat exchangers are commonly used on gas transmission pipelines to reduce the temperature of hot gases at the discharge of compressor stations using ambient air. The basic components of these exchangers include:

- One or more fans, which force air through the finned tubes (cross-flow operation)
- Fan drivers
- Motors
- Fan speed controllers
- Headers
- Supports

When assembled together, this set of equipment forms the body of a gas cooler, which is normally called a "bay." When two or more bays are added together they form a "unit." Putting two or more units together forms a "bank" of coolers. Air-cooled heat exchangers are categorized as "forced draft" or "induced-draft," depending on the location of the fans. In forced-draft coolers, the tube bundles are located on the discharge side of the fans, whereas in induced-draft coolers, the tube bundle is located on the suction side of the fans (see Figures 4-39, 4-40 and 4-41).

The American Petroleum Institute's Standard API-661, "Air-Cooled Heat Exchangers for General Refining Services," describes the minimum requirements for the design and testing of air-cooled heat exchangers. Although the standards are set for petroleum refinery air coolers, they can be generalized to cover both Petrochemical and Gas Industry services.

The tube bundle is one of the most important components of an air-cooler. The extremely low heat-transfer coefficient of air makes the use of extended surface tubes (finned tubes) a necessity. The tube itself is generally made of carbon or stainless steel, and the fins are normally made of aluminum, due to the high thermal conductivity of aluminum and its light weight. It is common to position 10 to 11 fins per inch in industrial air-coolers. These fins will introduce an extended surface area almost 20 times that of the surface area of the tubes. The tube diameter range-between 5/8 inch and 1 3/4 inches OD, whereas fins are within the 1/2 inch to 1 inch high range. Due to the very low heat-transfer coefficient and specific heat capacity of air, large quantities of air must be forced across the tube bundles to achieve the required cooling load. This is accomplished by using large-diameter fan blades (3 feet to 28 feet in diameter), rotated at high speed, which produces high noise levels. Due to the noise problem, it is common practice to ensure that no more than the maximum blade diameter is chosen (between 14 and 16 feet) to maintain acceptable noise and vibration levels (GPSA 1994; Mukherjee 1997).

COOLER HEAT TRANSFER EQUATIONS

The general procedure to derive the cooler design equations involves performing heat and material balances for the air and gas sides of the exchangers. As part of these balances, calculation of the overall heat-transfer coefficient (U) and log mean temperature difference (LMTD) correction factors are also required.

The heat dissipated by the hot gas is represented as follows:

$$q_g = \dot{m}_g \cdot C_{pg} \cdot (T_1 - T_2) \qquad (4-85)$$

where q_g = heat dissipated by the gas, BTU/hr
$\dot{m}_g$ = gas mass flow rate, lb$_m$/hr
C_{pg} = specific heat of gas at flowing conditions, BTU/lb$_m$·°F
T_1 = gas inlet temperature, °F
T_2 = gas exit temperature, °F

The gas exit temperature T_2 is the combined gas temperature at the outlet of the cooler. For coolers equipped with variable speed drive fans, the temperature of the combined stream from all bays will be the same as that from the individual bays. For coolers equipped with single speed drives, such would be the case only when all fans in all bays comprising the coolers are operational. If some of the fans are not operating, the temperature of the gas from each bay may be different.

The heat absorbed by the ambient air can be expressed as:

$$q_a = \dot{m}_a \cdot C_{pa} \cdot N_{\text{Fan}} \cdot (t_2 - t_1) \qquad (4-86)$$

Typical Side Elevations of Air-Coolers

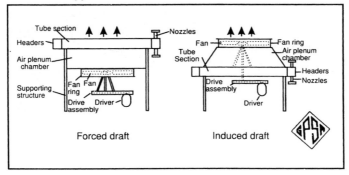

Figure 4-39. Typical side elevation of air coolers [Courtesy GPSA, Engineering Data Book, 1994]

where q_a = heat absorbed by air, BTU/hr

       $\dot{m}_a$ = air mass flow rate per fan, lb_m/hr

       C_{pa} = specific heat of air at the ambient pressure and temperature conditions, BTU/ lb_m, °F

    N_{Fan} = number of fans, dimensionless

       t_1 = inlet ambient air temperature, °F

       t_2 = exit air temperature, °F.

Typical Plan Views of Air-Coolers

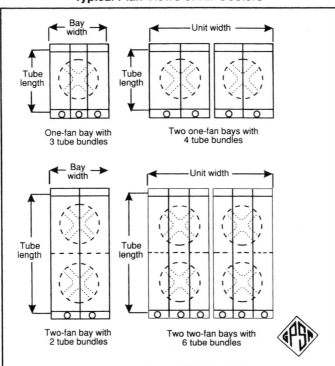

Figure 4-40. Typical plan views of air coolers [Courtesy GPSA, Engineering Data Book, 1994]

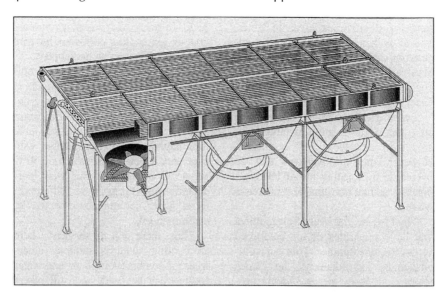

Figure 4-41. An aerial cooler with fans and tube bundle [Mukherjee, R., 1987, Chemical Engineering Progress]

In addition to Equations (4-85) and (4-86), the basic equation to calculate heat transfer for all heat exchangers is written as follows:

$$q = U \cdot A \cdot F \cdot \text{LMTD} \qquad (4-87)$$

where q = heat transfer rate, BTU/hr
   U = overall heat-transfer coefficient, (BTU/hr·ft^2·°F)
   A = heat transfer area, ft^2
   F = temperature correction factor, dimensionless
   LMTD = log mean temperature difference, °F.

Equation (4-87) can also be written as

$$q = U_b \cdot A_b \cdot N_{\text{Bays}} \cdot F \cdot \text{LMTD} \qquad (4-88)$$

where U_b = bare overall heat transfer coefficient, (BTU/hr·ft^2·°F)
   A_b = bare outside heat transfer area per bay, ft^2
   N_{Bays} = number of bays in service, dimensionless.

The heat dissipated by the gas, the heat absorbed by the air, and the heat transferred from the gas to the air should all be equal, so

$$q = q_g = q_a \qquad (4-89)$$

or

$$\dot{m}_g \cdot C_{pg} \cdot (T_1 - T_2) = \dot{m}_a \cdot C_{pa} \cdot N_{\text{Fan}} \cdot (t_2 - t_1) = U_b \cdot A_b \cdot N_{\text{Bays}} \cdot F \cdot \text{LMTD}$$
$$(4-90)$$

In Equation (4-88) the values of LMTD (ΔT_m), F, and U are computed as follows: The log mean temperature difference is:

$$\text{LMTD} = \frac{(T_1 - t_2) - (T_2 - t_1)}{\ln \frac{T_1 - t_2}{T_2 - t_1}} \qquad (4-91)$$

The temperature correction factor F may be obtained from Figure 4-42 for a single-pass cross-flow cooler, or from Figure 4-43 for a double-pass cross-flow cooler (GPSA 1994). To use these figures, the temperature-dependent functions P and R need to be computed as:

$$P = \frac{t_2 - t_1}{T_1 - t_1} \qquad (4-92)$$

$$R = \frac{T_1 - T_2}{t_2 - t_1} \qquad (4-93)$$

To obtain the value of the overall heat-transfer coefficient U, the following equation can be used (Holman 1997):

$$U = \frac{1}{\frac{1}{h_{io}} + r_f + r_m + \frac{1}{h_o}} \qquad (4-94)$$

where U = overall heat-transfer coefficient based on outside surface area, (BTU/hr·ft²·°F)
  h_{io} = inside heat-transfer coefficient (gas side) based on tube outside surface area, (BTU/hr·ft²·°F)
  r_f = combined fouling resistance, (hr·ft²·°F/BTU)
  r_m = metal resistance, (hr·ft²·°F/BTU)
  h_o = outside heat-transfer coefficient (air side), (BTU/hr·ft²·°F)

In Equation (4-94), the value of $(1/h_{io})$ is:

$$\frac{1}{h_{io}} = \frac{A_o}{A_i h_i} = \frac{r_o}{r_i h_i} = \frac{d_o}{d_i h_i} \qquad (4-95)$$

where A_o = tube outside surface area, ft²
  A_i = tube inside surface area, ft²
  r_o = tube outside radius, ft
  r_i = tube inside radius, ft
  d_o = tube outside diameter, ft
  d_i = tube inside diameter, ft
  h_i = inside (tube side) heat-transfer coefficient, (BTU/hr·ft²·°F)

In Equation (4-95), the value of h_i or inside heat-transfer coefficient can be easily computed using one of the common correlations (Dittus and Boelter or Seider and Tate), for tube flow heat transfer, see (Holman 1997). The value of h_o (air-side heat-transfer coefficient) can be calculated using an appropriate cross-flow equation (Holman 1997). The metal resistance or r_m is:

$$r_m = \frac{A_o \ln r_o/r_i}{2 \pi K L} = \frac{r_o}{K} \ln \frac{r_o}{r_i} \qquad (4-96)$$

where K = tube metal thermal conductivity, (BTU/hr·ft·°F)

The value of r_f, or fouling resistance, is normally a constant value and can be obtained from tables listed for gas cooling operations. Upon the substitution of all parameters into Equation (4-94), the value of the overall heat-transfer coefficient (U) can be calculated.

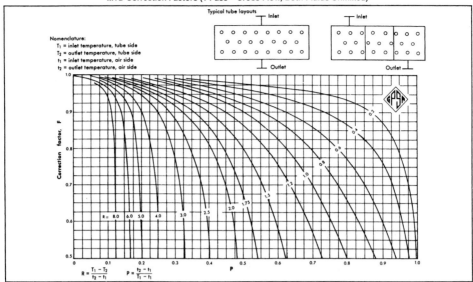

Figure 4-42. MTD Correction factors (1 Pass - Cross-Flow, Both Fluids Unmixed) [Courtesy GPSA 1994]

Generally, all the parameters required for the operating condition of the gas cooler can be computed from the design parameters supplied by the manufacturer as follows:

$$h_{io} = h_{iod} \left(\frac{N_{\text{Bays}d}}{N_{\text{Bay}}} \cdot \frac{\dot{m}_g}{\dot{m}_{gd}} \right)^{0.8} \qquad (4-97)$$

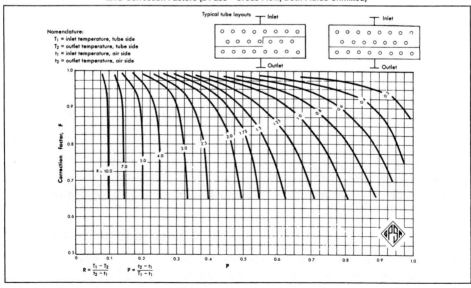

Figure 4-43. MTD correction factors (2 pass - cross-flow, both fluids unmixed) [Courtesy GPSA 1994]

where h_{io} = inside heat-transfer coefficient under simulation conditions, (BTU/hr·ft²·°F)
h_{iod} = inside heat-transfer coefficient under cooler design conditions, (BTU/hr·ft²·°F)
N_{Bayd} = number of bays under cooler design conditions, dimensionless
N_{Bay} = number of bays under simulation conditions, dimensionless
$\dot{m}_g$ = gas mass flow rate under simulation conditions, lb$_m$/hr
$\dot{m}_{gd}$ = gas mass flow rate under cooler design conditions, lb$_m$/hr.

The air-side heat-transfer coefficient is sensitive to both the rate of airflow across the tubes and the temperature of ambient air, as shown in Equation (4-98).

$$ h_o = [h_{od10} - F_t(t_1 - t_{10})]\left(\frac{\dot{m}_a}{\dot{m}_{a10}}\right)^{0.6} \qquad (4-98) $$

where h_o = outside heat transfer coefficient under simulation conditions, (BTU/hr·ft²·°F)
h_{od10} = outside heat transfer coefficient at 10°C (50°F), (BTU/hr·ft²·°F)
F_t = temperature correction factor, dimensionless
t_1 = ambient air temperature under simulation conditions, °F
t_{10} = ambient air temperature of 10°C (50°F), °F
$\dot{m}_a$ = fan air mass flow under simulation conditions, lb$_m$/hr
$\dot{m}_{a10}$ = fan air mass flow at 10°C (50°F), (lb$_m$/hr).

It should be noted that the air temperature under cooler design conditions is 10°C (50°F).

FAN AIR MASS FLOW RATE

Provided the coolers have a fixed pitch for the fan blade, which is the case for most gas coolers, the air mass flow rate per fan ($\dot{m}_a$) can be computed as follows:

$$ \dot{m}_a = \dot{m}_{ad} \frac{N}{N_d} \cdot \frac{t_{1d}}{t_1} \qquad (4-99) $$

where $\dot{m}_a$ = fan air mass flow rate under simulation conditions, lb$_m$/hr
$\dot{m}_{ad}$ = fan air mass flow rate under cooler design conditions, lb$_m$/hr
N = fan speed under simulation conditions, RPM
N_d = fan speed under cooler design conditions, RPM
t_{1d} = ambient air temperature under cooler design conditions, °R
t_1 = ambient air temperature under simulation conditions, °R.

REQUIRED FAN POWER

The required power of a fan motor in a gas cooling system can be achieved by specifying the degree of cooling, and it is calculated as follows:

$$ HP = HP_d\left(\frac{N}{N_d}\right)^3 \cdot \frac{t_{1d}}{t_1} \qquad (4-100) $$

where   HP = required fan power under simulation conditions, HP

HP_d = required fan power under cooler design conditions, HP

N = fan speed under simulation conditions, RPM

N_d = fan speed under cooler design conditions, RPM

t_{1d} = ambient air temperature under cooler design conditions, °R

t_1 = ambient air temperature under simulation conditions, °R.

GAS PRESSURE DROP IN COOLERS

The pressure drop in the gas flowing through a cooler can be expressed as

$$\Delta P = \Delta P_d \left(\frac{N_{\text{Bay }d}}{N_{\text{Bay}}}\right)^2 \left(\frac{\dot{m}_g}{\dot{m}_{gd}}\right)^2 \left(\frac{P_{1d}}{P_1}\right) \left(\frac{T_1 + T_2}{T_{1d} + T_{2d}}\right) \tag{4 - 101}$$

where ΔP = pressure drop under simulation conditions, psia

ΔP_d = pressure drop under cooler design conditions, psia

$N_{\text{Bay}d}$ = number of bays under cooler design conditions, dimensionless

N_{Bay} = number of bays under simulation conditions, dimensionless

$\dot{m}_g$ = gas mass flow rate under simulation conditions, $(\text{lb}_\text{m}/\text{hr})$

$\dot{m}_{gd}$ = gas mass flow rate under cooler design conditions, $(\text{lb}_\text{m}/\text{hr})$

P_{1d} = gas inlet pressure to the cooler under design conditions, psia

P_1 = gas inlet pressure to the cooler under simulation conditions, psia

T_1 = gas inlet temperature to the cooler under simulation conditions, °R

T_2 = gas exit temperature from the cooler under simulation conditions, °R

T_{1d} = gas inlet temperature to the cooler under design conditions, °R

T_{2d} = gas exit temperature from the cooler under design conditions, °R.

Some cooler assemblies bypass part of the gas to keep the pressure losses and gas velocities through the coolers down to an acceptable level. The gas mass flow rate through the tubes $(\dot{m}_{g \cdot \text{cooler}})$ can be found as follows:

$$\dot{m}_{g \cdot \text{cooler}} = \dot{m}_{g \cdot \text{total}} \cdot \left(1 - K_{\text{bypass}}\right) \tag{4 - 102}$$

where $\dot{m}_{g \cdot \text{cooler}}$ = gas mass flow rate through the cooler under simulation conditions, $(\text{lb}_\text{m}/\text{hr})$

$\dot{m}_{g \cdot \text{total}}$ = total gas mass flow rate under simulation conditions, $(\text{lb}_\text{m}/\text{hr})$

K_{bypass} = fraction of flow bypassing the cooler, dimensionless

When gas coolers are operating, it is generally necessary to set the gas discharge temperature. On the TransCanada System, this set point is 3°C above the ambient temperature. A minimum allowable set point is usually imposed on the gas discharge temperature. The TransCanada System set point is 10°C. Cooling will be stopped to prevent gas temperature from falling below this minimum value (Yoshikai 1994).

Design tables are normally supplied by the manufacturer and give a complete list of cooler and gas parameters at the design condition. This list of parameters usually includes:

- Cooler ID
- Cooler bypass
- Number of passes

TABLE 4-5. Typical overall heat-transfer coefficient for air coolers (Courtesy GPSA, Engineering Data Book, 1994)

	1 in. Fintube			
	1/2 in. by 9		5/8 in. by 10	
Service	U_b	U_x	U_b	U_x
1. Water & Water Solution				
Engine jacket water ($r_d = 0.001$)	110	7.5	130	6.1
Process water ($r_d = 0.002$)	95	6.5	110	5.2
50-50 ethylene glycol-water ($r_d = 0.001$)	90	6.2	105	4.9
50-50 ethylene glycol-water ($r_d = 0.002$)	80	5.5	95	4.4
2. Hydrocarbon liquid coolers				
Viscosity, cp, at avg. temp.				
0.2	85	5.9	100	4.7
0.5	75	5.2	90	4.2
1.0	65	4.5	75	3.5
2.5	45	3.1	55	2.6
4.0	30	2.1	35	1.6
6.0	20	1.4	25	1.2
10.0	10	0.7	13	0.6
3. Hydrocarbon gas coolers				
Pressure, psig				
50	30	2.1	35	1.6
100	35	2.4	40	1.9
300	45	3.1	55	2.6
500	55	3.8	65	3.0
750	65	4.5	75	3.5
1,000	75	5.2	90	4.2
4. Air and flue-gas coolers				
Use one-half of value given for hydrocarbon gas coolers				
5. Steam Condensers (Atmospheric pressure & above)				
Pure Steam ($r_d = 0.005$)	125	8.6	145	6.8
Steam with non-condensibles	60	4.1	70	3.3
6. HC Condensers				
Condensing Range, °F				
0° range	85	5.9	100	4.7
10° range	80	5.5	95	4.4
25° range	75	5.2	90	4.2
60° range	65	4.5	75	3.5
100° & over range	60	4.1	70	3.3
7. Other Condensers				
Ammonia	110	7.6	130	6.1
Freon 12	65	4.5	75	3.5

Notes: U_b is the overall rate based on bare tube area, and U_x is overall rate based on extended surface area.
(Based on approximate air face mass velocities, between 2,600 and 2,800 lb/hr·ft² of face area).
*Condensing range = hydrocarbon inlet temperature to condensing zone minus hydrocarbon outlet temperature from condensing zone.

TABLE 4-6. Fintube data for one-inch OD tubes (Courtesy GPSA, Engineering Data Book, 1994)

Fin Height by Fins/inch	1/2 in. by 9		5/8 in. by 10		
APF, sq ft/ft	3.80		5.58		
AR, sq ft/ft	14.5		21.4		
Tube Pitch	2 in. △	2 1/4 in. △	2 1/4 in. △	2 3/8 in. △	2 1/2 in. △
APSF (3 rows)	68.4	60.6	89.1	84.8	80.4
APSF (4 rows)	91.2	80.8	118.8	113.0	107.2
APSF (5 rows)	114.0	101.0	148.5	141.3	134.0
APSF (6 rows)	136.8	121.2	178.2	169.6	160.8

Notes: APF is total external area/ft of fin tube in sq ft/ft. AR is the area ratio of fin tube compared to the exterior area of 1 in. OD bare tube, which has 0.262 sq ft/ft. APSF is the external area in sq ft/sq ft of bundle face area.

- Number of bays
- Surface area per bay
- Number of fans per bay
- Fan drive type (single drive, two-drive, variable drive)
- Maximum power per fan
- Rated fan RPM
- H_{iod}
- $\dot{m}_{ad}$
- h_{o10C}
- $\dot{m}_{ad}$ per fan
- h_{od}
- t_1 (ambient temperature)
- HP_d
- HP_{motor}
- Pressure drop
- Standard flow
- Mass flow
- Molecular weight of gas
- Gas inlet pressure
- T_{1d}
- T_{2d}
- C_p
- Metal resistance
- Fouling factor
- Bypass %, summer
- Bypass %, winter
- Altitude

It was discussed earlier that in gas coolers, the use of finned tubes or extended surface tubes is necessary to increase the surface area for heat-transfer. The finned tubes will in turn reduce the heat-transfer coefficient within the tubes as they block the flow of air across the tubes. In Tables 4-5 and 4-6, typical overall heat-transfer coefficients for air coolers (both bare and extended tubes) and fin tube data for one-inch OD tubes are given (GPSA 1994).

ITERATIVE PROCEDURE FOR CALCULATIONS BASED ON UNKNOWN T_2

There are different approaches to solving gas cooler problems using iterative solutions. One such approach is to determine the gas temperature at the outletside of the cooler when the user has specified this temperature as being unknown. The solution depends on whether the cooler bays are equipped with single or variable speed drive fans. The steps comprising the procedure for each of these cases are slightly different, but the solution technique and equations used are almost identical for different types of gas coolers with different speed drive fans. The procedure to solve a gas cooler problem with a variable speed fan when the gas outlet temperature is specified is outlined below. For further details see (GPSA 1994) and (Yoshikai 1994).

Iterative Solution of a Gas Cooler with Variable Speed Drives by Specifying Gas Outlet Temperature

1. Assume a value for the gas cooler outlet temperature (T_2).
2. Calculate cooler duty q_g (gas side) using Equation (4-85).
3. Calculate air mass flow rate $\dot{m}_a$ using Equation (4-99).
4. Compute t_2 using Equation (4-86) ($q_a = q_g$).
5. Calculate LMTD using Equation (4-91).
6. Calculate factors P and R using Equations (4-92) and (4-93).
7. Use either Figure 4-42 or Figure 4-43 to find the temperature correction factor (F).
8. Calculate the corrected LMTD (F · LMTD).
9. Calculate both gas and air side heat-transfer coefficients using Equations (4-97) and (4-98), and the overall heat-transfer coefficient using Equation (4-94).
10. Calculate heat transferred (q) using Equation (4-88).
11. If q does not match q_g in Step 2, change gas cooler outlet temperature, T_2, until q and q_g are equal (with an allowable tolerance limit).
12. Compute pressure drop using Equation (4-101).

T_2 must not exceed the maximum allowable discharge temperature. This maximum value depends on different factors, particularly the pipeline coating, which will be damaged at temperatures exceeding 65°C. The maximum discharge temperature of the gas coolers in the TransCanada System is set at 45°C, so all coolers are designed to meet this maximum. In areas with an extremely cold climate, a minimum allowable discharge temperature may be specified. This set point is important because the contractions of the pipeline and its coating at very low temperatures will cause decay and dew point problems.

REFERENCES

Asante, B., 1994, "Justification for Internal Coating of Natural Gas Pipelines, Design Methods & Technical/ Facilities Planning Department," *NOVA Gas Transmission Limited (NGTL)* Internal Report, Calgary, Alberta, Canada.

Asante, B., 1996, "Design Criteria and Parameters Impact Study, System Design Department," *NOVA Gas Transmission Limited (NGTL)*, Internal Report, Calgary, Alberta, Canada.

Beggs, Dale, H., 1991, *Gas Production Operations*, OGCI Publication, Oil & Gas Consultants International Inc., Tulsa.

Campbell, J. M., Hubbard, R. A., and Maddox, R. N., 1994, *Gas Conditioning and Processing, Volume 1.*

Canadian Standard Association (CSA), 1999, *Z662-94 Oil and Gas Pipeline Systems*, Oil and Gas Industry System.

Gas Processors Suppliers Association (GPSA), 1994, *Engineering Data Book*, (Volume I), Tulsa, OK.

Golshan, H., and Narsing, M., 1994(a), "Impact of Different Gas & Pipeline Parameters on Three Major Gas Flow Equations," *Facilities Planning, NOVA Gas Transmission Limited (NGTL)*, Calgary, Alberta, Canada.

_____., 1994(b), "Study of Pipeline Deterioration Due to Age (Phase I)," *Design Methods & Technology/ Facilities Planning*, NOVA Gas Transmission Limited Internal Reports, Calgary, Alberta, Canada.

Holman, J. P., 1997, *Heat Transfer, 8th Edition*, McGraw-Hill Publications.

Hughes, T., 1993, "Optimum Pressure Drop Project," *Facilities Planning Department Internal Reports*, NOVA Gas Transmission Limited (NGTL), Calgary, Alberta, Canada.

Hydraulic Analysis & Resources, Tool (HART), 1995, "Users Guide Manual," System Design Department, NOVA Gas Transmission Limited, Calgary, Alberta, Canada.

Katz, Donald, L., et al., 1959, *Handbook of Natural Gas Engineering*, McGaw-Hill Book Company.

Kern, Donald, Q., 1997, *Process Heat Transfer*, Tata McGraw-Hill Edition.

Mohitpour, M., 1991, "Temperature Computations in Fluid Transmission Pipelines," ASME, New York, NY.

Mukherjee, R., 1997, "Effectively Design Air-Cooled Heat Exchangers," Chemical Engineering Progress.

NOVA Gas Transmission Limited, 1995, "Gas Turbine Overview," Technical Training Department, Airdrie Service Center, Alberta, Canada.

Segeler, C. G., 1965, *Gas Engineers Handbook, First Edition*, American Gas Association, Industrial Press Inc., New York, NY.

Smith, J. M., and Van Ness, H. C., 1959, *Introduction to Chemical Engineering Thermodynamics, 2nd Edition*, McGraw-Hill.

Starling, K. E., and Savidge, J. L., 1994, *Compressibility Factors of Natural Gas and Other Related Hydrocarbon Gases, AGA Transmission Measurement Committee Report*, No. 8.

Wilson, G. G., Ellington, R. T., and Farwalter, J., 1991, *Gas Distribution Home Study Course*, Institute of Gas Technology Education program,.

Weiss, M., 1998(a), "Nova Research and Technology Corporation (NRTC)," Internal Reports, NOVA Chemicals.

_____, 1998(b), "Mixtures for Storage and Pipeline Transport of Gases," *NOVA Research & Technology Corporation (NRTC)*, Internal Report, NOVA Chemicals, Calgary, Alberta, Canada.

Yoshikai, H., 1994, "Specifications for Representation of Coolers in HART," Internal Report, NOVA Gas Transmissions Limited.

Chapter *5*

LIQUID FLOW AND PUMPS

INTRODUCTION

This chapter outlines some of the major flow equations used for hydraulic calculations in pipeline transportation of liquids. It also provides a detailed review of centrifugal pumps. These pumps are most commonly used in liquid pipeline transmission systems. The objective is to briefly discuss the flow equations and their applications in order to relate them to the analysis of the centrifugal pumps used in liquid transportation.

In Chapters 3 and 4, gas equations were presented in Imperial Units. In this chapter, liquid flow equations are presented in both SI and Imperial Units.

FULLY DEVELOPED LAMINAR FLOW IN A PIPE

The velocity profile for a fully developed laminar flow in a pipe is given by

$$u = \frac{\gamma^2}{4\mu}\left(\frac{dP}{dx}\right) + \frac{C_1}{\mu}\ln \gamma + C_2 \qquad (5-1)$$

where
u = fluid velocity
γ = distance from the pipe centerline
μ = fluid dynamic or absolute viscosity
dP/dx = pressure drop over the length of the pipe
C_1 = constant
C_2 = constant.

The constants C_1 and C_2 of Equation (5-1) can be obtained using the appropriate boundary conditions. The velocity u at the pipe centerline for a laminar flow system must have some finite value. However, substituting $\gamma=0$ gives an ambiguous second term, since $\ln(0)$ is undefined. To eliminate this term, set $C_1 = 0$. Now

$$u = \frac{\gamma^2}{4\mu}\left(\frac{dP}{dx}\right) + C_2 \qquad (5-2)$$

The velocity at the pipe wall, when $\gamma = R$, is $u = 0$. Substituting these values into Equation (5-2) to solve for the C_2 gives

$$C_2 = \frac{-R^2}{4\mu}\left(\frac{dP}{dx}\right) \qquad (5-3)$$

or

$$u = \frac{1}{4\mu} \left(\frac{dP}{dx} \right) \left(\gamma^2 - R^2 \right) \qquad (5-4)$$

Equation (5-4) can be further simplified to

$$u = \frac{R^2}{4\mu} \left(\frac{dP}{dx} \right) \left[\left(\frac{\gamma}{R} \right)^2 - 1 \right] \qquad (5-5)$$

Using the velocity profile for fully developed laminar flow in a pipe, the volumetric flow rate can be calculated as follows:

$$Q = \int u \cdot dA \qquad (5-6)$$

or

$$Q = \int_o^R u \cdot 2\pi\gamma d\gamma \qquad (5-7)$$

Substituting for u (velocity profile) from Equation (5-4)

$$Q = \int_o^R \frac{1}{4\mu} \left(\frac{dP}{dx} \right) \left(\gamma^2 - R^2 \right) \cdot 2\pi\gamma d\gamma \qquad (5-8)$$

and after integration and substitution:

$$Q = \frac{-\pi R^4}{8\mu} \cdot \left(\frac{dP}{dx} \right) \qquad (5-9)$$

The pressure gradient, dP/dx, is constant for fully developed laminar flow, so $dP/dx = (P_2 - P_1)/L = \Delta P/L$, where L is the length of the pipe. This gives

$$Q = \frac{-\pi R^4}{8\mu} \left[\frac{\Delta P}{L} \right] = \frac{\pi D^4}{128\mu} \cdot \left[\frac{P_1 - P_2}{L} \right] \qquad (5-10)$$

Equation (5-10) represents steady-state isothermal laminar flow in a horizontal pipe. It can be rearranged and written for the pressure drop:

$$\Delta P = P_1 - P_2 = \frac{128\,\mu \cdot L \cdot Q}{\pi \cdot D^4} \qquad (5-11)$$

Substituting for $Q = \bar{u} \cdot A$, where $\bar{u}$ is the average liquid velocity in the pipe, and $A = (\pi \cdot D^2/4)$ (pipe cross-sectional area), the pressure drop Equation (5-11) is

$$P_1 - P_2 = \frac{32\mu \cdot L \cdot \bar{u}}{D^2} \qquad (5-12)$$

If the right hand side of Equation (5-12) is multiplied and divided by $\rho \cdot D \cdot \bar{u}$, the resulting equation will be:

$$P_1 - P_2 = 32 \cdot \left(\frac{\mu}{\rho \cdot D \cdot \bar{u}} \right) \cdot \left(\frac{\bar{u}^2 \cdot L \cdot D \cdot \rho}{D^2} \right) \qquad (5-13)$$

After simplification, Equation (5-13) becomes

$$P_1 - P_2 = 32 \cdot \left(\frac{\mu}{\rho \cdot D \cdot \bar{u}} \right) \cdot \left(\frac{\bar{u}^2 \cdot L}{D} \right) \cdot \rho \qquad (5-14)$$

Adding the effect of elevation changes (ΔP_E) and substituting for $R_e = \left(\frac{\rho \cdot D \cdot \bar{u}}{\mu} \right)$ (Reynolds number), Equation (5-14) can be rewritten as

$$\frac{P_1 - P_2}{\rho} - \Delta P_E = \frac{64}{R_e} \cdot \frac{L}{D} \cdot \frac{\bar{u}^2}{2} \qquad (5-15)$$

In Equation (5-15), $\frac{P_1 - P_2}{\rho} - \Delta P_E$ the term is the head loss considering the elevation changes. This head loss is inversely proportional to the Reynolds Number (R_e), and directly proportional to the L/D ratio and the square of the average liquid velocity $\bar{u}$ (i.e., kinetic energy of the fluid).

Equation (5-15) can be written in the form of head loss, as

$$\text{head loss} = \frac{64}{R_e} \cdot \frac{L}{D} \cdot \frac{\bar{u}^2}{2} \qquad (5-16)$$

Using a dimensional analysis, the above equation can also be written as

$$\text{head loss} = f \cdot \frac{L}{D} \cdot \frac{\bar{u}^2}{2} \qquad (5-17)$$

where f is the pipe friction factor.

Comparing Equation (5-16) and (5-17), the friction factor f in a fully developed laminar flow is

$$f = \frac{64}{R_e} \qquad (5-18)$$

Equation (5-18) shows that in laminar flow, friction factor (f) is only a function of the Reynolds number, and is totally independent of pipe roughness. For further details, refer to (Fox and McDonald 1985).

Using kinematic viscosity $\upsilon = \mu/\rho$, and substituting $\bar{u} = Q/A = Q/\pi D^2/4$, the expression for R_e can be rewritten as

$$R_e = \frac{4}{\pi} \cdot \frac{Q}{D \cdot \upsilon} \qquad (5-19)$$

or

$$R_e = C_1 \cdot \frac{Q}{D \cdot \upsilon} \text{ when } C_1 = \frac{4}{\pi} \qquad (5-20)$$

It can be further shown that the friction factor in laminar flow, as defined in Equation (5-18), can be expressed as

$$\sqrt{\frac{1}{f}} = C_2 \cdot \left[\frac{Q}{D \cdot \upsilon}\right]^{0.5} \qquad (5-21)$$

Laminar flow occurs when $R_e < 2,000$. The corresponding values of C_1 and C_2 in metric units are:

$$C_1 = 353,680$$
$$C_2 = 148.67$$

For laminar flow ($R_e < 2,000$), head loss can be calculated either by using Equation (5-17) or by using Equations (5-22) and (5-23), as shown:

$$Q = C_3 \cdot D^4 \cdot \frac{(P_1 - P_2 - \Delta P_E)}{L \cdot G \cdot \upsilon} \qquad (5-22)$$

Equation (5-22) is a rearrangement of Equation (5-10). The following equation defines the head loss, ΔP_E, to the elevation change:

$$\Delta P_E = C_E \cdot G \cdot (H_2 - H_1) \qquad (5-23)$$

where $C_E = 0.0999$ kPa/m
Q = liquid flow rate, m^3/hr
$C_3 = 8.616 \times 10^{-6}$
D = internal pipeline diameter, millimeters
P_1 = upstream pressure, kPa
P_2 = downstream pressure, kPa
ΔP_E = elevation head, kPa
L = length of pipeline, kilometers
G = specific gravity of liquid relative to water, dimensionless
υ = kinematic viscosity at the average flowing temperature, centistokes
H_1 = upstream elevation, meters
H_2 = downstream elevation, meters.

Equation (5-22) can be further rearranged to include the pipeline friction factor. The equation (for laminar liquid flow, assuming no elevation changes) in Imperial Units becomes:

$$P_1 - P_2 = 0.0134 \times \frac{f \cdot L \cdot G \cdot Q^2}{D^5} \qquad (5-24)$$

where P_1 = upstream pressure, psia
P_2 = downstream pressure, psia

f = friction factor, dimensionless
L = length of pipe, ft
G = specific gravity of liquid with respect to water, dimensionless
Q = liquid flow, gpm
D = pipeline inside diameter, inches.

In Imperial units, the Reynolds number is:

$$Re = \frac{3,160 \times Q}{D \cdot v} \qquad (5-25)$$

where $\quad Q$ = liquid flow, gpm
D = pipeline inside diameter, inches
v = fluid kinematic viscosity, centistokes

The R_e range between 2,000 and 4,000 is termed the critical zone. In general, the transmission factor $\left(\sqrt{1/f}\right)$ in this range is determined by experimentation once the pipeline is in operation. However, for a new pipeline, a transmission factor of 10.3 can be used for pressure drop/flow calculations with the understanding that the flow may be in error by ±10%. Consequently, the corresponding pressure drop may be in error by ±20%. Usually, the error is of little consequence in short pipelines (under 12–13 miles or 20 km). In longer lines, the error may result in over- or underdesign of pumping facilities.

TURBULENT FLOW

There is no mathematical relationship to describe the velocity profile for turbulent flow, so it is impossible to derive an analytical equation for head loss when the flow regime is fully turbulent. Therefore, effort is made to obtain experimental relationships to correlate friction factor, R_e, and pipe roughness, and to introduce these relationships into the original pressure drop/flow equation (Equation 5-17), which was based on dimensional analysis.

Colebrook–White Equation

The general equation for steady-state isothermal flow for liquids in a pipe was given in Equation (5-11). This equation can be further rearranged to introduce the transmission factor, as follows:

$$Q = C_4 \cdot \sqrt{\frac{1}{f}} \cdot D^{2.5} \cdot \left[\frac{P_1 - P_2 - \Delta P_E}{L \cdot G}\right]^{0.5} \qquad (5-26)$$

or in terms of pressure drop ΔP:

$$\Delta P = P_1 - P_2 - \Delta P_E = \frac{f \cdot L \cdot G \cdot Q^2}{C_4^2 \cdot D^5} \qquad (5-27)$$

Table 5-1. Reynolds number R_e vs. transmission factor $\sqrt{1/f} = F$
$K_e = 0.0007$ inches $= (0.0178$ mm) (ASCE 1975)

Pipe Size NPS 3 (ID = 3.068 inches, 77.9 mm)

F	0.0	0.1	0.2	0.3	0.4	0.5	0.6	0.7	0.8	0.9
10	4.6	4.9	5.2	5.6	6.0	6.4	6.9	7.4	7.9	8.5
11	9.0	9.7	10.4	11.1	11.9	12.8	13.7	14.7	15.7	16.9
12	18.1	19.4	20.8	22.3	23.9	25.7	25.6	29.6	31.8	34.1
13	36.7	39.5	42.4	45.7	49.2	53.1	57.1	61.6	66.6	71.9
14	77.7	84.2	91.2	98.9	107.4	116.8	127.3	138.9	151.8	166.3
15	182.6	201.1	222.1	246.1	273.8	306.1	343.8	388.8	442.9	509.1
16	591.7	697.0	837.4	1,032.1	1,313.8	1,769.9	2,611.5	4,667.9	18,235.5	—

Pipe Size NPS 4 (ID = 4.000 inches, 101.6 mm)

	0.0	0.1	0.2	0.3	0.4	0.5	0.6	0.7	0.8	0.9
10	4.5	4.9	5.2	5.6	6.0	6.4	6.8	7.3	7.8	8.4
11	9.0	9.6	10.3	11.0	11.8	12.6	13.5	14.5	15.5	16.6
12	17.8	19.1	20.4	21.9	23.5	25.1	27.0	28.9	31.0	33.2
13	35.7	38.3	41.1	44.1	47.4	51.1	54.8	58.9	63.4	68.3
14	73.6	79.3	85.5	92.3	99.8	107.9	116.8	126.6	137.4	149.3
15	162.4	176.9	193.2	211.3	231.6	254.5	280.5	310.1	344.0	383.3
16	429.0	482.7	547.3	625.8	721.3	842.8	999.3	1,207.3	1,503.2	1,947.9

Pipe Size NPS 6 (ID = 6.065 inches, 154.1 mm)

	0.0	0.1	0.2	0.3	0.4	0.5	0.6	0.7	0.8	0.9
10	4.5	4.8	5.2	5.5	5.9	6.3	6.8	7.3	7.8	8.3
11	8.9	9.5	10.2	10.9	11.7	12.5	13.4	14.3	15.3	16.4
12	17.5	18.7	20.0	21.5	23.0	24.6	26.3	28.2	30.2	32.3
13	34.6	37.1	39.7	42.5	45.6	49.0	52.4	56.2	60.3	64.6
14	69.4	74.5	80.1	86.0	92.5	99.5	107.1	115.3	124.2	134.0
15	144.5	156.1	168.8	182.5	197.7	214.4	232.8	253.1	275.7	300.7
16	328.6	359.8	395.2	435.4	480.5	532.8	592.9	662.1	744.5	842.2
17	959.4	1,104.7	1,279.5	1,516.4	1,827.0	2,272.4	2,890.6	3,923.7	5,893.2	11,020.1

Pipe Size NPS 8 (ID = 7.981 inches, 202.7 mm)

	0.0	0.1	0.2	0.3	0.4	0.5	0.6	0.7	0.8	0.9
10	4.5	4.8	5.2	5.5	5.9	6.3	6.8	7.2	7.7	8.3
11	8.9	9.5	10.1	10.8	11.6	12.4	13.3	14.2	15.2	16.2
12	17.4	18.6	19.9	21.3	22.7	24.3	26.0	27.8	29.8	31.9
13	34.1	36.5	39.1	41.8	44.8	48.0	51.3	55.0	58.9	63.1
14	67.6	72.5	77.7	83.3	89.4	96.0	103.0	110.7	118.9	127.8
15	137.5	147.9	159.3	171.6	185.0	199.6	215.6	233.0	252.1	273.1
16	296.1	321.4	349.6	381.0	415.4	454.3	497.7	546.3	601.8	664.9
17	736.9	821.0	915.3	1,032.3	1,169.9	1,340.9	1,538.7	1,795.0	2,126.7	2,566.8
18	3,187.8	4,122.6	5,667.1	8,727.4	17,659.3	—	—	—	—	—

Pipe Size NPS 10 (ID = 10.020 inches, 254.5 mm)

	0.0	0.1	0.2	0.3	0.4	0.5	0.6	0.7	0.8	0.9
10	4.5	4.8	5.1	5.5	5.9	6.3	6.7	7.2	7.7	8.3
11	8.8	9.4	10.1	10.8	11.6	12.4	13.2	14.1	15.1	16.2
12	17.3	18.5	19.8	21.1	22.6	24.2	25.8	27.6	29.6	31.6
13	33.8	36.2	38.7	41.4	44.3	47.3	50.7	54.2	58.0	62.1
14	66.5	71.3	76.3	81.7	87.6	93.9	100.7	108.0	115.8	124.3
15	133.4	143.2	153.9	165.4	177.9	191.4	206.1	222.0	239.3	258.2
16	278.8	301.3	326.0	353.3	383.0	416.0	452.4	492.4	537.4	587.6
17	643.6	707.4	776.9	860.6	955.1	1,067.3	1,190.5	1,340.3	1,519.4	1,734.7
18	2,001.8	2,339.8	2,775.7	3,362.6	4,193.9	5,453.3	7,581.9	11,941.6	25,738.1	—

Pipe Size NP 12 (ID = 12.000 inches, 304.8 mm)

	0.0	0.1	0.2	0.3	0.4	0.5	0.6	0.7	0.8	0.9
10	4.5	4.8	5.1	5.5	5.9	6.3	6.7	7.2	7.7	8.2
11	8.8	9.4	10.1	10.8	11.5	12.3	13.2	14.1	15.1	16.1
12	17.2	18.4	19.7	21.1	22.5	24.1	25.7	27.5	29.4	31.4
13	33.6	35.9	38.4	41.1	43.9	47.0	50.3	53.8	57.5	61.5
14	65.8	70.5	75.4	80.7	86.4	92.6	99.1	106.2	113.8	122.0
15	130.8	140.2	150.5	161.5	173.4	186.3	200.2	215.2	231.5	249.2

Table 5-1. (Continued)

16	268.4	289.2	312.0	337.0	364.0	393.9	426.4	462.0	501.6	545.3
17	593.5	647.6	705.8	774.6	850.8	939.5	1,034.4	1,146.6	1,276.3	1,426.2
18	1,603.8	1,815.9	2,070.9	2,384.7	2,779.9	3,289.4	3,970.2	4,924.2	6,346.8	8,701.0
19	13,307.5	26,378.4	59,128.4	—	—	—	—	—	—	—

Pipe Size NPS 16 (ID = 15.250 inches, 347.4 mm)

10	4.5	4.8	5.1	5.5	5.9	6.3	6.7	7.2	7.7	8.2
11	8.8	9.4	10.1	10.8	11.5	12.3	13.1	14.1	15.0	16.1
12	17.2	18.4	19.6	21.0	22.4	23.9	25.6	27.4	29.2	31.3
13	33.4	35.7	38.2	40.8	43.6	46.6	49.8	53.3	57.0	60.9
14	65.1	69.6	74.5	79.7	85.2	91.1	97.6	104.4	111.8	119.7
15	128.1	137.2	147.0	157.5	168.9	181.1	194.2	208.4	223.7	240.2
16	258.1	277.4	298.4	321.2	345.8	372.7	401.9	433.5	468.4	506.4
17	548.0	594.1	643.0	700.0	762.2	833.1	907.5	993.5	1,090.4	1,199.0
18	1,323.4	1,466.1	1,629.9	1,820.6	2,045.1	2,311.9	2,633.5	3,028.0	3,520.5	4,153.7
19	4,992.4	6,157.2	7,883.5	10,666.1	15,939.8	29,630.0	—	—	—	—

Pipe Size NPS 18 (ID = 17.250 inches, 438.2 mm)

10	4.5	4.8	5.1	5.5	5.9	6.3	6.7	7.2	7.7	8.2
11	8.8	9.4	10.1	10.8	11.5	12.3	13.1	14.1	15.0	16.1
12	17.1	18.3	19.6	20.9	22.4	23.9	25.5	27.3	29.2	31.2
13	33.3	35.6	38.0	40.7	43.5	46.4	49.6	53.1	56.7	60.6
14	64.8	69.3	74.1	79.2	84.7	90.6	96.9	103.7	110.9	118.7
15	127.0	136.0	145.6	155.9	167.0	179.0	191.8	205.6	220.5	236.6
16	253.9	272.6	292.9	314.9	338.5	364.3	392.2	422.3	455.4	491.4
17	530.6	573.8	619.5	672.3	729.7	794.6	862.3	939.8	1,026.4	1,122.5
18	1,231.2	1,354.4	1,493.7	1,653.1	1,837.2	2,051.0	2,301.7	2,599.5	2,956.9	3,394.3

Pipe Size NPS 22 (ID = 21.250 inches, 539.8 mm)

10	4.5	4.8	5.1	5.5	5.9	6.3	6.7	7.2	7.7	8.2
11	8.8	9.4	10.1	10.8	11.5	12.3	13.1	14.1	15.0	16.1
12	17.1	18.3	19.6	20.9	22.4	23.8	25.5	27.3	29.1	31.1
13	33.2	35.5	37.9	40.5	43.3	46.2	49.4	52.8	56.4	60.2
14	64.3	68.8	73.5	78.6	84.0	89.7	96.0	120.6	109.7	117.3
15	125.5	134.2	143.6	153.6	164.4	176.0	188.4	201.8	216.1	231.6
16	248.2	266.0	285.3	306.3	328.6	352.9	379.1	407.3	438.1	471.4
17	507.4	547.0	588.5	636.2	687.5	745.2	804.7	872.1	946.6	1,028.3
18	1,119.4	1,220.9	1,333.8	1,460.4	1,603.3	1,765.1	1,949.3	2,160.7	2,404.3	2,688.7
19	3,023.1	3,422.2	3,906.7	4,500.3	5,248.6	6,217.5	7,513.2	9,336.5	12,079.8	16,652.4
20	25,773.4	52,658.5	—	—	—	—	—	—	—	—

Pipe Size NPS 24 (ID = 23.250 inches, 590.6 mm)

10	4.5	4.8	5.1	5.5	5.9	6.3	6.7	7.2	7.7	8.2
11	8.8	9.4	10.1	10.8	11.5	12.3	13.1	14.1	15.0	16.1
12	17.1	18.3	19.6	20.9	22.4	23.8	25.5	27.2	29.0	31.0
13	33.1	35.4	37.8	40.4	43.2	46.1	49.3	52.6	56.2	60.1
14	64.2	68.6	73.3	78.3	83.7	89.4	95.6	102.2	109.3	116.8
15	124.9	133.6	142.9	152.8	163.5	175.0	187.2	200.4	214.5	229.8
16	246.1	263.6	282.6	303.2	325.1	349.4	374.5	401.9	432.0	464.4
17	499.3	537.6	577.8	623.8	673.1	730.3	785.2	849.5	920.1	977.2
18	1,082.9	1,177.9	1,282.8	1,399.8	1,531.0	1,678.3	1,844.5	2,033.3	2,248.5	2,496.3
19	2,723.1	3,119.3	3,518.7	3,995.3	4,577.4	5,301.3	6,220.7	7,428.3	9,077.9	11,456.1
20	15,174.2	21,759.6	36,828.3	49,635.6	—	—	—	—	—	—

Pipe Size NPS 26 (ID = 25.000 inches, 635.0 mm)

10	4.5	4.8	5.1	5.5	5.9	6.3	6.7	7.2	7.7	8.2
11	8.8	9.4	10.1	10.8	11.5	12.3	13.1	14.1	15.0	16.1
12	17.1	18.3	19.6	20.8	22.3	23.8	25.4	27.2	29.0	31.0
13	33.1	35.4	37.8	40.4	43.1	46.1	49.2	52.6	56.2	60.0

Table 5-1. (Continued)

14	64.1	68.5	73.2	78.2	83.5	89.3	95.4	101.9	108.9	116.5
15	124.5	133.1	142.3	152.2	162.8	174.1	186.3	199.4	213.4	228.4
16	244.6	261.9	280.7	300.9	322.5	345.9	371.1	398.1	427.5	459.3
17	493.5	530.8	570.1	614.8	662.8	716.3	771.3	833.3	901.3	975.3
18	1,057.2	1,147.8	1,247.4	1,357.9	1,481.3	1,619.1	1,773.6	1,947.9	2,145.0	2,370.1
19	2,627.9	2,926.6	3,276.6	3,687.6	4,179.8	4,777.7	5,515.0	6,448.1	7,661.8	9,298.2
20	11,620.1	15,148.5	21,226.7	34,049.3	76,907.1	—	—	—	—	—

Pipe Size NPS 30 (ID = 29.000 inches, 736.6 mm)

10	4.5	4.8	5.1	5.5	5.9	6.3	6.7	7.2	7.7	8.2
11	8.8	9.4	10.1	10.8	11.5	12.2	13.1	14.1	15.0	16.1
12	17.1	18.3	19.6	20.8	22.3	23.8	25.4	27.1	29.0	30.9
13	33.0	35.3	37.7	40.3	43.0	45.9	49.1	52.4	56.0	59.8
14	63.8	68.2	72.9	77.9	83.2	88.8	94.9	101.4	108.4	115.8
15	123.7	132.2	141.3	151.1	161.5	172.5	184.7	197.5	211.3	226.0
16	241.9	258.8	277.1	296.9	317.9	340.7	365.1	391.2	419.6	450.3
17	483.1	519.0	556.4	599.1	644.7	695.3	747.2	805.3	868.8	937.6
18	1,013.2	1,096.4	1,187.2	1,287.2	1,398.2	1,520.7	1,656.9	1,808.3	1,978.3	2,169.1
19	2,384.3	2,629.0	2,909.0	3,231.5	3,605.8	4,045.5	4,565.9	5,192.5	5,958.4	6,912.2
20	8,130.3	9,730.7	11,949.9	15,206.5	20,318.3	29,797.2	52,650.0	82,395.4	—	—

where $\Delta P_E = C_E \cdot G \cdot (H_2 - H_1)$, as previously defined. The transmission factor $F = \sqrt{1/f}$ can be described according to the following equation:

$$\sqrt{\frac{1}{f}} = -4 \log \left[\frac{K_e}{3.7D} + \frac{1.413\sqrt{\frac{1}{f}}}{R_e} \right] \qquad (5-28)$$

Equation (5-28) is referred to as the Colebrook–White equation, where K_e is the pipe internal roughness (mm), and D is the internal pipe diameter (mm).

The Colebrook–White equation is utilized when R_e exceeds 4,000 (turbulent range). For turbulent flow, either Equation (5-26) or (5-27) can be used to calculate pressure drop or liquid flow volumes. In metric units, the value of the constant C_4 (in Equation 5-27) is 19.8072×10^{-6}. For Imperial Units, Equation (5-24) can be utilized with the appropriate constants. In this case, units of measurement for K_e and D in Equation (5-28) will be in inches. The values of the Reynolds number and transmission factor $\sqrt{1/f}$ (Equation $5-28$) for various pipeline diameters are provided in Table 5-1(ASCE 1975).

Hazen–Williams Equation

The Hazen–Williams Equation is used only for selected materials such as refined oil products. The main drawback is that the Hazen–Williams coefficient, C_h, depends largely on the users' experience with experimental results for the viscosity of the product (ASCE 1975). The transmission factor is given by the following equation:

$$\sqrt{\frac{1}{f}} = C_f \frac{Q^{.074} \cdot C_h^{.926}}{D^{.065}} \qquad (5-29)$$

where $C_f = 0.1507$ for metric units
C_h = Hazen–Williams coefficient (Table 5-2).

Table 5-2. Hazen-Williams coefficient, C_h (ASCE 1975)

	At 45°F (7.2°C)			At 60°F (15.6°C)		
	G	v	C_h	G	v	C_h
Propane	0.527	0.230	165.0	0.510	0.22	165.4
Butane	0.587	0.320	160.6	0.580	0.30	161.4
Natural Gasoline	0.670	0.475	155.8	0.660	0.44	156.6
Gasoline	0.764	0.750	150.0	0.757	0.68	151.0
Power Fuel	0.790	1.400	142.8	0.780	1.28	144.4
Kerosene	0.815	3.000	134.4	0.810	2.45	36.5
Furnace Oil #1	0.815	3.800	131.8	0.810	3.20	133.6
Furnace Oil #2	0.845	4.550	130.0	0.840	3.70	132.0

Values for kinematic viscosity (v) in centistokes and specific gravity (G) for several liquid hydrocarbons.

Substituting Equation (5-29) into Equation (5-26), the Hazen–Williams equation can be rewritten as follows:

$$Q = C_5 \cdot C_h \cdot D^{2.63} \left[\frac{P_1 - P_2 - \Delta P_E}{L \cdot G} \right]^{0.54} \tag{5 - 30}$$

where $C_5 = 1.0745 \times 10^{-6}$ for metric units

Equation (5-30) is applicable for Reynolds numbers ranging from 4,000 to 1,000,000. The units are in metric.

In Imperial Units, the Hazen–Williams Equation (Equation 5-30) is expressed as follows:

$$Q = 0.148 \times C_h \cdot D^{2.63} \left[\frac{P_1 - P_2 - \Delta P_E}{L \cdot G} \right]^{0.54} \tag{5 - 31}$$

where Q = liquid flow, barrels per day
C_h = Hazen–Williams coefficient (from Table 5-2)
D = inside diameter of pipe, inch
P_1 = upstream pressure, psia
P_2 = downstream pressure, psia
ΔP_E = elevation changes correction, psia
L = pipeline length, mile
G = specific gravity of liquid with respect to water, dimensionless

For a nonisothermal situation when the temperature gradient between the pipe and the surrounding environment is significant, heat transfer between the pipe and environment needs to be considered. Refer to Chapter 3 for detailed temperature calculations. For further information, see (McAllister 1988).

Classification of Pumps

Pumping equipment is generally divided into two major categories: kinetic and positive displacement. General classes in each category are shown in Figure 5-1.

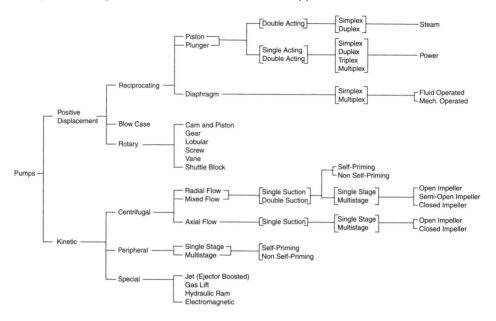

Figure 5-1. Classification of pumps

Centrifugal Pumps

Centrifugal pumps are the most common type of kinetic displacement pumps. They are further classified as radial, mixed, and axial flow.

Radial Flow

Radial flow pumps are those in which pressure is developed principally by the action of a centrifugal force. Generally, such pumps with a single-inlet impeller have a specific speed below 4,200, and pumps with double-suction impellers have a specific speed below 600. Liquid enters the impeller at the hub and flows radially to the periphery.

Mixed Flow

Pumps in which head is developed partly by centrifugal force and partly by the lift of the vanes on the liquid are classified as mixed flow pumps. This type of pump has a single-inlet impeller, with the flow entering axially and discharging in both axial and radial directions. Specific speeds generally range from 4,200 to 9,000.

Axial Flow

Axial flow pumps are sometimes called propeller pumps and develop most of their head by the propelling or lifting action of vanes on the liquid. These pumps have single-inlet impellers, with flow entering axially and discharging nearly axially. Specific speed of this type of pump is usually above 9,000.

Further Classification

- **Single Stage:** Pumps in which total head is developed by one impeller.
- **Multistage:** Pumps having two or more impellers acting in series in one casing.

- **Volute Pump:** Pumps having a casing made in the form of a spiral or volute.
- **Circular Casing Pump:** Pumps having a casing of constant cross-section concentric with the impeller.
- **Diffuser Pump:** Pumps with diffusion vanes.
- **Horizontal Pump:** Pumps with the shaft normally in a horizontal position.
- **Vertical Pump (Dry-Pit Type):** Vertical shaft-type pumps located in dry walls.
- **Single-Suction Pump:** Pumps equipped with one or more single-suction impellers.
- **Double-Suction Pump:** Pumps equipped with one or more double-suction impeller.

Rotation

Centrifugal pumps are designated as rotating either clockwise (CW) or counterclockwise (CCW). To determine the type of rotation for horizontal pumps, stand at the driving end facing the pump and note the type of rotation. For vertical pumps, determine the rotation by looking at their tops.

Rotary Pumps

A rotary pump is a positive displacement pump that consists of a fixed casing containing gears, cams, screws, vanes, plungers or similar elements actuated by rotation of the drive shaft. Rotary pumps generally fall into the following basic types:

Cam and Piston Pumps

Cam and piston pumps consist of an eccentrically bored cam, rotated by a shaft concentric in a cylindrical bored casing. They possess an abutment or follower arranged so that, during each rotation of the drive shaft, a positive quantity of liquid is displaced from the space between the cam follower and the pump casing.

Gear Pumps

Gear pumps consist of two or more gears operating in a closely fitted casing. These gears are arranged such that when the gear teeth unmesh on one side, liquid fills the space between the gear teeth, and is carried around in the tooth space to the opposite side. The liquid is then subsequently displaced as the teeth again mesh. There are two types:

1. **External gear pumps** have all gear rotors cut externally; these may be spur, single helical, or double helical teeth.
2. **Internal gear pumps** have one rotor with internally cut gear teeth, which mesh with an externally cut gear idler. They may be made with or without a crescent-shaped partition. Teeth may be of the spur or single helical type.

Lobular Pumps

Lobular pumps resemble gear-type pumps in action, but consist of two or more rotors cut with two, three, four, or more lobes on each rotor.

Screw Pumps

Screw pumps consist of one, two, or three screw rotors arranged so that as the rotors turn, liquid fills the space between screw threads and is axially displaced as the rotor threads mesh.

Vane Pumps

Vane pumps consist of one rotor in a casing that is machined eccentrically to the driveshaft. The rotor is fitted with a series of vanes, blades, or buckets that follow the bore of the casing, displacing liquid with each revolution of the driveshaft. There are two types of vane pumps:

1. **Swinging vane pumps** contain a series of hinged vanes that swing out as the rotor turns.
2. **Sliding vane pumps** consist of a series of vanes in a rotor that move outward and follow the casing bore. Vanes are usually flat with parallel sides (some modifications use vanes shaped like blocks or rollers).

Shuttle Block Pump

Shuttle block pumps consist of a cylindrically shaped rotor in a concentric casing. The rotor consists of a shuttle block and piston, which reciprocate within the rotor by means of an eccentrically located idler pin.

Reciprocating Pumps

Reciprocating pumps are generally classified into three major divisions: direct acting, crank and flywheel, and power pumps.

Direct Acting

A direct acting pump is a reciprocating, steam-powered pump, in which the steam piston is directly connected to the liquid piston or plunger through a piston rod. The length of the stroke is determined by the action of the steam in the steam cylinder. Pumps of this kind can be either simplex or duplex.

Crank and Flywheel

The crank and flywheel pump is also generally steam-powered, with a crankshaft on which a flywheel is mounted for storing energy during the early part of the stroke. The flywheel then imparts the stored energy to the liquid piston or plunger during the latter part of the stroke, after the steam is cut off in the steam cylinder. The stroke length is determined by the throw of the main crank.

Power

Power pumps are driven by power from an outside source applied to the crankshaft of the pump. The three types are as follows:

1. **Simplex pump:** A reciprocating pump having one liquid piston or its equivalent single- or double-acting plungers.
2. **Duplex pump:** A reciprocating pump having two liquid pistons or their equivalent single- or double-acting plungers.
3. **Triplex pump:** A reciprocating pump having three pistons or their equivalent single- or double-acting plungers.

Pumps for Trunkline Pipeline Stations

The most commonly used pumps for trunkline pipeline stations are either centrifugal or positive displacement. As these pumps have the most prevalent application in pipeline transmission, their theory and application are detailed later in the chapter. Centrifugal pumps are generally high-speed, high-volume units connected through speed increasers to internal combustion engines or directly to electric motors. In large stations, such pumps are generally connected in series, allowing each pump to handle total flow, and each adds an increase in pressure (head) to the liquid being moved.

Centrifugal pumps offer certain distinct advantages; chief among these is the fact that the flow of liquid from them is relatively even and smooth, with very few pulsations. When properly installed and operated, little or no vibration results from their use. These pumps can be used outside or in small buildings, they need only light foundations, and are easily kept clean. In addition, these pumps are of comparatively low cost, are simple to construct and easy to operate, and require a comparatively small space.

As discussed earlier, positive displacement pumps are divided into two general classes: reciprocating and rotary. Reciprocating pumps are further divided into two types: plunger and piston.

Plunger pumps always have a plunger longer than their stroke and only one end is available for pumping. Double-acting plunger pumps require at least two cylinders and plungers. Although not as adaptable as piston-type pumps, plunger pumps can be made to provide higher pressures and greater capacities. Nearly all large, high-pressure, reciprocating pumps used in trunkline stations are of the plunger type.

Pistons are usually shorter than their stroke and both ends are available for pumping with the same cylinder. Piston-type pumps are used extensively at gathering stations because of their flexibility in capacity, range in pressure, and portability. They can serve either as suction or discharge pumps. Most field pumps are piston pumps. Capacity and pressure rating are varied easily by changing the diameter of the piston and cylinder liners.

Reciprocating pumps are single construction, have a long service life, and are highly efficient. Disadvantages include their large weight, greater space requirements than other types, higher initial cost, pulsations, and sometimes the need for speed reducers in the case of very-slow-speed reciprocating pumps. Vertical, multicylinder, high-speed plunger pumps used in recent installations have helped overcome the disadvantages of the horizontal types. Vertical arrangement of the cylinder reduces the amount of floor space required; the higher speed range allows direct connection (without reduction of gearing), and the increased number of cylinders reduces the effects of surges (pulsations).

CENTRIFUGAL PUMPS

As discussed previously, centrifugal pumps have prevalent application in liquid pipeline transmission systems as they are capable of handling variable heads and flow rates (Labanoff and Ron 1987). These pumps can handle multiproducts and other liquids over a wide range of viscosities and other properties (Karassik 1976). Therefore, the remainder of this chapter will deal with the operation and theory of centrifugal pumps only.

Pump Operation

A typical pipeline centrifugal pump is shown in Figure 5-2. The centrifugal action within the pump takes place as follows:

1. The pump shaft is rotated by an external source (e.g., an electrical motor).
2. The impeller rotates inside the stationary casing.
3. Liquid flows into the impeller from the suction piping.

Figure 5-2. Typical pipeline centrifugal multistage pump [Courtesy David Brown Union Pumps Canada Ltd.]

4. The impeller imparts velocity energy to the fluid.
5. The liquid velocity energy is converted into pressure energy as it passes through the volute or diffuser.

Pump Characteristics

Centrifugal pumps are generally characterized by the following:

- Most common and most preferred for pipeline applications.
- Have minimal discharge pulsation
- Capable of efficient performance over a wide range of pressures and capacities (a requirement for most pipeline applications)
- Capable of high and variable throughputs
- Their discharge pressure, and hence head, are mostly a function of fluid density
- Their relatively small devices are less costly than other types of pumps
- High reliability
- Can be used with viscosities up to 300 centipoise (CP), depending on pump size and speed before efficiency losses begin to be economically significant
- Can be multistaged for higher heads

Centrifugal Pump Performance Curves
An individual set of pump performance curves is derived from a "family" set of pump curves as illustrated in Figure 5-3. The following information can usually be obtained:

- Pump model and size
- Rated speed

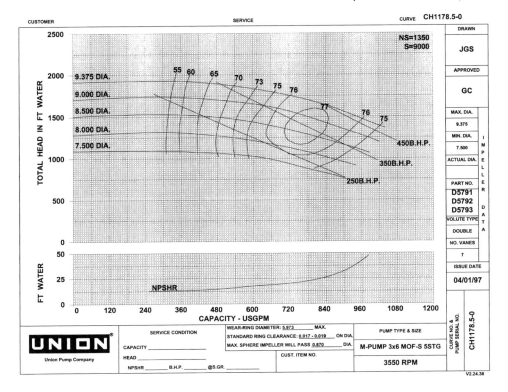

Figure 5-3. Typical centrifugal multistage pump performance curve [Courtesy David Brown Union Pumps Canada Ltd.]

- Curve identification number
- Impeller type and eye area
- Wear ring clearances
- Range of impeller diameters available
- Capacity versus head developed for the different impeller diameters (H–Q curves)
- Power to drive pump (pressure head, flow P–Q) curves, based on water, specific gravity (SG = 1.0)
- Pump efficiency (efficiency – Q curves)
- Net positive suction head required (NPSHR)
- Pump specific speed, N_S
- Pump suction specific speed, N_{SS} or S

In a double-suction pump, the area of the impeller eye is the area of the impeller eyes of both impellers. Pump shut-off is the head developed at zero (0) capacity. Pump run-out is the pump capacity above which the pump should not be operated due to instability, excessive NPSHR, vibration, and a dramatic reduction in head. This point is usually described as 120 percent of the best efficiency point.

Power and Efficiency

Brake horsepower (BHP) is the actual power delivered to the pump shaft. It is expressed as:

$$\text{BHP} = \frac{Q \times H \times G \times 100}{3,960 \times \text{ Pump Efficiency}} \qquad (5-32)$$

Hydraulic horsepower (HYD HP) is the liquid power developed by the pump. It is expressed as:

$$\text{HYD HP} = \frac{Q \times H \times G}{3,960} \qquad (5-33)$$

The pump efficiency is the ratio of hydraulic horsepower and brake horsepower:

$$\text{Pump Efficiency} = \frac{\text{HYD HP} \times 100}{\text{BHP}} \qquad (5-34)$$

where
Q = flow capacity, USGPM
H = total developed head, feet
G = specific gravity at the temperature of the liquid pumped.
Pump efficiency = pump efficiency in percent (%)

Impeller Theory

The work of a centrifugal impeller (H_t) can be derived by applying the principle of angular momentum to the mass of fluid going through the impeller passages. This principle is applied to the inlet and exit conditions of a frictionless fluid in an impeller to drive the impeller head H_t.

Using the following nomenclature:
U = peripheral velocity of impeller
V_r = relative velocity of flow
V = absolute velocity of flow
V_m = radial velocity
α, β = vane angle
ω = angular velocity
γ = density
g = gravitational acceleration
Q = flow
W = weight (mass) flow
1 = inlet subscript
2 = discharge subscript.

Refer to Figure 5-4 for a sketch of an impeller inlet and discharge velocity diagram. The torque (T) is defined as the difference of the peripheral components of impulse forces:

$$T = (\text{ outlet impulse} \times \cos \alpha_2) - (\text{ inlet impulse} \times \cos \alpha_1) \qquad (5-35)$$

$$T = \left[\left(\frac{\gamma}{g}\right) Q R_2 V_2 \cos \alpha_2\right] - \left[\left(\frac{\gamma}{g}\right) Q R_1 V_1 \cos \alpha_1\right] \qquad (5-36)$$

$$T = \left(\frac{\gamma}{g}\right) Q(R_2 V_{2_u} - R_1 V_{1_u}) \qquad (5-37)$$

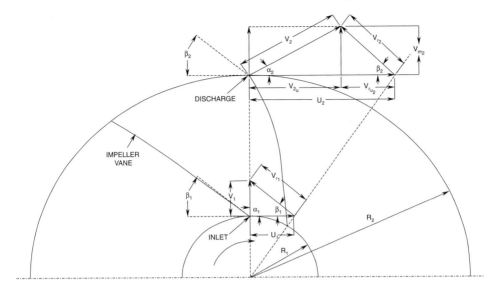

Figure 5-4. Impeller inlet and discharge velocity diagram

Considering Euler's theory, the following can be derived for the work done by the impeller (i.e., with theoretical head H_t achieved):

$$H_t = \frac{\text{energy input}}{\text{weight (mass) flow}} \tag{5-38}$$

$$H_t = \text{torque}\left(\frac{\omega}{W}\right) \tag{5-39}$$

Substituting:

$$H_t = \frac{T\omega}{W} = \frac{T\omega}{\gamma Q} = \left(\frac{1}{g}\right)(R_2\,\omega\,V_{2_u} - R_1\,\omega\,V_{1_u}) \tag{5-40}$$

The Euler equation for head H_t is thus:

$$H_t = \left(\frac{1}{g}\right)(V_{2_u}\,U_2 - V_{1_u}\,U_1) \tag{5-41}$$

Assuming the fluid enters the impeller without a tangential component (radial inlet), then $V_{1_u} = 0$ and Euler's equation (5-41) can be reduced to:

$$H_t = \frac{V_{2_u}\,U_2}{g} \quad \text{where } V_{2_u} = U_2 - \frac{V_{m_2}}{\tan \beta_2} \tag{5-42}$$

$$H_t = \frac{U_2^2}{g}\left[1 - \frac{V_{m_2}}{\tan \beta_2}\right] \tag{5-43}$$

For any given impeller size, V_{m_2} will vary directly with flow, and if the rotative speed is held constant, U_2 will be constant and H_t will vary linearly with flow. For an angle β_2 less than 90 degrees (backward curved vanes), $V_{m_2}/\tan \beta_2$ will increase with flow and cause H_t to

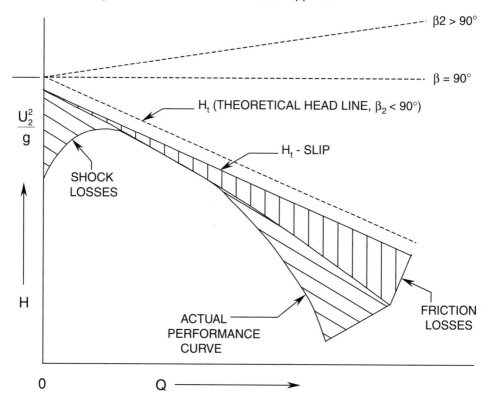

Figure 5-5. Development of the head capacity curve

decrease with flow. This will result in a theoretical curve, as shown in Figure 5-5. If shock losses, slip factors, and friction losses are added to the theoretical curve, the typical pump characteristic curve (as in Figure 5-5) is generated. It is these types of curves that pipeline engineers need to be familiar with in order to select pumps for a particular service application.

Specific Speed

Specific speed (N_S) is used to predict pump characteristics for the purpose of classifying pump impellers according to type, proportions, and performance. It is expressed as:

$$N_S = \frac{N\sqrt{Q}}{H^{\frac{3}{4}}} \qquad (5-44)$$

where
N_S = pump specific speed, dimensionless
N = pump speed, RPM
Q = capacity at best efficiency point, USGPM
H = total head per stage at the best efficiency point, feet.

For double-suction impellers, one half the flow is used to calculate the specific speed. The specific speed (N_S) determines the general shape or class of the impeller. As the specific speed increases, the ratio of the impeller outlet diameter, D_2, to the inlet or eye diameter, D_1, decreases. This ratio becomes 1.0 for a true axial flow impeller.

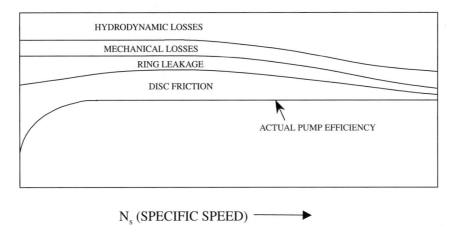

Figure 5-6. Specific speed and efficiency

Radial flow impellers develop head mainly through centrifugal force. Pumps of higher specific speeds develop head partially by centrifugal force and partially by axial force. A pump with a higher specific speed generates head more by axial forces and less by centrifugal forces. An axial flow or propeller pump with a specific speed of 10,000 or greater generates its head exclusively through axial forces.

The specific speed of an impeller can provide a wide variety of information about its performance:

- **Impeller Type:** Impellers with low specific speed are long and thin and are used for low-flow, high-head applications. Impellers with high specific speed are short and stubby and are used for high-flow, low-head applications.
- **Efficiency:** Efficiency is arrived at by considering the losses through the pump impeller and driver system. Pump efficiency is affected by losses due to pump impeller friction, ring leakage, and mechanical losses, as well as losses incurred by movement of the liquid within the pump, referred to as hydrodynamic losses. Specific speed affects pump efficiency (see Figure 5-6). The lower the specific speed, the lower the efficiency. The reason is that a higher percentage of energy is lost to overcome the impeller disc friction that is necessary to generate the high heads (Figure 5-6).
- **Shape of curve:** Once an impeller is designed for a certain specific speed, it will produce a typical head capacity curve and efficiency curve shape. A low specific speed impeller has a flat curve with a wide efficiency range. A high specific speed impeller produces a steep curve with a narrow efficiency range (Figure 5-7).

Impeller Curve Characteristics

There are a number of pump head-capacity (H–Q) curve shapes that are shown in Figures 5-8 and 5-9. These characteristic curves are:

- Rising
- Drooping
- Steep
- Flat

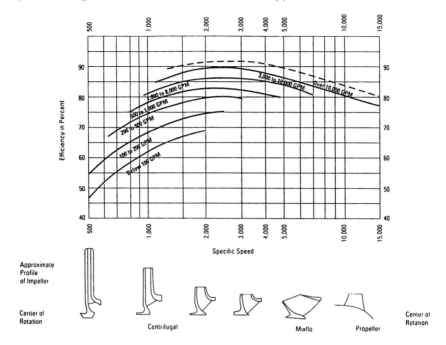

Figure 5-7. Specific speed and performance [Darby, R., 1996, reproduced with permission from Marcell Dekker]

- Stable
- Unstable
- Nonoverloading
- Overloading

Affinity Laws

The flow and head of a centrifugal pump may be changed by varying the pump speed or changing the impeller diameter. This results in a change to the impeller tip speed or velocity of its vanes, which causes a change in the velocity at which the liquid leaves the impeller. Usually, impellers can be cut down to 80 percent of their original diameter without lowering their efficiency significantly.

For centrifugal pumps with radial impellers, the relationships are approximately as follows:

For diameter change only:

$$Q_2 = Q_1\left(\frac{D_2}{D_1}\right), \; H_2 = H_1\left(\frac{D_2}{D_1}\right)^2, \; \text{BHP}_2 = \text{BHP}_1\left(\frac{D_2}{D_1}\right)^3 \qquad (5-45)$$

For speed change only:

$$Q_2 = Q_1\left(\frac{N_2}{N_1}\right), \; H_2 = H_1\left(\frac{N_2}{N_1}\right)^2, \; \text{BHP}_2 = \text{BHP}_1\left(\frac{N_2}{N_1}\right)^3 \qquad (5-46)$$

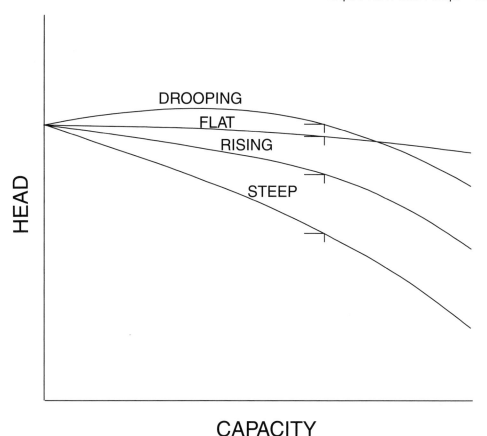

Figure 5-8. Various types of pump characteristic curves

For diameter and speed change:

$$Q_2 = Q_1\left(\frac{D_2}{D_1} \times \frac{N_2}{N_1}\right), \ H_2 = H_1\left(\frac{D_2}{D_1} \times \frac{N_2}{N_1}\right)^2, \ \mathrm{BHP}_2 = \ \mathrm{BHP}_1\left(\frac{D_2}{D_1} \times \frac{N_2}{N_1}\right)^3 \quad (5-47)$$

where D = impeller diameter, in inches
H = head, in feet or meters
Q = capacity, in USGPM
N = speed, in RPM
BHP = brake horsepower
1 = original conditions subscript
2 = new design conditions subscript.

Pump Head-Flow (H–Q) Curve and System Head Curve

A pump H–Q curve overlaid on the pipeline system hydraulic system head curve is shown in Figure 5-10. The pump is designed to operate at point H_1 unless the pump or the system curve is changed.

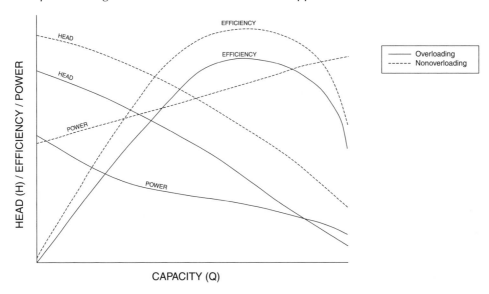

Figure 5-9. Typical characteristic curves for overloading and non-overloading

The pump includes a throttle valve on the discharge side. If such a throttle valve is partially closed, friction will be added to the system and the pump will be forced to operate back on the curve at point H_2. When the system head curve is changed by throttling, it is called throttle control.

If the speed of the pump is reduced from N_1 to N_2 or the impeller decliner is reduced from D_1 to D_2, then the $H–Q$ curve is changed as shown in Figure 5-10. The pump would now operate at point H_3.

Sizing and Selection

When sizing and selecting centrifugal pumps, it is most economic to use the physically smallest pump that will perform the service. Pumps are sized on the following basis:

- **Suction and discharge nozzles:** Suction nozzles are the same size or larger than the size of the discharge nozzles. The discharge nozzle is never larger than the suction nozzle. The larger the nozzle size the higher the flow capacity of the pump.
- **Impeller diameter:** The larger the impeller diameter the higher the head (and hence the discharge pressure). The head is proportional to the square of the impeller diameter [refer to Equation (5-45).]. However, due to the effect of inertia, impeller size is limited by speed. Typical speeds are 1,800 RPM for 26 inches impeller and 3,600 RPM for 12 inches impeller chambers.
- **Speed:** The flow rate varies linearly with speed—the head varies as the square and the power varies as the cube of the speed change. Centrifugal pumps are generally not used below 1,200 RPM because of the low head developed.
- **Suction limitations:** The suction conditions quite often are the limiting factors that affect speed, size, and capacity. Requirements are determined by the NPSHR, which will be defined later in this chapter.

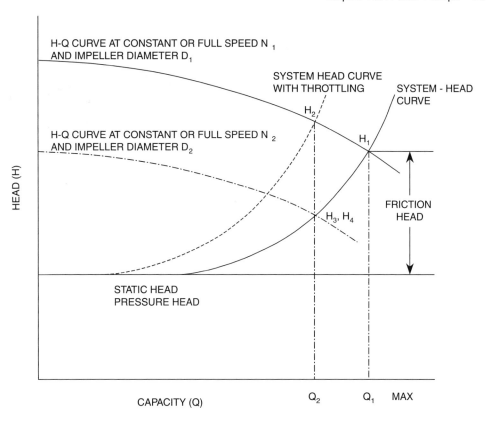

Figure 5-10. Capacity change with throttling

Performance in Pipeline Applications

The performance of pipeline pumps often needs to be altered to accommodate varying liquid transmission conditions. The following are generally considered by the pipeline industry:

- **Impeller change:** In order to change the specific speed, impeller size may be changed to meet the new demand on performance. Pump manufacturers can usually offer several different impeller diameters and vanes that will fit the pump casing without any further internal modifications. The effect of changing the impeller characteristics are illustrated in Figure 5-11.
- **Restaging:** Pumps with multistaging capabilities can be restaged (up or down) to meet the change in pressure, head, or flow requirements. For example, if an entire pressure range is not needed for a particular period of time, a number of impellers can be removed to meet the required conditions. Manufacturers provide destaging kits to block off the unused pump impeller areas to maintain efficiency (see Figure 5-12).
- **Impeller underfiling and overfiling:** This is undertaken to alter the performance of a pump. It involves modifying the flow area of the impeller by grinding metal off the impeller outlet vanes (Figure 5-13).

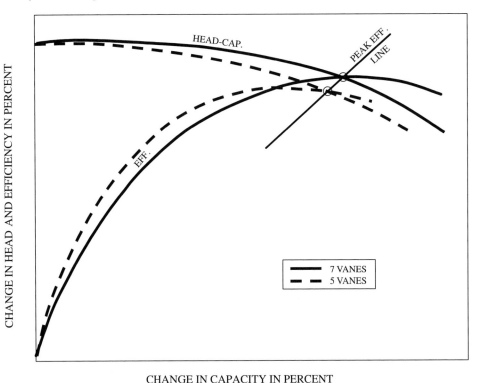

Figure 5-11. Changing performance by impeller design

- **Impeller volute chipping:** This is a technique that is used to alter the outlet flow area of the pump casing in order to modify performance (Figure 5-14). Again, caution is needed to avoid making costly mistakes.
- **Impeller volute inserts:** This technique (Figure 5-15) involves inserting special removable volutes into the pump to allow for a wider performance range. It also allows for more accurate and close control over the performance.

Cavitation

Cavitation control is a very important consideration in any liquid system and can be understood by considering the following:

- A local pressure drop causes an increase in flow velocity and hence an acceleration.
- A liquid will boil when the local pressure falls below its vapor pressure.
- A vapor occupies many more times the volume of a liquid.

It is important to avoid cavitation-induced conditions when operating centrifugal pumps. If a liquid is accelerated in such a manner that the local pressure falls below the liquid vapor pressure, the liquid will transform into the vapor phase, which results in the formation of bubbles. If the local pressure recovers, the vapor bubbles will transform

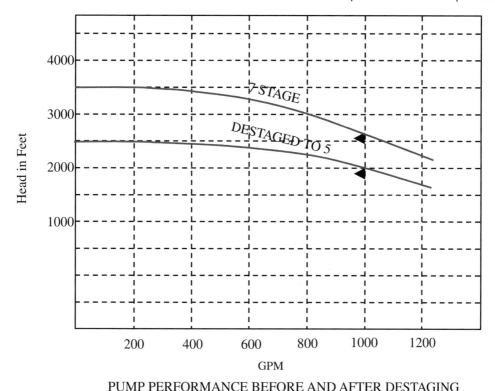

PUMP PERFORMANCE BEFORE AND AFTER DESTAGING

Figure 5-12. Changing performance by destaging

themselves back into a liquid. There is a tremendous volume change during transformation, because collapsing bubbles release a large amount of energy. Because the bubbles are very small, the resulting impact loads on the surrounding metal can be significant. This can result in physical damage to the metal and, thus, the creation of high noise levels.

Some liquids (such as water) are more difficult to handle from a cavitation point of view. When the vapor pressure of a homogenous fluid such as water is reached, the entire fluid begins to change phase, resulting in the formation of a large number of damage-causing bubbles. For a nonhomogeneous fluid such as a hydrocarbon, only the light ends (such as condensates) with low specific volume are affected.

In a centrifugal pump, the fluid is accelerated by the impeller. The area of lowest pressure in the pump suction system, as shown in Figure 5-16, is the eye of the impeller at cross-section A-A. If the pressure falls below the vapor pressure of the liquid, vapor bubbles form. As the mixture of liquid and bubbles continue through the pump, the pressure increases and the bubbles return to the liquid state. Damage to the impeller occurs where the bubbles collapse as shown at cross-section B-B. This location varies for different impellers and different suction conditions:

The effects of cavitation include:

- Noise and vibration
- Pump damage (e.g., pitting of the impeller)
- Fall-off of pump performance and efficiency

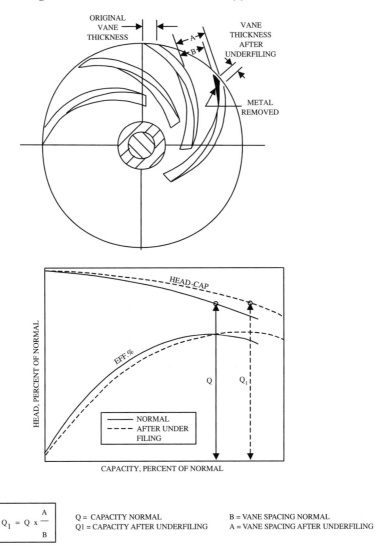

Figure 5-13. Changing performance with underfiling

Cavitation in centrifugal pumps can be recognized by a characteristic noise, which sounds just like it is trying to pump gravel. A typical break-off in the performance curve of a pump due to cavitation is shown in Figure 5-17.

Net Positive Suction Head

Net positive suction head (NPSH) is the total absolute suction pressure at the impeller eye less the absolute vapor pressure of the liquid pumped. NPSH must be of a magnitude to avoid vapor formation in the liquid and hence cavitation. Another way to describe NPSH is that it represents the amount of head required to push the fluid into the pump to suppress cavitation. Available and required NPSH are calculated as shown below.

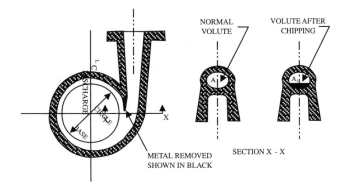

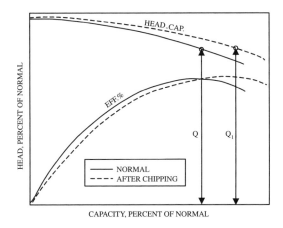

$$Q_1 = \sqrt{\frac{A_1}{A}} \times Q$$

Q = NORMAL CAPACITY   A = NORMAL VOLUTE AREA
Q_1 = CAPACITY AFTER A_1 = VOLUTE AREA AFTER
    VOLUTE CHIPPING   CHIPPING

Figure 5-14. Changing performance with volute chipping

Net Positive Suction Head Available (NPSHA)

The NPSHA is calculated by the following expression:

$$\text{NPSHA} = h_a - h_{VP} + h_{st} - h_{fs} \qquad (5 - 48a)$$

or

$$\text{NPSHA} = h_{st} + (P_a - P_{VP})\frac{2.31}{G} - h_{fs} \qquad (5 - 48b)$$

where P_a = absolute pressure at the surface of the liquid supply level (for suction from an open tank, this is the barometric pressure)

          P_{VP} = vapor pressure of the liquid at the temperature it is being pumped

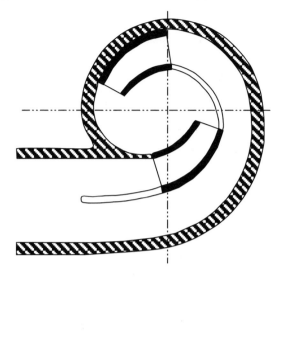

VOLUTE INSERTS CONVERT HIGH -CAPACITY PUMP TO LOW CAPACITY

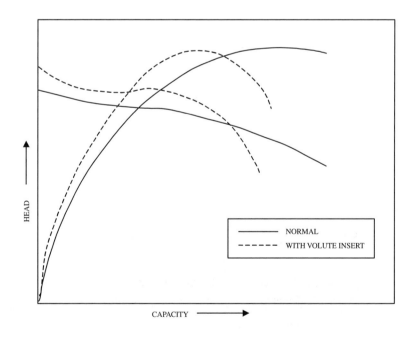

Figure 5-15. Changing performance with volute inserts

$$h_a = P_a \text{ expressed in equivalent head}$$
$$h_{VP} = P_{VP} \text{ expressed in equivalent head}$$

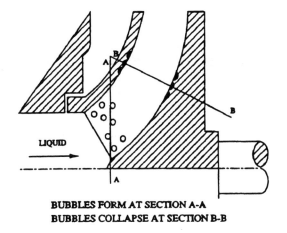

BUBBLES FORM AT SECTION A-A
BUBBLES COLLAPSE AT SECTION B-B

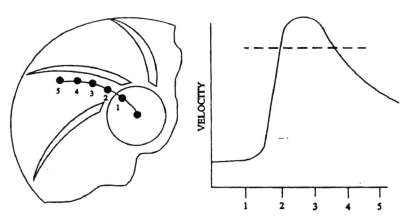

POINT 2 - VAP▼BELOW VP = BUBBLES FORM
POINT 3 - VAP▲ABOVE VP = BUBBLE COLLAPSES
                                    RELEASES ENERGY
WHERE:
V = VELOCITY   P = PRESSURE   VP = VAPOR PRESSURE

Figure 5-16. Cavitation

h_{st} = static elevation of the liquid supply above or below the pump inlet centerline. This value is negative for suction lift.

h_{fs} = suction line losses, including entrance losses and friction of the piping

G = specific gravity, dimensionless

The NPSHA must always be equal to or exceed the net positive suction head required (NPSHR) to prevent cavitation.

Net Positive Suction Head Required (NPSHR)

All pumps accelerate fluid and possess a corresponding internal friction loss. Therefore, every pump requires a certain amount of positive suction pressure in order to avoid cavitation. The pump NPSHR is determined by actual tests conducted by the

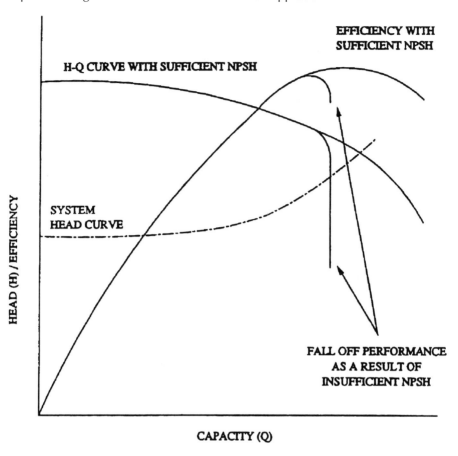

Figure 5-17. Result of cavitation on pump

pump manufacturer with procedures established by the Hydraulic Institute (ANSI/HI 2000). It must be understood that the NPSHR quoted by the pump manufacturer corresponds to the point at which the pump is fully cavitating. Therefore, it is important to allow a margin between NPSHA and NPSHR. At least three feet (one meter) of margin is generally required by the industry.

Correcting for Inadequate Suction Conditions
The following can be undertaken to correct for inadequate suction conditions:

Increase NPSHA.
- Raise liquid level (e.g., increase tank elevation, increase liquid level).
- Lower the pump (e.g., use vertical can pump).
- Reduce friction losses in suction piping (e.g., increase suction pipe diameter, reduce piping length).
- Use booster pump, prime pump.
- Reduce the vapor pressure of the liquid (e.g., subcool the liquid).

Reduce NPSHR.

- Use slower speeds (may not be economic).
- Use double-suction impellers (e.g., by doubling the impeller area the NPSHR can be reduced by up to 27 percent).
- Use a large impeller eye area (increase N_S). Caution: This may cause problems for lower flows.
- Use an inducer ahead of a conventional impeller. Caution: This reduces NPSHR, but also reduces the operating range of the pump.
- Use several smaller pumps in parallel (e.g., a smaller pump may require less NPSHR)

Viscous Liquids

The performance of centrifugal pumps is affected by fluids with higher viscosities, such as non-Newtonian fluids. A significant increase in brake horsepower, a reduction in head and efficiency, and some reduction in capacity result when handling moderate and high viscosities. A performance correction factor, as depicted in Figure 5-18, can be utilized to determine the actual performance of a pump handling viscous flow.

Minimum Flow

Minimum flow limits for a pump are established to limit or accommodate the following:

- Temperature rise
- Radial thrust
- Internal recirculation
- Reduced brake horsepower for pumps with high specific speed and overloading curves

Temperature Rise

The difference between BHP and HYD HP is the power loss within the pump. Most of the power loss is transferred and used to heat the liquid passing through the pump. The heat loss through the casing and other pump components is small. It is important that such a temperature rise does not cause vapor pressure formation, which would create abnormal operation, particularly for a single-stage pump.

The following equations can be utilized to determine the temperature rise across a pump impeller:

$$\text{Power Loss} = \text{BHP} - \text{HYD HP} \qquad (5-49)$$

$$T_{\text{Rise}}(°F) = \frac{(\text{BHP - HYD HP}) \times 2,545 \, \frac{\text{Btu}}{\text{hr} \times \text{hp}}}{(Q) \, \text{lb/ hr} \times C_p} \qquad (5-50)$$

where C_p = specific heat at constant pressure, BTU/lb.$\dot{\text{F}}$
Q = main flow rate, lb/hr.

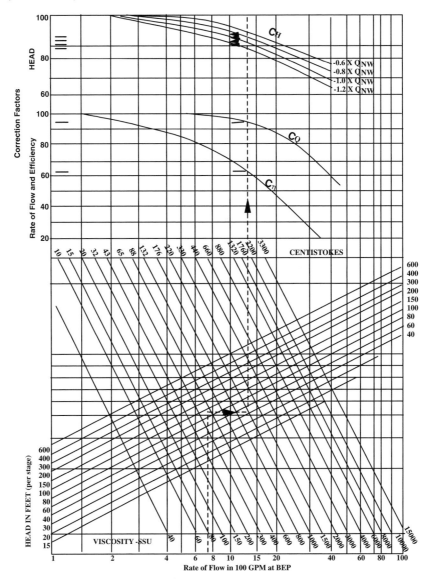

Figure 1.65B - Performance correction chart for viscous liquids (US units)
Courtesy of Hydraulic Institute

Figure 5-18. Performance correction chart for viscous liquids [ANSI/HI (2000), Courtesy Hydraulics Institute]

RETROFITTING FOR CENTRIFUGAL PUMPS (RADIAL-FLOW)

Very often, changes in pipeline throughput cause pumps to operate at below design capacity.

Reduced Pipeline Throughput

When pipeline throughput is reduced, the pumps in effect become technically oversized, and are therefore operating inefficiently. These pumps then become subject to problems that are associated with low flows, such as vibration or seal and bearing failure. There are a number of retrofit solutions to this problem: change the gear ratio to adapt to the desired output speed; exchange a low-capacity impeller with the existing one; install volute inserts; or combine a number of the above solutions.

Increased Pipeline Throughput

When pipeline throughput is increased, the existing pumps in effect become technically undersized, and either cannot handle the increased throughput or start to operate inefficiently. Retrofit solutions to this problem include:

- Increasing the gear ratio to adapt to the desired output speed (for a relatively small increase in flow),
- Adding a parallel pump unit capable of delivering the same pump head to the existing pump or pump series (for a relatively large increase in flow),
- Initially installing volute inserts with low-capacity impellers in a designed high-capacity pump. Through a combination of volute chipping, impeller underfiling, and ultimate installation of a high-capacity impeller, it is possible to meet a wide variety of conditions, all at optimum efficiency. Note: This option is possible only if it is known that the pumps will be required for an initial low-capacity, low-pressure start-up with ultimate change to high-capacity, high-pressure through field modification.

PUMP STATION CONTROL

Basic Control Operations

To properly control a pumping station (or compressor station), four basic control operations are involved to provide safety:

1. Control of prime mover (engine, motor or turbine)
2. Control of line valves (positioning)
3. Alarm notification (abnormal conditions)
4. Shutdown protection (to prevent damage to equipment and piping).

The first two control operations can be combined and interlocked to affect automatic sequence starting. Since some abnormalities (such as discharge valve closed, zero flow, etc.) can be tolerated for a short period of time and possibly corrected without requiring a shutdown, alarm notifications of abnormal conditions must be provided. However, other abnormalities (such as station overpressure, excessive pump vibration, etc.) require immediate shutdown to prevent unnecessary damage to equipment.

Sequence Starting

A typical starting control sequence for a motor-centrifugal pump unit on a liquid line is as follows (all station piping must be primed with the shipping liquid before a sequence start):

1. Position line valves [station discharge valve closed; flow control valve fully opened; suction valve(s) opened].
2. Start motor and bring to speed.
3. Open station discharge valve.
4. Adjust flow control valve for desired flow rate.

The starting sequence, similar to the sequence for starting gas compressors, is checked by a sequence timing relay. If the unit is not started within the given time, it is shut down, an alarm is set off, and the sequence begins again either by manual or automatic initiation.

Alarm and Shutdown Circuits

Similar to gas compressor stations, alarm and shutdown circuits are required for the protection of equipment.

Sensing switches for alarm condition (temperatures, pressures, liquid levels, etc.) are installed where required. The most basic alarm circuits include a sensing device, a relay to key a horn, a light to indicate the particular malfunction, and a relay to acknowledge and silence the horn. The light indication (and sometimes a buzzer or other audible signal) are also provided at the remote console in the dispatch office.

Switches for shutdown conditions of temperature, pressure levels, etc., are also installed where necessary. These switches are used to provide immediate protection against malfunctions that could cause serious damage to equipment. The switches generally include alarms. Operation of any of these devices will initiate a shutdown — stop the engine and proper positioning of line and unit valves.

Shutdown Sequence Control

Just as a pump unit must be started automatically on a set sequence, it must also be shut down in the same manner. Means for normal and emergency shutdowns must provide for the positioning of line and unit valves, a tripping motor, and de-energizing the fuel-gas control solenoid (on gas-fueled engines) after a cooling period.

Flow Control

The operational control of compressor stations in gas transmission lines and pumping stations in crude oil and products pipelines have much in common. However, there are basic operational differences that greatly affect the design of control systems for the two types of stations. In both cases, it is necessary to have on-off and continuous control of valve settings. Gas transmission lines are primarily controlled by gas pressure, which is propelled by compressors; liquid lines are controlled by liquid flow-rate, which is propelled by pumps.

Flow control of pumping stations can be accomplished by several methods:

- Changing throttle valve settings
- Using fluid drives

- Utilizing load controllers that automatically gather flow rate information and, through feedback, control engine or motor speeds. Load set points can be fed to these controllers as flow-rate changes are required, which is usually accomplished, by remote, from the dispatching office. Controllers can also be used to start, run, adjust, and stop units based on preset conditions.

PUMP STATION PIPING DESIGN

Piping arrangement within a pump station must be designed to enable safe, economic and functional objectives in terms of their ability to contain and convey a fluid under specified temperatures and pressures. For high-vapor-pressure (HVP) liquids, particular attention must be directed toward providing adequate suction and flow to the pump in order to avoid cavitation problems. As mentioned earlier, the minimum absolute suction pressure required at the suction of the inlet pump must exceed the vapor pressure of the liquid or the NPSH of the pump, whichever is larger. The alignment of the suction pumping to the pump must be arranged so as to minimize any rotational element with uniform velocity.

Failure of the suction piping to deliver the liquid to the pump under the above-mentioned condition will lead to noisy operation, random axial oscillation of the rotor, premature bearing failure, and cavitation damage to the impeller and inlet portion of the pump casing. The chances of any or all of these events occurring increases with pump size. These problems are also a function of the liquid being handled.

REFERENCES

American National Standards Institute (ANSI), and Hydraulic Institute (HI), 2000, "Centrifugal Pumps for Design and Application," ANSI/HI Standard 1.3 - Figure 1.65B, Parsippany, NJ.

American Society of Civil Engineers (ASCE), 1975, "Pipeline Design for Hydrocarbon Gases and Liquids," "Report, Test Com or Eng Direct in The Design of Pipelines," New York, NY.

Colebrook C. F., 1938–39, *Turbulent Flow in Pipes with Particular Reference to Transition Regions Between the Smooth and Rough Pipe Laws*, Jr. Inst. Civil Eng., London, England, Vol. II.

Darby, R., 1996, *Chemical Engineering Fluid Mechanics*, Marcel Dekker, Inc., New York, NY, p. 239.

Fox, R. T., and McDonald, A. T., 1985, *Introduction to Fluid Mechanics*, 3rd Edition, Wiley, New York, NY.

Karassik, I. J. (ed), 1976, *Pump Handbook*, McGraw Hill, New York, NY.

Labanoff, V. S. and Ron, R. M. R., 1987, *Centrifugal Pumps: Design and Application*, Gulf Publishing Co., Houston, TX.

McAllister, E. W., 1988, *Pipeline Rules of Thumb Handbook*, 2nd Edition, Gulf Publishing Co, Houston, TX.

Chapter 6

TRANSIENT FLOW IN LIQUID AND GAS PIPELINES

PURPOSE OF TRANSIENT ANALYSIS

Traditionally, pipeline transmission systems have been designed using steady-state simulations, which are sufficient for optimizing a pipeline when supply/demand scenarios are relatively stable. In the case of a gas pipeline, it is also important that flows in and out of the system or storage are not highly variable. In general, steady-state simulations provide the designer with a reasonable level of confidence when the system is not subject to radical changes in flow rates or operating conditions.

However, situations that require more than a conventional steady-state analysis do arise. These situations include large load factors, surges in mass/volume flow rates, and the loss of facilities and facility commissioning and operation (e.g., air purging and loading of pipelines; pigging operations). In these and other instances, the designer will want to perform a dynamic (or transient) analysis to test the capability of the system, choose its components, and maintain the appropriate level of safety.

This chapter illustrates the importance of transient simulations when designing transmission systems subject to unstable conditions. Example scenarios, from some major projects undertaken by the authors, are used to depict a diverse range of dynamic problems. The examples help identify the need for a transient analysis, and exemplify the downfalls in a system when the analysis is not employed during the optimization and design process.

(Note: Extracted from papers by M. Mohitpour, et al., 2nd ASME–OMAE Conferences, 1996 and 1999).

BACKGROUND

There are many important factors that need to be considered before a pipeline can be designed and constructed. However, arguably, the single most important factor is the flow requirements. Therefore, before the pipeline diameter, pump, or compressor unit size, etc. can be chosen, a great deal of consideration must be given to the supply and demand flow buildup and their variations and, hence, the hydraulics of a pipeline system.

Since the 1800s, a great deal of work has been performed to develop equations that accurately predict flow behavior of gases and liquids. Major contributions to the development of the flow equations have been made by people like Darcy, Reynolds, Fanning, Stanton, Pannel, Prandtl, Nikurasdse, Bernoulli, Weymouth, Colebrook, Moody, White (Uhl 1965). More recently, the Bureau of Mines and the American Gas Association (AGA) have played a part in the further development of gas flow equations. With the exception of two- and three-phase flows, the development of flow equations have evolved to a stage where, in most applications, the error inherent in the flow equation is insignificant in comparison to the error caused by the inaccuracy of the actual inputs into these equations (such as roughness or fluid properties). For example, research into the steady-state hydraulic behavior of gas pipelines has shown that the AGA gas equation can predict pressure drop to within 3%. In comparison, the parameters used as input to the flow equations, such as flow, operating temperature, operating pressure, and internal roughness generally involve a great deal of estimation and have a far greater impact on the overall error (Asante 1995; Price et al. 1996).

In general, steady-state hydraulics have been used almost exclusively in the facility selection process. This process involves reviewing flows and pressure drops and determining capacity, pipeline diameters, pipeline loop lengths, and overall pump or compressor station power requirements.

The facility selection process begins with the creation of a demand and supply forecast. These forecasts usually involve examining all existing and potential customers along the pipeline route, and creating a projection of their demand requirements over a predetermined time frame. The next step is the creation of a hydraulic computer model of the pipeline system to which the forecast for supply and demand is added. Steady-state simulations are then run to determine where the flow is restricted. Additional facilities, which may mean either loops, pumps or compression facilities, are added to the model to eliminate the restrictions. Usually, several alternate facility buildups are created over the forecast period. Ultimately, these alternate facility buildups will be compared on an economic and technical basis in order to choose the best alternative (Mohitpour and McManus 1995).

Steady-state analysis is also used to determine system design flow capacity and to provide a "best guess" of future pipeline operation so that detailed design can be performed on new facilities. To a lesser extent, it is also used in determining operation problems. However, steady-state modeling is limited in providing a realistic system response even under simple flow fluctuations. Steady-state simulations cannot be applied at all to abnormal operational situations such as customer delivery or supply switching, pipeline pigging, purge and loading, leak/line breaks, compression or pumping unit loss, liquid surge, or partial operations.

To appreciate the differences between the modeling of steady-state and transient-state flows, a basic understanding of compressible and incompressible flows must first be obtained. In reality, all fluids are compressible to varying degrees; however, gases are much more compressible than liquids. This fact is reflected in pressure drop and flow calculations. In order to accurately predict the pressure drop, the properties of the fluid (i.e., density, enthalpy, etc.) must be calculated along the pipeline length to reflect the changes with pressure and temperature (Mohitpour 1995). With most liquids, properties vary little with pressure and, therefore, there is less of a need to calculate fluid properties along the length of the pipeline.

Variance in fluid properties is not the only observable effect compressibility has on flow. Under transient flow conditions, where inlet (supply) and outlet (demand) flows vary with time, pressure, and compressibility-related phenomena such as linepacking and pressure surges occur. Linepacking is a term used to describe the storage effect of gas pipelines. The linepack can easily be comprehended by comparing a gas pipeline to an air compressor pressure tank. Inside an air compressor, gas is compressed into a pressure tank for storage and later use. As gas is compressed into the pressure tank, the tank pressure slowly increases with time until gas is removed from the tank. This is similar to what occurs when a pipeline is linepacked.

Pressure surge, on the other hand, is a term used to describe a relatively rapid process that occurs with "almost" incompressible fluids. With an incompressible fluid such as water, there is no storage capability obtained by pressurizing the fluid. In addition, when an attempt is made to compress an incompressible fluid, the pressure of the fluid will rise rapidly throughout the system. Phenomena are very important during the design of liquid pipeline systems.

Unlike transient modeling, steady-state hydraulics modeling requires that inlet and outlet flows balance and it does not reflect the time dependency of system flow. Steady-state hydraulics do not permit the modeling of either linepack or pressure surge phenomena. This creates a significant departure from real life applications, like those illustrated by examples later in this chapter.

THEORETICAL FUNDAMENTALS AND TRANSIENT SOLUTION TECHNIQUE

The following information provides an outline of theoretical correlations and solution techniques that are generally utilized for computer-based hydraulic modeling of transient flow.

The time-dependent, one-dimensional flow of a fluid in an inclined conical conduit, as shown in Figure 6-1, can be described by three equations representing conservation of mass, momentum, and energy (Wylie and Streeter 1990).

Conservation of mass (continuity):

$$\frac{\partial \rho}{\partial t} + \frac{\partial}{\partial x}(\rho u) = 0 \qquad (6-1)$$

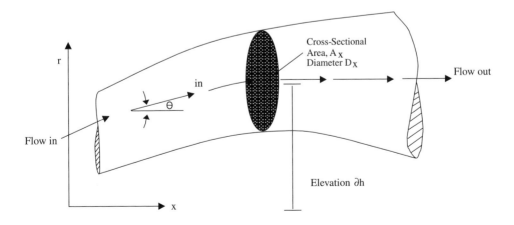

Figure 6-1. Flow of fluid through a cylindrical nonuniform pipe segment

Conservation of momentum:

$$\frac{\partial \rho}{\partial t}(\rho u) + \frac{\partial}{\partial x}\left(\rho u^2 + P\right) + \rho g \frac{dy}{dx} + \frac{4 f u |u|}{2D} = 0 \tag{6-2}$$

Conservation of energy:

$$\frac{\partial}{\partial t}\left[\rho\left(h - \frac{P}{\rho} + \frac{u^2}{2}\right)\right] + \frac{\partial}{\partial x}\left[\rho u\left(h + \frac{u^2}{2} + gy\right)\right] + q = 0 \tag{6-3}$$

where D = pipe diameter
f = friction factor
g = acceleration due to gravity
h = specific enthalpy of gas
ρ = gas density
P = gas pressure
q = heat transfer rate through the pipe wall into gas, per unit volume
t = time
u = gas flow velocity
x = distance along the axis of the pipe
y = elevation from a defined datum (positive in the opposite direction of gravity).

Solution Technique

For nonisothermal situations, such as the case of an operating pipeline, Equations (6-1) and (6-2) are usually solved simultaneously (Stuchly and Schmalz 1984). To consider heat transfer to and from the fluid and environment, separate heat transfer equations can be utilized to solve for temperature variation along the pipeline, if required (Mohitpour 1991). Therefore, the following equations are applicable:

$$\frac{\partial p}{\partial t} + \frac{ZRT}{A_x}\frac{\partial M}{\partial x} = 0 \tag{6-4}$$

$$\frac{1}{2}\frac{\partial p^2}{\partial t^2} + \frac{p^2}{ZRT}g.\sin\theta + \frac{p}{A_x}\frac{\partial M}{\partial t} + \frac{ZRT}{A_x^2}\frac{\partial M^2}{\partial x^2} + \frac{ZRTf|M|M}{2A^2D} = 0 \tag{6-5}$$

where A_x = pipe cross-sectional area at distance x
D_x = pipe internal diameter at distance x
M = mass flow rate
R = gas constant
T = gas temperature
Z = gas compressibility factor
$\sin\theta = dy/dx$.

Since Equations (6-4) and (6-5) are nonlinear, a numerical technique is usually utilized to solve memo. The method used here is generally referred to as the "implicit method," and it is a slightly modified version of the one given by (Wylie and Streeter 1990).

The implicit method consists of transforming Equations (6-4) and (6-5) from partial differential to algebraic equations, by using finite-difference approximations for the partial derivatives. The resulting nonlinear algebraic equations can then be solved by standard numerical techniques.

Let

$$P = p(x, t) \tag{6-6}$$

$$M = M(x, t) \tag{6-7}$$

$$PT = p(x, t + \Delta t) \tag{6-8}$$

$$MT = M(x, t + \Delta t) \tag{6-9}$$

$$PX = p(x + \Delta x, t) \tag{6-10}$$

$$MX = M(x + \Delta x, t) \tag{6-11}$$

$$PXT = P(x + \Delta x, t + \Delta t) \tag{6-12a}$$

$$MXT = M(x + \Delta x, t + \Delta t) \tag{6-12b}$$

then

$$\frac{\partial p}{\partial t} = \frac{PT + PXT - (P + PX)}{2\Delta t} \tag{6-13}$$

$$\frac{\partial M}{\partial x} = \frac{MX + MXT - (M + MT)}{2\Delta x} \tag{6-14}$$

$$\frac{\partial p^2}{\partial x^2} = \frac{(PX + PXT)^2 - (P + PT)^2}{4\Delta x} \tag{6-15}$$

$$\frac{\partial M}{\partial t} = \frac{MT + MXT - (M + MX)}{2\Delta t} \tag{6-16}$$

$$\frac{\partial M^2}{\partial x} = \frac{(MX + MXT)^2 - (M + MT)^2}{4\Delta x} \tag{6-17}$$

For a constant diameter pipe (i.e., when $A_x = A$ and $D_x = D$), Equations (6-4) and (6-5) can now be written as follows:

$$F_1 = \frac{PT + PXT - (P + PX)}{2\Delta t} + \frac{ZRT}{A}\left[\frac{MX + MXT - (M + MT)}{2\Delta x}\right] = 0 \tag{6-18}$$

$$F_2 = \frac{1}{2} + \frac{(PX + PXT)^2 - (P + PT)^2}{4\Delta x} + \frac{(P + PX + PT + PXT)^2 g.\sin\theta}{16ZRT}$$

$$+ \frac{(P + PX + PT + PXT)}{4A}\left[\frac{(MT + MXT) - (M + MX)}{2\Delta t}\right]$$

$$+ \frac{ZRT}{A^2}\left[\frac{(MX + MXT)^2 - (M + MT)^2}{4\Delta x}\right] + \frac{ZRTf}{A^2 D}$$

$$\frac{|(M + MX + MT + MXT)|(M + MX + MT + MXT)}{32} = 0 \tag{6-19}$$

For a horizontal pipe, the second term in Equation (6-19) can be ignored.

The Newton-Raphson method can be used to solve the equations. A similar technique has been used by (Mohitpour 1972) and described by (Vagra 1962), for solution of transient heat transfer problems. The procedure starts with trial values for all the unknowns and then calculates successive approximations until a set of solution values has been obtained to the desired degree of accuracy. These successive approximations are obtained from the solution of the following linear equations:

$$\frac{\partial f_1}{\partial PT}(\Delta PT) + \frac{\partial f_1}{\partial PXT}(\Delta PXT) + \frac{\partial f_1}{\partial MT}(\Delta MT) + \frac{\partial f_1}{\partial MT}(\Delta MXT) = -f_1 \qquad (6-20)$$

$$\frac{\partial f_2}{\partial PT}(\Delta PT) + \frac{\partial f_2}{\partial PXT}(\Delta PXT) + \frac{\partial f_2}{\partial MT}(\Delta MT) + \frac{\partial f_2}{\partial MT}(\Delta MXT) = -f_2 \qquad (6-21)$$

The delta (Δ) quantities are the corrections to be applied to the values of PT, PXT, MT, and MXT assumed at the beginning of the iteration. These assumed values are used to evaluate both values F_1 and F_2 [using Equations (6-18) and (6-19)] and the matrix of partial derivatives. When the delta quantities are zero, F_1 and F_2 are satisfied and the correct values for the unknowns have been obtained.

The above solution techniques are generally utilized in most computer programs that are used by the industry for hydraulic simulation and design for transient flow modeling. There are a number of software packages available—such as Pipeline Studio of LIC Energy (1999), Stoner & Gregg Transient Models, TransCanada's TFlow and PTRANS—that can undertake this type of simulation.

APPLICATIONS

Liquid Pipelines

Application of transient modeling for the design of liquid (in particular, water and crude oil) pipeline systems has been commonly used over the past several decades. Technical journals and other literature give abundant examples of cases where transient analysis has been successfully employed to ascertain safety of pipeline operation resulting from uncontrolled pressure transient (Burnet 1960; Goldberg 1985; Cohn and Nalley 1980; and Drivas et al. 1983). In particular, transient analysis has been cited for the following applications: **Pipeline construction economics:**

- Use of surge control system versus higher wall thickness pipe.

Water hammer control (water pipeline system design):

- Size and locate air vacuum and positive pressure surge relief valve

Pipeline pump station design and operation:

- Station control valve design for pressure override logic to protect pipeline from overpressurization
- Hydraulic transients created due to pump trip in water or oil pump stations
- Effect of rapid valve closures
- Effect of rapid flow (supply/demand) fluctuations

- Pressure surge relief system design
- Line break control
- Pipeline leak/failure
- Waxy crude transportation

Consider a transient analysis for a 250 km NPS 12 pipeline carrying non-Newtonian fluid—a mixture of bitumen and condensate (Mohitpour and McManus 1995)—as shown in Figure 6-2. In this case, various valve closing times were imposed on a mainline block valve located near the end of the pipeline downstream of a steep, hilly terrain and upstream of an environmentally sensitive region, including streams/rivers.

As can be inferred from Figure 6-2, the analysis indicates that while wall thicknesses selected for the pipeline during the preliminary design were adequate for steady operation, the pipeline would not have safely withstood the high surge pressures resulting from the rapid valve operation. For this example, it was also assumed that the upstream pump station control valve would fail and thus the pumps would continue operating. While installation of a surge control device would have mitigated operational concerns, it would not have adequately covered the environmental protection requirements. As a result, higher wall thicknesses were selected for the pipeline throughout the affected areas.

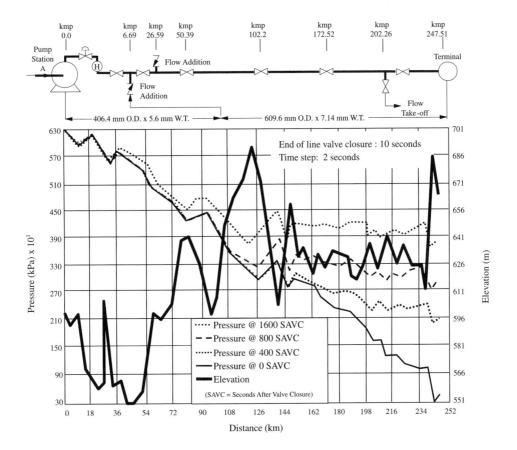

Figure 6-2. Pressure profile in a buried liquid pipeline subjected to end valve closure

Gas Pipelines

Unlike liquid pipeline systems, limited information is available on the application of transient analysis to the design of gas pipelines. What information exists covers mostly areas such as gas pipeline blowdowns (Jungowski et al. 1989), linebreak control leak detection systems, pipeline system response due to compression and flow increase steps, flow diversions, opening and closing valves, linepack determination, and flow-generated pulsation. The following case studies highlight the importance of applying transient state analysis to natural gas system design and operation.

Case Study 1—Capability Determination/Facility Location Optimization

This example is based on a study of the GasAndes Transmission System, which was inaugurated in August 1998. The pipeline transports natural gas from Argentina to Chile through the Andes Mountain Range (Figure 6-3). The system consists of approximately 500 km of NPS 24 pipe, and eventually will have up to four compressor stations and a number of loops installed along its length. The pipeline system ties into the Transportadora Gas del Norte (TGN) System near the town of La Mora, Argentina, and terminates at Santiago, Chile. For clarity, the natural gas pipeline systems in the southern cone of South America are shown in Figure 6-4.

During the preliminary design, transient state analysis was one of the approaches used to meet the objective of minimizing system costs while meeting customer requirements.

Prior to any design work, a forecast of natural gas demand in Santiago was estimated. Due to strict Chilean environmental regulations, the conversion of many of the thermoelectric plants to natural gas have been envisioned. The majority of natural gas demand in Santiago will thus stem from power generation. A typical cyclic gas demand profile for a thermoelectric power plant is shown in Figure 6-5. The minimum delivery pressure to a power station is usually around 35 bar. In many cases, the cyclic nature of the demand makes transient analysis mandatory for the gas transmission system design to ensure adequate pipeline capacity.

To minimize the complexity of illustrating the benefits of transient simulation, the example here is based on only one year of forecast data and one compression station at the inlet to the pipeline.

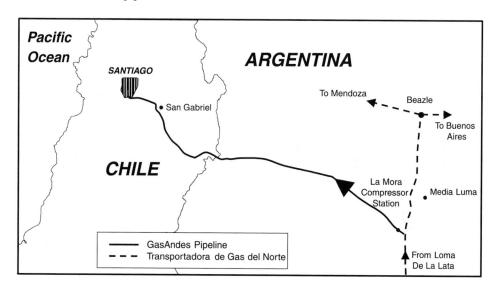

Figure 6-3. GasAndes pipeline project

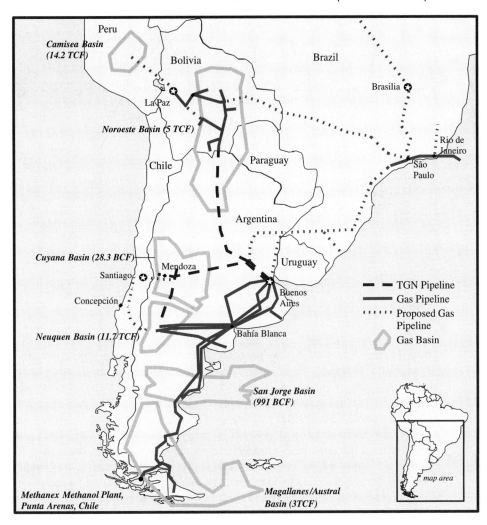

Figure 6-4. South American gas transmission network

Figure 6-5 illustrates the effects of compressible flow and transient analysis. This graph plots the flows at the inlet and outlet to the pipeline against time instead of distance. The buffering effect of linepack is clearly illustrated by the differences in the two supply and demand curves. The receipt flow or gas supplied to the pipeline remains relatively stable, between 309 kscm/hr (262 MMSCFD) and 355 kscm/hr (300 MMSCFD), whereas the delivery varies between 181 kscm/hr (153 MMSCFD) and 470 kscm/hr (398 MMSCFD). The other noticeable difference is the time lag of 11 hours between the outlet and inlet flows. If the fluid and the pipeline was entirely incompressible, the inlet flow would track the outlet flow almost exactly, and only a steady-state simulation would be required in the design.

A review of Figure 6-5 also indicates that with steady-state analysis, there remains the dilemma of choosing a design flow. Again, the flow at Santiago can range between 181 kscm/hr and 470 kscm/hr (153 MMSCFD and 398 MMSCFD).

In steady-state analysis, often daily flows are used instead of hourly flows. In many instances this can lead to underdesign of the system, as can be seen by comparing data

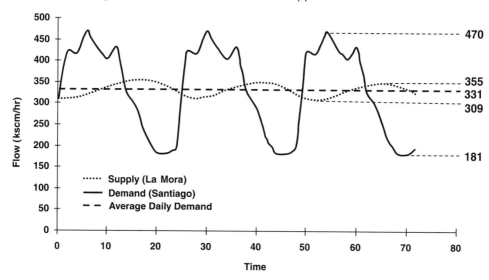

Figure 6-5. Supply and demand variations

from Figures 6-5 and 6-6. To create the daily flow profile in Figure 6-6, hourly flows from Figure 6-5 were totaled, and then this total was used in the steady-state pressure profile depicted in Figure 6-6. The steady-state daily flow results thus obtained show that the delivery pressure reaches a minimum of 53 bar (768.5 psia) for an average daily flow of 331 kscm/hr (281 MMSCFD).

To demonstrate the inadequacy of steady-state simulation, a transient analysis was performed on the same pipeline system using the hourly demand flows from Figure 6-5. Figure 6-7 illustrates the outcome of this analysis in addition to results from two other transient simulations. In this comparison, the transient profile for the "without loop" case is first considered; the other two curves will be discussed subsequently. From Figure 6-7, the delivery pressure for the "without loop" case reaches a minimum of 35 bar (507 psia), far below the 53 bar (768.5 psia) predicted by the steady-state simulation in Figure 6-6.

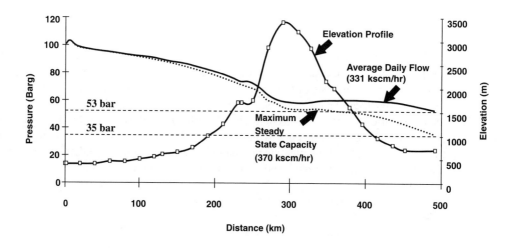

Figure 6-6. Steady-state pressure profile (without loop) NPS 24

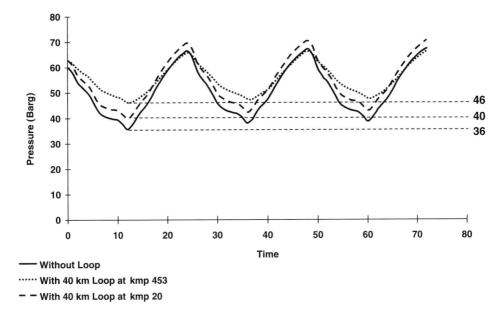

Figure 6-7. Transient state simulation - flow time plots

Obviously, applying steady-state simulation in this manner will lead to serious underdesign of the system.

Another alternative to using average daily flow is to use the peak hourly flow in the steady-state design. From Figure 6-5, the peak hourly flow at Santiago is 470 kscm/hr (398 MMSCFD). Figure 6-6 shows a pressure profile for the maximum steady-state capacity of the pipeline, assuming a minimum delivery pressure of 35 bar (507 psia). This maximum steady-state capacity corresponds to a flow of 370 kscm/hr (313 MMSCFD). In order to meet the peak hourly flow from Figure 6-5 under steady-state conditions, many additional facilities would be required. However, no additional facilities are required to meet this flow under transient conditions. If steady-state hydraulics were applied in this manner, the system would become overdesigned.

As can be seen from Figures 6-7 and 6-8, another important design consideration is the placement of additional loop on a pipeline system. If only steady-state hydraulics were used in design, the hydraulics would tend to favor placing the loop immediately downstream of a compressor station. This is because the additional pipeline volume immediately downstream of the compressor station slows gas movement and permits greater heat transfer to occur. The cooler gas then has a greater flowing efficiency, and therefore has less of a pressure drop. This is reflected in Figure 6-8. Although the same amount of facilities and the same flow rates were used to create the two curves in Figure 6-8, placing the loop closer to the compressor station leads to a slightly higher delivery pressure.

Contrary to the curves in Figure 6-8, the transient profiles in Figure 6-7 show that the minimum delivery pressure at the delivery point is higher for the downstream loop than for the upstream loop (46 bar and 40 bar or 667 psi and 580 psi, respectively). These phenomenon occur because of the buffering effect that linepack has on transient flows. Since the downstream pipeline is closer to the delivery point, it acts as a much better buffer or gas storage facility than the upstream facility.

If only steady-state hydraulics were considered in the design of the pipeline system, very erroneous results would have been obtained. Clearly, design using steady-state

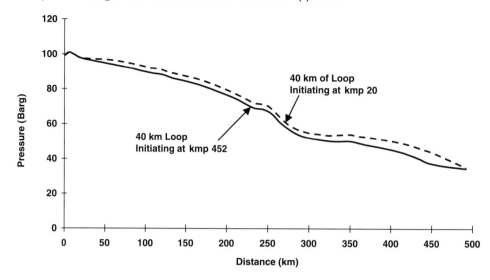

Figure 6-8. Steady-state pressure profile (with loop) NPS 24

analysis exclusively may lead to very different results and sometimes wrong conclusions. These errors would be much more pronounced if the demand is very cyclic.

Case Study 2—Pipeline Availability Design

This example studies the Mayakan Gas Pipeline, a 690 KM (430 miles), 8,274 kPa (1200 psia) natural gas pipeline supplying gas to Mexico's Yucatan Peninsula. The pipeline (shown in Figure 6-9) will initially provide 6.21 MMm³/day (219.3 MMSCFD) of gas from Pemex City to a number of delivery points along the Yucatan Peninsula to Valladolid. For clarity the natural gas pipeline systems of Mexico are shown in Figure 6-10.

Facilities' buildup for various phases of the Mayakan Pipeline System is shown in Figure 6-11. The "Phase 1" configuration consists of 264 km (164 miles) of NPS 24

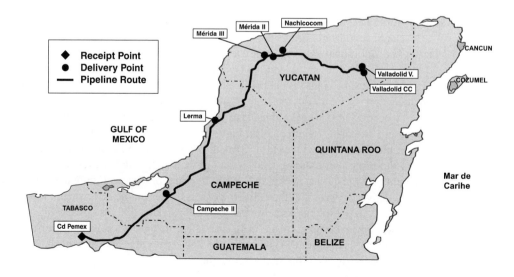

Figure 6-9. Yucatan Peninsula gas pipeline project—Mexico

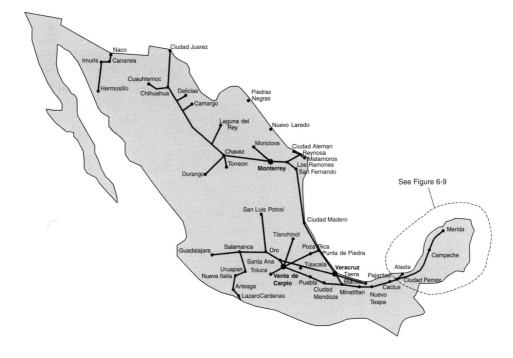

Figure 6-10. Existing Mexico gas pipeline systems

linepipe reducing to NPS 22 for 275 km (171 miles), and finally to NPS 16 for the last 153 km (95 miles). In addition, laterals provide gas into Merida and Campeche. The pipeline will subsequently be increased in capacity to 7.63 MM m³/day (269.4 MMSCFD) during Phase 2 in year 2004, and finally to 8.81 MM m³/day (311 MMSCFD) during Phase 3 in 2008 (311 MMSCFD). Capacity increases will be achieved through the addition of compression facilities along the mainline as well as laterals at Nachicocom and Lerma.

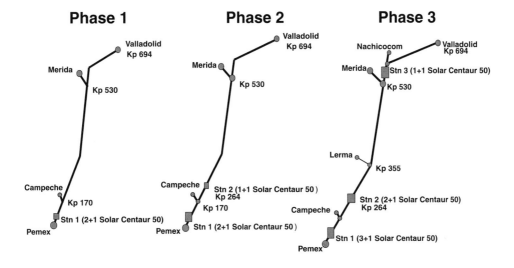

Figure 6-11. Mayakan gas pipeline system facilities buildup

Phased expansion of the pipeline is based on predicted gas requirements as new customers, primarily power generation plants, become available. Diurnal demand (i.e., hourly demand for any 24-hr period) for each phase of the pipiline's operation is shown in Figure 6-12.

Stringent requirements for availability of gas supply to delivery points were stipulated, since the power generation facilities generally have no backup fuel source.

Transient analysis was therefore undertaken to review system capability in meeting the availability requirements, which specified not only the overall availability of the pipeline, but also the maximum duration for which gas could not be delivered during any 1-year, 5-year, 10-year, or 26-year (project life) window over the operating life of the pipeline. The criteria depicting the reliability of the system included:

1. Unscheduled shutdown to be restricted to the following cumulative duration:
   - 1 day in any continuous 1-year operational period (99.7% availability)
   - 2 days in any continuous 5-year operational period (99.9% availability)

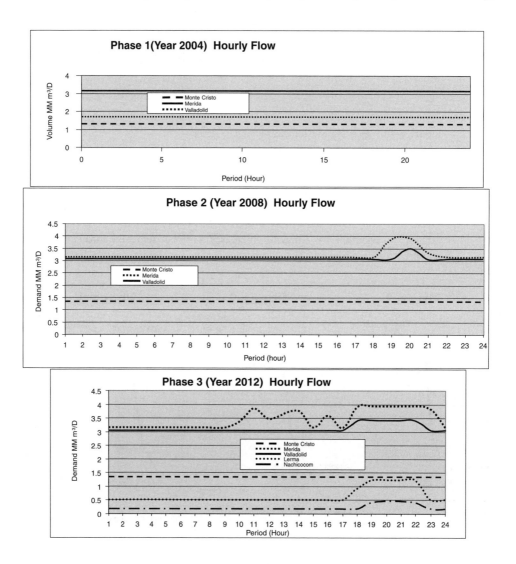

Figure 6-12. Mayakan pipeline hourly demand (phases 1 to 3)

- 3 days in any continuous 10-year operational period (99.9% availability)
- 5 days in any continuous 26-year operational period (99.95% availability)

2. Maximum gas velocity in mainline and laterals to be below 20 m/s.

Transient analysis of pipeline hydraulics, along with a reliability model (BS 5760 1994) of pipeline operation, was used to design the system to meet availability requirements without overbuilding. System reliability was determined separately for each phase of the pipeline operation by means of:

- construction of a reliability block diagram (RBD, see Figure 6-13)
- running a Monte Carlo simulation (see below)
- post-processing information (see below)

The main cause for lack of gas at delivery points on a single-line pipeline, such as Mayakan, is unavailability of one or more of the gas-turbine-driven compressor units. Since availability of these units is crucial, installation of a spare at each compressor station was justified. However, the cases of either two compressor unit outages simultaneously (i.e., due to repairs), or failure of an operating unit with the spare on outage for maintenance, are still possible. While the likelihood of these cases may seem extremely remote, the strict operating requirements necessitated their consideration.

Transient analysis was therefore used to determine the time response for supply pressure at each delivery point for each phase described above. This response varied considerably depending on where the unit outage occurred (i.e., which compressor station was down) and which phase of operation was affected. Ample graphical presentation of delivery presentation variation for Phase III is provided in Figure 6-14. At each delivery point, curves defining delivery pressure decline and recovery during unit outage and restart were generated for each station and for each phase.

The transient analysis results were used in combination with a pipeline availability model. This model was built in Visual Basic on a Microsoft Excel© spreadsheet using Crystal Ball, a commercially available Excel add-in for Monte Carlo simulation, a technique by which events are predicted based on randomly sampling statistical distributions of those events (Stoll 1989). Time between failure (TBF), time to repair

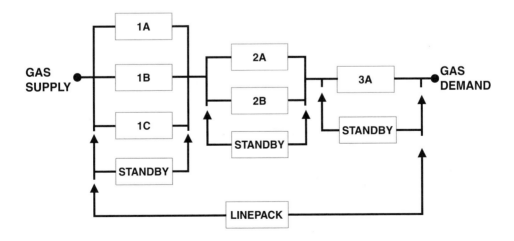

Figure 6-13. Reliability block diagram, phase 3

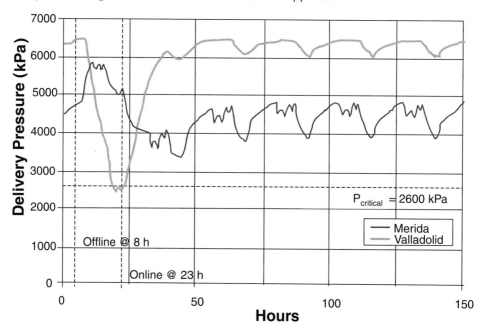

Figure 6-14. Pressure transients: phase 3, station 3, unit loss

(TTR), and compressor starting reliability were input into the model based on failure and repair data taken from operating logs of a mainline transmission system. The availability model simulated operation of the pipeline over time, including scheduled maintenance outages and unscheduled outages for compressor units. Durations that are typically considered for compression unit maintenance are:

- Hot gas pass inspection every 4,000 hours (10-hour duration)
- Unit overhaul every 30,000 hours (48-hour duration)
- Unit soak/wash every day (4-hour duration)

The unscheduled outages were generated randomly by the Monte Carlo method described above. When outage of the spare and an operating unit occurred simultaneously, the availability model utilized the transient response curves to determine the effect on pipeline delivery pressure. System failure hours were determined by tracking the pressure at delivery points and determining the length of time pressure fell below the minimum allowable. The program finally performed a calculation to determine the maximum outage duration over any 1-, 5-, 10-, or 26-year window during the life of the pipeline.

The results of the analysis verified that spare compressor units were necessary at each station. By utilizing transient analysis, in conjunction with a probabilistic analysis of pipeline operation, it was determined that the percentage confidence in the system's ability to meet the availability requirements, and thus prove adequate levels of equipment redundancy, were present.

Case Study 3—Pipeline Commissioning (Purge & Load)

After a pipeline is constructed and hydrostatically tested and dried, it is usually filled with air. In some instances, depending on availability, it can be filled with nitrogen.

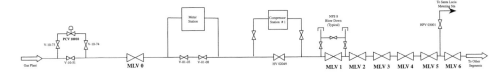

Figure 6-15. Mayakan pipeline segments (pemex gas plant to mainline block valve MLV 6)

When commissioning a gas pipeline, the air is purged at relatively low pressure, 200 kPag (29 psig), depending on diameter and segment length. The pipe segment is then pressured up, usually to a pressure lower than the maximum or normal operating pressure of the pipeline, to approximately 1,000 kPag (145 psig).

When purging a gas pipeline, purge velocity needs to be high enough to avoid natural gas/air stratification (refer to Chapter 9, purge calculations) and to ensure any dust accumulated in the pipeline is removed. Industry practice is to attain an average velocity of about 15 m/s for this latter purpose. Linepacking to 1,000 kPag (145 psig) is also a common practice, as it will ensure lowest gas inventory cost prior to commissioning the pipeline for its daily operation.

Purge and loading of the Mayakan pipeline described in Case 2 is considered for this case study. The pipeline has 26 sectionalalizing valves along its route (11 on the NPS 24 section, 9 on NPS 22 section, and 6 on the last NPS 16 section).

LIC Energy TGNET, Gas Pipeline Network simulator, was used to simulate the transient nature of this entire network for the prediction of time to purge, pressure, pack, and equalize successive segments between block valves.

Each pipeline valve has an NPS 8 blowdown on either side. For purging purposes, pipeline mainline valves downstream of each section were closed and the NPS 8 blowdown was opened (see Figure 6-15 for pipeline segments).

Natural gas composition entering the pipeline is given in Table 6-1.

For purging purposes, the methane component of the gas, which makes up the largest percentage of the gas composition, was tracked down each pipeline segment, from entry into the segment to exit from the segment through the NPS 8 blowdown. When the methane arrived at the exit, the blowdown valve was closed and the line was pressured up to 200 kPag (29 psig). The line was subsequently packed to 1,000 kPag (145 psig).

The results of the transient simulation study for velocity calculation are provided in Table 6-2. This table gives the time required to pressure to 200 kPag (29 psig) and linepack to 1,000 kpag (145 psig).

Once a segment was linepacked to 1,000 kPag (145 psig), through the blowdown bypass, the next segment was purged and pressured up to 200 kPag and subsequently

TABLE 6-1. Mayakan pipeline—gas composition at 98.064 kPa, 20°C

Component		Natural Condition (Mol %)
Methane	C_1	80.04
Ethane	C_2	17.44
Propane	C_3	1.88
ISO-butane	IC_4	0.03
n-butane	NC_4	0.05
Nitrogen	N_2	0.55
Carbon dioxide	CO_2	0.01

TABLE 6-2. Mayakan pipeline—purge times, times to pressure to 200 kPa and to 1,000 kPa

Pipeline Section	Length (km)	Cumulative Length (km)	Time to Purge Section (T1) (minutes)	Time to Pressure to 200 kPa (T2) (minutes)	Equalization between Sections Time (minutes)	Equalization between Sections Pressure (kPa)	Time to Pressure to 1,000 kPa (T3) (minutes)	Total Time per Section T1+T2+T3 (minutes)
NPS 24								—
V-10-52 to V-01-03/08	1.87	1.87	10*	5*	—	—	5	20
V-01-03/08 to HV02049	5.07	6.94	15*	10*	—	—	12	37
HV 02049 to MLV 1	26.55	33.49	22	20	35	999	43	85
MLV 1 to MLV 2	31.1	64.59	33	25	70	992	83	141
MLV 2 to MLV 3	31.66	96.25	33	25	74	972	100	158
MLV 3 to MLV 4	31.29	127.54	33	25	65	930	122	180
MLV 4 to MLV 5	31.04	158.58	33	25	60	900	148	206
MLV 5 to MLV 6	30.55	189.13	33	25	63	900	180	238
MLV 6 to MLV 7	29.84	218.97	33	25	62	900	204	262
MLV 7 to MLV 8	21.04	240.01	20	15	39	900	192	227
MLV 8 to HV 04049	29.09	269.10	33	25	60	900	255	313
NPS 22								
HV 04049 to MLV 9	31.07	300.17	33	25	55	900	268	326
MLV 9 to MLV 10	30.07	330.24	33	25	56	900	290	348
MLV 10 to MLV 11	31.67	361.91	33	25	64	900	330	388
MLV 11 to MLV 12	30.25	392.16	33	25	64	900	353	411
MLV 12 to MLV 13	31.43	423.59	33	25	70	900	385	443
MLV 13 to MLV 14	31.71	455.30	33	25	72	900	408	466
MLV 14 to MLV 15	29.50	484.80	33	25	65	900	435	493
MLV 15 to MLV 16	26.71	511.51	25	20	66	900	465	510
MLV 16 to HV 07049	23.50	535.01	22	17	48	900	437	476
NPS 16								
HV 07049 to MLV 17	23.65	558.66	25	20	23	900	330	375
MLV 17 to MLV 18	29.36	588.02	37	24	42	900	383	444
MLV 18 to MLV 19	28.63	616.65	37	24	55	900	395	456
MLV 19 to MLV 20	26.64	643.29	30	22	60	900	397	449
MLV 20 to MLV 21	30.68	673.97	37	24	75	900	452	513
MLV 21 to Vallodolid	21.61	695.58	22	17	48	900	392	431

brought to the 1,000 kPag linepack. However, when pressuring the subsequent segment from 200 kPa to 1,000 kPa through the NPS 8 blowdown, the associated mainline block valve was opened as soon as pressure equalization upstream and downstream of the valve was achieved.

The time to pressure (load) each pipeline segment to 1,000 kPa following initial pressurizing to 200 kPa was determined to assist in planning pipeline commissioning. Each pipe section was simulated to determine pressure buildup times required to reach a constant pressure of 1,000 kPa, that is, under fully packed conditions. Results are summarized in Table 6-3. This table also contains the actual purge and pressure times for each segment of pipeline to 200 kPa. The total time per segment to perform the commissioning operation is also included; it provides the maximum time required to complete commissioning operations for each segment of pipeline.

It should be noted that pressuring up successive segments of pipeline could commence prior to each previous segment of pipe attaining a pressure of 1,000 kPa. Once the pressure upstream and downstream of a sectionalizing block valve is equalized, commission (purge and load) of the next section downstream can commence.

A typical graph depicting pressures upstream and downstream of a sectionalizing valve is shown in Figure 6-16. In real practice, the blowdown bypass valve at the sectionalizing block valve will be used to throttle gas flow between pipeline segments until the pressure on each side of the valve is equalized, which would then allow the mainline sectionalizing valve to be opened. In effect, the graph provides equalization information for each sectionalizing valve of the pipeline. The segments of pipeline upstream and downstream of such a valve are equalized when the pressures for each of the pipe segments are equal. The graph shows that initially the pressure upstream of a valve is 1,000 kPa, while the pressure downstream is 200 kPa. These are steady-state conditions in the pipeline segments. At an arbitrary time of 25 minutes, the transient calculations commence to determine pressure buildup in each segment of pipeline to 1,000 kPa.

The time elapsed between the commencement of pressuring up to the time when the upstream and downstream pressure curves intersect is considered as the duration required for achieving equalization for a particular valve before it can be opened. After this period of time the two segments are pressured up to 1,000 kPa. These durations are depicted in Figure 6-16.

TABLE 6-3. Mayakan pipeline—purge velocity, purge time and time to pressure to 200 kPa

Size	Length (km)	Pipeline (pinlet kPag)	Pipeline (poutlet kPag)	Pipeline (velocity at inlet m/s)	Pipeline (velocity at outlet m/s)	Volume (MMscfd)	Purge Pipeline (Minutes)	Estimated Time to Pressure to 200 kPa
NPS 24	21	200	0.6895	13.9	22.4	31.1	20.2	15
NPS 24	31.6	200	0.6895	11.9	21.6	27.3	32.6	25
—	—	—	—	—	—	—	—	—
NPS 22	23.5	200	0.6895	13.4	25.3	26.0	21.7	17
NPS 22	31.7	200	0.6895	11.8	24.1	23.2	32.6	25
—	—	—	—	—	—	—	—	—
NPS 16	21.6	160	0.6895	12.1	27.7	11.1	21.7	17
NPS 16	30.7	160	0.6895	10.1	23.9	9.3	37.2	24

Notes:
1. Blowdown pipe diameters assumed to be NPS 8.
2. Actual results may vary depending on field conditions, e.g., temperature, at time work is performed.
(Minimum and maximum length between block valves for each pipe size).

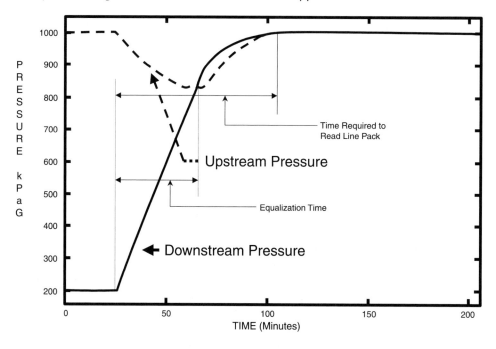

Figure 6-16. Time to pressurize section MLV 1 to MLV 2

COMPUTER APPLICATIONS

The following provides a number of solved problems using computer techniques for both liquid and gas pipeline systems.

Example 6.1: Oil Pipeline System

The example of an oil pipeline system is given in Figure 6-17. Determine the liquid hydraulic design summary for this system if the following data is available:

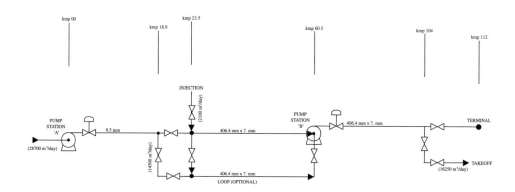

Figure 6-17. Facilities diagram, oil pipeline system

A. Pipeline Route

kmp	Elevation (m)
0	823
8	762
16	841
19	857
23	814
25	797
29	807
31	799
32	760
45	753
48	799
53	736
55	738
61	723
66	713
68	713
88	701
91	710
97	670
103	692
104	684
106	686
107	680
112	682

B. Operational Limitations

Minimum operating pressure	= 350 kPa-g
Maximum operating pressure	= 9,930 kPa-g
Minimum operating temperature	= 0.0 °C
Maximum operating temperature	= 40 °C
Initiating station inlet pressure	= 350 kPa-g
Initiating station discharge pressure	= 5,787 kPa-g
Default station discharge pressure	= 8,895 kPa-g
Pump/cooler/heater efficiencies	= 85%
Cooler/heater outlet temps	= 40 °C
Pump station suction/discharge losses	= 50 kPa

C. Oil Properties

Oil temperature coefficient of density	= −.88 kg/m^3 °C
Density of oil at 15°C	= 830.52 kg/m^3
Oil bulk modulus	= 1,400,000 kPa
Oil viscosity	= 19.8 mm^2/s @ 0 °C
	= 8.0 mm^2/s @ 19.8°C

D. *Pipeline Mechanical Configuration Information*

KMP 0-16	Diameter = 406.4 mm
	WT = 9.53 mm
KMP 17-112	Diameter = 406.4 mm
	WT = 7.00 mm
Depth of cover	1 m
Pipe roughness	0.0457 mm (1,800 micro inch)
Insulation	None
Ground temperature	0°C

Solution

The solution obtained using microcomputer hydraulic software is shown in Tables 6-4 and 6-5 and Figures 6-18 and 6-19.

Example 6.2: Water Pipeline System

The following water pipeline, as shown in Figure 6-20, is available. Provide a complete hydraulic design simulation for the cases detailed below. Additional data is as follows:

For the pipeline configuration shown in Figure 6-20, the following hydraulics simulations are required.

1. Study the hydraulics of the pipeline for the year 2000 and determine what additional facilities will be required for the year 2005.
2. For the year 2010, when future pipeline facilities are required, determine acceptable pipe diameter for the additional pipelines and any additional pumping facility, including its location, to take care of future demands at locations 6 and 7

Profile Summary
A. *Pipeline Route*

Location (km)	Elevation (m)
0.00	513.00
4.00	571.00
6.00	526.00
10.00	641.50
12.50	664.00
15.00	637.00
20.00	648.00
25.00	657.00
25.50	664.00
27.25	604.00
29.00	655.00
30.00	646.00
34.00	647.00
35.00	647.00
40.00	619.00
45.00	595.00
50.00	610.00
55.00	560.00

57.50	583.00
60.00	548.00
65.00	548.00
70.00	548.00
72.00	503.00
75.00	562.00
76.20	538.00
80.00	556.00
84.00	579.00
85.00	579.00
90.00	624.00
95.00	610.00
100.00	617.00
105.00	632.00
106.00	632.00
110.00	655.00
113.00	655.00
140.00	655.00
172.00	655.00

B. Operational Limitation

Minimum operating pressure	= 0.00 kPa-g
Maximum operating pressure	= 9,930.00 kPa-g
Minimum operating temperature	= −5.00 °C
Maximum operating temperature	= 25.00 °C
Adiabatic efficiency of pump	= 80.00%
Adiabatic efficiency of cooler	= 80.00%
Delivery pressure	= 675 kPa
Pump inlet loss	= 50.00 kPa
Pump outlet loss	= 50.00 kPa
Default pump discharge pressure	= 9,930.00 kPa-g
Default cooler outlet temperature	= 21.10 °C
Default heater discharge temperature	= 21.10 °C
Adiabatic efficiency of heater	= 80.00%
Design wind velocity	= 0.00 km/hr

C. Water Properties

1. Density of liquid at 15.0 Deg C	= 1,000.00 kg/m**3
2. Liquid bulk modulus	= 2,343,961.00 kPa
3. Inlet viscosity at 0.00 Deg C	= 1.79 (mm**2/s)
4. Inlet viscosity at 21.10 Deg C	= 0.98 (mm**2/S)
5. Liquid temp coefficient of density	= −0.08200 kg(m**3*°C)
6. Minimum pressure of density table	= −3,000.00 kPa
7. Minimum temperature of density table	= −20.00 °C
8. Maximum pressure of density table	= 10,000.00 kPa
9. Maximum temperature of density table	= 60.00 °C

Solution

The results of one of the computer hydraulic simulation for the water pipeline example are shown in Tables 6-6 and 6-7.

TABLE 6-4. Oil pipeline hydraulics design summary

Number	Facility Type	Inlet Location (km)	Inlet Elevation (m)	Flow (m**3/d)	Velocity (m/s)	Reynolds Number	Friction Factor	Inlet Temp (°C)	Skin Temp (°C)	Inlet Pressure (kPa-gauge)	Power (kw)
1	Pump	0.000	822.96	28,700.0	—	—	—	5.00	—	350.00	2,141.81
2	Segment	0.000	822.96	28,700.0	2.7907	5,405,651	0.003165	5.60	5.74	5,787.00	—
3	Segment	8.670	762.00	28,700.0	2.7915	5,407,070	0.003165	5.88	5.99	5,363.15	—
4	Segment	15.790	841.25	28,700.0	2.7919	5,407,952	0.003165	6.10	6.14	3,951.15	—
5	Flow change	18.890	857.10	28,700.0	—	—	—	—	—	—	—
6	Diameter/weight change	18.890	857.10	14,350.0	—	—	—	—	—	—	—
7	Segment	18.890	857.10	14,350.0	1.3605	2,669,322	0.003227	6.19	6.08	3,489.79	—
8	Segment	23.210	813.82	14,350.0	1.3603	2,668,985	0.003227	5.96	5.95	3,735.47	—
9	Flow change	23.540	811.27	14,350.0	—	—	—	—	—	—	—
10	Segment	23.540	811.27	16,250.0	1.5404	3,022,227	0.003212	5.94	5.91	3,748.02	—
11	Segment	25.380	797.05	16,250.0	1.5402	3,021,920	0.003212	5.87	5.81	3,805.06	—
12	Segment	28.510	807.72	16,250.0	1.5401	3,021,390	0.003212	5.75	5.71	3,615.34	—
13	Segment	30.680	798.58	16,250.0	1.5400	3,020,586	0.003212	5.67	5.64	3,619.87	—
14	Segment	32.120	759.53	16,250.0	1.5395	3,019,807	0.003212	5.62	5.40	3,894.29	—
15	Segment	44.640	752.86	16,250.0	1.5391	3,019,414	0.003212	5.19	5.15	3,541.60	—
16	Segment	47.530	749.81	16,250.0	1.5389	3,019,085	0.003212	5.10	5.02	3,472.65	—
17	Segment	52.590	736.09	16,250.0	1.5388	3,018,718	0.003212	4.95	4.92	3,420.95	—
18	Segment	54.520	737.62	16,250.0	1.5386		0.003212	4.89	4.80	3,345.55	—
19	Flow change	60.540	723.90	16,250.0	—	—	—	—	—	—	—
20	Pump	60.540	723.90	32,500.0	—	—	—	4.72	—	3,262.68	2,510.24
21	Segment	60.540	723.90	32,500.0	3.0793	6,041,669	0.003151	5.33	5.47	8,895.00	—
22	Segment	65.800	713.23	32,500.0	3.0800	6,042,886	0.003151	5.60	5.66	8,310.85	—
23	Segment	68.260	713.23	32,500.0	3.0817	6,046,334	0.003151	5.72	6.18	7,996.53	—
24	Segment	87.910	701.04	32,500.0	3.0864	6,049,549	0.003151	6.64	6.71	5,584.61	—
25	Segment	91.070	710.19	32,500.0	3.0842	6,051,188	0.003151	6.78	6.96	5,105.21	—
26	Segment	99.660	679.99	32,500.0	3.0850	6,052,736	0.003151	7.15	7.21	4,254.30	—
27	Segment	102.650	691.90	32,500.0	3.0853	6,053,315	0.003151	7.27	7.30	3,773.81	—
28	Flow change	104.050	683.67	32,500.0	—	—	—	—	—	—	—
29	Segment	104.050	683.67	16,250.0	1.5426	3,026,573	0.003212	7.33	7.28	3,662.22	—
30	Segment	106.130	685.80	16,250.0	1.5425	3,026,401	0.003212	7.22	7.19	3,576.89	—
31	Segment	107.530	679.71	16,250.0	1.5422	3,025,818	0.003212	7.15	7.04	3,581.25	—
32	End point	111.830	682.14	—	—	—	—	6.93	—	3,421.07	—

TABLE 6-5. Oil pipeline profile summary

Location (km)	Elevation (m)	Depth of Cover (m)	Outer Diameter (mM)	Wall Thickness (mm)	Pipe Roughness (mm)	Ground/Air Temperature (°C)	Ground Conductivity [W/(m °C)]	Insulation Thickness (mm)	Insulation Conductivity [W/(m °C)]
0.00	822.96	1.000	406.400	9.5000	0.0457	0.00	1.2000	0.00	0.00000
8.67	762.00	1.000	406.400	9.5000	0.0457	0.00	1.2000	0.00	0.00000
15.79	841.25	1.000	406.400	9.5000	0.0457	0.00	1.2000	0.00	0.00000
18.89	857.10	1.000	406.400	7.0000	0.0457	0.00	1.2000	0.00	0.00000
23.21	813.82	1.000	406.400	7.0000	0.0457	0.00	1.2000	0.00	0.00000
25.38	797.05	1.000	406.400	7.0000	0.0457	0.00	1.2000	0.00	0.00000
28.51	807.72	1.000	406.400	7.0000	0.0457	0.00	1.2000	0.00	0.00000
30.68	798.58	1.000	406.400	7.0000	0.0457	0.00	1.2000	0.00	0.00000
32.12	759.53	1.000	406.400	7.0000	0.0457	0.00	1.2000	0.00	0.00000
44.64	752.86	1.000	406.400	7.0000	0.0457	0.00	1.2000	0.00	0.00000
47.53	749.81	1.000	406.400	7.0000	0.0457	0.00	1.2000	0.00	0.00000
52.59	736.09	1.000	406.400	7.0000	0.0457	0.00	1.2000	0.00	0.00000
54.52	737.62	1.000	406.400	7.0000	0.0457	0.00	1.2000	0.00	0.00000
60.54	723.90	1.000	406.400	7.0000	0.0457	0.00	1.2000	0.00	0.00000
65.80	713.23	1.000	406.400	7.0000	0.0457	0.00	1.2000	0.00	0.00000
68.26	713.23	1.000	406.400	7.0000	0.0457	0.00	1.2000	0.00	0.00000
87.91	701.04	1.000	406.400	7.0000	0.0457	0.00	1.2000	0.00	0.00000
91.07	710.19	1.000	406.400	7.0000	0.0457	0.00	1.2000	0.00	0.00000
99.66	679.99	1.000	406.400	7.0000	0.0457	0.00	1.2000	0.00	0.00000
102.65	691.90	1.000	406.400	7.0000	0.0457	0.00	1.2000	0.00	0.00000
104.05	683.67	1.000	406.400	7.0000	0.0457	0.00	1.2000	0.00	0.00000
106.13	685.80	1.000	406.400	7.0000	0.0457	0.00	1.2000	0.00	0.00000
107.53	679.71	1.000	406.400	7.0000	0.0457	0.00	1.2000	0.00	0.00000
111.83	682.14	1.000	406.400	7.0000	0.0457	0.00	1.2000	0.00	0.00000

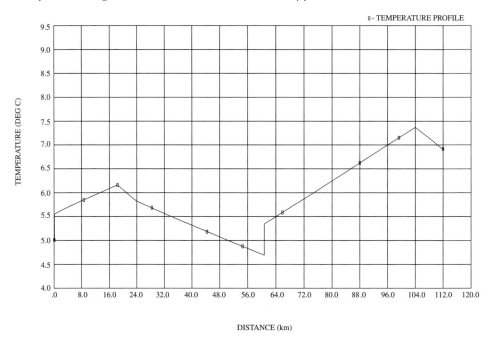

Figure 6-18. Oil pipeline temperature profile

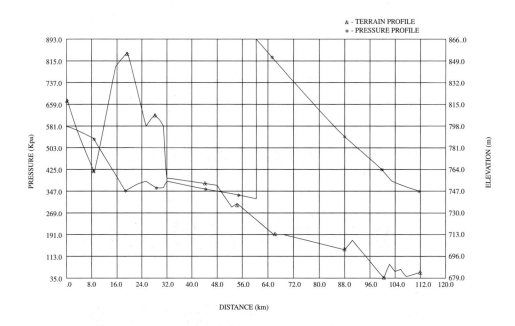

Figure 6-19. Oil pipeline hydraulic results

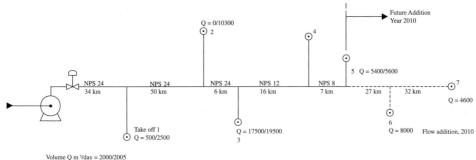

Figure 6-20. Facilities diagram: water pipeline

Example 6.3: Gas Pipeline System

Consider the gas pipeline system shown in Figure 6-21. In this diagram, the amount of flow added or taken off at each kilometer is clearly shown, together with the location of compressors and coolers. Provide a complete hydraulic design summary for this system. Undertake transient analysis using one of the transient simulation models for closing the end valve in 5 seconds. Simulate for a total time of 200,000 seconds. Additional data is given in the following:

Gas Composition

Methane	= 91.90%	N-heptane	= 0.00%
Ethane	= 5.00%	N-octane	= 0.00%
Propane	= 2.00%	Ethylene	= 0.00%
N-butane	= 1.00%	Carbon dioxide	= 0.00%
I-butane	= 0.00%	Hydrogen sulfide	= 0.00%
N-pentane	= 0.10%	Hydrogen	= 0.00%
I-pentane	= 0.00%	Nitrogen	= 0.00%
N-hexane	= 0.00%		

Gas Properties

Mixture molecular mass	=	17.7694
Base temperature	=	0.000 °C
Base pressure	=	102.620 kPa-abs
Density at base conditions	=	0.797 kg/m³
Net heating value	=	37.1983 MJ/m³
Gross heating value	=	41.1893 MJ/m³

TABLE 6-6. Water pipeline hydraulics design summary

Number	Facility Type	Inlet Location (km)	Inlet Elevation (m)	Flow (m**3/d)	Velocity (m/s)	Reynolds Number	Friction Factor	Inlet Temp (°C)	Skin Temp (°C)	Inlet Pressure (kPa-gauge)	Power (kw)
1	Pump	0.000	513.00	62,150.0	—	—	—	1.00	—	10.00	6,637.18
2	Segment	0.000	513.00	6,2150.0	2.6140	907,709	0.003165	1.44	1.48	7,300.00	6,637.18
3	Segment	4.000	571.00	62,150.0	2.6140	909,248	0.003164	1.51	1.53	6,437.64	6,637.18
4	Segment	6.000	526.00	62,150.0	2.6140	910,784	0.003164	1.54	1.58	6,733.23	6,637.18
5	Segment	10.000	641.50	62,150.0	2.6140	912,444	0.003163	1.61	1.63	5,306.29	6,637.18
6	Segment	12.500	664.00	62,150.0	2.6140	913,718	0.003162	1.65	1.68	4,902.47	6,637.18
7	Segment	15.000	637.00	62,150.0	2.6140	915,632	0.003161	1.70	1.74	4,984.81	6,637.18
8	Segment	20.000	648.00	62,150.0	2.6140	918,169	0.003160	1.78	1.82	4,511.23	6,637.18
9	Segment	25.000	657.00	62,150.0	2.6140	919,558	0.003160	1.86	1.87	4,057.41	6,637.18
10	Segment	25.500	664.00	62,150.0	2.6141	920,125	0.003160	1.87	1.89	3,952.13	6,637.18
11	Segment	27.250	604.00	62,150.0	2.6141	921,013	0.003159	1.90	1.92	4,413.50	6,637.18
12	Segment	29.000	655.00	62,150.0	2.6141	921,704	0.003159	1.93	1.94	3,784.80	6,637.18
13	Segment	30.000	646.00	62,150.0	2.6141	922,963	0.003158	1.95	1.98	3,800.13	6,637.18
14	Flow change	34.000	647.00	62,150.0	—	—	—	—	—	—	6,637.18
15	Segment	34.000	647.00	59,400.00	2.4984	883,306	0.003176	2.01	2.02	3,498.15	6,637.18
16	Segment	35.000	647.00	59,400.00	2.4984	884,617	0.003175	2.03	2.07	3,431.07	6,637.18
17	Segment	40.000	619.00	59,400.00	2.4984	886,798	0.003174	2.10	2.14	3,370.66	6,637.18
18	Segment	45.000	595.00	59,400.00	2.4985	888,973	0.003173	2.18	2.22	3,271.08	6,637.18
19	Segment	50.000	610.00	59,400.00	2.4985	891,134	0.003172	2.25	2.29	2,788.60	6,637.18
20	Segment	55.000	560.00	59,400.00	2.4985	892,751	0.003171	2.33	2.35	2,944.54	6,637.18
21	Segment	57.500	583.00	59,400.00	2.4985	893,827	0.3003171	2.36	2.38	2,551.18	6,637.18
22	Segment	60.000	548.00	59,400.00	2.4985	895,443	0.003170	2.40	2.44	2,727.42	6,637.18
23	Segment	65.000	548.00	59,400.00	2.4985	897,581	0.003169	2.48	2.51	2,392.54	6,637.18

24	Segment	70.000	548.00	59,400.00	2.4985	899,072	0.003169	2.55	2.56	2,057.77	6,637.18
25	Segment	72.000	503.00	59,400.00	2.4985	900,135	0.003168	2.58	2.60	2,365.79	6,637.18
26	Segment	75.000	562.00	59,400.00	2.4985	901,030	0.003168	2.62	2.63	1,585.62	6,637.18
27	Segment	76.200	538.00	59,400.00	2.4985	902,090	0.003167	2.64	2.67	1,740.99	
28	Pump	80.000	556.00	59,400.00				2.69		1,309.96	5,228.35
29	Segment	80.000	556.00	59,400.00	2.4986	914,414	0.003162	3.06	3.09	7,300.00	6,637.18
30	Flow change	84.000	579.00	59,400.00							6,637.18
31	Segment	84.000	579.00	48,070.0	2.0220	740,765	0.003249	3.11	3.12	6,806.94	6,637.18
32	Segment	85.000	579.00	48,070.0	2.0221	741,362	0.003249	3.12	3.14	6,761.99	6,637.18
33	Flow change	90.000	624.00	48,070.0							
34	Diameter/weight Change	90.000	624.00	26,620.0							6,637.18
35	Segment	90.000	624.00	26,620.0	4.0081	782,198	0.003368	3.16	3.36	6,095.35	6,637.18
36	Segment	95.000	610.00	26,620.0	4.0082	792,061	0.003364	3.56	3.75	4,500.81	6,637.18
37	Segment	100.000	617.00	26,620.0	4.0083	801,879	0.003359	3.95	4.14	2,702.33	6,637.18
38	Pump	105.000	632.00	26,620.0				4.33		827.50	2,028.81
39	Segment	105.000	632.00	26,620.0	4.0085	815,768	0.003353	4.65	4.69	6,000.00	6,637.18
40	Flow change	106.000	632.00	26,620.0							
41	Segment	106.000	632.00	20,020.0	3.0147	615,774	0.003460	4.72	4.80	5,655.10	6,637.18
42	Segment	110.000	655.00	20,020.0	3.0147	618,429	0.003459	4.88	4.94	4,624.05	6,637.18
43	Flow change	113.000	655.00	20,020.0							
44	Segment	113.000	655.00	13,860.0	2.0871	430,942	0.003620	5.00	5.14	4,020.43	6,637.18
45	Flow change	140.000	655.00	13,860.0							
46	Segment	140.000	655.00	5,060.0	0.7619	155,304	0.004245	5.29	4.81	1,294.83	6,637.18
47	End point	172.000	655.00					4.34		789.86	6,637.18

TABLE 6-7. Water pipeline profile summary

Location (km)	Elevation (m)	Depth of Cover (m)	Outer Diameter (mm)	Wall Thickness (mm)	Pipe Roughness (mm)	Ground/Air Temperature (°C)	Ground Conductivity [W/(m °C)]	Insulation Thickness	Insulation Conductivity [W/(m °C)]
0.00	513.00	2.500	610.00	9.2000	0.0250	1.00	1.2110	0.00	0.00000
4.00	571.00	2.500	610.00	9.2000	0.0250	1.00	1.2110	0.00	0.00000
6.00	526.00	2.500	610.00	9.2000	0.0250	1.00	1.2110	0.00	0.00000
10.00	641.50	2.500	610.00	9.2000	0.0250	1.00	1.2110	0.00	0.00000
12.50	664.00	2.500	610.00	9.2000	0.0250	1.00	1.2110	0.00	0.00000
15.00	637.00	2.500	610.00	9.2000	0.0250	1.00	1.2110	0.00	0.00000
20.00	648.00	2.500	610.00	9.2000	0.0250	1.00	1.2110	0.00	0.00000
25.00	657.00	2.500	610.00	9.2000	0.0250	1.00	1.2110	0.00	0.00000
25.50	664.00	2.500	610.00	9.2000	0.0250	1.00	1.2110	0.00	0.00000
27.25	604.00	2.500	610.00	9.2000	0.0250	1.00	1.2110	0.00	0.00000
29.00	655.00	2.500	610.00	9.2000	0.0250	1.00	1.2110	0.00	0.00000
30.00	646.00	2.500	610.00	9.2000	0.0250	1.00	1.2110	0.00	0.00000
34.00	647.00	2.500	610.00	9.2000	0.0250	1.00	1.2110	0.00	0.00000
35.00	647.00	2.500	610.00	9.2000	0.0250	1.00	1.2110	0.00	0.00000
40.00	619.00	2.500	610.00	9.2000	0.0250	1.00	1.2110	0.00	0.00000
45.00	595.00	2.500	610.00	9.2000	0.0250	1.00	1.2110	0.00	0.00000
50.00	610.00	2.500	610.00	9.2000	0.0250	1.00	1.2110	0.00	0.00000
55.00	560.00	2.500	610.00	9.2000	0.0250	1.00	1.2110	0.00	0.00000
57.50	583.00	2.500	610.00	9.2000	0.0250	1.00	1.2110	0.00	0.00000
60.00	548.00	2.500	610.00	9.2000	0.0250	1.00	1.2110	0.00	0.00000
65.00	548.00	2.500	610.00	9.2000	0.0250	1.00	1.2110	0.00	0.00000
70.00	548.00	2.500	610.00	9.2000	0.0250	1.00	1.2110	0.00	0.00000
72.00	503.00	2.500	610.00	9.2000	0.0250	1.00	1.2110	0.00	0.00000
75.00	562.00	2.500	610.00	9.2000	0.0250	1.00	1.2110	0.00	0.00000
76.50	538.00	2.500	610.00	9.2000	0.0250	1.00	1.2110	0.00	0.00000
80.00	556.00	2.500	610.00	9.2000	0.0250	1.00	1.2110	0.00	0.00000
84.00	579.00	2.500	610.00	9.2000	0.0250	1.00	1.2110	0.00	0.00000
85.00	579.00	2.500	610.00	9.2000	0.0250	1.00	1.2110	0.00	0.00000
90.00	624.00	2.500	323.900	5.6000	0.0250	1.00	1.2110	0.00	0.00000
95.00	610.00	2.500	323.900	5.6000	0.0250	1.00	1.2110	0.00	0.00000
100.00	617.00	2.500	323.900	5.6000	0.0250	1.00	1.2110	0.00	0.00000
105.00	632.00	2.500	323.900	5.6000	0.0250	1.00	1.2110	0.00	0.00000
106.00	632.00	2.500	323.900	5.6000	0.0250	1.00	1.2110	0.00	0.00000
110.00	655.00	2.500	323.900	5.6000	0.0250	1.00	1.2110	0.00	0.00000
113.00	655.00	2.500	323.900	5.6000	0.0250	1.00	1.2110	0.00	0.00000
140.00	655.00	2.500	323.900	5.6000	0.0250	1.00	1.2110	0.00	0.00000
172.00	655.00						1.2140		

Gas Pipeline System Description
Pipeline

- Diameter   = 863.66 mm
- Wall thickness   = 9.52 mm
- Maximum operating pressure   = 5,572 kPa
- Maximum operating temperature   = 60 °C
- Minimum operating pressure   = 1,520 kPa
- Minimum operating temperature   = 5 °C
- Pipe roughness   = 0.05 MM
- Drag factor   = 1.0

Compression / Cooling
Initiating Station:

- Suction pressure   1,367 kPa
- Suction temperature   35 °C
- Suction losses   88 kPa
- Discharge losses   118 kPa
- Cooler outlet   50 °C

Default Station

- Suction pressure   1,520 kPa
- Discharge pressure   2,000 kPa
- Cooler/heater temperature 50 °C

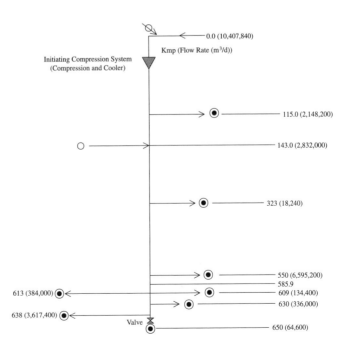

Figure 6-21. Gas pipeline system facilities diagram

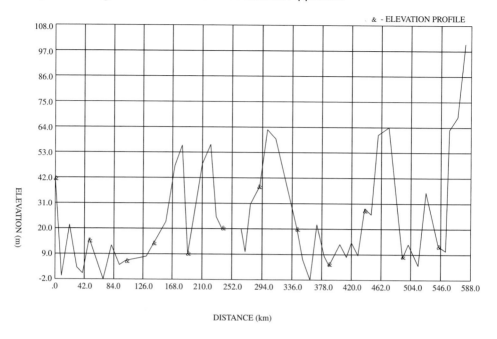

Figure 6-22. Gas pipeline elevation profile

Solution

The results of a steady-state microcomputer hydraulic simulation for the gas pipeline problem are shown in Tables 6-8 through 6-10 and Figures 6-22 and 6-23.

A transient analysis was then performed by closing the valve (See Figure 6-21) at the end of the pipeline in 5 seconds. The results of the transient analyses for various times after closing the end valve are shown in Figures 6-24 through 6-27.

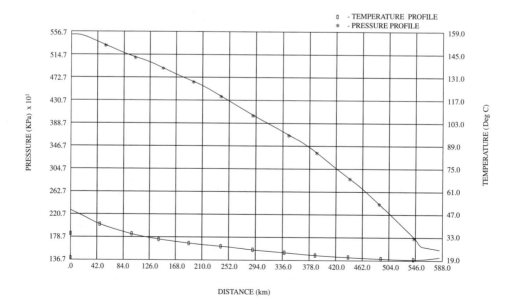

Figure 6-23. Gas pipeline hydraulics profile

TABLE 6-8. Gas pipeline hydraulic design summary (steady-state)

Facility Type	Inlet Location (km)	Inlet Elevation (m)	Inlet Flow (m**3/d)	Average Velocity (m/s)	Average Density (kg/m**3)	Reynolds Number	Friction Factor	Inlet Temp (°C)	Inlet Pressure (kPa-gauge)	Pressure
Compressor	0.000	42.00	10,407,840	—	—	—	—	35.00	1,367.55	26,905.4
Cooler	0.000	42.00	10,407,840	—	—	—	—	150.09	5,551.24	33,649.4
Segment	0.000	42.00	10,407,840	4.1724	41.063	11,297,330	0.002717	50.00	5,551.24	
Segment	10.000	0.00	10,407,840	4.1754	41.033	11,353,070	0.002717	48.26	5,522.16	
Segment	20.000	22.00	10,407,840	4.1838	40.951	11,409,470	0.002717	46.32	5,467.27	
Segment	30.000	4.00	10,407,840	4.1907	40.884	11,461,000	0.002717	44.72	5,428.37	
Segment	40.000	0.00	10,407,840	4.2035	40.759	11,514,760	0.002717	43.16	5,383.77	
Segment	50.000	16.00	10,407,840	4.2177	40.623	11,565,950	0.002717	41.60	5,331.02	
Segment	60.000	7.00	10,407,840	4.2302	40.504	11,611,390	0.002717	40.28	5,288.11	
Segment	70.000	−2.00	10,407,840	4.2467	40.345	11,655,950	0.002717	39.05	5,245.04	
Segment	80.000	14.00	10,407,840	4.2648	40.175	11,698,710	0.002717	37.78	5,191.89	
Segment	90.000	4.00	10,407,840	4.2827	40.008	11,738,340	0.002717	36.73	5,148.81	
Segment	100.000	6.00	10,407,840	4.3081	39.767	11,787,700	0.002717	35.69	5,100.80	
Flow change	115.000	7.00	10,407,840	—	—	—	—	34.27	5,029.16	
Segment	115.000	7.00	8,259,640	3.4354	39.558	9,398,065	0.002717	34.27	5,029.16	
Segment	130.000	8.00	8,259,640	3.4512	39.396	9,436,309	0.002717	32.87	4,983.69	
Flow change	143.000	14.93	8,259,640	—	—	—	—	31.77	4,941.75	
Segment	143.000	14.93	11,091,640	4.6876	38.950	12,729,340	0.002717	31.77	4,941.75	
Segment	160.000	24.00	11,091,640	4.7579	38.379	12,784,440	0.002717	30.58	4,844.65	
Segment	170.000	47.00	11,091,640	4.8122	37.946	12,823,750	0.002717	29.84	4,780.09	
Segment	180.000	56.00	11,091,640	4.8576	37.591	12,851,850	0.002717	29.22	4,720.19	
Segment	190.000	9.00	11,091,640	4.9050	37.228	12,880,600	0.002717	28.93	4,680.45	
Segment	200.000	26.00	11,091,640	4.9629	36.794	12,917,310	0.002717	28.33	4,616.61	
Segment	210.000	45.00	11,091,640	5.0213	36.366	12,952,150	0.002717	27.74	4,551.43	
Segment	220.000	56.00	11,091,640	5.0748	35.983	12,979,700	0.002717	27.24	4,488.51	
Segment	230.000	28.00	11,091,640	5.1266	35.597	13,004,220	0.002717	26.97	4,438.76	
Segment	240.000	20.00	11,091,640	5.2001	35.116	13,034,720	0.002717	26.61	4,381.32	
Segment	250.000	20.00	11,091,640	5.2802	34.583	13,066,230	0.002717	26.23	4,320.22	
Segment	260.000	20.00	11,091,640	5.3603	34.066	13,095,570	0.002717	25.87	4,258.18	
Segment	270.000	10.00	11,091,640	5.4454	33.534	13,127,120	0.002717	25.58	4,198.54	
Segment	280.000	30.00	11,091,640	5.5354	32.989	13,160,240	0.002717	25.14	4,127.97	
Segment	290.000	30.00	11,091,640	5.6278	32.448	13,193,270	0.002717	24.80	4,060.34	
Segment	300.000	38.00	11,091,640	5.7204	31.923	13,223,820	0.002717	24.38	3,986.26	
Segment	310.000	36.00	11,091,640	5.8216	31.367	13,252,030	0.002717	24.13	3,919.99	
Flow Change	323.000	60.00	11,091,640	—	—	—	—	23.90	3,836.39	
Segment	323.000	42.67	11,091,640	5.9547	30.566	13,265,540	0.002717	23.90	3,836.39	

TABLE 6-8. (Continued)

Facility Type	Inlet Location (km)	Inlet Elevation (m)	Inlet Flow (m**3/d)	Average Velocity (m/s)	Average Density (kg/m**3)	Reynolds Number	Friction Factor	Inlet Temp (°C)	Inlet Pressure (kPa-gauge)	Pressure
Segment	340.000	42.67	11,073,400	6.1260	29.760	13,301,310	0.002717	23.61	3,724.45	
Segment	350.000	20.00	11,073,400	6.2577	29.134	13,328,250	0.002717	23.44	3,656.39	
Segment	360.000	7.00	11,073,400	6.3995	28.488	13,359,510	0.002717	23.25	3,584.98	
Segment	370.000	0.00	11,073,400	6.5435	27.861	13,388,930	0.002717	22.91	3,503.76	
Segment	380.000	22.00	11,073,400	6.6860	27.267	13,414,730	0.002717	22.78	3,430.83	
Segment	390.000	8.00	11,073,400	6.8357	26.670	13,442,830	0.002717	22.58	3,352.93	
Segment	400.000	6.00	11,073,400	6.9884	26.072	13,469,910	0.002717	22.34	3,270.91	
Segment	410.000	13.00	11,073,400	7.1897	25.358	13,500,260	0.002717	22.17	3,190.21	
Segment	420.000	8.00	11,073,400	7.4253	24.553	13,532,170	0.002717	21.95	3,104.38	
Segment	430.000	14.00	11,073,400	7.6718	23.764	13,565,400	0.002717	21.79	3,018.72	
Segment	440.000	8.00	11,073,400	7.9274	22.998	13,598,260	0.002717	21.50	2,924.07	
Segment	450.000	28.00	11,073,400	8.1938	22.251	13,632,370	0.002717	21.32	2,837.52	
Segment	460.000	26.00	11,073,400	8.4702	21.524	13,666,210	0.002717	20.96	2,727.98	
Segment	470.000	60.00	11,073,400	8.7747	20.724	13,693,480	0.002717	20.76	2,627.78	
Segment	480.000	37.00	11,073,400	9.2124	19.789	13,723,040	0.002717	20.71	2,530.35	
Segment	490.000	8.00	11,073,400	9.6850	18.825	13,756,800	0.002717	20.66	2,427.91	
Segment	500.000	14.00	11,073,400	10.1889	17.893	13,792,510	0.002717	20.40	2,313.19	
Segment	510.000	6.00	11,073,400	10.7240	17.001	13,830,540	0.002717	20.21	2,195.07	
Segment	520.000	35.00	11,073,400	11.2857	15.999	13,866,550	0.002717	19.82	2,064.44	
Segment	530.000	25.00	11,073,400	12.3685	14.769	13,911,520	0.002717	19.61	1,933.63	
Segment	540.000	13.00	11,073,400	13.5462	13.486	13,955,990	0.002717	19.37	1,790.28	
Flow change	550.000	10.00	11,073,400					19.04	1,631.77	
Segment	550.000	10.00	4,478,200	5.7904	12.735	5,647,661	0.002717	19.04	1,631.77	
Segment	560.000	60.00	4,478,200	5.9077	12.482	5,637,655	0.002717	19.74	1,598.05	
Segment	570.000	68.00	4,478,200	6.0207	12.247	5,628,839	0.002717	20.51	1,569.05	
End point	580.000	100.00						21.03	1,536.64	

TABLE 6-9. Gas pipeline flow characteristics summary (steady-state)

Facility Number	Facility Type	Inlet Location (km)	Inlet Velocity (m/s)	Inlet Density (kg/m**3)
1	Compressor	0.000	16.5478	40.355
2	Cooler	0.000	6.0153	28.486
3	Segment	0.000	4.1760	41.033
4	Segment	10.000	4.1698	41.094
5	Segment	20.000	4.1822	40.972
6	Segment	30.000	4.1864	40.931
7	Segment	40.000	4.1959	40.838
8	Segment	50.000	4.2122	40.680
9	Segment	60.000	4.2240	40.566
10	Segment	70.000	4.2370	40.442
11	Segment	80.000	4.2573	40.249
12	Segment	90.000	4.2731	40.100
13	Segment	100.000	4.2930	39.915
14	Flow change	115.000	4.3249	39.620
15	Segment	115.000	3.4322	39.620
16	Segment	130.000	3.4430	39.496
17	Flow change	143.000	3.4606	39.296
18	Segment	143.000	4.6471	39.296
19	Segment	160.000	4.7303	38.605
20	Segment	170.000	4.7862	38.154
21	Segment	180.000	4.8390	37.737
22	Segment	190.000	4.8768	37.445
23	Segment	200.000	4.9338	37.012
24	Segment	210.000	4.9927	36.576
25	Segment	220.000	5.0506	36.156
26	Segment	230.000	5.0994	35.810
27	Segment	240.000	5.1608	35.384
28	Segment	250.000	5.2402	34.848
29	Segment	260.000	5.3212	34.318
30	Segment	270.000	5.4003	33.815
31	Segment	280.000	5.4914	33.254
32	Segment	290.000	5.5802	32.725
33	Segment	300.000	5.6763	32.171
34	Segment	310.000	5.7653	61.674
35	Flow change	323.000	5.8794	31.059
36	Segment	323.000	5.8697	31.059
37	Segment	340.000	6.0624	30.072
38	Segment	350.000	6.1912	29.447
39	Segment	360.000	6.3257	28.821
40	Segment	370.000	6.4750	28.156
41	Segment	380.000	6.6137	27.566
42	Segment	390.000	6.7601	26.969
43	Segment	400.000	6.9131	26.372
44	Segment	410.000	7.0740	25.772
45	Segment	420.000	7.3091	24.943
46	Segment	430.000	7.5453	24.162
47	Segment	440.000	7.8022	23.367
48	Segment	450.000	8.0566	22.629
49	Segment	460.000	8.3352	21.872
50	Segment	470.000	8.6095	21.175
51	Segment	480.000	8.9932	20.272
52	Segment	490.000	9.4430	19.306
53	Segment	500.000	9.9389	18.343
54	Segment	510.000	10.4514	17.444
55	Segment	520.000	11.0096	16.559
56	Segment	530.000	11.0807	15.439
57	Segment	540.000	12.9301	14.100
58	Flow change	550.000	14.1641	12.871
59	Segment	550.000	5.7281	12.871
60	Segment	560.000	5.8519	12.599
61	Segment	570.000	5.9624	12.366
62	End point	580.000	6.0786	12.129

TABLE 6-10. Gas pipeline profile summary (transient)

Location (km)	Elevation (m)	Depth of Cover (m)	Outer Diameter (mm)	Wall Thickness (mm)	Pipe Roughness (mm)	Drag Factor	Ground Temperature (°C)	Ground Conductivity [W/(m °C)]	Insulation Thickness (mm)	Insulation Conductivity [W/(m °C)]
0.000	42.00	2.000	863.660	9.5200	0.05000	1.0000	25.00	0.7000	0.00	0.00000
10.000	0.00	2.000	863.660	9.5200	0.05000	1.0000	25.00	0.7000	0.00	0.00000
20.000	22.00	2.000	863.660	9.5200	0.05000	1.0000	25.00	0.7000	0.00	0.00000
30.000	4.00	2.000	863.660	9.5200	0.05000	1.0000	25.00	0.7000	0.00	0.00000
40.000	0.00	2.000	863.660	9.5200	0.05000	1.0000	25.00	0.7000	0.00	0.00000
50.000	16.00	2.000	863.660	9.5200	0.05000	1.0000	25.00	0.7000	0.00	0.00000
60.000	7.00	2.000	863.660	9.5200	0.05000	1.0000	25.00	0.7000	0.00	0.00000
70.000	−2.00	2.000	863.660	9.5200	0.05000	1.0000	25.00	0.7000	0.00	0.00000
80.000	14.00	2.000	863.660	9.5200	0.05000	1.0000	25.00	0.7000	0.00	0.00000
90.000	4.00	2.000	863.660	9.5200	0.05000	1.0000	25.00	0.7000	0.00	0.00000
100.000	6.00	2.000	863.660	9.5200	0.05000	1.0000	25.00	0.7000	0.00	0.00000
130.000	8.00	2.000	863.660	9.5200	0.05000	1.0000	25.00	0.7000	0.00	0.00000
160.000	24.00	2.000	863.660	9.5200	0.05000	1.0000	25.00	0.7000	0.00	0.00000
170.000	47.00	2.000	863.660	9.5200	0.05000	1.0000	25.00	0.7000	0.00	0.00000
180.000	56.00	2.000	863.660	9.5200	0.05000	1.0000	25.00	0.7000	0.00	0.00000
190.000	9.00	2.000	863.660	9.5200	0.05000	1.0000	25.00	0.7000	0.00	0.00000
200.000	26.00	2.000	863.660	9.5200	0.05000	1.0000	25.00	0.7000	0.00	0.00000
210.000	45.00	2.000	863.660	9.5200	0.05000	1.0000	25.00	0.7000	0.00	0.00000
220.000	56.00	2.000	863.660	9.5200	0.05000	1.0000	25.00	0.7000	0.00	0.00000
230.000	28.00	2.000	863.660	9.5200	0.05000	1.0000	25.00	0.7000	0.00	0.00000
240.000	20.00	2.000	863.660	9.5200	0.05000	1.0000	25.00	0.7000	0.00	0.00000
250.000	20.00	2.000	863.660	9.5200	0.05000	1.0000	25.00	0.7000	0.00	0.00000
260.000	20.00	2.000	863.660	9.5200	0.05000	1.0000	25.00	0.7000	0.00	0.00000

270.000	10.00	2.000	863.660	9.5200	0.05000	1.0000	25.00	0.7000	0.00	0.00000
280.000	30.00	2.000	863.660	9.5200	0.05000	1.0000	25.00	0.7000	0.00	0.00000
290.000	38.00	2.000	863.660	9.5200	0.05000	1.0000	25.00	0.7000	0.00	0.00000
300.000	63.00	2.000	863.660	9.5200	0.05000	1.0000	25.00	0.7000	0.00	0.00000
310.000	60.00	2.000	863.660	9.5200	0.05000	1.0000	25.00	0.7000	0.00	0.00000
320.000	20.00	2.000	863.660	9.5200	0.05000	1.0000	25.00	0.7000	0.00	0.00000
330.000	7.00	2.000	863.660	9.5200	0.05000	1.0000	25.00	0.7000	0.00	0.00000
340.000	0.00	2.000	863.660	9.5200	0.05000	1.0000	25.00	0.7000	0.00	0.00000
350.000	22.00	2.000	863.660	9.5200	0.05000	1.0000	25.00	0.7000	0.00	0.00000
360.000	8.00	2.000	863.660	9.5200	0.05000	1.0000	25.00	0.7000	0.00	0.00000
370.000	6.00	2.000	863.660	9.5200	0.05000	1.0000	25.00	0.7000	0.00	0.00000
380.000	13.00	2.000	863.660	9.5200	0.05000	1.0000	25.00	0.7000	0.00	0.00000
390.000	8.00	2.000	863.660	9.5200	0.05000	1.0000	25.00	0.7000	0.00	0.00000
400.000	14.00	2.000	863.660	9.5200	0.05000	1.0000	25.00	0.7000	0.00	0.00000
410.000	8.00	2.000	863.660	9.5200	0.05000	1.0000	25.00	0.7000	0.00	0.00000
420.000	28.00	2.000	863.660	9.5200	0.05000	1.0000	25.00	0.7000	0.00	0.00000
430.000	26.00	2.000	863.660	9.5200	0.05000	1.0000	25.00	0.7000	0.00	0.00000
440.000	60.00	2.000	863.660	9.5200	0.05000	1.0000	25.00	0.7000	0.00	0.00000
450.000	64.00	2.000	863.660	9.5200	0.05000	1.0000	25.00	0.7000	0.00	0.00000
460.000	37.00	2.000	863.660	9.5200	0.05000	1.0000	25.00	0.7000	0.00	0.00000
470.000	8.00	2.000	863.660	9.5200	0.05000	1.0000	25.00	0.7000	0.00	0.00000
480.000	14.00	2.000	863.660	9.5200	0.05000	1.0000	25.00	0.7000	0.00	0.00000
490.000	6.00	2.000	863.660	9.5200	0.05000	1.0000	25.00	0.7000	0.00	0.00000
500.000	35.00	2.000	863.660	9.5200	0.05000	1.0000	25.00	0.7000	0.00	0.00000
510.000	25.00	2.000	863.660	9.5200	0.05000	1.0000	25.00	0.7000	0.00	0.00000
520.000	13.00	2.000	863.660	9.5200	0.05000	1.0000	25.00	0.7000	0.00	0.00000
530.000	10.00	2.000	863.660	9.5200	0.05000	1.0000	25.00	0.7000	0.00	0.00000
540.000	60.00	2.000	863.660	9.5200	0.05000	1.0000	25.00	0.7000	0.00	0.00000
550.000	68.00	2.000	863.660	9.5200	0.05000	1.0000	25.00	0.7000	0.00	0.00000
560.000	100.00	2.000	863.660	9.5200	0.05000	1.0000	25.00	0.7000	0.00	0.00000

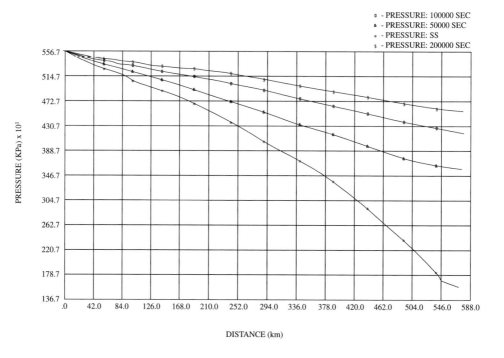

Figure 6-24. Gas pipeline transients, pressure vs. distance

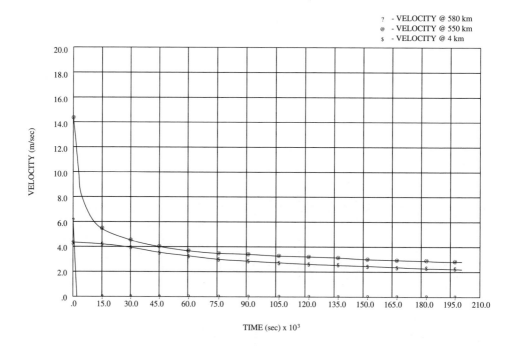

Figure 6-25. Gas pipeline transients, velocity vs. time

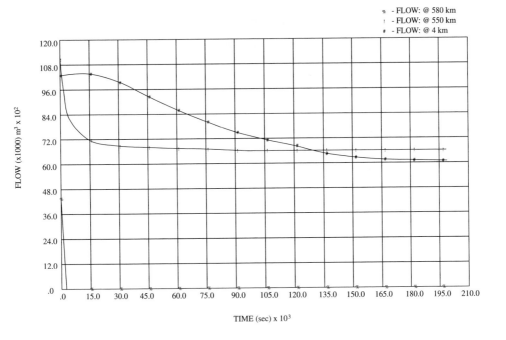

Figure 6-26. Gas pipeline transients, flow vs. time

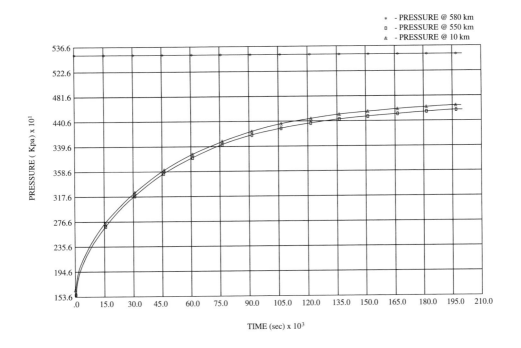

Figure 6-27. Gas pipeline transients, pressure vs. time

Example 6.4: Non-Newtonian Fluids

Consider the non-Newtonian pipeline system shown in Figure 6-28.

Perform a transient hydraulic design simulation (in closing the pipeline terminal valve in 5 seconds) for this system using the following data:

Pipeline Route

Location (km)	Elevation (m)
0	600
6.09	591
10.59	599
15.09	568
20.40	557
26.59	562
31.59	603
34.09	560
37.72	563
43.09	551
47.09	551
52.87	556
58.83	597
64.09	595
71.83	605
76.44	639.6
81.95	641
88.68	626
93.57	597
98.39	608
106.83	645
111.69	663
117.27	672
123.44	690
128.023	673
133.34	644
140.59	601
146.79	658
150.23	627
154.99	631
162.68	619
166.72	629
171.09	623
176.87	632
183.35	628
187.44	620
191.32	619
196.78	625
202.26	653
207.74	638
214.29	656
219.99	640
223.94	640
230.39	640
234.40	640
238.89	626
244.72	685
247.51	666

Operational Limitations

Minimum operating pressure	=	300.00 kPa-g
Maximum operating pressure	=	8,395.00 kPa-g
Minimum operating temperature	=	0.00 °C
Maximum operating temperature	=	71.00 °C
Non-Newtonian transition temperature	=	15.00 °C
Adiabatic effect of initial pump station	=	70.00%
Adiabatic efficiency of cooler	=	75.00%
Pump inlet loss	=	40.00 kPam
Pump outlet loss	=	40.00 kPa
Default pump discharge pressure	=	8,395.00 kPa-g
Default cooler outlet temperature	=	71.00 °C
Default heater discharge temperature	=	0.00 °C
Adiabatic effect of booster pump station	=	70.00%
Adiabatic efficiency of heater	=	80.00%

Newtonian Fluid Properties

Liquid temperature coefficent of density	=	-0.617kg/(m**3°C)
Minimum pressure of density table	=	50.00 kPa
Minimum temperature of density table	=	0.00 °C
Density of liquid at 15.00 °C	=	926.80 kg/m**3
Liquid bulk modulus	=	1,530,000.00 kPa
Maximum pressure of density table	=	9,000.00 kPa
Maximum temperature of density table	=	71.00 °C
Viscosity at 20°C	=	140 mm^2/s
Viscosity at 60°C	=	39 mm^2/s

Non-Newtonian Fluid Properties

SHEAR RATE (s^{-1}) = 3.96	APPARENT VISCOSITY (mm^2/s) = 480.00	TEMPERATURE (°C) = 0.0
SHEAR RATE (s^{-1}) = 3.96	APPARENT VISCOSITY (mm^2/s) = 387.18	TEMPERATURE (°C) = 0.0
SHEAR RATE (s^{-1}) = 3.96	APPARENT VISCOSITY (mm^2/s) = 367.62	TEMPERATURE (°C) = 0.0
SHEAR RATE (s^{-1}) = 3.96	APPARENT VISCOSITY (mm^2/s) = 296.47	TEMPERATURE (°C) = 0.0
SHEAR RATE (s^{-1}) = 3.96	APPARENT VISCOSITY (mm^2/s) = 211.41	TEMPERATURE (°C) = 0.0
SHEAR RATE (s^{-1}) = 3.96	APPARENT VISCOSITY (mm^2/s) = 146.32	TEMPERATURE (°C) = 0.0
SHEAR RATE (s^{-1}) = 3.96	APPARENT VISCOSITY (mm^2/s) = 106.51	TEMPERATURE (°C) = 0.0
SHEAR RATE (s^{-1}) = 3.96	APPARENT VISCOSITY (mm^2/s) = 78.78	TEMPERATURE (°C) = 0.0
SHEAR RATE (s^{-1}) = 3.96	APPARENT VISCOSITY (mm^2/s) = 60.43	TEMPERATURE (°C) = 0.0
SHEAR RATE (s^{-1}) = 7.92	APPARENT VISCOSITY (mm^2/s) = 480.00	TEMPERATURE (°C) = 0.0
SHEAR RATE (s^{-1}) = 7.92	APPARENT VISCOSITY (mm^2/s) = 378.18	TEMPERATURE (°C) = 0.0
SHEAR RATE (s^{-1}) = 7.92	APPARENT VISCOSITY (mm^2/s) = 359.07	TEMPERATURE (°C) = 0.0
SHEAR RATE (s^{-1}) = 7.92	APPARENT VISCOSITY (mm^2/s) = 294.73	TEMPERATURE (°C) = 0.0
SHEAR RATE (s^{-1}) = 7.92	APPARENT VISCOSITY (mm^2/s) = 204.16	TEMPERATURE (°C) = 0.0
SHEAR RATE (s^{-1}) = 7.92	APPARENT VISCOSITY (mm^2/s) = 133.79	TEMPERATURE (°C) = 0.0
SHEAR RATE (s^{-1}) = 7.92	APPARENT VISCOSITY (mm^2/s) = 92.72	TEMPERATURE (°C) = 0.0
SHEAR RATE (s^{-1}) = 7.92	APPARENT VISCOSITY (mm^2/s) = 65.61	TEMPERATURE (°C) = 0.0
SHEAR RATE (s^{-1}) = 7.92	APPARENT VISCOSITY (mm^2/s) = 48.98	TEMPERATURE (°C) = 0.0

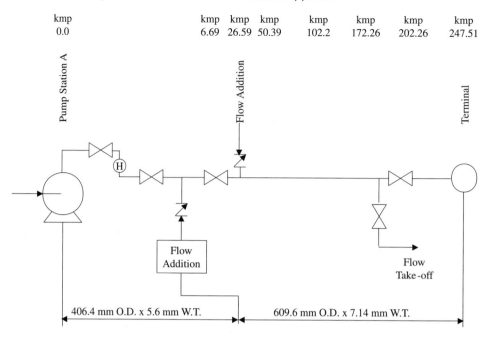

Figure 6-28. Facilities diagram: non-Newtonian pipeline

Solution

Sample results of the hydraulic simulation for the non-Newtonian pipeline example are shown in Tables 6-11 and 6-12, and in Figures 6-29 to 6-33.

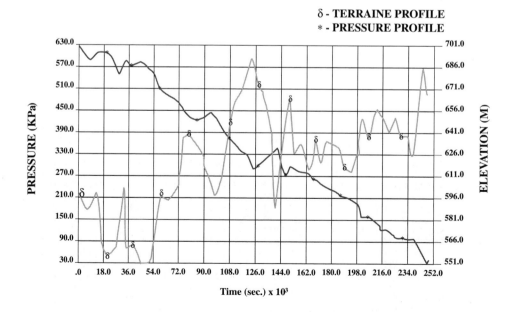

Figure 6-29. Steady-state non-Newtonian liquid pipeline hydraulics, pressure, and elevation vs. time

TABLE 6-11. Non-Newtonian liquid hydraulic design summary

Number	Facility Type	Inlet Location (km)	Inlet Elevation (m)	Flow (m**3/d)	Velocity (m/s)	Reynolds Number	Darcy-Weisbach Friction Factor	Inlet Temperature (°C)	Skin Temperature (°C)	Inlet Pressure (kPa-gauge)	Equivalent Viscosity (mm**2/s)	Power
1	Pump	0.000	600.00	12,000.0				38.50		300.00		1,212.01
2	Segment	0.000	600.00	12,000.0	1.1452	6,265	0.035224	39.97	38.64	6,230.00	72.41	
3	Flow change	6.090	591.00	12,000.0								
4	Diameter/weight Change	6.090	591.00	16,200.0								
5	Segment	6.090	591.00	16,200.0	0.6831	5,142	0.037187	36.72	35.90	5,986.56	79.09	
6	Segment	10.590	599.00	16,200.0	0.6824	4,870	0.037769	35.08	34.30	5,854.89	83.42	
7	Segment	15.090	568.00	16,200.0	0.6816	4,598	0.038401	33.52	32.64	6,072.31	88.26	
8	Segment	20.400	557.00	16,200.0	0.6808	4,308	0.039134	31.76	30.80	6,098.27	94.08	
9	Flow change	26.590	562.00	16,200.0								
10	Segment	26.590	562.00	17,800.0	0.7465	4,260	0.039261	28.50	27.86	5,966.82	104.32	
11	Segment	31.590	603.00	17,800.0	0.7461	4,116	0.039658	27.22	26.92	5,512.81	107.90	
12	Segment	34.090	560.00	17,800.0	0.7456	3,986	0.040035	26.61	26.06	5,858.05	111.36	
13	Segment	38.720	563.00	17,800.0	0.7451	3,833	0.040501	25.51	25.01	5,751.35	115.74	
14	Segment	43.090	551.00	17,800.0	0.7446	3,699	0.040929	24.51	24.07	5,783.74	119.83	
15	Segment	47.090	551.00	17,800.0	0.7441	3,553	0.041423	23.64	23.03	5,713.51	124.66	
16	Segment	52.870	556.00	17,800.0	0.7435	3,397	0.041986	22.43	21.87	5,565.65	130.30	
17	Segment	58.530	597.00	17,800.0	0.7430	3,257	0.042525	21.31	20.79	5,092.78	135.83	
18	Segment	64.090	595.00	17,800.0	0.7424	3,104	0.043150	20.27	19.59	5,009.69	142.40	
19	Segment	71.830	605.00	17,800.0	0.7419	2,977	0.043183	18.92	18.54	4,776.19	148.37	
20	Segment	76.440	639.60	17,800.0	0.7415	2,880	0.042117	18.16	17.72	4,377.26	153.29	
21	Segment	81.950	641.00	17,800.0	0.7410	2,771	0.040915	17.29	16.79	4,265.42	159.20	
22	Segment	88.680	626.00	17,800.0	0.7406	2,677	0.039864	16.29	15.95	4,284.08	164.70	
23	Segment	93.570	597.00	17,800.0	0.7403	2,604	0.039038	15.61	15.28	4,464.41	169.27	
24	Segment	98.390	608.00	17,800.0	0.7399	1,331	0.048098	14.96	14.44	4,284.16	331.01	
25	Segment	106.830	645.00	17,800.0	0.7395	1,310	0.048863	13.92	13.64	3,774.59	336.10	
26	Segment	111.690	663.00	17,800.0	0.7392	1,295	0.049436	13.36	13.06	3,509.62	339.91	
27	Segment	117.270	672.00	17,800.0	0.7389	1,279	0.050046	12.75	12.44	3,310.21	343.96	
28	Segment	123.440	690.00	17,800.0	0.7386	1,266	0.050569	12.12	11.90	3,014.83	347.44	
29	Segment	128.023	673.00	17,800.0	0.7384	1,254	0.051030	11.68	11.43	3,071.09	350.49	
30	Segment	133.340	644.00	17,800.0	0.7381	1,241	0.051583	11.19	10.88	3,219.96	354.16	
31	Segment	140.590	601.00	17,800.0	0.7378	1,228	0.052132	10.56	10.31	3,452.95	357.80	
32	Segment	146.791	658.40	17,800.0	0.7377	1,220	0.052474	10.06	9.93	2,792.03	360.06	
33	Segment	150.230	627.00	17,800.0	0.7375	1,215	0.052657	9.80	9.63	3,001.79	361.24	
34	Segment	154.990	631.00	17,800.0	0.7373	1,209	0.052922	9.45	9.19	2,858.79	362.94	

TABLE 6-11. (Continued)

Number	Facility Type	Inlet Location (km)	Inlet Elevation (m)	Flow (m**3/d)	Velocity (m/s)	Reynolds Number	Darcy-Weisbach Friction Factor	Inlet Temperature (°C)	Skin Temperature (°C)	Inlet Pressure (kPa-gauge)	Equivalent Viscosity (mm**2/s)	Power
35	Segment	162.680	619.00	17,800.0	0.7371	1,204	0.053151	8.93	8.79	2,795.44	364.43	
36	Segment	166.720	629.00	17,800.0	0.7370	1,201	0.053310	8.66	8.53	2,612.96	365.45	
37	Segment	171.090	623.00	17,800.0	0.7368	1,196	0.053492	8.39	8.22	2,568.83	366.62	
38	Segment	176.870	632.00	17,800.0	0.7366	1,192	0.053699	8.06	7.88	2,355.39	367.96	
39	Segment	183.350	628.00	17,800.0	0.7365	1,188	0.053866	7.70	7.59	2,244.25	369.03	
40	Segment	187.440	620.00	17,800.0	0.7364	1,185	0.053987	7.49	7.39	2,223.86	369.81	
41	Segment	191.320	619.00	17,800.0	0.7363	1,182	0.054123	7.29	7.16	2,144.12	370.69	
42	Segment	196.780	625.00	17,800.0	0.7362	1,179	0.054273	7.04	6.91	1,963.93	371.65	
43	Segment	202.260	653.00	17,800.0	0.7361	1,176	0.054415	6.79	6.67	1,581.84	372.57	
44	Segment	207.740	638.00	17,800.0	0.7359	1,173	0.054562	6.56	6.43	1,592.52	373.51	
45	Segment	214.290	656.00	17,800.0	0.7358	1,170	0.054702	6.30	6.19	1,276.40	374.41	
46	Segment	219.990	640.00	17,800.0	0.7357	1,168	0.054806	6.09	6.02	1,290.54	375.08	
47	Segment	223.940	651.00	17,800.0	0.7357	1,165	0.054913	5.95	5.84	1,098.17	375.77	
48	Segment	230.390	640.00	17,800.0	0.7356	1,163	0.055014	5.73	5.67	1,048.67	376.42	
49	Segment	234.400	640.00	17,800.0	0.7355	1,162	0.055092	5.61	5.54	955.18	376.92	
50	Segment	238.890	626.00	17,800.0	0.7354	1,160	0.055183	5.47	5.54	978.45	377.50	
51	Segment	244.720	685.00	17,800.0	0.7354	1,158	0.055255	5.30	5.39	302.28	377.96	
52	End point	247.510	666.00						5.23	5.27	410.83	

TABLE 6-12. Non-Newtonian liquid pipeline profile summary

Location (km)	Elevation (m)	Depth of Cover (m)	Outer Diameter (mm)	Wall Thickness (mm)	Pipe Roughness (mm)	Ground/Air Temperature (°C)	Ground Conductivity [W/(m °C)]	Insulation Thickness	Insulation Conductivity [W/(m °C)]
0.00	600.00	1.000	406.400	5.1600	0.0457	1.35	1.2110	0.00	0.00000
6.09	591.00	1.000	609.600	7.1400	0.0457	1.35	1.2110	0.00	0.00000
10.59	599.00	1.000	609.600	7.1400	0.0457	1.35	1.2110	0.00	0.00000
15.09	568.00	1.000	609.600	7.1400	0.0457	1.35	1.2110	0.00	0.00000
20.40	557.00	1.000	609.600	7.1400	0.0457	1.35	1.2110	0.00	0.00000
26.59	562.00	1.000	609.600	7.1400	0.0457	1.35	1.2110	0.00	0.00000
31.59	603.00	1.000	609.600	7.1400	0.0457	1.35	1.2110	0.00	0.00000
34.09	560.00	1.000	609.600	7.1400	0.0457	1.35	1.2110	0.00	0.00000
38.72	563.00	1.000	609.600	7.1400	0.0457	1.35	1.2110	0.00	0.00000
43.09	551.00	1.000	609.600	7.1400	0.0457	1.35	1.2110	0.00	0.00000
47.09	551.00	1.000	609.600	7.1400	0.0457	1.35	1.2110	0.00	0.00000
52.87	556.00	1.000	609.600	7.1400	0.0457	1.35	1.2110	0.00	0.00000
58.53	597.00	1.000	609.600	7.1400	0.0457	1.35	1.2110	0.00	0.00000
64.09	595.00	1.000	609.600	7.1400	0.0457	1.35	1.2110	0.00	0.00000
71.83	605.00	1.000	609.600	7.1400	0.0457	1.35	1.2110	0.00	0.00000
76.44	639.00	1.000	609.600	7.1400	0.0457	1.35	1.2110	0.00	0.00000
81.95	641.00	1.000	609.600	7.1400	0.0457	1.35	1.2110	0.00	0.00000
88.68	626.00	1.000	609.600	7.1400	0.0457	1.35	1.2110	0.00	0.00000
93.57	597.00	1.000	609.600	7.1400	0.0457	1.35	1.2110	0.00	0.00000
98.39	608.00	1.000	609.600	7.1400	0.0457	1.35	1.2110	0.00	0.00000
106.83	645.00	1.000	609.600	7.1400	0.0457	1.35	1.2110	0.00	0.00000
111.69	663.00	1.000	609.600	7.1400	0.0457	1.35	1.2110	0.00	0.00000
117.27	672.00	1.000	609.600	7.1400	0.0457	1.35	1.2110	0.00	0.00000
123.44	690.00	1.000	609.600	7.1400	0.0457	1.35	1.2110	0.00	0.00000
128.02	673.00	1.000	609.600	7.1400	0.0457	1.35	1.2110	0.00	0.00000
133.34	644.00	1.000	609.600	7.1400	0.0457	1.35	1.2110	0.00	0.00000
140.59	601.00	1.000	609.600	7.1400	0.0457	1.35	1.2110	0.00	0.00000
146.79	658.40	1.000	609.600	7.1400	0.0457	1.35	1.2110	0.00	0.00000
150.23	627.00	1.000	609.600	7.1400	0.0457	1.35	1.2110	0.00	0.00000
154.99	631.00	1.000	609.600	7.1400	0.0457	1.35	1.2110	0.00	0.00000
162.68	619.00	1.000	609.600	7.1400	0.0457	1.35	1.2110	0.00	0.00000
166.72	629.00	1.000	609.600	7.1400	0.0457	1.35	1.2110	0.00	0.00000

TABLE 6-12. (Continued)

Location (km)	Elevation (m)	Depth of Cover (m)	Outer Diameter (mm)	Wall Thickness (mm)	Pipe Roughness (mm)	Ground/Air Temperature (°C)	Ground Conductivity [W/(m °C)]	Insulation Thickness	Insulation Conductivity [W/(m °C)]
171.09	623.00	1.000	609.600	7.1400	0.0457	1.35	1.2110	0.00	0.00000
176.87	632.00	1.000	609.600	7.1400	0.0457	1.35	1.2110	0.00	0.00000
183.35	628.00	1.000	609.600	7.1400	0.0457	1.35	1.2110	0.00	0.00000
187.44	620.00	1.000	609.600	7.1400	0.0457	1.35	1.2110	0.00	0.00000
191.32	619.00	1.000	609.600	7.1400	0.0457	1.35	1.2110	0.00	0.00000
196.78	625.00	1.000	609.600	7.1400	0.0457	1.35	1.2110	0.00	0.00000
202.26	653.00	1.000	609.600	7.1400	0.0457	1.35	1.2110	0.00	0.00000
207.74	638.00	1.000	609.600	7.1400	0.0457	1.35	1.2110	0.00	0.00000
214.29	656.00	1.000	609.600	7.1400	0.0457	1.35	1.2110	0.00	0.00000
219.99	640.00	1.000	609.600	7.1400	0.0457	1.35	1.2110	0.00	0.00000
223.94	651.00	1.000	609.600	7.1400	0.0457	1.35	1.2110	0.00	0.00000
230.39	640.00	1.000	609.600	7.1400	0.0457	1.35	1.2110	0.00	0.00000
234.40	640.00	1.000	609.600	7.1400	0.0457	1.35	1.2110	0.00	0.00000
238.89	626.00	1.000	609.600	7.1400	#	#	#	#	#
244.72	685.00	1.000	609.600	7.1400	#	#	#	#	#
247.51	666.00	#	#	#	#	#	#	#	#

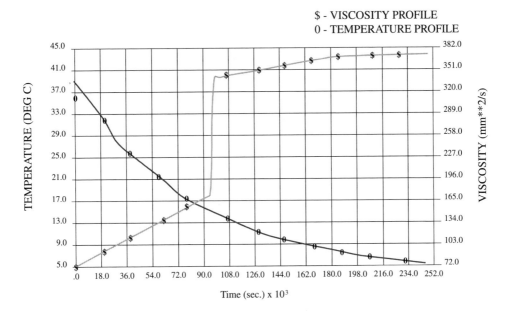

Figure 6-30. Non-Newtonian liquid pipeline hydraulics, temperature, and viscosity vs. time

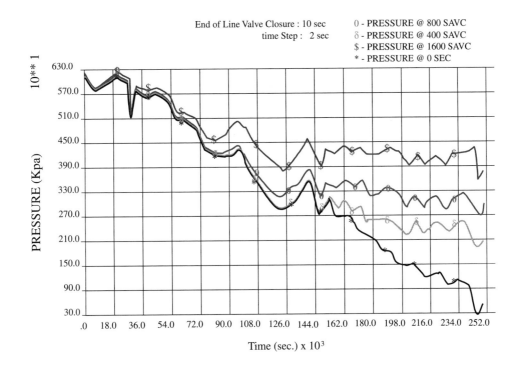

Figure 6-31. Non-Newtonian liquid pipeline transients, pressure vs. time

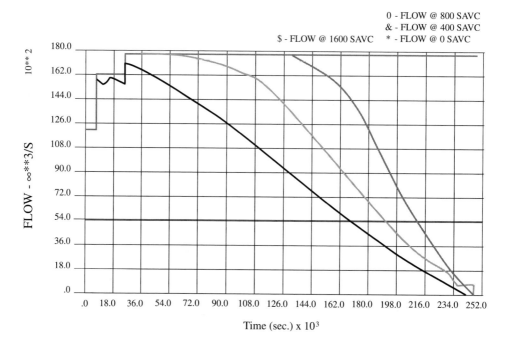

Figure 6-32. Non-Newtonian liquid pipeline transients, flow vs. time

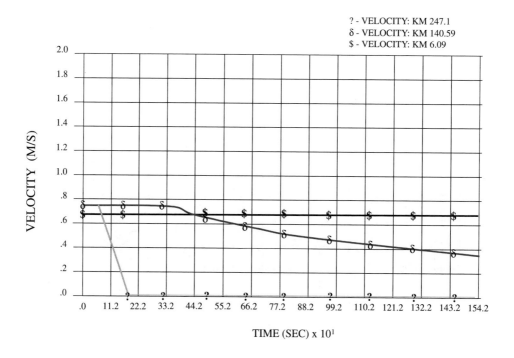

Figure 6-33. Non-Newtonian liquid pipeline transients, velocity vs. time

REFERENCES

Asante, B., 1995, "Justification for Internal Coating of Natural Gas Pipelines," Proc. ASME 14th OMAE Conf., Vol. V, Pipeline Technology, pp. 241–248.

British Standard BS 5760, 1994, "Reliability of Systems, Equipment and Component," (Part 2, B 25, 1992, 99:1992).

Burnett, R. R., 1960, "Predicting and Controlling Transient Pressure in Long Pipelines," Oil and Gas Journal, Vol. 58, No. 18, p. 153.

Cohn, A. R., and Nalley, R. R., 1980, "Using Regulators for Line Pressure Relief," Pipeline Industry 45.

Colebrook, C. F., 1939, "Turbulent Flow in Pipes with Particular Reference to the Transition Region Between the Smooth and Rough Pipe Laws," J. Institution Civil Eng.

Drivas, P. J., Salbris J., and Teuscher L. H., 1983, "Model Simulates Pipeline, Tank Storage Failures," Oil Gas Journal, p. 162.

Goldberg, D. E., 1985, "Quick Stroking," Pipeline & Gas Journal, p. 36.

Jungowski, W. M., Botros, K. K., and Weiss, M. H., 1989, "Simulation of Gas Pipeline Blowdowns," ASME ETC&E Conf. Pipeline Eng. Symp. P. D., Vol. 20, p. 75.

Kung, P., and Mohitpour, M., 1986, "Non-Newtonian Liquid Pipeline Hydraulics Design and Simulation Using Microcomputers," Proc. ETC&E Conf., Pipeline Eng. Symp., Vol. 3, pp. 73–78.

LIC Energy, 1999, "Pipeline Studio-transient Gas Network Pipeline Simulation (TGNET)," Users Manual Version 1.1. Houston, TX.

Mohitpour, M., 1972, "Effects of Strain Rate, Friction and Temperature Distribution in High Speed Axisymmetric Upsetting," Ph.D. Thesis, University of London, UK.

Mohitpour, M., 1995, "Pipeline Design and Construction Training Manual," Ed., NOVA Gas International Ltd., Calgary, Alberta, Canada.

Mohitpour, M., 1991, "Temperature Computations in Fluid Transmission Pipeline," ETC&E Conf. Pipeline Eng. Symp. P. D., Vol. 34, pp. 79–85.

Mohitpour, M., and Kung, P., 1986, "Gas Pipeline Hydraulics Design and Simulation," A Microcomputer Application, Proc. Conf. Continuous Sys. Simulation Language, Society of Compt. Sim. (SCS), St. Diego, CA, pp. 121–126.

Mohitpour, M., and McManus, M., 1995, "Pipeline System Design, Construction and Operation Rationalization," Proc. ASME 14th OMAE Conf., Vol. V, Pipeline Technology, pp. 459–467.

Mohitpour, M., Thompson W., and Asante, B., 1996, "The Importance of Dynamic Simulation on the Design and Optimization of Pipeline Transmission Systems," *Proc. ASME – OMAE 1st International Pipeline Conference*, Vol. 2, pp. 1183–1187.

Mohitpour, M., Kazakoff, J., Braun, A., and Brittin, R., 1999, "Gas Pipeline Design for Operational Reliability," 19th OMAE Conference, St. Johns, Newfoundland, Canada, Vol. 11–15.

Price, G. R., McBrien, R. K., Golsham, H., and Rizopoulos, S., 1996, "Improved Evaluation of Pipeline Friction Factors and Heat Transfer Coefficient Based on Transient Flow Simulation," Proc. ASME IPC Conf., Calgary, Canada.

Stoll, H. G., 1989, "Least-cost Electric Utility Planning," John Wiley & Sons, New York, NY.

Stuchly, J. M., and Schmatz R. A., 1984, "Enhancing the Versatility, Accuracy and Speed of Pipeline Transient Modeling," Proc. CETCE Conf. New Orleans, Feb. 11–14.

Uhl, A. E., 1965, "Steady Flow in Gas Pipelines," I.G.T.T. Report No. 10, American Gas Association (AGA).

Vagra, R. S., 1962, "Matrix Iterative Analysis," Prentice Hall, London, UK.

Wylie, E. B., and Streeter, V. L., 1990, "Fluid Transient," FEB PRESS Ann Arbor, MI.

Chapter *7*

PIPELINE MECHANICAL DESIGN

INTRODUCTION

This chapter applies specifically to the design of steel pipelines for transmission of gas and liquid petroleum products. It is not intended to be applied to other types of lines, such as low pressure, plastic, or distribution lines.

CODES AND STANDARDS

The design, material selection, and construction of pipeline facilities are governed by codes and standards that prescribe minimum requirements. The purpose of the codes and standards are to ensure that the completed structure will be safe to operate under the conditions to be used. Some examples of the codes and standards that are commonly used during the design and construction of pipelines have been cited in Chapter 1, Table 1-1.

Besides these, other regional, national, or international standards may be used in specific situations, while local regulations may also apply.

Individual companies may elect to develop in-house standards or specifications, which are more stringent than the minimum code requirements. Such standards are generally developed based on past experience and are intended to complement design and operating procedures.

In all cases, familiarity with applicable codes, standards, or specifications is required before the design commences.

LOCATION CLASSIFICATION

The most significant factor contributing to the failure of a gas pipeline is damage to the line caused by human activity. Pipeline damage generally occurs during construction of other services. These services may include utilities, sewage systems or road construction and will increase in frequency with larger populations living in the vicinity of the pipeline. To account for the risk of damage, the designer determines a

TABLE 7-1. Location classification for North American codes (Ref CSA Z662, 2007)

Location Classification	Canadian Code Z662-07 (Liquid & Gas)	American Code ASME B31.4 (Liquid)	ASME B31.8 (Gas)
Class 1	Dwelling units $\geq$ 10	Not applicable	Dwelling units $\geq$ 10
Class 2	10 < dwelling units < 46 Building or area occupied by more than 20 people. Industrial installation susceptible to environmentally hazardous conditions.	Not applicable	10 < dwelling units < 46.
Class 3	Dwelling units $\geq$ 46 Institutes that are difficult to evacuate.	Not applicable	Dwelling units $\geq$ 46
Class 4	Dwelling $\geq$ 4 stories	Not applicable	Building $\geq$ 4 stories. Dense traffic or underground facilities.

location classification based predominantly on population concentrations. Canadian and American codes differ marginally in the requirements for determining a specific location classification. Table 7-1 reflects the requirements of the relevant North American codes.

PIPELINE DESIGN FORMULA

The widely used formulae for determining the circumferential and axial stresses in a pressurized thin-walled pipe can be developed quite easily by considering the vertical and horizontal force equilibrium in Figures 7-1 and 7-2, respectively.

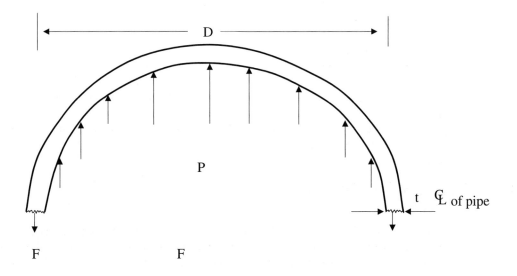

Figure 7-1. Force equilibrium in a pressurized thin pipe

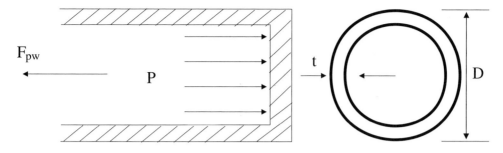

Figure 7-2. Longitudinal force equilibrium

In Figure 7-1, a unit tangential force F is created in the pipe wall due to the application of an internal pressure, P, assumed to act on a unit length of pipe. The resultant vertical force due to this pressure is PD, so the equilibrating force F in the pipe wall must be $PD\ell/2$. This force acts on an area of pipe wall, A, given by the product of the wall thickness t and unit depth (ℓ) of pipe. The tangential or hoop stress S_T in the pipe wall is F/A, or

$$S_T = \frac{PD}{2t} \qquad (7-1)$$

This is the hoop stress equation, which can be transposed to yield the familiar Barlow equation for the wall thickness of the pipe.

A similar consideration of the horizontal equilibrium of forces as shown in Figure 7-2 enables an expression for development of longitudinal stress in the cylinder.

The longitudinal force acting on the pipe end and caused by the internal pressure is approximately $F = P(\pi D^2/4)$, which is equilibrated by the force (F_{pw}) in the pipe wall. This force F_{pw} acts on an area approximated by πDt, so the axial stress S_x is

$$S_x = \frac{F}{A} \cong \frac{P\pi D^2}{4\pi Dt} = \frac{PD}{4t} \qquad (7-2)$$

A more accurate representation of the axial stress would be found by using the actual area of pipe wall resisting the longitudinal pressure force, that is,

$$S_x = \frac{F}{\dfrac{\pi(D^2-d^2)}{4}} = \frac{Pd^2}{D^2-d^2} \qquad (7-3)$$

where D and d are the outer and inner diameters, respectively. This is slightly less conservative than using the thin-wall approximation. In a similar fashion, the bending section modulus Z of a thin-walled cylinder can be approximated as $Z \cong \pi r^2 t$, where r is the mean radius, or in exact terms: $Z = \pi/32[(D^4-d^4)/D]$.

By way of comparison, consider a NPS 10 schedule 40 pipe subjected to an internal pressure of 1,000 psi. The longitudinal stresses for the exact and the thin-wall approximation are as follows:

Exact: $\quad S_x = \dfrac{Pd^2}{D^2 - d^2} = \dfrac{1{,}000 \times 10.02^2}{10.75^2 - 10.02^2} = 6{,}622$ psi

Thin wall: $\quad S_x = \dfrac{PD}{4t} = \dfrac{1{,}000 \times 10.75}{4 \times 0.365} = 7{,}363$ psi

or an 11.2% variation.

Similarly the variation in section modulus for the same pipe is

Exact: $\quad z = \dfrac{\pi}{32}\left(\dfrac{D^4 - d^4}{D}\right) = \dfrac{\pi}{32}\left(\dfrac{10.75^4 - 10.02^4}{10.75}\right) = 29.9$ in^3

Thin wall: $\quad Z = \pi r^2 t = \pi[(10.75 + 10.02)/4]^2 \times 0.365 = 30.9$ in^3

or a 3.4% difference.

The percentage variations on NPS10 schedule 80 pipe for longitudinal stress and section modulus are 16% and 5%, respectively. The degree of conservatism in the thin-wall approximation is worth noting when further consideration is given to the formulations in the various codes.

Note: The foregoing is applicable only to thin-wall pipe where it is reasonable to assume that the stress is uniform throughout the wall thickness. As a rule of thumb such an assumption is invalid for D/t ratios less than 16. In thicker-walled pipe (commonly found in offshore applications), the radial stress varies from a maximum at the inner surface to a minimum at the outer surface and an alternative formulation to Equation (7-1) is required.

Figure 7-3 illustrates an element in a thick-walled pipe for the general case of an internally applied pressure, p_i, and external pressure, p_e Timoshenko (1965), through the use of classical elasticity theory, has shown that the following expressions for radial stress S_r and tangential stress S_t apply to the element at any radius r.

$$S_t = \frac{p_i r_i^2 \left(r_0^2 + r^2\right) + p_e r_o^2 \left(r^2 - r_i^2\right)}{r^2 \left(r_o^2 - r_i^2\right)} \tag{7 – 4a}$$

and

$$S_r = \frac{-p_i r_i^2 \left(r_0^2 - r^2\right) - p_e r_o^2 \left(r^2 - r_i^2\right)}{r^2 \left(r_0^2 - r_i^2\right)} \tag{7 – 4b}$$

Since the numerator in Equation (7-4a) will always exceed that of Equation (7-4b), it follows that the tangential stress will always be larger than the radial stress.

Considering first the case where the pipe is subjected only to internal pressure, Equation (7-4a) becomes for ($r = r_i$).

$$S_t = \frac{p_i \left(r_o^2 + r_i^2\right)}{r_o^2 - r_i^2} \tag{7 – 5}$$

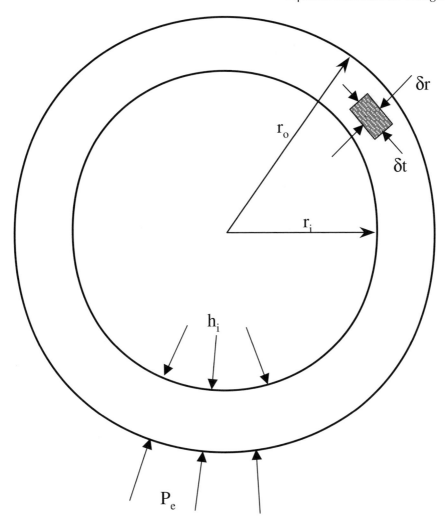

Figure 7-3. Tangential and radial stress in a thick-wall pipe

By substituting for r_o in terms of r_i and t and rearranging the terms, the following quadratic equation is obtained:

$$(S_t - p_i)t^2 + 2r_i(S_t - p_i)t - 2p_ir_i^2 = 0 \qquad (7-6a)$$

Solving for the wall thickness, t gives:

$$t = \frac{D}{2}\left(\left(\frac{S_t + p_i}{S_t - p_i}\right)^{0.5} - 1\right) \qquad (7-6b)$$

Equations (7-5) or (7-6b) may be rearranged to determine the bursting pressure for the pipe, which occurs when the tangential stress equals the ultimate tensile stress.

TABLE 7-2. Comparison of design factors in North American codes (Ref CSA Z662, 2007 and ASME B31.8, 2003)

Application	CSA Z662-07 (Gas & Liquid) Design Factor × Location Factor F × L				ASME B31.8-03 (Gas) and ASME B31.4-02 (Liquid) Design Factor, F			
	Class 1	Class 2	Class 3	Class 4	Class 1	Class 2	Class 3	Class 4
Gas (nonsour)								
General & cased crossings	0.80	0.72	0.56	0.44	0.80	0.60	0.50	0.40
Roads	0.60	0.50	0.50	0.40	0.60	0.50	0.50	0.40
Railways	0.50	0.50	0.50	0.40	0.60	0.50	0.50	0.40
Stations	0.50	0.50	0.50	0.40	0.50	0.50	0.50	0.40
Gas (Sour service)								
General & cased crossings	0.72	0.60	0.50	0.40	0.72	0.60	0.50	0.40
Roads	0.60	0.50	0.50	0.40	0.60	0.50	0.50	0.40
Railways	0.50	0.50	0.50	0.40	0.60	0.50	0.50	0.40
Stations	0.50	0.50	0.50	0.40	0.50	0.50	0.50	0.40
High Vapor Pressure Liquid General & cased crossings		0.64	0.64	0.64	0.72	0.72	0.72	0.72
Roads		0.64	0.64	0.64	0.72	0.72	0.72	0.72
Railways		0.50	0.50	0.50	0.72	0.72	0.72	0.72
Stations		0.64	0.64	0.64	0.72	0.72	0.72	0.72
Low vapor pressure liquid All but uncased RR crossings		0.80	0.80	0.80	0.72	0.72	0.72	0.72
Uncased railroad crossings		0.80	0.50	0.50	0.72	0.72	0.72	0.72

ASME B31.4-02 does not define a design factor, but opts to factor the specified minimum yield strength by 0.72. This factor has been incorporated into this table for comparison purposes only.

The following design formulae are used to calculate the wall thickness of a pipeline:

$$P = \frac{2St}{D} FLJT \ (Ref \ CSA \ Z662 - 07) \qquad (7-7)$$

where P = design pressure
S = specified minimum yield strength, SMYS
t = wall thickness
D = outside diameter of pipe
F = design factor = 0.80 (refer to Table 7-2)
L = location factor (refer to Table 7-2)
J = longitudinal joint factor (refer to Table 7-3)
T = temperature derating factor (refer to Table 7-4).

From (ASME B31.8 2003):

$$P = \frac{2St}{D}FET \tag{7 - 8}$$

where P, S, t, and D are defined as above and
   F = design factor (based on the location classification; refer to Table 7-2)
   E = longitudinal joint factor (refer to Table 7-3)
   T = temperature derating factor (refer to Table 7-4).

And from (ASME B31.4 2002):

$$P = \frac{2St}{D}FE \tag{7 - 9}$$

where F = design factor = 0.72 (refer to Table 7-2)
   E = longitudinal joint factor (refer to Table 7-3).

The temperature derating factor is 1.0 provided:

$$-30^{\circ}C \leq T_{\text{pipe}} \leq 120^{\circ}C \text{ (refer to Table } 7 - 4).$$

where T_{pipe} = pipe temperature.

The maximum operating pressure (MOP) is used as the design pressure, P. The specified minimum yield strength S is chosen either from what is stockpiled or from what is available from the pipe mills. Commonly available yield strengths are 241, 289, 317, 414, 448, and 483 Mpa, which correspond to grades X35, X40, X50, X60, X65, and X70. The number following the X indicates the specified minimum yield strength of the material. The higher strengths are generally utilized on large-diameter pipelines.

The pipeline diameter, D, is established during the system planning stage (refer to Chapter 3).

The location factor, L, in CSA Z662, and the design factor, F, in ASME B31.8, depend on the location classification of the pipeline. No equivalent factor is defined in ASME B31.4, where an allowable stress is defined as 0.72 SMYS. Table 7-2 outlines the differences between the location-based factors for the North American codes. Code requirements should be reviewed in detail for specific exceptions and clarification.

TABLE 7-3. Comparison of longitudinal joint factors in North American codes (Ref CSA Z662, 2007 and ASME B31.8, 2003)

Pipe Type	Longitudinal Joint Factor, J (CSA Z662-07) or E (ASME B31.4-02 and B31.8-03)	
	CSA Z662-07	ASME B31.4 and ASME B31.8
Seamless	1.00	1.00
Electric welded	1.00	
Resistance welded		1.00
Induction/flash welded		1.00
Fusion arc welded		0.80
Fusion welded		0.80–1.00[1]
Submerged arc welded	1.00	1.00
Spiral welded[2]		0.80
Furnace butt welded	0.60	0.60

[1] Value of E is dependent on the ASTM material classification.
[2] ASME B31.8 only.

TABLE 7-4. Comparison of temperature derating factors for North American codes (steel pipe) (Ref CSA Z662, 2007 and ASME B31.8, 2003)

Temperature (°C)	Temperature Derating Factor, *T*	
	CSA Z662-07 ASME B31.8	ASME B31.4
> 120	1.00	
150	0.97	
180	0.93	
200	0.91	
130	0.87	
> 30		1.00
< 120		1.00

The longitudinal joint factor *J* in CSA Z662 and *E* in the ASME codes are generally 1.0 for commonly used pipe types. However there are exceptions. Table 7-3 outlines the variations in the joint factor for the North American Codes.

The temperature derating factor, *T*, is defined only in CSA Z662-07 and ASME B31.8. ASME B31.4 defines a range of temperatures for which Equation (7-9) is applicable. The range of temperatures defined by B31.4 is not as broad as that in the other North American codes. However, the requirements of B31.4 are consistent with the requirements of the other codes. Table 7-4 outlines the variations in the temperature-derating factor for the North American codes.

Both CSA and ASME codes specify minimum allowable wall thicknesses for various pipe diameters. The intent is to minimize pipe damage during normal manufacturing, handling, construction, and operation.

EXPANSION AND FLEXIBILITY

Piping has to be designed to have sufficient flexibility to prevent thermal expansion or contraction from causing excessive stresses in the piping material or imposing excessive forces or moments on equipment or supports. In many cases, allowable forces and moments on equipment may be less than those for the connecting piping.

If expansion is not absorbed by direct axial compression of the pipe, flexibility should be provided by the use of bends, loops, offsets, and, in rare instances, by mechanical joints or couplings.

In simple terms, a flexibility analysis determines a suitable piping layout so as to minimize pipe stresses. Such an analysis evaluates the range of stresses the piping system will encounter while undergoing cyclic loading. The most common of these ranges is the thermal expansion stress range caused by system start-up and shut-down conditions. Often the stress range encompassed by the consideration of the cold (start-up) and hot (shut-down) conditions will be sufficient to cover other thermal expansion conditions such as standby, or variations in operating envelopes. However, any cyclic condition that could result in a significant stress range, irrespective of whether or not the source is thermal expansion, will require evaluation. Hence, any event that could cause relative displacements between anchor points, such as settlement or seismic disturbance, would fall into this category. Some worked examples of sizing expansion loops are given later in this chapter. After determining that the piping layout has adequate flexibility, the designer is free to use judgement, span tables, or simple formulae to locate suitable positions for intermediate supports to carry weight and other loads. The nature of these supports will be dependent upon the magnitude of the likely pipe displacements at their proposed locations.

For example, if these are very small then a rigid support may be used. Springs or gapped supports may be used to accommodate vertical or lateral displacements. The overriding consideration is that the support type must offer minimal restraint to the system as it undergoes cyclic loading, in order that the stress range is not unduly increased. Detailed consideration of support spacings and anchoring can be found later in this chapter.

There are fundamental differences in loading conditions for buried or similarly restrained portions of piping and the above-ground portions not subjected to substantial axial restraint. Therefore, different limits on allowable expansion stresses are necessary. Note that the ASME B31.8 code does not clearly differentiate between restrained or unrestrained lines. Rather, it is the operator's responsibility to define the type of restraint present, determine the loads causing axial stress and limit them appropriately. The assumption is made in both B31.4 and B31.8 that a buried pipeline is restrained and an above-ground pipeline is unconstrained, but this is not always an appropriate assumption.

Restrained Lines

Expansion calculations are necessary for buried lines if significant temperature changes are expected. Thermal expansion of buried lines may cause movement at points where the line terminates, changes direction, or changes size.

The hoop stress due to design pressure is determined in accordance with Equation (7-1):

$$S_h = \frac{PD}{2t}$$

For a pressurized pipe, the radial growth of the pipe will induce an opposite longitudinal (Poisson) effect to that caused by thermal effects. This will reduce the compressive stress such that

$$S_L = \nu S_h - E_c \alpha (T_2 - T_1) \qquad (7-10)$$

where S_L = restrained longitudinal compression stress [given by Equation (7-2)]
ν = Poisson's ratio
E_c = cold modulus of elasticity of steel
α = linear coefficient of thermal expansion
T_2 = maximum operating temperature
T_1 = ambient temperature at time of restraint (e.g. during backfilling)
t = wall thickness.

In order to limit the combined equivalent stress on the pipe, various codes require that:

$$S_h - S_L \leq 0.9\, S \times T \qquad (7-11)$$

where S = specified minimum yield strength
T = temperature derating factor

Note that this formula does not apply if S_L is positive (i.e. tension). Its derivation lies in considering the bi-axial state of stress in the pipe wall caused by the circumferential hoop stress and the longitudinal stress. The Tresca theory of failure states that failure will occur when the maximum shearing stress reaches the yield stess of the pipe material. The maximum shearing stress is the absolute value of the difference of the principal stresses in the pipe wall. To arrive at a maximum allowable shearing stress, $S_h - S_L$ is reduced by a design factor, in this case 0.9.

For those portions of restrained pipelines that are freely spanning or supported above ground, the combined stress is limited in accordance with the following formula:

$$S_h - S_L \pm S_B \leq 0.9\,S \times T \qquad (7-12)$$

where S_B = absolute resultant value of beam bending stresses caused by dead and live loads acting in and out of plane on the pipe, and is given as follows:

$$S_B = \left[(0.75\,i_i\,M_i)^2 + (0.75\,i_o\,M_o)^2 + M_t^2 \right]^{0.5} \Big/ z \qquad (7-13)$$

Note that M_i, M_o, and M_t are the in plane, out of plane, and torsional moments, respectively, acting on the pipe and Z is the section modulus. The terms i_i and i_o refer to the in and out of plane stress intensification factors that have been identified for various critical piping components, usually by experimental means. They can be found for a variety of components such as elbows, mitres, and fabricated tees in Appendix 'E' of ASME B31.8 (2003).

Equations (7-10) through (7-13) apply for both CSA Z662 and ASME B31.4, with one exception: ASME B31.4 does not include a temperature derating factor, but limits the range of applicable temperatures to under 120°C. It should also be noted that ASME B31:8 does not distinguish between restrained and unrestrained lines.

Unrestrained Lines

Expansion calculations for above-ground lines have to account for thermal changes as well as beam bending and the possible elastic instability of the pipe and its supports (due to longitudinal compressive forces).

CSA Z662, ASME B31.8, and ASME B31.4 segregate requirements for combining stresses in terms of unrestrained lines. However, the requirements vary slightly between the two codes and will therefore be addressed separately.

In both CSA Z662 and B31.8, the stresses due to thermal expansion for those portions of pipeline systems without axial restraints are combined and limited in accordance with the following formulae:

For bending

$$S_b = \frac{iM_b}{Z} \qquad (7-14)$$

where S_b = resultant bending stress
  i = stress intensification factor
  M_b = resultant bending moment
  Z = section modulus of pipe.

For twisting

$$S_t = \frac{M_t}{2Z} \qquad (7-15)$$

where S_t is the torsional stress and M_t is the twisting moment.

The combined thermal expansion stresses can be combined as follows:

$$S_E = \left(S_b^2 + 4S_t^2\right)^{\frac{1}{2}} \tag{7 – 16}$$

where S_E = combined thermal expansion stress.

However, the combined thermal expansion stress is not permitted to exceed 0.72 of the specified minimum yield stress times the temperature derating factor.

In ASME B31.4, the bending stresses due to thermal expansion for those portions of pipeline systems without axial restraint have to be combined and limited in accordance with Equation (7-17) (for temperatures less than 120°C):

$$S_B = \left\{ (i_i M_i)^2 + (i_o M_o)^2 \right\}^{0.5} \Big/ z \tag{7 – 17}$$

where S_B = resultant bending stress
$\quad\quad\quad M_i$ = the in-plane bending moment
$\quad\quad\quad\ i_i$ = the stress intensification factor for bending in the plane of the member
$\quad\quad\quad M_o$ = bending moment out of, or transverse to, the plane of the member
$\quad\quad\quad\ i_o$ = stress intensification factor for out-of-plane bending.

The following limiting cases must also be satisfied;

$$0.5\, S_H + S_{B(D)} \leq 0.54\, S \tag{7 – 18}$$

where $S_{B(D)}$ = absolute value of beam bending compression stress resulting from dead loads

A further limitation is caused by the addition of live loading so that

$$S_E + S_L + S_{B(D + L)} \leq S \tag{7 – 19}$$

where S_L = longitudinal stress $PD/4t$ and $S_L + S_{B(D + L)} \leq 0.85S$
$\quad\quad S_{B(D + L)}$ = absolute value of longitudinal bending stress from both dead and live loading.

Both CSA and ASME codes specify minimum allowable wall thickness for various pipe diameters. The intent is to minimize pipe damage during normal manufacturing, handling, construction, and operation.

Sizing an Expansion Loop

Before the widespread use of computer-based methods of structural analysis, designers had to rely on hand calculations and a number of simplifying assumptions in order to estimate pipe flexibility. The ASME B.31 code only deemed such calculations, or model testing, to be necessary in those circumstances where some doubt existed as to the adequate flexibility of the proposed system. The codes indicate that adequate flexibility may be assumed if a piping system is:

- of uniform size
- has two anchor points

- has intermediate restraints and satisfies the following approximate criterion:

$$\frac{DY}{(L-U)^2} \leq \frac{30\,S_A}{E_C} \qquad (7-20)$$

For ferrous metals this simplifies to:

$$\frac{DY}{(L-U)^2} \leq 0.03\ (\text{Imperial})\ \text{ or } \leq 208.3\ (\text{SI units}) \qquad (7-21)$$

where
$S_A = f\,(1.25\,S_c + 0.25\,S_H)$
D = nominal pipe size in inches (mm)
Y = thermal expansion of the pipe, inches (mm)
S_c = 0.67 ST at the minimum temperature
S = specified minimum yield stress
L = developed (total) length of pipe, feet (m)
U = distance between anchors, feet (m)
E_C = cold modulus of elasticity, psi (MPa).
f = a stress range factor to accomodate thermal cycling fatigue effects

Cycles, N	Factor, f
7,000 and less	1.0
> 7,000 to 14,000	0.9
> 14,000 to 22,000	0.8
> 22,000 to 45,000	0.7
> 45,000 to 100,000	0.6
> 100,000 to 200,000	0.5
> 200,000 to 700,000	0.4
> 700,000 to 2,000,000	0.3

Figure 7-4. shows a simple layout illustrating these parameters. Use of the above formula may be unconservative for a pipe having a large D/t ratio and should also be checked to ensure that the anchor reactions are satisfactory.

Consider the sizing of an expansion loop required for a NPS 16 pipe of API 5LX60 material supported between anchor points 60 m apart (Figure 7-4). It is assumed that the installation temperature is 15°C and the operating temperature is 80°C.

For a carbon or low alloy steel the rate of thermal expansion for the piping between 15°C and 80°C is approximately 75 mm/100 m. The restrained anchor movement y is therefore

$$y = U \times \frac{75}{100} = \frac{60 \times 75}{100} = 45 \text{ mm}$$

Substituting D, L, U, and y into Equation (7-21) and noting that $L = U + 2\,h$.

$$\frac{406(45)}{(60+2\,h-60)^2} \leq 208.3$$
$$87.79 = 4h^2$$
$$h = 4.68 \text{ m}$$

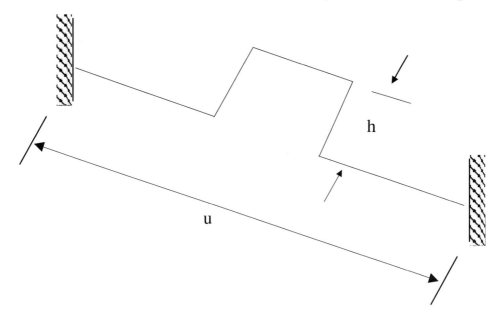

Figure 7-4. Expansion loop

An alternative method of calculation involves treating each leg of the expansion loop as guided cantilevers of length h and assuming that each cantilever absorbs half the thermal expansion between the anchors (Figure 7-5).

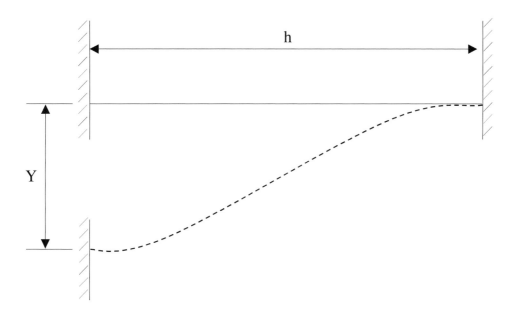

Figure 7-5. Guided cantilever end displacement

TABLE 7-5. Summary of end forces for high pressure pipelines (Schnockenberg, 2002)

Pipe (API 5LX 60)	P (psig)	T_o (°F)	F_s (lbf/ft.)	σ_H (psi)	σ_{LA} (psi)	σ_{LB} (psi)	L (ft.)	Δ(in.)	F (lbf.)
16 inches OD × 0.312 inches w.t.	1,150	162	142.2	28,334	−6,957	14,167	2,285	9.81	324,890
20 inches OD × 0.344 inches w.t.	1,000	162	222.2	28,070	−7,036	14,035	2,014	8.62	447,550
16 inches OD × 0.281 inches w.t.	1,000	162	142.2	27,470	−7,216	13,735	2,045	8.71	290,800
16 inches OD × 0.250 inches w.t.	880	162	142.2	27,280	−7,273	13,640	1,819	7.73	258,690
16 inches OD × 0.688 inches w.t.	2,800	150	142.2	29,758	−4,268	14,879	4,457	17.66	633,770
10 inches NB × 0.500 inches w.t.	2,800	150	64.2	28,700	−4,585	14,350	4,728	18.60	304,850
6 inches NB × 0.432 inches w.t.	2,800	150	24.4	18,670	−7,594	9,335	5,828	20.41	142,200

OD = Outside Diameter, NB = Nominal Bore.

The flexibility stress induced in the cantilever due to an offset Y is found in Roark and Young (1975) to be

$$S_F = \frac{DYE_C}{48h^2} \leq S_A \text{ (Imperial) or } \frac{3DYE_C}{h^2} \leq S_A \text{ (S I)} \qquad (7-22)$$

where

S_A = allowable stress range = $f(1.25\, S_C + 0.25\, S_H)$
D = nominal pipe size, inches (mm)
Y = thermal displacement of the anchor, inches (mm)
$S_C = 0.67\, S\, T$ at the minimum temperature
$S_H = 0.67\, S\, T$ at the maximum temperature
S = specified minimum yield stress of material (60,000 psi (4.14 MPa) in the example)
T = stress derating factor due to temperature (Table 7-4)
h = leg length of the loop, feet (mm)
f = 1.0 for a system with less than 7,000 cycles

Since $T = 1.0$ for this steel over the temperature range of 15–80°C, $S_C = S_H$ hence

$$S_A = 1.0\{(1.25 \times 0.67 \times 60,000 \times 1.0) + (0.25 \times 0.67 \times 60,000 \times 1.0)\}$$
$$= 60,000 \text{ psi or } 4.14\, \text{MPA}.$$

Note that the restraint movement will be half that computed previously (i.e., 22.5 mm). Substituting in Equation (7-22) and solving for h,

$$\frac{3 \times 406 \times 22.5 \times 208 \times 10^3}{h^2} \leq 414$$

from which $h = 3.69$ m.

For a fully restrained line the longitudinal stress in the pipe is made up of two components: a tensile stress due to the Poisson effect from the internal pressure and a compressive stress caused by temperature change. It can be determined from the following expression:

$$S_L = \frac{\nu P D_i}{2t} - E \alpha \Delta T \qquad (7-23)$$

where

ν = Poisson's ratio (0.3 for steel)
E = modulus of elasticity, psi (MPa) (200×10^3 MPa for steel pipe)
α = linear coefficient of thermal expansion (11.7×10^{-6} mm/mm °C)
ΔT = temperature change, °C.

P, D and t are the internal pressure, the inside diameter, and the wall thickness of the pipe, respectively. A negative sign for S_L indicates that the pipe is in compression, which will generally be the case for a pipe operating at a temperature greater than that prevailing at the time of installation. The maximum effective stress S_{EFF} on the pipe occurs when the absolute largest value is obtained from the following equation:

$$S_{EFF} = S_H - S_L \qquad (7-24)$$

where

S_H = the hoop stress in the pipe. As noted this requirement corresponds to a maximum shear stress criterion.

Anchoring and Support

Reinforced-concrete blocks are often used to serve as anchors. However, depending on soil conditions, other types of anchoring, such as steel pile and bracket, may be employed.

All above-ground piping, barrels, and valves must be adequately supported. Supports usually bear on footings, slabs, or piles. The piping must be restrained from lateral movement. The supports must be insulated from the piping to isolate the facility electrically from ground faults.

Stresses and deflections occur in pipelines at the transition from the below-ground (fully restrained) to the above-ground (unrestrained) condition. An analysis of the stresses and deflections in transition areas, resulting from internal pressure/temperature changes, is necessary to determine anchor block requirements and pipe size. Longitudinal deflections are used to check if an anchor block is required. The forces required to maintain the pipe in a fully restrained condition are then used to size the anchor block.

Case 1 – No Anchor

Consider a length of pipe capped at C, as shown in Figure 7-6. The section up to Point A is fully restrained, while the portion immediately to the right of B is unrestrained. Between A and B there is a transition from being restrained to unrestrained. Note that fully restrained is taken to mean a condition of zero longitudinal strain.

As this line is pressurized and heated the pipe will elongate and end B will move to the right an amount ΔL from some point of complete fixity, A, which is at a distance L_1 from B. This distance, L_1, to the point where the line can be considered fixed is a function of all of the parameters in Equation (7-23) plus the soil friction or resistance to pipe movement.

In the unrestrained portion of the pipe, B-C, the longitudinal stress caused by the internal pressure will be half the hoop stress, less that due to the Poisson effect. The strain in the longitudinal direction due to internal pressure is therefore

$$\varepsilon_{PR} = \frac{S_H}{2E} - \nu \frac{S_H}{2E} \qquad (7-25)$$

while the strain caused by any change in temperature ΔT is given by

$$\varepsilon_{TH} = \alpha \Delta T$$

The net longitudinal strain at Point B will therefore be

$$\varepsilon_B = \alpha \Delta T + \frac{S_H}{2E}(1 - \nu) \qquad (7-26)$$

The longitudinal strain at point A is of course zero.

The transition of stress and strain between Points A and B is assumed to vary as a linear function of length, L, as shown in Figure 7-7.

In order to establish the length, L, over which the transition occurs, the longitudinal resistance of the soil must be known. A simplifying assumption is to consider the soil constraint to be constant per unit length so that any tendency to move will be counteracted by a constant and opposite soil force. Wilbur (1983) has recommended a design value for the resistance of average soils (in SI units):

$$F_S = 0.0813 \, D_o^2$$

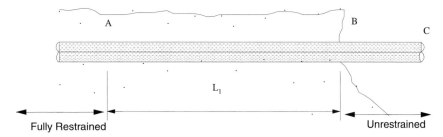

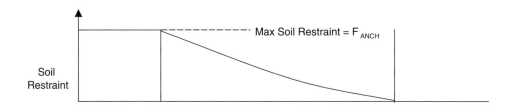

Figure 7-6. Transition from full to unrestrained

where F_S = soil resistance/unit length of pipe, kN/m
D_o = outside pipe diameter, m.

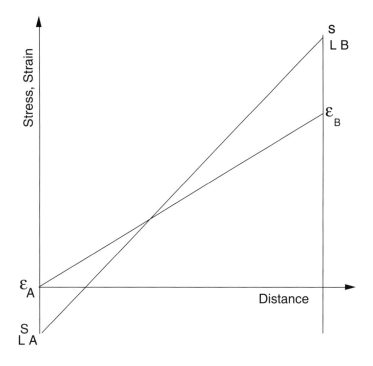

Figure 7-7. Stress and strain between points A and B

Between A and B, equilibrium of forces exists and can be described as

$$F_S L = A_P (S_{LB} - S_{LA})\qquad(7-27)$$

or

$$L = A_P \frac{(S_{LB} - S_{LA})}{F_S}\qquad(7-28)$$

where the longitudinal stress at B, caused only by internal pressure is $S_H/2$ while $S_{LA} = \nu S_H - E\,\alpha\,\Delta T$

where A_p = cross-sectional area of the pipe
         L = length of pipe between fully restrained and unrestrained regions.

The total movement at B will be the average strain between A and B over length L, given by $\delta = \varepsilon_B/2$, which is exactly half what the movement at B would have been had the soil restraint been zero.

Case II — With Anchor

When an anchor is used to restrict longitudinal deflections, the stress distribution will be that shown in Figure 7-8.

The transition from fully restrained to unrestrained occurs at the anchor. The resultant force on the anchor is simply the difference in stress on each side multiplied by the cross section area of the pipe.

The anchor force is thus given by

$$F = (S_{LB} - S_{LA})A_P\qquad(7-29)$$

which can be written for the case of a capped end pipe

$$F = -\frac{P\left(\pi D_i^2\right)}{4} + \left(\nu\frac{PD}{2t} - E\alpha\,\Delta T\right)A_P\qquad(7-30)$$

In the case of an increase in wall thickness beyond the anchor block, the result is essentially the same. The decrease in stress is compensated by an increase in pipe metal area.

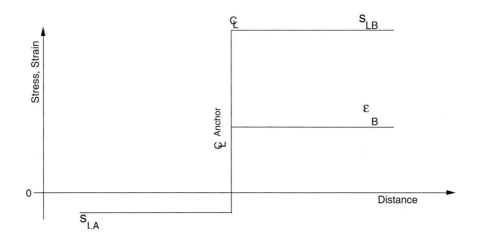

Figure 7-8. Distribution at anchor location

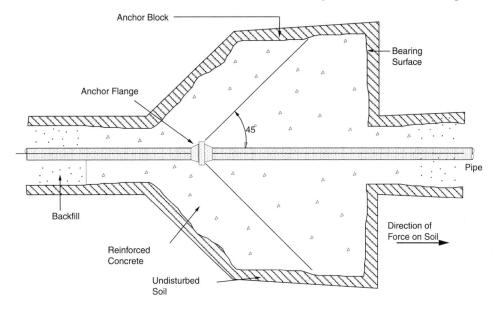

Figure 7-9. Plan view of anchor block (Schnockenberg, 2002)

Table 7-5 provides a sample summary of end forces for high pressure pipelines. As shown in the table, anchor block forces can be quite large. Therefore, careful attention must be paid to the design of the anchors. In order to minimize the size of the block, upper limits of lateral soil bearing pressures should be considered. Friction between the block and the soil can be used to reduce the size of the block.

The block should be reinforced concrete, cast against undisturbed soil. Figures 7-9 and 7-10 depict a typical anchor block.

Care should be taken to ensure that connected surface piping has sufficient flexibility to absorb a degree of lateral anchor movement. Scraper traps (see later), should be installed such that they move with the piping, rather than being rigidly attached to blocks. Instances of traps, together with their support blocks, being displaced a few centimeters are not uncommon.

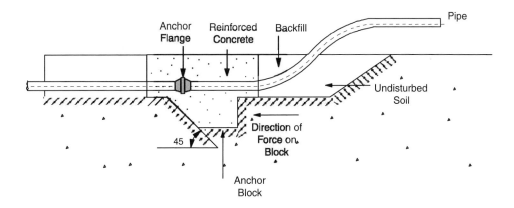

Figure 7-10. Cross-section through anchor block

Support Spacing

The various codes provide little guidance when selecting the spacing of above- or below-ground supports. For example, CAN/CSA Z184-M92 simply states "supports shall be designed to support the pipe without causing excessive local stresses in the pipe." The piping is thus identified as being the weaker member since, in general, the support can be designed to withstand practically any vertical load. The structural bearing capacity of the pipe will therefore govern the maximum allowable support load. TransCanada PipeLines' practice (Kormann and Zhou 1995) has been to use a spacing equation derived from considerations of localized stresses in the pipe and the principles contained in the ASME Boiler & Pressure Vessel Code (1992). Roark and Young (1975) reference the following relationships with respect to a cylindrical shell subject to a line load over a very short length in the longitudinal direction. The membrane and bending stress components at the loading "point" are, respectively:

$$S_M = -0.13\, BPR^{0.75}\, b^{-1.5}\, t^{-1.25} \tag{7-31}$$

and

$$S_B = -B^{-1}\, PR^{0.25}\, b^{-0.5}\, t^{-1.75} \tag{7-32}$$

where R = pipe radius, mm
t = wall thickness, mm
P = reaction force, or applied load, N
$2b$ = width of the applied load, which corresponds to the length of the support, usually $3R$
$B = \lfloor 12(1-\nu^2)\rfloor^{0.125} = 1.348273$ for steel with $\nu = 0.3$.

The applied load P can be generalized by assuming a backfill density, ρ_M, of 15 kN/m^3, and a steel density ρ_S, of 77 kN/m^3. In the case of a gaseous fluid, the weight of the internal contents of the pipe is negligibly small, but for liquids it is necessary to add the weight of fluid to the self-weight of the pipe. Hence for a pipe on continuous supports spaced L distance apart, the applied load P is given by P = weight of overburden of pipe and contents (if applicable).

For a gas:

$$P = 15\, DLH + 77\, \pi DtL \cong 15\, DLH \times 10^{-6} + 240\, DtL \times 10^{-6}$$
$$P = 15\, DL\,(H + 16t) \times 10^{-6} \tag{7-33}$$

where H is the depth of soil cover (mm), L is the support spacing (mm), and D is the pipe diameter (mm).

Substituting for P and B in Equations (7-31) and (7-32) gives:

$$S_M = -2.41 \times 10^{-6}\, L(H + 16t)D^{0.25}\, t^{-1.25} \tag{7-34}$$

and

$$S_B = -10.803 \times 10^{-6}\, L(H + 16t)D^{0.75}\, t^{-1.75} \tag{7-35}$$

The circumferential stress in the pipe can be written as:

$$S_{\text{CIRC}} = S_H - S_B + S_M \tag{7-36}$$

The code requirements limits the hoop stress S_H in piping at compressor and meter stations to 0.5 SMYS. Hence,

$$S_{\text{CIRC}} = 0.5 \text{ SMYS} + 10.803 \times 10^{-6} L(H + 16t)D^{0.75} t^{-1.75} \left[1 - 0.223(t/D)^{0.5}\right]$$

$$(7-37)$$

For practical D/t ratios, the last term is negligible so

$$S_{\text{CIRC}} = 0.5 \text{ SMYS} + 10.803 \times 10^{-6} L(H + 16t)D^{0.75} t^{-1.75} \qquad (7-38)$$

Similarly the allowable or limiting value of the circumferential stress is set by the codes. In the case of combined bending and membrane stresses, the ASME Boiler and Pressure Vessel Code sets the design stress as

$$S_{\text{ALLOWABLE}} = 1.5\,S_M \qquad (7-39)$$

For compressor and meter station piping, the limiting value for S_M is given in CAN/CSA Z184-92 as,

$$S_M = 0.8 \times 0.625 \times \text{ SMYS} = 0.5 \text{ SMYS} \qquad (7-40)$$

The limit for the total circumferential stress is therefore from Equations (7-38) and (7-39).

$$S_{\text{CIRC}} = S_{\text{ALLOWABLE}} = 1.5 \times 0.5 \text{ SMYS} = 0.75 \text{ SMYS} \qquad (7-41)$$

Inserting this value into Equation (7-38) it can be rearranged to give,

$$L = 23,142 \frac{SD}{(H + 16t)}(D/t)^{-1.75} \qquad (7-42)$$

where S is the specified minimum yield stress (SMYS) in MPa. Note that S_{CIRC} could be tensile or compressive.

For above-ground supports it is necessary to omit the effects of overburden in Equation (7-34) and rework the various equations to yield,

$$L = 23,142 \frac{SD}{16t}(D/t)^{-1.75} \qquad (7-43)$$

There are of course alternative formulations to Equation (7-43). One commonly used formula for estimating free span is:

$$L = 0.123\sqrt{SD} \qquad (7-44)$$

where L = support spacing, m
   S = specified minimum yield stress (SMYS), MPa
   D = pipe outside diameter mm.

In its derivation this equation disregards any localized stress effects, assumes zero axial constraint, and combines the primary membrane and bending streses such that they meet Tresca's (maximum shear stress) yield criterion.

$$S_{\text{TRESCA}} = S_{\text{MAX PRINCIPAL}} - S_{\text{MIN PRINCIPAL}} = S_H - S_L \qquad (7-45)$$

As mentioned above, the various codes limit S_H to 50% SMYS in the vicinity of compressor and meter stations while the longitudinal stress is the sum of the bending stress in the pipe due to its weight plus the membrane stress caused by the pressure acting on the end cap closure:

$$S_L = \frac{M}{I}D/2 + 0.5\,S_H \qquad (7-46)$$

where I = second moment of area of the pipe $\cong \pi D^3 t/8$.

Since the pipe is assumed to be continuously supported on a uniform span spacing of L the maximum bending moment M will be

$$M = 0.107\,w\,L^2 \qquad (7-47)$$

where w is the weight/unit length of the pipe and contents (if liquid). Neglecting the weight of its contents the maximum moment on the pipe will be,

$$M = 0.107 \times \rho_S\,\pi\,DtL^2 = 0.107 \times 77 \times \pi DtL^2 = 25.88\,DtL^2 \times 10^{-6} \qquad (7-48)$$

The longitudinal stress becomes

$$S_L = \frac{32.9 \times 10^{-6}L^2}{D} + 0.5 \times 0.5\,S \qquad (7-49)$$

Following the same argument used to develop Equation (7-43), the allowable stress in Tresca's yield criterion is limited to 1.5 times 50% SMYS or 0.75 SMYS. Equation (7-45) can then be written as

$$0.75\,S = 0.5\,S - \left\{0.25\,S - 32.9\frac{L^2}{D} \times 10^{-6}\right\}$$

Solving for L,

$$L = \sqrt{\frac{SD}{65.8}} \times 10^3 \text{ mm} = 0.123\sqrt{SD} \text{ m,}$$

as per Equation (7-44).

Above-ground supports have two additional functions besides supporting the weight of the piping: First, they provide protection to the termination points of compressors and vessels by restricting lateral and sometimes axial movement of the piping. Second, they help prevent excessive vibration of the piping. This is accomplished by choosing span lengths such that the natural vibration frequencies of the piping system is well removed from that of any source of excitation. Tuning the fundamental frequency of the piping to be greater than 30 Hz is a useful "rule of thumb," since excitation at frequencies above this value rarely has sufficient energy to create large displacements.

TABLE 7-6. Comparison of support spacings equations

Pipe	D (mm)	t (mm)	S (MPa)	Spacing Eq. (7-43) (m)	Spacing Eq. (7-44) (m)	Spacing Eq. (7-53) (m)
1	610	18.3	290	30.2	51.7	8.2
2	610	11.0	483	33.4	62.9	8.2
3	914	19.2	414	32.98	75.6	10.1
4	914	16.5	483	34.37	81.7	10.1

The fundamental frequency of an individual span of a continuously supported piping system can be estimated by assuming that the end supports are fixed.

$$\omega_1 = \left(\frac{4.73}{L}\right)^2 \sqrt{\frac{EI}{\rho A}} \qquad (7-50)$$

where ω = the fundamental angular frequency (rad/s), A is the cross-sectional area of the pipe $\cong \pi D t$, I is the second moment of area $\cong \pi D^3 t/8$ (mm^4), and the other symbols are as defined earlier in this chapter. Substituting for E, ρ, A and I in Equation (7-50) yields

$$f = \frac{\omega}{2\pi} = 6.37 \times 10^6 \frac{D}{L^2} \qquad (7-51)$$

By limiting frequency f to greater than 30 Hz, Equation (7-51) can be solved for L:

$$L = \frac{D^{0.5} \times 10^3}{2.17} \qquad (7-52)$$

If in Equation (7-52) D and L are expressed in units of millimeters and meters, respectively, and the denominater is replaced by the more easily remembered 3 (which also serves to increase the frequency of the system), a simple design formula based on vibration avoidance would be

$$L = \frac{D^{0.5}}{3} \text{ m} \qquad (7-53)$$

Table 7-6 illustrates the variation in pipe spans for a few typical pipe sizes using the various equations developed above, and clearly shows that large pipe spans result if localized stresses and vibration considerations are neglected and only the Tresca criterion is used.

JOINT DESIGN FOR PIPES OF UNEQUAL WALL THICKNESS

Background

Situations arise where the joining of two pipelines of equal diameters but of unequal wall thickness is required. The usual industrial practice is to specify a particular length of counterbore that will ensure that local moments at the joint (caused by the interaction of longitudinal stress) and the centroidal eccentricity of dissimilar wall thicknesses are reduced to a satisfactory level at the joint location. It has been shown (Colquhoun and Cantenwalla 1984) that the local moment caused by centroidal eccentricity falls off quite rapidly prior to reaching the heat-affected zone of butt weld joints. The rate of decay is a function of the diameter of the pipe and the diameter to wall thickness ratio (D/t). The moments and corresponding local stresses dissipate more rapidly in thinner, smaller pipe diameters.

Design Procedure

If the difference between wall thicknesses is equal to or less than 1.0 mm, no transition is required. In such cases the stress concentration generated by such a small discontinuity is not large enough to pose a problem; however, differences greater than 1 mm have been shown by industry experience to result in failure in large diameter pipe.

Pipes Operating with High Hoop Stresses

For pipes operating at hoop stresses of 80% of the specified minimum yield strength (SMYS), the following procedures are applicable.

First, whenever $(t_1 - t_2) \rangle 0.3\ t_2$, the thicker wall pipe is counterbored and tapered with a length of counterbore (L) greater than or equal to L_o, but not less than 50 mm, where:

$$L_o = \frac{0.85D}{\sqrt{(D/t_2)}}\text{mm} \qquad (7-54)$$

This technique ensures a reduction in the local stress of over 75% at the weld joint.

In the case where $(t_1 - t_2) \leq 0.3\ t_2$, the thicker wall pipe may be backbevelled as per Figure 7-11.

Another consideration is a transition occurring at a bend location. The transition welds are not typically located in the immediate region of high bending moments (such as may be generated by a field overbend or sag bend). In these situations the transition weld is generally located at a distance of not less than X_o from the point of high bending moment, where

$$X_o = \frac{\sqrt[4]{\left(D^2 t\right)}}{10} \qquad (7-55)$$

for $t = t_1$, or t_2, as appropriate.

As an example, consider the case when $D = 914$ mm and $t = 12.7$ mm. Then the minimum distance is

$$X_o = \frac{\sqrt[4]{914^2 \times 12.7}}{10} = 5.7\ \text{m}$$

Fitting Pipes Operating at Lower Stress Levels

For pipes where the transition joint is designed to operate at hoop stresses equal or less than 60% SMYS, joint connections may be formed by a backbevel as shown in Figure 7-11. Alternatively the thicker wall pipe may be counterbored and a tapered transition provided as per the above requirement, detailed in Figure 7-12.

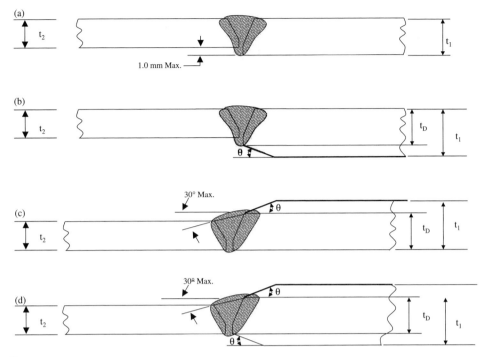

Notes:
1. t_1 & t_2 = nominal wall thickness
   t_D = the design wall thickness, where t_D is less than or equal to 1.5 t_2
   θ - 30° max., 14° min.. No minimum where materials being joined have equal specified minimum yield strength.
2. Provided all the requirements of Figure 7-11 are met, and wall thicknesses t_2 and t_D meet the pressure design requirements of CSA Standard Z662 - 96 the parts to be joined need not have the same SMYS when using the above backbeveled details.

Figure 7-11. Backbeveled joint design for piping designed to operate at a hoop stress of 60% SMYS or less

Design of Branched Connections

Some typical branch connections are shown in Figure 7-14; of these, the extruded or welding tee is preferred in the various codes, from a design viewpoint, since the absence

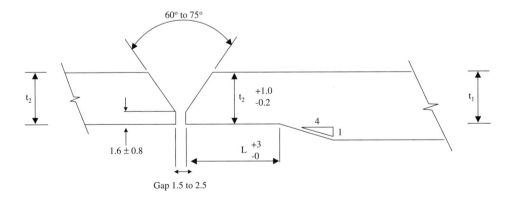

Figure 7-12. Counterbore and taper joint design for transition welds joining pipe of unequal wall thickness. Tolerances in mm.

a. Definition of Transition

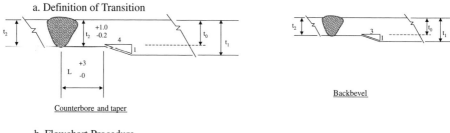

Counterbore and taper

Backbevel

b. Flowchart Procedure

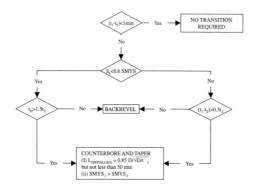

NOTES:

1. All machining + 1.0 mm/-0.2 mm
   (need only apply over 95% of circumference)
2. Subscript '1' refers to thicker pipe; '2' refers to thinner pipe
3. S_h - Hoop stress, check for both 1_1 and t_2
4. For backbevel, $SMYS_1$ may be less than $SMYS_2$
5. Fittings shall be joined to pipe using backbevel
6. Except for fittings, a counterbore and taper may replace a backbevel
7. D = Pipe outside diameter (mm)
8. SMYS = specified minimum yield strength

Figure 7-13. Flowchart summary of procedure for selection of transition

of sharp edges reduces pressure losses as well as eliminates potential stress raisers. The underlying concept, whether for single or multiple openings, is that of area reinforcement. Figure 7-15 (Adapted from ASME B31.8) shows that the area removed from the carrier pipe has to be compensated and this may involve the addition of full encirclement reinforcing material.

Generally reinforcement is not required if the opening is less than 2.5 inches (60.3 mm) outside diameter; although, since these types of openings are often subject to pulsating vibration, appropriate support should be provided. Table 7-7 shows the special requirements needed for branch connections. Fabricated tees are often used and the following worked example illustrates the necessary steps to ensure an adequate connection.

Example

An NPS 8 branch connection is to be welded to an NPS 16 carrier pipe. The carrier pipe material is API 5L X46 with a wall thickness of 0.406 inch. The outlet material is API 5L Grade B (Seamless) Schedule 40 with a wall thickness of 0.322 inch. The working pressure is 1,450 psig and the working temperature is 125°F. The pipe is being used in a Class 1 Division 2 location.

1. Design factors: $F = 0.72$, $T = 1.0$, and $E = 1.0$ (from Tables 7-2, 7-3).
2. Specify the branch pipe wall thickness
   For Grade B, $S_y = 35,000$ psi

$$T_b = \frac{PD}{2SFET} = \frac{1,450 \times 8.635}{2 \times 35,000 \times 0.72 \times 1 \times 1} = 0.248 \text{ inch sufficient}$$

Note: Schedule 40 provides a wall thickness of 0.322 inch (i.e. T_b)

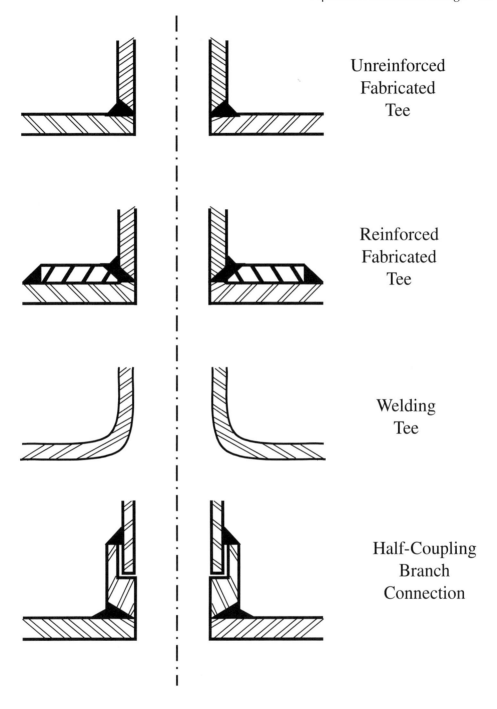

Figure 7-14. Typical branch connections

3. Specify the carrier pipe wall thickness

i.e. $T_{mh} = \dfrac{PD}{2SFET} = \dfrac{1,450 \times 16}{2 \times 46,000 \times 0.72 \times 1 \times 1} = 0.35\,\text{inch} < 0.406\,\text{inch specified (i. e. } T_p)$

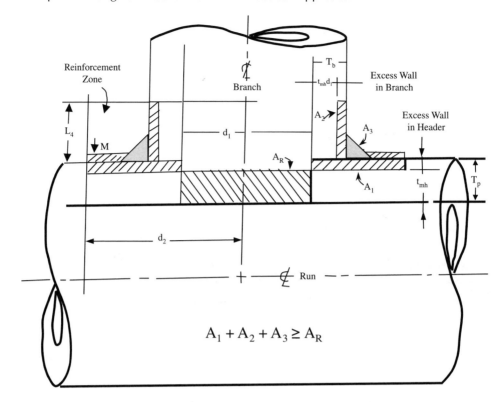

d_1 = inside diameter of branch
d_2 = 1/2 width of reinforcing zone = d_1
L_4 = height of reinforcement zone
A_R = required area of reinforcement
  = $t_{mh}d_1$

Figure 7-15. Concept of area reinforcement

4. Required reinforcement Area

$$A_R = d \times t_{mh}$$
$$d = 8.625 - 2 \times 0.322 = 7.981 \text{ inches}$$
$$\therefore A_R = 7.981 \times 0.35 = 2.763 \text{ in}^2$$

5. Extra area in carrier pipe wall - A_1 (see Figure 7-15)

$$A_1 = (T_p - t_{mh}) \times d$$

Where T_p is the specified wall thickness

$$A_1 = (0.406 - 0.35) \times 7.981$$
$$= 0.447 \text{ in}^2$$

TABLE 7-7. Special requirements, branch connections (taken from ASME B31.8, Clause 831.42)

Ratio of Hoop Stress to SMYS in Run	Ratio of Branch Diameter to Run Diameter		
	1/4 or less	Greater than 1/4 but 1/2 or less	Greater than 1/2
20% or less	Reinforcement not mandatory, but may be required for special cases.	Reinforcement not mandatory, but may be required for special cases.	If reinforcement required, it must be full encirclement if reinforcement extends more than halfway around pipe.
Greater than 20 % but 50% or less.	Calculation not required for openings 2.5 in. and smaller. If reinforcement required, any type of reinforcement permitted with special welds.	If reinforcement required, any type of reinforcement permitted with special welds.	If reinforcement required, it must be full encirclement with special welds if reinforcement extends more than halfway around pipe.
Greater than 50%	Calculation not required for openings 2.5 in. and smaller. If reinforcement required, any type of reinforcement permitted with special welds not thicker than run.	Welding tee preferred, full encirclement should be used, but pad, saddle, or weld-on fitting permitted. Special welds required.	Welding tee preferred, otherwise full encirclement is required with special welds, opening inside corner radiused, and thick pad tapered at ends.

Note: A welding tee may always be used instead of any fabricated branch connection.

6. Extra area in branch wall - A_2

$$A_2 = 2(T_B - t_b)L_4$$

Where T_B is the specified branch wall thickness

L_4 is the smaller of $2 - Tp/2$ or $T_b/2 + M$

and M is the actual or nominal thickness of added reinforcement

$$L_4 = 2 - T_p/2$$
$$L_4 = 2 - T_p/2 = 2-0.406/2 = 1.797 \text{ inches}$$
$$\text{or } L_4 = 2 - 0.322\big/2 + M = (1.839 + M) \text{ inches}$$
$$\text{use } L_4 = 1.797 \text{ inches to determine } A_2$$

Hence use $A_1 = 2(0.322 - 0.248) \times 1.797$
$$A_2 = 0.266 \text{ in}^2$$

7. Additional reinforcement area required - A_3

$$A_3 = A_R - (A_1 + A_2)$$
$$= 2.763 - (0.442 + 0.266)$$
$$= 2.055 \text{ in}^2$$

For a class 1 location, the ASME B31-8 requirement (clause 831.42) calls for the reinforcement to be of the full encirclement type.

8. Determine the thickness of the reinforcement - "M"
The weld leg length l_w is the smallest of T_B, M or 3/8 inch
$l_w < 0.322$ inch, M or 0.375 inch, say 0.322 inch since M is unknown

$$l_r = d - \left(\frac{d}{2} + T_B\right)$$

$$= 7.981 - \left(\frac{7.981}{2} + 0.322\right)$$

$$l_r = 3.669 \text{ inches}$$

$$A_3 = 2\left(1/2\, l_w^2 + Ml_r\right)$$

$$2.055 = 2\left(1/2 \times 0.322^2 \times 3.66M\right)$$

$$M = 0.266 \text{ inch}$$

Since the specified carrier and branch pipe thickness are substantially larger than needed for the operating pressure, a reinforcing ring of 0.27 inch will be adequate.

It is important to note that when the ratio of the size of opening to the diameter of the parent pipe becomes large the current methodology of area replacement specified by the various codes can lead to lower factors of safety than anticipated (Horsley et al. 1999). This is particularly the case with hot tapping, where the branch pipe is "tapped" into an existing pressurized or "hot" main line. Hot tapping has the obvious benefit of not requiring the main carrier pipe to be empty or depressurized, thus there is no loss of product nor interruption to production. (Many hot taps are performed on pipelines at full line pressure and full flow conditions). These demanding conditions however require careful measures to ensure high integrity welds. The flowing product carries away heat from the weld zone, hence high cooling rates are experienced and it is impossible to maintain a level of preheat in the carrier pipe. Often hot tap connections are made with an older pipe, which may have a high carbon equivalent content. It is the susceptibility to hydrogen cracking caused by the combination of high carbon equivalent material and high cooling rates that can cause problems, particularly if the thick-walled branch connection is used. For this reason Horsely et al. (1999) recommends that the material in the branch should be of as high a grade as is practical so that the pipe wall is as thin as possible.

Piping Vibration
The minimum recommended mechanical natural frequency for piping components is 30 hertz. In addition, the predicted mechanical natural frequencies should not lie within 20% of primary excitation frequencies.

Typical primary excitation frequencies at reciprocating compressor sites include:

1. First and second harmonics of the reciprocating equipment speed range for piping close to the equipment.

2. First harmonic of the rotating equipment run speed range for piping close to the equipment.
3. First harmonic of a reciprocating compressor run speed range for piping exposed to compressor pulsations when only single-acting load steps are part of operating conditions. First and second harmonics of reciprocating compressor run speed range when double-acting load steps are present.

Note that the term "close to the equipment" refers to piping for which significant transmissibility exists between the piping and mechanical excitation.

Acoustical Guidelines

Allowable Pressure Pulsation

Pulsation levels in piping should follow the guideline (API 618 - Section 3.9.2.7):

$$P_p = \frac{1.255 \times P_1}{(P_1 \times \text{ID} \times f)^{0.5}} \qquad (7-56)$$

where P_p = lineside piping guideline pulsation level, in kPa (peak to peak). About 1% of line pressure
P_1 = average absolute line pressure, kPa
ID = inside pipe diameter, meters
f = pulsation frequency, hertz.

Pulsation levels at reciprocating compressor valves should meet the following guideline (API 618 - 3.9.2.2.3):

$$P_{rv} = 0.03 \times P_1 \times R \qquad (7-57)$$

where P_{rv} = reciprocating compressor valve guideline pulsation level, kPa (peak to peak)
P_1 = average absolute line pressure, kPa
R = ratio of absolute discharge pressure over absolute suction pressure for the stage to which the valve belongs.

Allowable Acoustical Shaking Forces

Shaking forces in above-ground piping should meet the following industry accepted guideline:

$$F_p = \frac{7,874 \times \text{ID}}{(f/30)^{0.5}} \qquad (7-58)$$

where F_p = piping shaking force guideline in Newtons peak (up to 2,000 Newtons peak maximum)
ID = inside pipe diameter, meters (if greater than 0.254 m then use 0.254 m)
f = shaking force frequency, hertz (if less than 30 Hz, then use 30 Hz).

Shaking forces in vessels should meet the following guideline:

$$F_v = \frac{2,000}{(f/30)^{0.5}} \qquad (7-59)$$

F_v = vessel shaking force guideline, Newtons peak (up to 2,000 Newtons peak maximum)
f = shaking force frequency, hertz.

Allowable Pulsation-Induced Meter Error

Meter errors caused by pulsations are calculated for orifice and turbine types of meters. The maximum predicted metering error should be reduced to one percent or lower.

Orifice plate metering error is evaluated as follows (ISO 3313):

$$E_0 = \left[\left(1 + \left(V^2/2Q^2 \right) \right)^{0.5} - 1 \right] \times 100 \tag{7-60}$$

where E_0 = orifice plate metering error, percent
   V = volume velocity at the orifice, cubic meters per second peak
   Q = actual mean flow rate through the orifice, cubic meters per second.

Performance Guidelines

Allowable Pressure Drop

Reciprocating compressor pulsation suppression devices should meet the following guideline (API 618 - 3.9.2.2.4):

$$P_d = \frac{1.67(R-1)}{R} \tag{7-61}$$

where P_d = maximum pressure drop based on steady flow through a pulse suppression device expressed as a percentage of the absolute average line pressure at the inlet of the device; the minimum guideline pressure drop is 0.25%
   R = ratio of absolute discharge pressure over absolute suction pressure for the stage to which the pulsation suppression device belongs.

Pipe Soil Interaction

From the earlier discussions on codes and standards in Chapter 1, it is apparent that in order to ensure the structural integrity of a pipeline system, the codes must also recognize design conditions other than those of pressure and temperature. In geotechnical design slopes, fault movements are important considerations while seismic-related earth movements, thaw settlements, and frost heave may also bear consideration. All of the foregoing are imposed deformations with regard to the pipeline and would ordinarily result in a localized, nonlinear response, which is not specifically covered in the codes. Rather, the designer is required to determine what supplemental design criteria will be needed to provide an adequate response to these imposed deformations. Longitudinal strain has often been chosen as the quantity upon which to establish the governing criteria for the following reasons:

1. The loads ensuing from imposed deformations are self-limiting in nature, because they are developed by the constraints of adjacent material or by the self-constraint of the pipeline. That is to say, secondary stresses are developed, which satisfy strain displacement compatibility within the pipe or between the pipe and the supporting medium. Local yielding and minor distortions may result from these secondary stresses but they will generally not cause failure in and of themselves. Under these conditions, deformations or strains, rather than loads, are the quantities that control the response of the pipe.

2. All of the types of ground movement noted above will usually be sufficiently large as to cause the pipe to deform into the elastoplastic regime of the material. Here, the almost flat stress-strain characteristic of a pipeline steel means that strain is the sensitive quantity when considering the equilibrium state of the pipeline.

3. It is essential to note that it is strain and not stress that causes structural failures. If the longitudinal strain is sufficiently large in compression, then local buckling of the pipe wall will result. Conversely, there is also a limit on the magnitude of the tensile longitudinal strain since in combinations with the hoop strain in a pressurized line, the material could readily become inelastic. A commonly used criterion for the maximum tensile strain in the longitudinal direction is 0.5%, though in general, this is a conservative value. Usually, the maximum acceptable value of compressive longitudinal strain is that which will bring about the onset of local buckling of the pipe wall, even though there are a number of documented instances where severe corrugation of the pipe has taken place without the initiation of a rupture (Dykes 1996). The conservatism here is due to the unstable postbuckling behavior of cylindrical thin shell structures. Generally the load-carrying capacity of such shells falls off dramatically once local buckling is initiated. However, for a buried pipeline load-carrying capacity is not a significant factor, hence a comparatively large amount of deformation can occur before the pipeline will lose its capacity to carry internal pressure.

Pipe Lowering Stresses

On occasion, such as when there has been severe topsoil erosion or when a line needs to be relocated at new road or rail crossings, it becomes necessary to lower the initially straight pipe. This action will cause two new stresses on the pipe; the first a bending stress due to the change in direction, and the second an extension stress caused by the longer path length it needs to follow.

Conceptually the line lowering profile consists of a series of uniform-radius bends. Considering the segment shown in Figure 7-16, it can be deduced from the principle of intersecting arcs that:

$$(2R - h/2)\, h/2 = (\ell/4 \cdot \ell/4) \qquad\qquad (7-62)$$

where R = radius of curvature, m
        h = is the lowering depth ℓ is the span length over which lowering occurs.

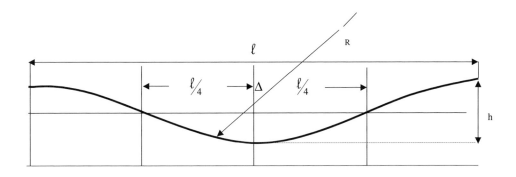

Figure 7-16. Line lowering profile

From which

$$R = \frac{1}{h}\left[\left(\frac{\ell}{4}\right)^2 + \left(\frac{h}{2}\right)^2\right] \qquad (7\text{—}63)$$

The bending stress S_B induced is found from

$$S_B = \frac{E.r}{R}$$

where r is the external radius of the pipe, and E Young's modulus of elasticity.

The change in length of the pipe due to axial straining can be shown to be approximately:

$$\Delta\ell = \ell\left(\frac{8}{3}\frac{h^2}{\ell^2} - \frac{32}{5}\frac{h^4}{\ell^4}\right) \qquad (7-64)$$

and since the sag is a small fraction of the lowered length, the second term can be neglected. The axial stress S_x in the pipe can then be written:

$$S_x = \frac{8E}{3}\left(\frac{h}{\ell}\right)^2 \qquad (7-65)$$

The combined longitudinal stress S_{comb} due to line lowering is the sum of S_x and S_B, that is:

$$S_{comb} = \frac{Er}{R} + \frac{8E}{3}\left(\frac{h}{\ell}\right)^2 \qquad (7-66)$$

Allowable Differential Settlement — Simplified Method

A simple, first approximation of the allowable differential pipe settlement in soil can be developed by assuming the bottom of the pipe remains in contact with the soil. The expected soil deformation may be represented as a sine curve, in which case Figure 7-17a illustrates the ditch bottom profile. Although the differential settlement is derived using an elastic analysis, the model is also instructive for an inelastic case.

From the Engineer's Theory of Simple Bending:

$$EI\frac{d^2y}{dx^2} = M \text{ or } \frac{d^2y}{dx^2} = \frac{M}{EI} = \frac{1}{R} \qquad (7-67)$$

where M = applied bending moments
E = Young's modulus of elasticity
R = radius of curvature
I = second moment of area.

For a sine curve

$$y = a \, \sin\frac{\pi x}{\ell} \qquad (7-68)$$

where ℓ = half-span or half-wave length
a = deflection amplitude
$\Delta = 2a$ = total pipe deflection.

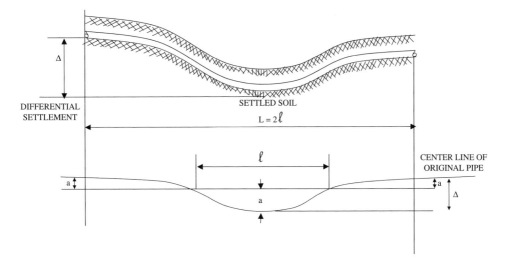

Figure 7-17a.   Differential settlement

the curvature d^2y/dx^2 at midspan is readily deduced to be:

$$\frac{d^2y}{dx^2} = -a\frac{\pi^2}{\ell^2} = \frac{1}{R} \qquad (7-69)$$

hence $a = -\ell^2/\pi^2 R$.

However, $\Delta = 2a$ and $L = 2\ell$, so the total pipe deflection Δ becomes:

$$\Delta = \frac{-L^2}{2\pi^2 R} \qquad (7-70)$$

Differential Settlement Analysis

A more accurate, though complex means of determining whether the critical "pipe" curvature has been exceeded can be obtained by modeling the pipe as a continum of "finite elements," supported by a nonlinear soil foundation. In the analysis, the pipe profile is permitted to deviate from the ditch bottom profile for two reasons:

1. The soil under the pipe deforms under its weight and the overburden.
2. The pipe will span short depressions in the soil profile.

The key step in the analysis is to represent the pipe soil interaction by a finite-element model in which the pipe is discretized into a number of one-dimensional beam elements interconnected at their end points (nodes).

Pipe stiffness, transverse loads, axial loads, and foundation "spring" stiffness are aggregated at these node points to form a set of algebraic equations, which can easily be solved on a computer. The model is subject to the usual assumptions of beam theory, that is: the deflections are small, plane sections remain plane, shear deflections are ignored, etc. The interaction of the beam model and the soil is defined at each node point using a load deflection curve, such as that shown in Figure 7-17b. In this figure δ represents the

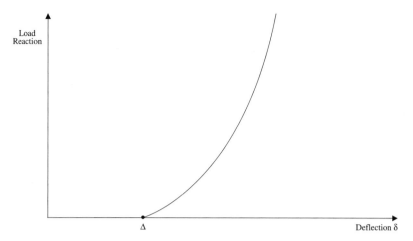

Figure 7-17b. Load deflection curve

deflection of the pipe and Q is the soil reaction, defined as a force per unit length of pipe. In the figure, the soil reaction is zero for deflections less than the soil settlement Δ. A deflection δ larger than Δ will cause an upward reaction Q as determined by the curve. While the shape of the Q - δ curve is the same at all nodes, except for the value of the settlement Δ, it can vary if soil properties change locally. Because of the nonlinear soil response defined by Figure 7-17b, an iterative procedure must be used to solve the algebraic equations and so determine the final deflected shape. The calculated deflections will be correct for the model to within the tolerance set for the iteration process; errors in the analysis are introduced only by the approximation of the true pipe soil interaction as a discretized model. This error can be reduced by increasing the number of elements in the model — at the expense of increasing solution time.

Geotechnical Considerations

In addition to the need to consider pipe soil interaction, discussed earlier in this chapter, there are two further geotechnical considerations that warrant attention: slope stability and seismicity. There is a considerable body of literature on both topics and space limitations here will permit a discussion of only the principal issues rather than analysis. The attention of the reader is therefore directed to the cited references for a full understanding of the issues. On slope stability: Bowles (1997), CGL (1984), Rajani and Morgenstern (1993), Wroth and Wood (1978); and for seismicity: O'Rouke and Lane (1989), Cornell (1969), Bazzurro and Cornell (1994), Campbell, (1991), and Mashaly and Datta (1989).

Slope Stablility

For those pipelines that are required to pass through hilly or even mountainous areas, slope instability can be a critical design issue. In many instances even gentle slopes exhibit slow soil movements of the order of $10^{-3} - 10^{-2}$ m/year (Bughi et al. 1996), while Rizkalla et al. (1996) cite ground movements as rapid as 6 cm/year on steep slopes. When such movements interfere with a pipeline, they can induce unacceptably high strains in the pipe wall over a period of time (1 to 5 years). They can in fact trigger failure modes in the pipe as evidenced by statistics collected for pipelines crossing active slopes (EGIG 1993). Hence there is a need for acceptable design criteria and appropriate monitoring systems when lines go into service. Current techniques for producing ground movements are based on traditional geotechnical analysis (deterministic or probabalistic), and require extensive geotechnical data that is often hard to obtain for large pipeline networks. Hence the identification of slide mechanisms, and of the controlling parameters of ground movement, is a difficult task. Landslides, or slips, occur whenever an unstable mass of material slides along a specific surface or zone known as the failure surface. Many of the steeper slopes along a pipeline route occur on the approaches to river crossings. In order to maintain the integrity of the pipeline and minimize slope disturbances, all such sites have to be investigated and, if necessary, stabilized prior to commissioning the line. The analysis of the slow movement of slopes is invariably based on continuum mechanics and usually consider the following aspects:

1. A limit state for the slope (see later section on "Reliability-Based Design").
2. Continuous deformation in the soil when a state of limit equilibrium is reached.

The first type of model involves examining the equilibrium conditions along a number of potential failure surfaces. The ratio of the available forces providing equilibrium (the shear strength of the material along the failure surface) to the forces tending to cause movement (gravity/horizontal seismic forces) is defined as the factor of safety. A condition of incipient failure is postulated along a continuous slip surface of known, or assumed, shape. A quantitative estimate of the factor of safety of the slope, with respect to the shear strength of the soil to that required to maintain equilibrium, is obtained by examining the soil equilibrium above the potential rupture surface.

This exercise is repeated for each potential failure surface, and the failure surface resulting in the lowest factor of safety is considered the most critical. Normally, for ease of computation, circular shape failure surfaces are analyzed; however, where geologic conditions indicate the possibility of a significantly noncircular critical failure surface, such failure surfaces are also analyzed. An example of the latter case would be that of a specific weak plane in otherwise more competent soil.

A frequently used model (Bughi et al. 1996) in the case of limit equilibrium analysis (aspect 2, above), uses the Coulomb Mohr failure criterion

$$\tau_{\mathrm{appl}} \leq \tau_{\mathrm{lim}} = c^1 + (\sigma - \mu_w) \tan \varphi \qquad (7-71)$$

where τ_{appl} = value of the applied shear stress due to load
τ_{lim} = shear strength of the material
c^1 and φ = cohesion and friction angles of the soil
μ_w = pore pressure
$(\sigma - \mu_w)$ = current value of the normal effective stress.

A limit equilibrum condition at a point in the soil may occur whenever there is an increase in the shear stresses τ_{appl} due to additional external loads, or through a decrease in the soil shear strength τ_{lim}. Depending on the constitutive law describing the stress-strain characteristic of the soil, deformation will occur as the stress state in the soil approaches the failure condition.

For shallow types of failures, there is a general understanding of the failure mechanism, which enables the stability of the slopes to be analyzed. Shallow failures on the right-of-way or in the ditch backfill above the pipe will not directly threaten the integrity of the pipeline. However, they may initiate surface water erosion, which could ultimately affect pipe stability.

Deep-seated failure can be analyzed using conventional slope stability analysis provided sufficient field data is available. The necessary data are: the slope geometry, soil conditions, ground-water conditions, stratigraphy, shear strength properties of the soil, and thermal properties of the soil, if the location is such that soil freezing can occur. A failure involving soil movement to a depth greater than 2–3 m (8–10 feet) is considered to be deep-seated. In order to determine which slopes along the proposed pipeline route may be susceptible to a deep-seated failure, a close inspection of aerial photography and site inspection (see Chapter 2) is required. This will enable the general slope and ground conditions to be assessed, particularly in the vicinity of those areas where this type of failure has occurred in the past.

Where slopes on or adjacent to the right-of-way are considered to be marginally stable, and where a slope failure may affect the integrity of the pipe, or cause unacceptable erosion, some form of slope stabilization becomes necessary. Common methods for stabilizing slopes are based on one or a combination of the following techniques:

- Reduce the loads causing failure
- Increase the internal soil resistance through reduction of the pore-water pressure
- Increase the resisting forces by toe loading
- Increase the resisting forces by structural means

Generally speaking, preventive measures involving gravity drainage earthworks will prove to be the most suitable and cost-effective, though it may be necessary to resort to sheet piling and grading, which are both more expensive.

Reducing Loads Causing Failure
Flattening a marginally stable slope by cutting or grading it to a flatter angle is one means of stabilizing to prevent either shallow or deep-seated failure. However, it should not be used in areas of permafrost because the removal of the top layer of ice-rich soil will also remove the insulation it provides. Thus thawing will be greatly accelerated and this could lead to shallow slumping.

Lowering the Ground-Water Level
This method is particularly effective in preventing deep-seated failures. It involves lowering the ground-water level in the "driving" or active earth pressure area of a potential or existing slide, thus reducing the hydrostatic loading and increasing the soil shearing resistance. It is relatively inexpensive to achieve where drainage can be accomplished by a gravity system.

It is generally used in conjunction with a diversion system for the control of ground water (see Chapter 9).

Reduction of Pore-Water Pressures
Improving Surface Drainage
The improvement of surface drainage by reducing the flow of water on or near the surface of slopes, increases the soil-shearing resistance by reducing the pore-water- pressures as well as the load-causing failure. The method referred to in Lowering the Ground-Water Level is applicable to improvement of surface drainage. In addition, the use of shallow trench drains down the right-of-way slope will control downslope drainage by channeling the flow in a gravel-filled trench. These methods are most effective in stabilizing slopes susceptible to shallow-type failure.

Internal Drainage
Internal drains such as horizontal bored drains, blanket drains, and vertical sand drains are often used to reduce excess pore-water pressure and thereby increase the soil shearing resistance. It is important to know the pore pressure regime in the soil since it is the variation in this regime that ultimately causes deformation. For instance, negative pressures in the pore fluid (suctions) produce an increase in the soil's shear strength (Bughi et al. 1996). This can be seen in Equation (7-71).

Revegetation

Revegetation is a favored method for slope erosion control, but normally it becomes fully effective several years after construction.

Toe Loading

The equilibrium of a stable slope can be upset by the addition of new overloads at the top of a slope or else by erosion at the toe. Both of these effects will alter the value of the applied shear stress. In cases where a slope is deemed to be marginally stable, and particularly where there is evidence of active erosion at the toe, toe loading or earth buttresses may be used to provide stabilization. Compacted earth buttresses placed on a slope, or berms placed on the lower portion of a potential slide area and beyond the toe, should consist of free draining material so as to encourage drainage from the slope. This method applies equally well to both shallow and deep-seated types of failure. When earth fills are used to stabilize slopes at river crossings, they have to be suitably armored to provide protection against erosive action from river flows.

Structural Methods

Stabilization of a potential or existing slide area may be accomplished by structural methods such as the use of retaining walls, earth anchors, vertical piles, or sheeting. Since these methods require that resistance to mass movement be obtained from below the slide area, it must be determined where a potential or existing failure plane is located. This will usually require detailed field investigation in order to determine the ground conditions so that the extent and type of slope stabilization can be determined.

Grading of soil or rock slopes is occasionally undertaken to increase soil strength or to decrease compressibility or permeability. However, it is generally quite costly and only justified when the more convential, simpler procedures described previously are unsuccessful.

Seismic Effects

Pipelines in seismic regions are vulnerable to earthquake loads, which could cause significant damage. Most earthquakes have their origin in tectonics, that is the energy causing the motion is produced from the tearing and grinding of the material associated with a slippage movement within an active fault system. A consideration of the geology along the pipeline route, is therefore pertinent to an assessment of potential earthquake activity. It would be appropriate, for example, to consider that the maximum magnitude of earthquake would occur either on an existing fault system or relatively close to it, where supplementary faulting might develop. However, as one moves away from such a region it is reasonable to assume that somewhat smaller magnitudes of earthquake are likely to be developed than along the main fault system. Associated with the magnitude is the local intensity felt at the pipeline, of an earthquake centered deep in the earth beneath the pipeline, or some distance away horizontally but at the same focal depth. In general, large earthquakes occur at focal depths of 70 to 80 km or greater, while those smaller in magnitude have a depth of focus of approximately 50 km. The ground wave attenuates with distance; Esteva (1970) concluded that the analysis of several strong motion accelerograph records "indicates that the predominant part of strong earthquake ground motions is represented by surface waves." Hence, in order to take account of earthquakes centered some distance away, either directly below the pipeline or offset from it, an appropriate attenuation relationship for the region should be used. While not mandated, the Canadian Code for Oil and Gas Pipeline Systems Z662-03 (2007), in an Appendix, states that the determination of earthquake loads and pipeline displacements from ground motion data shall be based on the response spectrum method, the time history method, or equivalent methods. ASME B31.8 while addressing the subject of soil liquefaction (where the soil behaves like a fluid having lost its shear strength due to intense dynamic cyclic loading), states that the seismic design conditions used to predict liquefaction shall have the same recurrence interval as that used to determine the operating design strength. On this basis the available historical data, the recurrence rates of earthquakes and their intensity can be assembled for number of predetermined sections or "links" of the pipeline. The seismic risk along the pipeline route can be described in statistical terms as a Poisson process, and the attenuation law for assessing the ground motion intensity at a given location of the pipeline is written as a function of the earthquake magnitude, source to site distance, and the local soil conditions.

The effect of earthquakes on the pipeline is multifaceted. The shaking from the earthquake induces additional stress into the pipe and additional forces into the anchors in the above-ground structures. In addition, the earthquake waves traveling through the ground impose bending, tension, and compression stresses in buried pipe. Shaking of the ground can cause liquefaction or compaction of some granular materials. The ground accelerations during an earthquake also add to the gravity and ground-water seepage forces, which contribute to slope failures.

The shaking of the ground also causes dynamic movements in the above-ground pipe and supports. Shaking perpendicular to the pipe causes the pipe to sway back and forth, thereby inducing bending stresses in the pipe between anchor supports. In addition, the longitudinal shaking of the pipe induces additional forces on the anchors. The stresses and forces become greater as the slope along the pipeline becomes greater.

The traveling seismic wave will induce stresses within the buried pipeline as it moves with the ground. The pipe though is sufficiently flexible in bending to accommodate the large radius of curvature that the ground movements will produce; thus, the induced bending stresses will be quite small. For example, the curvature of the ground caused by an earthquake is approximately the ratio of the maximum ground acceleration to the square of the wave propagation velocity of the surface waves (Newmark 1970). Suppose during a severe earthquake the ground acceleration has a value of 0.33g while the surface wave velocity is 760 m/s (2,500 ft/s). The ground curvature will be of the order of 0.0000016 ft^{-1} and the induced bending stress in a 48 inches diameter pipe would be approximately 100 psi, which is negligible. However, considerably larger compression and tensile stresses can be produced by direct ground strains during the earthquake since the pipe is very rigid with respect to longitudinal strains.

At the surface of the soil, compressional waves are of negligible amplitude. However, the surface waves can have appreciable amplitudes. These waves will produce strains within the pipeline when their direction of propagation is at an angle to the pipeline. The maximum effect in the longitudinal direction of the pipe occurs when the wave is propagating at an angle of 45° to it, with shearing motion occurring in a direction at right angles to the propagation direction. Under these conditions the longitudinal strain in the pipe is approximately given by

$$\varepsilon_L = \frac{\text{maximum ground velocity}}{2 \times \text{wave propagation velocity}}$$

For a pipe velocity of say 3 ft/s, and a wave velocity of 2,500 ft/s the longitudinal strain in the pipe will be 0.0006, which corresponds to a longitudinal stress of 18000 psi. In reality the pipe will not move entirely in sympathy with the soil, that is the soil in all likelihood will move faster than the pipe, and hence the longitudinal stress calculated above will be conservative. Nevertheless, it must be included in addition to other effects that cause stresses in the pipe in order to ensure its structural adequacy. A reasonable criteria for permissible deformation to avoid rupture is to limit strain to 2% at any section except at stress concentrations where it could double.

Seismic liquefaction: This occurs most commonly in fine-grained loose granular materials that are saturated. In this situation, the grains of soil are loosely stacked and all of the void spaces are filled with water. Upon occurrence of a seismic event, shaking causes the grains of soil to lose contact in an attempt to densify. During that period of time when the water is attempting to drain out so as to allow the densification of the soil grains, the soil mass takes on the characteristics of a dense viscous liquid. When the soil at the base of a slope liquefies the slope will become unstable, and in places where this has the potential to occur, the pipeline should either be rerouted if possible, or the soil suitably stabilized to reduce the risk of slumping.

With above-ground piping the dynamic motion is applied to the base of the anchor blocks and supports and hence their flexibility and freedom to rotate or tilt must be considered. The ground motion can be expressed either as a time history that enables maximum accelerations and velocities to be determined or as response spectra. This information, in either form, can be used as the forcing function in a dynamic analysis of a finite-element representation of the above-ground pipework. Finally, consideration must also be given, for both buried and above-ground pipelines, to the relative motion at faults crossing the pipeline. It is not uncommon to have vertical or horizontal displacements of several feet where faulting occurs. These of themselves will not cause the pipeline to rupture (Dykes 1996) if it has been designed to accommodate such movements. Relative movement of 4 to 5 feet can be sustained without failure by a properly supported above-ground pipeline, although one or two supports may lose contact with the pipe. With buried pipelines such displacements might cause severe distortions or misalignment but not necessarily rupture or collapse. Very useful guidance for the seismic design of water pipelines and hydrocarbon pipelines may be found respectively in the American Lifelines Alliance (2005) and Honegger (2005).

Reliability-Based Design

The purpose of mechanical engineering design is to ensure the safety and performance of a given pipeline, over a given period of time and under a specified loading condition. Once it is accepted that the loading and the structural strength of the pipe are known only in an approximate manner, it becomes clear that absolute safety is an untenable and impractical objective because of the number of uncertainties involved. These uncertainties may be due to randomness of loadings (both internal and external to the pipe), simplifying assumptions in the strength analysis, material properties, dimensions, etc. (Guedes 1997). However, by expressing these uncertainties in statistical (probabilistic) terms, a rational measurement of safety can be obtained from the estimated probability that some predetermined failure criterion has been exceeded (that is, some prescribed limiting condition for the pipe has been violated). In this manner we can limit the risk of unacceptable consequences. The major benefit of a reliability-based design approach is that the designer can create a pipeline structure that is both efficient and safe to the level specified. Currently five different pipeline codes permit the use of probabalistic approach to design, namely:

Norsok & DNV(2002)	— Norway/Offshore
CSA Z662-07	— Canada
NEN 3650	— Netherlands
prEN 1594, Annex G	— Europe (preliminary)
ISB 16708/13623	

In order to better understand the discussion of reliability-based design later in the chapter, a few concepts will be outlined and some terminology introduced at this point.

In general, the structural reliability of pipeline can be considered as a supply and demand problem. Failure occurs whenever the resistance or strength of the pipe (supply) is exceeded by the loading placed upon it (the demand). Wardenier (1982) has identified four different levels of safety analyses, which can be categorized in accordance with the nature and extent of information one has about the problem:

> Level 0 - Working stress approach
> Level 1 - Partial safety factor approach [also known as Load Resistance Factored Design (LRFD)]
> Level 2 - A semiprobabilistic method
> Level 3 - A "fully" probabilistic method

Generally speaking, this hierarchy represents the increasing amount of knowledge (the less uncertainty) one has about the problem. In essence, the level 0 approach is that taken earlier in the chapter and is both purely deterministic and routinely used. The uncertainty that exists in Equation (7.1) concerning the internal pressure load on the pipe and its ability to withstand it, is embodied in the "safety factor" represented by the factors F, E, and T, all of which have been assessed solely on the basis of experience and engineering judgment. The safety equation in such an approach is written simply as:

Design strength = Maximum lifetime load $\times$ factor of safety

With such an approach, the actual level of safety is unknown.

Load Resistance Factored Design

In recent years, there has been a significant increase in knowledge related to both the statistical descriptions of the distributions of the variables influencing pipeline strength and in the methods available for incorporating these uncertainties, in a rational manner, into a partial factors of safety approach. This approach is generally referred to as a Level 1 or load resistance factored design (LRFD) method. The term "partial" refers to the fact that more than one factor is used in the definition of the safety equation.

Thus uncertainties are grouped together as either partial load or material factors. These partial safety factors are associated with "characteristic" load and resistance effects, the definition of the characteristic values being certain percentiles of the load and resistance probability functions. The characteristic values of load parameters are generally chosen to be on the high side, that is, say a 98% probability of being exceeded, whereas on the strength side the characteristic values are usually those with a 2–5% chance of being exceeded. The values of the partial factors are developed (calibrated) using the higher order Level 2 or 3 approaches with the objective of providing a consistent level of safety for closely similar designs. (See later section on calibrating LRFD factors.)

The selected design format should be a simplified representation of the actual limiting condition under consideration and contain the most significant variables. Equation (7-72) expresses a representative load resistance factored design format,

$$\gamma_E \, S_{CE} + \gamma_F \, S_{CF} \leq \frac{R_C}{\gamma_R} \qquad (7-72)$$

where S_C and R_C are the characteristic load effect and resistance for the failure mode being considered. The subscripted γ's are the partial safety factors to be calibrated, and the subscripts E and F denote environmental loads and functional loads, respectively (note that these can both act in combination).

Equation (7-72) is generalized in the codes to take account of a number of other factors including a safety class factor, so that, for example, in the Canadian Limits States Design approach (CSA Z662-07), one finds:

$$-\emptyset R \geq \gamma \left(\alpha_G \, G + \alpha_Q \, Q + \alpha_E \, E + \alpha_A \, A \right) \qquad (7-73)$$

Where
$\emptyset$ = resistance (strength) partial safety factor (0.9)
R = characteristic resistance or strength
γ = safety class factor
$\alpha_G, \alpha_Q, \alpha_E, \alpha_A$ = load partial safety factors for G, Q, E, and A load effects
G, Q, E, A = permanent, operational, environmental, and accidental load effects respectively.

Figure 7-18 depicts the limit states design methodology used in the Canadian Code (Appendix C of CSA Z662-07). The safety class factor γ is given in Table 7-8 and the partial load factors in Table 7-9.

Safety class factors are related to the consequence of failure or human exposure to the hazard presented. Their various values are set to ensure that higher consequence failures have a lower probability of occurrence. For this reason the class factors are based upon both the location of the pipeline and its contents. The safety class factor can therefore vary along the length of the pipeline. Zimmerman et al., 1996 in a discussion of CSA Z–662 Appendix C presents in pictorial form (Figure 7-19) a possible relationship between safety class factor, human exposure, and

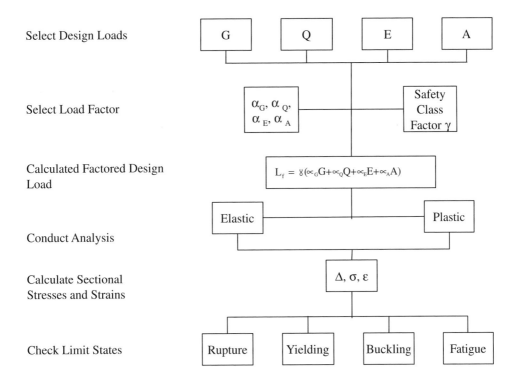

Figure 7-18. LSD method [Ref CSA Z662, 2007]

TABLE 7-8. Safety class factors for ultimate limit states (Ref CSA Z662, 2007)

Class Location	Gas Sour Service	Gas (non-sour)	HVP and CO_2	LVP
1	1.1	1.0	1.0	1.0
2	1.3	1.1	1.2	1.0
3	1.6	1.4	1.2	1.0
4	2.0	1.8	1.2	1.0

target values of annual reliability. Similarly Sotberg et al. (1997) have related acceptable failure probabilities to safety classes and the various generic types of limit states discussed next.

In order to perform a structural reliability analysis, a mathematical model relating the load and resistance needs to be devised. This relationship is expressed in the form of a limit-state function, which can be theoretical, semiempirical, or fully empirical in nature. It does however predict the onset of failure of the pipeline. For pipelines, limit states are defined in accordance with three specific design requirements (Dinovitzer et al. 1999):

The ultimate limit state

The serviceability limit state

The service limit state

The ultimate limit state (ULS) is the state at which the pipe cannot contain the fluid it is carrying. This limit state clearly has safety and environmental implications. Examples of this limit state are leaks and ruptures (bursting). The ultimate limit state is generally associated with failure modes involving defects. The serviceability limit state (SLS) is the state at which the pipeline no longer meets the full design requirements but is still able to contain the fluid. In short, it is not fit for purpose, e.g., it cannot pass sufficient fluid/inline maintenance inspection tools. This limit state has no direct safety or environmental implications. Examples of this limit state include permanent deformation due to denting or yielding. The serviceability limit state is generally associated with failure modes of defect free pipe.

The service limit states are those that develop slowly during the operating life of the structure, such as fatigue failure.

Limit states design requires that all potential failure modes be investigated, though in the case of pipeline design it is possible to conceive of some limit states where the state of knowledge is so incomplete (e.g., stress corrosion cracking) that they are currently not suited for inclusion in such a design format. Table 7-10 contains a list of pertinent pipeline design limit states, the first of which, radial yielding, will be used as an example of the LRFD approach.

TABLE 7-9. Load and resistance factor values of (Ref CSA Z662, 2007) Appendix C

Load Combinations	Load Factors				
	α_G	α_Q		α_E	α_A
		Pressure	Other		
ULS 1: Max operating	1.25	1.13	1.25	1.07	0
ULS 2: Max environmental	1.05	1.05	1.05	1.35	0
ULS 3: Accidental	1.0	1.0	1.0	0	1.0
ULS 4: Fatigue	1.0	1.0	1.0	1.0	0
SLS	1.0	1.0	1.0	1.0	0

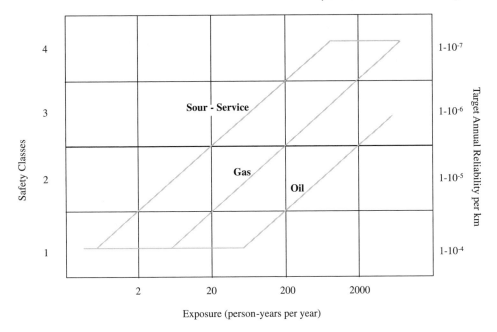

Figure 7-19. Safety class factor related to target annual reliability [Zimmerman et al., 1996]

Example:

The limit state function Z for radial yielding can be expressed explicitly in terms of the four basic quantities D, t, P, and S_y as:

$$Z = S_y - \frac{PD}{2t} < 0 \qquad (7-74)$$

TABLE 7-10. Pipeline design limit states

- Strength limit states/ultimate
- Radial pressure-yielding bursting, fracture, fatigue
- Axial loading-plastic collapse, fracture, buckling
- Flexural load-yielding, fracture, local buckling, wrinkling
- Combination of above with surcharge or thermal effects

Deformation Limit States/Serviceability

- Excessive cross-section deformation - ovalization
- Excessive beam-column deformation - curvature
- Resonance at the natural frequency with imposed loading

Progressive Degradation Limit States/Service

- Fatigue failure due to fluctuating internal loads
- Fatigue failure due to external (surcharge) loads
- Failure due to corrosion fatigue
- Failure due to crack propagation
- Corrosion-induced failure (leakage)

where P is the internal pressure, D the mean pipe diameter, t is the actual wall thickness at any point in the pipe, and S_y is the actual value of the yield stress at the same point.

The four-dimensional basic quantity "space" is split into two regions by the equation $Z = 0$. That part where $Z < 0$ is called the "failure region," while the portion defined by $Z > 0$ is denoted the safe region. The surface defined by $Z = 0$ is referred to as the failure surface or failure boundary. Introducing the quantity $S_H = PD/2t$, that is the applied hoop stress, the failure function can be rewritten as:

$$Z(S_y, S_H) = S_H - S_y \qquad (7 - 75)$$

The problem is now reduced to two quantities, one describing the load (S_H) and the other representing the pipe strength or resistance (S_y). A pictorial representation of the concept of the failure region, failure surface, and safe region is given in Figure 7-20.

Both the stress caused by the applied load S_H and the pipe strength S_y in Equation (7-75) are represented by statistical distributions. The probability density function (PDF) for the yield strength is found by a statistical analysis of measured test values from pipe mill certification records. Similarly, the mean value and variability of the wall thickness and diameter are found from pipe delivery records. Since the pipe is

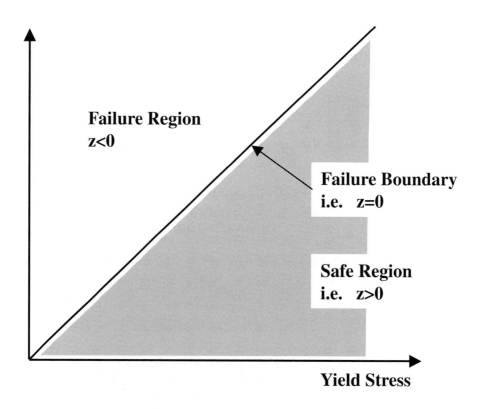

Figure 7-20. Pictorial definition of limit sate concept

made to a specification this limits the amount of variability present in the delivered pipe. In Level 1, load resistance factored design terms, the limit state described in Equation (7-74) can be written as

$$\gamma \, \alpha_Q \, S_H \leq \phi_y \, S_y \qquad (7-76)$$

The interplay between the load [the left-hand side of Equation (7-76)] and resistance is shown in Figure 7-21. By altering the values of α/ϕ, the amount of overlap of the tails of the two distributions can be altered. As will be described later in the description of the level 2 methods of analysis, the convolution integral of the joint probability density functions in these overlapping tails is directly related to the probability of failure.

The choice of partial load factors in the LRFD approach should be such as to be in accord with previous engineering experience. In the case of radial yielding, we know from a deterministic analysis (working stress design) that:

$$\frac{PD}{2t} \leq 0.8 \, S_y$$

while from Equation (7-76) the LRFD format is

$$\gamma \alpha_Q \frac{PD}{2t} \leq \phi \, S_y \qquad (7-77)$$

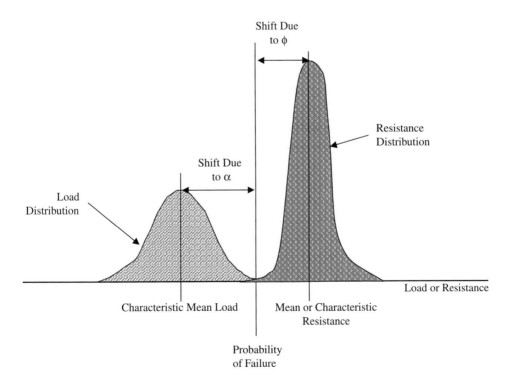

Figures 7-21. Partial factors in LSD

Since the partial safety factor in the Canadian Code is 0.9 and for nonsour gas in a class location 1 the safety class factor γ is 1.0 (from Table 7-8), it follows that:

$$\frac{\phi}{\gamma\,\alpha_Q} = 0.8$$

or

$$\alpha_Q = \frac{\phi}{0.8 \times \gamma} = \frac{0.9}{0.8 \times 1} = 1.13$$

which is the partial load factor found in Table 7-9. for the ultimate limit state due to pressure.

Calibrating LRFD Partial Load Factors

The use of the LRFD approach to pipeline design enables the designer to include probabilistic information without having to be deeply versed in probability theory. To do so with confidence though means that the various partial safety factors contained in Tables 7-8 and Table 7-9 have to be determined such that their use does not lead to designs that are drastically different from existing general practice. The calibration of the partial factors to ensure predetermined probabilities of failure is performed using Level 2/Level 3 methods and follows the steps outlined in Figure 7-22. It will be sufficient here to limit the argument

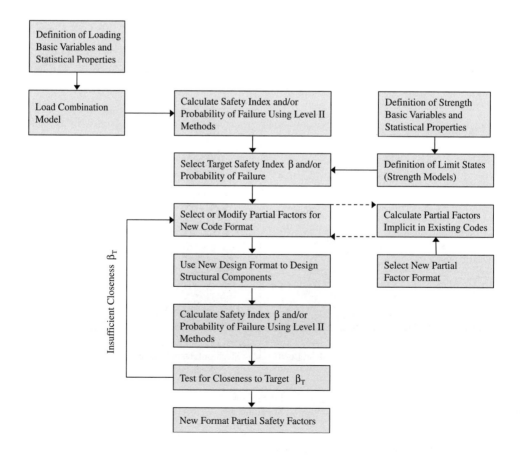

Figure 7-22. Flowchart for selection of partial safety factors of a probability-based design code

to the Level 2 approach, though the interested reader is referred to Thoft-Christensen and Baker (1982), Chen and Nessim (1994), and Ellinas et al. (1986) for a description of the Level 3 approach.

Level 2 Method

The previous example on radial yielding can be written in more general terms as follows. Suppose that the random loading in the pipe is denoted by L and its resistance or capacity is denoted by R, then the safety margin Z of the pipe is defined as:

$$Z = R - L \qquad (7-78)$$

Since R and L are random variables, Z must also be a random variable with a corresponding probability density function $f_Z(z)$.

In this case, and as shown in Figure 7-20, failure is clearly the event ($Z \leq 0$), so the probability of failure is:

$$P_f = \int_{-\infty}^{0} f_z(z)dz = F_{z(0)} \qquad (7-79)$$

Graphically, this is represented by the area under $f_Z(z)$ to the left of 0, as shown in Figure 7-23.

If R and L are assumed to be normally distributed with mean values μ_R and μ_L, respectively, and variances σ_Z^2 and σ_R^2, then the function $Z = R - L$ is also normal, based on

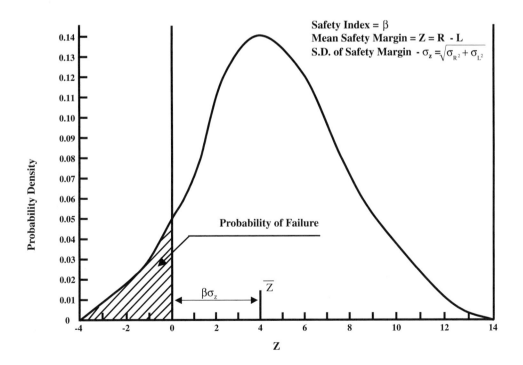

Figure 7-23. Distribution of safety margin

the assumption of independence. $F_{z(0)}$ is the value of the cumulative density function (*cdf*) corresponding to $Z = 0$. Thus for normal distributions one can write:

$$\mu_Z = \mu_R - \mu_L \text{ and } \sigma_Z^2 = \sigma_R^2 + \sigma_L^2$$

Furthermore, the standardized safety margin $(Z - \mu_z)/\sigma_Z$ is normally distributed with a zero mean and a standard deviation equal to one. The probability of failure is computed as follows:

$$P_f = F_Z(0) = \Phi(-\mu_z/\sigma_Z)$$

where Φ is the standard normal CDF. It can be seen that the probability of failure is a function of the ratio: $\beta = \mu_z/\sigma_Z$, which is usually called the Cornell *safety index*. Thus

$$\beta = \frac{\mu_R - \mu_L}{\sqrt{\sigma_R^2 + \sigma_L^2}}$$

Hence if one knows these quantities for the load and resistance distributions, then the safety index is readily found.

For lognormal quantities (which often are used to describe the variables that make up L and R), it is easier to consider

$$Z = R/L \tag{7 - 80}$$

Let the means and coefficients of variation be denoted μ_R, μ_L and δ_R, δ_L, respectively, where $\delta_R = \mu_R/\sigma_L$, that is the ratio of the mean value and the standard deviation of the distribution. Then the logarithm of Z is normally distributed with a mean and variance equal to

$$\mu_{\ell nZ} = \ell n \left\{ \frac{\mu_R \left(1 + \delta_L^2\right)^{0.5}}{\mu_L \left(1 + \delta_R^2\right)^{0.5}} \right\} \tag{7 - 81}$$

and

$$\sigma_{\ell nZ}^2 = \ell n\left(1 + \delta R^2\right)\left(1 + \delta L^2\right) \tag{7 - 82}$$

Consequently, the probability of failure is equal to

$$P_f = \text{Prob.}(R/L < 1) = \text{Prob.}(\ell nZ < 0) \tag{7 - 83}$$

which is the value of the normal cumulative frequency distribution at $\ell nZ=0$. The normal and log normal distributions adequately describe the variability in the parameters of interest in pipeline design so that Equations (7-78) and (7-83) are most often used to determine the probability that a limit state has been violated, i.e., the probability of failure. The safety or reliability index β and the probability of failure are uniquely related and since β is a quantity defining the safety of the pipeline structure, it is often used as the target measure of structural performance. Table 7-11 provides some corresponding values of P_f and β, other values may be found from the standard table for the cumulative normal distribution.

TABLE 7-11. Corresponding values of the failure probability P_f and the reliability index β

P_f	β	P_f	β	P_f	β
0.5	0.000	0.04	1.751	10^{-4}	3.719
0.4	0.253	0.03	1.881	10^{-5}	4.265
0.3	0.524	0.02	2.054	10^{-6}	4.753
0.2	0.842	0.01	2.326	10^{-7}	5.199
0.1	1.282	0.005	2.576	10^{-8}	5.612
0.05	1.645	0.001	3.090	10^{-9}	5.998

Having determined a target reliability index say $\beta = 3.719$ with a corresponding probability of failure of 10^{-4}, the necessary values of the particular factors are found by the trial-and-error-loop depicted in Figure 7-22.

VALVE ASSEMBLIES

Block Valves
Purpose
Block valve assemblies are used to isolate sections of mainline or long laterals when isolation is required in the event of a line break or if maintenance in a section of the line is necessary. Since their fuction is to provide a leak tight seal, it is important that they do not experience undue deflection. For this reason, they are substantially stiffer than the adjacent pipe and their stress levels are about half of the pipe.
Required Components
The following are the main components required for a block valve assembly:

- A gate or ball valve the size of the mainline to allow passage of pigs.
- Two blowdowns (gas only), either remote from or directly connected to the mainline, interconnected for equalizing the pressure on both sides of the block valve.
- A riser on each side of the block valve to provide a power supply for a hydraulic/pneumatic operator, or for taking fluid samples, connecting pressure gauges or performing flow tests.

Location
Code requirements for maximum block valve spacing vary with class location as shown in Table 7-12:

Ease of access and site conditions should always be evaluated when selecting a location for a valve assembly.

A detailed review of valve location requirements and need for valve automation to reduce oil spill in the event of a rupture in liquid lines is given by Mohitpour et al. (2003).

Side Valves
Purpose
The side valve assembly is required to isolate a lateral from the mainline in situations where a line break may occur or when maintenance of the lateral may be necessary.

TABLE 7-12. Code requirements for maximum block valve spacing (Ref CSA Z662, 2007 and ASME B31.8, 2003)

Class Location	CAN/CSA Z662			ANSI B31.4	ANSI B31.8
	Gas Pipelines	HVP Pipelines	LVP Pipelines		
1	NR	NR	NR	12 km	32 km
2	25	15	NR	12 km	24 km
3	13	15	NR	12 km	16 km
4	8	15	NR	12 km	8 km

Required Components

A side valve assembly consists of the following components:

- A gate or ball valve the size of the lateral
- A check valve and bypass line (for receipt laterals)
- A blowdown with appropriate valving (gas only)
- A flange and insulation set to separate the lateral electrically from the mainline
- Test leads from the mainline and the lateral

Note that the purpose of a check valve in the assembly is to prevent reverse flow. It will also prevent flow from the mainline into the inflowing lateral when the pressure in the lateral is less than that in the mainline. Check valves are not required on sales (offtake) laterals.

Location

These assemblies are located on the lateral immediately adjacent to the mainline.

Compressor Station Tie-Ins

Purpose

The main purpose of compressor station tie-ins is to direct the flow either through the compressor station (by opening the side valves and closing the block valve) or past the station (by closing the side valves and opening the block valve).

These valves allow the station to be isolated (in an emergency or for maintenance) without stopping the flow of fluids.

Required Components

The main components required for side valves and a block valve for a compressor station are:

- A mainline block valve
- A suction and a discharge valve (ball or gate)
- Power operators, normally tied into the station automation

Valve Operators

Operators are chosen according to ease of operation and economics. Typical operator configurations include:

- Direct handwheel or wrench-operated valves for NPS 4 valves and smaller
- Handwheel gear -operated for NPS 6 to NPS 12 valves

- Power operators for all NPS 16 valves and larger, and for smaller valves in meter and compressor stations that are designed for remote or automatically controlled operation

Possible modes of operation include:

- Pneumatic controls with a choice of high pressure, low pressure or rate of pressure drop line break controls
- Remote electronic signal operation, is generally used:
  1. To shut-in a meter station when sour gas is detected.
  2. When a meter station is remotely operated by a central gas control facility.

Blowdowns (Gas Only)

Blowdowns are used to vent gas to atmosphere, to expel air during purging, and to hook up a pull-down compressor (used to conserve the gas rather than venting to atmosphere).

When locating a valve assembly with a blowdown, it is important to choose an area that has no buildings immediately downwind, contains no source for igniting vapours, and is easily accessible.

All blowdowns and valve assemblies should be enclosed in a fenced area to protect them from vandalism or damage.

The blowdown size is governed by the time available to depressurize the section of line. Operation requirements for the particular line will govern the time requirements. Common practice involves a 30–60 minute timeline when two blowdowns are open on an isolated section.

There are numerous methods of calculating blowdown sizes. A quick formula for calculating the blowdown time is:

$$T_m = \frac{0.0588 P_1^{1/3} \, G^{1/2} \, D^2 \, L F_c}{d^2}$$

where
T_m = blowdown time, minutes
P_1 = initial trunkline pressure, psig
G = specific gravity of gas
D = ID of trunkline, inches
L = length of trunkline being blown down, miles
d = ID of blowdown stack, inches
F_c = choke factor:

- ideal nozzle = 1.0
- full-port gate valve = 1.6
- reduced port gate valve = 1.8
- regular plug valve = 2.0
- Venturi plug valve = 3.2.

A better formulation for blowdown time calculation or blowdown facilities sizing is provided in the following section.

Blowdown Analysis

The nomenclature used herein is as follows:

A = cross-sectional flow area, square inches

A_v = smallest cross-sectional area of blowdown assembly (including valve, riser, etc.), square inches

C_d = coefficient of discharge to convert A to effective flow area, dimensionless

g_c = acceleration of gravity, 32.2, ft/s^2

k = ratio of specific heats, Cp/Cv, dimensionless

m = mass, lb

M = molecular weight

P = pressure of internal gas in system being blown down, psia

R = gas constant, 1,545 ft·lb/lb·mol °R

S_g = specific gravity of gas, dimensionless

T = temperature of gas in system being blown down, °R

t = blowdown time, sec.

u = point velocity, ft/sec.

V = system volume involved in blowdown, cu. ft.

W = mass flow rate out of system during blowdown, lb./s.

Z = compressibility factor of gas, dimensionless

ρ = density, lb./cu. ft.

$_s$ = property at point where gas reaches sonic velocity

$_1$ = initial system condition for time interval being considered

$_2$ = final system condition for time interval being considered.

The following analysis develops a blowdown time formulation for blowing down a section of gas pipeline with volume V (Figure 7-24).

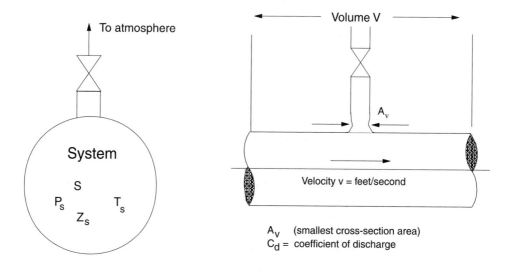

Figure 7-24. Blowdown pressure-time computation

Mass flow rate:

$$W = \rho_s \, A_s \, u = \rho_s \, C_d \, A_v \, u \qquad (7-84)$$

From Real Gas law of thermodynamics:

$$PV = m Z R T$$
$$\therefore P = \frac{m}{V} Z R T$$
$$P = \rho Z R T \ \text{ or } \rho = \frac{P}{ZRT} \qquad (7-85)$$

Therefore, the mass flow rate is:

$$W = \frac{P_s}{Z_s \, R \, T_s} \times C_d \, A_v \times V_s \qquad (7-86)$$

where $\quad V_s$ = sonic velocity

As long as the initial pressure is much higher than atmospheric pressure, critical flow can be assumed to exist through the blowdown valve so that

$$V_s = \sqrt{g_c \, k \, Z_s \, R \, T_s} \qquad (7-87)$$

Substituting for V_s in Equation (7-86):

$$W = \frac{P_s}{Z_s \, R \, T_s} C_d \, A_v \, \sqrt{g_c \, k \, Z_s \, R \, T_s} \qquad (7-88)$$

but

$$Z_s \, T_s = ZT \left(\frac{2}{k+1} \right)$$

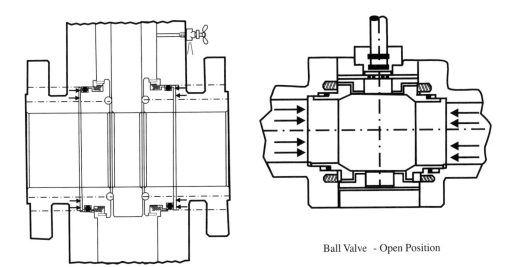

Ball Valve - Open Position

Gate Valve - Open Position

Figure 7-25.   Gate valve and ball valve [Courtesy Daniel Industries Canada]

and

$$\frac{P_s}{P} = \left(\frac{2}{k+1}\right)^{\frac{k}{k-1}}$$

or

$$P_s = P\left(\frac{2}{k+1}\right)^{\frac{k}{k-1}}$$

Substituting into Equation (7-88), the above expressions for Z_S, T_S, and P_S one obtains

$$W = \frac{P\left(\frac{2}{k+1}\right)^{\frac{k}{k-1}}}{ZTR\left(\frac{2}{k+1}\right)} C_d\, A_v\, \sqrt{g_c\, k\, ZTR\left(\frac{2}{k+1}\right)}$$

$$W = \frac{P}{ZTR} C_d\, A_v\, \left(\frac{2}{k+1}\right)^{\frac{k+1}{2(k-1)}} \sqrt{g_c\, kZTR} \qquad (7-89)$$

If flow is constant,

$$Wt = V\,\Delta\rho$$

or

$$t = \frac{V\,\Delta\rho}{W} \qquad (7-90)$$

but from Equation (7-85)

$$\Delta\rho = \frac{\Delta P}{ZRT}$$

$$t = \frac{V}{W} 144 \frac{\Delta P}{ZRT} = \frac{144V}{WZRT}(P_1 - P_2) \qquad (7-91)$$

or

$$dt = \frac{144V}{WZRT}\, dP$$

Equation (7-89) can be rewritten as,

$$\frac{WZRT}{144} = P\left[C_d A_v\left(\frac{2}{k+1}\right)^{\frac{k+1}{2(k-1)}} \sqrt{g_c\, k\, ZRT}\,\right] \qquad (7-92)$$

Assume the expression in square brackets = X, then $dt = V/PXdP$.

If the adiabatic index $k = C_p/C_v = 1.13$ (average value for natural gas), then the bracketed term X can be written as:

$$X = 0.182\, C_d\, A_v \sqrt{\frac{ZT}{S_g}} \qquad (7-93)$$

$$\text{Integrating } dt = \frac{VdP}{PX}$$

$$\int_O^t dt = \frac{V}{X} \int_{P_1}^{P_2} \frac{dp}{P}$$

$$t = \frac{V}{X} \ln \frac{P_2}{P_1} + C$$

at $t = 0, P_1 = P_2, \therefore C = 0$

so

$$t = \frac{V}{X} \ln \frac{P_2}{P_1} = \frac{5.5\, V\,(\text{ft}^3)}{A_v\,(\text{sq.in})\,C_d} \sqrt{\frac{S_g}{ZT}} \ln \frac{P_2(\text{psi})}{P_1} \quad (\text{seconds}) \qquad (7-94)$$

Note

$$S_g = \frac{M\,\text{gas}}{M\ \text{air}}$$

Problem:

Determine the blowdown stack size required to evacuate/blowdown in one hour a section of NPS 36 Grade 483 ($\times$ 70) pipeline designed to ANSI 600# flange rating. The line will be normally operating at 7,000 kPa (1,000 psi) and is designed for class 1 location (i.e., distance between blowdown valves will be 32 km). The following gas composition is applicable:

- Methane — 96.8%
- Ethane — 1.1%
- Propane — 0.0%
- Nitrogen — 2.0%
- Carbon dioxide — 0.01%
- Assumption: $C_d = 0.6$
- Pipeline gas temperature = ground temperature = 20°C

Valve Selection for Pipeline Application

Mainline isolation valves fulfill three basic functions: sectionalizing, diverting, and segregating. Sectionalizing, or dividing up the pipeline into smaller segments that can be isolated, is required to minimize and contain the environmental effects of a line rupture. Where pipelines are interconnected, valves are required to divert product flow to meet production needs. Finally, valves provide the means to segregate or isolate individual process equipment such as scraper traps and, on a broader level, entire plants for safety, maintenance, or operating reasons. Mainline valves must be of a through-conduit or through-bore design to accommodate the scrapers. Gate and ball valves are generally utilized for fluid transmission pipeline applications (Figure 7-32).

There is no single valve and actuator combination that is correct for every pipeline or every application. Variables that must be considered and specifically evaluated for each valve installation include the following:

- Operating characteristics
- Function
- Location
- Fluid service
- Materials options
- Space available
- Maintenance
- Repair capability
- Delivery schedule
- Costs.

Operating Characteristics

The majority of regular slab-gate and ball valves are specified with soft seats that are capable of providing a bubble-tight shutoff. Their temperature limit is generally around 130°C. Both types of valves have two seats and are considered to be "double block and bleed" by the loose definition that their tightness can be verified by opening the vent between the seats. Metal-seated valves are also available for use in abrasive/high temperature applications.

In a ball valve, the ball is mounted on fixed trunnions and the seats are free to move. At very low pressures, the spring force provides the seal and as line pressure increases it forces the seat against the ball proportionally. In most designs, if the upstream seat leaks, the downstream seat is pushed back by line pressure and fluid leaks past the valve. In a few designs, line pressure is also made to act on the downstream seat, thus providing a dual seating capability. However, a body relief valve must be provided to protect against overpressuring.

In the gate valve, seats, as well as the gate are free to move. Both seats can seal at the same time because the upstream seat seals against the gate and the gate seals against the downstream seal. If the upstream seat leaks, there is a possibility that the downstream seat may hold. Also, because line pressure acts on the entire exposed gate area, the sealing load is much higher than when compared to a ball valve.

Function and Location

Most pipeline valves are remotely located and therefore are either manually operated or fluid powered. Generally, the geared handwheel is the most cost-effective. Valves that are infrequently operated and those that do not need to be operated immediately or at high speed are in this category. The use of a gate valve is also recommended in low usage situations in order to utilize cheaper assets for large valves; in situations where high speed operation is essential, high gear ratios are required. In these cases, a two-speed gearbox and a portable driver such as a small gasoline engine are utilized for this purpose.

Due to their tight shutoff capabilities, the gate and ball valve are employed as tie-in hot tap isolation valves when additional interconnections have to be made to an existing pipeline.

Fluid Service and Materials Options

There is no evidence to show that either type of valve has an advantage in any particular fluid service. Either type should perform well in gas or liquid applications. Both gate and ball valves are available in various materials to suit the service conditions. For relatively noncorrosive applications, a carbon steel body with a nickel-plated trim are used. For medium corrosive services, stainless steels or alloy trims in the full solid form or as overlays are generally specified.

Space Availability

Due to its greater height, the gate valve is at a distinct disadvantage in locations where above-ground space is limited or in buried applications where construction costs are at a premium. In above-ground applications, the gate valve will require more elaborate access platforms. If the valve is to be buried, the longer length of the gate valve below the bottom of the pipe means that additional excavation is needed in the immediate area of the valve.

Maintenance

A major problem encountered with ball vales is spring and behind-the-seat body corrosion. Its primary cause is the lack of proper draining during the initial phases of operation or subsequent to hydrotesting. This usually results in sluggish operation of the valve. In some cases, corrosion has caused the valve to simply "freeze."

The gate valve is usually prone to operational problems due to an accumulation of particulates in the cavity. This occurs if the valve is not properly seated in the fully open or fully closed position for long periods of time. The debris will tend to solidify and thus prevent the gate from closing fully. Serious damage may result if the operator overrides the actuator limit switches or exerts excessive force on the handwheel. Regular draining is required to remove debris if such occurrences are frequent. Although not always successful, the best results can be obtained by flushing the cavity through two drain ports located 180 degrees apart.

A problem experienced in sandy environments is seizing between the stem nut and the stem due to entrapment of wind-borne matter when the stem is exposed. Stems must be adequately protected and cleaned on a regular basis to prevent this occurrence.

Actuator maintenance checklists should address the following:

- Verification of adequate hydraulic fluid levels
- Lubrication
- Pneumatic supply
- Absence of leaks
- Corrosion of electrical terminations and contacts
- Mechanical linkage problems

Repair Capability

While all gate valves are considered to be repairable in-place, this is only true for the top entry style ball valve. The split-body and all-welded-body ball valves must be removed from the line for repair of internals. Repair of the gate, however, is much simpler than repairing the ball because of its shape. This is a major consideration in remote areas or in locations where sophisticated machining facilities are unavailable and the ball can only be temporarily patched.

Delivery and Costs

Given the infrequent rate of failure of a ball valve the extra expense of a split body is rarely justified. Little difference is expected in delivery times between the gate and ball valves. Valve costs are also similar. Actuator cost also forms a small percentage of the valve cost.

TABLE 7-13. Sealing characteristics: gate valves versus ball valves

Factor	Gate Valve	Ball Valve
Sealing	✓	
Infrequent operation	✓	
Fail-safe		✓
Hot tap isolation		✓
High speed		✓
Repairability	✓	
Cost	✓	
Space limitations		✓
Hydraulically operated		✓

Note: ✓ indicates the better choice

Selection of the type of actuator will influence the total cost of the valve, since the more complex actuators cost proportionally more. Table 7-13 provides a general guideline when deciding which type of valve to buy for a project.

SCRAPER TRAPS

Scrapers, commonly referred to as pigs, are used in the daily operation and construction of pipelines. They are used for a variety of reasons:

- To clean a pipeline, thereby increasing the line's efficiency.
- To gauge or survey any objectionable restrictions or pipe deformations such as dents and buckles.
- To remove water after hydrostatic testing of newly constructed pipelines.
- To separate product batches.
- To inspect the pipeline internally, to detect any loss-of-metal defects (caused by internal or external corrosion).

Pig traps should be designed and installed to meet the requirements of the intended usages, such as cleaning or inspection. The relevant design codes must be followed to ensure overall compatibility and safety. The manufacturer should be consulted for any dimensional and configurational requirements for the trap.

Location

There is currently no formula for determining the maximum length of a pig run or the location of pig traps. The location of an intermediate pig launch and receipt assembly depends upon the quality of the pig, the quality of construction, the pig velocity, the pig design, the interior line conditions (rough, semirough, smooth), and the medium in which the pig is running. However, there are some general guidelines (Table 7-14).

The recommended spacing between traps is not necessarily the maximum distance that a pig may travel in a single run. This will vary considerably from pig to pig, and from pipeline to pipeline.

TABLE 7-14. Guidelines for choosing pig trap locations (Ref ASME/ANSI B31.8, 1993)

Pipeline Category	Recommended Spacing (km)
Newly constructed gas lines	100–200
Newly constructed product lines	200–300
Newly constructed oil lines	300–500

Major Components of Scraper Trap Assembly

The major components of a scraper trap assembly (Figure 7-26) include:

- Pig barrel and end closure
- Isolation valve
- Kickoff valve
- Bypass valve
- Mainline isolation valve (on mainline bypass)
- Blowdown stack and valve (for gas service only)
- Pig barrel drain valve (mainly for liquids)
- Pipe bends leading to the assembly
- Anchors and supports.

Pig Barrel and End Closure

A pig launch barrel (Figure 7-27) should be at least 1½ times as long as the longest pig (cleaning or pig inspection). An exception to this rule would be for products lines where it is necessary to launch two or more pigs in succession to separate buffer batches (batching pig).

The launch barrel diameter should be one to two NPS sizes larger than the line pipe. The launch barrel must be equipped with connections (usually flanged) for the kickoff line and, if drainage is required, drain valves. The kickoff connection should be located on the

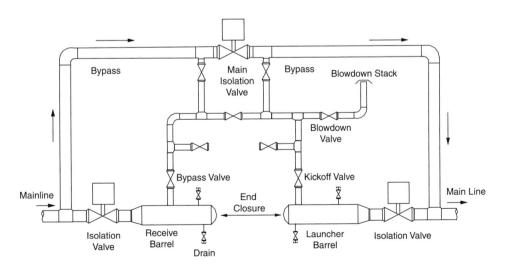

Figure 7-26. Scraper launcher/receiver

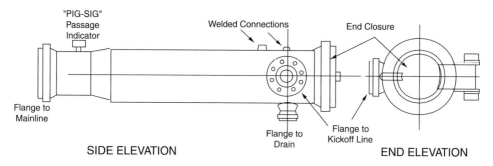

Figure 7-27. Typical pig launcher

side of the trap close to the closure end. The drain outlets should be located on the bottom of the barrel, also close to the end closure.

For gas lines, a blowdown stack and valve are required. Welded connections for gauging, purging, and blowdown should be included on the barrel and located on the top, close to the end closure.

A pig signal detector should be installed downstream from the launcher to indicate the passage of the pig into the main line.

Most operating companies prefer the use of seamless pipe for the scraper barrel. Low temperature materials must be used. Welds must be thermal stress relieved for low ambient temperature operation.

The pig receipt barrel should be at least 2½ times as long as the longest pig. Consideration must be given to the possibility of running long inspection pigs or, for product lines, a series of pigs for batching. When cleaning pigs are considered, attention must be given to the amount of debris expected.

The receipt barrel must be equipped with flanged connections for a bypass line and drainage outlets. The bypass connection should be located near the mainline connection. The drain outlets should be located near the end closure. The barrel diameter should be one to two NPS larger than the line pipe.

Welded connections should be provided for gauging, purging, and blowdown. Seamless material is recommended. Low temperature materials and thermal stress-relief welds are required for low ambient temperature applications.

The receipt barrel should also contain a connection for a pig passage indicator. This connection should be located just upstream of the reducer (Figure 7-28).

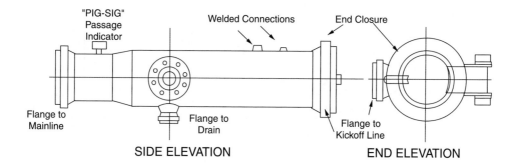

Figure 7-28. Typical pig receiver

A quick open end closure will facilitate easy access. A typical end closure is shown in Figure 7-29. Quick open closures should come equipped with a safety pressure locking device to prevent access while the barrel is under pressure.

Isolation Valve

A valve is required to isolate the scraper barrel from the main pipeline. The isolation valve should be a double block and bleed valve. This will ensure a bubble-tight seal to effectively isolate the trap from the mainline prior to opening the end closure. The valve must be a full-bore, through-conduit type to ensure pig passage. The valve trim and body should be designed for the intended service and ambient temperature conditions.

Depending on the size of the valve, the valve operator can be manually operated or equipped with a power operator. The requirement for automated valve operation will depend on the operating and control requirements. An insulated flange set is usually installed on the mainline side of the valve to electrically separate the mainline from the scraper barrel and valve (cathodic protection).

Kickoff Valve

The kickoff valve is used to launch the pig. The valve should be between 1/4 and 1/2 of the line size. Depending on the valve size, it can be either a bevel-geared or vertical-stem type. Normally, this valve is manually controlled to launch the pig slowly into the mainline.

The valve should have double block and bleed features. A reduced port valve is adequate. The valve body and trim must be compatible with the intended service and ambient temperature conditions.

Bypass Valve and Line

The bypass valve is identical to the kickoff valve with the exception that it is connected to the receipt barrel at a point near the reducer or valve end of the barrel. This location allows the pig to pass the isolation valve and then decreases the flow behind the pig to reduce its speed once it enters the trap.

Mainline Isolation Valve

In an emergency situation, the mainline isolation valve is used to isolate the upstream section of the mainline from the downstream section.

Blowdown Stack and Valve

A blowdown assembly is required for gas services only. The stack and the valve should be designed for low temperature construction. The blowdown size should be sized according to the formulae presented earlier in this chapter (Equation 7-94).

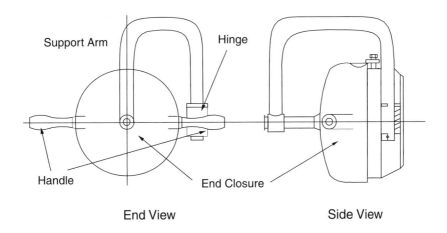

Figure 7-29. Threaded closure with hinge

Drain Valve

Even though drain valves are mainly required for liquid pipelines, they are common in gas services as well. The valve size is usually NPS 2, although NPS 4 valves are found on larger diameter barrels. The materials and design must be compatible with the design and operation conditions set out for the barrel. Drain valves are located at the bottom of the barrel, and are usually piped to a tank or include connections for attachment to tanker trucks.

Pipe Bends Leading to the Pig Trap Assembly

The radius of the bend leading to the trap must meet the pig traverse requirements set out by the pig manufacturer. For normal scraper and batch pigs, a minimum radius equal to three times the pipe outside diameter is usually adequate. However, for electronic instrument pigs, a much longer bend radius may be required.

General Launching Procedures

Launching Procedures — Gas

The following procedure applies to gas kickoff service and the launching assembly shown in Figure 7-30.

1. The isolation valve and the kickoff valve must be closed.
2. Open the blowdown valve to vent pressure from the launcher. (WARNING: Do not attempt to open the end closure until the launcher is completely blown down to atmospheric pressure.)
3. Open the end closure and insert the pig until the front cup reaches the reducer and forms a tight fit against the reducer.
4. Close the end closure.
5. Open the kickoff valve slightly to purge air from launcher.
6. Close the blowdown valve and slowly bring the launcher up to line pressure.
7. Close the kickoff valve. (CAUTION: Pig damage can occur if the kickoff valve is not closed when the isolation valve is opened. The pig may try to leave the launcher before the isolation valve is fully opened.)

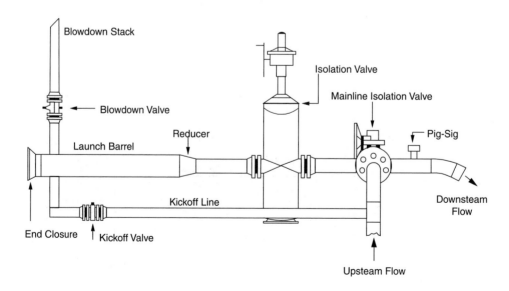

Figure 7-30. Typical gas line launching trap

8. Open the isolation valve.
9. Open the kickoff valve. [NOTE: If the pig does not leave the launcher immediately, slowly close the mainline isolation valve (partially) until the pig does leave the launcher.]
10. Open the mainline isolation valve.
11. Close the kickoff and isolation valves.

Launching Procedure — Liquid

The following procedure applies to the liquid service launching shown in Figure 7-31.

1. The isolation valve and the kickoff valve must be closed.
2. Open the drain valve and the vent valve to allow liquid to drain from the launcher. (WARNING: Do not attempt to open the end closure until the launcher is empty and at atmospheric pressure.)
3. Open the end closure and insert the pig until the rear cup passes the kickoff line connection and the front cup of the pig fits tight against the reducer.
4. Close the end closure and drain valve.
5. Open the kickoff valve slightly to purge air from the launcher.
6. Close the vent valve and slowly bring the launcher to line pressure.
7. Close the kickoff valve. (CAUTION: Pig damage may occur if the kickoff valve is not closed when the isolation valve is opened. The pig may try to leave the launcher before the isolation valve is fully opened.)
8. Open the isolation valve.
9. Open the kickoff valve.
10. Slowly close the mainline isolation valve (partially) until the flow through the kickoff line forces the pig into the mainline.
11. Open the mainline isolation valve.
12. Close the isolation valve and the kickoff valve.

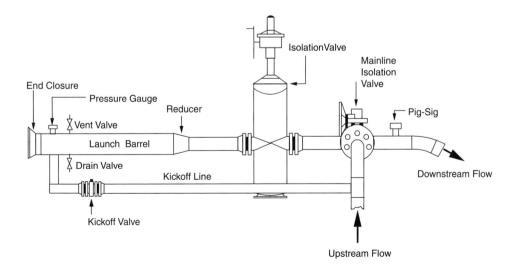

Figure 7-31. Typical liquid line launching trap

General Procedures for Pig Receipt

Receiving pigs are also different for liquid versus gas lines.

Receiving Procedures — Gas

The following procedure applies to the receiver shown in Figure 7-32:

1. The blowdown valve and end closure must be closed.
2. Before the pig arrives, open the bypass valve and then open the pig trap valve. (NOTE: If the pig does not enter the trap, slowly close the station suction valve until the pig is forced into the trap.)
3. Once the pig is in the trap, open the station suction valve.
4. Close the pig trap valve and bypass valve.
5. Open the blowdown valve to vent pressure from the receiver. (WARNING: Do not attempt to open end closure until receiver is completely blown down to atmospheric pressure.)
6. Open end closure and remove pig(s). (CAUTION: Some internal pipeline residues may smolder and ignite when exposed to the atmosphere. These residues should be buried or put in a pit where burning can be controlled.)
7. Close end closure.
8. Open bypass valve slightly to purge air from the barrel.
9. Close blowdown valve and slowly bring receiver up to line pressure.
10. Close bypass valve. (The bypass valve and trap valve may be opened at this time to be ready for the next pig.)

Receiving Procedures — Liquid

The following procedures apply to Figure 7-33:

1. Close the drain and vent valves and the end closure. Ensure the trap has been purged with air and filled with line fill.

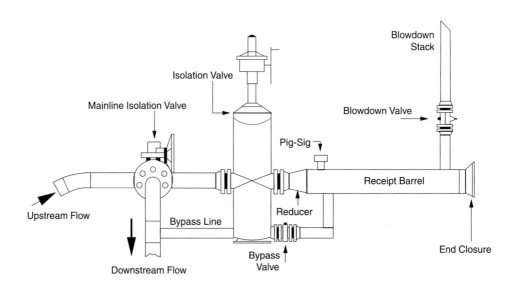

Figure 7-32.   Typical pig-receiving gas-trap line

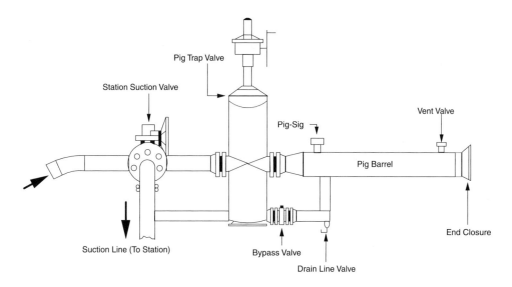

Figure 7-33. Typical pig receiver for liquid lines

2. Open the bypass valve before the pig arrives, and then open the bypass valve, and then the pig trap valve. (NOTE: If the pig does not enter the trap, slowly close the station suction valve until the pig is forced into the trap.)
3. Once the pig is in the trap, open the station suction valve.
4. Close pig trap valve and bypass valve.
5. Open the drain valve and vent valve to allow liquid to drain from the receiver. (WARNING: Do not attempt to open the end closure until receiver is empty and at atmospheric pressure.)
6. Open the end closure and remove pig(s).
7. Close the end closure.
8. Close the drain line valve.
9. Open the bypass valve slightly to purge air from receiver.
10. Close the vent valve and slowly bring receiver to line pressure.
11. Close the bypass valve. (The bypass valve and trap valve may be opened at this time to be ready to receive the next pig.)

Valves and Operators

As mentioned earlier, valves can be either gate or ball type. However, where it is necessary to open and close valves where there are high differential pressures across the valves, a ball-type valve should be used.

Valves must be lubricated regularly to ensure quality performance. Lubrication every six months is common practice, but the frequency may be adjusted based on local operating experience.

The selection of valve operators is based on ease and convenience of operations and economics. For manual valve operation, common practice is to operate valves of NPS 4 and smaller, via direct handwheel, and to operate NPS 6 and larger by means of handwheel gear.

Piping

The choice of the steel grade and wall thickness of the pipe and barrel is based on many considerations, such as:

- Availability of material
- Design pressure and combined stress requirement
- Adequate ring stiffness
- Design operating and ambient temperature
- Code requirements
- Economics

Coating

Valve assemblies, valves, and piping must be properly primed and painted to meet operating and ambient temperature conditions. Gaps between mating flanges should be taped to prevent entry of any foreign materials and moisture.

BUOYANCY CONTROL

Pipelines are subjected to buoyant forces when they encounter freestanding or flowing water, and when buried in saturated soils. The buoyant forces are counteracted by the addition of weight, such as concrete swamp weights(saddles), river weights (bolt-on) or continuous concrete coating. In this way, the pipe can be backfilled and kept in place once it has been installed. Two other methods that can be used to keep the pipe down are the installation of mechanical anchoring devices or backfilling and utilizing the mass of a select material to counteract the buoyant forces.

The final selection of the type and extent of buoyancy control measures is usually made on a site-specific basis. For example, swamp weights are good for wet areas where the pipe will stay on or close to the bottom of the ditch and the weights will not slide off during installation. If the ditch is filled with water, a river weight or continuous concrete coating may be required to keep the pipe down. The choice between these two is usually based on which is the more economic.

The required amount of weight, or the spacing of standard-size weights, is calculated by equating the forces due to the mass of the pipe and weights with the buoyant forces due to the fluid displaced by the pipe and weights. In this calculation, a factor of safety called a negative buoyancy factor is introduced to ensure that the pipe will stay down. Negative buoyancy factors between 5% and 20% are common, depending upon the fluid density of the soil/water mixture encountered and the construction methods used.

Methods of Buoyancy Control

Buoyancy control may be achieved in three ways: (1) by a mechanical anchoring system, (2) by backfill, and (3) by a density anchoring system.

Mechanical Anchors

Mechanical anchors (Figure 7-33) are designed to maintain a minimum holddown force on the pipeline by utilizing the shear strength of the underlying soil. They consist of individual or pairs of anchors attached to the pipeline by holddown clamps or straps. A mechanical anchor involves the use of a steel anchor rod with an auger at the lower tip. They are not commonly utilized for large-diameter pipelines.

Backfill

Backfill, using either native or borrowed select material for buoyancy control, relies on the mass of the backfill over the pipe to counteract the buoyancy forces. Native backfill may be considered if it consists of stable, ice-free soil that is capable of achieving a reasonable level of strength. If the native backfill is not adequate, select backfill can be used. Select backfill should be of coarse-grained, free-draining materials exhibiting sufficient shear strength when thawed or mixed with water. Although gravel is undoubtedly the best material, other materials such as a mixture of gravel, sand, clay, and silt can be used.

The stability of the ditch wall is another important factor to consider in this method of buoyancy control. Unstable ditch walls will not provide sufficient support for the select backfill to keep it over the pipe. This method should not be used if the ditch contains water during backfilling. The ditch must be dewatered prior to and during the placement of any backfill. The backfill can then be placed in a controlled manner.

This method can be used when a ditch is dry during construction but may be subjected to buoyant forces after backfilling is complete. Subsequent buoyant forces could be due to thawing, flooding, or a rising water table. Select backfill can then be used in place of weights in this situation. It can also be used in relatively impervious soils where the ditch excavation has penetrated below the water table.

Density Anchors

Density anchors are a system of weights added to the pipeline. The types of anchors are usually in the form of concrete and include swamp weights (saddle or set-on), river weights (bolt-on), and continuous concrete coating.

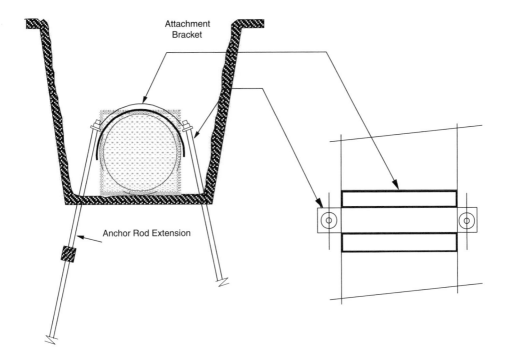

Figure 7-34. Mechanical anchor

Swamp weights (Figure 7-35) are shaped like an inverted "U" and are placed over the pipeline after the pipe is in the ditch. They are constructed by pouring concrete into prefabricated molds or forms and are reinforced with steel reinforcing bars or wire mesh. They can be manufactured at central batch sites or at a specific field site. The choice of location for manufacture depends primarily on the availability and proximity of materials and on transportation costs. Rockshield or felt lining is attached to the inside of the weight to prevent damage to the pipe coating. Lifting hooks are built into each weight for handling. Swamp weights are the most economical density anchor for pipelines primarily because they are the easiest to handle and install. They are usually used for small stream crossings, drainage courses, river floodplains, muskeg (bogs), and swamps.

River weights (Figure 7-36) are constructed in two halves and are designed to be clamped to the pipe. High tensile corrosion-resistant steel bolts are used to fasten the two halves together. The weights are manufactured by pouring concrete into prefabricated molds or forms and are reinforced with steel wire mesh/reinforcing bars. River weights can be manufactured at central batch sites or at a specific field site. The choice of location depends primarily on the availability and proximity of materials and transportation costs. Lifting hooks are built into each weight for handling. As with swamp weights, the inside corners of the weight are chamfered and the inside of the weight is lined with rockshield or felt to prevent damage to the pipe coating. The ends are chamfered to prevent snagging in cases where the pipeline has to be pulled into place.

Wood lagging and slats, strapped to the pipe along its length, are normally used between the river weights to prevent movement of the weights, thereby maintaining the required spacing and providing mechanical protection for the pipeline. If used, they should be taken into consideration in the spacing calculations because of their buoyant effect. River weights are used primarily at minor river and creek crossings (double-sag crossings) where the construction technique consists of welding up the pipeline in a staging area, installing the weights, and then walking the pipeline into place. For large-diameter pipelines, river weights may be used in muskeg or swamp areas, or any other

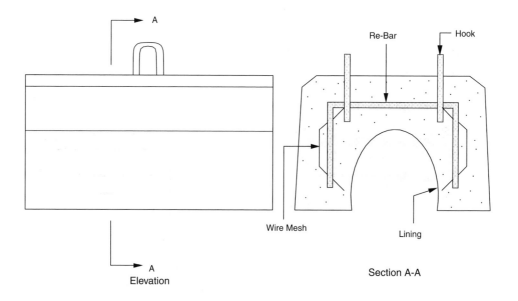

Figure 7-35. Swamp weight

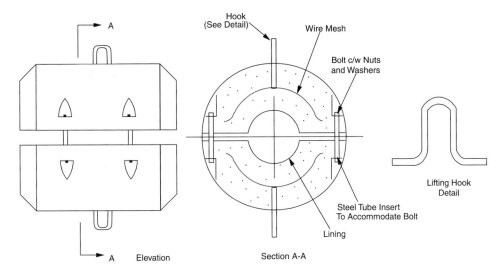

Figure 7-36.   River weight

areas where swamp weights may have been proposed but cannot be used because the ditch water cannot be evacuated.

Continuous concrete coating (Figure 7-37) consists of a coating that completely surrounds the pipe. It provides excellent mechanical protection for the pipeline, particularly in fast-flowing water or in rocky areas. It is reinforced with steel wire mesh or reinforcing bars.

The concrete is usually applied by forming and pouring. The coating may be applied at a central batching plant or at a specific field site. The choice of location depends primarily on the availability of materials, transportation costs and portability of manufacturing equipment.

Some of the other methods used to apply the concrete to the pipe include guniting, where the concrete is manually sprayed onto the pipe; impingement, where the concrete is sprayed onto the pipe as it is rotated past; and extrusion, where the concrete is applied to the pipe as the pipe passes through a machine. At the ends, where the joints are to be welded, the pipe is left bare for some distance, with reinforcement protruding from the concrete. After the sections are welded together, the reinforcement is extended across the bare areas and concrete is applied.

Concrete coating is used for crossing the main active channels of major rivers where the pipeline installation will be performed by pulling. It is also used at large lakes and possibly in large floodplains, swamps, or bogs where the ditch is expected to be filled with water and cannot be dewatered.

Selection, Design, and Verification of Buoyancy Control Systems

Final selection of the type and extent of buoyancy control measures should be made on a site-specific basis, taking the following into consideration: type of terrain, type of soil, ditch

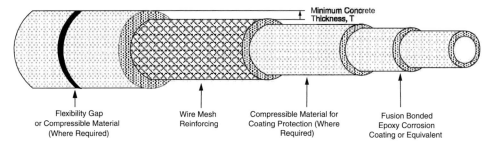

Notes:
- Compressible may be required for protection of pipe coating depending on how concrete is applied.
- Flexibility gap required for severe bends (e.g. offshore pipeline).
- This design to be used mainly at river crossings.

Figure 7-37. Continuous concrete coated pipe

conditions (dry or wet), construction season, cost (economics), availability of materials, access to site, ease of handling during transport and construction, and limitations of equipment. The following values are typically incorporated into design calculations of buoyancy control in Canada:

- For river floodplains, small streams, drainage courses, swamp, muskeg, small lakes, and local depressions where water will be encountered in the ditch during construction:
  Negative buoyancy: 5%
  Fluid density: 1,040 kg/m^3
- For main river channels and areas where flowing or moving water will be encountered during construction:
  Negative buoyancy: 10%
  Fluid density: 1,000 kg/m^3

The maximum allowable swamp weight spacing is calculated from the following design formula:

$$L = \frac{M_c \times g - B_c}{B_p - M_p \times g} \qquad (7-95)$$

where L = center-to-center weight spacing, m
M_c = mass of concrete weight, kg
B_c = buoyant force due to fluid displaced by weight, N
B_p = buoyant force per unit length of pipe due to fluid displaced by pipe, N/m
M_p = mass per unit length of steel pipe, kg/m
g = acceleration due to gravity, m/s^2.

See Figures 7-38 and 7-39, as well as the following table, for further details.

W_p = weight of pipe per unit length of pipe
W_c = weight of concrete.

For system equilibrium:

$$(W_p - B_p)L + (W_c - B_c) = 0$$
$$(M_p g - B_p)L + (M_c g - B_c) = 0$$

Therefore:

$$L_{\max} = \frac{M_c \times g - B_c}{B_p - M_p \times g} \qquad (7-96)$$

Note that the weight of the pipe corrosion coating has been considered negligible.

The following considerations must be taken into account for typical swamp weight design:

- The weights should not impose excessive stresses that could cause rolling or buckling of the pipe.
- The center of gravity of the weights should be kept as low as possible to prevent the weights from rolling off in cases where there are pipe movements and the ditch wall cannot hold them in place.
- The upper limit of the mass of the weight should take into consideration the maximum lifting capability at maximum boom of the side booms likely to be used in the handling and installation of the weights.

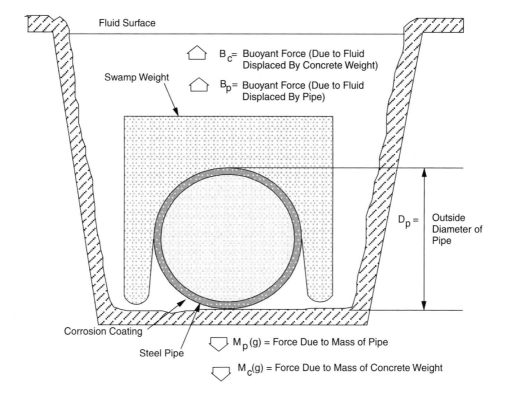

Figure 7-38. Free body diagram forces action on submerged pipeline swamp weight configuration

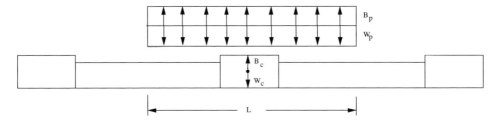

Figure 7-39. Free body of swamp weight spacing formula

Assuming the use of wood lagging, the maximum allowable river weight spacing is calculated from the following design formula:

$$L = \frac{M_c \times g - B_c + (Bw - Mw \times g)R}{B_p - M_p \times g + Bw - Mw \times g} \qquad (7-97)$$

where L = center-to-center weight spacing, m
M_c = mass of concrete weight, kg
B_c = buoyant force due to fluid displaced by weight, N
B_p = buoyant force per unit length of pipe due to fluid displaced by pipe, N/m
M_p = mass per unit length of steel pipe, kg/m
Mw = mass per unit length of wood lagging, kg/m
Bw = buoyant force per unit length of wood lagging due to fluid displaced by wood lagging, N/m
g = acceleration due to gravity, m/s^2
R = concrete weight length, m.
Wp = weight per unit length of pipe
Ww = weight per unit length of wood
Wc = weight of concrete.

See Figures 7-40 and 7-41, as well as the Table 7-15, for further details.

To derive the formula [Equation (7-97)] for the spacing of river weight consider the system equilibrium as shown in Figure 7-41.

$$(W_p - B_p)L + (W_c - B_c) + (W_w - B_w)(L - R) = 0$$
$$W_c - B_c + (B_w - W_w)R = (B_p - W_p + B_w - W_w)L$$

Therefore:

$$L = \frac{W_c - B_c + (B_w - W_w)R}{B_p - W_p + B_w - W_w} \qquad (7-98)$$

or

$$L_{\max} = \frac{M_c \times g - B_c + (Bw - Mw \times g)R}{B_p - M_p \times g + B_w - M_w \times g} \qquad (7-99)$$

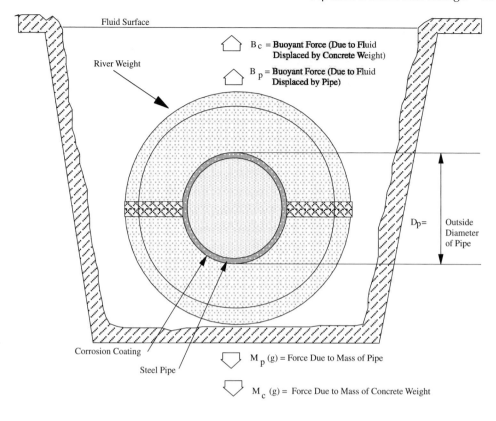

Figure 7-40. Free body diagram forces action on submerged pipeline river weight configuration

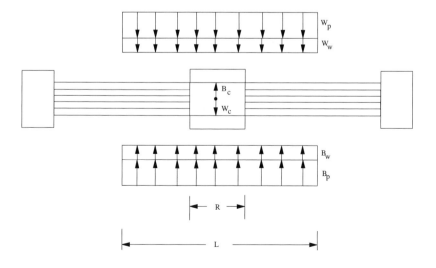

Figure 7-41. Free body diagram river weight spacing formula

TABLE 7-15. Sizing table for swamp weights (Ref ASME/ANSI B31.8, 2002)

		Sizing Table for Swamp Weights			
Nominal Pipe Size (mm)	Mass (kg)	Wall Thickness of Pipe (mm)	Maximum Weight Spacing (mm)	Minimum Ditch Width at Weight (mm)	Minimum Ditch Depth at Weight (mm)
114.3	100	3.2	20.0	600	1,160
168.3	200	3.2	9.0	710	1,230
		4.8	20.0		
219.1	300	4.0	7.7	830	1,320
		4.8	9.7		
273.1	500	4.8	8.0	900	1,380
		5.6	9.4		
323.9	700	5.6	7.8	1000	1,450
		6.4	9.0		
355.6	900	5.6	7.7	1,040	1,490
		6.4	8.6		
406.4	1,150	6.4	7.5	1,100	1,550
		7.1	8.3		

Note that the weight of the pipe corrosion coating was again considered negligible.

The minimum allowable concrete coating thickness is calculated from the following design formula:

$$Tc = \frac{1}{2}\left\{\left[\frac{D^2\gamma_c - \left[D^2 - (D-2t)^2\right]\gamma_s}{\gamma_c - \gamma_F\left(1 + \frac{N}{100}\right)}\right]^{1/2} - D\right\} \qquad (7-100)$$

where T_c = concrete coating thickness, mm
  D = pipe outside diameter, mm
  t = Pipe wall thickness, mm
  γ_c = concrete density, kg/m^3
  γ_s = steel density, kg/m^3
  γ_F = fluid density, kg/m^3
  N = negative buoyancy, %.

See Figures 7-42 and 7-43 for further details.

Let
  M_p = mass of pipe, kg/m
  M_c = mass of concrete, kg/m
  B_p = buoyant force acting on pipe, N/m
  B_c = buoyant force acting on concrete, N/m.

Then for neutral buoyancy or equilibrium:
  $B_p + B_c = (M_p + M_c)\,g$

Negative buoyancy N is the incremental downward force required in excess of the force required for neutral buoyancy (i.e., it is a safety factor):

$$N = (M_p + M_c)g - (B_p + B_c)$$

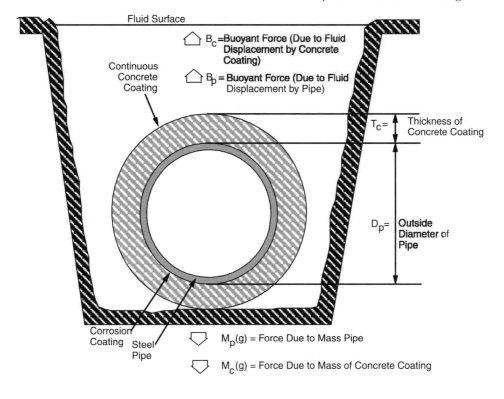

Figure 7-42. Free body diagram forces action on submerged pipeline continuous concrete coating

As a percentage:

$$\% N = \frac{(M_p + M_c)g - (B_p + B_c)}{B_p + B_c} \times 100 \qquad (7-101)$$

If

$$M_p = \left[D^2 - (D - 2t)^2\right] \frac{\gamma_s \pi}{4}$$

$$M_c = \left[(D + 2T_c)^2 - D^2\right] \frac{\gamma_c \pi}{4}$$

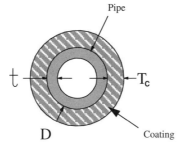

Figure 7-43. Parameters in continuous concrete coating thickness formula

$$B_p + B_c = (D + 2T_c)^2 \frac{g\gamma_F \pi}{4}$$

Then substituting these values in Equation (7-101) gives:

$$(D + 2T_c)^2 = \frac{\left[D^2 - (D - 2t)^2\right]\gamma_s - D^2\gamma_c}{\gamma_F\left(1 + \frac{N}{100}\right) - \gamma_c}$$

Therefore,

$$T_c = \frac{1}{2}\left\{\left[\frac{D^2\gamma_c - \left[D^2 - (D - 2t)^2\right]\gamma_s}{\gamma_c - \gamma_F\left(1 + \frac{N}{100}\right)}\right]^{1/2} - D\right\} \qquad (7 - 102)$$

Again the weight of the pipe corrosion coating was considered negligible.

The need for river and swamp weights and their spacing can be verified during construction on a site-specific basis, as illustrated in Figure 7-44 and in Table 7-15. A certain depth of mineral soil is required to counteract buoyancy forces. If this depth is not available, weights are required. Another option is to excavate a deeper ditch to obtain the minimum thickness of mineral soil, thus reducing or eliminating the need for weights. Economics and the cost of transporting excess swamp or river weights plays a major role in the field verification of weight requirements.

Buoyancy & Specific Gravity

Since weight always acts downward it follows that the reactive force (Buoyancy) must always act up and is positive in that direction. The term "negative" buoyancy has been coined to indicate that in circumstances where the body's own weight is insufficient to keep it immersed, additional weight must be added. A completely immersed body will experience a buoyant up thrust equal to the weight of the fluid it has displaced. It can only remain immersed if (1) its own weight exceeds this buoyant force or (2) an external force or weight is applied (negative buoyancy).

Negative buoyancy can be related to specific gravity since the concept of specific gravity is to compare the density of a given material to that of water. Consider a block of wood having a density of 481 kg/m³, placed in water with a density of 1,000 kg/m³. When the block is completely immersed in water it will require a net down ward force of 5091N

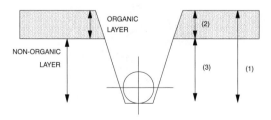

Figure 7-44. Pipeline ditch inspection recording diagram

$(519 \text{ kg} \times 9.81 \text{ m/s}^2)$ to keep it submerged. The specific gravity of the wood will thus be $481/1000 = 0.481$. The additional force required to keep the wood immersed is $(1/0.481 \times$ the water density). i.e an additional weight of 2.08 times its original weight. By comparison, a concrete block with a density of 2240 kg/m^3 has a specific gravity of 2.24 ensuring that it will sink in water.

Consider an NPS 36 Pipe $\times$ 0.622" wall thickness. The weight of the pipe is 236.1 lb/ft, (or 351.36 kg/m linear). If a linear foot of this pipe is submerged, the weight of water displaced is 441.8 lbs (657.7kg per metre), obtained from water density $\times$ pipe volume. The specific gravity of the pipe is then the ratio of the weight of the pipe in air to the weight of water displaced i.e. $236.1/441.8 = 0.54$. Thus the pipe will float almost half out of the water or more precisely with a free board of 17 inches (Fig 7. 44a). The weight needed to sink the pipe in water, (the required negative buoyancy) is then 441- 236.1 or 203.9 lbs per linear foot (or 305 kg/m).

Assume now that the pipeline is submerged in a saturated soil whose density is 123.7 lb / ft^3 (1981 kg/m^3) or a specific gravity of 1.981. The buoyancy of the pipe in such soil is given by the weight of soil displaced minus the pipe weight in air, that is, $123.7 \times \pi \times 3^2/4 - 236.1 \text{ lb/ft} = 873.1 - 236.1 = 637 \text{ lb/ft}$.

Hence under the circumstance of the soil behaving like a liquid, (a fine mud), the pipe, at 637 lb/foot, will have more than triple the positive buoyancy up thrust than previously and will now protrude 26 (66 cm) above the free surface. When the pipe is completely buried in a saturated soil, the only forces that will act on the pipe to keep it submerged will be the weight of the soil over burden and the shearing forces between the ditch wall and the overburden, assuming that the soil has a shearing strength (see Figure 7.44a).

If it is assumed that the soil has sufficient shearing and cohesion strength to act as a plug over the pipe, then a buoyancy calculation can be made as follows. In order for the pipe to move upward under the action of the positive buoyancy force, the soil plug above it must slide upward too, opposed by the shearing action between the plug and the sides of the ditch wall as well as the weight of the plug. Suppose the soil has a shearing strength of 100 lb/ft^2 and a cohesion strength of 60 lb/ft^2, then for a 5' depth of cover, the shearing force per unit length of pipe will be: $2 \times 100 \times 5 = 1000$ lbs/ft.

The weight of the soil overburden is found from : Soil density $\times$ { (depth of cover) $\times$ pipe diameter minus the semi circular cross sectional area of pipe}- $= 123.7 \times \{5 \times 3 - \pi \times 3^2/8\} = 1419$ lb/ft. The total buoyancy resistance is therefore $1000 + 1419 = 2419$ lb/ft. This is sufficient to provide a negative buoyancy of $2419 - 637 = 1782$ lb/ft. The pipe should therefore be stable. Should the soil above the pipe become cohesionless like sand i.e. have very little or no resistance, then the pipe will need to be weighted down to keep it submerged.

TABLE 7-16. Verification of weight requirements during construction

Station 50 m Intervals	Alignment Sheet (AS) No.	Swamp Weights on as Yes * or No	(1) Ditch Depth (m)	(2) Depth Organic Layer (m)	(3) Difference (1) – (2) (m)	Are Swamp Weights Required: Yes or No	Remarks
88 + 780	33	No	1.10	0.30	0.80	NO	
88 + 830	33	No	1.12	0.40	0.72	NO	Begin Weights at Station Surveyors
88 + 880	33	Yes	1.10	0.45	0.65	YES	are to adjust start of
88 + 930	33	Yes	1.15	0.80	0.35	YES	weights

* Assumes ditch can be dewatered prior to backfill.
** If (3) is less than 0.7m swamp weights are required.

For reference purposes, the results of various buoyancy calculations neglecting the weight of overburden on the pipe are presented in Table 7-14. These sample calculations are provided for pipe diameters of 30" and 24" with continuous concrete coating and for concrete saddle weights. For the latter case different water densities have been assumed to demonstrate its effect on negative buoyancy.

Buoyancy Control Using Membrane/ Geotextile Weights

Membrane or geotextile fabric weights, Figure 7-44b are made from very tough, geotextile polypropylene fabrics made into sacks that, when filled with local gravel ballast, are set onto a pipeline to achieve buoyancy control. The fabrics are not biodegradable, cannot be attacked by acidic soil, are permeable to water and support cathodic protection systems.

Geotextile membranes have been used for several years in such applications as dike lining. When used for pipeline weighting, the filled membrane weights have compelling advantages over cast concrete weights.

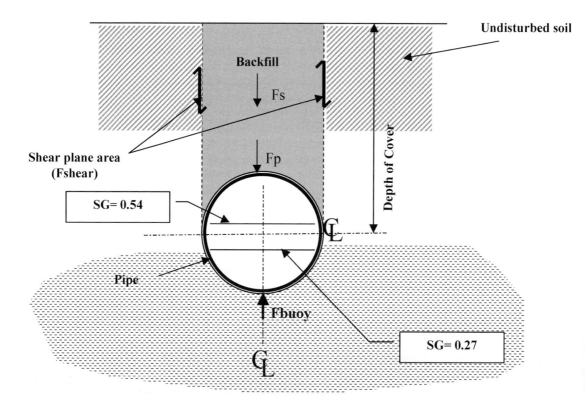

Fbuoy = buoyant force on the pipe
Fp = downward force of pipe and coating
Fs = downward force of soil mass on pipe
Fshear = backfill shear stress

Figure 7-44a. Free-body Diagram of a Buried Pipeline in Saturated Soil

TABLE 7-17. Sample weight calculation results

1- Continuous Concrete Coated Pipe	Fresh Water Density 62.366 LBM/ft³	
Pipe Diameter Outside inches	30.000	24.000
Pipe Wall Thickness inches	0.500	0.500
Concrete coat thickness inches	3.50	2.50
Concrete coating DENSITY lb/ft³	140	140
WEIGHT OF PIPE lb/ft	157.53	125.49
WEIGHT OF COATING lb/ft	358.17	202.38
WT OF PIPE AND COATING lb/ft	515.70	327.87
WEIGHT OF WATER DISPLACED PIPE lb/ft	306.18	195.95
WEIGHT OF WATER DISPLACED COATING lb/ft	159.55	90.15
WEIGHT OF WATER DISPLACED lb/ft	465.73	286.11
NET BUOYANCY lb/ft	49.97	41.76
SPECIFIC GRAVITY =	1.11	1.15

2 - Saddle on Weight	30" Pipe		24" Pipe		
Medium Specific Gravity =	1	1	1.1	1.15	1.2
Concrete weight lbs	6800	6800	6800	6800	6800
Concrete Volume ft³	48.6	48.6	48.6	48.6	48.6
AVERAGE DISTANCE Between weight ft	20.0	26.0	26.0	26.0	26.0
WATER DISPLACED BY PIPE IN AVERAGE lbs	6124	5095	5604	5839	6114
WEIGHT OF PIPE IN AVERAGE lbs	3151	3263	3263	3263	3263
Submerged Concrete weight lbs	3771	3771	3468	3316	3165
WEIGHT OF CONCRETE AND PIPE Lbs	6921	7034	6731	6579	6428
NEGATIVE BUOYANCY lb/ft	39.89	74.57	43.34	27.68	12.08
SPECIFIC GRAVITY*	1.087	1.24	1.13	1.076	1.03

Note: Specific gravity is equal to the weight of the pipe and coating divided by the weight of the water/mud that it displaces.

- No extra trench depth or width is required. Unlike concrete weights which have most of their weight on top of the pipe, geotextile/membrane bag weights have their weight distributed along the sides of the pipeline.
- Geotextile/membrane bag weights easily conform to the bottom of the ditch decreasing the need to dewater the trench. Also in mountainous areas no rock shield is needed between the pipe and the bag.

Figure 7-44b. Geotextile/membrane weights (Courtesy PipeSaks, Ontario, Canada)

- Construction personnel are generally not required to be in the trench to release fabric weights, increasing work place safety.
- No bolts are needed installing or aligning as occurs with a bolt-on concrete weight.
- Multiple lines can be installed in one ditch without the need to place sand bags between the lines.
- Geotextile/membrane bag weights do not extend above the pipe and thus less ditch depth is required than with set-on type concrete weights
- Movement of the pipeline will not result in damage to corrosion coating.
- In remote areas where concrete is not readily available, geotextile/membrane bags can be filled using locally available stones rather than transporting a ready mix plant and cement to the site.
- Geotextile/membrane bags can be filled and applied to the pipe immediately without any cure time.
- No concern exists with producing and applying the geotextile/membrane bags at any ambient temperature, while freezing can destroy the structural integrity of concrete.

Perhaps most significant is that for almost all ballast materials, filled membrane weights require fewer weights to achieve the same buoyancy control as concrete weights. The reasons for this phenomenon can be found in examining the buoyancy properties of the ballast materials that can be used to fill the geotextile membrane weights, McGill, 2002.

For example crushed limestone weighs approximately 1682 kg/m^3. The difference between the 2450 kg/m^3. for solid limestone and the crushed density of 1682 kg/m^3. is due to the amount of void volume in the crushed stone sample. Using 2450 kg/m^3. for solid density and 1682 kg/m^3 for dry crushed density, the void volume will be = 100% $\times$ [1 − (1682/2450)] = 31.4%.

The issue in buoyancy control is not the weight in air but rather the weight in water or mud. A concrete weight with a volume of 1 m^3 will displace 1 m^3 of fresh water (density 1000 kg/m^3), thus causing an upward buoyant force of 9810N. However, a filled geotextile/membrane bag with a volume of 1 m^3 will displace less than 1 m^3 of water due to the void space between the ballast/gravel particles. In the above example, the upward force is 6730N as only 0.686 m^3 of water are displaced.

It can thus be deduced that equal dry weights of concrete and crushed stone will result in a greater submerged weight for the crushed stone than the concrete and allow wider spacing of the weights.

Buoyancy Calculations for Membrane/ Geotextile Weights

The use of geotextile/membrane weights can generally decrease the number of overall weights necessary for buoyancy control. Since the stones/gravel ballast are used to fill geotextile sacks, the resulting weight generally has a higher specific gravity than concrete. Therefore geotextile weights tend to provide a more positive specific gravity for pipeline buoyancy control than solid concrete weights. The following calculations (Connors, 2003) demonstrates the reasons for such a better buoyancy control for a 30" (762 mm), 0.562 " (14.3 mm) wall thickness pipe buried in a trench subjected to continuous flooding. The following assumptions are applicable

- Negative buoyancy 10%
- Concrete weight 2240 kg/m^3
- Geotextile sack filled with stone having density of 2560 kg/m^3 (dry bulk density is 1522 kg/m^3). Sack weight = 3200 kg
- Pipe weight = 176.6 kg/m
- Mud water density 1150 kg/m^3

As per equation 7-95 Saddle weight spacing = Weight of submerged saddle weight (kg) / Total weight required (kg/m).

Therefore, total weight required for buoyancy control = (pipe buoyancy − pipe weight) × Negative buoyancy.

Pipe buoyancy = weight of mud water displaced by the pipe volume = 3.14/4 × (762/1000)^2 × 1150 = 0.456 m^3/m × 1150 kg/m^3 = 524.2 kg/m.

Thus total weight required = (524.2 kg/m − 176.6 kg/m) × 1.10 = 382.4 kg/m.

Weight of submerged geotextile (or concrete) weight = Volume of geotextile (or concrete) weight (V) × submerged density.

For geotextile/membrane weights = V × (density of stone − density of mud water) = (3175 kg/ 2560 kg/m^3) × (2560 kg/m^3 −1150 kg/m^3) = 1748.7 kg. For concrete weights = V × (density of concrete − density of mud water) = (3175 kg / 2242 kg/m^3) × (2242 kg/m^3 − 1150 kg/m^3) = 1546.4 kg.

Calculating the Center-to-Center Spacing = Weight of submerged geotexile (or concrete) weight / total weight required, which indicates:

Geotextile/membrane Weight Spacing = 1748.7 kg / 382.4 kg/m = 4.57 m. Concrete Weight Spacing = 1546.4 kg / 400 kg/m = 4 m

Therefore, for this example, almost 14% fewer geotextile weights will be required than solid concrete weights.

CROSSINGS

When pipelines are required to cross other facilities, structures, or watercourses, it is usually necessary to produce individual designs, including drawings. It is also necessary to obtain individual approval for such crossings from either the owner company of the facility being crossed or from the appropriate regulatory authority having jurisdiction. Usually these are required for the crossing of roads, railways, other pipelines, buried and overhead utilities (electrical, telephone), rivers, and watercourses.

In addition to the general characteristics of the operating pipeline, design considerations and information generally required for appropriate approval are outlined below.

Roads

Information required for approval to cross roads includes:

- Geometry and location of crossing
- Minimum depth of cover to ditch bottom and to road surface
- Construction methods (bored, punched, or open-cut)
- Pipeline specifications (size, wall thickness, operating pressure, class location, etc.)
- Cross-sectional profile
- Warning signs

Railroads

Information required for approval to cross railroads includes:

- Geometry and location of crossing
- Casing details if crossing is cased
- Construction method
- Cross-sectional details
- Minimum depth of cover to ditch bottom and to the rails.
- Pipeline specifications
- Warning signs

Pipeline and Utility Crossings

Information required for approval to cross other pipelines and utilities includes:

- Geometry and location of crossing
- Relative position and clearance between the two
- Pipeline specification
- Cathodic protection (test leads)

River Crossings

Information required for approval to cross rivers includes:

- Geometry and location of crossings
- Scour considerations
- Depth of burial
- Sag bend location
- Environmental considerations
- Timing of construction
- Bank stabilization methods
- Buoyancy control methods
- Construction techniques
- Pipeline specifications
- Warning signs

Design of Uncased Crossings of Highways

The following describes a method for designing uncased pipeline crossings of highways without compromising the structural integrity of the pipeline.

Pipeline codes (specifically CSA Z662-96) usually permit uncased pipeline crossings of highways, provided that the pipeline is installed at a great enough depth (usually a minimum of 1.2 metres) and that a thicker wall pipe is used within the right-of-way.

Historical Background

During the past two decades, the state-of-the-art of pipeline design has advanced significantly. The design of buried pipelines under vehicular crossings is rather unique in that it also involves soil-structure interaction. Casing of pipeline crossings under roads and railroads was a common practice in the past. Technological improvements in the manufacturing process, construction, and the protection of pipelines have reduced the need for casing. The current practice is to utilize uncased crossings whenever possible.

On the basis of studies sponsored by the American Society of Civil Engineers, and the independent research conducted by M.G. Spangler, a design procedure was developed and introduced in a paper presented on June 3, 1964, at the annual conference of the American Water Works Association. The Spangler method, often referred to as the "Iowa Formula," has become the most widely accepted procedure for the design of uncased pipeline vehicular crossings.

Design Procedures

The design of an uncased pipeline crossing must be adequate to provide the following:

1. Sufficient support for the soil, the road surface, and the live loads over the pipe.
2. Sufficient rigidity so that excessive flattening of the pipe does not occur.
3. Sufficient strength so that combinations of internal and external forces do not cause failure of the pipe.

The adopted basic theories are commonly those credited to Marston and Spangler. Marston's equation has been proven to be conservative. (All design equations are provided at the end of this section.) The analysis of pipeline crossings can be broken down into two main categories: the determination of forces acting on the pipe and determination of the stress and deformations in the pipe section due to the combinations of the forces acting on the pipe. These are detailed below.

Determination of Forces Acting on the Pipe

In this category, the forces include the pipe's internal pressure and the external pressure acting on the pipe due to the weight of the soil above it, as well as the wheel loads.

(a) Internal Pressure

The internal pressure is normally known for an operating pipeline or can be calculated using the expression given in the codes for design pressure [see Equation (7-7)]. Included in this expression is a design factor known as the class location factor which limits the internal stress level on the basis of population density adjacent to a pipeline.

(b) Soil Load

To calculate the soil load, Marston's theory can be used. Included in Marston's formula is a design parameter called the load coefficient, C_d. It is a function of the ratio of the height of the backfill or earth above the pipe to the width of the ditch or diameter of the bored hole. It is also a function of the internal friction of the soil backfill and the coefficient of friction between the backfill and the sides of the ditch. Marston recognized five different classes of soil in the development of his original formula. The generally accepted factor used today in design involving highway subsoil material is the soil class, which Marston labelled "ordinary maximum for clay (thoroughly wet)." Values of the load coefficient can be found in Figure 7-45.

(c) Wheel Loads

Vehicle wheel loads are considered as concentrated loads applied at the roadway surface and the load on the buried pipe, including impact loading, is calculated using the Boussinesq Point Load Formula. An influence coefficient, C_T, is included in this formula, which represents the fractional part of the wheel load that is transmitted through the soil to the buried pipe, and is based on Holl's integration of the Boussinesq equation. This influence coefficient is dependent upon the length and width of the section of pipe under consideration, its depth below the roadway surface, and the position of the point of application of the wheel load with respect to the area in plan of the pipe section. The area on which the load is calculated is a projection of the pipe section on a horizontal plane through the top of the pipe. The influence coefficient can be found in Figure 7-46. Impact factors for vehicles operating on unpaved roads or those paved with a flexible-type surface range from 1.5 to 2.0; for rigid pavements it is taken to be 1.0.

Determination of Stresses and Deformations

A buried pipe under a road crossing is subjected to the hoop stress caused by internal pressure and circumferential bending stress due to the external static and dynamic loads. The bending stresses are assumed to be algebraically additive to the tensile hoop stress due to internal pressure. Also, flexible pipes are characterized by their ability to deform extensively without rupture of the pipe wall. They could fail by excessive deflection rather than by rupture. Therefore, design procedures are also directed toward predicting the pipe deflection under load. These stresses are as follows:

(a) Hoop Stress (S_h)

The hoop stress in the pipe is computed using Barlow's Formula (S_h) [see Equation (7-1)], which establishes a relationship between the tensile strength, the internal pressure, and the nominal dimensions of a pipe.

(b) Circumferential Bending Stress (S_b)

Stresses in the pipe caused by external loads can be computed using Spangler's Formula. Included in this formula are bending and deflection parameters, K_b and K_z which are dependent upon the distribution of load over the top half of the pipe and the resultant distribution of the bottom reaction. The load distribution over the top half of the pipe may be considered as uniform. The bottom reaction, however, depends largely upon the extent to which the pipe settles into, and is supported by, the soil at the bottom of the trench or bored hole. For bored installations, the bottom reaction may be considered to occur over an arc of 90°. For an open trench installation, the bottom reaction is generally assumed to occur over an arc of 30°.

(c) Deformations

Pipe deflections may be calculated by using the Iowa Formula (Spangler's), and ignoring the support of the soil. The pipe, therefore, acts as an elastic ring having no effective lateral soil support and the deflection is controlled entirely by the elastic resistance to bending of the pipe wall.

Design Equations

Marston's Formula for soil load on an assumed rigid pipe:

$$W_d = c_d \gamma B_d^2 \qquad (7-103)$$

Boussinesq's Point Load Formula for wheel load:

$$W_L = \frac{c_T\, I}{L} P \qquad\qquad (7-104)$$

Coefficient C_d (graph on left)

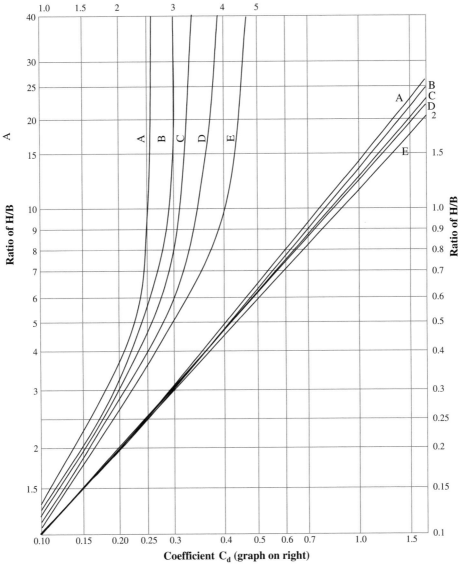

Coefficient C_d (graph on right)

A = 0.1924 for granular materials without cohesion   D = 0.3 max for Ordinary soil
B = 0.165 max for sand and gravel   E = 0.11 max for Saturated Clay
C = 0.15 max for saturated top soil

Figure 7-45. Values of load coefficient C_d (Trench Fill) [Spangler et al., 1964]

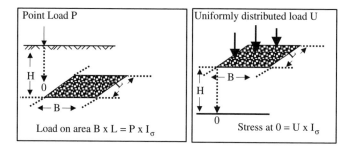

$$m = L/H \qquad n = B/H$$

m and n are interchangeable

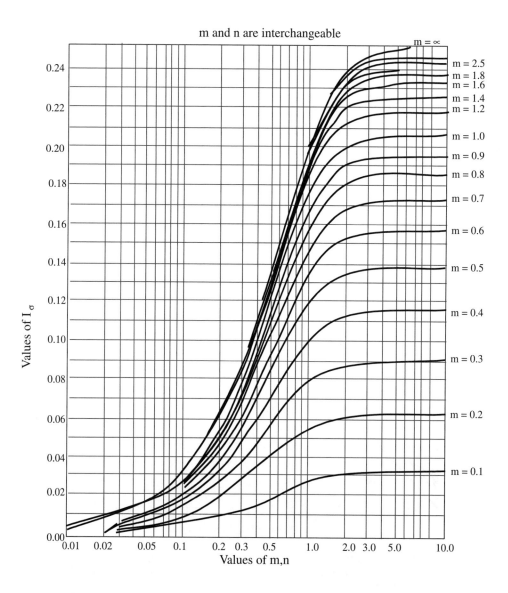

Figure 7-46. Influence value I_σ for concentrated or uniform surcharge of limited extent

Barlow's Formula for hoop stress:

$$S_h = \frac{P_i R}{t} \qquad (7-105)$$

Spangler's Formula for circumferential bending stress:

$$S_b = \frac{6\,K_b\,W_e REt}{Et^3 + 24K_z P_i R^3} \qquad (7-106)$$

IOWA Formula (Spangler's) for pipe deflections:

$$\Delta = \frac{12\,K_z\,W_e R^3}{Et^3 + 24\,K_z P_i R^3} \qquad (7-107)$$

Nomenclature:

B = pipe diameter, m
B_d = width of trench at the top of a pipe (i.e., effective width), m
C_d = load coefficient for fill load
C_T = Influence coefficient for a single concentrated load = $4I_\sigma$
E = Young's Modulus, MPa
S_b = circumferential bending stress in a thin steel pipe wall, MPa
S_h = tensile hoop stress in a steel pipe, MPa
I = impact factor for live loads, varies as function of buried depth. API RP1102, recommends 1.75 for railroads and 1.5 for highways each decreasing by 0.3 per foot of depth below five (5) feet until the factor equals 1.0
I_σ = Influence value
K_b = bending parameter
K_z = deflection parameter
L = length of pipe (taken as 0.91 m), m
P = wheel load, kg
P_i = internal pressure in pipeline, MPa
R = outside radius of pipe, mm
t = pipe wall thickness, mm
W_d = soil dead load on the pipe, N/m
W_e = total external load on the pipe, N/m (includes the soil dead load and the vehicle wheel live load)
W_L = wheel live load on the pipe, N/m
Δ = maximum vertical deflection in pipe, mm
γ = unit weight of soil, kg/m^3

Sample Calculation

Given:

- Pipe diameter = 610 mm
- Pipe wall thickness = 12.7 mm
- Pipe grade = 359 MPa
- Class location = 1
- Internal pressure = 8450 kPa
- Depth of cover = 1.2 m (min. per code)
- Wheel load = 9,072 kg
- Unit weight of soil = 1,922 kg/m^3
- Young's modulus = 200 × 10^3 MPa

Determine the combined circumferential stress and vertical deflection in the pipe due to internal pressure and external loads.

(a) Calculate the soil load:

- Trench width, assume $B_d = 0.61 + 0.34 = 0.95$ m
- Depth of cover, $H = 1.2$ m
- $H/B_d = 1.2/0.95 = 1.26$

Then, using Figure 7-45, $C_d = 1.07$, from curve D for "ordinary maximum for clay." Therefore, using Equation (7-103), the soil dead load on the pipe is:

$$W_d = 1.07 \times 1{,}922 \times 9.81 \times 0.95^2 = 18{,}208 \text{ N/m}$$

(b) Calculate the wheel load on the pipe. (See Figure 7-47). Consider the load to be acting on an area:

$$B \times L = 0.61 \text{ m} \times 0.91 \text{ m}$$

Using Figure 7-46:

$$m = \frac{B}{2H} = \frac{0.61}{2 \times 1.2} = 0.254$$

$$n = \frac{L}{2H} = \frac{0.91}{2 \times 1.2} = 0.379$$

which gives $I_\sigma = 0.038$ and $C_T = 4 \times I_\sigma = 0.152$.

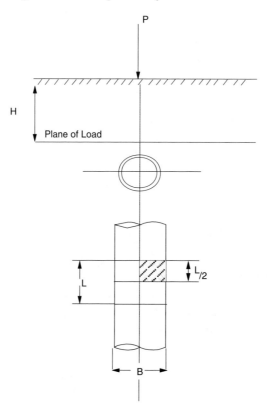

Figure 7-47. Wheel load on the pipe

Therefore, using an impact factor of 1.5 within Equation (7-104), the wheel live load on the pipe is:

$$W_L = \frac{0.152 \times 1.5 \times 9,072 \times 9.81}{0.91} = 22,298 \, N/M$$

(c) Calculate the hoop stress due to internal pressure and the circumferential bending stress due to external loads to give the combined stress in the pipe:

$$S_h = \frac{8,450 \times 10^{-3} \times 305}{12.7} = 203 \, MPa$$

This is lower than the maximum limit of 0.6 SMYS (= 215 MPa) required for uncased crossings in Class 1 location.

The circumferential bending stress is given by Equation (7-106) where (see below), for an open trench installation, the bending and deflection parameter K_b and K_z are 0.235 and 0.108, respectively.

From Equation (7-106), S_b when calculated will be found to be 43 MPa.

Width of Uniform Reaction (Degrees)	Crossing Construction	Parameters Deflection K_z	Parameters Moment K_b
0	Consolidated rock	0.110	0.294
30	Open trench	0.108	0.235
90	Bored	0.096	0.157

Therefore, the combined stress is:

$$S_h + S_b = 203 + 43 = 246 \, MPa = 0.69 \, SMYS$$

This is lower than the maximum limit of 0.72 SMYS permitted in a Class 1 location for the pipe sections adjacent to the crossing. This ensures that the uncased pipeline within the crossing operates at a lower stress level and has a higher factor of safety than the pipe sections adjacent to the crossing.

(d) Calculate the vertical deflection (with no internal pressure to give the worst loading condition):

Using Equation (7-107), the vertical deflection of the pipe is:

$$\Delta = \frac{12 \times 0.108 \times (18,208 + 22,298) \times 10^3 \times 305^3}{200 \times 10^6 \times 12.7^3 + 24 \times 0.108 \times 8.45 \times 10^6 \times 305^3} = 1.44 \, mm$$
$$= 0.24\% \text{ of the nominal pipe diameter}$$

This is within the limit of 3% of nominal pipe diameter for vertical deflection as recommended by Spangler.

DEPTH OF COVER

To protect pipe from general surface activity (e.g., farming, traffic), minimum cover is typically specified. Normal standards are 0.8 m of cover from top of the pipeline to ground surface for pipeline right-of-way, 2.0 m under railroad crossings with 1.0 m minimum at the ditch low point, 1.4 m in the ditch of highway crossings, and 1.1 m in the ditch of local road crossings. At installations and assemblies, 0.9 m of cover is normal. Depth of cover may be reduced in bedrock areas or may be increased in highly unstable soil areas.

AERIAL MARKINGS

Aerial markings are usually installed every 10 km along a pipeline. Through their identification numbers they provide location on the pipeline system to assist in aerial survey/reconnaissance.

WARNING SIGNS

Warning signs are required at all road (including trails, survey, and seismic lines in undeveloped country), railroad, and watercourse crossings. They are there to warn the public of the installation, and should be laid out as follows:

- There is to be one on each side of a crossing.
- They are not to be placed within the right-of-way of a road or railroad, and are to be a maximum of 0.3 m outside the fence line.
- They are to face the crossing.
- There are to be no obstructions in front of them (e.g., brush) and they are to be maintained in a legible condition.

Warning signs should say "WARNING PIPELINE" or "DANGER HIGH PRESSURE PIPELINE" with the company name and phone number. Sizes and other details for such signs are usually established by local regulatory authorities.

REFERENCES

American Lifelines Alliance (2005), "Seismic Guidelines for Water Pipelines" National Institute of Building Sciences Washington D.C.

American Petroleum Institute (API) RP1102, 1993, "Steel Pipeline Crossing Railroads and Highway," 6th Edition.

ASME Boiler and Pressure Vessel Code, 1992, "Section VII, Division 2," ASME, New York, NY.

ASME B31.8, 2003, ASME Code for Pressure Piping, ASME, New York, NY.

Bazzurro, P., and Cornell A., 1994, "Seismic Hazard Analysis of Non Linear Structures - Methodology," Journal of Structural Engineering, ASCE, Vol. 120, No 11, pp. 3320–3344.

Bowles, J. E., 1997, *Foundation Analysis and Design*, McGraw Hill, New York, NY, XIV + 750p.

Bughi, S., Alesetti, P., and Bruschi, R., 1996, Slow Movements of Slopes Interfering with Pipelines, "Modeling and Monitoring," Proc 15th ASME Int. Conference Offshore Mechanics and Arctic Engineering, Florence, Italy, Vol. 5, pp. 363–372.

Campbell, K. W., 1991, Analysis of Ground Shaking Hazard and Risk for Life Line Systems, *Lifeline Earthquake Engineering*, pp. 581–590.

CAN/CSA, Z184-M92, 1992, "Gas Pipeline System," CSA, Rexdale Ontario Canada.

Chen, Q., and Nessim, M. A., 1994, "Risk Based Calibration of Reliability Levels for Limit States Design of Pipelines," CFER Report to National Energy Board of Canada.

Conners, G.W, 2003 'The Use of Geotextile Fabric, Anti-Buoyancy Weights for Buried Pipelines" Prented at GAIL Gas Industry O&M Conference at Mumbai June 18th - 19th, http://pipesak.com/news/gailpaper.htm

Cornell, C. A., 1968, "Engineering Seismic Risk Analysis," Bull Seismological Society of America, Vol. 58, No 5, pp. 1583–1606.

Colquhoun, I., and Cantenwalla, M., 1984, "Welding Pipe of Unequal Wall Thickness - Determination of Minimum Counterbore Length," Internal Review Memo, NOVA, Calgary, Alberta, Canada.

CSA., Z662-07, 2007, Oil and Gas Pipeline Systems, Canadian Standards Association, Etobicoke, Ontario, Canada.

Committee on Gas and Liquid Fuel Lines (CGL), 1984, "Guidelines for the Seismic Design of Oil and Gas Pipeline Systems," ASCE, New York, NY, 471.

DNV, 1996, "Rules for Submarine Pipelines," Det Norske Veritas Oslo, Norway.

Dykes, S. A. M., 1996, "Pipeline Earthquake Survival, Tennant Creek, Australia, 1988," Paper 2–15, Banff Symposium on Pipeline Reliability, CANMET, Ottawa, Canada.

EGIG, 1993, European Gas Pipeline Incident Data - 1993 Update. European Gas Pipeline Incident Data Graph: British Gas, Dansk Gas Teknisk Centre, Enigas, Gaz de France, Gas Unie Ruhrgas AG, Distrigas and SNAM.

Ellinas, C., Raven, P. W. J., Walker, A. C., and Davies, P., 1986, "Limit State Philosophy in Pipeline Design," Journal Energy Resources Technology, Trans., ASME, January.

Esteva, L., 1970, "Seismic Risk and Seismic Design Decisions," in *Seismic Design for Nuclear Power Plants*, ed R.J. Hansen, MIT, Cambridge, pp. 142–182.

Guedes, S. C., 1997, "Quantification of Modeling Uncertainty in Structural Reliability," in *Probabilistic Methods for Structural Design*, ed. G. Guedes Soares, Klauwer Academic Publishers.

Hingoraney, R., and Goto, D. N., 1995, "Isolation Valve Selection Play Important Role in Pipelining," Pipeline and Gas Journal, August, pp. 38–41.

Honegger, D. G and Nyman, D. J. "Guidelines for the Seismic Design and Assessment of Natural Gas and Liquid Hydrocarbon Pipelines" Pipeline Research Council International Catalog No L51427.

International Standards Organisation. (2006) "Petroleum and natural gas industries – Pipeline transportation systems – Reliability-based limit state methods" ISO 16708-2006. 57pp.

International Standards Organisation. (2000) "Petroleum and natural gas industries – Pipeline transportation systems". ISO 13623-2000.

Kormann, P., and Zhou, Z. J., 1995, "Support Spacing of Buried and Above Ground Piping," Proc. 2nd Int. Conf. Advances in Underground Pipeline Engineering, Bellevue, WA, Eds J.K. and M. Jeyapalan, ASCE, New York, NY.

Mashaly, E. A., and Datta, P. K., 1989, "Seismic Risk Analysis of Buried Pipelines," Journal of Transporation Engineering, ASME, Vol. 115, No. 3, pp. 232–252.

McGill, J. C., 2002 "Water Engineering & Management " April Vol: 149 Number: 4

McGill 2002 and Connors (2003) American Lifelines Alliance (2005), "Seismic Guidelines for Water Pipelines" National Institute of Building Sciences Washington D.C.

Mohitpour, M., Trefaneuks, B., Tolmanquin, S. T., and Kossatz, H., 2003, "Oil Pipeline Valve Automation for Spill Reduction," Proceedings of Rio Pipeline Conference, Oct 2003.

Newmark, M. M., 1970, "Seismic Design Criteria for Alyeska Pipline Serric Company," Appendix A-3. 1051, Alyeska Pipeline Design Submission.

NORSOK Standard Y-002, 1997, "Reliability Based Limit State Principles for Pipeline Design."

O'Rouke, T., and Lane, P. A., 1984, "Liquidfaction Hazards and their Effects on Buried Pipelines," Material Center for Earthquake Engineering Research, University of California, Berkeley, Technical Report, NCEER - 89 - 007, February.

Paulin, M. J., Philipps, R., and Boivin, R., 1996, "An Experimental Investigation into Lateral Pipeline Soil

Interaction," Proc 15th ASME International Conference, Offshore Mechanics and Arctic Engineering, Florence, Italy, Vol. 5, pp. 313–324.

Rajani, B. and Morgenstern, N.R., 1993, "A Simplified Design Method for Pipelines subjected to Transverse Soil Movements," Proc 12th ASME Int. Conference, Offshore Mechanics and Arctic Engineering, Glasgow, Scotland, Vol. 5 pp. 137–165.

Rizkalla, M., Trigg, A., and Simmonds, G., 1996, "Recent Advances in the Modeling of Longitudinal Pipeline Soil Interaction for Cohesive Soils," Proc 15th ASME International Conference Offshore Mechanics and Arctic Engineering, Florence, Italy, Vol. 5, pp. 325–332.

Roark, R. J. and Young, W. C., 1975, *Formulas for Stress and Strain, 5th Edition*, McGraw-Hill Book Company, New York, NY.

Schnockenberg, P. J., 2002, "How to Calculate Stress in Above/Below Ground Transition," Pipeline Rules of Thumb Handbook, 4th Edition.

Sotberg, T., Moan, T., Bruschi, R., Jiao, Y., and Mork, K. J., 1997, "The SUBERB Project: Recommended Target Safety Levels for Limit State Based Design of Offshore Pipelines," Proc 16th Int. ASME Conference Offshore Mechanics and Arctic Engineering, Florence, Italy.

Spangler, M. G., and Marston, 1964, Proc. Annual Conf. American Water Work Association, June.

Taylor, R. N. (ed), 1995, *Geotechnical Centrifuge Technology*, Blackie Academic and Professional, London, England.

Thoft-Christensen, P., and Baker, M. J., 1982, *Structural Reliability Theory and its Applications*, Springer Verlag, Berlin, Germany.

Timoshenko, S., and Young, D. H., 1965, *Theory of Structures*, PWS Publishing, Boston.

Venzi, S., Malacarne, C., and Cuscuna, S., 1993, "Development of an Expert System to Manage the Safety of Pipelines in Unstable Slopes." Proc 12th ASME Int. Conference Offshore Mechanics and Arctic Engineering, Glasgow, Scotland, Vol. 5, pp. 127–134.

Wanttand, G. M., O'Neill, M. W., Reese, L. G., and Kalajian, E. H., 1979, "Lateral Stability of Pipes in Clay," Proc. 11th Offshore Technology Conference, Houston, TX, pp. 1025–1034.

Wardenier, J., 1982, *Hollow Section Joints*, Delft University Press, Delft, The Netherlands.

Wilbur, W. E., 1983, Pipe Line Industry, February.

Wroth, G. P., and Wood, D. M., 1978, "The Accumulation of Index Properties with Some Basic Engineering Properties of Soils," Canadian Geotechnical Journal, Vol. 15, No. 2, pp. 137–145.

Zhou, Z., and Murray, D. W., 1993(a) "Numerical Structural Analysis of Buried Pipelines," Structural Engineering Report 181, Department of Civil Engineering, University of Alberta, Edmonton, Alberta, Canada.

Zhou, Z., and Murray, D. W., 1993(b), "Behavior of Buried Pipelines Subjected to Imposed Deformations," Proc. 12th ASME Int. Conference, Offshore Mechanics and Arctic Engineering, Glasgow, Scotland, Vol. 5, pp. 115–122.

Zimmerman, T. J. E., Chen, Q., and Pandy, M. D., 1996, "Target Reliability for Pipeline Limit State Design," pp. 111–120, Vol. 1, Proc. 1st ASME IPC Conference, Calgary Alberta.

Chapter 8

MATERIALS SELECTION AND QUALITY MANAGEMENT

INTRODUCTION

This chapter will provide an overview of the processes involved in the selection of materials and in quality management for oil and gas transmission systems. To understand the complexities of these processes, it is first beneficial to identify the key activities involved in building pipeline facilities, and to outline the framework in which the selected materials will function. In this chapter, emphasis will be placed on natural gas transmission systems though the process and techniques described may also be used more generally.

ELEMENTS OF DESIGN

A pipeline system consists of three major segments: gathering, transmission, and distribution. The transmission system is usually different from gathering and distribution since it combines high operating pressures (5,500 to 15,000 kPa) with large diameter pipelines (NPS 16 to NPS 42).

The high level of energy stored in compressed gas requires a gas transmission system that provides an acceptable level of protection against leaks and ruptures. For an oil transmission system, emphasis is usually placed on leak prevention. In high vapor pressure systems, the potential for ruptures is the most important issue to be addressed. Figure 8.1, taken from Nara (1983), shows the relationship between the design and operating requirements for a pipeline and those of the line pipe material.

To transport large volumes of gas economically over long distances, the system design requires a particular combination of pipe size and operating pressure. The system operating pressure, pipe diameter, wall thickness, and pipe grade specifications are based on the detailed design of the facility. The severity of the external environment and the nature of the fluid being transported, e.g., corrosive/abrasive are important factors, too. Once these specifications have been established, material selection can take place.

The purpose of this task is to ensure that the material selected will perform safely, reliably, and efficiently. This is achieved by determining performance criteria that effectively address:

- Resistance to fracture initiation and propagation
- Material strength
- Good weldability in both shop and field conditions
- Fit-up requirements
- Acceptable defect size

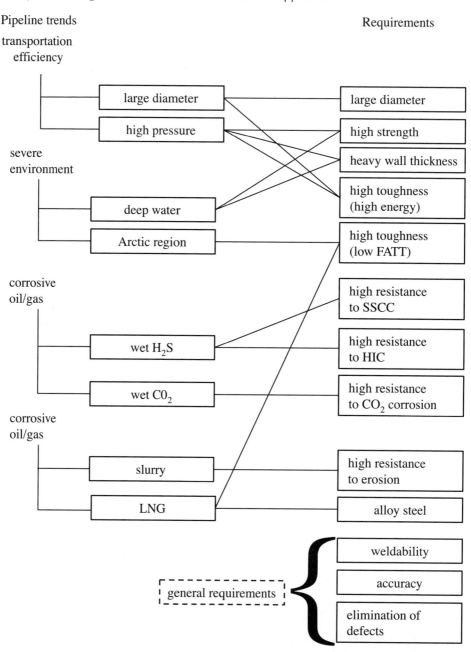

Figure 8-1. Relationship between pipeline service and line pipe requirements [Nara 1983]

The result of this exercise is that each major pressure containing component in the system (e.g., pipe, valves and fittings) is specified to have the key material characteristics required to meet the identified performance criteria at optimum cost. The toughness level is specified to reduce the risk of fracture initiation and propagation while sufficient strength is required to safely withstand the design pressure.

The material performance criteria are considered by addressing or specifying the following:

- The toughness level is specified to reduce the risk of fracture initiation and propagation,
- Sufficient strength is selected to safely withstand the design pressure,
- Restrictions on carbon and chemical composition are specified to help ensure good weldability,
- Risk of field fit-up problems is reduced by placing tight limits on dimensional tolerances,
- Inspection procedures are specified to ensure that components are free of defects and that sound workmanship and manufacturing standards are being followed.

While the quality control function is most evident during the materials selection process and the construction of the facilities, the implementation of quality management systems extends from the initial design stages through to operation.

Material Design and Selection

Material selection is carried out during the design stage of the pipeline project. The engineer has to select and consequently designate the materials for the line pipe and associated equipment.

It is often beneficial to standardize the design parameters and operating requirements. Standardizing these requirements allows the engineer to designate design standards for material selection.

A material designation standard covers the requirements of the selection of materials for use in pipelines, compressor or pump stations, and meter stations, according to approved company and industry standards. It also allows for the designation of items such as design drawings, purchase orders, and the quality procedures to be followed.

In selecting materials, due consideration is given to their safe and reliable performance under the anticipated in-service conditions over the lifetime of the component.

When developing the materials designation standard, components up to a certain size, typically NPS 16, are normally specified to an applicable industry standard.

For components exceeding this size, the material performance criteria may change significantly. There are also a number of components, such as end closures and orifice fittings, for which there is no applicable industry standard.

In both of these cases, the pipeline owner/operator may wish to develop proprietary specifications. These proprietary specifications supplement the applicable industry standards, where available, by providing details of additional requirements. This is particularly important for larger components, which may not normally be available as stock items. In some cases, the proprietary specification provides a basic, de facto standard where no industry standard exists.

Restrictions are placed on composition and dimensional tolerances to minimize welding and construction difficulties. Toughness specifications are adjusted to provide a higher degree of mechanical integrity consistent with design conditions. Strength requirements are modified to reflect the manufacturing process involved in producing a given component. Finally, proprietary specifications may include clarification of the industry standards requirements that are listed as "subject to agreement."

It is important to recognize that materials not listed in the Materials Designation Standard may be acceptable under a different set of design and operating conditions. Such changes may be implemented on a project basis after "due consideration" is given to the specific requirements of a given project.

A separate group, the quality systems group, is responsible for developing and administering quality management programs. These programs typically cover quality during design, procurement, and construction.

It is important to establish a quality assurance system to ensure that the materials specified are purchased and correctly installed. Without this control system, problems can arise. For example, consider a case where a material with a high degree of weldability has been specified to facilitate welding under difficult field conditions. (Weldability is a term used to describe a metal's capability of being welded to form a sound structure.) If the vendor ships material with poor weldability and a quality control system either has not been established, or is ineffective in preventing installation of the shipped material, then extra, costly work will be required to remedy the problem and ensure quality welds.

Deviations from the material specifications need to be closely monitored by individuals in the quality systems group to ensure that design criteria are met.

Where small diameter components, less than NPS 16, are required, the available industry standards are generally used. The design and operating stresses associated with these components are usually relatively low (less than 175 MPa), therefore the strength and toughness requirements are also low. The required strength (241 MPa) and toughness (27 J) levels can be readily achieved. Past experience and the consequences of failure would indicate that the industry standards provide a sufficient degree of safety and reliability.

Large diameter components (NPS 16 and larger), generally operate at higher stress levels and contain a greater volume of stored energy (which is much more dangerous in gas systems). The results of failure can be extreme. The required material properties (strength and toughness) to prevent or limit failures may not be readily achieved without special attention. At the same time, the costs of the component can be substantial and its supply limited. Often the more stringent the proprietary requirements the higher the cost will be. Therefore, it is important to determine whether the increased safety/reliability above the industry standard threshold is value-added.

The primary performance criteria that are reviewed and considered are:

- Fracture control capabilities
- Material strength
- Chemical composition/weldability
- Dimensional tolerances
- Inspection requirements

Fracture Control Design

Thomas (1988) has defined fracture control as, "... the rigorous application of those branches of engineering, management, manufacturing, and operations technology dealing with the understanding and prevention of crack initiation and propagation leading to catastrophic failure."

Hence one of the main design aspects considered in the selection of materials or the development of any specification is the fracture control requirement. It must be recognized that while leaks in a transmission system cannot be entirely eliminated, fractures initiating from a leak can be controlled. Leaks caused by mechanical damage, corrosion, defects in the material, or any other reason must remain stable and should not initiate a fracture in the component involved. Fracture initiation control is obtained by specifying material that has sufficient notch toughness or resistance to fracture.

The notch toughness requirements, and hence the fracture control requirements, are also affected by the component design stresses. Stress levels must therefore be considered in the determination of the toughness requirements. Items with high design stress levels require the specified materials to have sufficiently high toughness for adequate fracture control. Conversely, items with low design stress levels have low toughness requirements.

Valve designs, for example, require the body section to be fairly rigid to allow for proper valve operation. As such, valves are designed with thick sections such that the resultant stresses are less than 50% of Specified Minimum Yield Strength (SMYS). Due to the low stresses, the toughness requirements are also relatively low.

Pipe, however, forms a major part of the transmission system and wall thicknesses are usually minimized to achieve optimal costs. The thin wall design results in stress levels up to 80% SMYS for natural gas pipelines, considerably higher than that for valves. The fracture control toughness requirements are therefore higher.

Rothwell (1997) has tabulated (Table 8-1) a summary of the main requirements for fracture control found in a number of major international standards and recommendations. He cautions though that it is not possible to show every detail of the standards in such a format, so a full understanding can be obtained only by referring to the actual documents. Although Table 8-1 is far from comprehensive, it does cover North American practice, the United Kingdom, and the EPRG recommendations for ductile fracture arrest toughness, which has been incorporated into the ISO Standard 31821 for line pipe. Some interesting observations and comparisons can be obtained from the table; for example, the Australian Code is the only one of those presented that requires a formalized approved fracture control plan such as that shown in Figure 8-2 (courtesy of Venton and Dietsch 1997). (Though individual regulators often require such a plan before approving highly visible projects.) Besides, most, if not all, of the elements contained in this flow diagram can be found in the formal approach taken to fracture control design by many pipeline companies— and outlined later in this chapter.

All of the codes and recommendations state requirements for the control of both brittle and ductile fracture. There are some differences in the size of pipe to which the requirement is applied as well in the requirements themselves. The interested reader is referred to Rothwell's (1997) paper for a very full discussion of the code comparisons, and indeed the proceedings of the entire seminar on "Fracture Control in Gas Pipelines" in which it appears. In the Canadian code (see Chapter 1 for "Material Code"), the general requirements for fracture control have been delineated according to service category as shown in Table 8-2.

Category III is provided for short pipe runs where resistance to initiation is more critical than crack propagation. Arrest requirements are particularly critical for gas pipelines and are also important for oil pipelines containing high vapor pressure liquids. The use of the drop weight tear test, with a defined shear area (SA) requirement, ensures that the pipe material will fail in a ductile mode, which, as will be seen, propagates at a slower velocity than a brittle (cleavage) mode. It also ensures crack arrest in a pipeline carrying low vapor pressure oil. Before looking at the required pipe toughness levels and the methods to determine these levels, the basis of the initiation control method should be reviewed.

TABLE 8-1. Examples of fracture control requirements in international standards and recommendations (Rothwell 1997)

	AS 2885 - 1997	ASME B31.8	CSA Z662	IGE TD/1 (and referenced documents)	EPRG recommended
Fracture control plan	Required	Not required	Not required	Not required	
Fracture initiation toughness:					
Pipe body	No	No	Yes	Yes	
Seam weld	No	No	Yes (design temperature <−5°C)	Yes	
Fracture propagation toughness:					
Brittle	DWTT 85% SA	DWTT 40% SA (Charpy 50%)	DWTT 60% SA any lot, 85% all heat average	DWTT 75% SA	—
Ductile	Charpy energy	Charpy energy or crack arrestors	Charpy energy or crack arrestors	Charpy energy	Charpy energy
Arrest toughness calculation:					
Method	Battelle equation	Choice of Battelle, AISI, BGC, BSC	AISI referenced, higher values may be needed	BGC equation	AISI equation (X70 or less) Battelle equation (X80)
Limitations	CH₄, MAOP 15.3 MPa or less, X70 or less	CH₄	Lean gas, OD < 1067, MAOP < 8 MPa	(Lean gas implied)	(Lean gas)
Specification approach	Statistical - minimum Charpy >75% of arrest value (for default fracture length)	All-heat average > arrest value	All-heat average > arrest value	Not specified. In BGC practice, statistical based on specified fracture length	Minimum 75–90% AISI value (X70 or less), 100% Battelle value (X80)
Arrest length specified	Yes, default to two pipes each direction	No	No	In practice, 2 or 3 pipe lengths at 95% probability	—
Exemptions	Stress at MAOP < 30% SMYS (brittle)	For OD 406 mm and larger, stress at MAOP < 40% SMYS	(Initiation) Stress < 50 MPa, diameter < 60.3 mm, wall thickness 5 mm or less, design stress 225 MPa or less with design temperature > − 30°C (Propagation) Design stress lower than threshold stress based on AISI formula	Stress at MAOP, <30% SMYS	—
	MAOP < 50% SMYS (ductile)	For OD < 406 mm, stress at MAOP 72% SMYS or less			
	Diameter < DN 300 or wall thickness < 5 mm				
Risk management context	Yes	No (may be applied in practice)	No	No (is applied in practice)	Some national regulations
Design factors based on Class Locations	No	Yes	Yes	Yes (two only)	Two design factors used in tabulations

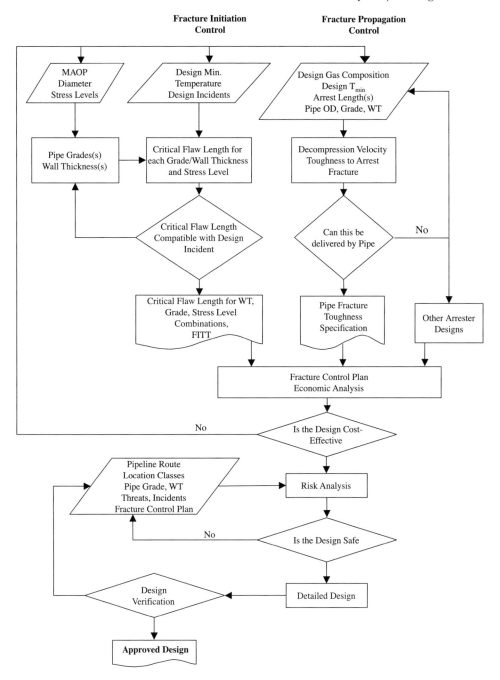

Figure 8-2. Fracture control design process [Venton and Dietsch, 1997, Permission from Welding Technology Institute of Australia]

One of the key requirements in pipe design is the prevention of brittle failure. For brittle fracture to occur there have to be three preconditions:

- The presence of tensile stress, (though this can be very low).
- A notch, defect, or stress concentration.
- A temperature at which the steel is "notch brittle."

TABLE 8-2. Line pipe category and application (Ref CSA Z662, 2007)

Pipe Category	Requirements	Typical Application
I	No notch toughness requirements	Low vapor pressure fluids e.g., water/crude oil)
II	Notch toughness in form of energy absorption and fracture appearance. CVN 27J, OD < 457 mm, 40J, ID > 457 mm, DWTT shear area 60% average	Buried and above ground pipe line (−5C to −45C)
III	Proven notch toughness in form of energy absorption only CVN as for Cat II No shear area requirement	High Vapor Pressure (HVP) Above-ground piping (e.g., compressor station for pipe length < 50m)

Note: DWTT refers to a drop weight tear test described later.

Brittle fracture can spread at very high speed; indeed the speed of a brittle fracture in a gas pipeline is greater than the gas decompression velocity. Thus, the crack tip sees a constant pressure and the length of a failure can be very extensive, in fact it stops only when it encounters something tougher. The failure surface exhibits chevron markings pointing toward the origin of the crack. One of two tests can be specified to measure resistance to fracture propagation by a shear area requirement, i.e., either a Charpy test, or a drop weight tear test. The drop weight tear test is the better of the two because it allows a larger energy of absorption to be imparted to the test specimens. As most pipe materials exhibit a ductile to brittle transformation with decreasing temperatures, the testing is generally conducted at the minimum design temperature to ensure that the results are representative of the material at operating temperatures. A typical transition curve is shown in Figure 8-3. The location of the transition region depends on the material. For some materials it is highly variable, with steel microstructure and chemistry. Transition may occur at +150°C, while in other materials it can be −75°C.

The Charpy V notch (CVN) test is conducted using either a standard full size specimen, 10 mm × 10 mm with a 2-mm notch located at the midpoint of its 55-mm length, or a sub size, e.g., 2/3 or 6.7 mm × 6.7 mm section with no change in length or depth relative to the notch. The drop weight tear test uses a plate specimen 12 inches long × 3 inches wide with a 0.2 inch chevron through the thickness notch machined at its midpoint.

After the brittle fracture test is completed, the fracture surface of the test specimen is examined to determine the percentage of the surface that has failed in a ductile or tearing manner. A dull grey surface is indicative of ductile failure. To prevent brittle fracture in line pipe, acceptance is based on a minimum of 60% of the surface exhibiting ductile behavior. Note that notched impact tests measure resistance to fracture propagation - a dynamic event.

All the commonly used pipeline steels are strain rate sensitive, the effect of increasing strain rate being to increase the yield stress of the material. This makes yielding more difficult and hence the material behaves as if it is more brittle at high strain rates (impact loading). High loading rates have the effect of shifting the transition temperature approximately 60°F in line pipe steel.

Once resistance to brittle fracture is established, fracture initiation control can be achieved by specifying that the pipe has sufficient notch toughness. The toughness is determined through Charpy impact testing, the amount of energy absorbed by the material during fracture being used as the basis for acceptance.

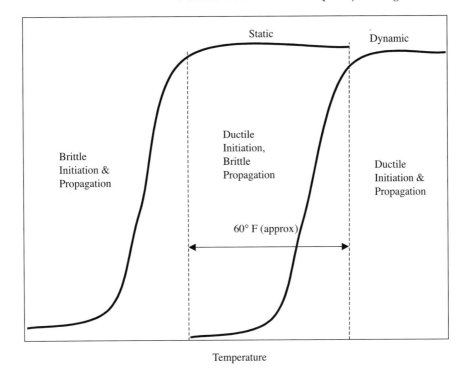

Figure 8-3. Typical ductile to brittle transition curve

The method used to determine the minimum toughness requirements is based on the Battelle/AGA relationship for ductile initiation developed by Maxey (1974) from work originally described by Burdekin and Stone (1966). They, in turn, used the Dugdale (1960) elastic plastic strip yield model to derive the relationship between the crack opening displacement and the applied stress in a biaxial stress field (Figure 8-4). They added the Folias factor M_T and the use of flow stress to account for the strain hardening of typical pipeline steels. The Battelle relationship correlates Charpy toughness to the critical throughwall defect size for fracture initiation. That is, the critical defect size required to initiate a fracture can be calculated for various Charpy toughness levels using the following formula:

$$\frac{K_c^2 \pi}{8c\bar{\sigma}^2} = \ln \, \sec \frac{\pi}{2} \left[\frac{M_T \, \sigma_T}{\bar{\sigma}} \right] \qquad (8-1)$$

where

$$K_c^2 = \frac{12 \, C_v \, E}{A_c}$$

σ_T = hoop stress at failure (PR/t)
P = internal pressure level at failure, MPa
R = outside radius of the pipe, mm
t = wall thickness, mm
$\bar{\sigma}$ = flow stress of the material (yield strength, σ_{y_s} + 68.95 MPa)
2_c = length of the throughwall flaw, mm
C_v = Charpy, V notch shelf energy, J

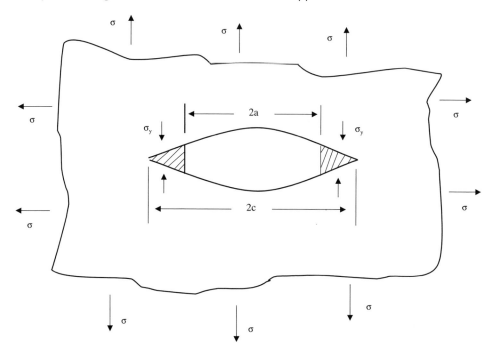

Figure 8-4. Dugdale strip yield model

A_c = area of fracture surface of a Charpy V notch specimen, mm^2
E = Young's modulus, MPa (207,000 MPa for steel)
M_T = "Folias" correction, which takes account of the effects of bulging.

$$M_T = \left[1 + 1.255\frac{c^2}{Rt} - 0.0135\frac{c^4}{R^2t^2}\right]^{0.5} \qquad (8-2)$$

A number of groups have developed formulae to predict the required Charpy V notch energy required for a propagating failure to be arrested within one pipe length in each direction from the initiating defect. Some of the more frequently used are listed below and can be found in clause 941.1 of ASME B31.8:

(a) $\qquad CVN = 0.0108\, \sigma^2 R^{0.33} t^{0.33}$ (Battelle)

(b) $\qquad CVN = 0.0345\, \sigma^{1.5} R^{0.5}$ (AISI)

(c) $\qquad CVN = 0.0315\, \sigma R/t^{0.5}$ (BGC)

(d) $\qquad CVN = 0.00119\, \sigma^2 R$ (BSC)

where CVN = full size Charpy V notch absorbed energy, ft.lb
σ = hoop stress, psi
R = pipe radius, in.
t = wall thickness, in.

The sources of the above equations are:

Battelle - Battelle Memorial Institute
AISI - American Iron and Steel Institute
BGC - British Gas Council
BSC - British Steel Corporation

Another representation based on 2/3 Charpy values has been given by Cabral and Kimber (1997) as follows:

Battelle Memorial Institute
$$C_{v2/3} = 2.382 \times 10^{-5} \, \sigma^2 (Rt)^{1/3}$$

AISI
$$C_{v2/3} = 2.377 \times 10^{-4} \, \sigma^{3/2} (2R)^{1/2}$$

British Gas
$$C_{v2/3} = \sigma \left(\frac{2.08R}{t^{1/2}} - \frac{v \, R^{1.25}}{t^{3/4}} \right) \times 10^{-3}$$

Japan
$$C_{v2/3} = 2.498 \times 10^{-4} \, \sigma^{2.33} (2R)^{0.3} t^{0.47}$$

CSM (Italy)
$$C_{v2/3} = -0.627t - 6.8 \times 10^{-8} \left(HR^2 / t \right) + 2.52$$
$$\times 10^{-4} R\sigma + 1.254 \times 10^{-5} \left(Rt\sigma^2 / H \right)$$

Mannesmann (Germany)
$$C_{v2/3} = 19.99 \times e^{0.287 \times 10^{-8}} \left(\sigma^{1.76} (2R)^{1.09} t^{0.585} \right)$$

where the additional parameters are defined as

ν = a constant = 0.396 for natural gas
H = backfill depth mm.

When these formulae were applied to a buried pipe 752 mm in diameter, with 8.1 mm wall thickness, X65 grade, buried to a depth of 750 mm, the resulting 2/3 Charpy values were plotted against stress level in Figure 8-5 for a range of and pipe diameters shown in Figure 8-6. There are quite noticeable differences between the formulae with the British Gas results being the most conservative.

For a given design stress level, the critical defect size increases with increasing Charpy toughness. However, even with infinite toughness, the critical defect size is limited by the flow stress of the pipe. The theoretical relationship is shown in Figure 8-7. Figure 8-8 presents an example for the situation where a NPS 24 × 6.7 mm Grade, 483 pipe operates at a design pressure of 8,450 KPa.

Assuming the stress in the pipe wall remains constant, the toughness required to prevent fracture initiation also increases with the pipe diameter. This is caused by the bulging of the pipe as described by the Folias factor.

The minimum toughness specified by some companies to control fracture initiation ensures that the critical throughwall defect size is at least 90% of the maximum theoretical defect size at the maximum operating pressure of the pipeline. The effectiveness of this approach was demonstrated when an NPS 16 pipeline, operating at a pressure of 5,930 kPa, was punctured by a ditching machine. The 50-mm-long by 60-mm-wide hole resulted in a leak, but fracture initiation did not occur. The pipe had a specified Charpy energy of 27 Joules and, using the Battelle/AGA relationship, the calculated critical defect size was 100 mm. Thus, fracture initiation was successfully controlled by having a greater critical defect size than the actual defect.

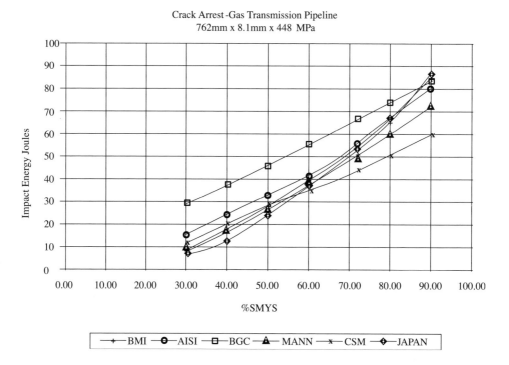

Figure 8-5. Comparisons of various formulae for determining minimum toughness for crack arrest [Cabral and Kimber, 1997, Permission from Welding Technology Institute of Australia]

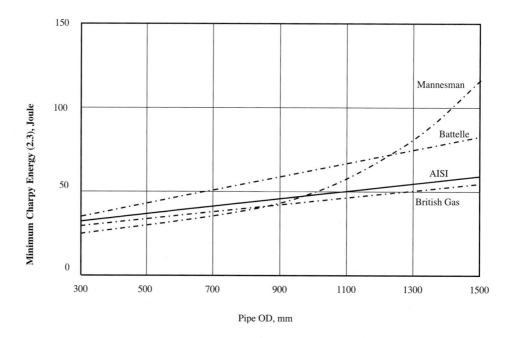

Figure 8-6. Minimum charpy value predicted by four empirical equations Class 900–X70 pipe at 72% SMYS [Piper and Morrison, 1997, Permission from Welding Technology Institute of Australia]

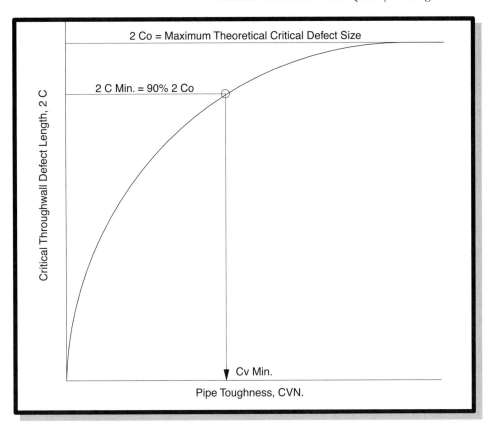

Figure 8-7. Theoretical relationship between pipe toughness and critical throughwall defect size

For gas pipeline systems with a diameter of NPS 36 or larger and high pressure (7,000 kPa), supplemental requirements for ductile fracture propagation control may be required. Fracture propagation can occur if the defect exceeds the critical size. Under some conditions it is possible for a crack in a gas pipeline to propagate through the pipe in a ductile manner for a very long distance. This propagation occurs when the fracture velocity in the pipe equals or exceeds the gas decompression velocity. In this situation, the pressure at the crack tip would provide sufficient energy for propagation to continue. The potential for ductile fracture propagation increases for rich gas compositions because the decompression velocity for "rich" gas can be quite slow.

Ductile fracture propagation is normally controlled through the specification of notch toughness. While it is unlikely that ductile fracture propagation will occur, the toughness requirements chosen provide a high probability that any failure would be limited to an acceptable length. Experience has shown that ductile fractures are often arrested by factors other than toughness (e.g., by a girth weld breaking ahead of the crack tip). Another method to control ductile fractures is to install a fracture arrestor, such as a sleeve, on the pipe. Fracture arrestors are frequently installed on carbon dioxide pipelines.

The required toughness for fracture arrest can be determined by using a hypothesis developed by Battelle/AGA. Where necessary, this value is specified as an all-heat average. The result is that a sufficient quantity of pipe is obtained with the required

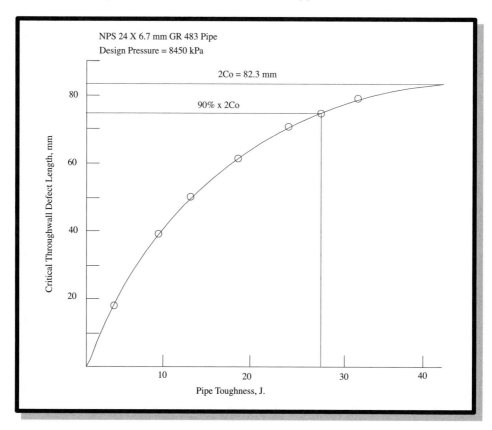

Figure 8-8. Critical throughwall defect length vs. pipe toughness

arrest toughness to be randomly distributed in the system and limit failures to a statistically acceptable length (average of approximately 75 m). A worked example of the technique is given in the next section.

Fracture propagation in an oil pipeline is normally not a concern due to the very rapid decompression of the product. However, if an oil pipeline were to be pressure-tested using a gas such as air, or if high vapor pressures are involved, the control of fracture propagation should be considered.

For components, it is normal to consider only fracture initiation because the length of a failure is limited by the size of the components (Figure 8-9). Critical defect size can be related to Charpy toughness using a fracture mechanics approach. The following formula is frequently used to determine the required Charpy toughness for components:

$$C_v = \left[\frac{4.5045 \, \sigma^2 \, \pi \, 2c}{2E}\right]^{2/3} \tag{8-3}$$

where C_v = Charpy energy, J
                σ = Applied stress at failure, MPa
                $2c$ = Length of the throughwall flaw, mm
                E = Young's modulus, MPa (207,000 MPa for steel).

Operating Stress

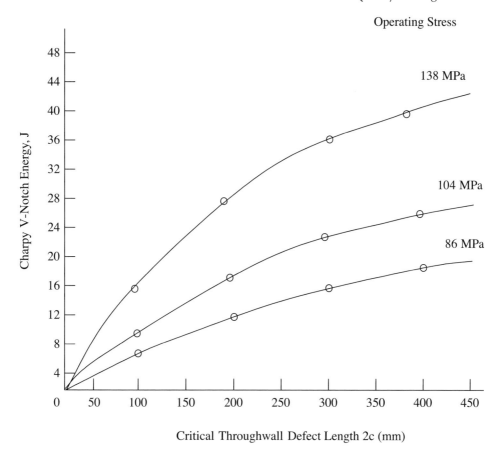

Figure 8-9. Critical throughwall defect vs. Charpy toughness for components

Fracture Velocity

The toughness regime determines the fracture velocity in the pipe. For example, one might expect a brittle fracture to propagate at a rate between 1,500 and 3,000 ft/sec (457 m/s and 914 m/s), whereas a ductile fracture is significantly slower — 400 to 700 ft/sec (121 m/sec to 213 m/sec) being typical speeds though values as high as 1,100 ft/sec (355 m/sec) have been observed.

The gas decompression velocity is equivalent to the acoustic velocity for the mixture with wave speeds of between 1,240 and 1,450 ft/sec (381 m/sec and 442 m/sec) typical for a dry lean gas.

Example

Determine the all-heat toughness for a 42" diameter API 5L × 70 grade line pipe operating at 1,000 psig if the requirement is to ensure fracture arrest within two joint lengths with a 90% probability of success.

Step 1: Determine the strength requirements.

Since the pipe line likely passes through a number of locations the required wall thickness will vary. By taking into account the location factor "F" (see Chapter 7, Table 7-2). The required wall thicknesses are determined from Chapter 7, Equation (7-1).

Assuming a Class 1 Location

$$t = \frac{PD}{2SFLJT}$$

For an operating temperature less than 250°F, the derating factor $T=1.0$, while the longitudinal factor E may also be taken as 1. The Table 8-3 contains the results of the various computations similar to the following:

$$t = \frac{1000 \times 42}{2 \times 70000 \times 0.8 \times 1 \times 1 \times 1} = 0.375 \text{ inches} \qquad D/t = 112$$

Step 2: Determine the required toughness

There is a requirement in ASME B31.8 Clause 841.11(c) to specify a fracture toughness for cases where $F \geq 0.4$ and $D \geq 16$ NPS, i.e., all of the above cases. This clause also provides four equations that may be used to calculate the full-size Charpy absorbed energy requirements.

(a) $\qquad CVN = 0.0108 \, S^2 R^{0.33} t^{0.33}$ $\qquad$ (Battelle)

(b) $\qquad CVN = 0.0345 \, S^{1.5} R^{0.5}$ $\qquad$ (AISI)

(c) $\qquad CVN = 0.0315 \, S \, R/t^{0.5}$ $\qquad$ (British Gas)

(d) $\qquad CVN = 0.00119 \, S^2 R$ $\qquad$ (British Steel)

where $\quad$ CVN = full-size Charpy V notch absorbed energy (ft lb)
$\qquad S_H$ = hoop stress (psi) = $S \times F \times E \times T$
$\qquad R$ = pipe radius, in
$\qquad t$ = nominal wall thickness, in.

Selecting the wall thickness corresponding to the Class 1 Division 1 location, all four CVN values and their average are computed as follows.

$$S_H = S_y \times F \times E \times T = 70,000 \times 0.8 \times 1 \times 1$$

$$S_H = 56,000 \text{ psi}$$

(a) $\qquad CVN = 0.0108 \times 56^2 \times 21^{.33} \times 0.375^{.33}$ $\qquad$ $= 67.4$

(b) $\qquad CVN = 0.0345 \times 56^{1.5} \times 21^{0.5}$ $\qquad$ $= 66.2$

(c) $\qquad CVN = 0.0315 \times 56 \times 21 / 0.375^{0.5}$ $\qquad$ $= 60.92$

(d) $\qquad CVN = 0.00119 \times 56^2 \times 21$ $\qquad$ $= 78.4$

Average $\qquad\qquad\qquad\qquad\qquad\qquad\qquad\qquad\qquad\qquad\qquad\qquad$ $= 68.2$ ft.lb
and for all classes (See Table 8–4)
Step 3: Determine the percent arrest

TABLE 8-3. Pressure design requirement

Location	F	t	D/t
Class 1, Division 1	0.8	0.375 (in inches)	112
Class 1, Division 2	0.72	0.417 (in inches)	101
Class 2	0.6	0.5 (in inches)	84
Class 3	0.5	0.6 (in inches)	70
Class 4	0.4	0.75 (in inches)	56

From Figure 8-10 (Eiber and Maxey 1979), the percentage of pipe in the total supply capable of arresting a fracture within two joint lengths (80 feet) with a probability of success of 90% is approximately 90%. Using this number in Figure 8-11 and assuming that the quality control at the pipe mill is such that the coefficient of variation (the ratio of the standard deviation to the mean) for the all-heat toughness is, say, 20%, then the ratio

$$\frac{\text{arrest toughness}}{\text{all-heat average}} = 0.75$$

Hence the required mill all-heat average test results for Class 1 Division 1 pipe must not be less than;

$$\text{all-heat average} = \frac{68.2}{0.75} = 90.92 \qquad \text{use 92 ft.lb}$$

If quality control is tighter and the coefficient of variation is, say, 5% then from Figure 8-11 (Eiber and Maxey 1979) the arrest level/all-heat average will be 0.94 and the required mill heat average will be 68/0.94 = 72.3, say 75ft.lb.

In this manner one can link the fracture control requirements to the pipe supplier's production quality.

Strength Requirements

While the required strength for a material is determined by the system design and the applicable design codes, the available strength limits are governed by the material selected. The strength levels that can be reliably achieved using current manufacturing processes for a given material should be regularly reviewed and guidelines established to assist in the system design function.

In discussing the strength of a material, reference is usually made to the yield strength, which is the point at which plastic deformation begins to occur as the applied stress is increased. The specified minimum value of yield strength is commonly known as the grade of the material; thus a Grade 414 material has a specified minimum yield strength of 414 MPa.

TABLE 8-4. Notch toughness requirement class

Equation/cPass	1-1	1-2	2	3	4
(a)	67	56.5	41.7	30.8	21.2
(b)	66	56.6	43	32.7	23.4
(c)	61	51.6	39.3	29.9	21.3
(d)	78	63.5	44	31	20
Average	68	57	42	31	22

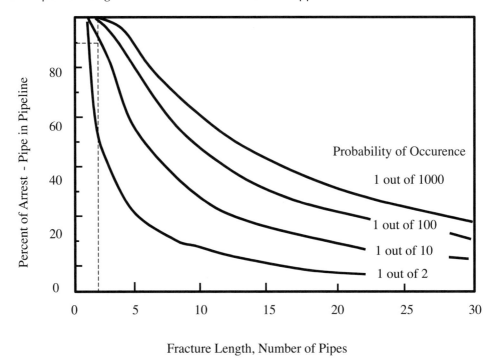

Figure 8-10. Percentage of arrest pipes in a population of pipe lengths [Eiber and Maxey, 1979] (Used with permission of Pipeline Research Council International, Inc. (PRCI). Reproduction is prohibited without the written consent of PRCI. Use of this (these) figure(s) in this publication grants no rights to its (their) subsequent use.)

In general, it is desirable to use the highest strength or grade of material possible. With a higher strength material, the amount of material required is reduced, assuming the component size and operating pressure are constant. This reduction in the quantity can result in a significant cost saving. For example, an NPS 42 pipeline constructed from Grade 483 pipe, as compared to Grade 414, would require 49 tonnes per kilometer less steel for the same operating pressure. The Grade 483 pipe is more expensive per tonne, but because fewer tonnes are required there is a net cost saving in material of about $25,000/km. The reduced wall thickness also translates into lower costs for transportation, welding, and nondestructive inspection.

Sanderson et al. (1999) have provided a very comprehensive account of the potential savings that could be realized using X100 grade line pipe - a material that is still at the developmental stage of production. Their study suggests that the use of this very high strength steel for large diameter gas pipelines, over long distances, in remote environments where social and environmental risks are low,can provide present value cost savings of up to 8% when compared to the use of a X70 grade line pipe over a 30-year life.

Although the use of the thinnest and highest strength material can result in a cost saving, some practical considerations do exist. A thin component may be more susceptible to construction and handling damage. The compressive loads applied during transportation or field fabrication (i.e., during bending) can cause denting or buckling.

Most companies establish guidelines based on past experience to limit the minimum allowable thickness for each pipe diameter. CSA standard Z662 also imposes minimum requirements.

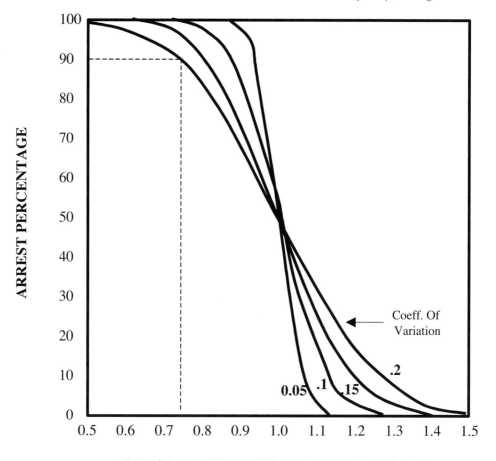

Figure 8-11. Fracture length as related to probability and percent of arrest pipe [Eiber and Maxey, 1979] (Used with permission of Pipeline Research Council International, Inc. (PRCI). Reproduction is prohibited without the written consent of PRCI. Use of this (these) figure(s) in this publication grants no rights to its (their) subsequent use.)

Chemical Composition

The chemical composition of a material will affect its weldability as well as its strength and toughness. When considering any restrictions on the chemical composition to improve weldability, the strength and toughness requirements must also be reviewed.

Where the system operating conditions allow for the use of materials with strength and toughness levels that are considered to be normal, weldability is controlled by limiting the carbon equivalent. Numerous attempts have been made to relate weldability to chemical composition, historically, the carbon equivalent (CE) was calculated using the International Institute of Welding (IIW) formula Boniszewski (1979):

$$CE = C + \frac{Mn}{0.6} + \frac{Cr + Mo + V}{5} + \frac{Ni + Cu}{15} \qquad (8-4)$$

TABLE 8-5. Compliance factor (F) - carbon equivalent formula

Carbon (%)	Compliance Factor	Carbon (%)	Compliance Factor	Carbon (%)	Compliance Factor
<0.06	0.53	0.11	0.70	0.17	0.94
0.06	0.54	0.12	0.75	0.18	0.96
0.07	0.56	0.13	0.80	0.19	0.97
0.08	0.58	0.14	0.85	0.20	0.98
0.09	0.62	0.15	0.88	0.21	0.99
0.10	0.66	0.16	0.92	>0.21	1.00

More recently, the carbon equivalent has been calculated using the Yurioka formula, which is supposed to give a better indication of weldability, particularly for very low carbon steels:

$$CE = C + F\left[\frac{Mn}{0.6} + \frac{Si}{24} + \frac{Cu}{15} + \frac{Ni}{20} + \frac{Cr + Mo + V}{5} + 5B\right] \qquad (8-5)$$

Where F is a compliance factor that is dependent on carbon content.

The carbon equivalent formula controls the chemical composition by including the effects of the primary alloying elements (C, Mn, Si, Cr, Mo, V, Cb, B, Ni, Cu) on weldability.

To ensure that the material is weldable, the maximum allowable carbon equivalent is specified. The international standards for pipeline materials specify a maximum carbon equivalent. Individual companies may specify a tighter limit.

The need to have a suitable carbon equivalent requirement was demonstrated when a problem was encountered with the welding of pipe to valves. In this case, the valve material met the strength requirements but the carbon equivalent (0.70%) was well in excess of that allowed (0.50% maximum). The consequence was that excessive cracking occurred in the heat-affected zone during welding.

Dimensional Tolerances

While tighter dimensional tolerances, such as diameter and ovality, can reduce field fit-up and construction difficulties, the effects on the cost of the component must also be considered. The tighter tolerances may reduce construction costs but may increase the component cost. The end result may be that the solution is more expensive than the problem.

This was demonstrated in an internal TransCanada study conducted to determine what advantages could be gained by reducing the allowable plus-tolerance on the diameter of a fitting from 2.4 mm to 0.0 mm. By reducing this tolerance, savings approximately equal to 5% of the fitting cost during field construction could be realized due to improved fit-up, reduced welding times, and ease of inspection. However, there would also be an estimated increase of approximately 25% in the cost of the fitting. In this case, the overall result was that the tighter diameter tolerance was not economically justified.

Inspection Requirements

Inspection requirements primarily serve as a quality control tool. Inspection of the component in the manufacturer's plant ensures that the product is free of defects and suitable for its intended service.

A secondary function of any inspection requirement is to provide some means of process control. With this control, manufacturing methods can be adjusted to eliminate or drastically reduce problems that are encountered on a continual basis. An example of this is in the inspection of castings. If an inspection reveals that the same type of defect continually occurs in the same area on the casting, steps can be taken to eliminate the defect and reduce the overall costs by modifying the casting design, the sand mixture, etc.

When reviewing inspection requirements or where additional requirements are considered, the component manufacturing method and the inspection technique used must also be considered. The manufacturing method will determine what type of defects may be present and the inspection technique must detect those defects. At the same time, criteria to select and reject defects must be established only to consider imperfections that could cause a problem.

For example, castings are subject to void type defects. As such, radiographic inspection is a suitable technique. The acceptance criteria (MSS SP-54) recognizes that small slag inclusions or minor amounts of porosity, which are inherent in castings, are not injurious. The cost required to remove and repair these defects would exceed any benefit gained.

MATERIALS DESIGNATION STANDARDS

As previously stated, a material designation standard can be used as a tool when selecting materials. The materials and specifications listed in the Materials Designation Standard have been reviewed based on their ability to meet the required materials performance criteria, as well as any code and regulatory requirements. Again, it must be emphasized that this standard lists only materials suitable for the design and operating conditions most commonly used.

When using the Materials Designation Standard, it is necessary to have a completed and detailed system design. System size, design pressure, component type, and design temperature all influence the materials to be selected. The following steps outline the process for determining the correct materials designation from one company's Materials Designation Standard. Table 8-6 is one page from such a Standard.

Step 1: Locate the appropriate table for the application (Pipeline & Compressor Station - Sweet Gas Services) shown here as Table 8-6.

Step 2: Find the listing in Column 1 for the component type and size (NPS 24 blind flange)

Step 3: Applicable designations are given in Columns 2 and 4 for below-and above-ground designs, respectively

Step 4: Material strength is listed in Columns 3 and 5

Step 5: Column 6 indicates whether a specification sheet (data sheet) is required for components that have a proprietary specification

Pipelines Containing Sour or Corrosive Substances

In many instances interest extends beyond dry, sweet gas pipelines, to include wet/sour service for both oil and gas. For the most part, the aim in selecting a suitable material for a pipeline to handle corrosive products (untreated oil, gas, and condensate containing amounts of CO_2, H_2S, and chlorides) is to determine which material will most reliably and

TABLE 8-6. Example of a Material Design Standard

	Below-Ground Service Design Temperature $-5°C$			Above-ground Service Design Temperature $-45°$		
Blind flanges:						
NPS 16 and Larger	CSA CAN3-Z245.12	248		CSA CAN3-Z245 12	248	No
Smaller than NPS 16	CSA CAN3-Z245.12	248		CSA CAN3-Z245, 12	248	No
Swaged nipples:						
NPS 2 and Larger	ASTM A333 Gr. 6 Seamless, Schedule 80	241		ASTM A333 Gr. 6 Seamless, Sched. 80	241	No
Smaller than NPS 2	ASTM A105	248		ASTM A105	248	No
Solid bull plugs and pipe plugs (hex or round):						
NPS 2 and Smaller	ASTM A105	248		ASTM A105	248	No
Socket welding and threaded fittings:						
Elbows, Tees, Caps						
Unions, and Couplings						
NPS 2	ASTM A350 LF2	248		ASTM A350 LF2	248	No
Smaller than NPS 2	ASTM A105	248		ASTM A105	248	No
Nipples						
NPS 2	ASTM A333 Gr. 6 Seamless, Sched. 80	241		ASTM A333 Gr. 6	241	No
Smaller than NPS 2	ASTM A106 Gr. B Sched 80	241		ASTM A106 Gr. B Sched 80	241	No
Hex Head Bushings						
NPS2 and Smaller	ASTM A105	248		ASTM A105	248	No

cost-effectively carry the product for the required duration of service. Materials have to meet not only the corrosion resistance criteria but also mechanical limitations such as weldablility.

The generally accepted starting point in the selection process is to evaluate the use of a carbon steel having some corrosion allowance, with corrosion protection being provided by means of the injection of corrosion-inhibiting chemicals.

However, if the fluid stream corrositivity is high due to factors such as high flow rates, high CO_2, H_2S, and chlorides, it is more common to use corrosion-resistant alloys like duplex stainless steels and clad pipes.

When more than one alternative meets the technical requirements, say the use of carbon steel plus inhibitors, or the use of corrosion-resistant alloys, one has to make recourse to an economic assessment, preferably based on a life-cycle cost (LCC) approach.

Evaluating the Risk of Water Wetting of the Surface

If the environment contains no free water or if the water present is prevented from actually wetting the walls because of the flow regime, then there is no risk of internal corrosion and so materials selection is then determined by the mechanical (strength and toughness) and weldability requirements. When water wetting does occur, however, the corrosive effect of the CO_2 and any H_2S present have to be considered. An analysis of the flow regime is also critical in determining whether inhibition will be a possible option for effectively controlling corrosion (see previous chapter on turbulent flow regimes).

In gas lines, water condensation occurs whenever the temperature drops below the dew point ($-38°F$ is a commonly accepted value).

In oil lines, "free" water may be contained within an oil emulsion and will not give rise to corrosion as long as the flow rate is sufficient to entrain the water and provide a continuous oil film on the contact surface. The amount of water that can be entrained can be determined experimentally and depends upon the type, viscosity, and temperature of the oil. In extreme cases, oils that are capable of carrying more than 90% water in an oil emulsion have been found. In general, it has been found that most oils can safely entrain a water cut of up to 20% as long as the flow velocity is above the critical level of about 1 m/s. Stagnant regions (e.g., low spots on the pipeline) where water may accumulate below the oil layer are at risk even at very low water cuts (of just a few percent). Light gas condensate does not offer the same protection as oil and, in general, does not entrain water, so water wetting is likely even at very low water cuts.

In multiphase (gas-liquid) conditions, the wetting behavior depends strongly on the flow regime, derived either from computer flow simulations or estimated using flow regime maps. When multiphase pipelines are operating in the stratified flow regime there will be significantly different corrosion conditions on the top and bottom surfaces of the pipeline, with the top-of-the line pipe being largely unprotected by any liquid-carried inhibitors.

It should be remembered that flow rates and regimes may vary considerably over the operational life, and this can have a significant impact on the possible corrosion risks during the life of a project.

CO$_2$ Corrosion

When considering the feasibility of using carbon steel, the potential general corrosion rate due to CO_2 has to be calculated. Laboratory work on CO_2 corrosion has continued to be very active throughout the past few years and has yielded interesting developments and refinements in the modeling of CO_2 corrosion. Several methods now exist for predicting the CO_2 corrosion rate of carbon and low alloy steels, and these have been brought together in one publication of the European Federation of Corrosion (EFC) (EFC 23, 1997). This volume covers the electrochemistry involved in CO_2 corrosion and also explores (in the various papers included) the influence of flow velocity, effects of minor alloying additions on CO_2 corrosion rates, and options for mitigating corrosion by altering the environment.

Environmental modifications, which are usually considered, are the injection of corrosion inhibitors/addition of glycols. Glycols are added in bulk and reduce the corrosion rate as well as prevent formation of gas hydrates in gas systems. The efficacy of these methods of corrosion control will depend on several factors, including the flow regime and water content. The selection of film-forming inhibitors is particularly critically dependent upon flow effects and can best be evaluated by testing in flow loops rather than simple exposure testing.

Once the anticipated wall thickness loss has been calculated, a decision can be made as to whether this can be accommodated in a corrosion allowance on carbon steel or if a corrosion-resistant alloy (CRA) is required. In either case, further consideration has to be given to the possible effect of H$_2$S on the performance of the chosen material.

Sour Service

In corrosive conditions containing hydrogen sulphide, metallic materials may suffer a variety of hydrogen-induced embrittlement and cracking problems, which can potentially cause catastrophic failure. Thus these risks have to be evaluated even in the presence of just traces of water.

The resistance of carbon and low-alloy steels to sulphide stress corrosion cracking (SSCC) has been shown to be dependent not only on the partial pressure of H_2S (p H_2S), but also on the pH of the environment (EFC 16, 1995).

For other forms of H_2S corrosion in carbon and low-alloy steels such as step-wise cracking (SWC) or stress orientated hydrogen induced cracking (SOHIC), the H_2S partial pressure of 3.5 mbar is taken as the separation between sour and nonsour conditions. This limit is based on the lowest partial pressure at which SWC has been reported in practice (AGA NG-18 Report No. 131 1982). The risk of SWC occurring does not depend the total pressure of the system. Requirements for materials to avoid SWC depend upon the product form; products made from rolled steel (e.g., longitudinally welded pipe) require tight restrictions on the chemistry of steel as well as qualification by SWC testing. Seamless pipe is generally regarded to be resistant to SWC if the sulphur content is below 0.01%.

For corrosion-resistant alloys (CRAs) that may fail in H_2S service due to a combination of mechanisms involving SSC and SCC, there is no simple cutoff in H_2S partial pressure, which can be used to define the limits of the risk of cracking. Each type of alloy has to be considered individually.

Application Limits for Corrosion-Resistant Alloys

While typical CRAs used in the oil and gas industry tend to be fairly resistant to corrosion in the presence of CO_2, they do limit the maximum temperature to which they can be exposed before localized pitting corrosion occurs in the presence of H_2S and chloride ions. There is also a risk of stress cracking beyond certain limits of H_2S, although this is chloride content, temperature- and pH-dependent for the different types of alloys.

Guidance on the application limits of various classes of materials is given in various references, including NACE standard MR0175 (NACE MR0175). In certain cases it would be preferable to establish the performance of a material according to a commended test protocol (EFC 14, 1995). Such a test protocol could be used to qualify a material for a specific set of field conditions, simulating the expected brine composition and pH where known, in order to optimize materials selection. It is particularly recommended to verify alloy performance by testing when conditions are extreme or when the alloy has only recently been developed.

In summary, the martensitic stainless steels (9 to 15Cr) may all be used generally up to 90°C (9Cr and 13Cr) or 150°C (Super-13Cr and 15 Cr). The amount of H_2S to which they can be exposed without cracking is rather low and is critically dependent on the pH of the environment and also on chloride content. To date these materials have seen rather limited field application for pipelines (with the exception of extensive use by Mobil Indonesia in the Arun field) but there is growing interest in them for future projects. Successful application of these alloys will depend on solving the challenge of welding them in an economic way, with or without postwelding heat treatment, as the various alloys demand.

The limiting levels of H_2S to which the 22Cr duplex stainless steels can safely be exposed is approximately 0.1 bar, but the 25Cr super duplex alloys can tolerate up to 0.7

bar, depending on the chloride content and pH-limiting service temperatures being around 200°C and dependent upon chloride content. There has been an extensive application of many grades of duplex stainless steels in pipelines.

Life-Cycle Costing

In choosing an appropriate material for a pipeline according to the steps outlined above, it is possible that a final choice has to be made—typically between a corrosion-resistant alloy (CRA) and carbon steel—that can have an impact on operations, inspection, and maintenance activities. Therefore, while it can initially appear to be the expensive option, the longer-term cost implications need to be evaluated. This evaluation can be made on economic grounds using life-cycle costing (LCC). A major benefit of LCC is that it demands that the materials selection be made while taking a global view of the wider impact and not simply solving one problem in isolation.

QUALITY MANAGEMENT

Crosby (1980) describes quality as "conformance to the defined requirement." A requirement is something that is needed for a particular design - fit for purpose. It can be a standard, a specification, a purchase order, or a drawing.

Quality is obtained by using and organizing competent personnel to work effectively in accordance with approved plans and procedures in order to meet the defined requirements. It also requires monitoring, recording, controlling, and correcting, in a timely manner, any deficiency or errors that occur. Further, the cause of each deficiency must be identified and remedied to prevent recurrence.

A noncompliance, or a nonconformance, is an indication of lack of quality and can occur at the design, procurement, fabrication, or construction stages of the project. The group establishing the requirements must evaluate the noncompliance. An accepted item with a noncompliance effectively amends the requirement.

How to Have a Quality Project

There are basically two reasons to have a quality project: (1) Provide for safe and reliable operation; and (2) give maximum cost benefit to the owner. Figure 8-12 illustrates an example of cost-effectiveness. The focus should be on "total life-cycle cost," the clear inference from the figure being that while extremely onerous specifications will increase cost, underspecifying requirements can lead to early and costly failure services.

There are four steps to building and maintaining a quality pipeline project.

1. Quality in design.
2. Quality in procurement.
3. Quality in manufacture.
4. Quality in construction.
5. Quality in operations and maintenance.

Quality in Design

The development of material specifications are part of the design process and must be applicable to the operating conditions. The specifications for a valve suitable for use in a wet, sour environment, for example, are different from the specifications for a valve suited

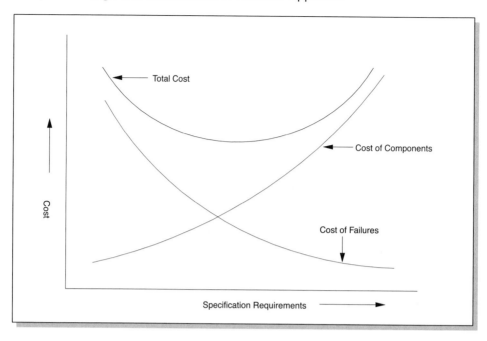

Figure 8-12. Example of cost-effectiveness

for a dry, sweet environment. Similarly service life should be taken into consideration, specifying requirements for a 25-year service life for a component with a much shorter life expectancy, and it will often prove to be expensive.

Design audits should be conducted to ensure the engineering materials are specified correctly. Typically, an audit would consist of a detailed check of the materials being specified, a report, and disposition of any noncompliance.

Quality in Procurement

For most pipeline projects the material cost will amount to about 40% of the total cost while construction will consume another 50%. The pipeline construction contractor will be dependent upon the timely arrival of quality product if his schedule is to be maintained and costly standby charges avoided. Vendor selection for the provision of line pipe, compression, and other components, as well as transportation logistics becomes extremely important. Since a large proportion of errors and nonconformances are caused by poor communications and undue haste, it is essential that the relationship between the internal project groups, the procurement team, and the supplier is effective. Figure 8-13 illustrates a report card that may be used to gauge continuous improvement between all three parties over the course of several orders. By quantifying the performance of each group and their adherence to the requirements, one is making use of a quality control process.

An integral part of assuming that goods and services are required in a timely quality manner involves selecting suitable vendors. One should not assume that every supplier is committed to providing quality service. Steele and Court (1997) stress the importance of understanding the current activity in the marketplace before selecting key vendors. They have illustrated the possible relationships shown in Figure 8-14. From a buying perspective it is essential to be able to identify and secure items that have a high criticality to the project, both in terms of delivery and cost. During times of feverish construction activity in the

Commodity:
Client (Department name / contact) :
Procurement (Contact) :

Supplier:
Date:

		Order 1	Order 2	Order 3	Order 4	Order 5
1	Procurement Assessment of Client Performance					
1 a	* clear definition of requirements					
1 b	* Timely advice of requirements					
1 c	* changes to requirements after order					
2	Client Assessment of Procurement performance					
2 a	* communicates clear expectations of client					
2b	* status reporting of materials progress / delivery					
2c	* materials provided on date as promised					
2d	* materials match requirements					
3	Supplier Assessment of Procurement Performance					
3 a	* clear scope /expectations provided					
3 b	* adequate lead time provided					
3 c	* timely issue of P.O., change notices					
3 d	* quality of P.O., change notices					
3 e	* changes after order					
3 f	* efficient payment					
4	Procurement Assessment of Supplier Performance					
4 a	* performance during bid / award stage					
4 b	* delivery performance - material					
4 c	* delivery performance documentation					
4 d	* quality performance					
4 e	* reporting performance					
4 f	* continuous improvement initiative					
4 g	* invoicing performance					

Figure 8-13. Procurement report card [Somerville 1999]

pipeline industry many items will have a long lead time. Besides understanding the potential vendors capability and capacity for meeting requirements, it is equally important to determine how they value the supplier relationship. Choosing a supplier for whom the volume of business is of little consequence may result in inattention that could impact the schedule—a very real "nonconformance." Ideally, the supplier will view you as a core customer and will be proactive in meeting the functional requirements (Coulson et al. 1999).

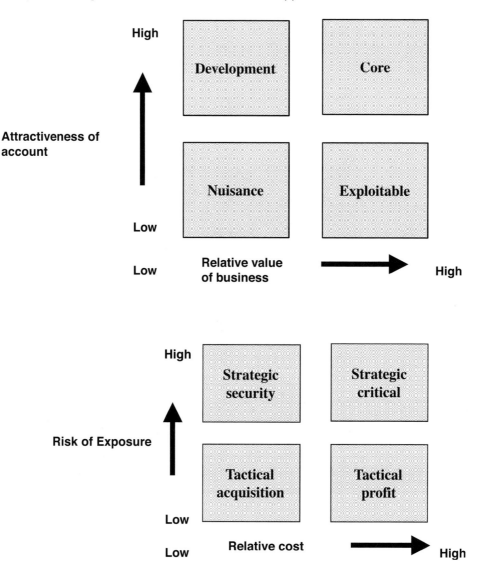

Figure 8-14. Supplier preference, supply positioning [Steel and Court, 1997, reproduced with permission from McGraw-Hill Co.]

Vendor Qualification

Qualification of a vendor provides assurance that the plant is capable of manufacturing items to the required specification using acceptable procedures. Therefore, from a technical viewpoint, each manufacturing location (or plant) should be qualified. Normally a more detailed evaluation is conducted for vendors supplying items to a proprietary or company specification than for items to an industry specification.

The following areas should be discussed and evaluated during a technical qualification for items supplied to a company specification:

- Capability
  - Specification requirements
  - Necessary equipment

- Quality
  - Quality organization
  - Measurement equipment
  - Documented procedures
  - Audit quality program
- General
  - Expertise
  - Required personnel
  - Competency
  - A copy of the appropriate regulatory registration if items are to be used under a particular jurisdiction

Once the above information is determined, it is advisable to have the vendor manufacture a few items on a trial order. The purpose of the trial order is to demonstrate how the vendor performs. His progress on further orders can be tracked using the report card shown in Figure 8-13. This process helps provide feedback for improvement or removal from the vendor list.

Most companies maintain a qualified vendors list, which indicates acceptable vendors, plant locations, size ranges, applicable specifications, any restrictions, etc.

Third-party Inspection Agencies

Third-party inspection agencies should also be qualified to ensure that conformance to design requirements can be obtained. Individual inspectors can be qualified by examination, depending on what is being inspected and how critical it is. Ideally one should deal with suppliers who have a very high standard of internal quality control. This will avoid the need for third-party inspection since quality will have been "built in" rather than be the result of costly inspection.

Contracts for inspection can be awarded by project, annually or semiannually, or in a variety of other ways. Contracts can be awarded to more than one agency (e.g., by factory, by geographic area, or by item to be manufactured). Most agencies have expertise in one or two areas. For example, one agency may have expertise in pipes, while another agency has expertise in pressure vessels.

It is very important that the agency be sent a copy of all applicable documents (the purchase order, specifications, any agreements, drawings, etc.).

The terms of reference should be established. In some cases, one may want a number of full-time inspectors (e.g., for pipe), whereas in other cases a release inspection may be all that is required (e.g., for flanges).

Audit of Purchase Orders

A purchase order is a legal document, so it is necessary to ensure that it is accurate. Purchase orders for important items should be audited for the following requirements:

- Correct specifications.
- Correct and complete description of items ordered, terms and conditions of payment.
- Correct reference to other documents, e.g., drawings, contracts.
- Other (e.g., identification of third-party inspection agency, requirement for drawings, customs broker, freight handler, shipping point).

Quality in Manufacture (Material Quality)

Material quality provides assurance that what was specified by the designer was ordered and received.

Review of Manufacturers' Drawings

Some companies review manufacturers' drawings to ensure that they comply with what was ordered and that the product will meet appropriate safety and environmental requirements. Although this review can be time-consuming, many errors can be detected and corrected before the item is manufactured (Coulson et al. 1999).

Pre-award/Pre-production Meetings

In some cases, it is advisable to hold a pre-award meeting before an order is placed, and to hold a pre-production meeting after an order is awarded. Typical cases where this would be advisable: for a new vendor, a new revision of the specification, a vendor that had problems on a previous order, a vendor that has not manufactured an item to the applicable specification for a long time, new personnel at the vendors, or a very large order. The specification should be reviewed in detail and any exceptions discussed. Hold points are established and notification for inspection are agreed upon. The third-party inspection agency should also attend the pre-production meeting.

Surveillance during Manufacture

Manufacturing surveillance is sometimes required and is normally undertaken on items with a large dollar value or some degree of complexity.

When surveillance is required, it is necessary to agree on either full- or part-time inspection. Full-time inspection is normal for the manufacture of pipe, the application of pipe coating, and for very large orders for other equipment. Part-time inspection is usually sufficient for components such as valves, tees, or elbows.

An important decision to be made is whether or not to use the owner's employees or a third-party inspection agency for inspection purposes. It is better to use the owner's employees for trial orders, new vendors, training, when serious problems are expected, and when the inspection would tie in with a visit for another reason (e.g., a periodic audit). However, the use of third-party inspection agencies can be more cost-effective if the owner's employees have far to travel.

An inspector's duties would normally consist of the following:

- Ensure the item meets the order requirements (e.g., dimensions)
- Check the material test reports
- Record all noncompliances and remedial actions
- Sign a release form authorizing shipment
- Advise head office of the release
- Stencil identification on the items inspected

Receiving Inspection

The receiving inspector will normally check the material shipped to ensure all items are received as listed on the manifest and that they are free from defects. If the item was inspected at the factory, the receiver does not need to inspect it again. The factory inspector will have identified the item in some way (e.g., by a stencil) so that the receiver knows it was inspected. If the item were not inspected at the factory, the receiver would do some additional checks (e.g., dimensions, marking, applicable specification, and approved manufacturer).

Resolve Noncompliances and Complaints

Noncompliances may be detected by the manufacturer, the plant inspector, the receiving inspector, and the end user, though the latter will prove costly.

Frequently, the item containing the noncompliance may be used to avoid delays in the project. In such cases, the group that established the requirements must evaluate the likely consequence of noncompliance on the functionality. Acceptance of the noncompliance means the requirements have been amended. If possible, the noncompliance should be corrected. (Mohitpour et al. 1993).

Complaints from the field should be investigated and documented. If the item met the specification but is still causing problems, the specification may need to be changed. If the item did not meet the specification but the noncompliance was not detected until it got to the field, the quality procedures may need to be amended. Periodic reports on noncompliances and complaints should be issued for control purposes.

Review Material Test Reports

Material test reports (MTRs) are documents completed by the manufacturer giving the results of tests. MTRs are not normally required for all items. When they are required, MTRs should be completely checked by the inspector. In addition, the owner may wish to spot-check some MTRs. The data on the MTR is checked against the specification requirements. A preprinted checklist will assist in this activity and help minimize errors.

Write Completion Summaries

A short report is written after the job is complete. This report summarizes the order, the applicable specification, and the inspections carried out. It lists each noncompliance and describes its disposition. Other comments may also be included in this report.

The completion report is used to give feedback on performance to the vendor, the third-party inspection agency, and the buyer.

The completion summary reports can also be used to provide input about a vendor performance assessment program. The vendor's performance should be assessed periodically. Some companies have relatively strict procedures for assessing the performance of vendors (e.g., number of noncompliances, delivery).

Maintain Material Identification

Marking requirements are contained in the applicable proprietary or industry specification. Marking normally includes specification number, manufacturer, size, grade, heat number, etc. Items are usually marked when they are received. Care must be taken to avoid removing markings. This is particularly true for pipes. A pipe is usually marked on both ends; however, in cutting short lengths of pipe, the two ends are usually used first to take advantage of the mill bevels. If the remaining pipe is not marked, the pipe will become useless for high pressure service.

Maintain Records of Technical Data

Technical data applicable to each item should be filed in case a problem develops with that item, or future modifications are required.

Quality in Construction

After materials meeting the design requirements have been acquired, they must be assembled in accordance with the construction requirements to complete the project (Table 8-7).

A number of companies utilize field inspectors during construction (quality control) as well as auditors from their head office (quality assurance).

During the construction of a pipeline, a large number of inspectors are used. The pipeline is assembled quite rapidly and the workforce is spread out over many kilometers. In the case of a pump station, a compressor station, or a meter station, the work occurs at one location and there are not as many activities involved, so one or two inspectors are usually sufficient.

Methods to obtain construction quality are discussed below. Many similarities will be noted between construction quality and material quality.

TABLE 8-7. Construction management guidelines

CMT			Construction Contractor
Contract/Site Administration	Quality Management	Safety Management	
• Administration	• Surveillence plan	• Health safety & environmental manual	• Safety requirements document
• Cost control	• Surveillence procedure	• Safety surveillence plan	• SIIP
• Construction progress	• NCR procedure	• Safety checklist	• Project description
• Auxillary contracts	• Turnover Requirements Document	• IMP	• Contract specifications
• Invoicing			• Quality assurance requirements document
• Schedule control • Change control • Drawing control • Material control • Reporting Requirements • Commissioning & turnover			• IIP

Contractor Qualification

Qualification of a contractor provides assurance that the contractor is capable of performing the required work using acceptable procedures. These will often include a credit rating check of the company and its cash flow disposition in order to ensure that it has the financial stablility to be able to perform the work to completion. It may be necessary to either require the contractor to post a performance bond, which can be costly, or else to subdivide the work into smaller amounts and spread it among several contractors. Since the cost of performance bonding will invariably be passed on from the contractor, there are significant savings to be had through working with only those contractors who are financially stable. Company reports, SEC filings, and Dun and Bradstreet reports are the usual authoritative sources of financial data. A qualified contractor list should be developed. This list would include any limitations for a particular contractor, such as maximum project size, diameter of pipe, etc.

Third-party Inspection Agencies

Third-party inspection agencies also need to be qualified. Individual inspectors can be qualified by examination. During construction, radiographic inspection of welds is almost always subcontracted. These inspectors must be qualified by the appropriate regulatory authority. However, the pipeline owner will sometimes require a sample radiograph that can be evaluated and used as astandard for comparison during construction. Contracts are normally awarded on a project basis. The radiographic inspection firm may be hired by the contractor or by the owner.

Preconstruction Meetings

Preconstruction meetings are held to review the technical requirements of the job. Special requirements or conditions are highlighted (e.g., the timing of a river crossing).

Inspection during Construction

Inspection during construction is a major activity. Each activity is monitored to ensure that all work is carried out in compliance with the pipeline specifications, contract documents, regulatory requirements, and any specific conditions stipulated in agreements with landowners.

Each inspector must prepare daily reports for their specific activity, which should record routine data, specific problems, general comments relating to conditions of work, equipment, and the contractor's attitude toward the work as it affects quality. A major emphasis is placed on safety.

A pipeline project will typically have inspectors for all of the following activities (refer to Chapter 9, which also contains detailed duties of the inspectors for each activity).

It is likely that an inspector will cover more than one activity. Each project will have a chief inspector to whom the other inspectors report. All the inspectors are trained to be environmentally sensitive. However, with more emphasis being placed on the environment, many projects also have an inspector whose sole responsibility is the environment. Inspectors may either be employees of the company or contract personnel.

Audits during Construction

Many companies have an audit team during the construction period. Audits are conducted to ensure that construction quality control activities are functioning effectively and that standards and codes are met. Audits are normally performed regarding material handling, bending, welding, nondestructive testing, coating, ditching, lowering-in, and backfilling to ensure that procedures are being followed and that the requirements of these procedures are appropriate.

Resolve Noncompliances

Noncompliances are identified and reported to the engineering and construction groups. Procedures should be in place for resolving noncompliances and periodic reports should be issued on noncompliances. The group establishing the requirement must be involved in resolving the noncompliance if it cannot be corrected.

Write Completion Summaries

Project completion reports are written by the auditor to identify the project, its size, main contractor, inspection staff, and information related to welding and radiography (e.g., number of weld repairs and the reasons for repairs).

Maintain Records of Technical Data

The technical data applicable to a project is filed for future reference.

Quality in Operations and Maintenance

After the project is completed, it must be operated and maintained according to established procedures and within its design parameters. Consider the following example: The maximum operating temperature has been established at the design stage as 40°C. A stringent specification for the pipe coating was developed and the coating was inspected and tested during application. The coating was again inspected as the pipe was being lowered into the ditch. The operations group however allows the temperature of the gas to reach 80°C, thus burning off the coating and destroying the protection provided against corrosion. All the work that went into design quality, manufacturing quality, and construction quality has now been negated.

SUMMARY

In the selection of materials for a fluid transmission system, the materials performance criteria should be considered for cost-effectiveness, safety, and reliability of the system. The materials performance criteria that are of primary concern include fracture control capabilities, chemical composition, material strength, dimensional tolerances, and inspection requirements. The Materials Designation Standard is a tool that can be used for materials selection.

Materials selection is one part of a complete quality management program. To ensure the safe, efficient, and reliable operation of a gas or oil transmission system, a coordinated quality program is required.

REFERENCES

American Gas Association (AGA) 1982 NG-18, Report No. 131.

Boniszewski, T., 1979, "Manual Metal Arc Welding - Old Process, New Development," Parts 1, 2, and 3, Metallurgy and Materials Technology, October, November, December.

Burdekin, J.M. and Stone, D.E.W., 1966, "The Crack Opening Displacement Approach to Fracture Mechanics," Journal of Strain Analysis 1, pp. 144–153.

Cabral, M.A. and Kimber M.J., 1997, "Pipeline Fracture Experiences in Australia and North America," Paper 1, Proc. Int. Seminar on Fracture Control in Gas Pipelines, Welding Technology Institute of Australia, Sydney.

Coulson, K.E.W., Quinton, D., and Slimmon, T., 1998, "Application of Material Standards and ISO Quality Management Systems," Proc. 2nd Int. Pipeline Conference, CEPA/ASME, Calgary, Alberta, Canada.

Coulson, K.E.W., Slimmon, T.C, Murray, A., and Puka, N., 1999, "How to Determine Your Best Suppliers of High Pressure Line Pipe and High Integrity Pipe Coatings," Oil and Gas Journal, November 5.

Crosby, P.H., 1980, Quality is Free, McGraw Hill Book Company, New York, NY, 270.

CSA Z662 2007, "Oil and Gas Pipeline Systems," Canadian Standard Assoc., Etobicoke, Ontario, Canada.

Dugdale, D.S., 1960, "Yielding of Steel Sheets Containing Slits," Journal of Mechanics and Physics of Solids, Vol 8, pp. 100–108.

Eiber, R. J., and Maxey, W. A., 1979, "Fracture Propagation Control Methods," Proc. 6th Symposium on Line Pipe Research Paper L1. AGA Cat. No. L30175 Houston, November.

European Federation of Corrosion, 1995, Publication No. 14, "Guidelines for Methods of Testing and Research on High Temperature Corrosion," Ed. Grabke, G. H., and Meadowcraft, D.B.

European Federation of Corrosion, 1997, Publication No. 23, "CO_2 Corrosion Control in Oil and Gas Production—Design Considerations,"

European Federation of Corrosion, 1995, Publication No. 16, "Guidelines on Materials Requirement for Carbon Steel and Low Alloy Steels for H_2S Containing Environments in Oil and Gas Production."

Fletcher, L. and Bilston, K.J., 1997, "Requirements of the Petroleum Pipeline Code AS2885.1–97 Fracture Control Plan," Paper 2, Proc. Int. Seminar on Fracture Control in Gas Pipelines, Welding Technology Institute of Australia, Sydney, June.

Folias, E.S., 1964, "The Stresses in a Cylindrical Shell Containing an Axial Crack," Aerospace Research Laboratories, ARL, pp. 64–174.

Maxey, W.A., 1974, "Fracture Initiation Propagation and Arrest," 5th Symposium on Line Pipe Research, Houston, TX.

Mohitpour, M., Oleksuk C., and Cumberland, R., 1993, "Quality Management for Consulting Engineering Services," Proc. 13th ASME International Conference, Offshore Mechanics and Arctic Engineering, Glasgow, Scotland.

NACE MR 0175/ISO 15156 "Petroleum and Natural Gas Industries - Materials for use in H2S-containing Environments in Oil and Gas Productions".

Nara, Y., 1983, "The Production of Line Pipe in Japan," Steels for Line Pipe and Pipeline Fittings, The Metal Society.

National Energy Board (NEB), 1996, "Stress Corrosion Cracking on Canadian Oil and Gas Pipelines," Report of Inquiry, MH-2-95, November, Calgary, Alberta, Canada, 139.

Piper, J. and Morrison, R., 1997, "The International Database of Full Scale Fracture Tests and its Applicability to Current Australian Pipeline Designs, Paper 3, Proc. Int. Seminar on Fracture Control in Gas Pipelines, Welding Technology Institute of Australia, Sydney, June.

Rothwell, A.B., 1997, "The International State of the Art in Pipeline Fracture Control and Fracture Risk Management," Paper 12, Proc. Int. Seminar on Fracture Control in Gas Pipelines, Welding Technology Institute of Australia, Sydney, June.

Sanderson, N., Ohm, R., and Jacobs, N., 1999, "Study of X-100 Line Pipe Costs Points to Potential Savings," Oil and Gas Journal, March 15, pp. 54–57.

Somerville, W.J.H., 1999, Private Communication.

Steele, P.T. and Court, B.H., 1997, *Profitable Purchasing Strategies*, McGraw Hill Publishing Company, New York, NY.

Thomas, J.M., 1988, "Fracture Control for Marine Structures," *Treatise on Materials Science and Technology*, Academic Press, New York, NY.

Venton, P. and Dietsch, A., 1997, "Design of Crack Arrestors," Paper 10, Proc. Int. Seminar on Fracture Control in Gas Pipelines, Welding Technology Institute of Australia, Sydney, June.

Chapter 9

PIPELINE CONSTRUCTION

INTRODUCTION

The material covered in this chapter highlights the activities involved in pipeline construction, testing and commissioning procedures. This discussion will focus on the necessary steps to ensure quality facilities are constructed and installed before handover to the operations group for daily operation. A typical pipeline spread is illustrated in Figure 9-1 and shows the various activities involved and their linear progression over the course of the project.

CONSTRUCTION

Pipeline construction usually takes place in relatively isolated areas. For this reason, it is often necessary to house workers in construction camps if little or no local temporary accommodation exists. Transportation, usually by bus, is provided to the worksite. Travel along the right-of-way is made using all-wheel-drive (4 × 4) pickup trucks or all-terrain vehicles (ATVs).

Typical pipeline construction activities include, in sequence:

- Route surveying
- Mobilizing equipment and personnel
- Preparing the right-of-way (ROW; clearing and grubbing the right of way)
- Transporting and storing pipe and materials
- Topsoil stripping (where appropriate)
- Grading
- Stringing (transport and laying of pipe on the ROW)
- Transporting other materials and weld equipment to site
- Welding, ultrasonic, and X-ray checking of welds
- Installing protective coating at pipe joints
- Jeeping pipe (testing for external coating integrity)
- Trenching
- Lowering pipe into the trench
- Installing watercourse crossings
- Installing block valves and terminus equipment
- Backfilling
- Leak testing (hydrotesting)
- Restoring the ROW
- Inspecting the route

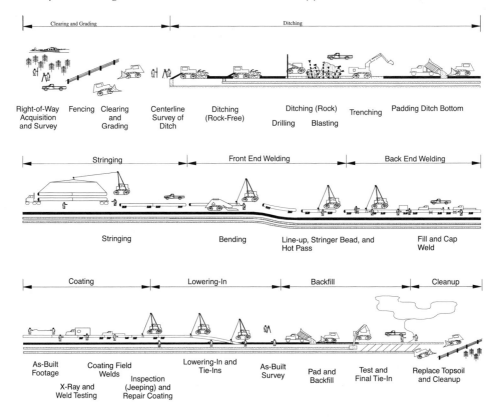

Figure 9-1. Typical pipeline spread

- Demobilizing equipment and personnel
- Installing cathodic protection

While the specifics of pipeline construction are dependent upon existing site conditions at the time of construction, the following sections describe typical pipeline construction techniques as well as those that are used in environmentally sensitive areas.

Construction Surveying

A construction survey is performed to gather sufficient detailed data to enable the design team to complete the design of all sections of the pipeline, including associated access roads and watercourse crossings. This activity may include the brush-cutting of centerlines and cross-section offsets of sufficient width to provide a clear line-of-sight for the surveyors. These centerlines may also be required to provide access for subsurface survey (e.g., soils testing) equipment. The width of line cut for surveying purposes should be kept to a minimum. Woodcutting in the surveying area will often leave merchantable timber in salvageable lengths of 4 feet or greater. Care has to be taken to ensure that no cut brush and trees are left in watercourses where they may form artificial barriers for water or fish. Figure 9-2 shows stacked, salvageable timber and, nearby, the burning of brush and scrub material.

Figure 9-2. Stacking salvageable timber and burning scrub

Pipeline construction workspace requirements are a function of pipe diameter, equipment size, slope conditions, bedrock, the location of construction (e.g., at road crossings or river crossings), pipeline crossovers, the method of construction (e.g., borings or open-cut construction), and the existing soil conditions during construction. As the size of the pipeline being installed increases, trench depths and widths increase, which creates additional spoil material. Larger equipment is needed to handle the heavier pipe, thereby increasing the amount of construction workspace required. In order to prevent the ROW from becoming too crowded, it becomes necessary on occasion to seek permission of the environmental regulator to establish minimum size and area requirements to promote worker safety.

Typical construction ROW widths using various pipe diameters are presented in Table 9-1, while Figure 9-3 shows a sectional view of the ROW during a normal (summer) construction season. In winter when snow is present, space is usually left to stockpile the

TABLE 9-1. Typical right-of-way construction width requirements

Pipe Size (mm)	Working Side(m)	Spoil Side (m)	ROW Width (m)
60.3–114.3	9	7	16
168.3–273.1	10	8	18
323.9–457	14	9	23
508–660	15	10	25
762–914	15	11	26
1,067	17	12	29
1,219	18	13	31

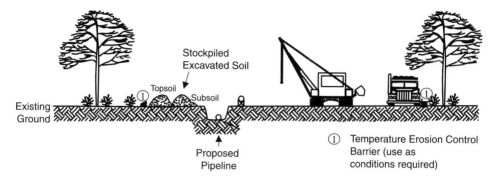

Figure 9-3. Typical right-of-way during normal (summer) construction season

snow. Expanded temporary workspace, in addition to the construction ROW, will be required at special locations to construct the pipeline where there are:

- Steep side slopes
- Road and railroad crossings
- Crossovers
- Additional topsoil segregation in agricultural areas
- Equipment and pipe staging areas
- Staging areas associated with wetland and water body crossings

As an example, for the construction of a new 24-inch (600-mm) diameter pipeline within a virgin ROW, a minimum 75-foot (25m) wide construction ROW will be required. The construction ROW will include the 15 m of proposed new permanent ROW, and 10 m of temporary workspace. For the construction of other pipelines ranging in diameter from 4 to 12 inches (100–300 mm), an 18-m-wide construction ROW is required. Figure 9-3 illustrates the use of the temporary workspace and ROW during typical construction of a pipeline.

Access to the construction ROW is often restricted to existing roads, and in some areas during critical times of the year, road bans are set up to prevent the movement of heavy equipment on them. Examples of types of access used are abandoned town roads, railroad ROWs, powerline access roads, logging roads, and farm roads. Permission for access to these roads can sometimes involve a fee, and invariably involves a commitment to repair any damage caused. An approximately 6-m (20 feet) width of road is needed to accommodate large pipeline equipment, such as tractor trailer pipe trucks and equipment transporters, excavators, sidebooms, and all other pipeline construction support vehicles. The width of the required access road is dependent on slope and road alignment. Because of the type of equipment traveling on the access roads, improvements may be required, such as grading, laying of gravel for stability, and improving stream crossings before they can be used. The same erosion control and restoration practices that apply to the construction ROW also applies to the access roads.

Clearing

The area of the right-of-way, all temporary working space, and access routes used for construction are completely cleared of trees and brush. Stumps are removed from the ditchline, spoil storage areas, and all areas to be graded. Logs are salvaged wherever possible (e.g., for use as rip rap or for eventual sale). Bulldozers with special attachments cut or knock down any remaining material. The material is then piled and burnt.

As clearing progresses, gates are installed on all fences. Underground utilities or structures are hand-exposed and protected with ramps or other means to prevent damage by equipment and vehicles. A typical clearing crew and equipment would include a foreman, a subforeman (for fencing and hand-exposing), several operators (including for night shifts), power saw laborers, laborers (for fencing and hand-exposing), and a bus driver. The necessary equipment consists of a dozer complete with cutting blade, a dozer with an angle blade and winch, a dozer complete with angle blade and ripper, dozers complete with brush rake, skidders, and a crew cab truck (for fencing operations).

Final equipment and manpower requirements will depend on the progress schedule and the type of terrain and vegetation. The clearing crew will do the minimum grading necessary to accommodate vehicles used to transport personnel, fuel, and equipment to the site.

Inspection staff are used to ensure that:

- Permanent right-of-way or temporary work area boundaries are clearly marked and clearings are confined within these boundaries.
- Temporary land-use agreements are executed prior to clearing rights-of-way and comply with each specific provision.
- Utilities are hand-exposed, marked, and protected.
- Drainage is not blocked.
- Cut timber/brush (slash) debris, and earth are not placed in watercourses.
- Burning permits are obtained and adequate measures are taken to prevent fires from spreading to adjoining areas.
- Sufficient area is cleared to ensure that slash or debris is not mixed with ditch or grade spoil and that spoil is not placed on uncleared ground.

Grading

Topsoil and organic surface material is stripped and stored away from ditchline areas, ditch spoil storage areas, and all areas to be graded. This material must be stored in such a way that it is not mixed with grade or ditch spoil and can be replaced during cleanup. The contractor must grade the right-of-way to provide a level work surface with sufficient width for the pipe-laying crews who follow (Figure 9-3). This grading will consist of the following areas:

- Topsoil storage area
- Ditch spoil area
- Ditch area
- Pipe area
- Pipe-laying equipment area
- Vehicle and equipment passing area

The passing lane may be eliminated on slopes that require heavy grading or are very rocky. On side slopes, the working side (pipe and equipment) may be constructed with fill.

The ditch must be excavated in previously undisturbed soil so that the pipe is left on undisturbed soil. As an option, side slopes may have the spoil, ditchline, and pipe on one level, and the equipment on another. The contractor will grade so that conventional vehicles can travel along most of the right-of-way. In steep, rugged terrain where this grading is not practical, it can be limited to the amount necessary to install the pipe, using towing equipment or winches. Vehicles are routed around these areas by using existing roads or constructing temporary roads near the site.

A typical grading crew would include a foreman, operators, and a bus driver. Typical equipment would include dozers complete with angle blade and ripper, dozers having an angle blade and winch, a grader cat, a bus, and 4×4 pickups.

Drilling and blasting with explosives will be required where the rock is too hard to break by ripping. Explosive charges are placed in holes made in the rock and then detonated. The explosives manufacturer will establish the spacing of holes and the size of the charge required to fracture the rock. This procedure needs to be approved by the appropriate regulatory authorities.

A typical drilling and blasting crew would include a foreman, drillers, drillers' helpers, powder men, a blaster, a powder truck driver, a bus driver, and operators.

The necessary equipment consists of an air track drill/compressor, a dozer complete with winch, backhoes, and a powder truck.

The size of the crew depends on the average daily progress required and the terrain condition. Crews may be split, or an additional crew mobilized, to work on difficult areas ahead of the main crew. Rock work may be done a season ahead of grading.

Inspectors are used to monitor and enforce the following:

- Topsoil or grade spoil is not placed on uncleared ground.
- Cuts and spoil storage piles are sloped for stability and do not present a hazard for the public, livestock, or wildlife, and that they are fully contained within the boundaries of the right-of-way or temporary workspace.
- Topsoil is placed in such a way that mixing with spoil will not occur.
- Soil is not placed in streams; bridges or culverts are used where necessary to maintain drainage in streams (Figure 9-4).
- Land-use agreements are obtained and approved prior to grading temporary access trails.
- Final right-of-way profile will allow the pipe to be bent and laid in undisturbed soil in accordance with specifications.
- Valve and meter station sites are graded level to the elevation specified on the construction drawings. Where cuts and fills are required, site boundaries must be adequately marked and staked to ensure that the final elevation is as specified by the contractor's surveyors.
- Where blasting is required, provision is made to ensure loose rock does not scatter over the right-of-way or adjacent land, causing damage to property or risk to workers and the public. The inspector should have the contractor pick up and dispose of any fly-rock. Mats or earth cover should be used in the vicinity of highways, houses, etc.
- The contractor has obtained all the required permits for the use of explosives, and met the terms of these permits. Only qualified drilling and blasting personnel are employed in the blasting operation.
- Proper notification is given to landowners and the public prior to blasting.
- All blasting materials are stored in an approved magazine.

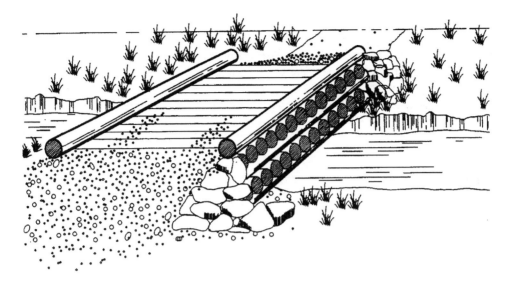

Figure 9-4. Temporary bridge structure

Loading, Hauling, and Stringing

The pipeline contractor usually receives the pipe in a stockpile away from the right-of-way. This stockpile may be located at the pipe manufacturer's plant, a coating plant, near a port of entry, or near the rail spur closest to the construction site. The contractor becomes responsible for the pipe as it is removed from the stockpile, and remains responsible for all loss and damage until the finished pipeline is accepted by the pipeline company. As each load is removed from the stockpile, a tally form is filled out that documents the length, wall thickness, and grade of each joint as well as notes any damage. Multiple copies are made for the contractor and the inspection staff. A crane is used to load the pipe on trucks. The trucks then haul the pipe to the right-of-way where it is off-loaded with a side boom and placed on wooden skids, end-to-end, on the right-of-way. The pipes are angled slightly to allow end hooks to be removed and to prevent damage to bevels.

Special precautions are required to prevent damage to the pipe or coating. These precautions include using rubber or other suitable material between the pipe and the tie-down chains, using padded bunks, and covering the pipe load with tarpaulins. Pipes will not be strung until blasting for ditch stock is completed.

A typical stringing crew would include a foreman, operators, laborers, and truck drivers.

Equipment consists of a crane, a side boom complete with a stringing boom, pipe transport trucks, a tow tractor, a 4×4 pickup, and crew cab trucks.

The number of trucks needed will vary depending on hauling distances and daily production requirements. Pipe may have to be loaded on special pipe carriers for transport in wet areas. Double-jointed pipes (lengths to 25 meters/80 feet) require self-steering trailers.

Inspectors monitor the stringing process to ensure that:

- Pipe is being handled so as to prevent damage to it or to the coating, and that any damage is clearly marked.
- Different pipe wall thickness/grades as marked on alignment sheets are correctly placed.

- Public roads and land are not being damaged by pipe hauling.
- Pipe is not placed on the ground or dragged over it.
- Temporary stockpiles are safe and properly supported.
- Proper end hooks are being used to load and unload pipe to prevent damage to bevels.
- Required gaps are made for access across the right-of-way for the landowner, livestock/wildlife.
- Tallies are accurate and distributed according to procedures.

Bending

The pipe will require bending to accommodate changes in direction and elevation of the trench. Bending can be done before or after trenching is complete. In terrain where the trench is expected to be stable, trenches may be dug prior to bending, with the bends made to fit the final ditch contour. To reduce or avoid ditch cave-ins in unstable soil areas and avoid delays, the pipe will be bent, welded, and the joints coated prior to trenching.

Large-diameter pipe must be precisely bent to avoid costs and delays due to cutting and rebending or redigging the trench during lowering-in. This is accomplished by first confirming the length of each joint and staking its proposed location along the completed trench or ditchline. The required bends are located and measured by the bending engineer and the exact location, type of bend (overbend, sag, sidebend, or combination), and degree of bends required are marked on each pipe joint. Sag bends must be measured and bent so that the pipe will rest on the bottom of the trench. Overbends must clear the high point of the ditch bottom.

Large-diameter pipe is bent using a hydraulic bending machine and an internal mandrel. Areas contacting the pipe will be padded with neoprene rubber when precoated pipe is used. The bending is accomplished by cold-stretching the pipe material. The contractor will use one or more side booms to place the pipe in the bending machine and to replace it on skids after bending. The bending engineer will allow for lapping where tie-ins are required (e.g., road crossings, side bends, test points). The bending crew will normally set up the start joint of each section to be welded by skidding it to the height required for welding.

In extreme terrain, a hot bend or prefabricated elbow segment will be used where grading cannot be done to a contour consistent with the allowable bending radius.

A typical bending crew would include a foreman, a bending engineer, measurement men, a roadman, laborers, operators, and a mandrel operator.

The required equipment consists of side booms, hydraulic bending machine complete with dies and mandrel, 4×4 pickups, crew cab truck, and a bus.

A second bending machine may be required, depending on daily production and the amount of bending required.

Inspectors will monitor to ensure that:

- Pipe is being handled in a manner that will prevent damage to it and to the coating.
- Maximum curvature of 1.5 degrees per diameter length is not exceeded.
- Bends are smooth and free from wrinkles, dents or flat spots; the maximum difference between minimum and maximum diameter does not exceed 5% of nominal pipe diameter.
- Bending is not done within 1.8 m (6 ft) of the end of the pipe or at a one-diameter length from a circumferential weld.
- Longitudinal welds are placed on the neutral axis.
- Gaps in the right-of-way for landowners, livestock/wildlife are maintained.

Welding

Welding is required for double-jointing, preparing sections for crossings, mainline welding, tie-ins, and fabrication. Welding processes that are commonly used for pipeline applications are gas submerged arc welding (restricted to double-jointing and fabrication), shielded metal arc welding (SMAW), or gas metal arc welding (GMAW). Either a mechanized welding technique or manual welding may be used for mainline welding; manual welding is used for all other requirements.

Gas metal arc welding (GMAW) is done using a consumable wire that melts when an arc is struck and maintained between the wire and the material being welded (Figures 9-5, 9-6, and 9-7). The welding wire is fed continuously into the arc during the welding process.

The arc and weld pool are shielded from the atmosphere by a concentric flow of gas. A number of gases/gas mixtures are used for shielding, with 100% CO_2, or 75% argon and 25% CO_2, being two of the most common.

If welding procedures are not followed or the operator technique is not correct, welding defects may result. The continuous nature of the wire electrode and the virtual absence of slag leads to high productivity and is ideally suited to mechanization. The mechanized GMAW process is invariably used in Canada and Europe for major large-diameter, cross-country pipeline construction, and has been successfully introduced in the United States on the Alliance Pipeline project.

Shielded metal arc welding (SMAW) is done using consumable stick electrodes that melt when an arc is struck and maintained between the electrode and material being welded (Figures 9-8 and 9-9). These electrodes are composed of two main parts, one being the core

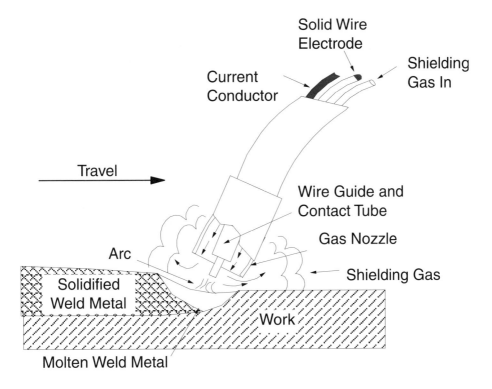

Figure 9-5. Gas metal arc welding

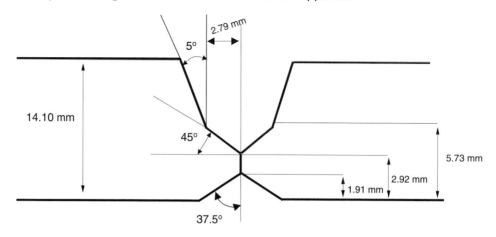

Figure 9-6. Typical joint design for gas metal arc welding

wire and the other the external flux coating. The core wire, and sometimes iron powder within the coating, provides the necessary metal to fill the weld joint.

The flux coating is required to shield the arc and molten metal from the atmosphere, add alloying elements to the weld metal, and provide a protective layer (slag) during and after solidification of the weld metal. This slag is subsequently removed between passes.

Shielded metal arc welding remains the most widely used and versatile process for general pipeline applications. With the correct choice of consumable and welding technique the SMAW process can be applied to all welding positions and will allow a wide range of mechanical property requirements to be met. However, the process is very dependent on the welders' manual skills for attainment of defect-free welds with acceptable properties. Production is inherently limited by its intermittent nature, and a duty cycle of 40% is

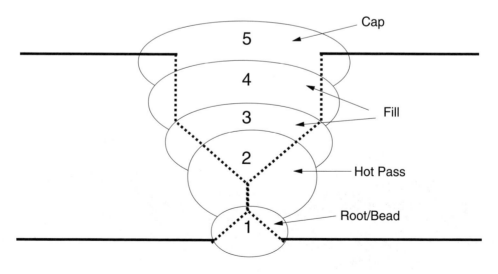

*Note: Number of passes will vary with pipe wall thickness

Figure 9-7. Typical number of passes for GMAW*

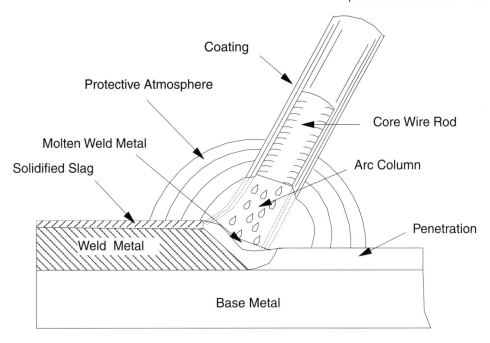

Figure 9-8. Shield metal arc welding

typical. Most of the developments in arc welding have been aimed at overcoming both these limitations.

Before commencing site welding, the contractor must develop or obtain qualified welding and repair procedures for any method proposed, and for all combinations of pipe

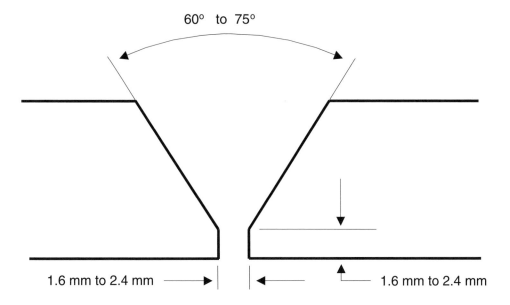

Figure 9-9. Shield metal arc welding joint preparation

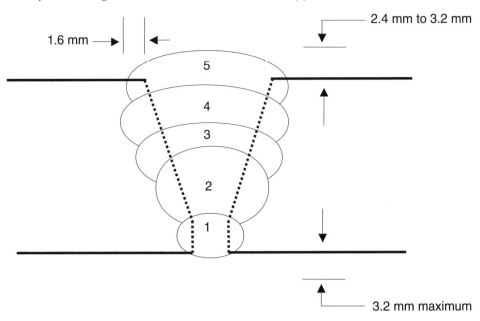

Figure 9-10.　Shield metal arc welding - typical welding passes

diameter, wall thickness, and grades involved. In order to be approved for work on the project, each welder is required to make a test weld using a qualified procedure on the project pipe (for example, see CSA Z662-07). Radiographs and destructive tests are done on the qualifying welds in accordance with the applicable code or standard as specified, for example API 1997. Each qualified welder is assigned a number to be used to identify each individual's work.

While some pipe mills and coating facilities have the capability of handling double joints of pipe (generally pipe is supplied in 12-meter average lengths), the contractor will often double-joint the pipe by welding two sections together to form a single 24-meter-long section. The contractor will usually do the double-jointing at the pipe stockpile site to reduce handling and transportation costs prior to hauling and stringing. The advantage of double-jointing the pipe is that half as many mainline welds are needed on the right-of-way, which results in lower weld defect rates and an increased daily production rate for the crew. However, the increased difficulty in handling and the special equipment needed to transport the longer pipe on roads and over rough terrain on the right-of-way may render this alternative impractical for some locations.

Multiple welding passes are required to complete each weld. The number of passes is dependent upon the pipe wall thickness and the welding process used. The first pass is called the stringer bead, which is followed by the second, called the hot pass. The required number of fill passes are then conducted, which are succeeded by a cap pass to complete the weld. In mainline welding, a crew called the pipe gang will make the stringer bead and hot pass, followed by welders who will complete the weld. The pipe gang's daily progress is normally used to size the other crews on the spread. An internal clamp, operated and propelled by compressed air, is used to hold the correct alignment during the placing of the stringer bead. On a large-diameter pipe, the stringer bead will be made using at least two and as many as four welders working simultaneously. The hot pass will also be done by one or two pairs of welders. The fill and cap passes will be completed by individual welders.

Mechanized welding makes the stringer bead internally, using gas metal arc equipment built into the lineup clamp. Subsequent passes are made using "bugs" that run on bands placed on the pipe at a preset distance from the weld bevel. These bugs are operated in pairs and require a shelter carried by a side boom to prevent the shielding gas from being blown away by prevailing winds. Mechanized welding has a consistently superior weld quality and welders can qualify with less training than manual welders. The disadvantages of mechanized welding include increased need for repair and maintenance of equipment, additional equipment requirements for welding shelters, and the need for end preparation machines.

A typical mechanized welding crew would include a pipe gang foreman, a welder foreman, a subforeman, welders, repair welders, welders' helpers, a journeyman (stabbers, spacers, end preparation.), laborers, operators, and mechanized welding technicians.

Equipment would include a quad welder, side booms (set-in), side booms (cut-outs), side booms complete with end preparation machines, side booms complete with generator and shelters, a bus, and 4×4 pickups.

A typical manual welding crew would include a pipe gang foreman, a welder foreman, subforemen, stringer bend welders, hot pass welders, backend welders, repair welders, welders' helpers, a journeyman (spacers, stabbers, clampmen), laborers, and operators.

Equipment for manual welding includes quad-welder tractors, side booms (set-in), side booms (cut-outs), internal clamp (plus one spare), welding rigs, 4×4 pickups, a bus, a buffing tractor, and a tow tractor.

Inspectors will monitor to ensure that pipe is being handled so as to avoid damage to the pipe and coating, that all parameters specified by the welding procedure are being met, and that all welders are qualified. Internal lineup clamps are not released until the stringer bead is complete. Welds are marked with welders' numbers, longitudinal seams are correctly located, (usually at the 10 o'clock and 2 o'clock positions offset at each joint by several inches), and pipe coating is protected from weld spatter damage. Weather conditions should be suitable for welding. Two welding passes are completed on all welds at the end of each day.

Welds that are to be left for more than 48 hours have one mechanized fill pass or two manual fill passes completed. All open ends of welded sections are securely night-capped at the end of each day, no waste material or debris, such as welding rod stubs, is left on the right-of-way or placed in the pipeline trench. All welds found to be unacceptable by nondestructive testing are cut out and replaced or repaired using a qualified procedure; gaps that are left in the right-of-way are necessary for landowners, livestock/wildlife.

Non-Destructive Testing
Radiographic Inspection
A radiographic inspection subcontractor working directly for the pipeline company will carry out inspection of welds using radiography.

The radiographic inspection subcontractor is required to provide a radiographic record of weld quality. This subcontractor will also carry out the interpretation of the radiographic films and report in writing to the welding inspector the existence and nature of any defects in the welds. Welds that fail to meet the acceptance requirements of API 1104, and any additional requirements detailed in the construction specifications, will be marked to indicate a need for repair or replacement.

Radiographic Procedure Qualification
A radiographic procedure developed by the radiographic inspection contractor is used to produce acceptable radiographs on each diameter and wall thickness of pipe in the project.

This procedure will detail which film and developing chemicals, screens, source strength, and exposure times should be used.

The radiographic procedure test consists of: A separate test radiograph to be completed for each procedure and technique to be used on the project (e.g., internal gamma-ray, single-wall image); and the use of equipment for the test radiographs similar to that used for production welds.

The welding inspector will invariably witness the radiographic inspection subcontractor's procedure test. The procedure test is performed either in conjunction with testing of welders or,if no welders are to be tested, as the first radiograph on the production welds. Further tests may be required before an acceptable result is produced. The radiographic inspection subcontractor will not be allowed to proceed with the work until satisfactory results are obtained and the procedure is approved.

After approval, the procedure test radiographs are retained on the job by the welding inspector for comparison with production radiographs. Copies of the approved procedure(s) are then distributed.

The radiographers employed by the radiographic inspection subcontractor must be qualified according to the relevant government specifications and regulations.

Inspector Subcontractor Responsibilities

A single senior radiographer designated by the radiographic inspection subcontractor will be responsible for all reports, interpretations, and evaluations. This senior radiographer is required to produce consistent, informative, and concise interpretations of the radiographs, and be responsible for the conduct and performance of all radiographic personnel, equipment, and supplies.

Equipment and Supplies

The radiographic inspection subcontractor is required to provide all the equipment and supplies necessary to perform the radiographic examination. These supplies include the following:

- Barricades and appropriate warning signs indicating the presence of a radioactive hazard
- Radiographic sources (either internal crawler or external units, gamma- or X-ray)
- Mobile facilities for processing and viewing film on the right-of-way
- Equipment and materials for developing and interpreting radiographs
- Sufficient radiographic sources, facilities, equipment, and materials to radiograph welds at the same rate at which the pipeline contractor is welding.

Radiographic Quality

The inspection contractor is required to produce radiographic film, which meets the minimum requirements of API 1104, and the contract specifications.

Radiography of Repair Welds

Whenever an initial radiograph indicates a repair to a weld is necessary, the inspection contractor will mark the position of the repair on the weld using an indelible marking pencil. All markers used for defect location are placed at intervals of not more than 150 mm. A survey ribbon will also be wrapped around the pipe at the location of the weld defect.

Marking of Welds

The inspection contractor is required to mark each field weld by applying a copper tag with a unique number. This number is die-stamped on the copper tag, which is then glued to the top of the pipe adjacent to and on the downstream side of the weld cap. This copper tag will appear as an image on the radiograph. Mainline welds are numbered consecutively, usually beginning at 1, for each diameter. Tie-in welds on the pipeline are given the prefix T and numbered consecutively (e.g., T1, T2).

If a weld has been repaired, it retains the original number followed by the letter R (e.g., T1 R1). If it is repaired a second time, the weld number will be followed by R2 (e.g., T1 R2), and so forth.

Film Identification Marking

The inspection contractor is required to put the following information on each radiograph. Its removal or modification after exposure would be cause for rejection of the radiograph:

- Project name or number
- Weld number
- Millimeter markers
- Proper penetrometers (as per API 1104, latest edition, and as specified).

On mainline welds, a station number is required to locate a weld at 200-meter intervals Each tie-in and transition weld made will be marked with a station number.

Film Cataloguing Procedure

The inspection contractor will ensure that all films are thoroughly washed and dried before interpretation and packaging. A copy of the radiographic evaluation form is packaged with the film.

Ultrasonic Inspection

On large diameter pipeline construction projects completed using mechanical welding, ultrasonic inspection is usually used to size and locate defects in welds, such as porosity, slag inclusion, lack of fusion, and lack of penetration. The results of the inspection are obtained much more quickly than radiography, enabling corrective actions to be taken sooner. Ultrasonic inspection is also used to detect laminations in the parent metal of the pipe or to confirm minimum wall thickness after grinding repairs.

A qualified and experienced ultrasonic operator will carry out the actual defect sizing and location on pipeline welds. Ultrasonic inspection for locating weld defects is usually required only on large-diameter projects where it is expensive to carry out repairs. By determining the exact location of the flaw, the repair may be avoided or minimized.

Magnetic Particle Inspection

Magnetic particle inspection is a technique used primarily to identify discontinuities in the parent metal of pipe or fittings. A fine-grained medium having magnetic attraction is applied to the surface to be tested, after or during induction of a magnetic field. This medium will show the magnetic field pattern, which will be distorted by any variations in material composition, such as voids or discontinuities.

Liquid Dye Penetrant Inspection

Liquid dye penetrant is a technique used primarily to identify cracks. After the surface is cleaned, a dye is applied on it, followed by a developer. A crack will show in color contrasting that of the surface color.

Trenching

Normally the trench will be excavated using wheel ditchers, supplemented by backhoes, to dig side bends and areas requiring extra depth (Figure 9-11). Wet areas and rock trenches are dug entirely by backhoes or equivalent machinery.

Minimum ditch dimensions are set to ensure that minimum cover requirements are met and that the backfill material will flow around the pipe and fill underneath the bottom quadrants of the pipe. Rocky areas have to be dug deeper to allow for bedding/padding (sand or fine-grained earth) to be placed beneath the pipe. Where the pipe has been bent ahead of trenching, extra care, including the use of cut stakes, may be necessary to ensure that the pipe fits the ditch.

Attention is required at crossings and other tie-in locations to dig the extra width and bell holes to allow welding to be done in the trench. A typical ditch crew and equipment would include a foreman, operators, oilers, and laborers. Ditch equipment would include a wheel trencher, hydraulic backhoes, a ripper tractor, a bus, and a 4×4 pickup.

In areas where the wheel trencher cannot work, additional backhoes will be substituted. Inspectors are used to monitor and enforce the following:

- Correct ditch dimensions, including extra depth where specified.
- Spoil placed separate from topsoil on stripped ground.
- Proper sloping of ditch sides.
- Hand-trenching near buried utilities.
- Proper alignment of trench.
- Gaps or plugs to allow for landowners, livestock, and wildlife crossings.
- Drainage is not blocked and spoil or debris is not placed in stream channels.

Coating

All weld areas require cleaning and coating after nondestructive testing has been completed. Precoated pipe and girth welds coated using shrink sleeves or fusion bond

Figure 9-11. Wheeled ditching machine

epoxy will be electrically checked for damage or "holidays." Repairs will be made as necessary while the pipe is still on skids. A typical joint coating crew (using shrink sleeves) would include a subforeman and laborers. Equipment would include a 4×4 pickup, a crew cab, propane torches, and a "holiday" detector.

A typical joint coating crew (using fusion bond epoxy) would include a subforeman, operators, and laborers.

Equipment would include a sandblast unit complete with an air compressor, an induction coil complete with a generator, an epoxy powder application unit, a "holiday" detector, a tow tractor, a 4×4 pickup, and crew cab trucks.

The accompanying inspection process ensures that:

- Welds are properly cleaned prior to coating.
- Correct preheat temperatures and application procedures are being followed.
- "Holiday" detectors are correctly calibrated.
- Shrink sleeves are free of dimples, bubbles, punctures, or burn holes.
- Fusion bond joint coatings meet minimum thickness specification.
- Weld and pipe number data are recorded prior to coating welds.

Lowering-In

The pipe is lowered into the trench using several side boom tractors working together. The pipe coating will be inspected again using a "holiday" detector and repairs will be made as necessary. Cradles (which roll along as the pipe is lifted) or wide nonabrasive belts are used for lowering the pipe, depending on the length of section and the terrain (Figure 9-12). Any weights required for buoyancy control will be installed by the lowering crew.

When rock is encountered, the bottom of the trench will require padding. A minimum of 150 mm of sand or fine-grained earth will be placed in the trench prior to the pipe. The pipe may also be supported on sandbags or a suitable alternative so that it does not touch the native soil bottom prior to its padding. In all cases, the trench bottom requires clearing of loose rock, roots, or any debris to avoid damage to the pipe coating.

Padding is also used to provide lateral support and longitudinal restraint to buried pipe. Lateral support along the length of the pipe prevents ovaling due to overburden loads, while at side bends it also prevents local overstressing of the pipe by restricting movement.

There are three main areas where there is a need to provide longitudinal restraint, the first of which is at the location of above-/below-ground transitions to prevent excessive axial pipe movements. It is also used on steep longitudinal slopes to prevent overstressing of the pipe due to an accumulation of compressive stresses. Its primary use, however, is in the prevention of overstressing due to temperature-change-induced expansions and contractions. Padding is generally compacted to eliminate large voids.

A typical lower-in crew would include a foreman, a subforeman, operators, laborers, and a skid truck driver. The necessary equipment would include side booms complete with belts, cradles, a skid truck, a tow tractor complete with a skid sled, skid racks, a clam, a backhoe (weights), side booms complete with ditch pumps, a "holiday" detector, a bus, and 4×4 pickups.

A typical ditch padding crew comprises a subforeman, operators, truck drivers, and laborers. The ditch padding crew uses a front end loader, a backhoe, a ditch dozer, dump trucks, a 4×4 pickup, and a bus.

Figure 9-12. Lowering in the line

Inspection procedures ensure that:

- Pipe is being handled to prevent damage to pipe and coating.
- Pipe coating is inspected with a properly calibrated "holiday" detector.
- An adequate number of side booms are spaced to prevent overstressing of the pipe and to allow it to be placed in the ditch without coming in contact with the ground or ditch walls.
- Ditch bottom is padded or pipe is placed on sandbags in rocky areas.
- Pipe fits the ditch and adequate slack is left.
- Minimum cover requirements are met.
- Weights, if required, are correctly installed with specified spacing.

Backfill

Backfilling involves replacing ditch spoil into the trench (Figure 9-13). Where soil conditions allow, backfilling is done directly behind the pipe as it is being lowered-in. If rocks are present in the ditch spoil to the extent that coating will be damaged, the pipe must be protected with rockshield or padding. The backfill material will be compacted, using heavy equipment. The ditchline will be left from 0.5 to 1 meter higher than normal grade to prevent erosion on the ditchline. Openings must be left to allow natural drainage across the pipeline.

A typical backfill crew would include a foreman, operators, and laborers. Equipment would include a dragline complete with a mormon board, dozers, a backhoe, a 4×4 pickup, and a bus.

Inspectors are used to continuously monitor and ensure that:

- Backfill material is suitable and is placed in the trench in such a way that the pipe and coating are not damaged.
- Rocks in excess of 0.2 m diameter are not used in backfilling.

Tie-Ins

Tie-in welds are those made in the trench to complete the pipeline. Tie-in welds are required at such places as crossings (road, railway, utility, streams), side bends, and gaps left for landowners, livestock/wildlife crossings. Large-diameter tie-ins are made by two welders who make all passes required on each side of the pipe. The backfill crew will leave 20 to 40 cm on each side of a tie-in to allow the pipe to be moved for measuring, cutting, and alignment for welding. The pipe is held for the stringer bead pass using an external lineup clamp. The tie-in crew will coat the completed welds after radiography and backfill the open trenches.

A typical tie-in crew would include a foreman, a subforeman, welders, welders' helpers, operators, laborers/swampers, and a bus driver. The crew uses side booms, a backhoe, a dozer complete with a winch, a ditch pump, a bus, and 4×4 pickups.

Inspectors will monitor and ensure that:

- Welding is done by qualified welders following approved procedures.
- Pipe is being handled so as to prevent damage to pipe and coating.
- Pipe is aligned and welded without excessive force, leaving sufficient slack after completion.
- Completed welds are radiographed and approved prior to backfilling.
- Weld areas are coated, exposed pipe coating checked, and any damage repaired prior to backfilling.
- Backfilling, including padding if necessary, is done according to the requirements of the backfill section.

Crossings (Road, Highways, Minor Streams, and Utilities)

A separate crew, working ahead of the tie-in crew, can bore through roads, highways, and railroad crossings to avoid interruptions in service and damage to the surface. However, ongoing maintenance and repair can be done in conjunction with tie-ins. The number of crossings, the terrain, and the water table conditions are major considerations when deciding whether to use a separate crew.

Utility crossings that require the pipe to be installed under an existing facility are usually completed by the tie-in crew. Stream crossings that can be excavated by backhoes are also installed by the tie-in crew. Additional equipment may be required for larger sections if they require the provision of weights for buoyancy control.

A typical boring crew would include a foreman, a subforeman, welders, welders' helpers, operators, laborers, and a bus driver. The necessary equipment would include side booms, a boring machine, backhoes, a ditch pump, 4×4 pickups, and a bus.

Inspectors will monitor and ensure that:

- Pipe handling, excavation, welding, coating, and backfilling requirements are met.
- Provisions of crossing agreements are met and representatives are notified and are on-site as required.
- Proper warning signs and barricades are installed and other provisions for public safety, such as flag men, are employed.
- Depths of cover and other dimensions specified on the drawings are met.
- As-built data are recorded prior to backfilling.
- Pipe is adequately supported by compacted fill or sandbags in areas that are over-excavated.

Figure 9-13. Backfilling the trench

Fabrication

Fabrication is the term used to describe assemblies consisting of pipe, valves, fittings, and instruments required to commission and operate the pipeline. Normally, it also includes the welding required to complete these assemblies to the extent that the pieces may be hauled to the final location. This welding may be carried out in a fabrication shop specifically set up by the contractor for the project, or may be done by a subcontractor using an established welding fabrication shop. The assembly will be completed and installed by a tie-in crew. Completed assemblies require a hydraulic pretest prior to welding to the pipeline and backfilling. The pretest will locate any leaks or faulty material that should be repaired prior to mainline testing.

Foundations are required to support the weight of these assemblies and to absorb the thrust from subsequent gas blowdowns. The type of foundation will depend on ground conditions and may consist of concrete blocks, concrete piles, or steel piles.

Inspectors ensure that:

- Shop welding is performed by qualified welders who are following approved procedures.
- Dimensions and configurations shown on drawings are correct.
- Pretests are correctly performed and documented.
- Installations, coating, and tie-ins are done correctly.
- Painting, site finishing, and fencing are completed to specification.

Cathodic Protection

Cathodic protection facilities, including test leads, sacrificial anodes, and ground beds with connection to rectifiers and the pipeline, are installed by the main contractor. Wire connections to the pipe will normally be made by the tie-in crew using a thermite welding process.

Cleanup

Cleanup involves restoration of the right-of-way, temporary workspace, and temporary access routes to their final state for operation of the pipeline. Specific activities include removal of gates; removal and disposal of rock, debris, and excess spoil; replacement of sloping cuts to the extent required; removal of temporary bridges and culverts; installation of erosion control measures; and replacement of topsoil.

Cleanup is done prior to hydrostatic testing to the maximum extent possible, leaving access to areas required by the testing crew.

A typical cleanup crew would include a foreman, a subforeman, operators, laborers, dump truck drivers, and a bus driver. Equipment would include dozers complete with ripper and awinch, backhoes complete with cleanup bucket, a front end loader, dump trucks, a grader, a 4×4 pickup, a bus, and a crew cab.

Inspectors will monitor and ensure that:

- Rock, debris, and excess spoil are removed and disposed of in approved locations.
- Approval of landowners and regulatory authorities has been obtained as required.
- Topsoil suitable for revegetation has been replaced.
- Erosion control measures are properly undertaken.
- Damaged or leaning trees are removed.
- Pipeline marker signs and aerial markers have been properly installed and fence posts have been painted.
- All surplus pipe and other construction materials have been removed.

Erosion Control and Revegetation

The completed pipeline must be protected from soil erosion throughout its operating life to prevent damage and possible failure. Failure can be caused by the removal of supports or the force of flowing water. The main methods of erosion control are revegetation, berming of the ditchline, and installation of diversion berms on slopes to control the downslope movement of surface water. Ditch plugs and subsurface drains will be installed to prevent the flow of ground water along the ditchline (see Chapter 2).

Suitable grass and legume species will be established on the right-of-way by planting or seeding to supplement or replace natural vegetation. Areas subject to high water flow are protected either with concrete or rock rip rap.

Major Water Crossings

Major water crossings, either large rivers or long swamps, must be constructed using special procedures and equipment. Pipelines across rivers may be installed by pulling the pipe under the river with a cable using a stationary or tractor-mounted winch. As many side booms as necessary will carry the concrete-coated pipe to the edge of the water on the side where the pipe section has been welded and concrete-coated. Intermediate welds in the section will be required, depending on workspace and equipment availability. The trench will be excavated using a large dragline, a backhoe, or a clam working on a sled or a barge. The depth and flow of water and the riverbed material will determine the method to be used to excavate and backfill the trench.

Long swampy areas will be crossed by pushing the concrete-coated pipe along the water-filled trench, buoyed with empty barrels or similar flotation that can be released after the pipe is in place. The individual joints of pipe are concrete-coated and welded together, and moved into position using a push station. The trench will be excavated and backfilled using a backhoe, dragline, or clam shells working from swamp mats. In situations where the excavation depths are greater than 30 feet below the mean water level, clam shells mounted on barges are used to excavate and then backfill the installed pipe.

TRENCHLESS CROSSINGS

Notwithstanding the widespread historical use of trenching as a means of crossing rivers, one of the most significant challenges facing the pipeline industry is compliance with increasingly stringent environmental regulations (Boivin et al; 1994). Regulatory expectations, particularly in North America, are becoming increasingly neutral to the cost of crossing installations and are demanding a more thorough assessment of a wider variety of river crossing options including trenchless technologies (Heinz et al; 2004). For example in Canada, a set of Watercourse Crossing Guidelines have been developed, and these have been incorporated in a best practices manual by Trow Engineering Consultants (1996) for PRC International. Table 9-2 lists a number of trenchless crossing techniques that are currently in use, indicating their respective advantages and disadvantages and their appropriateness of use. Perhaps the most commonly used is the directional drilling process, which has been accepted by most pipeline companies as proven technology.

Directional drilling has the following advantages over open trench excavation:

1. The riverbanks and the sensitive ecosystem along riverbanks are not damaged.
2. Since there is no excavation in the river itself or along its banks, there is no silt buildup or bank erosion. It is not necessary to take special measures to protect the marine life.
3. The excavation workspace is much smaller. One only needs to accommodate the fixed machinery at both ends plus a few backhoes and a crane and the water based mud pits. Most importantly the amount of material removed from the hole is many times less than would be excavated using normal open cut techniques.
4. The river can be crossed at almost anytime, (i.e. during fish spawning or high flows).

Directional drilling however has it limitations

1. If the soils are inconsistent, filled with till (small loose packed stones), sandy, or too water laden the directional drilling has a lower chance of being successful and is sometimes impossible.
2. It is a high cost operation, usually charged by the day, and if unsuccessful the crossing must then be open cut, increasing the cost even more (Hair; 1995).
3. It requires a high cost, wear resistant pipe coating for the pull through. One cannot be sure there has not been some coating damage.

There are two stages to direction drilling; the first stage involves drilling a small diameter pilot hole along a designated path, while the second stage involves enlarging the hole to accommodate the pipeline. Figure 9-14 illustrates the various stages in a typical horizontal directional drilled crossing. The pilot hole is kept on course by using a non rotating drill string with an asymmetrical leading edge (Trow 1996).

A steering bias is created by the asymmetry of the leading edge. If a change in direction is required, the drill string is rolled so that the direction of bias is the same as the desired change in direction. Drilling progress is sometimes achieved by hydraulic cutting action with a jet nozzle but is more generally accomplished using a guided drill bit (Figure 9-15).

Mechanical cutting action, when required, is provided by a downhole positive displacement mud motor.

The actual path of the pilot hole is monitored during drilling by taking periodic readings of the inclination and azimuth of the leading edge. These readings, along with measurements of the distance drilled since the last survey, are used to calculate the horizontal and vertical coordinates along the pilot hole, relative to the initial entry point on the surface.

The pilot hole is enlarged using a reaming process which can either form part of the pull back process or be completed prior to it. Ordinarily the reaming tool is attached to the drill pipe at the exit point. The drill pipe is then drawn toward the drilling rig, as the reamers rotate thus enlarging the pilot hole. Sections of drill pipe are continuously added as progress is made, to ensure that there is always a string of pipe in the hole. The reaming tools are comprised of a circular set of cutters and drilling fluid jets. The pressurized drilling fluid serves three purposes, to cool the cutting tools, support the reamed hole and to lubricate the trailing drill pipe. In soft soils with small diameter lines, pre-reaming is unnecessary and the final installation phase is undertaken when the pilot hole has been completed. In this circumstance the reaming assembly is attached to the actual pipeline pull section and sections of pipe are added as the reamer progresses towards the drilling rig. In order to minimize the torsion acting on the pipeline a swivel is used to attach the pull section to the leading reamers.

Route Selection and Assessment

The first stage in an HDD project is to determine its feasibility. Key factors which must be considered are:

(i) Sub surface conditions- certain ground conditions are more amenable to HDD than others, for example, the presence of extensive gravel, cobbles and boulders can be problematic.
(ii) Entry and exit site locations –Directional drilling is a surface launched process and does not require excavated pits. Suitable entry and exit sites however must be available which are compatible with both construction and operational requirements – relatively flat, so as to minimise grading, with sufficient space to allow movement of large equipment and the placement of sumps and retention ponds. For larger bores this can require a sizeable footprint and extensive land disturbance. The alignment of

TABLE 9-2. Trenchless crossing techniques

Method	Description	Environmental Factors		Construction/Engineering Factors		Appropriate Use
		Advantages	Disadvantages	Advantages	Disadvantages	
Boring.	• Bore under watercourse from bore pit on one side to bore pit on other with or without casing.	• No sedimentation.	• Bore pits may require dewatering onto surrounding land.	• Can be fast under the right conditions.	• Potential for the flow in the watercourse to enter the borehole.	• Fine textured impermeable soils.
	• Wet boring with pilot hole and reaming bit can also be performed.	• No disturbance of stream bed.	• Possibility of sump water causing sedimentation of watercourse.	• Minimizes clean-up of banks and bed.	• Can be slow or not feasible under adverse conditions.	• Low water table.
		• No bank disturbance.	• Deep bore pits cause terrain disturbance.	• Road boring equipment may be easily available.	• Difficult with till or coarse material.	• Where stream bed cannot be disturbed.
		• Maintains normal stream flow.		• Casing or pilot pipe prevents bore collapse.	• Potential for raveling at face.	• Used most often on irrigation ditches.
		• Maintains fish passage.			• Potential for bore pit cave-in.	• Where little or no approach slopes are present.
					• Potential to affect slope stability.	
					• Seepage into bore pit.	
					• Possible need for specialized equipment.	
					• Limited to = 100 m (328 ft).	
					• Casing may bind in hole if there is a pause in the boring.	
Punching/ramming.	• Ram or punch casing or pipe under watercourse.	• No sedimentation.	• Bore pits may require dewatering onto surrounding land.	• Can be quick under the right conditions.	• Can be slow under adverse conditions.	• Fine textured soils with low hydraulic conductivity.
	• Bore pits required as detailed.	• No disturbance of stream bed.	• Possibility of sump water causing sedimentation of watercourse.	• Minimizes clean-up of banks and bed.	• Potential bore-pit cave-in.	• Low water table.

Technique	Advantages	Disadvantages	Applications
	• No bank disturbance. • Maintains normal streamflow. • Maintains fish passage. • No cave-in of bore hole. • Can handle standard big inch pipe diameters.	• Bore pits cause terrain disturbance. • Potential to affect slope stability. • Seepage into bore pit. • Specialized equipment required. • Potential corrosion problems. • Relatively inaccurate. • Limited to ≈ 50 m (164 ft) in length. • Severely limited in the presence of cobbles or boulders.	• Irrigation ditches. • Where stream bed cannot be disturbed. • Can also be used in coarse textured substrate. • Where little or no approach slopes are present.
Directional drilling. • Set up drilling rig above approach slopes and drill to target above opposite approach slopes.	• No sedimentation. • No bank disturbance. • No stream bed disturbance. • No approach slope disturbance. • Maintains normal stream flow. • Maintains fish passage. • Eliminates clean-up and reclamation in valley.	• Disturbance of drilling and target area. • Disposal of drilling fluids. • Fractures in substrate may release pressurized drilling fluids into watercourse. • Potential for "blowouts". • Expensive. • Depends on substrate/bedrock. • Specialized equipment required. • Limited to stress-free radius of pipe.	• Large watercourse with sensitive habitat and no instream activity allowed. • Areas with very unstable approach slopes. • High aesthetic concerns (e.g., parks). • Large watercourse with extensive contaminated sediments.

Adapted from Canadian Watercourse Crossing Guidelines (1993).

the drill path relative to the pipe assembly corridors can sometimes be a determining factor.

(iii) Pipe Assemble/Lay down considerations. Ideally a lay down corridor, of sufficient length, to allow for the assembly of a single piece of pipe for the proposed crossing would be desirable. Otherwise the pipeline will need to be assembled in segments and the pulling operation halted in order to make the field welds. In order to minimise the installation stresses, it is highly desirable to assemble the pipe into no more than 2 or 3 segments prior to pulling the pipe into the reamed hole. On longer

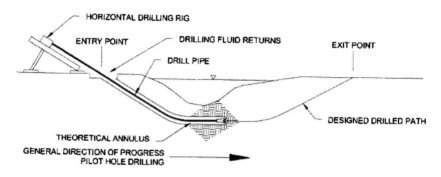

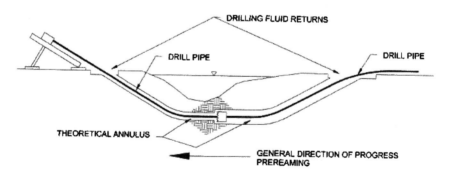

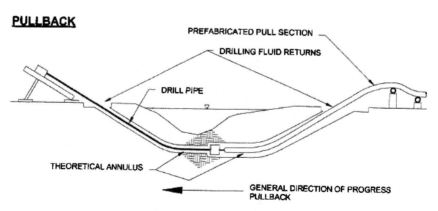

Figure 9-14. Stages in a Horizontal Directional Drill Crossing (CAPP 2005)

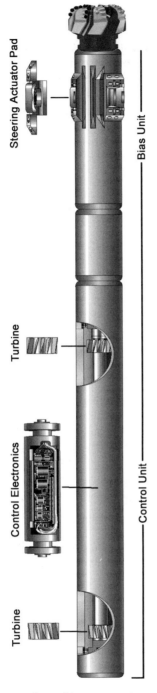

Figure 9-15. Steerable Drilling Head Assembly

crossings this can requires a corridor in excess of 500m. A straight assembly corridor is preferred but may not be possible, in which case, the corridor should be such that the calculated minimum radius of curvature for the pipe is not exceeded.

(iv) Structural considerations- the pipeline must be capable of withstanding the total combined stresses imposed during installation and ongoing operational conditions.

(v) Drilling considerations -limitations to the drilling process are assessed to ensure the optimum crossing solution is recommended.

An initial assessment of the geotechnical, hydrological and environmental setting is made based on any existing data for the site and the historical performance of any previous drilling experience in the area. It is usual to examine several conceptual drill paths and possible pipe assembly areas during the assessment process (Barlow et al; 1996).

In order to determine the drilling profile a schematic stratigraphic profile is prepared from data collected using a number of boreholes, (Figure 9-16); 3-D seismic data; resistivity sampling, and ground penetrating radar. The purpose of the boreholes is to provide information on:

The lithology (rock or soil types present);
The unconfined compressive strengths of the sub surface material;
Possible penetration rates.

The seismic and ground penetrating radar data is used to help identify:

Consistency of stratigraphy and the presence of such anomalies as boulders, caverns and loose gravel and density pockets.
Provide information on the size and extent of fractures in three dimensions.

Figure 9-17 illustrates schematically, a packer test assembly. This is used when drilling the boreholes to determine potential fluid loss zones and the maximum annular pressure for the drill path. The inflatable packer heads isolate the test section at the prescribed depth allowing the acceptable annular pressure to be determined. This is an important step in determining the potential for "frac outs" at various points along the possible drill path.

Having established one or more feasible drill path routings, the next steps in the directional drill design address the:

(i) Design of a drill path
(ii) Developing the drilling fluid design
(iii) Performing a drilling fluid pressure analysis
(iv) Assessing the pulling force and bending stresses during installation
(v) Investigating the need for surface casing installations

The design of the drill path involves determining the entry and exit angles, the required depth of cover, and for each segment the permissible arc radius. A detailed pipe stress analysis is completed using the methodology described later. It is prudent to take a conservative approach to ensure that unexpected conditions encountered during construction can be quickly assessed. A detailed site inspection will ensure that the pipe layout area, pipeline tie-ins, and other construction details are suitable. Later, once the optimal drill path is determined, the geotechnical, design, and construction drawings are prepared.

A drilling fluid is required to transfer power and lubricate the mud motor and bit, transport cuttings, and seal fractures in the formation. The properties of the drilling fluid are dependent upon the formation type and to ensure that the following key parameters are met; penetration rate, annular pressure control, and minimising the likelihood of formation fracture. The choice of fluid is determined by the type of soil / ground conditions likely to be encountered. These soils are characterised as being either coarse grained or fine grained, cohesive or cohesion less. A cohesive soil, like clay, is made up of fine grained particles which stick together through chemical bonding. As a consequence, the permeability of clay is very low. Sand or gravel, by comparison, are both coarse grained and cohesion less due to the weak chemical bonding of their particles. They are also highly permeable to water, which

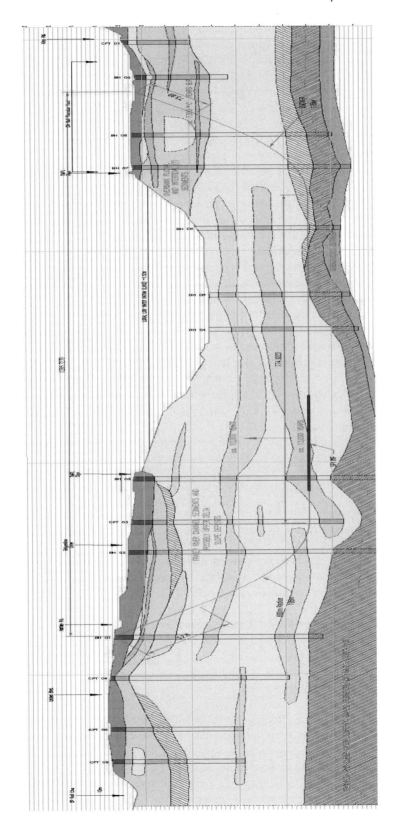

Figure 9-16. A Schematic Stratigraphic Profile of a River Crossing Showing Borehole Locations. Courtesy Entec Inc.

is why, when drilling through them, the drilling fluid generally used contains the natural clay lubricant sodium bentonite. Bentonite when mixed with water, the primary component of the drilling fluid, breaks down into a very large number of particles known as platelets. These platelets have a huge surface area relative to the volume of bentonite added. When the fluid is pumped into the bore hole, under pressure, the platelets attach themselves to the surface wall and begin to cake – effectively closing off the drilling fluid from migrating into the surrounding sand or gravel. To help the bentonite / water mixture flow better, long chain polymers can be added. Polymers are also added to the mixture when drilling in clay. Besides sealing the borehole, the bentonite performs another useful function, that of holding the drilling cuttings in suspension, which assists greatly in transporting them out of the hole. The cuttings, plus pressurised drilling fluid mixture, provide support for the bore hole until the pipeline is installed. When the slurry, containing spoil, flows from the borehole, either on the entry or exit side, it is indicative of an open bore path. It is highly beneficial to have good slurry flow during drilling and back reaming operations.

The purpose of the drilling fluid analysis is two fold; first to establish that sufficient fluid is available to allow the drilling and pulling operations to be completed smoothly and secondly to calculate the anticipated annular pressure acting on the formation during the drilling process. Foamed mud, or air mist systems can sometimes be used, and with their lower density they have the effect of lowering the annular pressure and thus minimising the chance of lost circulation and fracturing. In addition, foam has very good cuttings-carrying capacity, which assists in cleaning the borehole and in solids removal. Generally the volume of circulating fluid required is also less with foams than for conventional drilling fluids.

The amount of drilling fluid necessary is a function of the type of soil encountered. As a rule of thumb, equal or slightly greater, volumes of fluid are needed to remove a given volume of sand, whereas for clay type soils the amount will be three to four times the amount of soil removed. At least 30-35 lbs of high quality bentonite is needed per 100 gallons of drilling fluid to allow a sufficient margin of safety when drilling in most sands or gravel, but it could increase by 50% for other types.

A simple formula that it is useful in determining the amount of drilling fluid required is:
(Bit or reamer size in inches)^2 /24.5 = Gallons of soil/foot of borehole

The annular pressure will increase under the following conditions:

- Borehole instability causing sloughing
- Poor cuttings removal causing annulus loading
- Improper drilling fluid parameters

It will decrease under the following conditions:

- Fluid loss due to fracture
- Using low density drilling fluids
- Efficient hole cleaning

Onsite drilling fluid management, including monitoring and recording the annular pressure and formation reaction, are critical to prevent hydro-fracture during the drilling and reaming processes, and to ensure optimum performance. The pressure is continuously monitored and recorded at the start, middle, and end of each drill stem joint to ensure that drilling is proceeding as required. Drilling fluid volumes, hole cleaning efficiency and penetration and flow rates are closely monitored as is the pilot hole accuracy.

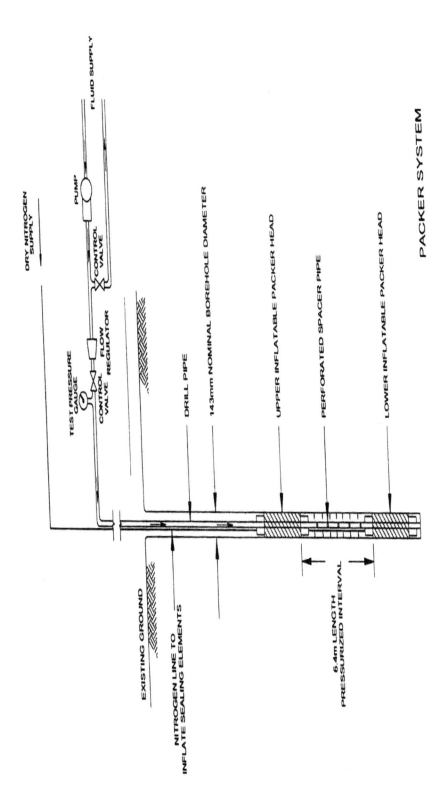

Figure 9-17. Packer Test System. Courtesy Entec Inc.

The provision of surface casing is needed in situations where the overlying material close to the entrance and exit of the drill path is weak, and could collapse during drilling or, form a sinkhole under service conditions (Baumgard et al; 2004). The necessary casing design therefore must be of adequate depth to provide a seal with competent strata, in order to mitigate the risk of fluid migration, or borehole collapse near the surface. The diameter of the casing should be sufficient to allow the passage of the final reaming tools. Casing can be installed on both sides of the crossing to allow for casing-to-casing drills. The casing is pile driven into position to the required depth.

Structural Considerations

Currently there is only one published standard which addresses specific requirements for directional drilling, the Dutch Code NEN 3651 (1994). In the absence of such standards, the design responsibility passes to the pipeline operator and the installation contractor, who would be wise to engage the services of an experienced drilling consultant. Fortunately, there are several excellent design and construction guidelines, among which, are the Canadian Association of Petroleum Producers (2004) publication, "Planning Horizontal Directional Drilling for Pipeline Construction," an ASCE recommended practice (2005), and the engineering guideline produced by the Pipeline Research Council International PRCI (Watson; 1995). The former is available as a free down load from the association's web site.

The PRCI document notes that, "load and stress analysis for an HDD pipeline is different from similar analyses of conventionally buried or laid pipelines because of the relatively high tension loads, and sometimes severe bending and external fluid pressures felt by the pipeline during the installation process." In many instances the installation loads will be substantially larger than the in service design loads, requiring the pipe to be sized accordingly.

Installation Loads

During installation, the pipeline experiences a tensile pulling load as it is being fed into the borehole; bending stresses as it is forced to conform to the curved profile, and external pressure due to the presence of the lubricating drilling mud in the annular space between pipe wall and borehole (unless the pipe is filled with drilling mud at the same pressure). The resulting stresses have to be combined in order to establish whether or not the pipe will fail. (Fowler and Langner (1991).)

In order to determine the tensile force required to pull the pipe through the reamed hole we need to consider four components:

The submerged weight of each pipe section;
The viscous force or fluid drag due to pulling the pipe through the drilling mud;
The sliding friction between the pipe and the surface of the bore hole; and,
The frictional force of the pipe section immediately outside the entrance as it moves over the support rollers.

The calculation process proceeds in a systematic manner from right hand end to the left, that is, from the pipe side to the rig side. Figure 9-18 illustrates the design drill path, which can be subdivided into a sufficient number of straight and curved segments, as to completely define its shape. For simplicity, straight segments may actually have a slight curvature, while a curved segment should have a constant radius of curvature. The individual loads acting on each segment are resolved to provide a resultant tensile load. The total axial load in the pipe during the last portion of the pull back process will be the sum of the tensile forces acting on each segment.

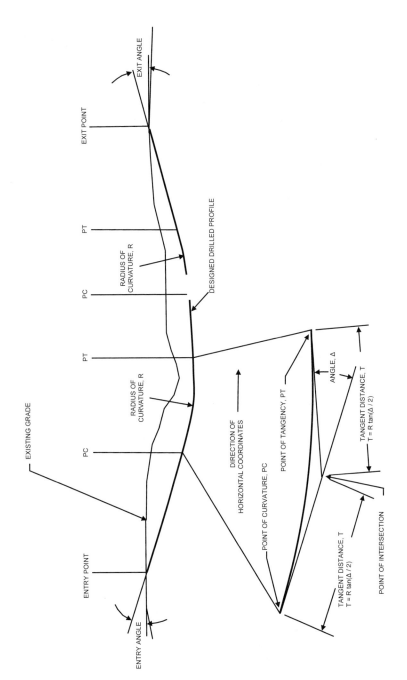

Figure 9-18. Design drill path

Frictional and fluid drag effects always increase the tensile load since they act in the opposite direction to the pipe movement (Figure 9-19). However, since the frictional and drag force components are caused by the shape of the hole, the axial tension at any point can be assumed to be linked to that particular location irrespective of which portion of pipe is passing over it during the pull back. This means that the worst case load at any point on the profile can be determined for the case when the pipe has just emerged from the hole.

The tension T_2 at the left hand end of the element is given by:

$$+/- T_2 = T_1 + | F_{frict} | + | F_{Drag} | + W_s \, LSin\theta \qquad (9-1)$$

The sign of the value for T_2 is dependent upon whether T_2 tends upward (+ve), downward (−ve) or has zero value, if the section is horizontal, that is, θ is zero.

T_1 is the tension at the right hand of the section, and may be zero if it is the first section entering the hole, or it could have a discrete value if there is friction present on an external section of pipe sliding on rollers.

F_{frict} = Friction force between the pipe and the bore hole surface
   $= |W_s \, LCos\theta \, \mu_{soil}|$

Ws = distributed submerged weight of pipe and its contents (if any)

F_{drag} = fluid drag between the moving pipe and the drilling fluid in the annulus
   $= |12 \, \pi \, DL \, \mu_{mud}|$ or $|\pi*L\mu*_{mud}|$

μ_{mud} = drag coefficient (recommended value of 0.05 psi (NEW3650), $\mu*_{mud} = 0.3447$ kPa)

μ_{soil} = friction coefficient between pipe and soil, recommended values between
      0.21-0.3 (Watson)

L = section length (ft or m)

D = outside diameter of the pipe (inches) or D* in m

θ = angle of inclination of the pipe section relative to the horizontal.

The absolute value of the friction and drag force is taken as it always acts in opposition to the direction of movement, and adds to T_2.

The forces acting on a curved pipe segment are as shown in Figure 9-20, and require a number of additional variables to be defined.

R = constant radius of curvature (ft or m)

α = included angle of the curved section (degrees)

   θ_1 = inclination from the horizontal of T_1, at the right hand end of the segment (degrees)

   θ_2 = inclination from the horizontal of T_2 at the left hand end of the segment (degrees)

   $\theta = (\theta_1 + \theta_2)$

   L_{arc} = section arc length = $R\alpha\pi/180$ (ft or m)

N_1, N, N_2 = forces normal to the pipe wall at the right, centre and left contact points

   fr_1, fr, fr_2 = frictional forces associated with the normal contact forces at the right, centre and left contact points.

The simplifying assumption is made that the distributed submerged weight of the pipe and any contents is concentrated to act at the centre of the segment as shown in Figure 9-20. In order to calculate the three normal contact forces, the segment is modeled as beam in 3 point bending as shown in Figure 9-21a. Since the bending moment over the segment is not constant, then strictly speaking it will not assume the shape of a circular arc. However, in order for the bent pipe section to fit into the curved drill path segment, it must deflect sufficiently so that its centre can touch a point matching the displacement h, of a circular arc of radius R, as shown in Figure 9-21b.

$$h = R(1 - Cos \, \alpha/2) \qquad (9-2)$$

Clearly there is a distinct possibility that any pipe section will touch the surface of the bore hole segment at more than three places, so this assumption and that of 3 point bending that follows from it, are simplifications. They are acceptable however, given that our primary interest lies in determining the frictional forces associated with the normal contact forces, and that our knowledge of the drag and friction coefficients is inexact. The solution for the value of the central contact force N that produces a central deflection of h is given by Roark (1965) as:

$$N = (12\ T_{ave}\ h - (Ws/12)Cos\theta \cdot Y)/X \qquad (9-3)$$

where, $X = 3L_{arc} - (j\ tanh\ (U/2))/2$
$Y = 18(L_{arc})^2 - j^2\ (1 - 1/cosh\ (U/2))$
$U = 12\ L_{arc}\ /\ j$
$j = (EI/\ T_{ave}\ )^{0.5}$

and E = Young's Modulus, I = Bending Inertia of the pipe section and tanh and cosh are respectively the hyperbolic tangent and cosine.

It can be seen that equation 9-3 requires an iterative solution since T_{ave}, the average tension $(T1 + T2)/2$ is implicit, being used to calculate both N and j. It is usual when performing hand calculations not to determine T_{ave} to very high accuracy, a few iterations being sufficient.

The central friction force is given by, $fr = \mu_{soil}\ N$.

From symmetry, the end reactions N_1 and N_2 are N/2 and the end friction forces fr_1, fr_2 are taken to be, fr/2. Figure 9-21b shows N acting downwards and if this is taken as a positive value, it means that the bending resistance and /or the buoyancy of the pipe section is sufficient to require a normal force acting against the top of the hole in order to enable the pipe to deflect by an amount, h. When N is negative, the submerged weight of the pipe section causes it to sink, coming to rest at the bottom of the curved segment where an upward normal force is felt at the point of contact. As noted previously, the absolute values of the friction and drag forces are taken, irrespective of the sign of N since these forces always oppose motion and serve to increase the tensile force T_2.

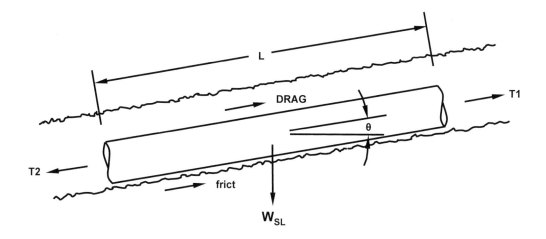

Figure 9-19. Forces acting on a straight pipe segment (Watson (1995))

If we assume that the pipe section behaves like a catenary, the forces acting upon it can be added hence,

$$+/- T_2 = T_1 + 2|F_{frict}| + |F_{Drag}| + W_s\,L_{arc}Sin\theta \qquad (9-4)$$

As with the straight section, the $+/-$ term is resolved by examining if T_2 tends upward ($+ve$), downward ($-ve$) or has zero value if the section is horizontal, that is θ is zero.

The tensile stress at any section on the pipe is given by the ratio of the local tension and the cross sectional area,

$$S_A = T_{local}/A$$

A commonly accepted value for the allowable tensile stress during installation is 90% of the specified minimum yield stress.

Bending Stresses

In addition to the tensile forces, the pipe will also experience bending stresses as it is forced to follow the curvature of the drill path. As a first approximation, the information required to calculate the various bend radii can be obtained from the as built centre line data of the pilot hole. By assuming the pipe is bent in the arc of a circle the radius of curvature R (in feet or m) of the drilled hole is given by,

$$R = 57.33\,(L_{arc}/\alpha)$$

where L_{arc} is the arc length of the drilled segment (feet or m) and α is the change in angle over the arc length L_{arc} (degrees)

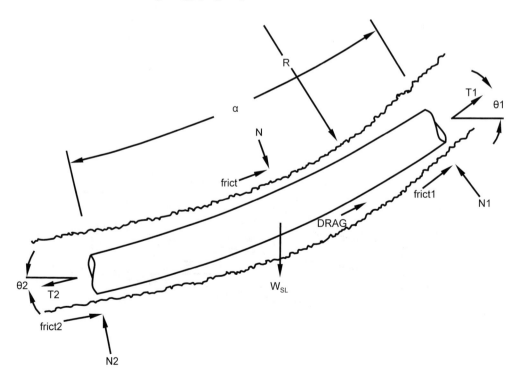

Figure 9-20. Forces acting on a curved pipe segment (Watson (1995))

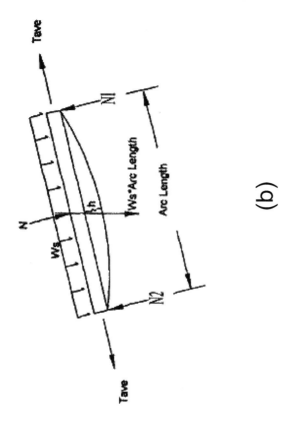

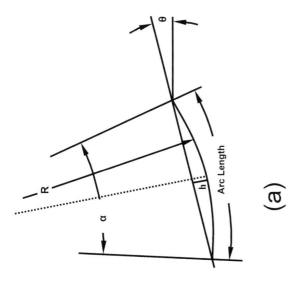

(b)

(a)

Figure 9-21. Three Point Bending Model

The bending stress is obtained from, $S_{BE} = \dfrac{ED}{24R}$ lbf/in^2 or $\dfrac{ED}{2R}$ kPa

where D is the pipe diameter in inches (m), E is Young's Modulus, lbf/in^2 or kPa.

The design guidance for the limiting value of bending stress, S_b , during pull back is based on the following formulations developed for the behaviour of thin walled tubular members in offshore structures (API Recommended Practice 2A; 1993).

$S_b = 0.75\,S_y$ for $D/t/</= 1{,}500{,}000/S_y$

$S_b = [0.84-(1.74\ S_yD)/Et]S_y$ for $1500{,}000/S_y\ <D/t/ \leq = 3{,}000{,}000/S_y$

$S_b = [0.72-(0.58\ S_yD)/Et]S_y$ for $D/t > 3{,}000{,}000/S_y$

Given the simplifying assumptions made about the drill profile, the idealized analytical models and the neglect of dynamic effects, it is good practice in order to provide a margin of safety, to keep the D/t ratio below 60.

As a further check against buckling failure or plastic collapse due to an external hoop stress, the following formula, proposed by Timoshenko, is used to determine the minimum wall thickness.

$$t = D(P_{ext}/2E)^{0.333}$$

The external hydrostatic pressure produced by the weight of the drilling mud column is given by

$$P_{ext} = W_mH_m/19.25 \text{ (psi) or } W_mH_m/101.94 \text{ (kPa)}$$

where

Wm is the mud density in pounds/gallon (kg/m^2) and H_m is the depth of mud column in feet (m)

The critical value of hoop buckling stress caused by external pressure on a tubular section can be checked using formulations contained in the API Recommended Practice 2A referenced above.

$$S_H = S_{HCritical}/1.5$$

where $S_{HCritical}$ is given in terms of the elastic hoop buckling stress S_{HE} as follows:

$S_{HE} = 0.88E(t/D)^2$ for long unstiffened cylinders

$S_{HCritical} = S_{HE}$ for $S_{HE} </= 0.55S_y$

$S_{HCritical} = 0.45\ S_y + 0.18\ S_{HE}$ for $0.55S_y < S_{HE} </= 1.6\ S_y$

$S_{HCritical} = 1.31\ S_y/[1.15 + S_y/S_{HE}]$ for $1.6S_y < S_{HE}/ = 6.2\ S_y$

$S_{HCritical} = S_y$ for $S_{HE} > 6.2\ S_y$

Using these formulae, the hoop stress due to external pressure should be limited to 67% of the critical hoop buckling stress.

Having determined that the pipe will not fail under the action of a single load condition, the effect of combined loading at discrete points along the pipe length, where high stresses are expected, should be checked. The first of these combined load cases is a unity check for combined bending and tension. That is,

$$S_A/(0.9 \ S_y) + S_{BE}/S_b \ </= 1.0$$

The unity check for the complete interaction of tension, bending and hoop stress is,

$$A^2 + B^2 + 2 \ \nu \ |A| \ B \ </= 1.0$$

where

$A = 1.25 \ (\ S_A + S_{BE} - 0.5 \times S_H \ ) / S_y$
$B = 1.5 \ S_{H/} \ S_{\ HCritical}$ and
ν = Poisson's ratio

A determination that the unity check is exceeded does not necessarily imply that a failure would result. Rather that extra care should be taken, perhaps to modify one of the load conditions. For example filling the pipe will alter the buoyancy effect and can reduce the hoop stress due to external pressure differential.

Example

An NPS10 Grade X52 product pipeline with a wall thickness of 0.25 inches is to be pulled through a directionally drilled crossing having the profile shown in Figure 9-22. The crossing of 1300ft projected length, can be assumed to lie in the plane of the paper. The entrance on the pipe side (right), and the exit on the rig side (left), are at the same vertical height. The profile is represented by three straight and two curved sections. The entrance angle is at 15° and the exit angle 12° to the horizontal. The rightmost curve has a 25 degree arc on a radius of curvature of 1100 feet, while the left curved section has a 15 degree arc on a radius of curvature of 1200 feet. The location of the horizontal mid section is 80 feet below the elevation of the entrance and exit datum points.

Determine the total force T_{tot} required to pull the entire pipeline through the reamed hole if the right side tension on the pipe, as it enters the hole, is zero and the pipe is installed empty. The density of the drilling mud can be taken to be 90.28 lb/ft³, the drag coefficient μ_{mud} = 0.05 psi, the coefficient of friction between pipe and the soil μ_{soil} = 0.3. Young's modulus for steel is 30×10^6 lbf /in².

The submerged weight of the pipe/unit length Ws = Weight of steel pipe Wp – Buoyancy B

$$Wp = 3\pi(10.75^2 - 10.25^2) \times 0.28 = 27.7 \ lbf/ft$$
$$B = \pi D^2 L \rho/4 = \pi 10.75^2 \ 1 \ 90.28/576 = 56.9 \ lbf/ft$$

$$Ws = 27.7 - 56.9 = -28.2lbf/ft$$

The depth of each intersection point on the drill path, the horizontal projection length and curved segment lengths are found from trigonometry. The drill path in Figure 9-22 has been subdivided into 5 segments whose properties and orientation from right to left are described as follows:

Section	Type	Angle	Length	Orientation of tensile force
A to B	Straight	$\theta = 15°$	69 ft	Down
B to C	Curved R = 1100 feet	$\theta = 7.5°, \alpha = 25°$	476.2 ft	Down
C to D	Straight	$\theta = 0°$	227.1 ft	
D to E	Curved R = 1200 ft	$\theta = 6°, \alpha = 15°$	314.2 ft	Up
E to F	Straight	$\theta = 12°$	227.5 ft	Up

The pull force on the downward inclined section A to B is found by considering the forces on the segment. The difference in tension between the segment ends is

$$\Delta T_{BA} = T_B - T_A$$
$$\Delta T_{BA} = |F_{frict}| + |F_{Drag}| - W_s \, L Sin\theta$$

$$F_{frict} = |W_s \, L Cos\theta \times \mu_{soil}|$$
$$= |-28.2 \times 68.97 \times Cos \, 15° \times 0.3|$$
$$= 583 \text{ lbf}$$

$$F_{drag} = |12 \, \pi D \, L\mu_{mud}|$$
$$= |12 \, \pi \times 10.75 \times 68.97 \times 0.05|$$
$$= 1397.6 \text{ lbf}$$

$$\text{Weight component} = WsL \, Sin\theta$$
$$= -28.2 \times 68.97 \times Sin \, 15°$$
$$= -503 \text{ lbf}$$

$$\Delta T_{BA} = 583 + 1397.6 - (-503) = 2483 \text{ lbf}$$
$$T_B = T_A + \Delta T_{BA}$$
$$T_B = 0 + 2483 = 2483 \text{ lbf}$$

Tension on Curved section at point C

The incremental tension is given by $\Delta T_{CB} = T_C - T_B$. In order to determine ΔT_{CB} we first need to determine the normal contact forces.

$$h = R \, (1 - Cos \, \alpha/2) = 1100 \times (1 - Cos \, 12.5°)$$
$$= 26.07 \text{ ft}$$

The second moment of area of the pipe I is given by $\pi \, t \, (D-t)^3/8$

$$I = \pi \times 0.25 \, (10.75 - 0.25)^3/8$$
$$I = 113.65 \text{ in}^4$$

Starting the iterative process to determine the segment tension we assume a trial value for the average tension in the segment $T_{ave} = 14000 \text{lbf}$.

$$j = (EI/T)^{0.5} = (30 \times 10^6 \times 113.65/14000)^{0.5}$$
$$j = 493.5 \text{ in}^2$$

$$U = 12 \times 476/493.5 = 11.67$$

$$X = 3L_{arc} - (j \tanh \, (U/2))/2$$

$$X = 3 \times 476 - (493.5 \tanh \, (11.67/2))/2$$

$$X = 1193.26$$

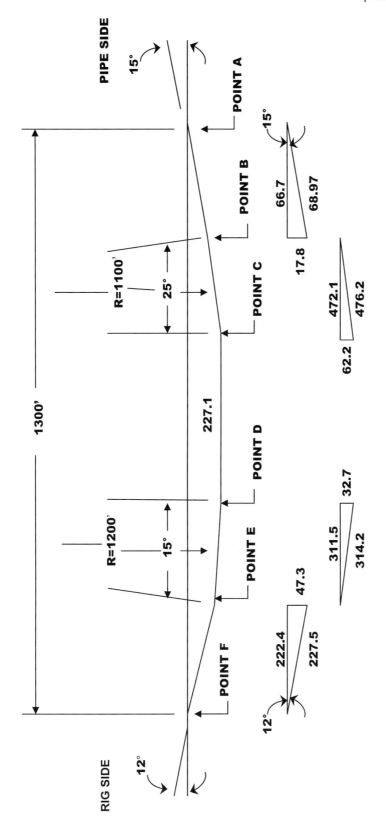

Figure 9-22. Dimensional Schematic of a Drilled Crossing

$$Y = 18(L_{arc})^2 - j^2(1 - 1/\cosh(U/2))$$

$$Y = 18 \times 476^2 - 493.5^2(1 - 1/\cosh(11.67/2))$$

$$Y = 3905088 \text{ in}^2$$

Substituting the above values in the equation for normal contact force

$$N = (12 \ T_{ave} \ h - (Ws/12)Cos\theta. \ Y)/X$$
$$N = (12 \times 14000 \times 26.07 - ((-28.2/12)Cos \ 7.5° \times 3905088)/1193.26$$

$$N = 11296 \text{ lbf}$$

The positive value for N indicates that the normal reaction force is acting downward to bend the pipe into the curve of the reamed hole.

The incremental tension force in the segment is $\Delta T_{CB} = T_C - T_B$

$$\Delta T_{CB} = 2|F_{frict}| + |F_{Drag}| - W_{sl}L_{arc}Sin\theta$$

The friction force, $F_{frict} = |\mu_{soil} \ N| = | \ 0.3 \times 11296| = 3388.8 \text{ lbf}$
The drag force $\quad F_{drag} = |12 \ \pi D \ L_{arc} \ \mu_{mud}| = |12 \ \pi \times 10.75 \times 476 \times 0.05$
$$= 9645 \text{ lbf}$$

Weight component $= WsL \ Sin \ \theta = (-28.2 \times 476 \times Sin \ 7.5°)$
$$= -1752 \text{ lbf}$$

$$\Delta T_{CB} = 2|3388.8| + |9645| - (-1752)$$
$$\Delta T_{CB} = 18170 \text{ lbf}$$

$$T_C = \Delta T_{CB} + T_B = 18170 + 8483 = 20654 \text{ lbf}$$

Checking $T_{ave} = (T_C + T_B)/2 = (20654 + 2483)/2 = 11568 \text{ lbf}$

This value is 18.5% adrift of the assumed initial value so the iteration process begins using the improved estimate for T_{ave} until the difference is within 5% . Convergence is obtained for a value of T_{ave} of 11245 lbf. The resulting value for ΔT_{CB} is 17923 lbf leading to a tensile force T_C of 20206 lbf.

Straight Section C-D
$T_C = 20206 \text{ lbf and } L = 227.1 \text{ feet, } \theta = 0°$

$$\Delta T_{DC} = T_D - T_C$$
$$\Delta T_{DC} = |F_{frict}| + |F_{Drag}| - W_s \ LSin\theta$$

$$F_{frict} = |W_s \ LCos\theta \times \mu_{soil)}|$$

$$= |-28.2 \times 227.1 \times Cos \ 0° \times 0.3|$$

$$= 1921.6 \text{ lbf}$$

$$F_{drag} = |12 \ \pi D \ L \ \mu_{mud}|$$

$$= |12 \ \pi \times 10.75 \times 227.1 \times 0.05|$$

$$= 4601.8 \text{ lbf}$$

Weight component $= \text{WsL Sin } \theta$

$$= -28.2 \times 227.1 \times \text{Sin } 0°$$

$$= 0 \text{ lbf}$$

$$\Delta T_{DC} = 1921.6 + 4601.8 = 6523.4 \text{ lbf}$$

$$T_D = T_C + \Delta T_{DC}$$

$$T_D = 20206 + 6523.4 = 26729 \text{ lbf}$$

Tension on Curved section at point E

T_D = 26729 lbf, L_{arc} = 314.2 feet, θ = 6°. The incremental tension is given by $\Delta T_{ED} = T_E - T_D$ as before, in order to determine ΔT_{ED} we first need to determine the normal contact forces.

$$h = R \ (1 - \text{Cos } \alpha/2) = 1200 \times (1 - \text{Cos } 7.5°)$$
$$= 10.27 \text{ ft}$$

To start the iterative process to determine the segment tension assume a trial value for the average tension in the segment T_{ave} = 32490 lbf.

$$j = (EI/T_{ave})^{0.5} = (30 \times 10^6 \times 113.65/32490)^{0.5}$$

$$j = 323.9 \text{ in}^2$$

$$U = 12 \times 314.2/323.9 = 11.64$$

$$X = 3L_{arc} - (j \tanh (U/2))/2$$

$$X = 3 \times 314.2 - (323.9 \tanh (11.64/2))/2$$

$$X = 780.63$$

$$Y = 18(L_{arc})^2 - j^2(1 - 1/\cosh (U/2))$$

$$Y = 18 \times 314.2^2 - 323.9^2(1 - 1/\cosh (11.64/2))$$

$$Y = 1672673 \text{ in}^2$$

Substituting the above values in the equation for normal contact force

$N = (12 \ T_{ave} \ h - (Ws/12)\text{Cos}\theta . \ Y)/X$
$N = (12 \times 32490 \times 10.27 - ((-28.2/12) \text{ Cos } 6° \times 1672673)/780.63$

$N = 10135 \text{ lbf}$

The positive value for N indicates that the normal reaction force is acting downward to bend the pipe into the curve of the reamed hole.

The incremental tension force in the segment is $\Delta T_{ED} = T_E - T_D$

$$\Delta T_{ED} = 2|F_{frict}| + |F_{Drag}| + W_{sl} \ L_{arc}\text{Sin}\theta$$

The friction force, $F_{frict} = |\mu_{soil}\ N| = |\ 0.3 \times 10135| = 3040$ lbf

The drag force $F_{drag} = |12\ \pi D\ L_{arc}\ \mu_{mud}| = |12\ \pi \times 10.75 \times 314.2 \times 0.05$
$$= 6366\ lbf$$

Weight component $= WsL\ Sin\ \theta = (-28.2 \times 314.2 \times Sin\ 6°)$
$$= -926 lbf$$

$$\Delta T_{ED} = 2|3040| + |6366| + (-926)$$
$$\Delta T_{ED} = 11521\ lbf$$

$$T_E = \Delta T_{ED} + T_D = 11521 + 26729 = 38250\ lbf$$

Checking $T_{ave} = (T_D + T_E)/2 = (26729 + 38250)/2 = 32489$ lbf

This value is identical to the estimate so there is no need to iterate further.

Straight Section E-F

$T_E = 38250$ lbf and $L = 227.5$ feet, $\theta = 12°$

$$\Delta T_{FE} = T_F - T_E$$
$$\Delta T_{FE} = |F_{frict}| + |F_{Drag}| + W_s\ LSin\theta$$

$$F_{frict} = |W_s\ LCos\theta \times \mu_{soil)}|$$
$$= |-28.2 \times 227.5 \times Cos\ 12° \times 0.3|$$
$$= 1882\ lbf$$

$$F_{drag} = |12\ \pi D\ L\ \mu_{mud}|$$
$$= |12\pi \times 10.75 \times 227.5 \times 0.05|$$
$$= 4610\ lbf$$

Weight component $= WsL\ Sin\theta$
$$= -28.2 \times 227.5 \times Sin\ 12° = -1334\ lbf$$
$$\Delta T_{FE} = 1882 + 4610 - 1334 = 5158\ lbf$$
$$T_F = T_D + \Delta T_{FE}$$
$$T_D = 38250 + 5158 = 43408\ lbf$$

Hence the total pulling load required to install the pipe is 43408 lbf. The axial stress on the pipe at point E is,

$$S_A = T_E/A = 38250/(\pi/4(10.75^2 - 10.25^2) = 4638\ lbf/in^2$$

Consider the stresses acting at the point E which is at a depth 47.3 feet. The bending stress at E is therefore,

$$S_{BE} = \frac{30 \times 10^6 \times 10.75 \text{lbf/in}^2}{24 \times 1200}$$
$$= 11978 \text{ lbf/in}^2$$

The D/t ratio for the pipe is 43 so the allowable bending stress is given by

$S_b = [0.84 - (1.74\ S_y\ D)/Et]\ S_y$ for $1,500,000/S_y <D/t/ \le= 3,000,000/\ S_y$

$S_b = [0.84 - (1.74 \times 52000 \times 10.75)/30,000,000 \times 0.25)] \times 52000$

$= 36936 \text{ lbf/in}^2$

So the actual bending stress is within the allowable limit.

The external hoop stress caused by the hydrostatic differential fluid pressure (29.38psi) exerted by the drilling mud at a depth of 47.3 feet, assuming no fluid within the pipe is,

$$S_H = \frac{\Delta PD}{2t} = \frac{29.38 \times 10.75}{2 \times 0.25}$$
$$S_H = 632 \text{ lbf/in}^2$$

The elastic hoop buckling stress S_{HE} is found using,

$S_{HE} = 0.88E(t/D)^2 = 0.88 \times 30,000,000 \times (0.25/10.75)^2$

$= 14278 \text{ lbf/in}^2$

$S_{HCritical} = S_{HE}$ for $S_{HE} \le= 0.55 \times Sy$

$= 14278 \text{ lbf/in}^2$

and $S_{HCritical}/1.5 = 14278/1.5 = 9518 \text{ lbf/in}^2$

Since $S_H = 632$ lbf/in^2 is less than $S_{HCritical}$ /1.5 the external hoop stress is well below the buckling threshold.

Considering the combined loading at Point E

(1) For tensile + bending

$$S_A/(0.9\ S_y) + S_{BE}/S_b \le= 1.0$$
$$4638/(0.9 \times 52000) + 11978/36936 = 0.423 <1.0$$

So the combined tensile and bending stresses at E are acceptable

(2) The unity check for the complete interaction of tension, bending and hoop stress is,

$$A^2 + B^2 + 2\ \nu\ |A|B \le= 1.0$$

where

$A = 1.25(S_A + S_{BE} - 0.5 \times S_H)/S_y = 1.25(4638 + 11978 - 0.5 \times 632)/52000 = 0.39$
$B = 1.5S_H/S_{HCritical} = 1.5 \times 632/14278 = 0.06639$

$$0.39^2 + 0.06639^2 + 2 \times 0.3 \times 0.06639 \times 0.39 = 0.172 \leq 1.0$$

so the combined stresses at point E are acceptable.

Contracting Considerations

Invariably directional drilling contracts are put out to bid which requires the preparation of a bid document package containing the detailed technical specifications, commercial terms and legal requirements. A suitable bid structure might be based on unit pricing, for example a price per foot; reimbursable costs plus allowed profit expressed as a day rate, or a fixed price (lump sum). If the crossing has been deemed technically feasible and contains detailed information on the expected ground conditions then the contract risk is lowered and a lump sum contract will be favoured by the pipeline operator. The other forms of contract are likely to be more appealing to a contractor. The technical specifications will include plans and profile drawings as well as the results of all hydrographic, topographic, seismic and geotechnical surveys. A performance specification will be provided which sets out the required tolerance on accuracy and the acceptable installation stresses. The environmental and health and safety guidelines and regulations to be followed are included in the bid documents. A compliant bid response, besides providing a detailed project schedule and cost estimate, will contain reporting commitments. For example it is reasonable to expect daily progress reports, frequent cost reporting and reports as needed on borehole survey data, drilling fluids and annular pressures.

The contractor should also commit to preparing an as-built drawing of the project as well as a project summary. In terms of staffing, the drilling contractor will supply on-site qualified drilling personnel as required which, as a minimum, would include a,

- Drilling Engineer
- General River Crossing Technician
- Drilling Fluid Technician
- Environmental Specialist
- Annular Pressure Technician

HYDROSTATIC TESTING

The completed pipeline will be subjected to a hydrostatic test to prove the integrity of the materials and identify any leaks. The pipeline is divided into test sections, (Figure 9-23) with a maximum elevation difference to allow maximum and minimum test pressures to be maintained during the duration of the test. An allowance for pressure changes during the duration of the test is also provided to account for environmental temperature variations.

During winter construction in cold climates, and at high elevations in some locations, the possibility of sub-zero ambient temperatures has to be considered. To enable hydrostatic testing to proceed under such conditions a water/methanol mixture is used as the test medium. Alternatively the test fluid may be heated. (See later section on hydrotesting at low temperatures).

A typical test head installation is shown in Figure 9-24a. Test heads with valved connections necessary for filling, pressuring, and instrument lines, are welded on at each end of the test section. The section is filled with water using pumps which have the capacity to overcome pressure due to hydrostatic head. Several sections may be filled together, using temporary piping connections between test heads. Air will be prevented from mixing with the fill water by the use of a sphere or bi-directional pig sized to fit the inside pipe diameter. Additional air may have to be pumped into the section to prevent the sphere from running downhill and creating a vacuum, which would cause air to bypass and be trapped in the section. This topic is discussed in more detail in a subsequent section on filling and dewatering.

After the section is filled and the temperature of the fill water is stabilized to ground temperature, the fill halves will be blind-flanged and additional water pumped in to increase the pressure to the level required. If the test pressure is to exceed 90% of the specific minimum yield strength (SMYS) of the pipe usually, and according to some codes, a yield plot must be done. The volume of water required for each increment of pressure (100 kPa) is plotted on a chart such as that shown in Figure 9-24b. If this plot deviates from the straight line established at lower pressures by an amount equal to 0.2% of the section's water volume (Table 9-3), pressuring ceases and the test is commenced. Otherwise, a maximum pressure will result in 110% of SMYS at the low elevation of the section. Deadweight pressure gauges, and circular chart pressure and temperature recorders are used to measure and record pressure, pipe surface temperature and ground temperature during the 4 hours pressure test. Any loss in pressure during the test must be linked to a temperature change or the test will not be accepted, Gray (1976). Test pressure will be maintained until the pressure stabilizes, or until a leak is evident which must be located and repaired. After dewatering, the line will be dried using dehydrated air to leave the completed pipeline filled with air usually with a dew point of −10°C or less. This is generally accomplished using multiple runs of foam pigs run by compressed air, which has been passed through a dessicant dryer to lower its dew point.

A typical hydrostatic test crew is made up of a foreman, a subforeman (night-shift), operators, laborers, a truck driver and a test supervisor. Their equipment would include; a fill pump, a pressure pump, a test instrument shelter, a compressors/pump, a winch truck, crew cab trucks, 4x4 pickups, deadweight gauges, and pressure and temperature recorders.

The test crew will be supported by a small tie-in crew to cut off and weld on test heads, install and remove pig catchers, and make tie-ins between test sections.

The inspection process involves ensuring that:

- Test sections have been cleaned and test head welds radiographed prior to filling;

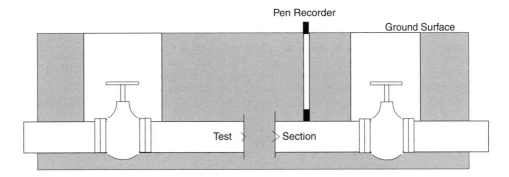

Figure 9-23. Pipe section subjected to hydrostatic testing

- Test heads and valves are designed for maximum test pressure;
- Instrumentation has been calibrated and is correctly installed and operating as necessary;
- The water source has sufficient volume, the water quality is acceptable, and screens and filters are used during filling to keep fish, silt, etc., out of the pipeline;
- Warning signs are placed at all public access points, and at all points where pipe or appurtenances are exposed;
- No heavy equipment is working on the right-of-way while the pipe is being tested;

Figure 9-24(a). Test Head Installation

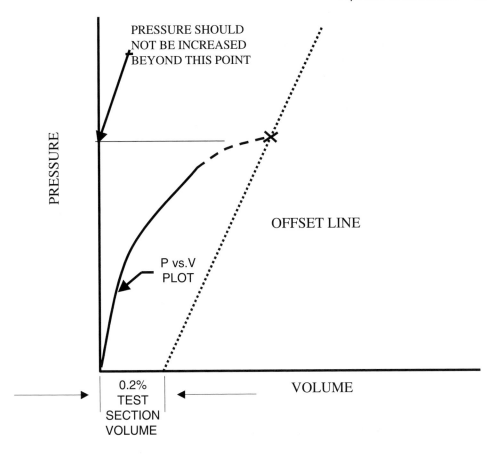

Figure 9-24b. Pressure vs. Volume Plot

- Yield plots are done by qualified personnel and documentation is completed correctly;
- Test pressure is safely released immediately after the test has been accepted;
- Dewatering is done in accordance with landowner agreements and regulations and in a manner which prevents erosion or damage to public or private property;
- Dry dewatering pig runs are made to remove all free water;
- Water is removed from valve bodies.

Fortunately, these days, relatively few newly constructed pipelines fail the hydrostatic strength test. It is not uncommon though for test sections to leak, and when the leak size is small, it can be problematic trying to find it. There is not always an immediate indication on the surface of its location. When the line remains full of water and under pressure, several methods are used to determine the leak location. In the first of these, the right of way is patrolled, whether by walking, driving or flying in an attempt to spot a wetted surface. Clearly such efforts will not work in wet weather or in wetlands. It may take time for the leak to become apparent on the surface, causing the project to be delayed. As a consequence, acoustical techniques have been developed and various claims have been made as to their capability. A third approach is to subdivide the test section into smaller segments, preferably without having to dewater and refill the now smaller test section.

Nitrogen can be applied to the outer surface of the pipeline at discrete locations so as to cause freeze plugs within the line, thus isolating the test section.

A costly approach is to section the line by emptying it and then cutting the test section in two before retesting each half in order to isolate the leak. The process is repeated to narrow down the leak location.

When air alone is used for test purposes, (see pneumatic testing) leaks are even more difficult to find and usually requires an odorant to be added to the test medium. The leaking fluid will permeate through the overburden (generally within fifteen to twenty minutes), and may be detected using trained sniffer dogs walking along the line. (Quaife and Moynihan(1991).

FILLING AND DEWATERING

Prior to starting the filling operation it is good practice to conduct a strength test on the fill line to ensure its integrity. This is normally done by conducting a hydrostatic test of the line at a pressure of around 3500 kPA (500 psig) for a 5 minute period. The maximum operating pressure of the fill line should not exceed this test pressure and the entire line should be monitored over its entire length throughout the filling operation. This is done so as to prevent any potential damage to the right of way or private property. As noted previously, the butt welds attaching the test heads must all pass a radiographic examination prior to hydrotesting. It is usual to install two pigs in the sending test head prior to starting a fill, with one pig remaining in the test head for subsequent use during dewatering.

It is essential prior to testing to prepare a fill and dewatering plan in order to determine the back pressures required to control both these operations without developing air locks in the pipeline. Generally the test section is filled by running a new, tight fitting, inflated spherical, or multi disc bi-directional urethane pig in front of the fill water, with adequate back pressure to prevent the pig and water from running ahead of the main water column while going down hill. The back pressure has to be sufficient to balance the weight of the pig plus the force due to the weight of water behind it when going down an incline. Should the pig run ahead, air can become entrapped in the system and the resulting bubble can prove to be very difficult to dislodge during dewatering. Unless the bubble is located at a

TABLE 9-3. Theoretical volumes of water and water methanol mixtures required to pressurize a 1.0 km long pipeline by 100 kPa

Pipe Size (NPS)	Pipe OD (mm)	Pipe WT (mm)	Water Volume L/l00 kPa/km	Water/Methanol Mixture L/100 kPa/km
4	114.3	3.2	0.56	0.7
6	168.3	3.2	1.43	1.75
8	219.1	3.8	2.52	3.06
10	273.1	4.7	3.93	4.77
12	323.4	4.8	5.92	7.12
14	355.6	5.3	7.12	8.56
16	406.4	6.0	9.34	11.22
18	457.2	5.6	12.95	15.35
20	508	5.6	16.87	19.86
24	610	6.4	24.98	29.29
30	914	7.3	40.96	47.71
36	914	8.2	61.13	70.87
42	1,067	9.0	86.32	99.62
48	1,422	10.2	113.19	130.56

high point in the line where a vent may be located, enabling the air to be bled off, it can only be moved using very high pressure air compressors which require time to mobilise. It is not uncommon to lose several weeks during the dewatering process as a result.

The rate of fill is measured using a calibrated flow meter and should not exceed the amount stated on the water permit. It is important for several reasons, to prevent air from entering the test section during filling. Besides causing problems during dewatering, it complicates the yield plotting, increases the sensitivity to temperature changes, can cause pressure surging, and increases the stored energy in the system causing a longer break in the line, if one were to occcur Air can enter the system through the fill pump though if it is drawing in air, a gurgling sound will be heard. If the water source is shallow, whirlpools sometimes can form and reach down to the suction inlet. To prevent this from happening a board can be placed on the water surface above the inlet.

EVALUATION OF SQUEEZE VOLUMES

The D/T ratio for on-shore pipelines is invariably sufficiently large such that stresses in the radial direction (Figure 9-25) can be neglected. In this case, thin wall cylinder behavior, as represented by the Barlow Equation, is applicable.

It can be further assumed that along the length of pipe to be considered:
1. Variations in pipe cross sectional diameter are negligible;
2. There is no variation in temperature across any cross section;
3. There is no variation in temperature along the pipe length, and that the temperature differential considered below is due to a variation of the environmental/ambient condition over time.

The following notation is used to derive the relationship between pressure, temperature and volume in a section of fluid filled pipeline. Since the pipe is under test, its ends are fully closed, so the internal volume of the pipe must always be equal to the volume of contained fluid. Thus, incremental volumes of fluid (δV_{fluid}) added to the line must equal the incremental volume change of the pipe (δV_{pipe}).

$$\delta V_{pipe} = \delta V_{fluid} \qquad (9-5)$$

Both the pipe and fluid are affected by changes in pressure (δP) and temperature (ΔT). From the definitions in the notation it follows that, by rearranging the expressions for both the bulk modulus and the coefficient of volume thermal expansion of the fluid, it can be written that

$$\delta V_{fluid} = BV\Delta T - \delta PV/K \qquad (9-6)$$

where B = coefficient of volumetric thermal expansion of the fluid and K = bulk modulus of fluid = $\frac{\delta P}{\delta V/V}$.

The determination of the incremental increase in volume of the pipe, δV_{pipe}, is dependent upon the degree to which the pipe is free to move. Consider first the case where the welded pipe has unrestrained movement such that it is free to expand in both the radial and longitudinal directions. The incremental volume increase in the pipe will be due to the

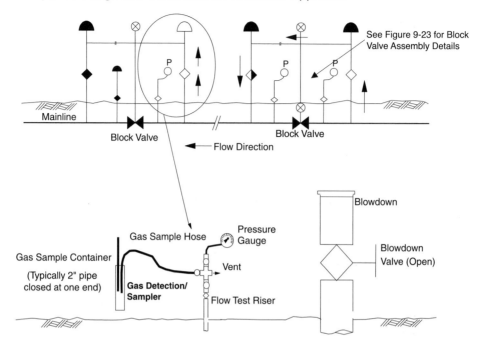

Figure 9-25. Purging a mainline

effects of both pressure and temperature. In the case of pressure change, the pipe wall experiences a state of biaxial stress, a circumferential or hoop stress S_θ, and an axial stress S_x. The respective hoop and axial strains are:

$$\varepsilon_\theta = \left(\frac{S_\theta - \nu S_x}{E}\right) = \frac{\delta PD}{4tE}(2 - \nu) \qquad (9-7)$$

and

$$\varepsilon_x = \left(\frac{S_x - \nu S_\theta}{E}\right) = \frac{\delta PD}{4tE}(1 - 2\nu)$$

where

ν = Poisson's ratio
S_x = axial stress = $\frac{\delta PD}{4t}$

S_θ = hoop stress = $\frac{\delta PD}{2t}$

E = modulus of elasticity
t = pipe wall thickness

and the new pipe length L^* and diameter D^* become

$$L^* = L(1 + \varepsilon_x) \qquad (9-8)$$
$$D^* = D(1 + \varepsilon_\theta)$$

The new volume of the pipe section is given by $\frac{\pi L^*(D^*)^2}{4}$. Upon substitution of the expressions for D^* and L^*, and ignoring the cross products of strain as being negligible, the new pipe volume V* is given by:

$$V^* = \frac{\pi D^2 L}{4}(1 + 2\varepsilon_\theta + \varepsilon_x) \qquad (9-9)$$

The incremental increase in volume, δV, due to incremental change in pressure, δP, is found by substituting the expressions for ε_θ and ε_x into Equation (9-9) and noting that $\delta V = V^* - V$, where V is the original volume of the pipe. Hence for fluid and pipe:

$$\delta V_{pipe} = \frac{V\delta PD}{4tE}(5 - 4\nu) \qquad (9-10)$$

The incremental volume change due to a temperature change ΔT is easily found using

$$\delta D = D\alpha\Delta T \text{ and } \delta L = L\alpha\Delta T \qquad (9-11)$$

Where α is the coefficient of linear expansion of pipe material, and δD and δL are changes in diameter D and length L respectively.

By neglecting the cross products of the coefficient of thermal expansion when evaluating the new volume, the incremental volume is:

$$\delta V = 3\alpha V\Delta T \qquad (9-12)$$

The combined effect of incremental pressure δP and temperature ΔT change is found by substituting Equations (9-6), (9-10) and (9-12) into (9-5):

$$V\left(\frac{\delta PD}{4tE}(5 - 4\nu) + 3\alpha\Delta T\right) = V\left(B\Delta T - \frac{\delta P}{K}\right) \qquad (9-13)$$

Rearranging the terms and eliminating V:

$$\delta P = \frac{(B - 3\alpha)\,\Delta T}{\frac{1}{K} + \frac{D(5-4\nu)}{4tE}} \qquad (9-14)$$

Equation (9-14) provides the change in pressure due to a temperature change for an unconstrained pipe with closed ends. It is interesting to note that, in this circumstance, it is independent of pipe length.

Buried pipelines behave differently, since the friction of the soil generally inhibits axial movement. In this case the axial strain, ε_x, is assumed to be zero. An inability to move longitudinally can impose considerable axial stress, and this is the essential difference between constrained and unconstrained fluid-filled pipes. The analysis of buried pipes closely follows that shown above, concentrating on diametral change in the pipe dimensions, since there is no change in length. Zero axial strain means that, for compatibility, the strain that would ordinarily be due to thermal expansion must be balanced by the (shortening) Poisson effect caused by the increase in diametral strain. The induced axial stress is therefore

$$S_x = E\alpha\Delta T - \nu\,S_\theta \qquad (9-15)$$

This is identical to the expression for axial stress in a restrained pipeline found in ASME ANSI B31.4 (1986).

The unit change in diametral strain due to changes in hoop stress and axial stress caused by an incremental temperature increase ΔT is given by:

$$\delta\varepsilon_\theta = \frac{\delta\,S_\theta}{E} + \nu\left(\alpha\Delta T - \nu\frac{\delta\,S_\theta}{E}\right) + \alpha\Delta T \qquad (9-16)$$

The increase in volume due to the increase in diameter $\delta D = D\delta\varepsilon_\theta$ is

$$\delta V_{pipe} = \frac{\pi\left[(D+\delta D)^2 - D^2\right]}{4}\cdot L \cong \frac{\pi D\delta D}{2}\cdot L = \frac{\pi D^2\delta\varepsilon_\theta}{2}\cdot L \qquad (9-17)$$

Substituting for $\delta\varepsilon_\theta$ from Equation (9-16) and simplifying,

$$\delta V_{pipe} = \frac{\pi D^2 L}{2}\left\{\frac{\delta\,S_\theta}{E}(1-\nu^2) + (1+\nu)\alpha\Delta T\right\} \qquad (9-18)$$

$$\delta V_{pipe} = V\left\{\frac{\delta PD}{tE}(1-\nu^2) + 2(1+\nu)\alpha\Delta T\right\} \qquad (9-19)$$

Substituting Equations (9-6) and (9-19) into (9-5):

$$V\left\{\frac{\delta PD}{tE}(1-\nu^2) + 2(1+\nu)\alpha\Delta T\right\} = BV\Delta T - \frac{\delta PV}{K} \qquad (9-20)$$

Equation (9-20) can be rearranged to provide an expression for the pressure increase in a restrained pipe due to an incremental change in temperature:

$$\delta P = \frac{[B - 2\alpha(1+\nu)]\Delta T}{D(1-\nu^2)/Et + 1/K} \qquad (9-21)$$

Gray (1976) neglected the effect of axial thermal stress on restrained pipelines capped at the ends. His formulation replaces the term expressing the apparent volume change $B - 2\alpha(1+\nu)$ by the simpler form $B - 2\alpha$. Since the coefficient of volume thermal expression of water is generally an order of magnitude larger than the coefficient of linear expansion for the pipe material, the effect of neglecting thermally induced axial stress is to over-estimate the pressure increase by at most 5%, compared with using Equation (9-21). Ignoring the temperature increase effects simplifies Equation (9-20) such that, for a restrained pipe, the theoretical volume required to pressure (squeeze) the pipeline section is given by:

$$\delta V = V\delta P\left(\frac{D}{Et}(1-\nu^2) + \frac{1}{K}\right) \qquad (9-22)$$

This is the equation used to derive the quantities shown in Table 9-3. Similar derivations have also been reported by Awoseyin (1985).

It is generally assumed that an increase in test water temperature will be accompanied by an increase in hydrostatic pressure; however, this is not always the case. At temperatures below 4°C, an increase in temperature will cause an increase in the density of the water which causes a decrease in pressure. Furthermore, the slight decrease in density with a temperature rise between 4°C and 5.5°C combined with the thermal expansion of the pipe will still cause a reduction in pressure. That this should be so can be seen from the numerator of Equation 9-21, where the term $B - 2\alpha(1+\nu)$ can become negative for some values of B, the coefficient of expansion of water.

To illustrate the effect of a change of temperature on the test pressure consider a pipe of outside diameter 114.3mm and wall thickness 3.18mm. As suggested by Gray we can generally neglect the effect of axial thermal stress on a restrained pipeline so that Equation 9-21 can be simplified to become:

$$\delta p = \frac{B - 2\alpha}{\frac{D}{Et}(1 - \nu^2) + C}$$

Where
$\alpha = 1.17 \times 10^{-5}$
$B = (-64.268 + 17.105T - 0.20396T^2 + 0.0016048T^3) \times 10^{-6}$
$E = 200 \times 10^6 \text{ kPa}$
$\nu = 0.3$
$C = 4.86 \times 10^{-7}$

Note that C, the reciprocal of the bulk modulus for the test fluid, water, is a function of the test pressure.

Assuming the test temperature T is 10°C then upon substituting, B has the value 87.072810^{-6} and the pressure increase due to a 1 degree C temperature change is:

$$\delta p = \frac{87.0728 \times 10^{-6} - 2 \times 1.17 \times 10^{-5}}{\frac{114.3}{200 \times 10^6 \times 12.7}(1 - 0.3^2) + 4.86 \times 10^{-7}}.$$

That is a pressure increase of 98.9 kPa.

In a similar manner we can compute the change in pressure due to a temperature increase over a range of temperatures. Consider a large restrained transmission pipeline whose outside diameter is 1219mm, having a wall thickness of 12.7mm.

Temp °C	0	1	2	3	4	5	6	7	8	9	10
Pressure Change(kPa)	−77	−59	−41	−24	−7	10	26	41	56	71	

Assume that the test water temperature increases from 1 to 3°C during the test period . The end result is a pressure decrease of 100 kPa, obtained by summing the pressure decreases in each interval, that is 59 kPa between 1 and 2°C , and 41 kPa between 2 and 3°C.

Note that had the temperature *decreased* from 3 to 1°C the pressure would have increased by 100 kPa.

HYDROSTATIC TEST SECTIONS CONTAINING TRAPPED AIR

As noted previously in this chapter, a yield plot or graph of pressure versus cumulative volume is developed by recording the squeezed volume for each 100 kPa pressure increment. In the case of a pipeline section containing only water or a water/methanol mixture, the graph will be a straight line until the elastic limit of the pipeline is reached. However, if the section contains a significant amount of entrapped air, the pressure-volume graph will be non-linear. As the pressure within the section increases, the air present will compress, thus requiring more fluid to be added.

The reduction in air volume per 100 kPa pressure is not constant, so a corrective procedure is required. First, the volume of air in the test section is determined in several stages. Assuming no air entrapment, the volume of fluid which would be required to raise the pressure by an amount δP can be found using Equation (9-22). Subtracting this quantity from the volume actually required to raise the pressure by the same amount δP gives the volume of fluid required to displace the air as it was being compressed, an amount V_a.

Applying Boyle's Law for gases,

$$\frac{P_1 V_1}{Z_1} = \frac{P_2 V_2}{Z_2} \qquad (9-23)$$

where V_1 and V_2 are the air volumes at pressure P_1 and P_2, respectively, and Z_1 and Z_2 are the compressibility factors at the respective pressures. Since $V_1 = V_2 + V_a$ then

$$\frac{P_1(V_1 + V_a)}{Z_1} = \frac{P_2 V_2}{Z_2} \qquad (9-24)$$

from which

$$V_2 = \frac{P_1 V_a}{\left(P_2 \frac{Z_1}{Z_2}\right) - P_1} \qquad (9-25)$$

so that the air volume at any pressure can be determined. In order to plot the points on the pressure volume graph, proceed as follows. The volume of air present at the pressure at which the yield plot is to start is obtained using Equation (9-23). The change in this volume as the pressure increases from P_1 to P_2 is found using Boyle's Law, Equation (9-23). Note that the air volume decreases, so $V_2 = V_1 - V_a$.

Substituting for V_2 in Equation (9-25) and rearranging

$$V_a = V_1 \left(1 - \frac{P_1 Z_2}{P_2 Z_1}\right) \qquad (9-26)$$

We need to add to this volume the volume required, V_R, to squeeze a fluid-packed line from P_1 to P_2 using Equation (9-22). The sum $V_R + V_a$ is the volume of fluid required to squeeze the line from an initial pressure P_1 to P_2, assuming no yielding occurs during the interval. This process is repeated for each new starting condition to obtain the modified reference curve. The 0.2% offset line is then drawn parallel to this reference curve.

Example

A test section of a pipe 1067mm outside diameter and 12mm wall thickness and 9410 m long contains entrapped air, calculate the 0.2% offset line for the yield plot of the test section. The squeeze pump has an output of 1 m^3 per 507.5 strokes and the on test pressure is 8140 kPa. (The pressure at the yield point is 7400 kPa). Consider two intermediate test pressures $P_1 = 5300$ kPa and $P_2 = 6300$ kPa.

Step 1 Determine the volume of air in the test section.

(a) $P_1 = 5300$ kPa and $P_2 = 6300$ kPa (Note pressures are absolute)

During the test it was found that,

$$V \text{ actual} = 5789 \text{ strokes} = \frac{5789}{507.5} \times 1m^3 = 11.407m^3$$

(b)
$$\frac{\Delta V}{\Delta P} = V\left(\frac{D}{Et}\left(1 - \nu^2\right) + C\right)$$

$V = 8039.855 m^3$ $D = 1067mm$ $E = 200 \times 10^6$ kPa $t = 12mm$
$\nu = 0.3$ $C = 4.525 \times 10^{-7}$ kPa^{-1}

Substituting the above values we obtain $\frac{\Delta V}{\Delta P} = 0.6935 m^3/100$ kPa

(c) $\Delta V_A = 11.407 - 10 \times 0.6935 = 4.472 m^3$

(d) $V_2 = \dfrac{P_1 \Delta V_A}{\dfrac{P_2 Z_1}{Z_2} - P_1}$ (Compressibility factors can be found from Tables)

$$= \frac{5300 \times 4.472}{\left[\dfrac{6300 \times 0.9785}{0.9735} - 5300\right]}$$

$$= 23.03 m^3 \text{ at } P = 6300 \text{ kPa}$$

Step 2 Calculate points on the Pressure Volume graph
(a) For $P_y = 7400$ kPa

$$\frac{P_y V_y}{Z_y} = \frac{P_2 V_2}{Z_2}$$

$$\frac{7300 V_y}{0.969} = \frac{6300 \times 23.03 V_2}{0.9735}$$

From which $V_y = 19.52 m^3$

(b) $\Delta V_A = 19.52 \left[\dfrac{1 - 7400 Z_i}{P_i \times 0.969}\right]$

$$= 19.52 - 149069\, \frac{Z_i}{P_i}$$

(c) $V_R = (P_i - P_y)\dfrac{\Delta V}{\Delta P}$

(d) $\quad = (P_i - 7400) \times 0.6935 \times 10^{-2}$

$$= 0.006935\, P_i - 51.319$$

Combining (b) and (c)
Total volume of water required to squeeze from P_y to P_i is,

$$V_T = 19.52 - 149069\, \frac{Z_i}{P_i} + 0.006935\, P_i - 51.319$$

$$V_T = 0.006935\, P_i - 149069\frac{Z_i}{P_i} + -31.799$$

TABLE 9-4. Methanol Washing Quantities

Pipe Diameter (NPS)	Liters of Pure Methanol per Kilometer of Pipe	US Gallons of Pure Methanol per Mile of Pipe
4	85	33.7
6	100	39.62
8	125	49.5
10	155	61.4
12	180	71.3
14	205	81.2
16	225	89.2
18	240	95.1
20	255	101
24	285	112.9
30	405	160.5
36	510	202
42	510	269.4
48		

Values for P_i between the leak test pressure and the strength test pressure are then chosen and the corresponding volumes calculated from the above formula. The resulting pairs of data can then be plotted and the 0.2% offset line drawn parallel to the calculated curve.

Internal Cleaning and Inspection

The pipeline will be internally inspected to ensure that it is free from debris and ovality, dents, or buckles. Compressed air, at a pressure of around 50 psi, is used to propel the foam cleaning pigs, and suitable launchers and receivers are required to contain the pigs and collect any debris.

For NPS 16 and smaller pipe, the completed pipeline will be inspected for ovality, dents and buckles by adding a sizing plate to the cleaning pig that has a diameter of 12.7 mm less than the inside diameter of the pipe. The sizing plate is a sectored aluminum disc which distorts whenever it encounters an out of roundness. Any deformation of this sizing plate will require the location of the cause to be found and repairs to be made to remove the source. Radioactive, magnetic, or radio-wave-transmitting devices are available to monitor the progress of the pig or to locate it, should it become jammed in the pipe.

Pipe larger than NPS 16 are internally inspected using a caliper pig. This inspection run is made after hydrotesting and the installation of all valve assemblies. These instrumented pigs mechanically measure the minimum inside diameter of the pipe and record the information on a strip chart. The runs are made with backpressure maintained in front of the pig to control its speed. The strip chart is calibrated by length so that indentations can readily be located and repaired as necessary.

HYDROSTATIC WATER CRITERIA

The pipeline codes do not provide criteria for the water quality to be used for hydrostatic testing. However, pipeline companies in their construction documents stipulate that contractors must use water that does not contain silt, suspended material, or harmful corrosive components, unless it can be treated satisfactorily by the use of filters or chemical additives.

Thus, it is at the owner's discretion to determine whether the water to be used by the contractor is suitable for hydrostatic testing or whether remedial measures are required to clean the water prior to use.

There is more concern regarding the quality of the water that will be returned to the environment after testing than with the quality of the water used for testing. Chemical additives in the water are usually specified so as not to contaminate the environment when discharged after testing. These water quality specifications are usually set by applicable Environmental Authorities.

Tests could be performed on the water used for testing to determine if there is a possibility of the pipeline being contaminated with iron or sulphur bacteria. The iron bacteria are capable of using the iron from the pipe as metabolic fuel, which could result in pitting of the pipe wall and/or a reduction in internal diameter via a deposition of biomass.

The sulphur bacteria are capable of reducing various sulphur species to hydrogen sulphide, causing corrosion problems and/or reduction in the internal pipe diameter via a biomass deposition.

One solution to the problem of bacterial contamination would be to chlorinate the water as it enters the pipeline.

Runoff water from agricultural areas can contain high levels of nitrates and phosphates. These components promote biological activity.

Salinity

Salt solutions support corrosion. All water sources must have a very low salinity level to be acceptable. Selection criterion is 10 mS (milliSiemens).

pH

Acidic solutions are more corrosive than saline solutions. A selection criterion of pH 5 is considered safe.

Suspended Solids

Suspended solids can leave deposits. An acceptable limit is < 50 mg/L. A settling tank or simple filter system might preclude putting unwanted suspended soils or stirred mud into the pipeline. Any injected water is required to be clear, not muddy or turbid.

Methanol Wash

All of the methanol solution displaced from the pipeline must be recovered and disposed of in a manner acceptable to the pipeline owner and the jurisdictional authorities. It is quite unacceptable for the solution to be discharged to the ground, where it could eventually make its way into the ground water.

It is good practice to measure the specific gravity of the recovered methanol solution in order to determine the likely amount of pure methanol recovered from the line. This is done by collecting a number of clear samples of the solution and using a hygrometer and a thermometer to determine its average specific gravity and temperature. Table 9-5 provides a range of specific gravities for various methanol solutions.

If the volume and content of the recovered solutions do not meet the specified criteria then the methanol wash is repeated until they are met. Upon completion of the methanol wash, if the section of pipe is not tied in immediately, the open ends of the section should be capped with a water tight seal.

Air Drying

Air drying may be used as an alternative method to a methanol wash. In this case the pipeline is considered to be dry when the atmospheric dew point of the air within the pipe reaches a specified minimum, usually around 45°C (−49°F).

In order to be effective the atmospheric dew point of the air entering the line must be considerably lower than −45°C, a value of −70°C (-94°F) is the norm. The dew point temperature of the air entering and exiting the pipe section is measured hourly until the specified value is reached. The line is then put under positive pressure with dry air having a dew point of −70°C (−94°F) or less and the section is left in that condition until it is tied into the system.

Vacuum Drying

Vacuum drying is another alternative method to air drying or methanol washing. Vacuum drying is done in four stages. In the first stage a number of dewatering pig runs are made to ensure that as much free water as possible has been removed. The line pressure is then drawn down below atmospheric pressure using a vacuum pump. This causes any remaining free water to boil and start to vaporize. Having ensured that there is no inward leaking of air, the third stage starts. This stage involves carefully controlling the vacuum pressure, holding it at the saturated vapor pressure until all the free water has vaporized. This condition is signaled by a noticeable decrease in the pressure within the section. The line is left to "soak" for about 12 hours before the final drying stage commences. During the fourth stage the line pressure is drawn down further until the dew point of the air remaining in the line has reached −45°C (−49°F). The vacuum drying equipment is then shut off and removed, and the line is placed under a positive pressure of 50 kPa by filling it with nitrogen.

HYDROSTATIC TESTING AT LOW AMBIENT TEMPERATURES

A major problem with using water to test in cold climates is the potential for it to freeze at low ambient temperatures. This is particularly true if the testing cannot be performed in summer, or if the pipeline lies in frozen soil or permafrost. The threat of freezing can be avoided by heating the water, or adding freeze depressants.

Warm water testing involves heating the water to about 10°C and pumping it through the test section until the temperature of the discharge water at the other end reaches between 2 and 4°C. Alternatively, water from a source may be mixed with much hotter water from pumps and flowed until the desired discharge temperature is reached. As the warm water flows through the pipe it thaws the surrounding soil and creates a "heat sink" around the pipe. The challenge is to maintain sufficient cirulation that will supply a large enough heat sink for the duration of the test period.

TABLE 9-5. Specific gravity of recovered methanol

Temperature °C	SG of 60% Methanol Solution	SG of 70% Methanol Solution	SG of 80% Methanol Solution	SG of 90% Methanol Solution
0°	0.9090	0.8869	0.8634	0.8374
10°	0.9018	0.8794	0.8551	0.8287
15°	0.8978	0.8751	0.8505	0.824
20°	0.8946	0.8715	0.8469	0.8202

When SG = specific gravity.
(source: Chemical Engineers Handbook, 5th Edition (Perry & Chilton).

Once the surrounding soil has been warmed, the test section can be pressure tested without freezing the aqueous medium. However, there is a maximum amount of time, called the "maximum shut-in time" after which there is a danger of freezing the test medium. Once the maximum shut-in time has been surpassed, it is necessary to either de-water the section or to resume circulation of warm water.

It is important to ensure that none of the test medium becomes frozen during the duration of the test. Freezing could occur due to inadequate water circulation or delays in de-watering, for example in the case of a leak. Although warming during the summer can subsequently melt any ice that forms in the pipe during the test, this is not true for permafrost regions. In areas where the ground temperature never rises above $0°C$, a portion of the frozen test medium may never melt, and could be very difficult to remove from the line. The remaining ice could create operating problems such as restrictions or blockages.

The disadvantage of testing with heated water is that it generally requires a far greater volume of water than is normally needed for regular hydrotesting. Instead of using a single volume of water to test all the sections in a spread, several times the fill volume is needed for each section. This amount cannot be simply passed on to the next section following a test. Large heaters and additional fuel quantities are also necessary. Moreover, there is the fear of thaw bulb formations in ice-rich insulated slopes, which could threaten ground stability around the pipe.

Nevertheless, warm-water testing has proven to be a reliable and economical means of hydrostatically teating pipelines during winter construction in cold climate, non-permafrost regions. Its has been successfully employed under such conditions for almost 40 years. However, there are many stability concerns with the ice-content of some permafrost backfills and there is the possibility of freezing the test medium at very cold ground temperatures. For this reason, warm-water testing, particularly in permafrost regions, must be very carefully planned.

Freeze Depressants

Another way to prevent the freezing of an aqueous test medium is to inject a freeze depressant that will lower the freeze point of the solution below $0°C$. There are a number of possible freeze depressants; among which are calcium chloride, methanol, ethanol, isopropanol, ethylene glycol and propylene glycol. Product availability must be considered with methanol generally found to be the most readily available as well as the most economical choice.

Methanol has a high volatility, which makes it easier to remove from the water after the test is completed, while it does not pose a threat of corrosion. By comparison, some of the other depressants require chemical additions, such as chromate inhibitors for calcium chloride or a denaturing agent for ethanol, which upon disposal could be harmful to animal life.

The advantage of adding a freeze depressant is that the freezing point of the solution can be brought down well below $0°C$ (about $-21°C$ using methanol). Moverover, the fill volume from one section can be moved forward to subsequent sections, with make-up from a tanker truck.

There are disadvantages to adding a freeze depressant to the test water. For example, there are hazards associated with transporting and storing large volumes of methanol. In the event of a line break, or leak, with associated fluid losses, additional quantities of methanol would be required for make-up solution. Furthermore, a failure or accidental spill would be hazardous to the environment, so substantial contingency planning is necessary. Once the testing is complete, there must also be an environmentally acceptable plan for disposal of the test solution; this will normally involve trying to separate the freeze depressant from the water as completely as possible.

PNEUMATIC TESTING OF PIPELINES

While water, or water with a freezing point depressant such as methanol, is commonly used for hydrotesting purposes, some codes (e.g. Canadian Code CSA Z662, 2007) permit the use, under special circumstances, of Low Vapor Pressure (LVP) liquids. However, Canadian Code CSA Z662 Clause 8.2.2.1.1 does state, "wherever practical water shall be used as the test medium."

All of the Codes and Standards listed in Table 9-6, with the exception of the DnV OS F101, permits pneumatic testing of *new* pipelines, although the conditions differ for each and are briefly reviewed below. The Canadian Standard permits air or gaseous testing for operating pressures at or below 700 kPa. For pressures exceeding 700 kPa, permissability depends upon whether the pipe has been shown to have sufficient notch toughness, has a longitudinal joint factor of 1.0 (see Chapter 7 Mechanical Design) and if one or more of the following conditions apply:

- The pipeline material requirements, particularly toughness are met
- The ambient temperature is, or is expected to be 0°C (32F), or lower
- Liquid is not available
- Removal of the liquid pressure-test medium would be impractical
- Elevation profile would require too many test sections if a liquid medium were to be used
- The strength test pressure will not cause a hoop stress above 80% SMYS at any location

Clause 12.8 of the Canadian standard permits gas distribution piping to be tested using a flammable, non toxic gas provided the strength test pressure does not cause a hoop stress exceeding 30% SMYS at any point in the test section.

In other jurisdictions the requirements for pneumatic testing vary considerably and the reader is cautioned to study the pertinent regulations carefully. For instance the ASME

TABLE 9-6. Pipeline Standards Permitting Pneumatic Testing

Country	Organisation	Standard	Title
Canada	Canadian Standards Association	CSA Z 662-07	"Oil and Gas Pipeline Systems"
Australia	Standards Association of Australia	AS 1978-1987	(1) Pipelines Gas and Liquid Petroleum Field Pressure Testing
United States	ASME	(i)ASME B31.4 (ii)ASME B31.8	(i)Transportation Systems for Liquid Hydrocarbons and Other Liquids (ii) Gas Transmission and Distribution Systems
United Kingdom	British Standards Institute	BS 8010-2.8	Code of Practice for Pipeline Part 2 Pipelines on land: Design Construction and Installation"
Russia	Ministry of Construction for the Oil and Gas Industries	VSN 011-88	Construction of Transmission and Field Pipelines Internal Cleaning and Testing
European Union	Comite Europeen de Normalisation (CEN)	CEN prEN 1594	Pipelines for Gas Transmission
International	International Standards Organisation (ISO)	ISO CD 13623	Pipeline Transportation for Petroleum and Natural Gas Industries

Standard Does Not Permit Pneumatic Testing

Norway	Det norske Veritas	OS-F101	Submarine Pipeline Systems

B31.4 code (2002) for liquids pipelines does not permit pneumatic testing for pipelines operating at a hoop stress exceeding 20% SMYS, such pipelines are to be hydrotested. Carbon dioxide pipeline systems are not permitted to be pneumatically tested at all. The ASME B31.8 code (2003) for gas pipeline systems allows for gaseous testing of pipelines operating at a hoop stress less than 30% SMYS. Above this threshold pipelines may only be pneumatically tested as follows:

1. Location Class 1 Division 2 (11-25 buildings) with an MOP causing a hoop stress not exceeding 72% SMYS (gas or air permitted).
2. Location Class 2.(36-45 buildings) (air permitted but not gas).
3. Location Class 3 (66 plus buildings) or four multi storey buildings (air permitted but not gas if:

   a. The ground temperature is 32°F or less or may reach this temperature during the test.
   b. A sufficient quantity of water of satisfactory quality is unavailable.
   c. The maximum hoop stress during the test is less than 50% SMYS in Location class 3 and less than 40% SMYS in Location class 4.
   d. The maximum operating pressure will not exceed 80% of the maximum field test pressure used and;
   e. The pipe to be tested is new pipe with a longitudinal joint factor of 1.0.

The British Standard BS 8010 (1994) is much more stringent. Dry, oil free air or nitrogen "may be considered", if the pipeline conveys non flammable substances that are gases at ambient temperature and pressure conditions. (That is, Catgory C substances such as oxygen, nitrogen, carbon dioxide and argon). Pneumatic testing is only permitted if the pipeline is designed to operate at a design factor not exceeding 0.3.

Water is the preferred test medium for pressure testing new gas pipelines within the European Community. However the CEN standard prEN 1594(1994); does permit testing using air, or an inert gas, provided appropriate safety precautions are taken in accordance with national legislation.

The ISO Pipeline Standard 13623 accepts the use of air or non toxic gas as a test medium only if hydrotesting is not possible. This is warranted under the commonly accepted circumstances of low ambient temperatures, an insuffucent supply of water of adequate quality, or the need to avoid contamination.

As noted above, air testing tends to be an exception and is used on occasion, but only if certain requirements are met Wight, (1983). It was used, for example to good effect on TransCanada's GasAndes (refer to Chapter 6) project which involved pipelining at high altitude (3500 m) and across the crater of an extinct volcano, scenarios where water was inaccessible or in short supply. The remoteness of the location negated one of the major drawbacks of air testing, namely safety. On one particularly difficult section, that ascended steeply from a river valley to cross a range of mountains, two air tests replaced 10 hydrostatic tests.

When considering using air as a test medium instead of water or a water-methanol mixture, several factors must be considered; some of the advantages and disadvantages are listed below.

Advantages

- Air is environmentally friendly. No environmental damage results from a rupture, as opposed to the damage caused from a water and methanol spill
- No need to draw water from a creek, river or dugout

- No water methanol mixture to dispose of upon completion
- No water and methanol mix to purchase and haul to site
- No storage tanks necessary (as is the case for hydrostatic testing in remote areas)
- No additional stresses within the pipe due to the mass of a liquid test medium
- Less methanol required for methanol wash
- No need to produce a Profile drawing of the line (i.e. no increase in pressures due to elevation heads)
- The point of pipe rupture can be easily located. However, to ensure the danger zone is clear of unauthorized personnel, test lengths are somewhat minimized.
- No chance of freezing during winter test conditions. As a result, there is no need to supply backup equipment or insulate exposed pipe to ensure the line does not freeze.
- Overall testing time may be less than with hydrostatic testing, since filling and displacing times are faster, however it can require a much longer time for the temperature to stabilise and for a measurable pressure drop to occur.
- Potential cost savings due to the above factors

Disadvantages

- Due to the compressibility of air, the potential danger for damage to property and injury to workers increases dramatically
- More time and effort is required to ensure proper placement of test equipment and to monitor the danger zone to eliminate unauthorized access
- Air testing is restricted to remote non-residential areas (Class 1 locations) where the proximity of dwellings and inhabited areas is minimized
- Air testing requires a minimum of 24 hours of test as opposed to a minimum of 8 hours in the case of hydrostatic testing (strength plus leak testing)
- It is very difficult to detect leaks during a air test. It is almost impossible to relate a small leak to a pressure drop because of the compressibility of the air and the volumetric changes which can occur as a result of a change in the temperature of the testing fluid over the period of the test.

Pipelines which cross roads or railways may also be air tested however there are stringent rules surrounding the length of the test sections. This length is determined by considering such factors as;

- The fracture propagation characteristics of the pipe material
- Pipe diameter
- Traffic density

Safety considerations are paramount when air testing since the stored energy in the line is extremely high and should the pipe rupture the potential for injury and property damage in the immediate vicinity is high. As a consequence pressure test warning signs are installed along all potential routes of public access. Roads in the proximity to the right of way are invariably closed for the duration of the test. Safety precautions established for the test section, also applies to the fill lines, instrument lines, and compression equipment.

COMMISSIONING

Commissioning is the process of checking the facilities to ensure they are capable of transporting volumes of various fluids, and preparing the facilities to perform their function.

It covers the period from the time that the construction of facilities is completed until the facility is on-stream. Completion of construction does not necessarily mean that final cleanup is completed, but only that the pipeline facilities are completed, tied-in, and ready for flow. Construction/Commissioning status definitions related to different activities are defined in Table 9-7.

Pipeline commissioning is complete only when the pipeline is on-stream and all measurement facilities have been calibrated and found to operate properly and accurately. All valving and piping facilities must be found to be operative and set for the operating mode.

Natural Gas Pipelines

Purging is done to remove all air or air-gas mixtures from natural gas pipeline facilities Mohitpour et. al., (2000). An air-gas mixture is a highly flammable and dangerous substance. It is therefore very important to ensure that air/gas mixtures within flammability limits are minimized. Purging of air is normally carried out using natural gas. A high gas flow velocity minimizes mixing at the air/gas interface. Adequate velocities are equated to minimum gas inlet pressures for various pipeline configurations.

Purging is done through bypass valves for safety and control (Figure 9-25). These smaller valves cause less vibration and allow for better control of flow and pressure. They are also easier to operate. Purging is performed on one section of the line at a time. On a mainline, purging is completed from one block valve to the next, continuing one section at a time, until the line is full of gas. On laterals, purging is completed from the side valve to the next valve assembly.

Normally, mainlines are purged in the direction of flow. However, they may be purged from either end, depending on the availability of gas supply.

Test equipment is attached to monitor pipeline conditions at the appropriate flow-test valves. A pressure gauge is installed on the upstream side of the upstream block valve to monitor existing line pressure (Figure 9-26). A pressure gauge is installed on the downstream side of the upstream block valve and on the upstream side of the downstream block valve to monitor purge section inlet pressures and outlet pressures, respectively. Pressure gauges are generally connected with a flexible hose to minimize the effects of vibration. Deadweights may be used instead of pressure gauges. A gas detector (sampler) is installed on the upstream side of the downstream block valve to monitor gas concentration. The detector should be calibrated to the gas stream used to purge the section. A second detector should be available in case of a malfunction (e.g. contamination of probe). All automatic control devices are deactivated for safe manual operation.

The upstream bypass is purged and pressured by slowly opening the upstream bypass valve and purging air through the downstream blowdown bleed valve. When a good gas flow is obtained, the bleed is tightened and the bypass allowed to pressure up (equalize). The upstream bypass valve should then be fully opened. Now the operator is ready to commence purging.

The upstream block valve assembly is purged by opening the downstream blowdown valve. A pressure reading is taken and the valve opened to obtain the required minimum

TABLE 9-7. Construction/Commissioning Definition

Activity	Definition
Fabrication/construction/as-built/clean-up/ integrity tests	Mechanical completion
Settings/calibrations/servicing/operability checks	Pre-commissioning completion
Purge/pressure/equalize/final calibration & operability checks	Commissioning activities and completion
Operation	On-stream

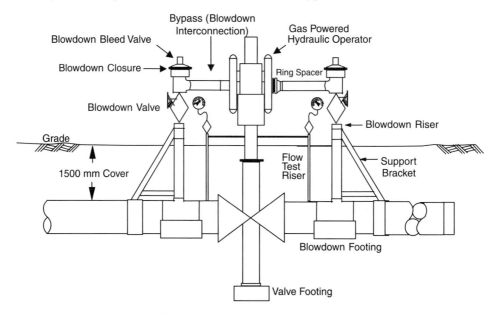

Figure 9-26. Mainline Block Valve Assembly

inlet pressure, which may be determined from blowdown minimum inlet pressure tables. This pressure must be maintained throughout the purge.

For purging the downstream block valve assembly, the upstream blowdown closure is opened while the activities listed previously are being performed at the upstream block valve. The upstream blowdown valve is opened, and the downstream blowdown valve and closure are closed. The gas detector is monitored for 100% gas flow. When gas flow reaches 100%, the purge is continued for five minutes. The 100% reading must be continuous during this period to ensure that no mixture is left in the line. After five minutes have passed, the upstream blowdown valve and closure are closed. Equalizing can then commence.

Laterals are purged from the side valve to the next valve assembly. Initially, a pressure gauge is installed at the side valve to monitor existing line pressure if a connection exists, and a second gauge is placed on the side valve blowdown or flow test connection to measure inlet purge pressure. A pressure gauge is installed at the outlet valve to monitor outlet purge pressure and buildup pressure. A gas detector is installed at the outlet blowdown to monitor gas concentration. Detectors should be calibrated to the gas stream used to purge the section. All automatic control devices are deactivated to allow for safe manual operation.

The operator is now ready to commence purging by opening the bypass valve. The pressure is read and the valve positioned to obtain the required minimum inlet pressure which may be determined from blowdown minimum inlet tables. This pressure has to be maintained throughout the purge.

When the gas flow reaches 100%, the purge is continued for five minutes. The 100% reading must be continuous for this period. At the end of five minutes, the blowdown valve and closure are closed. Equalizing then commences (refer to Equalizing section). Typical inlet pressures for the purging of various mainline sizes with natural gas are depicted in Figure 9-27.

Equalizing

Equalizing is done to pressurize the section of line that has been purged to the same pressure (operating pressure) as adjacent sections of line (Figure 9-28). It is usually performed in conjunction with purging, one section at a time. On mainlines, equalizing is done from block valve to block valve; on laterals, it is done from the side valve to the next valve assembly, or to the meter station.

Methods of Equalizing

Equalizing of the mainline begins upon completion of the purge, when the downstream valve assembly's upstream blowdown valve and closure are closed. The upstream valve assembly's downsteam blowdown valve is left in the open (purge) position and pressure builds in the purged line. As the pressure builds (differential decreases), this valve can be opened gradually until fully open. Vibration must be monitored closely while performing this operation.

Pressure at both assemblies is monitored and both assemblies are continuously checked for leaks. When the pressure is equalized (within 350 kPa of the upstream mode) all blowdown and bypass valving is closed (operating mode) and secured, all automatic controls are returned to service, and routine service (lubrication, etc.) is performed on the assemblies.

For laterals, the side valve's bypass valve is left in the open (purge) position and the pressure is allowed to build (upstream assembly closed). As the pressure in the purged line increases (differential decreases), the bypass valve can be opened gradually until fully open. Vibration must be closely monitored.

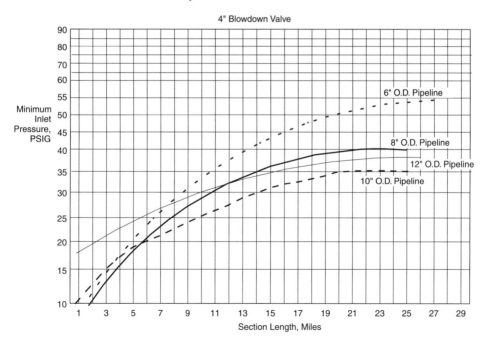

Figure 9-27. Typical Inlet Pressure for Purging Various Mainline Pipeline Segments

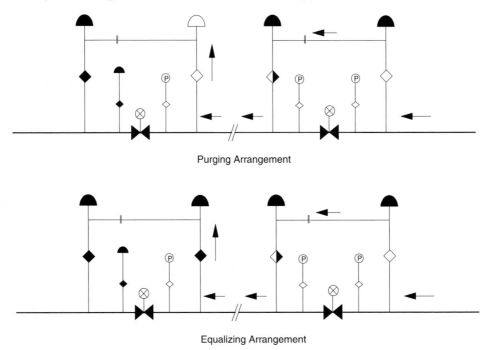

Figure 9-28. Equalizing

The pressure is monitored at the upstream and downstream assemblies and continuous checking for leaks is maintained. When the pressure has equalized and stabilized, all valves in the purged section are to be set for normal operation. All automatic controls are returned to service. Routine service is performed on the assemblies.

Commissioning of Meter Station Equipment

Facilities that are commissioned and made ready for onstream operation are as follows:

- Recorders and other equipment (such as meter station heaters)
- Instrumentation
- Controls
- Electrical equipment

Recorders and other equipment relate to all meter stations that are not automated or telemetered. Instrumentation, control and electrical equipment are for automated or telemetered stations where measurement data is transmitted to a Central Control Station/ Facility.

Purging Calculations

When purging gas pipelines (i.e. replacing air with natural gas) it is important to avoid natural gas/air stratification, which allows air to be trapped below the natural gas while the pipeline is purged. Calculations of the dimensionless Richardson's number (R#) is one way of determining whether stratification occurs during the purging process. A Richardson's number between one and five is generally accepted to assure air/gas mixing. The lower the number, the less likely stratification will occur. A complete background to purging computations is provided by Mohitpour et. al. (2000).

Richardson's Number is expressed as follows:

$$R_\sharp = \frac{g'd}{u_p^2} \qquad (9-27)$$

where

$$g' = g\frac{\Delta\rho}{\rho_{ave}}$$

and g = acceleration due to gravity = 9.81 m/sec^2
$\Delta\rho$ = difference between air and gas density, kg/m^3
ρ_{ave} = average density = $\frac{\rho_a+\rho_g}{2}$ kg/m^3
ρ_a = air density, kg/m^3
ρ_g = gas density, kg/m^3
d = internal pipe diameter, m
u_p = average velocity, m/s

Sample Purging Problem

A 24 km section of NPS 36 looped line (loop is also NPS 36) is required to be purged (Figure 9-29). Only 10% of the mainline gas flow of 0.8 BCFD is to be utilized for purging.

Calculate the following:

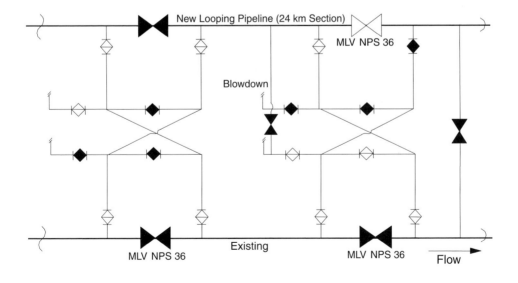

Figure 9-29. Mainline Loop Configuration

a) Whether 10% of the mainline gas flow would ensure full air/gas mixing;

b) The required size the orifice plate to ensure introduction of 10% of mainline flow to loop during purging; and

c) The time that it takes to fully purge this section of loop.

The following assumptions are applicable:

- Mainline operating pressure is 4100 kPa (g)
- Mainline operating temperature is 28°C
- Density of air at base condition is 1.225 kg/m^3
- Gas composition and properties assume the same properties as in Problem 2 Chapter 4

Solution

a) Richardson's number

$$g' = 9.81 \left[\frac{1.225 - 0.772}{\left(\frac{1.225 + 0.772}{2} \right)} \right] = 4.45 \text{ m/sec}$$

where

$$d = 36 \text{ inches} = 0.9144 \text{ m}$$
$$A = 0.6567 \text{ m}^2$$
$$Q = 0.8 \text{ BCFD} = 262.4 \text{ m}^3/\text{s}$$
$$\therefore 10\% \ Q = Q_p = 0.1 \times 262.4 = 26.24 \text{ m}^3/\text{s}$$

Referring to the table for Problem 2 in Chapter 4, the specific volume Vsg for natural gas at 28°C and 4100 KPa(g) can be calculated as follows:

$$V_{sg} = 0.02872 + [0.0314 - 0.02872] \frac{8}{(40 - 20)} = 0.0298 \text{ m}^3/\text{kg}$$

$$\rho_g = \frac{1}{V_{sg}} = \frac{1}{0.0298} = 33.56 \text{ kg/m}^3$$

$$\therefore u_p \text{ at pipeline conditions} = \frac{26.24}{0.6567} \times \frac{0.772}{33.56} = 0.92 \text{ m/s}$$
$$\therefore \text{ Richardson Number } R_\# = \frac{g' d}{u_p^2} = \frac{4.45 \times 0.9144}{(0.92)^2} = 4.8 \text{ which} < 5 \ \therefore \text{ OK}$$

b) Orifice Size

Gas sonic velocity $= V_s = \sqrt{g_c \ k \ Z_s \ T_s \ R^*}$ ft/sec

$$Z_s \ T_s = ZT \left(\frac{2}{k + 1} \right)$$

$$V_s = \sqrt{g_c \ Z \ T \left(\frac{2k}{k + 1} \right) R^*}$$

assume $R^* = 1543$ mole weight, gas constant ft lb/lb °R

$S_g = 0.6$ (gas specific gravity), Mw = 17.6

$Z = 0.91$

$k = 1.35 = $ Ratio of C_p/C_v

$g_c = 32.2$ ft/sec^2

$$T_s = 460 + [28 \times 1.8 + 32] = 542 \text{ R}$$

$$V_s = \sqrt{32.2 \times 0.91 \times 542 \times 1.15 \times 1543/17.6}$$
$$= 1266 \text{ ft/s}$$
$$= 386 \text{ m/s sonic velocity at line condition}$$

The above is an approximation, since the value taken for R (Gas Constant) is for a typical natural gas composition. Therefore, the actual sonic velocity V^s could be as low as 420 m/s.

Qp at line condition:

$$26.24 \times \frac{0.772}{33.56} = 0.6 \text{ m}^3/\text{s}$$

$$Q = A \times Vs$$

$$\therefore A = \frac{Q}{V_s} = \frac{0.6}{386} = \frac{\pi d_o^2}{4}$$

where do = orifice diameter

$$\therefore \text{ do orifice diameter} = \sqrt{\frac{4 \times 0.6}{\pi \times 386}} = 0.045$$

$$= 4.5 \text{ cm}$$
$$= \text{approximately 5 cm}$$

c) Time required for purging:

$$L = 24 \text{ km} = 24,000 \text{ m}$$
$$\text{Pipe Section Volume} = \frac{\pi}{4}(0.9144)^2 \times 24,000 = 15761 \text{ m}^3$$
$$\therefore t = \frac{15,761}{0.6} = 26,267 \text{ sec} = 7.31 \text{ hrs}.$$

Allow eight hours for purging.

Natural Gas Liquid Pipelines (Propane, Butane and Condensate Pipelines)

The commissioning of propane, butane and condensate pipelines requires additional activities and facilities to deal with pressure requirements to maintain the product in the liquid phase, and the "heavier than air" property of the gas phase.

The initial step in commissioning is to purge the air from the pipeline section with nitrogen. This will result in a nonflammable and non-explosive condition if mixed with hydrocarbon vapors. After the purge is completed, using a procedure as presented for a natural gas - air purge, the pressure in the section will be increased to a level which will prevent vaporization of the hydrocarbon. The pressure requirements due to elevation changes must be included in this calculation. The pipe section can then be filled with the hydrocarbon with a pig between the product and nitrogen. The nitrogen can be vented to atmosphere or used to purge and pressure additional sections. A flare pit must be utilized for venting so that any hydrocarbon released as the pig is landed can be burned. After the completed pipeline is full of hydrocarbon product, the pressure is raised to operating levels to complete the commissioning.

Requirement of all purging operations:

- Qualified personnel are present at all inlet and outlet locations
- Communications are maintained between sites
- A written procedure is prepared and understood by all personnel involved
- Sources of natural gas, nitrogen and hydrocarbon are available at the volumes and pressures required
- Instruments, including backups, are functional and have been calibrated
- Sources of ignition are not present when venting natural gas. Hydrocarbon liquids or vapor are vented to a facility where it will be flared safely

REFERENCES

Alberta Energy and Utilities Board, 2005 Directive 66 "Requirements and Procedures for Pipelines", December Calgary Alberta, Canada

American Petroleum Institute (2003). "Recommended Practice for Planning, Designing and Constructing Fixed Offshore platforms- Working Stress Design" 20th Edition. RP2A-WSD-03 Dallas, TX.

American Petroleum Institute (API), 2005, "Standard 1104 Welding Pipelines and Related Facilities"

American Society of Civil Engineers (2005) "Pipeline Design for Installation by Horizontal Directional Drilling." Manuals and Reports on Engineering Practice No.108

American Society of Mechanical Engineers 2002 Clause 437 "ASME B31.4-2002 Liquid Transportation Systems for Liquid Hydrocarbons and Other Liquids" New York, NY, USA

American Society for Testing of Materials (1999). "Standard Guide for Use of Maxi-Horizontal Directional Drilling of Polyurethane Pipe or Conduit Under Obstacles, Including River Crossings" F1962-99, West Conshohocken, PA.

American Society of Mechanical Engineers (ASME), 2002, "ASME/ANSI B31.4," *Liquid Transportation Systems for Hydrocarbons, Liquid Petroleum Gas, Anhydrous Ammonia and Alcohols*, Washington, D.C.

Awoseyin, R. S., 1985, "Pressure — Temperature Relationships in Buried Pipelines," *Proc. Instn. Mech. Engrs*. U.K., Vol. 199, No 41.

Barlow, J.P., and Cavers, D.S., (1996) "The role of Geotechnical Investigations for Directionally Drilled Crossings" Proc; 1st ASME International Pipeline Conference IPC96. June 9-13. Calgary. Alberta Vol.2 pp 1229-1235.

Baumgard, A., Savigny, K.W., and Cocciolo, P., "Post Installation Geotechnical Issues Associated with Large Scale HDD Crossings" Proc; 4th ASME International Pipeline Conference IPC2004. Oct. 4-8. Calgary, Alberta Vol.1. pp 275-282.

Boivin, R.P., Rizkalla, M., Marshall, R.G., and Simmonds, G.R., (1994) "Criteria for Evaluating Directionally Drilled Crossings" Pipelines and Gas Journal. June pp 12-22.

British Standards Institution 1992 "BS 8010-2.8: Code of Practice for Pipelines Part 2. Section 8 Pipelines on land: design, construction and installation" Including the reference IGE/TD/1 Edition 3, 1993, communication 1530: "Steel pipelines for high pressure gas transmission". London, England

Brower, W. B., Eisler, E., Filkorn, E. J., Gonenc, J., Plati, P., and Stagnitti, J., (1993) "On the compressible flow through an orifice," J. Fluids Eng. 115, pp 660–664.

Canadian Association of Petroleum Producers (2004) "Planning Horizontal Directional Drilling for Pipeline Construction." CAPP Publication 2004-0022. Calgary, Alberta. September.

Canadian Association of Petroleum Producers, (1993), "Watercourse Crossing Guidelines for the Pipeline Industry," Calgary, Alberta, Canada.

Canadian Standards Association, 2007 "Standard Z662-07," *Oil and Gas Pipeline Systems*, Etobicoke, Ontario, Canada.

Canadian Standards Association, 2007, Clauses 8 and 12 "CSA Z662-07; Oil and Gas Pipeline Systems", Mississauga, Ontario, Canada.

Comite Europeen de Normalisation, 1994, Section 7 and 11.4 "CEN prEN 1594 : Functional requirements for pipelines for gas supply systems for pressures over 16 Bar". Brussels, Belgium.

Dalton, P., Sobolevsky, J.T., Aynbinder, A., and Tabakman, Y., 1994 "FSU, US design codes differ on key points of pressure testing" Oil and Gas Journal Dec ,1994,Houston TX pp 52-55.

Det norske Veritas, 2000., "OS-F101: Submarine Pipeline Systems"Hovik, Norway.

Fowler, J.R., and Langner, C.G., (1991). "Performance Limits for Deepwater Pipelines "OTC 6757, 23rd Offshore Technology Conference Houston, TX May 6-9.

Gas Processors Suppliers Association(GPSA) 1998 *Engineering Data Handbook* Vol. 1 Section. 3 pp 3-17.

Gray, J. C., 1976, "How Temperature Affects Pipeline Hydrostatic Testing," *Pipeline and Gas Journal*, August, pp. 26–30.

Hair, J.D., (1995) "Design and Project Management Considerations Involved with Horizontal Directional Drilling." AGA Operations Conference Las Vegas. NV. May 7-10.

Heinz, H., Moore, T., and Cullum-Kenyon, S., (2004) "Geotechnical Assessments for Trenchless Water Crossings in Alberta" Proc; 4th ASME International Pipeline Conference IPC2004. Oct.4-8 Calgary.Alberta Vol.1. pp 595-600.

Hodge. B. K., and Koenig, K., 1995 *Compressible Fluid Dynamics*. Prentice Hall, Englewood Cliffs.

International Organisation for Standardisation, 2000. Clause 6.7 "ISO 13623:Petroleum and natural gas industries - Pipeline transportation systems." Geneva, Switzerland

Mohitpour, M., Kazakoff, K. J., Montemurro, D., 2000, "Planning for Purging and Conditioning of a Newly Constructed Gas Pipeline System Using a Pipeline Simulation", Proc, ASME, 3rd IPC 2000 Conference, October 1 – 5, Calgary, Alberta, Canada.

National Energy Board 1999. "Onshore Pipeline Regulations,1999" pursuant to sub section 48(2) National Energy Board Act, 23 June, Calgary, Alberta, Canada.

NEN 3651 (1994) "Supplementary Requirements for Steel Pipelines Crossing Major Public Works." Government Industry Standards Committee.343 20. The Netherlands. February.

Pick, A. R., and Smith, J. D., 1988 "Pressure testing with air works for Oil pipelines" Oil and Gas Journal Vol 86, No.21, June 13, Houston TX pp 44-47.

Quaife, L. R., and Moynihan K. J., 1991 "A new pipeline leak-locating technique utilizing a novel odorized test-fluid (patent pending) and trained domestic dogs" AAPG Bulletin (American Association of Petroleum Geologists); Vol/Issue: 75:3; Annual meeting of the American Association of Petroleum Geologists (AAPG); 7-10 Apr 1991; Dallas, TX

Roark, R. J., (1965). *Formulas for Stress and Strain*. Fifth Edition. McGraw Hill New York.

Robinson, C., 2000 "Pneumatic Testing" Internal Report August 2000, National Energy Board, Calgary, Alberta, Canada.

Standards Australia 2001 Section 2 "AS 2885.1-1997 Amendment 1-2001; Pipelines – Gas and Liquid Petroleum – Part 1 Design and Construction" North Sydney, Australia

Standards Australia 2002 Section 2 "AS 2885.5; Pipelines – Gas and Liquid Petroleum – Field Pressure Testing" North Sydney, Australia

Trow Engineering Consultants Inc, 1996, "Water Crossing Design and Installation Manual," Report PR-237-9428, *Offshore and Onshore Design Applications Supervisory Committee*, Pipelines Research Committee International.

Watson, P. D., "Installation of Pipelines by Horizontal Directional Drilling- An Engineering Design Guide. Prepared for Pipeline Research Council International Inc. by J.D. Hair and Associates, Louis J. Capozzi and Associates and Stress Engineering Services. Catalog No L51730e April 15th 1995.

Wight, R. S., 1983, "Developments in Hydrostatic Testing," *Proc. Int. Conference on Pipeline Inspection*, CANMET,Edmonton, Alberta.

Young, V. W., and Young, G.A., "Elementary Engineering Thermodynamics" 1947 McGraw Hill Book Company Inc New York, p 153-162.

Chapter 10

PIPELINE PROTECTION

INTRODUCTION

The most effective method of mitigating corrosion on the external surface of a buried or submerged facility utilizes a dual system of a protective coating supplemented by cathodic protection. Each of these means of protection will be discussed in detail in the sections below.

PIPELINE COATING

The primary function of an external coating is to establish a permanent barrier between a structure and its immediate environment.

The performance of any coating system is directly related to the conditions encountered during the installation and operational life of a facility. Therefore, before any coating selection is initiated, it is imperative that the environmental and construction conditions of a facility are well understood.

Based on an evaluation of the relevant chemical, electrical, environmental, geotechnical, and mechanical conditions experienced by a pipe coating system during its service life, an optimum external pipeline coating for buried service should exhibit the following properties Coulson and Temple (1983):

- Adhesion
- Chemical resistance
- Electrical resistance
- Compatibility with cathodic protection
- Flexibility
- Hardness/abrasion resistance
- Impact resistance
- Penetration resistance
- Soil stressing resistance
- Weathering resistance
- Compatible repair and girth weld coating material

Pipeline Coating Failure Modes

External corrosion and environmentally assisted cracking problems such as stress corrosion cracking (SCC) can threaten pipeline integrity. Most pipeline systems are protected from these problems by external coatings and impressed current cathodic protection. For a problem to develop, the coating must fail in such a way as to expose the pipe to a corrosive external environment and to prevent cathodic protection from reaching the steel surface. The mode of coating failure is the key to whether serious problems can occur.

Different coating systems under identical exposure conditions can fail in completely different ways. An example was recently encountered on a pipeline in Southern Alberta. The following photographs show coating conditions found at one site where the pipe was coated with polyethylene tape alongside fusion bond epoxy (FBE). The coatings were exposed to identical service conditions in identical soil environments. However, the way in which the coatings failed was completely different.

As can be seen in Figure 10-1, the polyethylene tape coating disbonded to form a wrinkle in the coating. By way of comparison, Figure 10-2 shows damage observed on the fusion bond epoxy section of pipe exposed to identical field conditions. In contrast to the wrinkling associated with the tape system, the FBE coating has failed through localized blistering.

Examination of the water trapped under these coating failures showed that beneath the tape disbondment the solution has a pH value of 6.5 to 7.0, whereas beneath the FBE blister the pH of the trapped water was in excess of 10. These observations indicated that the tape formed an electrically shielding disbondment, while the FBE allowed penetration of the cathodic protection and hence the generation of a high pH electrolyte next to the steel.

Figure 10-1. Disbonded tape coating showing longitudinal wrinkles where the tape has pulled away from the pipe

Figure 10-2. Blistered FBE coating

The failure mode of a pipeline coating has significant impact on whether corrosion and SCC will develop in the underlying metal. Due to the absence of effective cathodic protection, both localized corrosion and stress corrosion were observed beneath the disbonded tape coating, as shown in Figure 10-3. Cracks are grouped in colonies, often containing hundreds of individual cracks, and are usually found on the pipeline system in the body of the pipe. Research is ongoing to understand the mechanism and rate of crack growth.

In contrast, no localized corrosion or stress corrosion was found beneath the FBE blisters, suggesting that despite coating failure the mode of failure allowed protection of the pipe surface by cathodic protection. These and other field observations suggest that FBE coatings fail in such a way as to allow penetration of cathodic protection to the steel surface. For this reason, most pipeline companies now use FBE coatings.

Important Coating Properties for Operation
Adhesion
Permanent adhesion is essential for a material to perform as a satisfactory coating. A disbonded coating may permit moisture to accumulate between the structure and coating, thereby adversely influencing the effectiveness of the coating.

Therefore, it is imperative that the surface of the base pipe is properly prepared. This involves much more than simply removing surface dirt and rust. The presence of invisible contaminants such as soluble salts can have a very detrimental effect on the life of the coating (Neal 1999).

Figure 10-3. Corrosion and SCC observed in the steel surface beneath disbonded tape coating

In order to perform well, coatings need to maintain a strong adhesive bond throughout their working life. High initial adhesive strength will be obtained from having a suitably prepared surface, good coating flow, and wetting during application. The longevity of the bond is related to the properties of the coating, its resistance to moisture penetration, soil strength, and cathodic disbondment, all of which are discussed later in the chapter.

Cathodic disbondment, long-term water immersion, and soil stress tests are commonly employed to evaluate the adhesion of a coating system.

Chemical Resistance

A coating system should be resistant to all chemicals such as hydrocarbons, acids, alkalis, and biological agents likely to come into contact with the coating. A typical chemical resistance test exposes the coating to the liquid and vapor phase of a chemical for a specified period of time. At the end of a test period, the degree of coating degradation is evaluated.

Electrical Resistance

An effective underground coating must have good dielectric strength to assure high electrical resistance. This resistance should not change appreciably with time or with exposure to water.

Although coatings need not withstand continuous high voltage stresses normally associated with electrical insulation, they must exhibit good electrical resistance. For example, coatings are exposed to 1,500 to 7,500 volts during holiday detection, and 1 to 3 volts during the application of impressed current cathodic protection.

Compatibility

In areas where the coating has been damaged or is missing, cathodic protection is used to protect the metal from corrosion. However, cathodic protection can induce the evolution of hydrogen gas at the exposed surface and increase the transportation of water to the coating defect site. These two effects can singularly or in combination increase the loss of coating adhesion at the edge of a "holiday," that is at a failure point in the coating.

Cathodic disbondment tests are used to assess a coating's compatibility with cathodic protection. Typically a coating specimen with a predrilled holiday is immersed in a 3% salt solution at 20°C for 28 days. During the test, the coating samples are usually maintained at a nominal −1.5 volt potential.

After completion of the test period, a levering action is used to lift the coating around the holiday, until the coating demonstrates a definite resistance to the levering action. The radius of disbondment is measured and reported.

The test is performed at various temperatures to determine the operating temperature range of a coating.

Penetration Resistance

During storage and operation, coatings are subjected to concentrated pressures exerted by the weight of the structure/backfill. A coating system must be able to resist penetration or deformation under conditions of static load.

Resistance to penetration is evaluated by subjecting a known coating thickness to a constant static load. The percentage of coating thickness penetrated is evaluated.

Soil Stressing Resistance

Coatings are subject to deformation and damage by soil stresses. Cohesive clays around the coated structure expand with absorption of moisture and subsequently contract during drying. These clays exhibit sufficient gripping action to remove the coating during contraction. Coatings must be able to withstand this type of soil movement, which can also be exacerbated by temperature cycling of the pipeline.

After a predetermined number of cycles have elapsed, the clay is removed, and the coating is examined for adhesion loss.

Important Properties for Construction and Installation

Flexibility

Coatings must be sufficiently flexible to resist cracking or disbondment during construction. For example, plant-applied pipe coating requires sufficient flexibility for field bending. Coating flexibility is determined by bending a coated strap over a fixed radius mandrel corresponding to the maximum amount of curvature expected. Tests are performed over a wide temperature range to determine the bending limits of a coating system.

Hardness/Abrasion Resistance

By the nature of construction, coated materials undergo a great deal of handling from the time the coating is applied, until the coated material is lowered into the ditch and the backfilling operation is complete. A coating should have the ability to resist abrasion and any other damage during shipping, handling, backfilling, and in-service conditions. Soft coatings are more susceptible to damage than hard coatings.

Coating hardness is usually assessed using a Shore D Durometer.

Impact Resistance

It is important for coatings to resist mechanical damage during installation and backfilling. The method of assessing impact resistance consists of an indentor of fixed weight, which is dropped from varying heights to produce a "point impact" on the coating. The failure point is then related to the height from which the falling weight ruptures the coating film and exposes the underlying steel surface. Impact resistance is the amount of energy required to cause the failure.

Tests are done over a wide range of temperatures to anticipate the amount of damage that may occur at a given construction temperature.

Weathering Resistance

Coating may be applied well in advance of proposed construction in service dates. Therefore, coatings must be resistant to ultraviolet degradation and extreme temperature changes. Ultraviolet degradation and the effects of temperature are evaluated by placing coating compounds in areas where coating storage may occur, exposing them to the elements. After a specified time period, the effects of exposure on the coating are determined by visual examination and by performing cathodic disbondment, flexibility, and impact resistance tests. The weathered coating sample test results are compared to the test results obtained from an unweathered coating sample.

Compatible Repair and Girth Weld Coating Material

The girth weld coating system and repair coating should be compatible with the parent coating and possess properties that meet the construction and operating conditions.

Coating Selection

Long-term coating performance is essential to prevent deterioration of a pipeline. Errors in coating selection may lead to failure and remedial repair costs, which can easily exceed the increased initial cost of a higher integrity coating. Factors used in coating selection are:

- Operating temperature
- Backfill/terrain characteristics
- Coating availability
- Overall coating cost

Operating Temperature

The pipeline operating temperature downstream of compressor stations and producing facilities is one of the criteria used for coating selection. These are the regions of pipe exposed to the highest temperatures. Laboratory coating tests are undertaken to assess coating properties at the operating temperature envisaged at a facility. From the test results, a coating can be selected, which will resist disbonding, melting, sagging, and softening over the period of the design life.

Backfill/Terrain Characteristics

Backfill/terrain characteristics that need to be considered include:

- Rocky or frozen backfill—requires selecting a coating with excellent impact resistance; use of such a coating though relatively expensive, can obviate the need to import fine backfill/padding material for the pipe.
- Clay or glacial moraine—requires selecting a coating with excellent soil stressing resistance.
- Undulating terrain—for plant-applied systems, selecting a coating with flexibility sufficient for field bending under construction conditions is important.

Coating Availability

Consideration should be given to the geographic source of the pipe, the pipeline location, the availability and location of coating applicator facilities, and probable method of transportation of the coated pipe. There are occasions where the final analysis of these factors leads to the selection of a coating system that is applied in the field with portable equipment or at a portable coating facility.

Factors that may eliminate the choice of an over-the-ditch applied coating system (using portable equipment or facilities) include.

- Coating manufacturer application requirements and construction temperatures.
- The difficulty of providing and maintaining the optimum heat distribution in subzero conditions (e.g., variation of line heater speeds).
- The manufacturer's recommended application specification is difficult to meet for over-the-ditch systems during rainy periods and general winter construction.
- Pipeline terrain.
- Mountainous terrains may preclude the use of a field-applied coating system because it may be difficult to apply it properly under these conditions.

Specifications

Once a coating system is selected, comprehensive specifications are required for coating application, handling, storage, joint coating, and repairs. Inspection of all phases of the coating operation is essential to ensure a uniform, high-quality protective coating.

Field Joint Coatings

While there are numerous coating standards and specifications covering mainline pipe coatings, such is not the case for field joint coatings. Instead, individual users and contractors have developed their own specifications, often with little bearing on the operating performance. Surface cleanliness is the chief requirement in achieving a good joint coating, and in the sometimes difficult conditions found in the field, it can be achieved only by blast-cleaning. Pipe preheat is required prior to cleaning and again before coating application. This ensures that there is no condensed surface moisture. A good anchor profile is highly desirable to ensure the adhesion of coatings. Andrenacci et al. (1999) recommend a pattern at least 2.5 mm deep. They also have developed a weighted comparison of joint coatings (Table 10-1), which shows that by their reckoning a 2-layer mastic sleeve is superior to a number of other methods.

Another more subjective comparison of various types of field joint completion systems is found in Table 10-2.

Overall Coating Cost

There are four elements that make up the overall coating cost: the cost of the plant-applied coating, joint completion costs, field repair costs, and the expected life of the pipeline (amortization costs). The last point is important, since it is relatively easy to overdesign or overspecify a coating with consequent cost increase. The choice of coating should therefore fully reflect the design life of the pipe.

Summary

The steps involved in coating selection and evaluation are as follows:

1. Identify the coating properties required by a facility.
2. Design and carry out a testing program to assess coating properties (if not already established). Before committing to a major project, perform a plant/field trial to ensure that the coating can be applied satisfactorily and that the required properties are obtained.
3. Match a coating with known properties to the facility requirements.

TABLE 10-1. Comparison of joint coatings (Andrenacci et al. 1999)

Factors	Weighting factor	2 - Layer Mastic Sleeve R	W	2 - Layer H-Melt Sleeve R	W	3 - Layer Sleeve R	W	New Genera-tion Sleeve R	W	Cold-Applied Tapes R	W	Epoxy Polyure-thane R	W	FBE R	W
1. Surface preparation	1.2	1	1.2	4	4.8	3	3.6	3	3.6	1	1.2	5	6	5	6
2. Equipment requirements	1	1	1	3	3	4	4	3	3	3	3	2	2	5	5
3. Speed of installation	0.8	1	0.8	3	2.4	4	3.2	3.5	2.8	2	1.6	2	1.6	1	0.8
4. Operator skill set	1	2	2	3	3	4	4	3.5	3.5	1	1	2	2	5	5
5. Sensitivity to application	1.2	1	1.2	3	3.6	4	4.8	3.5	4.2	1	1.2	5	6	3.5	4.2
6. Crew size	1.2	1	1.2	2	2.4	4	4.8	3	3.6	3	3.6	2	2.4	5	6
7. Inspectability	0.7	2	1.4	2	1.4	2	1.4	2	1.4	4	2.8	5	3.5	5	3.5
8. Soil stress and mech. resistance	2	5	10	2	4	2	4	2	4	5	10	1	2	1	2
9. Shielding potential	0.8	3	2.4	4	3.2	3	2.4	3	2.4	5	4	1	0.8	1	0.8
10. Environmental concerns	0.7	1	0.7	1	0.7	4	2.8	3	2.1	4	2.8	5	3.5	4	2.8
11. Safety concerns	0.8	1	0.8	2	1.6	4	3.2	3	2.4	4	3.2	5	4	4	3.2
12. Cathodic disbondment	2	1	2	3	6	1	2	1	2	2	4	3	6	1	2
- low temp	2	5	10	5	10	3	6	3	6	5	10	3	6	2	4
- high temp.															
Best overall system rating X weighting factor			1		6		7		2		4		5		3

Note: Joint systems—important factors. R = rating; 1 best; 5 = worst; W = weighted. From Adrenacci, Worg et al. (1999).

CATHODIC PROTECTION

The prevention of external corrosion on underground structures is accomplished using a dual system of coatings and cathodic protection. To define corrosion, pipelines will be considered synonymous with all underground structures.

Definition of Corrosion

An understanding of cathodic protection cannot be achieved without first understanding the mechanisms of corrosion. A simplified definition of corrosion, in its most general sense, is the degradation of a material due to a reaction with its environment. The environment in pipeline applications includes the soil and its many constituents (water, air, salts, bacteria, etc.). For corrosion to occur, four conditions must be met:

1. There must be an anode and cathode.
2. There must be an electrical potential difference between the anode and the cathode.
3. There must be a metallic connection between the anode and cathode.
4. The anode and cathode must be immersed in a common electrically conductive medium (an electrolyte, e.g., the ground).

TABLE 10-2. Comparison of common joint completion systems (Andrenacci et al. 1999)

System	Advantages	Disadvantages
Cold applied tapes	— Straightforward to apply — Low skill required — Few sizes — Adheres to most coatings — Effective in cool, low soil stress applications — Relatively insensitive to ambient temperatures — Can tolerate a low degree of surface preparation	— Must use a primer (typically solvent-based) — Spiral voids — Bridging a major problem — Shields CP system — Very soft, pressure sensitive adhesives — Low soil stress resistance
Heat shrink sleeve 2-layer hot melt	— Simple & reliable installation — Low skill required — Excellent resistance to cathodic disbondment, (CD) and undercutting — Relatively insensitive to ambient temperatures — Good track record — Can tolerate a low degree of surface preparation	— Limited resistance to soil stresses — Requires a suitable torch — Installers must be trained
Heat shrink sleeve 2-layer hot melt	— Good resistance to soil stresses — Effective at higher operating temperatures — Excellent resistance to cathodic disbondment, water soak, and undercutting — Anticorrosion effectivess guaranteed if epoxy primer is properly applied and cured — Lower preheats than with 2-layer hot melts	— Requires personnel & facilities to store, mix, and dispense epoxy primer — Requires near-white surface preparation — Epoxy is sensitive to temperature for proper cure. force cure best in cold climates. — Expensive
Field-applied FBE	— Excellent soil stress resistance — Effective for high temperature uses — Excellent CD resistance and hot water soak if the surface is properly prepared and the FBE is properly cured	— Requires induction heating in the field — Difficult to control application and cure temperature in cold, windy conditions — Extremely sensitive to surface cleanliness — Only useable with FBE coatings
Epoxy-urethane	— Excellent soil stress resistance — Simple to apply (brush, spray, etc.) — Relatively inexpensive — Good CD resistance and hot water soak if the surface is properly prepared and the FBE is properly cured	— Extremely sensitive to surface cleanliness — Properties are very sensitive to cure temperature and surface dryness — Accurate mixing ratio and very good physical mixing are absolutely essential

Once all of these conditions have been met, corrosion will occur. The anode will corrode, producing metal ions and electrons. Corrosion occurs where current leaves the surface of the pipeline (anode) via the surrounding soil. Corrosion does not occur where current enters the pipeline (cathode).

Mechanism of Cathodic Protection

Corrosion prevention is affected by negating one of the four conditions listed above. In the case of cathodic protection, the electrode potential difference between the anode and the cathode is eliminated by making all of the pipeline cathodic.

Cathodic protection can simply be defined as the application of an external direct current source that causes current to flow to the structure being protected. In other words, an artificial potential gradient is created that will reverse the natural current flow. This causes the anode and cathode to exchange functions. An obvious problem is that something (the anode) will still be corroded. Cathodic protection provides an alternative (sacrificial) anode, which is intended to corrode instead of the pipeline (the cathode).

Cathodic protection can be applied by two methods—impressed current cathodic protection and galvanic current cathodic protection. Both systems function by causing current to flow from a sacrificial anode to the pipeline. The difference is the method by which the current is caused to flow.

Impressed Current System

An impressed current system typically utilizes an external AC power source, a rectifier (to convert AC current to DC current), a ground bed of anodes, a metallic connection from the anodes to the pipeline, and a common environment (the ground). These anodes are often made of high silicon iron or graphite. Table 10-3 describes their use and type of backfill required.

The materials allow a large amount of current to flow from them in proportion to their own corrosion or consumption rate. The anodes cannot function without an external power source. The power source forces current to flow from the rectifier along the anode header cable to the anodes. The current then exits the anodes and travels through the ground to the pipeline. The current enters the pipeline at holidays or voids in the pipeline coating and then moves to a thermite-welded cathode cable. To complete the electrical circuit, the current flows back to the rectifier. The rectifier contains meters to monitor the current output and the voltage output (Figure 10-4).

TABLE 10-3. Impressed current anodes

Type of Anode	Uses/Comments	Backfill
Sacrificial	Sacrificial anodes can be connected to an external power source and discharge current but their lifetime will be very short. Normally they are not used for this application.	75% Hydrated Gypsum 20% Bentonile 5% Sodium Sulfate
Steel	Any buried bare steel, such as old piping, or an automobile can be used.	None
High-silicon chromium iron	One of the most common anodes in use for pipelines and plants.	Coke breeze
Graphite	Good life in well-drained soil.	Coke breeze
Metal/metal oxide (ceramic)	Made from titanium and coated with a metal oxide. These anodes do not corrode away. They are designed to last a given time and stop working quite suddenly. Much lighter than most other anodes.	Coke breeze or seawater or fresh water
Anodflex (continuous anode)	Good for long, poorly coated structures, or very high soil resistivity locations.	Coke breeze

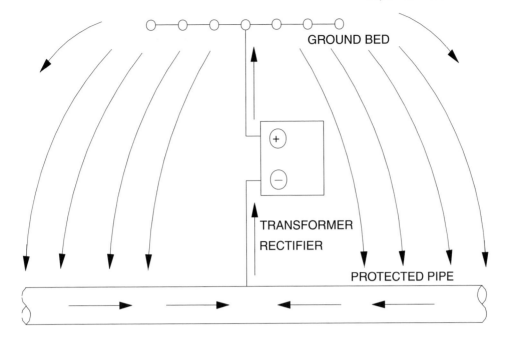

Figure 10-4. Impressed current cathodic protection system

The advantages and the limitations of an impressed current system are detailed below.
Advantages

- Can be designed for wide range of voltage and current
- High ampere year output available from a single ground bed
- Large areas can be protected by single installation
- Variable voltage and current output
- Applicable in high-resistivity environments
- Effective in protecting uncoated and poorly coated structures

Limitations

- Can cause cathodic interference problems with adjacent pipelines
- Subject to power failure and vandalism
- Requires periodic inspection and maintenance
- Requires external power, which involves costs
- Overprotection can cause coating damage

Galvanic Current System

A galvanic current system requires anodes and a wire joining the pipeline to the anodes. The current in this system is provided by a natural reaction—a galvanic reaction—between the anodic material and the pipeline. It utilizes the same concept that provides electrical power in batteries. The voltage created by the galvanic reaction prevents current from leaving the pipeline and thus provides corrosion. The anodes are constructed of magnesium, zinc, or sometimes aluminum.

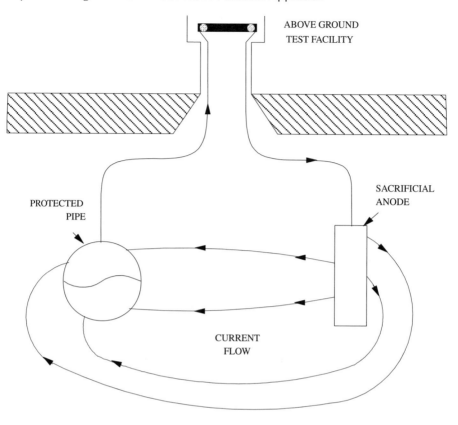

Figure 10-5. Sacrificial cathodic protection

The current flows from the anodes to the pipeline via the ground. The current mainly enters the pipeline surface at voids in the coating (holidays). The potential gradient caused by the current flow effectively prevents current from leaving the surface and hence prevents corrosion. Since all electrical circuits must be continuous (circuitous) to function, the current flows along the pipeline to the cathode cable connection—a thermite weld. The circuit is completed by joining the cathode and anode cables. The test lead is a convenient location for the joining to take place. It allows easy monitoring of the current flow (Figure 10-5).

Table 10-4 shows a galvanic series of metals in neutral soil referenced to a copper/copper sulfate reference cell. Basically, the table indicates the potential difference established between different metals. Current will flow from the more electronegative materials, or those materials with a more negative potential, to metals less electronegative. In other words, the current will flow from metals higher up on the table to those lower down on the table. For example, magnesium can be used to protect steel but graphite cannot. The greater the difference in the potentials the greater will be the driving force (voltage) that will allow cathodic protection to be affected in lower soil resistivities. It should be noted that the potential of a material is significantly affected by the addition of other elements (e.g., pure magnesium has a potential of -1.75 V while an alloy 91Mg-6A1-3Zn has a potential of only -1.55 V). Some typical sacrificial anodic materials and their uses are given in Table 10-5.

TABLE 10-4. Typical potentials for selected materials in neutral—soil referenced to a Cu/CuSO₄ electrode

Material	Potential Volts
Magnesium	-1.75
91MG-6A1-3Zn-0.15Mn	-1.55
Zinc	-1.1
95A1-5Zn	-1.05
Aluminum	-0.8
Steel (clean)	-0.5 to -0.8
Steel (rusted)	-0.2 to 0.5
Gray cast iron	-0.5
Lead	-0.5
Steel in concrete	-0.2
Copper, brass, bronze	-0.2
High-silicon cast iron	-0.2
Mill scale on steel	-0.2
Graphite (carbon)	$+0.3$

The following lists the advantages and limitations of galvanic systems.

Advantages

- No external power required
- No regulation required
- Easy to install
- Minimum of cathodic interference
- Anodes can be readily added
- Low maintenance
- Uniform distribution of current
- Installation (of anodes at time of construction) can be inexpensive
- Minimum right-of-way/easement costs
- Efficient use of protective current

Limitations

- Limited driving potential
- Low/limited current output/unit
- Installation (of anodes after construction) can be expensive
- Poorly coated structures require many anodes
- Can be ineffective in a high-resistivity environment

It might appear that galvanic systems would be the first choice for cathodic protection application since no external power sources or rectifiers are required.

TABLE 10-5. Sacrifical anodes

Types of Anode	Uses/Comments	Backfill
Magnesium	Soil with resistivity $< 10,000$ Ωcm	75% Hydrated gypsum
		20% Bentonite
		5% Sodium sulfate
Zinc	Soil with resistivity $< 2,000$ Ωcm	75% Hydrated gypsum
		20% Bentoite
		5% Sodium sulfate
Aluminum	Seawater	None

However, for extensive pipeline systems, the impressed current system is generally more practical. The major advantages of the impressed current system over the galvanic system are:

- Its wide range of voltage and current output
- The large area covered from a single location
- It is effective in protecting poorly coated pipelines

Cathodic Protection Design

Cathodic protection facilities may be required for new construction and the upgrading of old systems. In either case, the following similar factors must be considered prior to any design:

- Existing cathodic protection
- Length and diameter of existing/new facilities
- Future plans for expansion.

Existing cathodic protection in the area may have sufficient capacity to be able to protect the additional loads. This can be accomplished by simply resetting the rectifier output.

Current requirements for cathodic protection are calculated by using specific current densities and pipeline surface area (see Table 10-6). The specific current density varies with the pipeline coating quality and integrity, as well as the soil conditions (Table 10-7). Improved coating requires less current density. Knowing the length and diameter of the pipe allows surface areas to be calculated.

The need for gas supply or gas production is such that future facilities are continually being planned. If these plans are known in advance, cathodic protection facilities can be designed to accommodate them. It costs much less to install one large ground bed than to install two smaller ones.

Once the decision has been made as to the requirement for cathodic protection, a suitable location for the ground bed must be selected. Again, a number of factors must be considered. These are:

- Electrical power availability
- Land availability
- Soil resistivities
- Future plans for expansion.

TABLE 10-6. Typical current density requirements for cathodic protection of bare steel

Environment	Current Density (mA/m^2)
Neutral soil	4 to 16
Well-aerated neutral soil	22 to 32
Wet soil	22 to 65
Highly acidic soil	32 to 160
Soil with active sulfate reducing bacteria	65 to 450
Stationary fresh water	11 to 65
Moving fresh water	54 to 65
Turbulent fresh water with entrained oxygen	54 to 160
Seawater	54 to 270

TABLE 10-7. Typical average current density requirements for cathodic protection of pipelines in soil

Coating Type	Average Current Density (mA/m^2)
Asphalt/coal tar enamel	300 to 1,500
Thin mil tapes	25 to 200
Extruded polyethylene	10 to 75
Fusion bond epoxy	25 to 100

To install electrical power in rural areas it usually costs \$10,000–\$15,000 per kilometer. Since the average cost of ground-bed installation is also \$10,000–\$15,000 (including power), a poorly located ground bed can needlessly increase the costs. Alternately, expensive power sources such as thermal electric generators, solar power, wind power, or fuel-driven generators may have to be considered if no suitable locations with electrical power can be found. Table 10-8 lists the various elements that need to be considered in order to determine life-cycle costs.

Generally, the wetter the ground-bed location, the better the ground-bed output. Since surface conditions do not always indicate the below-ground conditions, soil-resistivity measurements are used to indicate moist soil conditions. The lower the soil resistivity, the lower will be the required driving force (voltage) of the ground bed to affect cathodic protection over a set distance. In other words, the lower the soil resistivity, the greater the coverage for cathodic protection one can obtain from a single ground-bed installation. Ponds, drainage ditches, or natural water runoff courses are common locations for ground beds due to the low resistivity of their soils. A ground bed radiates lines of equal potential (Figure 10-6), the potential decreasing with distance from the ground bed. Where there is a hole in the coating, the equal potential lines decrease as the current from the ground bed flows to the bare steel.

Just because a suitable site has been selected does not automatically mean the final ground-bed site has been found. Some landowners do not allow pipeline companies to install facilities on their land, especially without a large monetary payment. These landowners must be identified early before too much planning has been completed. In circumstances where there is limited land availability, or congested pipelines, a suitable alternative can be to install deep vertical ground beds (NACE 1985). The anodes, usually of high silicon iron, are installed within the right of way, at depths from 15m up to 100m in an "open hole", (aqueous electrolyte) or "closed hole" (surrounded by backfill).

TABLE 10-8. Cathodic protection life-cycle costs

Capital cost	Design of facilities
	Purchase of materials
	Surveying land
	Taking soil resistivity measurements
	Purchase of land/rights of access
	Paying for clearing of forested land
	Installation costs
	Cost of power (transmission line or generator)
Maintenance & monitoring costs	Cost to verify cathodic protection meets an accepted criteria and is operating properly on a regular basis
	Cost to repair cathodic protection facility when it fails to operate as required
Operating costs	Cost of power
	Annual payments for land access
	Cost to water groundbeds in dry soil

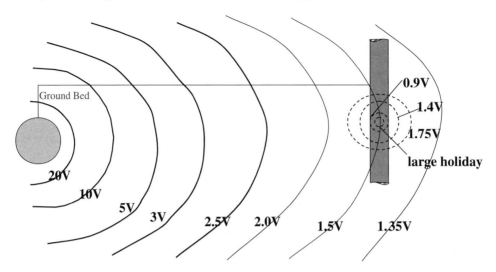

Figure 10-6. Equal potential lines radiating from a ground bed to a coated pipeline with a holiday [Wright, M. 1998]

Once the above information has been obtained, calculations are performed to determine the parameters for the required cathodic protection installation (ground bed). These calculations are iterative, repetitive, and very time-consuming, especially for a long pipeline. They can however be programmed for computer solution. The principles of cathodic protection calculations are detailed later in this chapter.

Design Example

The following is an example of the design for an impressed current cathodic protection facility. It is given that the estimated length of the pipeline is 38 km, the diameter is 500 mm, and the external coating will be fusion bond epoxy. Field experience and attenuation calculations confirm that in the area where the pipeline is to be installed, the maximum protective range is 44 km for the cathodic protection installation. If possible the ground bed should be located at km post 19. A suitable site for a cathodic protection installation was established at km post 20.8, based on soil resistance measurements. The soil resistivity at anode burial depth and below the anodes is less than 5,000 ohm-cm.

The current required after 20 years of service is estimated to be 0.00007 amps/m². Therefore the total current requirement for this 38-km section of pipe is 4.2 amps, which is calculated as follows,

$$\text{Total current requirement} = \text{current density} \times \text{total surface area } (\pi DL)$$
$$= 0.00007 \times \pi \times 0.5 \times 38000$$
$$= 4.2 \text{ amps}$$

The standard size of rectifier is 8 amps.

Silicon-chromium cast iron anodes will be used, each possessing an individual weight of 20 kg. Equation (10-1) can be used to determine the number of these anodes that will be required for a specified lifespan:

$$N = \frac{(\text{Life, years}) \times (\text{Size of Rectifier, amps}) \times (\text{Consumption Rate, kg/amp}-\text{year})}{(\% \text{ volume consumption}) \times (\text{Mass of anode, kg})}$$

$$(10 - 1)$$

In this case, a consumption rate of 0.45 kg per amp-year and a 60% volume consumption were assumed for the 20-year lifespan. Equation (10-1) indicates that six anodes will be required, i.e.,

$$N = \frac{20 \times 8 \times 0.45}{0.60 \times 20} = 6$$

The major components of the circuit resistance (R_T) are the anode-to-electrolyte resistance (R_A), the structure-to-electrolyte resistance (R_c) and the cable resistance (R_w).

There are various formulas available for determining R_c. The formula used is dependent on the ground-bed installation configuration. In this case, the anodes are installed horizontally at a depth of 2.5 m, the center-to-center spacing will be 4.5 m, and the canned anode diameter is 0.3 m. The equation for determining R_c in this example is as follows:

$$R_c = \frac{0.001588\rho}{L}\left[2.3\,ln\frac{4L^2 + 4L\sqrt{S^2 + L^2}}{DS} + \frac{S}{L} - \frac{\sqrt{S^2 + L^2}}{L} - 1\right] \qquad (10 - 2)$$

where
R_A = anode-to-electrolyte resistance, ohms (2.0 ohms)
R_c = structure to electrolyte resistance ohms
ρ = soil resistivity, ohm-cm (5,000 ohm-cm)
L = length of installation, m (6 anodes × 4.5m/anode = 27m)
D = canned anode diameter, m (0.3m)
S = twice the depth of anode, m (2.5m × 2 = 5.0m)
ln = natural logarithm

The structure-to-electroyte resistance (R_c) is estimated to be .05 ohms after 20 years.

$$R_c = \frac{0.001588}{27} \times 50\left[2.3\,ln\frac{4 \times 27^2 + 4 \times 27(5^2 + 27^2)^{0.5}}{0.3 \times 5} + \frac{5}{27} - \frac{(5^2 + 27^2)^{0.5}}{27} - 1\right]$$
$$= .05 \text{ ohms}$$

The cable resistance (R_w) is calculated by determining the length of positive and negative cable to be used, the wire size, and wire resistance. In this example 130 meters of positive and 50 meters of negative (wire size is AWG #4) is used. Therefore, R_w equals 0.15 ohms (180 m × 0.000833 ohms/m).

Adding R_A = 2.0 ohms, R_c = 0.05 ohms, and R_w = 0.15 ohms, the total resistance (R_T) in this example is found to be 2.2 ohms.

The rectifier voltage can be calculated using Ohm's Law, where V_{REC} = 2.2 ohms × 8 amps × 1.2 (design safety factor) = 21 volts. The manufacturer's standard size is 24 volts.

The easement length is determined as follows: The first anode needs to be remote from the structure, usually at least 100 m away. The installation was determined to require six anodes with a spacing of 4.5 m, so the total ground-bed length must be 127 m (round to 130 m).

The results of the design are:

- Size of rectifier (24 volts, 8 amps)
- Number of anodes required (6)
- Size of easement required (12 m × 130 m)
- Location of ground bed (km post 20.8)

Before the ground bed can be installed, additional work must be completed:

- A legal survey and land title search must be conducted
- The required easement must be requested
- The necessary electrical power has to be acquired

To ensure foreign companies, as well as the pipeline company, know where the ground bed is located, legal surveys and registered easements are essential. Their acquisition prevents many problems similar to those that have arisen in the past regarding new facilities being built through existing ground-bed facilities.

The rectifier must have power to function. To obtain electric power in the field takes upwards of three months or more. It is therefore imperative to apply for power as early as possible.

Cathodic Protection Installation

The installation and energizing of the ground bed is the final step. Again, a number of procedures must be followed.

Determining the Location of Easement

Often, the surveyor's stakes are still in the ground (see Chapter 9), which will make the exact determination of where the ground bed is to be installed very easy. If the stakes cannot be found, the legal survey plan of the ground bed must be consulted so that the correct location can be marked.

Laying Out of Materials and Excavation

The ground-bed materials are strung out along the proposed ditch to provide a marker for the backhoe operator as to the length of the ditch that is to be excavated. Anodes are usually buried at a minimum depth of 2.4 m and the cables are buried at a minimum depth of 0.6 m. The anode depths will vary according to the soil-resistivity measurements taken earlier. The cable depths must be deep enough to prevent damage by farmers working their fields.

Splicing the Anode to the Header Cable

The anodes are spliced to the main cable running from the ground bed to the rectifier by means of a mechanical clamp.

Recall from the earlier discussion that the current travels from the rectifier to the anodes via the anode header cable. Wherever the current leaves to go to the pipeline, corrosion occurs. If they are not electrically sealed, the splices would provide a spot for the current to leave; therefore, they must be well taped to ensure no current leakage.

Lowering Anodes into the Ditch and Backfilling

Safety regulations do not allow human entry into the deep anode ditches without extensive modifications, which would require extra time and money. Therefore, the anodes are lowered into the ditch by a method that does not require personnel to enter the ditch. Ropes are often used as guides and constraints for the lowering in. Once the anodes are positioned, the backfilling can commence.

Thermite-Welding the Negative Cable to the Pipeline

The negative cable running from the rectifier to the pipeline is connected to the pipe using mechanical clamps or a thermite-welding process. The finished thermite weld is very secure and requires a very substantial knock to remove it from the pipe. The welded connection is then taped to prevent damage.

Installing and Energizing the Rectifier

The last step to enable cathodic protection is to install the rectifier, connect the electrical power, and energize the ground bed. The cathodic protection will begin to work immediately. Potential readings are then taken along the pipeline and compared to the readings obtained before the cathodic protection was activated (the natural readings). Depending on the readings, the output of the rectifier can be increased or decreased. In order to properly conclude the ground-bed installation, the following requirements must be met: as-built drawings must be prepared, a detailed ground-bed description developed, and a detailed rectifier description prepared.

These records enable precise details of the installations to be recorded. This enables future location of the ground bed and a method of monitoring what type of materials are being used, and their effectiveness as sacrificial anodes.

Cathodic Protection Monitoring

Regularly excavating the pipeline is impractical, so corrosion technicians use special instruments to monitor the cathodic protection.

This is carried out by measuring the pipe-to-soil potential difference using a high-input impedance potentiometer with a reference cell (usually a saturated copper/copper sulfate reference cell) and some type of metallic connection to the pipeline—a test lead.

A widely accepted standard for cathodic protection states that the potential difference between the pipe and the soil shall be no more electropositive than -950 for a standard saturated $Cu/CuSO_4$ reference cell.

Also, a typical maximum allowable potential of -3,000 mV ($Cu/CuSO_4$) is set to prevent adverse effects on some coating types.

To assess the status of the cathodic protection along the pipeline, two wire test leads are placed at regular intervals, usually every 3 km, and at locations of possible concern—such as pipeline crossings, cased crossings, and river crossings. Two wires are attached as a safety precaution in case one of the wires becomes disconnected. The potential readings can be determined through an attachment to either wire.

Typically, a four-wire test lead is installed at insulation flanges, cased crossings, and pipeline crossings. Two wires are attached to each structure, which allow the monitoring of possible adverse effects from the foreign structure on the protected pipeline. If any effect is noticed, the test lead may be used to mitigate the problem.

Pipe-to-soil meters are permanently installed meters that monitor the cathodic protection. They are usually located at the ends of one pipeline system or in areas of high concern.

The meters allow nonspecialist technicians to easily monitor and record the cathodic protection status. All pipe-to-soil meters and rectifiers (volt and amp output) are monitored once a month.

All test leads are monitored at least once a year to determine the electrical pipeline potentials. This survey provides a more accurate determination of the status of the cathodic protection along the entire length of pipeline than the monthly pipe-to-soil meter readings.

Cathodic Protection Records

Copies of all surveys and checks are relayed to the corrosion prevention group and these records are reviewed to ensure no facilities are below the cathodic protection standards. Computer database systems have been developed to assist the review of these records by quickly highlighting the problem areas. Any such areas are then investigated.

Coating and Cathodic Protection

Transmission pipeline are generally protected from external corrosion by a dual system of cathodic protection and anticorrosion coating. The industry's practice is in some cases to specify and apply barrier coatings (such as urethane foams) over the anticorrosion coating at the coating mill or at the construction site.

Evaluation has shown that this practice i.e. the use of barrier coating on buried steel pipeline will not impede the performance of pipeline's cathodic protection system (Coulson et al. 1991).

Summary

Cathodic protection in conjunction with a coating system is a reliable and economic method of corrosion prevention. It functions by sacrificing a known material to prevent a buried structure (the pipeline) from corroding. Cathodic protection requires regular monitoring to ensure effective corrosion prevention. This review indicates that there are many factors to be considered when designing cathodic protection facilities.

CATHODIC PROTECTION CALCULATIONS FOR LAND PIPELINES

Attenuation

Symbols

R_s = linear pipe resistance, ohms per m
R_L = leakage (coating) resistance, ohms per m
α = attenuation constant, per kilometer
R_o = characteristic resistance, ohms
ρ = soil resistivity, micro ohm-cm
ρ^1 = steel resistivity, micro ohm-cm
g = coating conductivity, micro mhos per m^2
D = pipe diameter, mm
L = pipe length, kilometer
t = pipe wall thickness, mm
a = pipe cross-sectional area, square cm
A = surface area of pipe, m^2 per m
W = weight of pipe, kg per m
dE_O = change in potential at drain point, volts
dE_X = change in potential at point X, volts
dE_T = change in potential at end of line, volts
I_O = current at drain point, amps
I_X = current at point X on line, amps
I_T = current at end of line, amps.

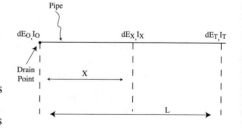

General Case

The equation for a general case can be written as follows:

(a) *Volts* $dE_X = dE_O \cdot \cosh(\alpha X) - I_A \cdot R_O \cdot \sinh(\alpha X)$

(b) *Amps.* $I_X = I \cdot \cosh(\alpha X) - \dfrac{dE_O}{R_O} \cdot \sinh(\alpha X)$

or

(c) *Volts* $dE_O = dE_X \cdot \cosh(\alpha X) + R_O \cdot I_X \cdot \sinh(\alpha X)$

(d) *Amps.* $I_A = I_X \cdot \cosh(\alpha X) - \dfrac{dE_X}{R_O} \cdot \sinh(\alpha X)$ $(10-3)$

Finite Lines

The equation for finite lines can be written as follows:

(a) *Volts* $dE_O = dE_T \cdot \cosh(\alpha L)$

(b) *Amps.* $I_O = \dfrac{dE_T}{R_O} \cdot \sinh(\alpha L)$

or

(c) *Volts* $dE_X = dE_T \cdot \cosh \alpha(L - X)$

(d) *Amps.* $I_X = I_O \cdot \dfrac{\sinh \alpha(L - X)}{\sinh(\alpha - L)}$ $(10-4)$

Infinite Lines

In almost all cathodic protection systems there are no infinite lines. However, the equation for infinite lines can be written as follows:

(a) *Volts* $dE_X = dE_O e^{-\alpha x}$

(b) *Amps.* $I_X = I_O e^{-\alpha x}$,   *or* $I_O = \dfrac{dE_o}{\alpha R_L} = \dfrac{dE_o}{R_o}$

(c) *Volts* $\text{Ln } dE_O = \text{Ln } dE_X + \alpha X$

(d) *Amps.* $\text{Ln } I_O = \text{Ln } I_X + \alpha X$ $(10-5)$

Linear Pipe Resistance

Steel resistivity varies from 16 to 23 micro ohms-cms. An average value is 18 micro ohms-cms. The equation for linear pipe resistance can be written as:

$$R_s = \rho^1 \frac{L}{A}(\text{ohm})/\text{m} \qquad (10-6)$$

Leakage (Coating Resistance)

For design purposes, the coating conductivity, g, of buried, nonfusion bonded epoxy coated landlines may be taken as 2,000 micro mhos per m^2. Fusion bonded epoxy coated lines can be considered to have a coating conductivity of 400 micro mhos per m^2. The equation for leakage (coating) resistance R_L can be written as follows:

$$R_L = \frac{1}{g\,A\,L}(\text{ohms per km}) \qquad (10-7)$$

Characteristic Resistance

The equation for characteristic resistance can be written as:

$$R_O = \sqrt{R_S \cdot R_L} \quad \text{ohms} \qquad (10-8)$$

Attenuation Constant

The equation for attenuation constant can be written as any of the following:

(a) $\alpha = \sqrt{\dfrac{R_S}{R_L}}$ per km

(b) $\alpha = \dfrac{R_O}{R_L}$

(c) $\alpha = \dfrac{R_S}{R_o}$ per km $\hspace{4cm}$ (10 − 9)

Galvanic Anodes Calculations

Symbols

I_a = anode current output, milliamperes
R_a = anode resistance to earth, ohms
R_C = cathode resistance to earth, ohms
E_a = open-circuit potential between anode and reference electrode, volts
E_c = open-circuit potential between cathode and reference electrode, volts
E_o = open-circuit potential between anode and cathode = E_a - E_c, volts
E_{cu} = closed circuit potential between cathode and reference electrode, volts
E_d = driving voltage between anode and cathode for protected system = E_a - E_{cu}, volts
W = weight per anode, kg
n = number of anodes
Q = weight of anode material consumed per unit current-time, kg per amp year
$I_{c.d.}$ = design current density for protection, milliamps per m^2
Y = design life of system, years
L = length of anode, cm
D = diameter of anode, cm
W_m = weight of anode material required, kg
A = surface area of structure to be protected, m^2
ρ = soil resistivity, ohm-cm.

Current Output

The equation for current output can be written as follows:

$$I_a = \frac{E_o}{R_a + R_c} = \frac{E_a - E_c}{R_a + R_c} \text{ milliamps} \hspace{2cm} (10 - 10)$$

For negligible cathode resistance, Equation (10-10) can be restated as:

$$Ia = \frac{E_d}{R_a} = \frac{E_a - E_{cu}}{R_a} \text{ milliamps}$$
$$= \frac{0.8}{R_a} \text{ for steel protected by Galvomag}$$
$$= \frac{0.6}{R_a} \text{ for steel protected by H - 1 alloy}$$

where

$$Ra = \frac{\rho}{2 \pi L}\left(\ln\frac{8L}{D} - 1\right)$$

Anode Life

The following equation assumes the anode will be replaced when it is 85% dissolved. It can be written as follows:

$$\text{Life} = \frac{71.4\ W_m}{I_a}\ \text{years} \qquad (10\text{-}11a)$$

For magnesium anodes the equation can be written as follows:

$$\text{Life} = \frac{108.7\ W_m}{I_a}\ \text{years} \qquad (10\text{-}11b)$$

Weight of Anode Material Required

The equation for the weight of anode material required is given by:

$$W_m = A \cdot I_{cd} \cdot Q \cdot Y\ (\text{kg}) \qquad (10-12)$$

where $I_{c.d.}$ = required current density, milliamps/m^2
Y = design life in years.

Number of Anodes Required

The equation for the number of anodes required can be written as follows:

$$n = \frac{W_m}{\text{Weight of Anode}} \qquad (10-13)$$

Impressed Current Anodes Calculations

Single Cylindrical Anode

The following formulae enable determination of the anode-to-earth resistance of a single cylindrical anode:

$$R_v = \frac{\rho}{2\pi L}\left(\ln\frac{8L}{d} - 1\right)$$

$$R_h = \frac{\rho}{2\pi L}\left(\ln\frac{4L^2 + 4L\sqrt{s^2 + L^2}}{ds} + \frac{s}{L} - \sqrt{\frac{s^2 + L^2}{L}} - 1\right) \qquad (10-14)$$

where R_V = resistance of single vertical anode, ohms
R_h = resistance of single horizontal anode, ohms
L = length of anode, including backfill, cm
d = diameter of anode, including backfill, cm
ρ = electrolyte resistivity, ohm-cm
s = twice the depth of anode, cm.

For a single vertical anode, the following simplified expression can be used:

$$R_V = \frac{\rho K}{L} \qquad (10-15)$$

TABLE 10-9. The shape function

L/d	K
5	0.0140
6	0.0150
7	0.0158
8	0.0165
9	0.0171
10	0.0177
12	0.0186
14	0.0194
16	0.0201
18	0.0207
20	0.0213
25	0.0224
30	0.0234
35	0.0242
40	0.0249
45	0.0255
50	0.0261
55	0.0266
60	0.0270

where R_V = resistance of single vertical anode, ohms
ρ = effective soil resistivity, ohm-cm
L = length of anode, cm
K = shape function, representing ratio of anode length/anode diameter (from Table 10-9)
L/d = ratio of length to diameter of anode.

Single-Row Vertical Anode Group

The total anode-to-earth resistance of a group of vertical anodes that are parallel and equally spaced in one row is given by the following equation:

$$R_n = \frac{1}{n}R_v + \frac{\rho}{S}P \qquad (10-16)$$

where R_n = total anode-to-earth resistance of vertical anodes in parallel, ohms
R_V = anode-to-earth resistance of a single vertical anode, from Equation (10-15), ohms
n = number of vertical anodes in parallel
ρ = soil resistivity, measured with pin spacing equal to S
P = paralleling factor, from Table 10-10
S = spacing between adjacent anodes, cm.

Multiple-Row Vertical Anode Group

An anode group composed of two or more rows of vertical anodes, separated by a distance substantially larger than that between the anodes within a single row, has a total resistance approximately equal to the total parallel resistance of all the rows. The formula generally used for paralleling resistance is:

$$\frac{1}{R} = \frac{1}{R_1} + \frac{1}{R_2} + \frac{1}{R_3} + \frac{1}{R_N} \qquad (10-17)$$

TABLE 10-10. Paralleling factor

n	P
2	0.00261
3	0.00289
4	0.00283
5	0.00268
6	0.00252
7	0.00237
8	0.00224
9	0.00212
10	0.00201
12	0.00182
14	0.00168
16	0.00155
18	0.00145
20	0.00135
22	0.00128
24	0.00121
26	0.00114
28	0.00109
30	0.00104

Cathodic Protection Calculation Example

Given the following information, for each two pipelines (of same length) constructed in parallel in the same trench, determine the cathodic protection current coverage requirement for Case A and B. Also, calculate the number of anodes required for a 20-year protection life and select the appropriate rectifier amperage plus voltage for each case.

Case A:
NPS 4, wall thickness 0.237 inch, length 4.5 miles
NPS 8, wall thickness 0.25 inch, length 4.5 miles
Case B:
NPS 6, wall thickness 0.156 inch, length 150 miles
NPS 12, wall thickness 0.250 inch, length 150 miles

Assumptions:
Case A:

1. Coating quality, resistivity = 500,000 ohm ft.^{2}=1/g
2. Drain point shift potential, E_o = 2.0 volt
3. Midpoint shift potential, E_x = 0.2 volt (1 rectifier)
4. Steel resistivity = 18×10^{-6} μohm/mile
5. Anode type = zinc

Case B:

1. Coating quality = 200,000 ohm ft.^{2}
2. Drain point shift potential, E_o = 2.0 volts
3. Midpoint shift potential, E_x = 0.15 volts
4. Steel resistivity = 18×10^{-6} μohm/mile
5. Anode type = zinc

Solution for Case A. NPS 12 line only, refer to Appendix E.

INTERNAL CORROSION

The presence of corrosive agents and conditions inside a pipeline can result in premature failure, loss of service, possible pollution, injury, property damage, or fatalities. Common causes of internal corrosion include:

- Hydrogen sulfide, which is a poisonous corrosive gas and forms sulfuric acid in the presence of water.
- Carbon dioxide, which forms carbonic acid in the presence of water.
- Water vapor present in a pipeline will condense to liquid water when cooled sufficiently. Liquid water in a pipeline will contain dissolved corrosive gases as acids and may also contain bacteria responsible for some corrosion reactions. High velocity flow conditions can remove protective oxide layers on the inside of a pipeline and accelerate the corrosion attack. Low velocity flow conditions can allow wetting of thepipe wall to occur as well as near-stagnant conditions, which are ideal for biological corrosion mechanisms.

Mitigating Internal Corrosion

The mitigation of internal corrosion involves the elimination of one or more of the ingredients essential to the corrosion reaction.

- Internal coatings and linings isolate the corrosive fluids from the pipe wall.
- Dehydration and separation of water from the process stream prevents the formation of acids by dissolved gases.
- The use of special chemical inhibitors results in a self-healing coating in the pipeline for protection from acids.
- Pipe materials resistant to the type of attack expected may be selected prior to construction.

Regular cleaning or pigging of the affected pipeline will remove any trapped condensed water and will break up any deposits that can harbor bacteria or produce stagnant conditions.

REFERENCES

Andrenacci, A., Wong, D., Mordaski, J.G., 1999, "New Developments in Joint Coating and Field Repair Technology," Materials Performance, February, pp. 33–43.

Coulson, K.E.W, Temple, D.G., 1983, "An Independent Laboratory Evaluation of External Pipeline Coatings," Paper as Presented at 5th Int. Conf on the Internal and External Protection of Pipes, Innsbruck, Austria October 25–27, BHRA Fluid Eng.

Coulson, K.E.W, Barlo, T.J., Werner, D.P., 1991, "Tests Show Barrier Coatings Do Not Block Cathodic Protection," Oil and Gas Journal, October 14, pp. 80–84.

National Association of Corrosion Engineers (1985), "Design, Installation, Operation and Maintenance of Impressed Current Deep Groundbeds". NACE RP-072-85.

Neal, D., 1999, "Good Pipe Coating Starts with Properly Prepared Steel Surface," Pipeline and Gas Industry, March 1999, pp. 38–43.

Wright, M., 1998, "Cathodic Protection Design Manual," Nova Gas International course notes.

Chapter *11*

PIPELINE INTEGRITY

INTRODUCTION

Pipelines continue to be the safest method of delivering gas and petroleum products across great distances and over all manner of terrain. The hundreds of thousands of major transmission pipelines throughout the world operating virtually around-the-clock, unseen to the general public, attest to this fact.

However, pipelines as an engineered facility do fail from time to time. The consequences of a catastrophic failure most certainly costs the operator additional money through repair and possibly loss of product throughput. Damage to the surrounding environment, especially in sensitive areas, can be very expensive. Most important, public safety must be maintained.

A safe, longlasting, and profitable pipeline is every operator's goal. Consequently, maintenance measures that are both cost-effective and prevent failures or high repair costs must be considered. In Canada alone there is an excess of 250,000 km of natural gas, crude oil, and petroleum product pipelines. In the world and up to 1999 there exists 1,700,000 km of such transmission pipeline Mohitpour (2000). In all of North America, over a half of the large-diameter pipeline system is more than 25 years old. With limited maintenance resources, it is essential that the available funds be spent where they are most effective in reducing the risks posed by pipeline failures to life, the environment, and financial assets.

Proactive engineering design will minimize but not completely eliminate the effect of these hazards.

Causes of Pipeline Failures

While the potential threats to pipeline integrity are numerous, they can generally be classified into five or six categories: general corrosion, contact damage, geotechnical, SCC, and other.

- General corrosion includes all forms of pipeline corrosion (internal, external, pitting, etc.)
- Contact damage is sometimes referred to as third-party damage (e.g., line strikes by earth moving equipment, etc.)

543

- Geotechnical failures refer to various types of ground or "right-of-way" movement (landslides, floods, etc.)
- SCC (stress corrosion cracking)
- Other causes of pipeline failure include material and equipment failure and operator error

Proactive engineering design and a vigilant right-of-way patrol program will minimize but cannot completely eliminate the effect of the hazards.

Reasons for an Integrity Program

Pipeline failures can have a significant impact on the public and the environment, which may lead to a serious effect on business. The cost of cleaning up after a spill and the cost of repairing the pipeline can be very high. Failures cause the pipeline to be shutdown, which halts delivery of the product. A shutdown may cause plants or manufacturing facilities at either end of the pipeline to close or use a more costly form of transportation, such as trucking.

Due to the public concern for safety and the impact on the environment, rigorous regulations are imposed on pipeline operators. The National Energy Board of Canada, for example, issued a strong recommendation regarding in-line inspection of pipelines following a number of pipeline ruptures across the country, and held a public inquiry into Stress Corrosion Cracking in Pipelines (1996). In the United States, Bill HR1489, the Pipeline Safety Act of 1991, states "all pipelines, gas and liquid, identified in environmentally sensitive and high density populated areas be inspected using smart pigs." A planned, responsible, integrity program may help avoid further stringent regulations being imposed on pipeline operators, and enable them to increase pipeline integrity without overly increasing pipeline operating and maintenance costs.

Planned pipeline integrity projects are required to minimize the impact of failure on the public and the environment and to avoid disruptions to operations.

Pipeline Integrity Management

Pipeline integrity management utilizes an engineering approach to develop programs to analyze, detect, evaluate, and reduce or eliminate the risks facing a pipeline.

The goal of a pipeline integrity management program is to prevent integrity problems from producing a significant impact on public safety, the environment, or business operations.

Pipeline integrity management is a four-phase program: pipeline assessment, inspection management, defect and repair assessment, and rehabilitation and maintenance management.

Pipeline Assessment

The foundation of a comprehensive integrity management program is the proper understanding of the hazards to which a pipeline is exposed as well as the associated business and safety consequences.

The hazards facing pipelines may be grouped into the following categories:

- Corrosion
- Stress corrosion cracking
- Soil type and instability
- Contact damage
- Material defects
- Construction practice
- Operating problems

The consequences of pipeline problems may be grouped into two categories; business and safety.

Business Consequences

System disruption with loss of:

- Revenue
- Product
- Reputation
- Prestige

Safety Consequences

- Fatalities
- Injuries
- Pollution
- Explosion
- Fire
- Property damage

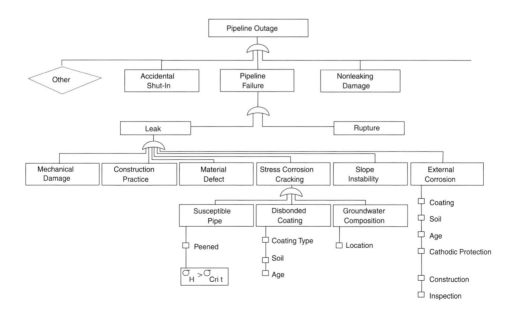

Figure 11-1. Partial fault tree for outage probability

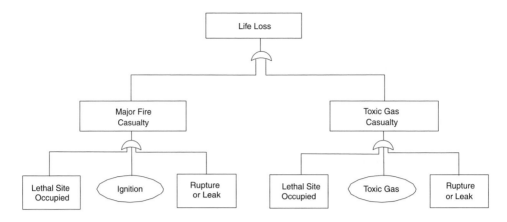

Figure 11-2. Schematic of safety risk assessment

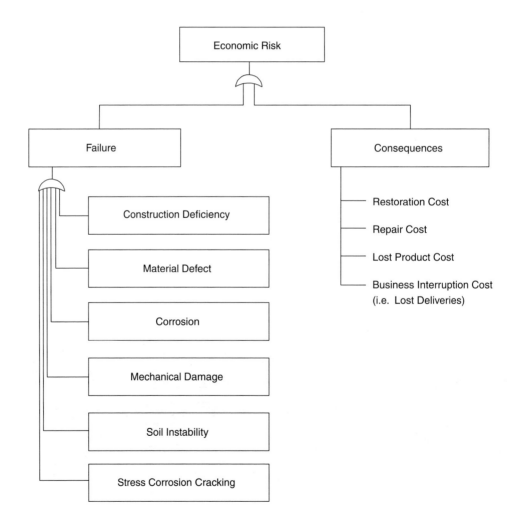

Figure 11-3. Schematic of economic risk assessment

Preventing hazards and reducing the consequences associated with pipeline failure is an active process and should be addressed during pipeline design, construction, and operation.

Risk assessment involves estimating both safety and economic risks. Fault tree analysis is a risk assessment tool used to estimate public safety and economic risks and to determine outage probability.

The probability of an outage caused by structural failure, which is one of the key factors required to assess safety and economic risks, is illustrated in Figure 11-1.

The data used to estimate outage probabilities are usually derived from the operating company's own engineering and operating records, data on pipeline characteristics, and failure statistics and are supplemented by industry data and experience. The level of detail in each branch of the fault tree may vary but should reflect an understanding of the deterioration process involved in structural failure. The fault tree method allows the contribution of each significant failure to the total outage probability to be estimated separately.

The probability of life loss due to a leak or rupture can be considered as:

Probability of life loss = (Probability of an ignited or toxic gas release) × (Probability of site being occupied)

The safety risk associated with deterioration of structural integrity depends largely on the physical location of the pipeline.

The fault tree used to estimate safety risks is illustrated schematically in Figure 11-2.

One of the most significant components included in the estimate of outage or failure consequences is the potential for business interruption costs due to failure to make deliveries.

The fault tree used to estimate economic risk is illustrated schematically in Figure 11-3.

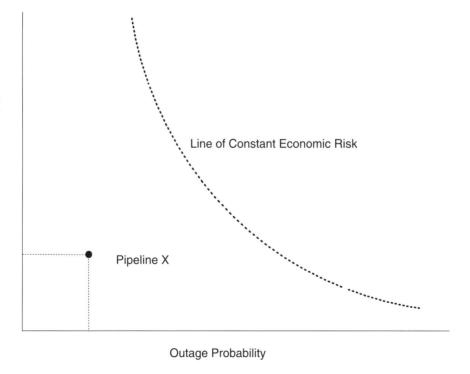

Figure 11-4. Economic risk components

The results of an economic risk assessment is illustrated in Figure 11-4, which shows that:

Economic Risk = (Outage Probability) × (Outage Consequence)

Inspection Management

There are a number of methods and tools available to determine the integrity of both new and existing pipelines. The most common in-line inspection tools, or "smart pigs," use conventional or high resolution magnetic flux leakage, or ultrasonic principles, to detect and locate structural anomalies in a pipeline. Other methods used to establish pipeline integrity include hydrotesting, coating surveys, magnetic particle inspection, and visual inspection. Geometry tools are also available to detect the presence of dents, ovality, and other diameter restrictions that would affect the integrity of the system and impede the passage of inspection tools.

Conventional Magnetic Flux Leakage Tools

The most widely used devices for corrosion detection are conventional magnetic flux leakage tools. These devices rely on magnetic field anomalies that are created by the presence of imperfections. A permanent magnet or a direct current electromagnet creates a longitudinally oriented flux field around the full circumference and throughout the wall thickness of the pipe. As the tool moves along the pipeline, a flux field is imposed on the pipe wall. The flux is contained within the pipe wall as long as it is free of imperfections. However, if the flux reaches an external or internal surface defect, the flux field is distorted outside the pipe wall. The distortion in the flux field induces an electric current or other signal in one or more groups of coils or detectors located between the poles of the magnet and arrayed around the circumference of the pipe (Figure 11-5). The resulting electrical analog signal and its location along the pipeline are recorded on magnetic tape. Later, a log of the analog signals of the pipeline is printed. This log is inspected and interpreted by the vendor and the indications are graded according to size.

The response of the conventional magnetic flux leakage tools to metal loss defects is highly dependent upon the depth of the defect (i.e., the degree of wall thickness penetration) and less dependent upon other dimensions of the defect. Therefore, a conventional magnetic flux leakage pig will not be able to detect the longitudinal length of an anomaly. Moreover, the shape of the defect is sometimes difficult to detect. A pit or gouge with a relatively sudden change in wall thickness will be easily detected, while one with a gradually changing profile is more difficult to detect. Therefore, uniform wall thickness reduction is not detectable except at the transition point.

The amplitude of the analog signal is used in direct correlation with the grading of the logs. The logs are usually graded using the following categories, according to body wall penetration:

Grade	Body Wall Penetration
Light	< 30%
Moderate	> 30% and < 50%
Severe	> 50%
Unknown	#

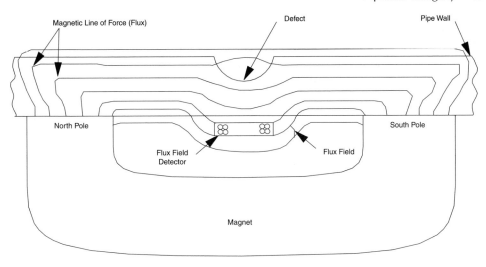

Figure 11-5. Magnetic flux leakage principle

The grading system is subjective and a distinction between internal and external indications cannot be made. Before the results are graded, the pipeline should be excavated and a number of indications located and measured to correlate the amplitude of the signal to the depth of the indication on the pipeline.

A grade of "unknown" is usually given to defects that are not due to corrosion but could consist of hard spots, expander marks, or magnetic material adjacent to the pipeline.

Conventional magnetic flux leakage tools do not have the capability to size the indications. Therefore, the indications have to be excavated and accurately measured prior to conducting defect assessment.

High Resolution Magnetic Flux Leakage Tools

High resolution magnetic flux leakage tools use the same principle of detection as the conventional tools, but offer a significantly improved ability to measure and represent the actual dimensions of metal-loss anomalies. This improved capability arises from the use of smaller sensors, a much larger number of sensors, a secondary bank of sensors to indicate which anomalies are internal or external, and a computer enhancement analysis of the data. These improvements enable accurate defect sizing by providing depth, length, width, and wall thickness profiles.

Excavation of the indication is not required for defect assessment or to correlate the signal. Defect assessment, including failure pressure calculations, can be performed based on the results of the high resolution tool. Results are usually presented in a digital format, versus the log format of the conventional magnetic flux leakage tools.

The cost of conducting a high resolution survey is approximately five to six times that of a conventional tool, although a portion of the added expense can be weighed against having to conduct correlation and assessment excavations.

Ultrasonic Tools

More recently, ultrasonic principles have been incorporated into in-line inspection equipment. The ultrasonic tool consists of a number of ultrasonic transducers mounted around its circumference. The ultrasonic transducer emits a high frequency pulse that is reflected from the inner and outer pipe walls. The time difference from when the signal leaves the transducer to when it reaches the inner and outer walls of the pipeline is used to calculate the sensor standoff distance from the wall and the wall thickness. The standoff indicates whether the corrosion is internal or external.

The actual wall thickness of the pipeline is obtained and the ultrasonic tool can detect wall thinning of as little as 10%. The tool must operate in a homogenous liquid environment to ensure that an acoustic couplant is present. In gas lines a liquid or gel slug can be put around the tool to act as the couplant. The tool must operate at relatively slow speeds (1 m/s maximum) to enable all data to be collected.

Geometry Tools

Geometry tools range from simple gauging plates mounted on cleaning pigs to detect the presence of dents or ovality to a predetermined threshold level without locating where they are, to tools that reliably size and locate dents and other diameter restrictions.

The most common geometry tool is the caliper pig, which contains a number of fingers around its circumference that deflect whenever they cross a dent or diameter restriction in the pipeline. This deflection is registered on a strip chart located in the tool. The amplitude of the deflection is used to size the diameter restriction. An odometer wheel is used to reference the location of the restriction from a known location. Other caliper tools that use ultrasonic transducers to detect diameter restrictions have been developed.

A tool has been developed recently that, as well as providing geometry information, will detect changes in the pipeline X-Y-Z position based on repeated runs using inertial guidance technology. This is very useful in monitoring pipelines in areas of unstable soils such as river valleys, permafrost areas, or in areas of mining subsidence.

Defect Assessment
Corrosion Damage Assessment

Corrosion damage in pipelines can be detected by various methods, but when it is discovered, the next concern is whether the pipeline is structurally sound enough to operate at the maximum allowable operating pressure (MAOP). Corrosion damage reduces the capacity of a pipe to contain internal pressure, and if the corrosion is allowed to proceed, the pipeline may eventually leak or rupture.

A number of analysis techniques have been developed to determine whether a defect will affect the pipeline's capability of operating at the MAOP. Prior to performing a defect assessment, the following data must be obtained:

Pipeline Data

- Diameter (nominal)
- Wall thickness (nominal or actual if available)
- Flow stress (pipe grade + 68,950 kPa)
- Maximum allowable operating pressure (MAOP)
- Zone or location class

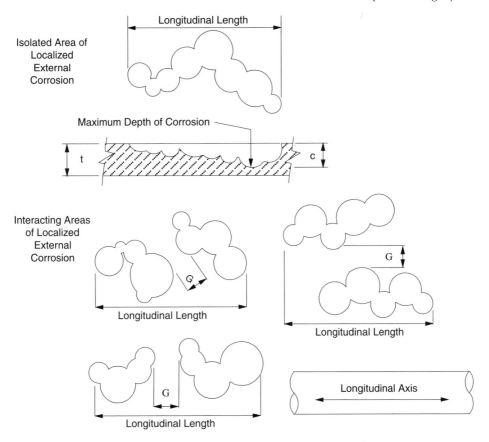

Figure 11-6. Method of deriving longitudinal length of localized external corrosion [CSA Z662, 2007] (Courtesy CSA International)

Corrosion Data

- Maximum depth or average depth depending on model used
- Length projected on longitudinal axis
- Distance between adjacent pits (required to test for interaction)

The Canadian Standards Association (CSA) Z662 states that: Localized external corrosion pitting to a depth of 80% of the nominal wall thickness of the pipe shall be permitted, provided that the longitudinal length of the corroded area, as derived from Figure 11-6, does not exceed the value given by the equation:

$$L = 1.12B \sqrt{Dt} \qquad (11-1)$$

where L = maximum allowable longitudinal length of the corroded area, mm
B = a value equal to 4.0 for maximum depths less than 17.5% of the nominal wall thickness, or a value obtained from Figure 11-7 for maximum depths between 17.5% and 80% of the nominal wall thickness
D = nominal outside diameter of the pipe, mm
t = nominal wall thickness of the pipe, mm
c = maximum depth of the corroded area, mm

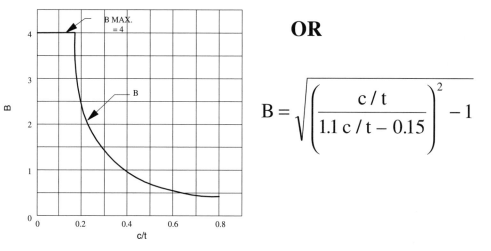

$$B = \sqrt{\left(\frac{c/t}{1.1\,c/t - 0.15}\right)^2 - 1}$$

Figure 11-7. Determination of B [CSA Z662, 2007] (Courtesy CSA International)

The corroded area must be thoroughly cleaned and the corrosion products removed so that the dimensions can be accurately measured.

Pipes with areas of localized external corrosion that exceeds the limits specified in the CSA Z662 equation must be repaired or replaced in accordance with the applicable clauses of the CSA or an equivalent standard.

NG-18 Surface Flaw Equation

In the early 1970s (Kiefner et al. 1973), the Battelle Institute was contracted by AGA to adapt equations derived by the NG-18 committee for use as a reliable method for evaluating corrosion damage. The basic form of the NG-18 surface flaw equation that evolved from the Battelle study is given as:

$$S_p = S \frac{(A_o - A)}{\left(A_o - AM^{-1}\right)} \tag{11-2}$$

where S_p = calculated hoop stress in undamaged pipe that causes failure in the corroded region, kPa

S = strength measure of the pipe called the "flow stress," kPa. (Yield strength + 68,950 kPa has been shown experimentally to be most representative.)

A = cross-sectional area of metal lost in the corroded region projected onto the longitudinal axis of the pipe, mm^2.

A_o = original uncorroded longitudinal cross-sectional area of the corroded region, mm^2.

M = Folias correction factor, a stress concentration or shape factor that accounts for the outward bulging that occurs at defects in thin-walled cylinders (Folias 1965). A good approximation of M is given in the following equations.

Three Term Folias Factor — For use when $L \leq 2R$

$$M = \sqrt{1 + 1.255 \frac{(L/2)^2}{RT} - 0.0135 \frac{(L/2)^4}{R^2 T^2}} \tag{11-3}$$

Rounded Corrosion - Parabolic Model

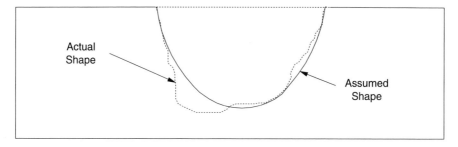

Irregular Corrosion - Rectangular Model

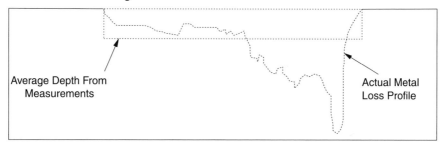

Figure 11-8. Corrosion models [Kiefner & Vieth, 1989]

Two Term Folias Factor — For use when $L > 2R$

$$M = \sqrt{1 + \frac{1.61(L/2)^2}{RT}} \qquad (11-4)$$

where T = pipe wall thickness, mm
              L = length of corroded region projected on the longitudinal axis, mm (Figure 11-6)
              R = pipe radius, mm.

The NG-18 surface flaw equation can be rearranged to solve for failure pressure.

$$P = \frac{ST}{R}\left(\frac{1 - A/A_o}{1 - (A/A_o)M^{-1}}\right) \qquad (11-5)$$

where P = failure pressure, kPa
              A = area of metal removed in pit, mm^2
              A_o = original area, mm^2.

The actual corrosion shape may be approximated by two methods (Figure 11-8). If the corrosion is rounded in shape, the metal loss may be approximated using the parabolic model equation:

$$P = \frac{ST}{R}\left(\frac{T - 2/3D}{T - 2/3DM^{-1}}\right) \qquad (11-6)$$

where P = failure pressure, kPa
          S = flow stress, kPa
          R = pipe radius, mm
          T = wall thickness, mm
          L = pit length, mm
          D = pit depth, mm
          M = Folias factor.

This model assumes the corrosion cross-section to be parabolic in shape, with a maximum depth equal to the measured maximum pit depth. Maximum pit depth and axial length are the corrosion measurements required for using the parabolic model.

If the corrosion is irregular in shape, the parabolic model will either overestimate or underestimate the actual amount of metal loss. In this case, the rectangular model equation should be used to obtain a more accurate assessment:

$$P = \frac{ST}{R}\left(\frac{T-D}{T-DM^{-1}}\right) \qquad (11-7)$$

This model assumes the corrosion cross-section to be a rectangular notch with a depth equal to the average depth of corrosion. The average depth must be determined by obtaining sufficient depth measurements inside the corroded region and assessing them to provide an accurate representation of the actual amount of metal loss. Depth measurements at five

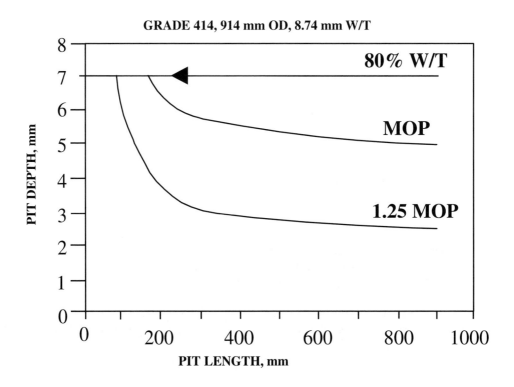

Figure 11-9.   Critical corrosion geometry curve, parabolic model

millimeter intervals along the corrosion length has been found to besatisfactory for this purpose. Average depth and axial length are the corrosion measurements required for using the rectangular model.

Calculation of maximum allowable pit depth provides a sensitivity check on the amount of additional metal loss that would be required before a failure will occur. The reasoning for using the parabolic or rectangular models in these calculations are the same as described above.

$$Y = 1.5T \left(\frac{P - \frac{SR}{R}}{\frac{P}{M} - \frac{ST}{R}} \right) \quad \text{for parabolic model} \qquad (11 - 8)$$

$$Z = T \left(\frac{P - \frac{ST}{R}}{\frac{P}{M} - \frac{ST}{R}} \right) \quad \text{for rectangular model} \qquad (11 - 9)$$

where
Y = maximum pit depth, mm
Z = maximum average pit depth, mm
P = test or operating pressure, kPa
S = flow stress, kPa = σ_y + 69, MPa
R = pipe radius, mm
T = pipe wall thickness, mm
M = Folias factor
σ_y = yield stress.

Using these analysis techniques, it is possible to generate critical corrosion geometry curves for any desired pressure.

Figure 11-9 shows critical corrosion geometry curves for a pipeline at the MAOP and the minimum hydrostatic test pressure (1.25 MAOP). As the pressure increases, the rejection criteria become more severe and the tolerable defect size decreases, shifting the curve down and to the right.

ASME B31G Criterion

The ASME B31G criterion for assessing corroded pipelines was also established on the basis of the state of the art in the early 1970s. It is well known to be overly conservative, except perhaps in cases of long shallow corrosion. As a result, rehabilitation efforts have led to the removal of large amounts of quite serviceable pipe. Kiefner and Veith (1989) have suggested that such wasteful effort can be avoided if attention is paid to the following factors, which cause the excess conservatism:

- The expression for flow stress (1.1 × SMYS in B31G)
- The approximation used for the Folias factor
- The parabolic representation of the metal loss (as used within the B31G limitations), primarily the limitation when applied to long areas of corrosion
- The inability to consider the strengthening effect of islands of full thickness or near full thickness pipe at the ends of or between arrays of corrosion pits

The Canadian code Z662, 03 (2003), uses as a value of flow stress for line pipe material the findings of Kiefner et al. (1973): yield stress + 69 MPa (10,000 psi). The addition of a third term to the Folias factor Equation (11–3) for use when $L \geq 2R$ or $L^2/Dt \leq 50$ is less conservative than the two-term factor given in Equation (11–4) and used in B31G. One should be careful when considering very long anomalies since the negative term in

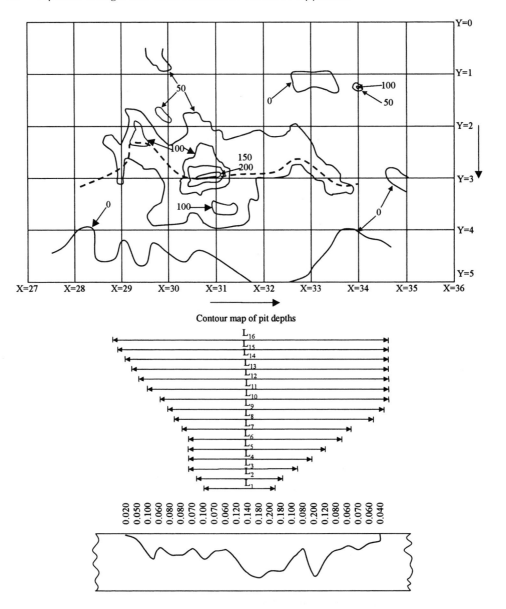

Figure 11-10. Contour map and profile of pit depths along "River-Bottom" path (dimensions in inches) [Kiefner and Vieth 1989]

Equation (11–3) will start to dominate and the equation will become invalid. Therefore, an alternative form is proposed for use when $L2/Dt \geq 50$;

$$M = 0.032\left(L^2/Dt\right) + 3.3$$

The third element of conservatism concerns the representation of metal loss as either being of a rectangular or parabolic shape. While the parabolic shape is the least conservative of the two, it also significantly underestimated the actual failure stress in 47 burst tests, as reported by Kiefner and Duffy (1971). Clearly more realistic representations of metal loss will be obtained if the contours of the pit profile were to be accurately measured. Figure 11-10 depicts both contour map and the "river bed" profile of a pitting fracture. By subdividing the depth of the fracture into 16 discrete intervals and measuring the length of the pit at each discrete depth, the area of metal loss can readily be found from summing the 16 trapezoidal areas. The average depth is found knowing the total metal loss area and the sum of the individual 16 lengths, L_{total}. This total length is used in Equation (11–3) to determine M which in turn is used in Equation (11–2) to determine the hoop stress at failure. Note that in Equation (11–2) the term A/A_0 becomes d_{avg}/t, where d_{avg} = pit depth average.

RSTRENG Technique

An even more accurate means of predicting the failure stress was the development of a computer-based tool, RSTRENG, as part of the PR3-805 project sponsored by the American Gas Association (Kiefner and Vieth (1989)). The basis for RSTRENG is the multiple evaluation of the predicted failure pressure based on an affected area rather than the total area as determined previously and given by $L_{total} \times d_{avg}$. Each calculation involves determining the area of metal loss beneath a particular length L_i and is slightly larger than what would be calculated using $L_i d_{avg}$ since at the end points the values for d are now zero. This approach is tedious and as an alternative Kiefner and Vieth (1989) suggest that an effective area based upon the maximum length, L, and maximum depth of pitting, d, be used. This area is given by

$$A = 0.85 \, dL$$

The above can be used in Equation (11–2) together with the modified Folias factor to calculate the predicted failure pressure. Tables 11-1 and 11-2 provide a comparison of flow stress and Folias factor for B31G and RSTRENG for several grades of pipe and flaw lengths (Haupt and Rosenfeld 1999).

The proposed modifications to the B31G criterion by Keifner and Vieth (1989) eliminate some of the conservatism in the criterion; however, a number of ambiguities remain. Metal loss is projected onto a longitudinal plane for evaluation. When corrosion is not oriented longitudinally, the B31G will tend to underestimate the remaining strength in the case of spiral corrosion by up to 50% (Mok et al. 1991). In addition, only circumferential stress due to pressure (perpendicular to the projection plane) is considered, making it difficult to evaluate the effect of longitudinal stresses due to endloads and bending. When corrosion pits are closely spaced, they must be treated either as contiguous or totally separate. There is no provision for the intermediate case where adjacent corrosion pits interact.

Corrosion geometries that fall below the 1.25 MAOP curve (hydrostatic test pressure curve) are not usually considered detrimental to pipeline integrity, and repair to the pipe coating is all that is required.

Corrosion geometries that fall above the hydrostatic test pressure curve are considered detrimental to the pipeline integrity and must be removed or repaired.

TABLE 11-1. Comparison of flow stress-B31G versus RSTRENG (Haupt and Rosenfeld 1999)

Pipe Grade API (SMYS)	B31G 1.1 SMYS (psi)	RSTRENG SMYS + 10,000 psi	Percent Increase from B31G to RSTRENG Definition (%)
5L GR.B (35,000 psi)	38,500 psi	45,000 psi	16.9%
X52 (52,000 psi)	57,200 psi	62,000 psi	8.4%
X60 (60,000 psi)	66,000 psi	70,000 psi	6.2%

TABLE 11-2. Comparison of folias factor-B31G versus RSTRENG 12.75 × 0.375, B (Haupt and Rosenfeld 1999)

Flaw Length (inches)	B31G Folias Factor	RSTRENG Folias Factor	Percent Decrease from B31G to RSTRENG Definition	Corresponding Increase in Predicted Failure Pressure (psi[1])
2	1.292	1.234	4.5%	13.7%
5	2.277	2.047	10.1%	7.7%
9	3.815	3.265	14.4%	4.6%
12	N/A	4.103	—	53.9%

[1] Based on a maximum depth of corrosion of 200 mils (53% wall loss).

The defect dimensions can be plotted on the critical corrosion geometry curves to determine which repair methods are required. Figure 11-11 also shows typical repair curves for a specific pipeline.

Mechanical Damage Assessment

Canadian Standards Association Code addresses the subject of dents and other mechanical damage in pipelines. Generally, the code requires that plain dents greater than 6% of the outside diameter (for pipes larger than 101.6 mm OD) be repaired by either cutting out or using full encirclement welded split sleeves. Dents that contain stress concentrators such as a gouge or arc burn, or dents that are deeper than 6 mm andlocated on a weld, must be repaired by either cutting out or using full encirclement welded split sleeves.

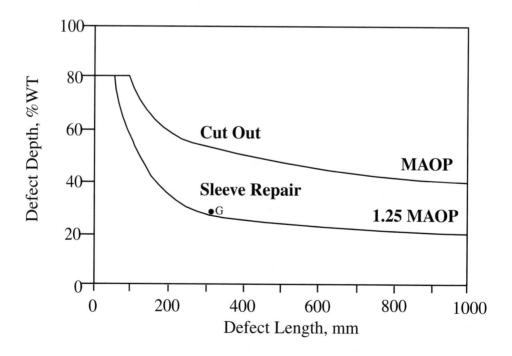

Figure 11-11. Defect assessment curve (NG-18)

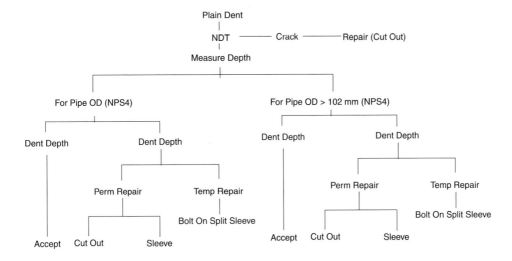

Figure 11-12. Plain dent damage assessment flowchart

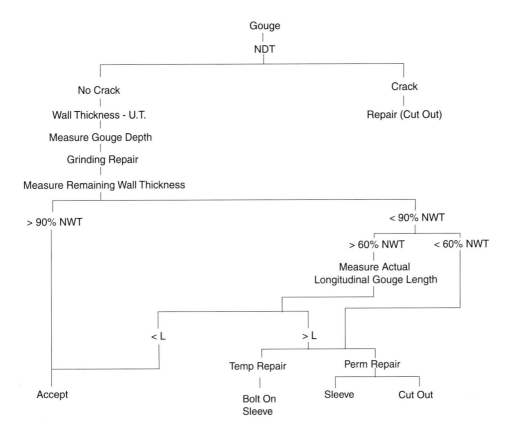

Figure 11-13. Gouge damage assessment flowchart

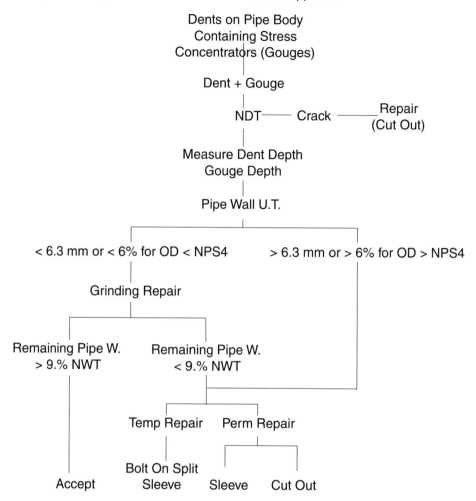

Figure 11-14. Dent on pipe body containing stress concentrators damage assessment flowchart

Figures 11-12 through 11-16 contain damage assessment flowcharts for dents and other mechanical damage.

Repair Assessment

There are basically three methods of repair used to fix corrosion damage and restore the integrity of the pipeline.

Coating Repair

Repairs to pipe coating should be undertaken to arrest the corrosion growth when the corrosion damage is of a minor nature and does not require sleeving or cut-out. The products of corrosion should be thoroughly removed from the pit and the old coating removed from the circumference of the pipe. New coating that is compatible with the existing coating should be reapplied to the pipeline. Repair will ensure that corrosion growth does not continue to the point that it becomes a major problem.

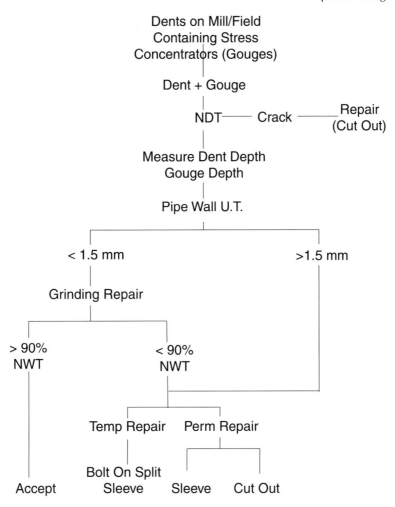

Figure 11-15. Dent on weld containing stress concentrators damage assessment flowchart

There are also methods to rehabilitate a pipeline by completely recoating it by using line travel equipment.

Sleeve Repair

Full encirclement sleeves are used to repair pipelines as an alternative to taking them out of service and cutting out the affected area. The sleeves are manufactured in two parts and are placed over the pipeline and welded to each other along the longitudinal seam. The defect should be cleaned of the corrosion products and a suitable filler material should be used to fill the void between the carrier pipe and the sleeve.

Epoxy-filled sleeves that allow epoxy to be injected under pressure into the space between the pipe and the sleeve have been developed. The ends of the sleeve are first sealed with a fast curing epoxy to contain the epoxy injected under pressure.

Full encirclement sleeves can be of the pressure-containing or non-pressure-containing type. The pressure-containing sleeves require that the ends of the sleeve bewelded to the pipeline around the circumference. Qualified welding procedures are required to ensure that

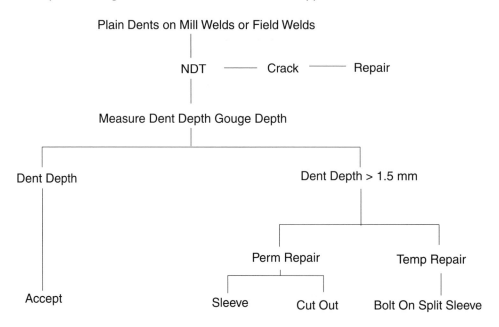

Figure 11-16. Plain dent on weld damage assessment flowchart

burn-through or hydrogen-induced cracking does not occur in the fillet weld or carrier pipe. The non-pressure-containing sleeves do not have their ends welded to the pipeline, but they do provide structural support to the affected area.

Cut-Outs

Corrosion that is greater than 80%, or corrosion that does not meet the requirements of applicable codes, and certain other defects such as laminations and cracks, require removal from the pipeline. The affected area must be removed as a cylinder and a new piece of pipe welded in place.

The pipeline section containing the defect must be depressured and the product either blown-down, displaced with an inert product, or drained. Qualified welding procedures are required and mud plugs, vapor plugging tools, or other methods should be used to ensure that no hydrocarbon vapors are present in the weld area. Pretested pipe must be used and the tie-in welds should be x-rayed to ensure they are satisfactorily completed.

Alternate methods of repair that eliminate the need for draining or blowing down the pipeline are also available. These include the use of stopples or remotely activated pipeline packers.

Examples of Corrosion Assessment
Pitting Corrosion Using NG-18 Surface Flaw Equations

NPS 16 gas pipeline: Outside diameter - 406 mm

- Wall thickness 6—mm
- Grade 359—MPa
- MAOP 8,453—kPa.

Three bell-hole locations were excavated on this pipeline. A single joint of pipe was exposed at each location. The coating was removed and the pipe sandblasted clean. The field supervisor has reported that pitting corrosion has been discovered at all three bell-hole locations, with none of the pits interacting.

A technician measuring the length and maximum depth of the most severe defect at each bell-hole location found the following:

	Defect Length	Maximum Depth
Bell-hole 1:	200 mm	1.5 mm
Bell-hole 2:	30 mm	4.6 mm
Bell-hole 3:	180 mm	4.5 mm

Determine the appropriate repair method to perform at each of these three exposed areas of pipe before they are backfilled.

Maximum Allowable Pit Depth

Suppose the defects described above were not detected from the bell-hole excavation and the corrosion engineer estimates that the corrosion growth rate for this pipeline has been conservatively estimated to be 0.4 mm per year. How many more years of active corrosion growth would be estimated using NG-18 surface flaw equations to reach a failure pressure equal to MAOP?

General Corrosion Using NG-18 Surface Flaw Equations

For the same NPS 16 pipeline, an additional bell-hole was excavated and a single joint of pipe was exposed. The coating was removed and the pipe sandblasted clean. The field supervisor reported that there is a 0.5-meter section of pipe that looks severely corroded. The corrosion is continuous and interactive for the entire 0.5-meter length.

A technician recorded the following measurements:

	Defect Length	Maximum Depth	Average Depth
Bell-hole 4:	520 mm	3.3 mm	2.1 mm

Determine the appropriate repair method

Damage Assessment Flowcharts

A dent was discovered on the body of a NPS 24 609 mm by 9-mm-wall-thickness pipeline. The depth of the dent was measured to be 25 mm. What should be the recommended repair method?

On the same NPS 24 pipeline, another dent measuring 60 mm was discovered. It was apparently caused by a backhoe. A gouge was also discovered that was 215 mm long, with a depth of 25% wall thickness. What should be the recommended repair method?

NDT testing of a weld revealed a small crack on the inside seam. What should be the recommended repair method?

A gouge was discovered on an NPS 12 pipeline. The maximum depth was 27% of the pipe wall and 25 mm long. What should be the recommended repair method?

REFERENCES

ASME, 1991, "B31G - Manual for Assessing Remaining Strength of Corroded Pipes," American Society of Mechanical Engineers, New York, NY.

CSA Z662 - 03, 2007, "Oil and Gas Pipeline System," Canadian Standards Association, Etobicoke, Ontario, Canada, p. 344.

Folias, E. S., 1965, "An Axial Crack in a Pressurised Cylindrical Shell," International Journal of Fracture Mechanics, 1, pp. 104–113.

Haupt, R., and Rosenfeld, M., 1999, "ASME B31-8 Gas Transmission and Distribution Piping Systems," Course Notes, ASME Professional Development Program, Houston, TX.

Keifner, J. F., and Duffy, A. R., 1971, "Summary of Research to Determine the Strength of Corroded Areas in Line Pipe," Presented at Public Hearing of U.S. Department of Transportation, July 20.

Kiefner, J. F., Maxey, W. A., Eiber, R. J., and Duffy, A. R., 1973, "Failure Stress Levels of Flaws in Pressurised Cylinders," Progress in Flaw Growth and Fracture Toughness Testing, ASTM STP 536, American Society for Testing and Materials, pp. 461–481.

Kiefner, J. F., and Vieth, P. H., 1989, "A Modified Criterion for Evaluating the Remaining Strength of Corroded Pipe," Final Report on Project PR3-805, Battelle Memorial Institute, Columbus, OH.

Kiefner, J. F., and Vieth, P. H., 1990, "New Method Concepts Criterion for Evaluating Corroded Pipe," Oil and Gas Journal.

Mohitpour, M., Dawson, J., Babuk, T., and Jenkins, A., 2000, "Concepts for Increased Natural Gas Supply — A Pipeline Perspective," Proc. 16th WPC, Calgary, Alberta, June 11–15.

Mok, D. H. B., Pick, R. J., Glover, A. G., and Hoff, R., 1991, "Bursting of Line Pipe with Long External Corrosion," International Journal of Pressure Vessels and Piping, 46, pp. 195–216.

National Energy Board (1996). Report of the Inquiry — Stress Corrosion Cracking on Canadian Oil and Gas Pipelines MH-2-95. November. Calgary, Alberta, Canada.

Chapter 12

<hr style="border-top: 4px solid black;" />

SPECIALTY FLUID TRANSMISSION

INTRODUCTION

This section discusses the technical considerations and facilities requirements for specialty fluid transmission. Pipeline design for specialty fluids such as carbon dioxide, hydrogen, high vapor pressure liquids, and slurries is not as common as pipeline design for oil and natural gas. This chapter will provide information specific to the design, construction, and operation of specialty fluid transmission systems and associated facilities.

Transportation of Slurry by Pipeline

Mixtures of solids and liquids have been transported in flumes and pipes for several centuries. The majority of applications over the past decades have involved in-plant transfers over relatively short distances. Only since 1957 has the slurry pipeline method been used to transport solids on a large scale over considerable distances. The first long-distance slurry pipeline was the 175 km coal slurry pipeline, which began operating between Cadiz and Cleveland, USA, in 1957. During the next 20 years, more than 1,500 km of pipelines capable of carrying 30 million tonnes of solids per year have been built.

Pipeline transportation for moving solid/liquid particulates and mixtures over long distances is now well-established (Guran 1985). The mining industry uses dredges to remove overburden and to process placer gold deposits. This industry also transports tailings through pipelines to disposal ponds or to backfill sites in mines.

A novel combination of directional drilling and pipeline slurry transportation was employed by Texas Industries when constructing a sand and gravel mine operation in Oklahoma (Davis 1999). The pipeline consists of a 4,000-foot-long continuous loop of 16 inches polyethylene pipe through which a gravel/water slurry is pumped at a rate of 15 feet per second in order to keep the gravel in suspension. The gravel mine is situated on a river bank but the market for the product lies on the other side of the river. The use of directional drilling to create two parallel 20 inches-diameter cased crossings reduces the distance between the mine and the distribution plants by approximately 42 miles. Considering that 12 million pounds of gravel must be transported per day, the pipeline alternative is also environmentally superior because it avoids more than 10,000 truck miles per day.

The construction industry uses dredges to recover and process sand and gravels. It also uses pumps and pipelines to transport sands, gravel and concrete to points of placement such as banks, dikes, embankments, and dams for large construction projects. Chemical, petroleum, metallurgical, pulp, and paper and other industrial plants use pipelines to transport raw and processed materials and finished products.

The following identifies important design parameters specific to slurry pipeline design. All other parameters can generally be based on single-phase liquid pipeline design parameters.

Classification

Slurry pipelines can be classified as follows:

1. In-plant lines
2. Short-distance pipelines
3. Long-distance transmission systems

In-plant lines are very short lines that move material from one processing stage to another either horizontally along the plant floor, vertically between floors, or from the bottom of one piece of equipment to the top of another (such as a tank). Usually, in-plant lines operate under fluctuating conditions and exhibit fairly wide operating capacities. Short-distance pipelines (less than 10 km in distance) are generally pipelines that carry materials to or from processing plants. They are usually limited to the distance over which a single pumping facility system is capable of conveying a slurry mixture. Long-distance pipeline transmission systems are lines that move material over distances of 10-20 km or greater and which require more than one pumping facility. An example of long-distance transportation of slurry is the transportation of a coal/water mixture, as shown in Figure 12-1, Wasp et al. (1977).

A coal-water mixture is probably the best known type of slurry because of its history and its immense potential. Slurry pipelining first drew attention in 1951 when Consolidation Coal, USA, began a study on moving coal by pipeline. The work culminated in 1957 with the construction of the celebrated 175 km, NPS 10 coal pipeline in Ohio from Cadiz to Cleveland. This pipeline operated successfully for six

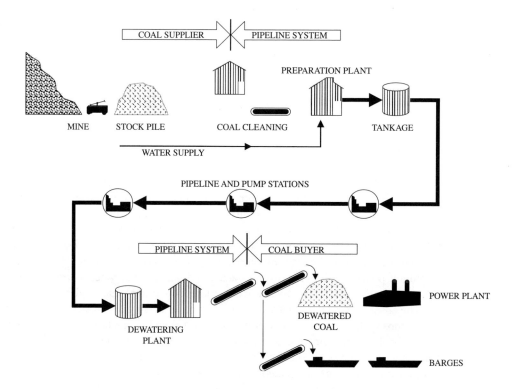

Figure 12-1. Schematic diagram of a long-distance slurry pipeline system [Wasp et al., 1977]

years and forced the railways to lower freight rates by one half, which eventually caused the mothballing of the pipeline itself.

The success of the Cadiz-Cleveland coal pipeline was followed by the construction of the Black Mesa coal slurry pipeline in 1970, carrying up to 4 million tonnes per year of coal over a distance of 440 km from coal mines in Arizona to the Mohave power station in Nevada.

The economic advantage of transporting coal through pipelines (in slurry form) created a tremendous global interest, such that now there are possibly over 4,000 km of coal pipeline in operation worldwide (refer to Table 12-1) (Wasp et al., 1977).

Advantages of Slurry Pipelines

Advantages that are connected with the transportation of slurry by pipelines are:

1. Commercial
   - High efficiency of transportation of large volumes over long distances
   - High reliability
   - Low inflation
   - Relatively strike immune
2. Geographical flexibility
   - Suitable for all terrain
   - Can be built at the location of a processing plant
   - Can take into account social considerations
3. Environmental protection
   - Buried
4. Little land used
   - Can take into account social considerations
5. Safety
   - Buried
   - Automated

TABLE 12-1. Selected world slurry pipelines (Wasp et al., 1977)

System	Slurry Material	Length (km)	Diameter (NPS)
Black Mesa	Coal	273	18
Calaveras	Limestone	17	7
Bougainville	Copper concentrate	17	6
West Irian	Copper concentrate	69	4
Pinto Valley	Copper concentrate	11	4
Tasmania	Magnetite concentrate	53	9
Waipipi (land)	Magnetite concentrate	4	8
Waipipi (offshore)	Magnetite concentrate	1.8	12
Pena Colorada	Magnetite concentrate	30	8
Nevada Power — Utah/Nevada	Coal	180	24
Energy Transportation System Inc. — Wyoming/Arkansas	Coal	1,036	38
Sierra Grande	Magnetite and Hematite	20	8
Brazil	Magnetite and Hematite	250	20
Mexico	Magnetite and Hematite	17	10
Brazil	Phosphate	71	10

Disadvantages and Limitations of Slurry Pipeline

The main technical limitations of long-distance slurry transmission pipelines are:

- Low turndown (i.e., flow rates cannot be varied widely)
- Fixed entry and delivery points
- Required water/liquid supply
- Required water/liquid cleanup, disposal, and return
- Limitations on the amount and type of materials conveyed
- High capital, low variable costs

Due to the nature of the flow regime in slurry pipelines, throughput cannot be varied widely. To avoid blockage, a certain minimum flow velocity has to be maintained to keep the solids in suspension. On the other hand, the higher the velocity, the greater the pumping power costs and the greater the erosion rates on the pipe walls. Thus, slurry pipelines usually have low turn-down ratios.

Compared to truck and rail transportation, slurry transmission pipelines have very limited distribution capability. Only a few intermittent delivery points can exist along a major line.

The supply of the liquid at the initiation of a pipeline system may be scarce and, in most cases, has to be separated from the slurry at the discharge point and treated before being disposed or returned. Such limitations can add to the cost of operation.

There are also limitations on the types of materials that can be transported as slurries through a pipeline. The material has to be compatible with the carrier fluid and amenable to separation from the carrier fluid at the pipeline terminal or somewhere in the subsequent processing or use of the material.

The following lists some of the products that are most suitable for water slurry:

- Coal
- Copper concentrates
- Copper-nickel concentrate
- Kaolin
- Mine tailings
- Gravel
- Tar sands
- Magnesium nodules
- Iron concentrates
- Copper ore
- Limestone
- Phosphate
- Sand
- Clay
- Wood pulp
- Food products

Design Considerations
Pipeline Flow Regime

The flow of a mixture of solids and liquids (slurry) in a horizontal pipe can be homogeneous, pseudohomogeneous, or heterogeneous as illustrated in Figure 12-2. In homogeneous flow, the solids are distributed uniformly over the pipe cross-section. Examples of such slurries are thick drilling mud or slurries of nonsettling fine clays. In pseudohomogeneous flow, the solids are not distributed uniformly over the pipe cross-section, but the flow still behaves as if it were. The third case involves segregated flow, where the solids settle to the bottom of the pipe and move along it as a sliding bed. Slurries of sand and gravel mixtures or coarse tailings can exhibit this behavior. There can also be a number of other flow conditions in between these limiting cases.

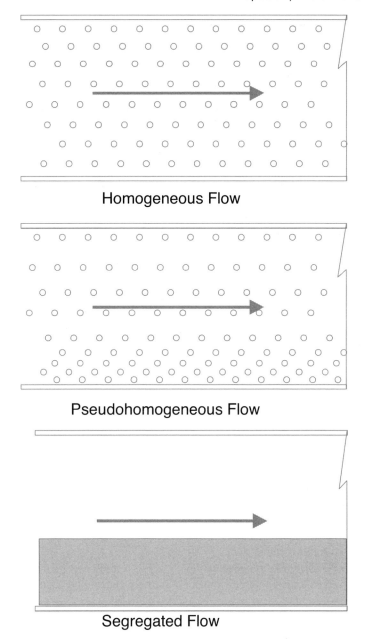

Homogeneous Flow

Pseudohomogeneous Flow

Segregated Flow

Figure 12-2. Three limiting cases of slurry flow in a pipeline

Slurry flow in pipelines can be altered by a change in flow parameters along the pipeline or by changes in the nature of the terrain, environment, etc. For example, if the flow rate or concentration is increased or the particle size is reduced, segregated flows can be changed to near-pseudohomogeneous flows. Conversely, if the flow rate is reduced sufficiently, a homogeneous flow could become a segregated flow. Further flow reduction can lead to a stationary bed and possible blockage of the pipeline.

The following considerations should be noted when designing slurry pipelines (Carleton and Cheng 1974):

- The minimum velocity in the pipeline at which solids just begin to settle out. (Typical value for coal water slurry is about 1.2–1.3 m/s)
- The maximum velocity that can be tolerated as far as erosion of the pipe wall or attrition of the solids are concerned. This velocity is about 5 m/s, depending on particulate size.
- The type of feeder, pump, valves, and instruments that are acceptable and compatible with the pipeline operations.
- The maximum slope that the pipeline can tolerate.

The flow of fully homogeneous mixtures can be laminar or turbulent because the mixture will not segregate under any circumstances. Pseudohomogeneous mixtures should be pumped at velocities that create turbulent flow to ensure the solids remain in suspension and the pseudohomogeneous character is maintained.

A pipeline flowing nonhomogeneous mixtures may contain more solids in the bottom half of the pipe than in the top half. Since it is not economic to operate a partially blocked line, segregated flow needs to be maintained such that all the solids are moving. Therefore, pipelines that transport heterogeneous mixtures must do so at velocities in excess of the deposit velocity to ensure that all solids are kept in suspension.

In a laboratory test setting where all the properties of the mixture are defined, the final design can become more accurate. The optimum combination of line size and flow velocity can be determined to transport the maximum concentration of solids for the existing properties.

Properties
Slurry properties important to design are:

- Density
- Viscosity
- Specific heat at constant volume and pressure
- Thermal conductivity
- Bulk modulus
- Degree of homogeneity
- Newtonian/non-Newtonian relationship (Kung and Mohitpour 1986)

Most of the slurry characteristics are dependent on one or more of the following:

- Properties of the solids
- Properties of the carrier fluid
- Solid concentration/suspension
- Particulate size and shape

Depending on the type of pipeline under consideration, some of the slurry properties can be varied, while in other cases they may be fixed.

Abrasiveness of the slurry may not be of primary concern when computing the hydraulic performance of a pipeline, but it strongly influences operating and maintenance aspects of the pipeline and facilities.

Density

The density of a slurry mixture (ρ_M) is a function of density of the solids (ρ_s), the density of the carrier fluid (ρ_ℓ), and the solids concentration. Solids concentration may be expressed as volume fraction (C_V) or weight fraction (C_W). While it is more convenient to use the weight fraction (C_W) for throughput calculations, it is the volume fraction (C_V), which influences the hydraulic characteristics of the slurry mixture. The following equations describe the method of calculating the density of a slurry mixture:

$$C_W = \frac{C_V \rho_s}{\rho_M} = \frac{\text{Weight of solids}}{\text{Weight of mixture}} \qquad (12-1)$$

and

$$C_V = \frac{C_W \rho_M}{\rho_s} \qquad (12-2)$$

Substitution of $\rho_M = C_V \rho_s + (1-C_V)\rho_\ell$ in Equation (12-1) gives:

$$C_W = \frac{C_V \rho_s}{C_V \rho_s + (1 - C_V)\rho_\ell} \qquad (12-3)$$

Equation (12-3) provides a relationship between mass (C_W) and volume (C_V) fractions when the densities of the solids and liquids are given. Equation (12-4) can be used to determine the density of the mixture when C_W, ρ_s, and ρ_ℓ are given, which can then be used in Equation (12-2) to determine C_V.

$$\rho_M = \frac{1}{\dfrac{C_W}{\rho_s} + \dfrac{(1-C_W)}{\rho_\ell}} \qquad (12-4)$$

Viscosity

The following equation predicts the apparent viscosity of mixtures of solids and liquids under the flow conditions that are generally used in the liquid slurry industry:

$$\frac{\mu_M}{\mu_o} = \left[1 + 2.5\, C_V + 10.05\, C_V^2 + 0.00273\, e^{(16.6 C_V)} \right] \qquad (12-5)$$

where
μ_M = viscosity of the mixture, mm^2/s
μ_o = viscosity of the liquid, mm^2/s)
C_V = solids concentration by volume, %.

Equation (12-5) gives reasonable results for values of C_v up to 40% and is good for initial design purposes. However, this correlation assumes the particles are spherical and of uniform size, a condition seldom encountered in practice.

To determine accurate values of slurry mixture viscosity, laboratory assessments are essential. Typical laboratory test data for a coal/water slurry are given in Table 12-2. Corresponding viscosity results are provided in Figure 12-3.

Although information from existing slurry systems and laboratory test data for similar materials can be useful, Equation (12-5) can be used to establish a minimum value for the apparent viscosity. However, operating data indicates that actual viscosities

TABLE 12-2. Typical laboratory test data sheet (coal/water slurry)

Coal Density 1,460 kg/m³, Ground to −8 Mesh in Water
Pipe Diameter 0.4953 m
Water Density 1,000 kg.m³, Mixture Density 1,161 kg/m³
Temperature 8.0°C
C_v=35.0%

Particle Size Distribution

Size Opening (mm)	Weight Passing (%)
0.044	11.5
0.074	13.4
0.105	15.7
0.149	19.3
0.210	24.4
0.297	31.3
0.420	40.9
0.595	51.3
0.841	64.9
1.190	81.8
1.680	95.8
2.380	99.5

Head Loss Measurement

Vel (m/s)	Head Loss (m_{water}/m)	Fanning Friction Factor	Apparent Viscosity (mm²/s)
3.05	0.0139	3.10×10^{-1}	2.20
2.75	0.0126	3.47×10^{-1}	3.65
2.44	0.0116	4.06×10^{-1}	7.4
2.14	0.0104	4.77×10^{-1}	14.2
1.80	0.00871	5.63×10^{-1}	25.0
1.65	0.00819	6.29×10^{-1}	36.5
1.52	0.00760	6.88×10^{-1}	48.0
1.43	0.00710	7.27×10^{-1}	55.5

Deposit velocity 1.3 m/s

of mixtures of irregularly shaped particles can be more than twice the values obtained from Equation (12-5).

Particle Size and Distribution

Slurry properties depend on the particulate sizes making up the slurry mixture and the distribution of different sizes of particles throughout the mixture. The viscosity and the degree of homogeneity are directly affected by the size of the particulates and their distribution. Larger particles tend to increase the deposit velocity and cause the mixture to be heterogeneous, but do not increase the viscosity. A mixture with smaller particles exhibits a lower deposit velocity. Such a mixture is likely to behave like a homogeneous mixture, and would have a higher apparent viscosity. As this mixture has a lower deposit velocity, it could be pumped at a lower velocity, resulting in lower pressure drops and energy requirements to transport the same volume of solids.

A mixture containing a variety of particle sizes could provide the advantage of being able to transport the larger particles at a lower velocity. This advantage is caused by the carrier fluid and the solids forming a homogeneous carrier fluid with higher density, thus allowing the larger particles to be carried at lower velocities.

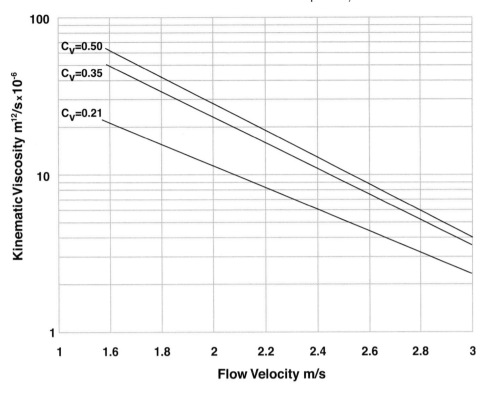

Figure 12-3. Apparent typical viscosities of coal/water slurry

Solid Concentration

The viscosity of a slurry mixture increases with the concentration of solids in the mixture. The viscosity is also dependent upon the particle size, shape, and distribution so it is difficult to examine the solids concentration in isolation. However, it can be clearly shown that in all cases, higher concentrations correspond to higher apparent viscosities. There is a clear economic incentive to try to transport as high a solid content as practical in the smallest amount of carrier fluid, but the mixture has to be capable of being pumped, so there is an upper limit to the solid concentration.

For mixtures where the particle sizes are given, laboratory tests can determine the practical limits of solids concentration. Too low a concentration would require an excessive amount of carrier liquid; too high a concentration would cause high viscosities, and hence excessive pumping power requirements.

Bulk Modulus

Figure 12-4 represents an idealized fluid-solid (slurry) system model. The effective bulk modulus, K_e, of this system is as follows (Kumano 1980):

$$K_e = \frac{p}{(\Delta V / V)} \qquad (12-6)$$

where p is the pressure, ΔV is the change in volume, and V is the total volume. The volume changes of the liquid (ΔV_ℓ) and solid (ΔV_s) are given by:

$$\Delta V_\ell = \frac{V_\ell p}{K_\ell} \quad \text{and} \quad \Delta V_s = \frac{V_s P}{K_s} \qquad (12-7)$$

where V_ℓ and V_s are volumes of the fluid and solid, and K_ℓ and K_s are the bulk moduli of the fluid and solid, respectively. Since $\Delta V = \Delta V_\ell + \Delta V_s$, Equation (12-6) may be written as:

$$K_e = \frac{K_\ell K_S}{c_\ell K_s + c_s K_\ell} \qquad (12-8)$$

where $c_\ell = V_\ell / V$ and $c_s = V_s / V$.

The ratio of change in volume to original volume V (i.e., $\Delta V / V$) can be expressed as a function of change in density over the original density ($\Delta \rho / \rho$). Equation (12-8) can be rewritten in terms of density. The change in density is a function of changes in temperature and pressure. Therefore, the density can be calculated from the bulk modulus and base density (at standard pressure and temperature) for slurry mixture densities at different temperatures and pressures.

Hydraulics Calculation

The following sections are applicable for performing hydraulic calculations for slurries.

Homogeneous and Pseudohomogeneous Mixtures

For a line moving a slurry that is a homogeneous mixture or that behaves as a pseudohomogeneous mixture, pressure drop calculations and pipeline sizing would follow the methods used for ordinary liquid pipelines (refer to Chapter 5). The success and accuracy of the calculation depends on the accuracy of the value used for apparent viscosity. However, if flow velocity is selected such that the flow regime is turbulent, the friction

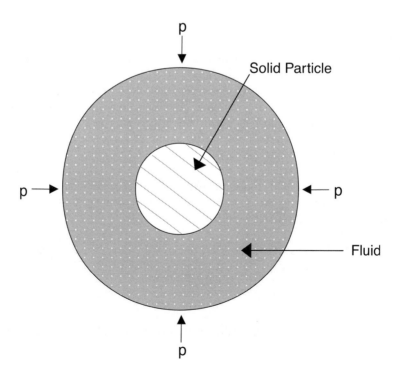

Figure 12-4. Fluid-solid system model for determination of the effective bulk modulus

factor (and hence pressure drop) is not very sensitive to changes in Reynolds number and thus the apparent viscosity.

Heterogeneous Mixtures

Calculating the pressure drop for the flow of heterogeneous mixtures is more complicated. For preliminary design purposes, it may be possible to use the same approach as for pseudohomogeneous mixtures providing the heterogeneous effect is small. For true heterogeneous flow or segregated flow, the solution is more complex. The treatment of such a flow regime is fully described by Wasp et al. (1977).

Pipe Roughness

The value commonly used for the absolute roughness in pipeline transportation of a slurry mixture in a steel pipe is 0.046 mm. For a new clean steel pipe, absolute roughness figures of 0.013 mm to 0.025 mm can be used. The sensitivity of the friction factor to changes in pipe roughness is very low, which diminishes the need for much refinement of the roughness value.

Pressure Transients in Slurry Pipeline

While the influence of pressure waves along fluid pipelines caused by fluid pressure surges (for example, by sudden valve closure) is well-defined, the effect of pressure transients (due to shock loading) in a slurry mixture cannot be treated the same way. Experimentation performed by Kumano (1980) compared the wave velocity of a water/granite mixture to the velocity of water alone: 1,293 m/s and 1,338 m/s, respectively. Kumano calculated that the addition of solid particles with a concentration of 13% contributed to a wave velocity reduction of about 3%. Therefore, it can be concluded that the effect of sudden pressure changes in a slurry pipeline is less than that experienced with a single (phase) fluid. The following provides a means of calculating wave velocity in slurry mixtures.

Equations of density (12-1 to 12-4) can be rewritten to express the effective density ρ_e, as follows:

$$\rho_e = \rho_s c_s \rho_\ell c_\ell \qquad (12-9)$$

where ρ_s and ρ_ℓ are the densities of solid and fluid, respectively. Then the wave velocity, a, for the fluid-solid system can be defined by:

$$a = \sqrt{\frac{K_e c_f}{\rho_e}} \qquad (12-10)$$

where c_f is a correction factor, which accounts for the effect of the elasticity of the pipe material (Streeter and Wylie 1967). Here, c_f is given by:

$$c_f = \frac{1}{1 + \frac{K_e}{E} \frac{D_m}{t} \left(1 - v^2\right)} \qquad (12-11)$$

where E = Young's modulus
  ν = Poisson's ratio
  D_m = mean diameter
  t = wall thickness of the pipe.

Computation of the wave velocity in a slurry mixture using Equation (12-11) typically underestimates actual wave velocity by about 3%.

Pumps and Drivers for Slurry Pipelines

The type, use and limitations of pumps for slurry transportation in pipelines are fully described by Wasp et al. (1977).

The use of centrifugal pumps for slurry pipeline systems is limited to low-pressure applications and for slurry transportation of fine, nonabrasive materials. Otherwise, excessive wear can result in pump impeller failure (Ahmed and Baker 1986). Positive displacement pumps are normally used for most high-pressure slurry applications. Pump drivers for slurry pipelines are generally electric motors. Where electric power is difficult or uneconomical to obtain, other drivers, such as diesel engines, have been considered.

Due to the abrasive nature of most slurries, flow control valves used on other liquid pipelines are not practical for slurry systems. It is thus essential that speed control of the pumps be incorporated into the drive system. This is achieved by utilizing any one of a number of available methods, such as solid-state electronic frequency converters, which can be used with standard induction motors.

Pump Station Equipment and Layout

When designing slurry pump stations or other liquid pipelines, special attention must be paid to the selection of valves and station piping configurations. Station piping should be designed so that there is sufficient fluid velocity to prevent solids from settling and to keep the velocity low enough to prevent excessive erosion. Sharp bends should be avoided where possible, otherwise piping should be designed such that it can be quickly and easily replaced when required. Dead pockets, which could become plugged with solids, should also be avoided.

Pump stations can be classified into two main categories: origin or source stations, and booster stations. At the origin station, the slurries are taken from the slurry preparation plant or storage facilities and pumped to the mainline pump suction. The storage facilities may or may not be part of pump station installations. At the origin station, it is advisable to install a time-delay loop through which the slurry passes and where its properties are monitored before it goes to the pipeline. If an "off-specification" slurry mixture material is received, it can be detected and diverted to a holding tank or pond before it reaches the pipeline.

Mainline booster stations for slurry systems may require tankage if the line cannot be shut down without the use of plugging. This requirement will vary with the type of material being handled.

Pump stations handling aqueous slurries should be designed to ensure that equipment and piping can be kept from freezing in cold weather. Freezing can be avoided by installing the pumps indoors, insulating, heat tracing/burying the pipe below the frost line in northern environments. The drain piping should be laid out so that all process piping can be drained or pumped out in case of winter emergency conditions.

Depending on the materials to be transported, it may be necessary to install special equipment to keep solids out of pumps, or to flush out pumps prior to shutdown and to refill them before start-up. The flushing medium is usually the same as the carrier fluid and requires a supply source/storage facilities and a means of getting the material to the station.

Defluidization System

The objective of defluidization of a liquid/solids slurry system is to return the solids to a condition in which they can be stored, reclaimed, and handled with the same ease as before they were slurried. Defluidization comprises a significant portion of capital and operating costs. It can also introduce complications when obtaining environmental regulatory approvals.

Defluidization is accomplished through natural drainage when dealing with coarse solids, such as 100% plus 6 mm up to the largest-size particle the pipeline system can handle. Intermediate sizes, 100% plus 0.5 mm (28 mesh) and minus 6 mm, usually require the application of mechanical forces to arrive at a moisture content amenable to handling and storing. Fine solids, 100% minus 1.4 mm (14 mesh), can be partially defluidized using mechanical forces, but if a product is essentially free of surface moistures, thermal dryers may be necessary.

Vibrating Screens

Vibrating screens can defluidize coarse solids sufficiently to satisfy most end use requirements. Surface moistures can be in the order of 4% to 5%.

Intermediate size particles can be defluidized on vibrating screens to between 8% and 14% surface moisture. This reduction is not generally satisfactory for end use and further treatment by centrifugal machines is necessary.

Fine sizes can be partially defluidized by high-speed vibrators but ultrafine sizes can be lost through the screen. If the final product is to be used to feed a thermal dryer, additional defluidization by centrifuging is necessary. Surface moisture from the high-speed vibrators will generally be in the range of 25–35%, depending on specific gravity and average particle size.

Centrifuges

Centrifuges are classified in several ways but are basically grouped according to whether they use screens. The number of G's developed is also a distinction, where G is the force of the acceleration of gravity. In the slurry industry, plus 0.5 mm solids with a top size of 9.5–12.5 mm are usually processed through a vibrating basket-type centrifuge, which gives a product surface moisture of 3–5%, depending on the average particle size of the minus 0.5 mm solids.

Fine solids are defluidized by basket-type or solid bowl-type centrifuges to a surface moisture of 14–18%. The centrifuges can be horizontal or vertical. Some ultrafines (minus 0.15 mm) would still pass through the screen and would have to be recollected and defluidized by other means. However, the solid bowl will retain most of the ultrafines.

The ultrafine solids are settled in a static thickener, usually together with 0.5 mm solids, and defluidized by vacuum or pressure filtration.

Filters

The most commonly used filter is the rotary disc vacuum filter. This filter type is best suited for defluidization of settled solids from a static thickener. The product moisture is mostly dependent on the size distribution and the vacuum pressure available. Speed of rotation is also a factor, as is the volume of air drawn through the solids and the filter cloth.

Tailings can be defluidized by belt pressure filters to a lower surface moisture than by vacuum drum or disc. A much lower surface moisture can be obtained for tailings by using a plate and frame pressure filter. The product is discharged as a large, tightly packed block and is easily handled.

Thickeners

A thickener is a device that uses gravitational forces to settle fine solids from a slurry and separate them from the bulk of the fluid. The product is not a defluidized mass of solid particles but a thickened mass in which the solids concentration is normally 25–60% by weight. The separation of solids and fluid is not absolute and the ultrafine solids, minus 0.044 mm, are usually difficult to settle out. Chemical reagents (flocculents) are used to agglomerate these sizes and assist gravitational settlings.

A static thickener can make a good separation of solids from fluid because of its longer retention time and virtual lack of disturbing turbulence.

High-Vapor Pressure and Multiproducts Pipelining

High-vapor pressure (HVP) pipelines are pipelines that transport hydrocarbons or hydrocarbon mixtures in the liquid or quasi-liquid state with a vapor pressure in excess of 240 kPa at 38°C. Examples of HVP liquids that are commonly transported through pipelines include ethane, ethylene, propane, and butane (condensate) in the dense liquid phase. These products are transported in the liquid phase to eliminate potential difficulties associated with two-phase flow and to avoid the danger of thermal effects associated with a phase change.

Ethylene, ethane, propane, and butane condensates can be transported through the same pipeline in a batch form. Other fluids that are commonly transported in a batch form are gasoline, kerosene, diesel, and jet fuel. A pipeline transporting liquid hydrocarbons in a batched form is commonly referred to as a "multiproducts pipeline".

This section highlights important design considerations specific to HVP and multiproducts pipelining.

Production and Properties of HVP Liquids
Sources and Uses

Hydrocarbon mixtures are available in varying amounts from all hydrocarbon wells. Production wells are connected through gathering lines to a processing plant. Generally, crude oil will be split into a lighter hydrocarbon gas consisting of methane (CH_4), ethylene (C_2H_4), and ethane (C_2H_6); into a heavier hydrocarbon liquid consisting of propane (C_3H_8) and butane (C_4H_{10}); and into the remaining liquid crude oil. Produced gas is usually split into a heavier hydrocarbon mixture of propane and butane, leaving the remaining methane with the ethane and ethylene. Water and sulphur are removed at the first processing point.

TABLE 12-3. Physical properties of some common high vapor pressure products (GPSA, 1994)

	Methane	Ethylene	Ethane	Propane	n-Butane	Isobutane
Formula	CH_4	C_2H_4	C_2H_6	C_3H_8	C_4H_{10}	C_4H_{10}
Molecular weight	16.043	28.054	30.070	44.097	58.124	58.124
Boiling point, °F at 14.696 psia	−258.69	−154.62	−127.48	−43.67	31.10	10.90
Vapor pressure, 100°F, psia	(5,000)	—	(800)	190	51.6	72.2
Freezing point at 14.696 psia	−296.46	−272.45	−297.89	−305.84	−217.05	−255.29
Critical pressure, psia	667.8	729.8	707.8	616.3	550.7	529.1
Critical temperature, °F	−116.63	48.58	98.09	206.01	305.65	274.98
Critical volume, cu. ft./lb.	0.0991	0.0737	0.0788	0.0737	0.0702	0.0724
Liquid specific gravity 60°F/60°F at 14.696 psia	0.3	—	0.3564	0.5077	0.5844	0.5631
Gas density 60°F, specific gravity air = 1	0.5539	0.9686	1.0382	1.5225	2.0068	2.0068
Specific heat 60°F at BTU/lb.°F	0.5266	0.3622	0.4097	0.3881	0.3867	0.3872
Net calorific value at 60°F BTU/ft.^{3}	909.1	1,499.0	1,617.8	2,316.1	3,010.4	3,001.1
Flammability limits, %	5.0–15.0	2.7–34.0	2.9–13.0	2.1–9.5	1.8–8.4	1.8–8.4

Ethane can be removed from the hydrocarbon mixture at any convenient location by a rather simple stripping operation. The ethane is then sent by a gathering system to an ethylene plant.

Ethylene is made by heating ethane until the required corresponding hydrogen atoms are split off, leaving ethylene.

Ethane is used primarily to make ethylene, which is the basic building block in the petrochemical and plastics industries. Such products as polyethylene (plastics), ethylene glycol (anti-freeze), and styrene are derived from ethylene.

Propane and butanes are used primarily for their heating value. There are a significant number of motor vehicles utilizing propane fuel; however, the trend has now moved toward compressed natural gas.

Physical Properties

The physical properties of HVP fluids are required for the solution of flow problems. Table 12-3 summarizes some of these properties.

High-Vapor Pressure Pipeline Design

There are a number of high-vapor pressure pipelines that are currently operational in the world. Table 12-4 lists some of these pipeline systems (Luk 1987).

System Design

High-vapor pressure products are gaseous under atmospheric conditions. However, it is generally more economic to transport these products above their critical pressure and temperature and in dense phase liquid form. The vapor pressures for various fluids that are commonly transported by pipelines are depicted in Figure 12-5.

Since HVP products are transported in liquid or quasi-liquid form (Rohleder 1972), they are governed by liquid pipeline codes such as ASME-ANSI B31.4 (refer to Table 1-1, Chapter 1). Such codes are utilized to guide the design, construction, and operation of HVP pipelines.

Operating Pressure Range

When transporting HVP fluids such as ethylene, ethane, propane, and butane, certain minimum operating pressures must be maintained. The minimum pressure that the various liquids may be exposed to depends on the operating temperatures that will be encountered

TABLE 12-4. Partial list of HVP pipeline systems [Luk 1987]

Pipeline System	Length (km)	Location	Operating Pressure (kPa)	Nominal Line Size (NPS)
AGEC Ethylene	241	Alberta, Canada	to 10,204	4 to 12
ARCO Ethylene	Unknown	Unknown	to 9,300	12
ARG GRID	993	Western Europe	4,275 to 10,204	10
Cochin	3,000	Canada/USA	6,200 to 10,204	12
Dow Chemicals	34	Texas, USA	9,300	4
ETEL	278	France	5,500 to 10,204	6 to 8
Ethane gathering	885	Alberta, Canada	to 10,204	6 to 12
French	30	France	5,500 to 10,204	6
ICI TransPenine	222	Britain	5,500 to 10,204	8
MAPCO Rocky Mountain Area	1,886	Western USA	to 11,000	10 and 12
MAPCO Seminole	1,000	Texas, USA	to 11,000	6 to 14
Novacorp, Porcupine Hills	160	Alberta, Canada	to 5,700	6
Russia	760	Bashkir	5,500 to 10,204	8
Shell	61	Texas, USA	8,600 to 9,300	6

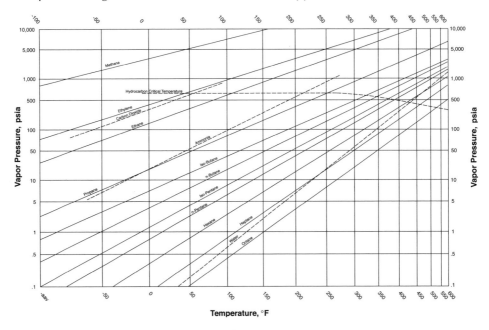

Figure 12-5. Vapor pressure chart [courtesy GPSA]

within the system and the vapor pressures of the product corresponding to these temperatures.

For ethane and heavier hydrocarbons such as propane and butane, the minimum pressure should generally be set such that the products are maintained in their liquid phase throughout the operating temperature range of the pipeline. The minimum pressure is selected using pressure-enthalpy diagrams and the product flowing temperature range on the upstream side of pump stations.

Pumping ethylene, which has a temperature of 48.58°F or 9.2°C (Table 12-3), will be somewhat different from pumping ethane and other heavier hydrocarbons that exhibit higher critical temperatures. Because the normal operating temperature of a pipeline is often higher than the critical temperature of ethylene, the minimum pressure will have to be set above the critical pressure of 729.8 psia (5,031 kPa). This means that the ethylene is often transported as a super-critical gas whenever the flowing temperature is higher than the critical temperature. A test program that was conducted by an ethylene pipeline company confirmed that transportation of ethylene as a super-critical gas is realistic and practical (Hanna and Bourbonnie 1978; Hanna 1978, 1979).

The *GPSA Handbook* and other such books typically contain a set of pressure-enthalpy diagrams that can be used to establish the operating temperature-pressure region for any pertinent HVP pipeline.

Figure 12-6 illustrates typically selected temperature-pressure regions for an ethane pipeline and an ethylene pipeline under the same product flowing temperature range of 30°F (−1°C) in winter to 65°F (18°C) in summer. A minimum pump suction pressure of 4,480 kPa (656 psia) should be considered when pumping ethane in order to maintain liquid phases at all times throughout the system. For ethylene, a minimum pump suction pressure of 6,200 kPa (900 psia) needs to be considered (refer to Chapter 5 for pump design).

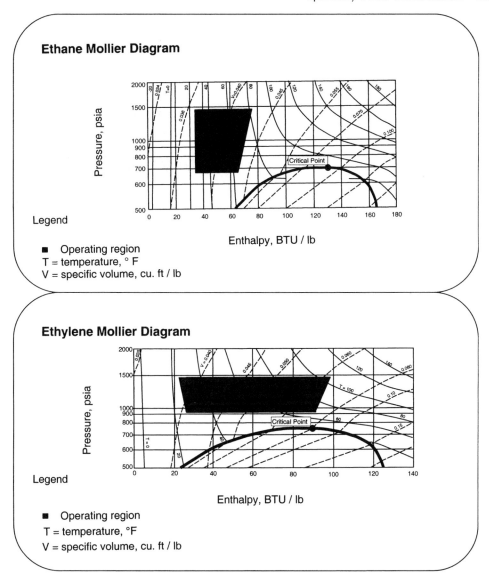

Figure 12-6. Typical operating range for ethane and ethylene liquids [courtesy *Oil and Gas Journal*, Hanna 1979]

For propane, butane, and other heavier hydrocarbon products, the minimum pressures under normal pipeline operating temperatures can be set much lower by investigating the pressure-enthalpy diagrams of the fluid. As an example, the minimum pressure for pumping propane between −1°C and 18°C flowing temperature, could be set at approximately 1,000 kPa (145 psia).

Maximum operating pressures are generally set at 1,480 psia (10,204 kPa) for most HVP pipelines. For propane and butane and other heavier hydrocarbons, the maximum operating pressure can be set lower, depending on the efficiency of the overall system. A typical maximum operating pressure (MOP) range for propane is

TABLE 12-5. Pipeline properties of some HVP liquids (kg/m^3)

Fluid	Density (kg/m^3)	Viscosity CP (mm^2/s)	Temperature (°C)	Reid Vapor Pressure (kPa) at 37.8°C
Raw condensate (at 977 kPa)	547.5	0.171	30	—
	572.3	0.199	15	
Refined condensate	653.8	0.456	44	83
	661.7	0.506	35	
	666.2	0.534	30	
Propane	446.9	0.166	52	1,379
	483.475	0.199	30	
	500.578	0.218	20	
Butane	529.0	0.212	52	483
	535.117	0.235	44	
	560.823	0.237	25	
Gasoline (leaded)	711.3	0.68	5	
		0.61	15	
Gasoline (unleaded)	699	0.7	5	
		0.63	15	
Jet fuel (A type)	744	8	20	
		1.5	0	
Water *	1,000	—	15	
		1.79	21.1	
		0.98	0	

Note: * Included for comparison purposes.

900 psia to 1,480 psia (6,200 kPa–10,204 kPa), while for butane it is 800 psia to 1,480 psia (5,500 kPa to 10,204 kPa).

HVP Pipeline Fluid Properties

Properties useful in pipeline hydraulic calculations are provided in Table 12-5 for a number of typical HVP fluids. Viscosity-temperature relationships for HVP liquids are provided in Figure 12-7.

Another property that is used for predicting flow densities at different pressures is the bulk modulus, K. The literature does not often provide values of the bulk modulus for various fluids. The following relationships are thus provided for determination of fluid densities at different pressure and temperatures and bulk modulus K.

Liquid density (ρ) at various pressures (P) and temperatures (T) is given by the following relationship:

$$\rho = \rho_b \left[\left(1 + \frac{P - P_b}{K} \right) - \alpha(T - T_b) \right] \qquad (12-12)$$

where
ρ_b = density at base condition
P_b = pressure at base condition
T_b = temperature at base condition
α = liquid temperature coefficient of density
K = bulk modulus.

For isobaric conditions (i.e., at constant pressure) $P-P_b/K = 0$, Equation (12-12) can be rewritten as:

$$-\alpha = \frac{\frac{\rho - \rho_b}{\rho_b}}{T - T_b} = \frac{\Delta \rho}{\rho_b \, \Delta T} \qquad (12-13)$$

where $\Delta\rho = \rho - \rho_b$ = change in density.

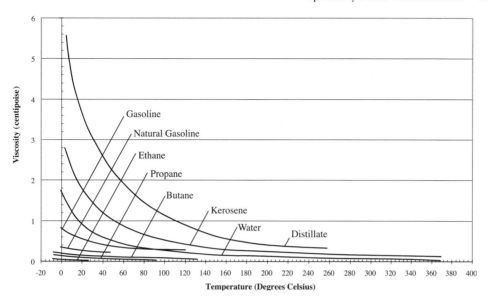

Figure 12-7. Viscosity of water and liquid petroleum products [Adapted from *Crane Handbook*]

If the liquid temperature coefficient of density is known, it is possible to compute liquid densities at different pressures.

For isothermal conditions (i.e., at constant temperature), $T - T_b = 0$, so Equation (12-12) can be rewritten as follows:

$$\rho = \rho_b \left[1 + \frac{P - P_b}{K} \right] \qquad (12-14)$$

or

$$K = \frac{\Delta P}{\Delta \rho} \rho_b \qquad (12-15)$$

where $\Delta P = P - P_b$ change in pressure.

TABLE 12-6. Industry accepted contamination levels

Contaminant	Product	Percentage of Contamination in Product
Butane	Gasoline	*
Premium gasoline	Regular gasoline	3
Regular gasoline	Premium gasoline	1
Jet fuel (kerosine)	Regular gasoline	1
Jet fuel (kerosine)	Premium gasoline	2
Jet fuel (kerosine)	Diesel	2
Any	Jet fuel	Nil

* Dependent on butane already added in refinery.

Example 12.1

Calculate the bulk modulus and liquid coefficient of density for liquid CO_2 if the pressure drop across a pipeline segment is 3,100 kPa. The inlet pressure is 13,100 kPa. The density at base pressure and temperature is 968.5 kg/m^3.

Solution

The density at inlet (13,100 kPa) is 1,073.5 kg/m^3. The density at the outlet (1,000 kPa) is 1,064 kg/m^3. Therefore, $\Delta\rho = 1{,}073.5 - 1{,}064.0 = 9.5$ kg/m^3.

Therefore,

$$K = \frac{\Delta P \rho_b}{\Delta \rho} = \frac{3,100 \times 968.5}{9.5} = 316,036 \text{ kPa}$$

Assume the following:

$T_1 = -20°C$ corresponding $\rho = 1{,}073.5$ kg/m^3
$T_2 = +20°C$ corresponding $\rho = 882.0$ kg/m^3
$\rho_b =$ (base density) at $15°C = 968.5$ kg/m^3

$$-\alpha = \frac{\Delta P}{\Delta \rho_b \Delta T} = \frac{1,073.5 - 882}{986.5 \times (40)} = -0.005 \ (\text{kg/m}^3 \ °C)$$

Liquid Batching

General

Sequential transportation of liquids in a batch form is commonly exercised by refineries and liquid pipeline companies to transport multitudes of products through a single pipeline. This form of transportation includes batch transportation of low- as well as high-vapor pressure fluids. Batching in the early days was achieved by injecting a liquid into the pipeline followed by a separation pig (usually a sphere) and then the second batch of another product/fluid. Batching of a multitude of products without a separation pig is more common today, and this chapter will provide the steps required to calculate the interface volume growth and batch cycle duration.

Minimum Batch Size

Minimum batch size can be determined from a knowledge of the allowable contamination of one product in another. Most product pipeline companies allow themselves a degradation in the quality of specification of the product. However, such a degradation allowance depends on legal requirements rather than practical necessity. From a legal point of view, a product (e.g., unleaded gasoline) has to be delivered to the market with contaminants not exceeding government-specified component limits. Contamination levels in batch product pipelines are related to the size and type of batches or tenders required for entrance to the pipeline.

Contamination usually varies from company to company and, as previously mentioned, is dependent on the specification of the final product and the quality of the products as run down by refineries.

Typical allowable contamination levels accepted by the industry are shown in Table 12-6.

Intervolume Predictions

Methods for prediction of the interface length and volume are described by Austin and Palfrey (1989). The following describes the steps involved in arriving at interface length and volumes.

Step 1

The blended viscosity for a 50/50 mix can be derived from the following equation:

$$v_B = 0.5 \ \log \ v_1 + 0.5 \ \log \ v_2 \qquad (12-16)$$

where v_B = blended viscosity, CS
   v_1 = product 1 viscosity, CS
   v_2 = product 2 viscosity, CS.

Step 2

The Reynolds number for the blended product is then:

$$\text{Re} = \frac{V \cdot d}{v_B} \times 1,000,000 \qquad (12-17)$$

where R_e = Reynolds number
   V = fluid velocity, m/s
   d = inside pipe diameter, m.

Step 3

The critical Reynolds number is found from

$$\text{Re}_c = 10,000^* \ \text{Exp} \left(2.75 \times \sqrt{d} \right) \qquad (12-18)$$

where Re_c = critical Reynolds number.

Step 4

The interface length (LC), if $R_e > \text{Re}_c$, is

$$LC = \frac{0.3716\sqrt{L \cdot d}}{\text{Re}^{0.1}} \qquad (12-19)$$

and if $R_e < \text{Re}_c$, then

$$LC = \frac{582.49\sqrt{L \cdot d} \ \text{Exp}\left(2.19 \ \sqrt{d}\right)}{\text{Re}^{0.9}} \qquad (12-20)$$

where LC = length of interface (50/50% mix), km
   L = length of travel, km.

Step 5

Finally the interface volume can be then found from

$$VC = 250 \quad \pi d^2 \cdot LC \qquad (12-21)$$

where VC = interface volume, m^3
   d = inside pipe diameter, m.

The above computation assumes that the product velocity and pipeline diameter remain constant throughout the pipeline. However, situations arise where products are delivered to different delivery (drop-off) points, and where the pipe diameter changes subsequent to drop-off. In this case, a new interface does not begin at each of the velocity change locations. Rather, the interface length continues to grow at a rate dependent on the new velocities. The following steps are thus taken to calculate interface lengths in situations where velocity changes in the pipeline as a function of diameter.

Step 6

At each velocity change point, a new Reynolds number and critical Reynolds number must be calculated. If $R_e > R_{e_c}$, then from Equation (12-19)

$$LC = \frac{0.3716 \sqrt{L \cdot d}}{Re^{0.1}}$$

Step 7

The next step is to calculate LC_1 for the end of the first section (with inside diameter d_1). The length LC_1 must be converted to an equivalent length that the interface will occupy in second section with inside diameter d_2.

$$LC_{(1-2)} = LC_1 \times \left(\frac{d_1}{d_2}\right)^2 \qquad (12-22)$$

Step 8

The length of the second section pipeline that would produce the equivalent length of interface received from Section 1 is then calculated by

$$L_{2\,equiv} = \frac{\left(\frac{LC_{(1-2)} \cdot Re_2^{0.1}}{0.3716}\right)^2}{d_2} \qquad (12-23)$$

Step 9

The total cumulative length of interface at the end of the second section is now.

$$LC_{2\,(Total\,Length)} = \frac{0.3716\sqrt{(L_{2\,equiv} + L_2) \cdot d_2}}{Re_2^{0.1}} \qquad (12-24)$$

Step 10

Steps 6 through 9 are repeated for all sections of the pipeline system where either velocity or pipe diameter changes occur.

Example 12.2

The following example provides a result of interface calculations for a pipeline transporting multiproducts from two refineries to market locations.

Design Data

- Pipeline schematic (Refer to Figure 12-8)
- Pipeline inlet pressure:
  - Refinery 1 = 300 kPa gauge
  - Refinery 2 = 300 kPa gauge
- System inlet temperature = 30°C
- Minimum operating pressure = 300 kPa gauge
- Maximum design pressure = 10,205 kPa gauge
- Maximum operating pressure = 9,184 kPa gauge
- Maximum design temperature = 50°C

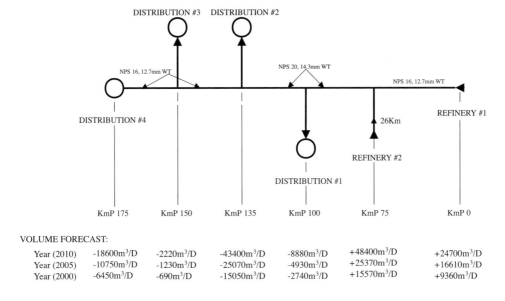

DISTRIBUTION #3   DISTRIBUTION #2

NPS 16, 12.7mm WT

NPS 20, 14.3mm WT

NPS 16, 12.7mm WT

DISTRIBUTION #4

26Km

REFINERY #1

REFINERY #2

DISTRIBUTION #1

| KmP 175 | KmP 150 | KmP 135 | KmP 100 | KmP 75 | KmP 0 |

VOLUME FORECAST:

Year (2010)	-18600m³/D	-2220m³/D	-43400m³/D	-8880m³/D	+48400m³/D	+24700m³/D
Year (2005)	-10750m³/D	-1230m³/D	-25070m³/D	-4930m³/D	+25370m³/D	+16610m³/D
Year (2000)	-6450m³/D	-690m³/D	-15050m³/D	-2740m³/D	+15570m³/D	+9360m³/D

NOTES: 1. Uplift of Jet A - 1 at Distribution #1 and Distribution #2 is in a ratio of 80:20
        2. Uplift of white product at Distribution #2 and Distribution #4 is in a ratio of 70:30
        3. Pipe Grade API 5L X 42

Figure 12-8. Multiproduct pipeline system schematic

- Minimum delivery pressure = 300 kPa gauge
- Pump station operating efficiency = 75%
- Pipeline depth of burial = 1 m
- Pipeline roughness = 0.045 mm
- Ground temperature = 27°C
- Ground conductivity = 1.14 W/m°C
- Pipeline route elevation = Refer to Table 12-8
- Corrosion allowance = 1.6 mm

Table 12-7 lists information regarding product type.
Table 12-8 provides an elevation profile for the multiproducts pipeline.
Contamination criteria from Table 12-6 was utilized in the prediction of interface lengths and volumes, as shown in Tables 12-9 through 12-11.

TABLE 12-7. Product information

	Viscosity (mm²/s)	Temperature (°C)	Density (Kg/m³)*
Diesel	6.86	5	847
	5.10	15	820
Gasoline (leaded)	0.68	5	711.3
	0.61	15	—
Gasoline (unleaded)	0.7	5	699
	0.63	15	690 (assumed)
Jet A fuel	8	29	774
	1.5	0	—

* Density at standard pressure (101.35 kPa) and temperature (15°C).

TABLE 12-8. Route elevation profile

Kilometer Position (kmp)	Elevation (m)
0	72
1	72
21	72
27	49
30	47
42	76
57	76
64	59
68	99
82	60
92	43
101	119
109	8
128	60
132	60
136	17
146	6
150	28
155	7
170	7
175	7

Results

The pipeline was designed for transporting the heaviest of the products (i.e., diesel). Pumping facilities were selected such that under the worst scenario, a single batch of diesel can be transported for the design conditions stipulated previously (refer to design data and Figure 12-8).

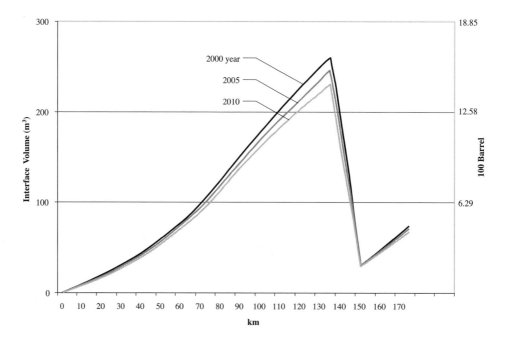

Figure 12-9. Interface volume trucking between diesel and Jet A-1

TABLE 12-9. Maximum interface mixture volume (cubic meters)

	Year 2000	Year 2005	Year 2010
Diesel/Jet A-1	259.37	245.71	234.08
Diesel/regular gasoline	250.28	237.95	226.69
Regular/unleaded gasoline	225.43	214.33	204.18
Unleaded/premium gasoline	219.93	209.09	199.20
Premium gasoline/diesel	244.17	232.14	221.15

Maximum interface mixture volumes and minimum batch size meeting the contamination criteria of Table 12-6 are provided in Tables 12-9 and 12-10, respectively. A sample calculation of interface volume and length is provided in Table 12-11 for regular and unleaded gasoline.

Typical interface volume accumulation tracking is shown in Figure 12-9 for the interface between diesel and Jet A-1. From this figure, it can be inferred that as velocity or flow rate increases, interface volume decreases.

BATCHED PRODUCTS PIPELINE DESIGN & OPERATIONAL ISSUES

In transportation of any batched products (LVP & HVP), there are a number of issues that will require specific attention to ensure that the products are transported and delivered with the appropriate quality/integrity, safety and least environmental impact in the event of a leak or pipeline failure. Design & operational issues that specifically affect batched products pipelines are summarized below:

- Technical design of facilities (Pipeline, pumps, measurements) including piping arrangements
- Pipeline valve spacing (for high vapor pressure product transportation)
- Dead leg impacts & remediation
- Parallel flow paths in piping systems
- Start/stop operations vs. continuous flow
- Batch Interfacial management
  - Batch sequencing
  - Contamination impacts
  - Cut-points
  - Degradation cost impacts
- Operational controls & procedures
- Quality testing
- Impacts of supply forecast
- Impacts of shared load pumps/meters – quality and degradation
- Post pipeline treatment for interfaces/contaminants
- Delivery considerations – filtration, interfacial cuts, contamination and degradation impacts, tankage residence times
- Safety, environmental, & risk
- Emergency response

Some of the operational issues are described in detail elsewhere, Mohitpour, Szabo & Van, Hardeveld, 2003.

CARBON DIOXIDE PIPELINE TRANSMISSION

Properties and Uses

Carbon dioxide (CO_2) is a colorless, odorless, nonflammable, nontoxic substance that may exist as a gas, liquid, solid, or in all three phases at its triple point. The critical pressure and temperature of CO_2 is 1,070 psia (7,377 kPa) and 88°F (31°C), respectively. It is present in earth's atmosphere at a concentration of approximately 330 ppm, although somewhat higher concentrations may occur in occupied buildings (L'Air 1976; Recht 1984). Air in the lungs contains approximately 5.5% (55,000 ppm) of CO_2. Although it is nontoxic, air containing 10–20% CO_2 concentrations by volume is immediately hazardous to life by causing unconsciousness, failure of respiratory muscles, and a change in the pH of the blood stream.

Carbon dioxide is generally used for:

- carbonated beverages
- aerosol propellants
- fire extinguishers
- enrichment of air in greenhouses
- fracturing and acidizing oil wells
- shielding gas for welding
- dry ice for refrigeration
- tertiary oil recovery, miscible flooding

Carbon dioxide can be successfully transported as a supercritical fluid through a pipeline, and designed and operated similarly to a natural gas pipeline with careful consideration given to specific differences in design and materials of construction.

In the oil recovery process, CO_2 is injected into the formation/reservoir where it dissolves in the oil to form a miscible fluid (Petroleum Frontiers 1984). The injection causes the oil to swell, which results in reduced viscosity, and increased density. Carbon dioxide presence exerts an acidic effect on the reservoir rock, and in some cases vaporizes some of the oil. As a rough rule of thumb, approximately 6–10 MCF (170–280 m^3) of CO_2 is required for recovery of one barrel of (0.16 m^3) oil. Carbon dioxide miscible flooding generally recovers about 10–15 percent of the oil remaining in oil reservoirs (Recht 1984, 1986).

TABLE 12-10. Minimum batch size

	Minimum Batch, Cubic Meters		
	Year 2000	Year 2005	Year 2010
Diesel/Jet A-1	12,968.50	12,285.50	11,704.00
Diesel/regular gasoline	25,028.00	23,795.00	22,669.00
Regular/unleaded gasoline	22,543.00	21,433.00	20,418.00
Unleaded/premium gasoline	7,331.00	6,969.67	6,640.00
Premium gasoline/diesel	24,417.00	23,214.00	22,115.00

TABLE 12-11. Interface between regular and unleaded gasoline (year 2000 volume)

Item	km 0–75	From REF #2 (26 km)	km 75–100	km 100–135	km 135–150	km 150–175
Blended viscosity cs	0.794	0.794	0.794	0.794	0.794	0.794
Reynolds number (Re)	456,206	456,206	965,715	965,715	481,561	481,534
Critical Reynolds number (Rec)	54,701	54,701	67,271	67,271	54,701	54,701
Interface length, LC, km	0.5397	0.317767	0.324272	0.383684	0.240059	0.309917
Total interface length, LC, km	N/A	N/A	N/A	N/A	N/A	N/A
Interface volume m^3	61.50	36.21	58.50	69.22	27.36	35.32
Cumulative interface volume m^3	61.50	97.71	156.21	255.43	27.36	62.67

Pipeline Transportation

CO_2 is transported in the supercritical phase (similar to a liquid form) above 700 psia (4,826 kPa) (Renfro 1979). Transmission in gaseous phase is not economical, as it can result in two-phase flow and thus high pressure losses, particularly in hilly terrain.

CO_2 Pipeline Transmission Network

In North America, notably the United States, there are three major CO_2 pipeline transmission systems (Alexander and Lee 1986, Cathro et al., 1986; Quarles 1983). The network extends between Texas through New Mexico and Colorado (Figure 12-10). These facilities deliver CO_2 in dense phase form (liquid) to customers in the Denver city area who inject CO_2 into old oil reservoirs (miscible flooding) and for enhanced recovery of previously unrecoverable oil. There are also other CO_2 pipelines in the USA, but not as extensive as those described here. For example, Chevron's 210 km NPS 16 CO_2 pipeline extends between the Exxon Rangely Field to Northwest Colorado (Tampkins and Johnson 1986).

The facilities that support the pipeline route illustrated in Figure 12-10 are:

1. Shell Cortez CO_2 Pipeline System (Cortez, Colorado, to Denver City, TX), which consists of:
   - Transmission pipeline (30 inches × 508 miles, MOP 2,140 psi, capacity 700 mmscfd)
   - Metering and dispatch facilities (orifice meters for custody transfer and vortex meter for system balance)
   - Booster facilities (compression for low pressure, pumps for high pressure)
   - Gathering and well cluster facilities (12 clusters total)
   - Central processing and compression facilities (three central facilities)
2. Amoco Bravo Pipeline System (Bravo Dome, New Mexico to Denver City, TX), which includes:
   - Transmission pipeline (24 inches × 219 miles, MOP 2,500 psi, capacity 600 mmscfd)
   - Meter stations (orifice meters for custody transfer)
   - Boosters (compression at low pressure and pumps at high pressure)
3. Arco Sheep Mountain Pipeline System (Gardner, Colorado, to Denver City, TX) (Warren 1989), consisting of:
   - Transmission pipeline (20 inches × 108 miles and 24 inches × 300 miles, MOP 2,625 psi, capacity 480 mmscfd)
   - Metering facilities (turbine meters for custody transfer and orifice meters for system balance)
   - Booster facilities (compressors at origin, no mainline boosters)

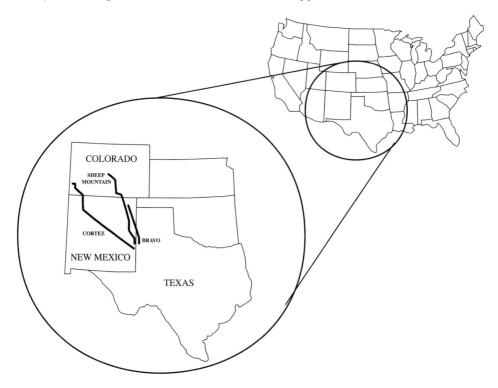

Figure 12-10. Major CO_2 pipeline network in USA

Design Considerations

The following is a list of considerations that must be evaluated for CO_2 pipeline transmission systems (Cathro et al., 1986; Recht 1984, 1986; King 1980, 1981).

Hydraulics

- Properties determination
- Viscosity determination
- Flow calculation methods
- Transient simulation
- Flow characteristics (how to prevent laminar flow)
- Typical CO_2 gas mixtures

Piping design

- Crack arrestors
- Blowdown assemblies (piping design and valve sizing)
- Blowdown rate determination and basis
- Necessity for linebreak controls
- Scraper traps
- Depth of cover

Material selection

- Pipeline/piping materials

- Carbon equivalent
- Hardness values
- Fracture strength
- Valve, fitting, and trim types and need for sour service specifications
- Experiences with Teflon, nylon, EPR-EPDM polymide, or semi-rigid polyurethane
- Valve actuators

Cleaning and strength testing

- Cleaning procedures
- Hydrostatic testing/air drying/dewatering

Construction techniques

- Corrosion monitoring and corrosion coupons
- External corrosion
- Crack arrestors
- Special construction and welding procedures
- Need for stress relieving

Pipeline operation

- Refrigeration effects during start-up/blowdown
- Method of start-up and shutdown
- Line pressuring/commissioning
- Requirements for blowdown mufflers
- Experience with pipeline ruptures and causes
- Environmental considerations
- Operational problems
- Operational safety (leaks, blowdowns, etc.)

Measurement

- Custody transfer methods: volumetric mass measurement, need for metering facilities (restrictions to be met)
- Moisture analyzers (water control measurement densitometers)

Salient technical features affecting the design of CO_2 pipeline facilities are:

- Excessive water content in CO_2 can cause formation of highly corrosive carbonic acid. Levels of between 18 and 30 lb/MMSCF (288–480 kg/m^3) are accepted by the industry for CO_2 transmission in carbon steel pipelines.
- Concentrated CO_2 causes deterioration and explosive decompression in petroleum-based sealing materials (O-rings, seals, valve seats, etc.) when they are taken from line pressure down to atmospheric pressure. Petroleum-based greases and many synthetic greases (for use in valves) are deteriorated by CO_2 such that they become hardened and ineffective. Specially selected inorganic sealing materials and grease must therefore be used in CO_2 service (Watkins 1983).
- Thermodynamic characteristics of CO_2 make pressure and temperature ranges critical in pipeline operation. Blowdowns and pipeline loading must be controlled over significantly longer times than in normal natural gas pipeline procedures, to prevent excessively low-temperature gradients.

- CO_2 is handled as a dense phase (supercritical) liquid in pipeline operation with pressures between 8,274 kPa (1,200 psi) and 17,500 kPa (2,500 psi) and temperatures below 30°C (80°F).
- Due to the sensitivity of CO_2 properties (such as compressibility and density) to temperature, ambient temperatures can have a significant effect on the operation of a CO_2 pipeline.
- Air or nitrogen leak tests do not confirm that a system will be leak-tight when commissioned with CO_2. All mechanical joints, seals, packings, etc., must be carefully designed, installed, and tightened to prevent leaking of CO_2.
- Pipeline pigging in CO_2 service is difficult without the aid of a precursor lubricant such as diesel because dry CO_2 has very poor lubricating characteristics.
- CO_2 pipeline operation involves a unique transient effect commonly referred to as "the slinky effect," which is a variation on the common "water hammer" experienced in pipelines flowing incompressible fluids such as water.
- Existing CO_2 pipeline facilities have been designed to meet current gas/oil pipeline systems codes and standards. No speciality fluid transmission code is applicable or available.
- The two primary safety considerations for CO_2 facilities are: (1) to avoid suffocation in areas where CO_2 may be blowing down, leaking, and displacing oxygen (especially in enclosed or low areas); and (2) to exercise extreme caution when operating or maintaining high-pressure CO_2 facilities due to the compressibility and potentially violent expansion (150 to 1) of CO_2 as it changes phase.

CO_2 Gathering, Processing, and Pipeline Transportation
General Description
A CO_2 system is generally comprised of wet and dry CO_2 gathering and transmission systems (Cathro et al., 1986). Two-phase CO_2 is initially pipelined from well clusters in the CO_2 fields to a central processing facility. At such a facility the CO_2 is processed, dried to reduce water content to acceptable levels, and compressed. The dry CO_2 is transported from a central processing facility to metering and dispatch stations. It is then transported through a pipeline consisting of various pumping and valving arrangements to delivery locations, usually for injection purposes.

A CO_2 dome field can be extensive; it can cover an area of approximately 250,000 acres (1,000 km²), extending 400 km (25 miles) long to 26 km (15 miles) wide. Wells are drilled into the CO_2 dome field to obtain CO_2 from a formation (usually 75–92 meters thick) generally consisting of limestone and dolomite at depths of 2,500 to 2,800 meters. An average well can generally produce in the order of 15 to 20 MMSCFD (425–566 Mm³D) of CO_2. Peak production from the field can be as high as 1,000 MMSCFD (28,300 Mm³D), with a production life of over 25 years.

At metering and dispatch stations CO_2 is measured for pressure, flow, density, dewpoint, and temperature. CO_2 in the dense phase state is then transported via a pipeline for miscible CO_2 flooding to extend the oil recoverable reserves by 25 to 30 years or more.

A typical configuration of a facility comprising the CO_2 source clusters, gathering, processing, and compression facilities is shown in Figure 12-11. CO_2 wells can be located up to seven or eight miles (11–13 km) away from the processing facility. CO_2 from the wells is first pumped into a well cluster separation facility for preliminary separation and removal of water. The two-phase CO_2 is then transported by a gathering pipeline to the processing facility for further processing and compression.

Well Cluster Facilities

A typical well-head piping arrangement consists of Duplex steel with stainless-steel valving. Most CO_2 wells consist of $7^{5/8}$ inches casing and $4^{1/2}$ inches tubing made from 13% chromium steel. The packer and all other equipment are made from 13% chromium steel.

The wells generally flow at about 750 psi (5,170 kPa). The well production flow temperature is controlled to a minimum of 12°C (54°F) to avoid hydrate formation.

CO_2 from the wells is then transported to a well cluster facility, typically through NPS (Nominal Pipe Size) 4, 316 stainless steel pipes. Each cluster facility can serve up to four or more wells and can handle between 70 and 100 MMSCFD (2–3 MMm^{3}D) of wet CO_2 gas.

At the cluster facilities produced CO_2 is generally 30% liquid and 70% vapor with approximately two barrels of free water per MMSCF on the average. Most of the free water is separated from the CO_2 and the CO_2 is injected with a hydrate inhibitor (50% diethylene glycol and 50% water). The two-phases of CO_2 are then metered prior to transmission to the processing facilities. Water is typically metered using turbine meters and the CO_2 using orifice meters. The collected free water is injected into disposal wells, usually located on the fringes of the CO_2 reservoir.

Typically, there is no hydrate formation protection between the wells and the cluster facilities, except for the maintenance of sufficient backpressure to keep the well-head temperature above critical hydrate formation temperature. This is achieved by controlling the well-head pressure.

Each cluster facility is controlled via a supervisory control and data acquisition (SCADA) system. Control includes regulation of flow from each well and actuation of the wells' safety control valves.

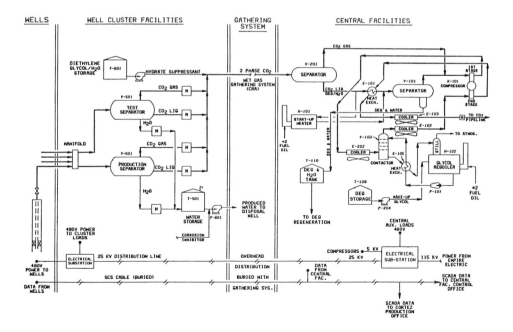

Figure 12-11. Typical CO_2 gathering and processing facilities

The cluster facility can be depressurized through a vent stack. Depressurization is typically undertaken in two stages. First, liquid is removed from the low points of vessels and equipment. The vapor is then vented through the vent stack. Depressurization is carefully controlled to avoid hydrate (dry ice) formation.

Processing and Compression Facilities

The two-phase CO_2 is usually transported from the well cluster facilities to the processing facilities via Duplex steel pipelines. Pressure at the well heads varies from 750 to 780 psi (5,170–5,380 kPa) and CO_2 is typically received at a processing plant at about 630 psi (4,340 kPa). Plant losses are usually about 15 psi (105 kPa).

Gathering pipelines feeding CO_2 into the processing facilities are typically equipped with scraper launchers and receivers, and are fitted with butterfly valves to control the CO_2 pressure reaching the processing plant.

The two-phase CO_2 received from a cluster facility is preheated in a heat exchanger. It is then processed at the processing facilities' main slug catcher, typically 48 (15m) feet long and 12 feet (3.7m) in diameter. Such a slug catcher is designed and operated between 600 psig (4,000 kPa) and 650 psig (4,500 kPa). The processing plant can typically handle approximately 300 MMSCFD (8.5 MMm^{3}D) of CO_2. The CO_2 from this main separator is further processed in the inlet separator prior to entering the first stage of compression. After the first stage of compression, the CO_2 gas is generally at 1,250 psi (8,620 kPa). It is then further dehydrated and cooled in a glycol contactor. CO_2 is then passed to a second stage compressor. The dry CO_2 gas from the second stage of compression at 2,500 psi (17,240 kPa) is then cooled via banks of fin-fan coolers, is changed to a dense-phase gas, and then enters a dry CO_2 pipeline for transportation to a metering and dispatch station. Inlet pressure and temperature to the pipeline is typically 2,140 psi (14,750 kPa) and 100°F (38°C). In the latter case, all valves, fittings, and flanges are designed to ASME ANSI Class 1,500 flange rating.

In order to avoid CO_2 liquid formation in the compressor cylinders, which could result in a blowout, the CO_2 gas is preheated in a heat exchanger to about 90°F (32°C) prior to the first stage of compression. Parallel separator/compression facilities, as required, process and compress the CO_2 for delivery into the transmission pipeline. Reciprocating compressors are typically utilized for compression of CO_2.

Processing facilities typically contain control and annunciator rooms, switch gear facilities, and an uninterrupted power supply (battery room and diesel generator units). Each compression unit is equipped with a local control panel, which can initiate a localized emergency shut down (ESD). If the plant is experiencing a major problem and it is picked up by the local panel, the entire system is shut down automatically once two compression units are tripped.

The control room is typically equipped with computerized facilities with a number of screens. Mechanical flow sheets and any information pertaining to each well facility can be displayed on a screen. Displays of the status of each system can be stepped down through various levels to show further details.

An important aspect of the processing plant operation and control is the unique instrumentation that is utilized. Unique and very accurate instruments are used to meet the stringent operating criteria. For example, CO_2 temperature in various processing units and at the well heads must be controlled to a fine degree, as undue variation from the set points can cause serious upsets if left unattended.

A simplified flow diagram of a model CO_2 gathering and processing facility is shown in Figures 12-12 and 12-13 (Cathro et al., 1986). In this system the CO_2 treatment unit comprises a test separator, vaporizer, superheater, and glycol contactor. Together, they process the CO_2 in order to lower the water content and ensure that all the CO_2 is in a vapor phase prior to being compressed. The compression equipment includes a two-stage reciprocating compressor. Each compression stage has separate scrubbers and pulsation bottles.

The efficiency of the facilities depicted in Figure 12-13 is reflected in its unique design features, which allow substantial removal of free water from the two-phase CO_2 received from the wells in the vaporizer. The CO_2 gas is further dried in the superheater prior to any compression.

The vaporizer's heat is generally received from the second-stage compression of CO_2 at about $1,400°F$ ($760°C$). In the vaporizer the CO_2 is separated from its constituent components (i.e., water and oil). The water from the test separator and the vaporizer is then stored in produced water tanks for subsequent reinjection. The oil from the vaporizer is stored in slop oil/water tanks and trucked out.

Gaseous CO_2 from the vaporizer at $60°F$ ($15°C$) is further superheated to $64°F$ ($18°C$) and passed through a glycol contactor prior to reaching the compressor suction scrubber. Leaving the glycol contactor, the CO_2 contains about 8–9 pounds of water per MMSCF ($1.1–1.2$ kg/MMsm^3) (Figure 12-14). For reboiler see Figure 12–15.

After the first stage of compression, CO_2 is cooled and further scrubbed by a second-stage scrubber. It is subsequently compressed and passed through a superheater and vaporizer for intermediate cooling. It is then passed through the compressor discharge cooler in order to change it to dense phase gas prior to transportation by pipeline at about $2,140$ psi ($14,750$ kPa).

CO_2 Pipeline Transmission System

Generally, pipelines carry all the CO_2 from the processing and compression facilities to the market, which may be an oil recovery field.

Table 12-12 shows a summary of current design and operating practices in the CO_2 pipeline transmission industry. Table 12-13 provides operating experience for CO_2

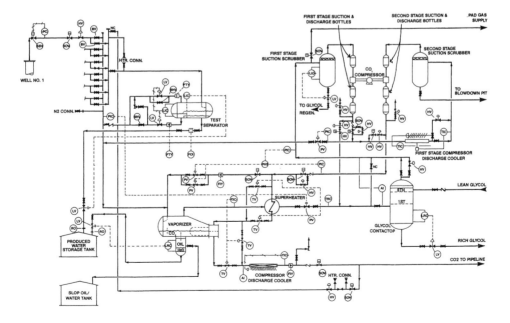

Figure 12-12. Simplified flow diagram CO_2 gas/liquid loop

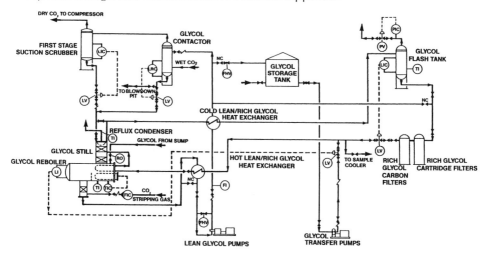

Figure 12-13. Simplified flow diagram of a gylcol dehydration loop [For reboiler, see Figure 12–15]

pumping and metering stations (Eyen 1986). Figure 12–16 to 12–21 are referred to in Tables 12–12 and 12–13.

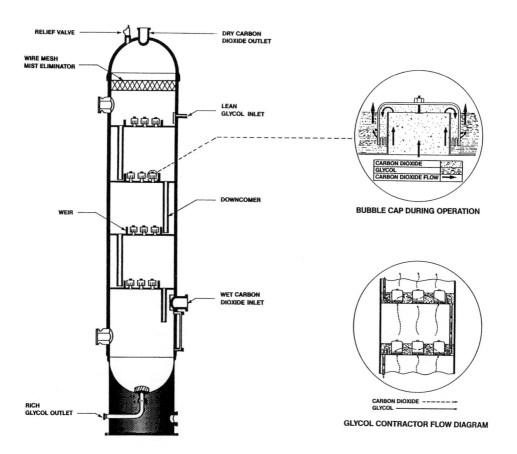

Figure 12-14. Cutaway of a typical glycol contractor

CO₂ Grade for Pipeline Transmission

The following compositions are typical of the CO_2 that is transported through pipelines:

- Carbon dioxide (CO_2): 98.372% or 98.350%
- Nitrogen (N_2): 1.521% or 0.136%
- Methane (CH_4): 0.107% or 1.514%
- Hydrogen sulphide (H_2S): 0.000

Pipeline Design Considerations

Design Pressure and Operating Ratings

Due to the inert characteristics of CO_2 and the requirement to keep this fluid above its critical pressure and temperature and in liquid form for pipeline transportation purposes (as well as for economic reasons), pipeline operating pressure and temperature ranges lie between 1,250 psi (8,619 kPa) at 4°C (40°F) (Farris 1983) and 2,220 psi (15,300 kPa) at 38°C (100°F). The upper limit is due to the ASME-ANSI 900# flange rating. The lower limit is set by winter ground conditions and is also set by the phase and behavior of CO_2 that is required to maintain supercritical conditions.

The phase diagram for pure CO_2 (Canjar et al., 1966) is depicted in Figure 12-22. This figure also illustrates typical cycles of compression and cooling for CO_2.

Usually CO_2, after being gathered from wells and prior to conditioning, is water-saturated, and a considerable amount of water is condensed during the first and second

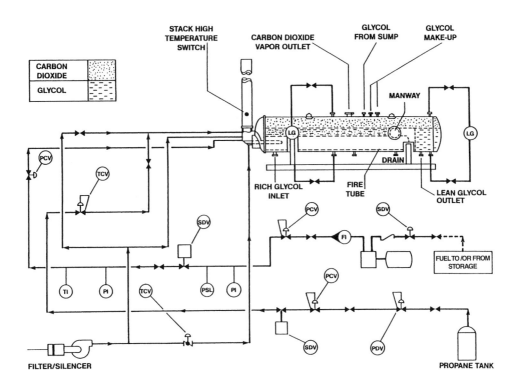

Figure 12-15. Cutaway of a glycol reboiler

TABLE 12-12. Summary of design and operating experience - CO_2 pipeline and metering (see bottom of Table)

Item	Description	Comments/Experience
1	Pipeline design	Material API 5LX65-X70, MOP 2,140 psi (14,750 kPa), ANSI, 900, flange rating, Depth of cover 4′ Most CO_2 pipelines are fitted with crack arrestors at 300–350 meters intervals (Fesmire 1983)
2	Pipeline welding	All welds are preheated and special welding rods such as 8018 C3 electrodes using a special welding procedure are utilized.
3	Pipeline coating and cathodic protection (CP)	A fusion bonded epoxy or polyurethane coating with full cathodic protection is utilized. Meter and control facilities usually provide the status of the CP. If alarmed, the CP system is checked.
4	Pipeline testing, commissioning and operation	For pipelines commissioned after they are filled with dry air or nitrogen at 100 to 500 psi (690–3,450 kPa). CO_2 displaces the dry air or nitrogen until the line is completely filled. The line is then pressurized. Temperature affects CO_2 much more than pressure. A change in temperature can cause an overrange in flow meters, but due to compressibility, pressure effects are negligible. CO_2 pipelines are hydrotested with water but test water is completely removed and then air dried to dewpoint of less than $-45°C$.
5	Relief valves	Anderson Greenwood or its equivalent is used and valves are tested with an nitrogen pilot. Valves must seal under high CO_2 pressure. Pressure relief arrangement must be installed on all exposed piping.
6	Valve operators	EIM and Limitorque operators or their equivalent are installed on all valves that use electric power.
7	Valve seat	Usually Viton as CO_2 eats Teflon. Other various elastometers have been used. EPDM with 85 hardness is the most suitable but cannot withstand hydrocarbon carryovers. It is good in decompression and dynamic works. Attention needs to be paid to the method of manufacturing elastometers. Extruded, machined, moulded, or formed elastometers all respond differently to dynamic conditions. For dynamic seals, the industry recommends using seals described by Watkins (1983). 4 RMS is recommended to be the best surface finish for valve seats. Dense chrome plating is best for hard seats. Anodized aluminum should be used if in contact with CO_2.
8	Valve sealant/lubricant	Chemflow or its equivalent is used. Petroleum-based products are not recommended as they react with CO_2 and decompose. Some sealants like Valtec 80 are not used as they harden when they are in contact with CO_2.
9	Valve gaskets	Flexitallic or its equivalent.
9a	Pressure regulator seats	60 HPRZ Polymide seats are typically seen in the industry.
10	Block valves/blowdowns	There are usually double-seating ball valves with Viton sealings; these seals vary. All block valves have blowndown assemblies. Blowdowns are sized for blowing down a 20 mile (32 km) section of pipeline in 6–8 hours to avoid dry ice formation. Industry specifies valves for sour service applications.
11	Valve operation	Slow opening of all blowdown valves is recommended. The industry does not work on pressurized piping. Pipeline segments and valve bodies always are depressurized prior to removal of valves. The industry never shuts in two valves on a short section/run (especially exposed piping) as high ambient temperatures can cause overpressuring of the pipe. Therefore, thermal reliefs or shut-in sections are used.
11a	Scraper traps	CO_2 pipeline are equipped with scraper traps. However, the industry has significant problems with pigging CO_2 pipelines when using rubber-type pig cups. These swell on contact with CO_2. Also, CO_2 has very low or no lubrication properties for a pig to move inside the pipeline. The industry experiences indicate that after pipeline commissioning, the use of scrapers is redundant as very little moisture dropout has been seen after pigging.
12	Fittings	Swagelok fittings are recommended by the industry.

TABLE 12-12. (Continued)

Item	Description	Comments/Experience
13	Meter station piping (Coating & CP)	All below-ground piping is epoxy-coated. Corrosion coupons are installed at every station.
14	Meters — orifice (Figure 12–16)	Typical orifice meter run design is shown in Figure 12-16. Flow computers are incorporated for custody transfer. They include differential pressure (DP) cells and temperature transmitters for measuring custody transfer flows.
	Meters — vortex (Figure 12–17)	These are used for flow calibration and line integrity checks. Industry recommends the use of a flow computer that does not require density measurements. Density can be computed from the monitored pressure and temperature readings. A typical design is depicted in Figure 12-17.
15	Quality control	H_2O, H_2S, N_2, and CO_2 composition are analyzed. Density measurements are usually at actual pipeline pressure and temperature conditions. (Figures 12-17 and 12-18).
16	Density measurement (Figure 12–18)	Industry's experience in use of electronic densitometers reflects that these instruments are very sensitive and drift on measurement and therefore are difficult to calibrate.
17	Flow controller	Programmable logic controller is used by the industry.
18	Hydrogen sulphide (H_2S) analyzer	Facilities must be equipped with an analyzer that can be easily utilized for measurement of H_2S. (Figure 12-17).
19	Carbon dioxide detection	Industry uses a mobile instrument, Furerite or its equivalent. The instrument needs to be housed in an enclosed building.
20	Moisture analyzer	Dupont analyzer or its equivalent with a desiccant column dryer is used by the industry. The analyzer compares dried CO_2 with samples from the pipeline and has a 10-minute sample cycle. Dupont 303 or its equivalent portable H_2O analyzer is also utilized for calibration or comparison purposes. The instrument is good only for 30–90 days based on the following H_2O contents: 100 lb/MMSCF (14 kg/MMsm^3) lasts 30 minutes. 30 lb/MMSCF (4.15 kg/MMsm^3) lasts 8–9 days. Less than 30 lb/MMSCF (4.15 kg/MMsm^3) lasts 30–90 days.
21	Instrument filters	Boston filters or their equivalent are used; these are generally changed once a week. The use of filters on all instruments is recommended to prevent problems.

* Refer to Figure 12–16 to 12–18.

compression cycles as shown in Figure 12-23 (Fesmire 1983). From Figure 12-23, it may be noted that the initial water content of CO_2 could be such that it would not be in a free water state prior to dehydration. This means that the selection of materials becomes less critical. However, it is important that the number of compression cycles be selected such that no water is formed after the final stage cooling.

Figure 12-24 illustrates the water content of CO_2 at various pressures and temperatures (Renfro 1979). Point one represents the typical water content of CO_2 at reservoir conditions. Typical separation processes would allow the saturated water content to drop to 1/4 of its original value, as shown at Point two.

Figure 12-25 illustrates the thermodynamic path that CO_2 goes through from initial reservoir conditions through well conditions, processing, compression, and heat recovery to pipeline operating condition. Both poor and good well conditions are illustrated. The pipeline is operating at 10,204 kPa (1,480 psia), the limit of ASME-ANSI 600# flange rating. As can be seen from Figure 12-25, in the case of a good well, the heat required for vaporization and superheating is approximately equal to the amount of heat available for recovery. Considering that there will be heat exchanger and transmission loss, supplemental heat will be required to compensate for these losses. In the case of a poor well, the heat required is small compared with the heat available for recovery, resulting in excess heat being available for the process. This would result in a reduction of supplemental heat

TABLE 12-13. Summary of design and operating experience — CO_2 pipeline pump and metering station [Cathro et al., 1986; Renfro 1979][*]

Item	Description	Comments/Experience
1	Booster pumps	The CO_2 pipeline industry usually uses centrifugal, single-stage, radial-split pumps. Inboard/outboard bearings: split sleeve with ball bearings. Pumps are checked for mechanical wear and efficiency every two years. The pump is manually started. Generally, there are no recycle lines on the pump. A typical booster pump arrangement is shown in Figure 12-19.
2	Pump/booster drivers	Generally, an electric driver, HZ 60, three-phase, 4,160 volts, operated between 1,500 and 2,000 RPM (single speed).
3	System lube oil	Stainless steel piping with three-stage lube oil pumping arrangement. Two centrifugal pumps are set to start up at about 2,000 psi (13,800 kPa) and stop at 2,500 psi (17,200 kPa). Each pump is fitted with a small orifice plate between its suction and discharge to measure the CO_2 for cooling the lube oil.
4	Booster seal arrangement	Borg Watner carbon face seals are typically used. Seals are oil-lubricated.
5	Booster strainer	An inlet strainer is installed on all pumps to remove contaminates.
6	Lube/seal oil	Petroleum-based, refrigeration-type, low-viscosity oil is used by the industry.
7	Station valves	Are generally Globe or Gate valves with body relief. Typical block valve arrangement is shown in Figure 12-20.
8	Valve grease	Camaflow 1,000 or equivalent body grease/sealant is used.
9	Pressure relief system	Pump stations typically have three pressure relief systems at three different relief settings (e.g., 2,140, 2,150 and 2,160 psi). The three settings (1,455 kPa, 1,462 kPa and 1,469 kPa, respectively) are due to the spongy effect of CO_2. Typical station piping relief system is schematically shown in Figures 12-19 and 12-21.
10	Meter run and provers	The industry usually insulates all meter runs, as a 1°F (0.5°C) temperature differential can swing pressure recordings up to 20 psi. Orifice plates (with a 0–400 inches pressure transmitter connected to a flow computer) are used for CO_2 flow measurements. Meter runs are typically equipped with differential pressure and temperature recorders as well as a densitometer. A pressure relief system is also added (see Figure 12-18).
11	Safety/inspection	Generally, a ball-type permanent meter prover is used for metering calibration. The provers are a piston-type ballistic prover. Proving is achieved for known CO_2 volume and pressure and ball speed. Orifice plates warp on rapid depressurization or pressure differentials greater than 30 psi (210 kPa). As a result, a 30-day inspection routine is recommended for the orifice plates. The industry recommends depressurizing CO_2 lines prior to any repair. CO_2 has an expansion factor of 150:1. Therefore, even a small pressure of 15 psi (100kPa) can cause serious damage if not depressurized.

[*] Refer to Figure 12–18 to 12–21.

requirements, depending upon the mix of production by well type. As previously noted, reservoir depletion increases the reservoir gas enthalpy, consequently reducing supplemental heat requirements. Eventually, in a typical CO_2 dome field, all wells will be produced into a lower pressure system and no supplemental heat will be required.

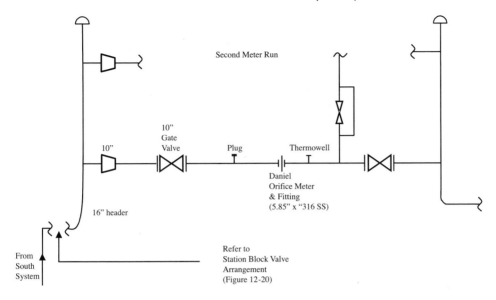

Figure 12-16. Typical orifice meter run (used for custody transfer) [Permission CRC Press]

CO₂ Properties Important to Pipeline Design

Some of the CO_2 properties that are useful in pipeline flow computations are depicted in Figures 12-26 to 12-31. These include specific heat (at constant pressure), density, viscosity, thermal conductivity, and enthalpy. These properties are typically commercially graded CO_2, produced and processed for pipeline transportation in western USA, and are provided as a guideline.

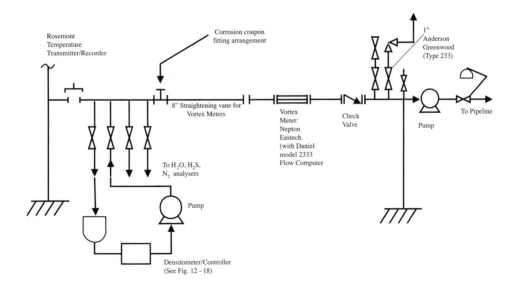

Notes: 1.) Densiometers are housed in a building and maintained at the same temperature as the CO_2 product in the pipeline.
2.) A pump is installed after the densitometer so that the density calculation is affected by the pump pressure returning CO_2 to the metering and pipeline.

Figure 12-17. Typical vortex meter run

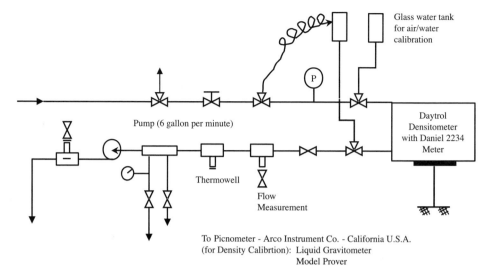

Figure 12-18. Densitometer piping arrangement

Phase behavior of CO_2 and CO_2-hydrocarbon and CO_2-sulfur (SO_2) mixtures that are important in enhanced oil recovery processes are detailed by Lee and Sigmond (1978), and Sayagh and Najman (1984). These mixtures are successfully processed, transported, and used for enhanced oil recovery by miscible flooding (Petroleum Frontiers 1984). The mixture attributes that contribute to improved oil displacements are:

- Viscosity reduction
- Oil swelling

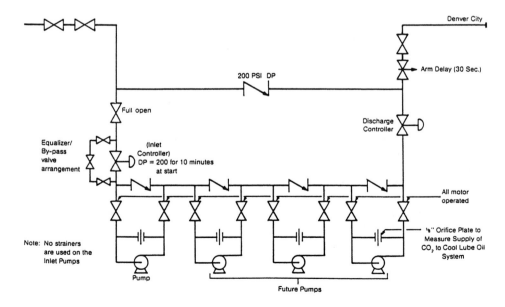

Figure 12-19. Booster pump piping arrangement

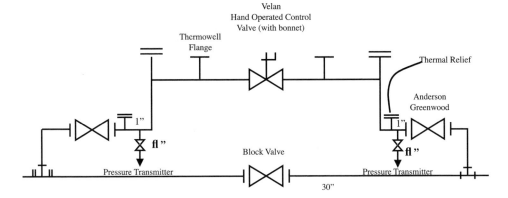

Figure 12-20. Station block valve arrangement (for sizes 4 inches and larger)

- Oil vaporization and extraction
- Condensation
- Permeability/porousity enhancement
- Blowdown recovery, etc.

CO_2 Impurities

Impurities influence the vapor pressure of carbon dioxide and affect pipeline fracture control/propagation properties. The impurities also affect pipeline capacity and hence system design.

Figure 12-32 illustrates the comparison of phase boundaries for pure CO_2 and for a CO_2/hydrogen mixture (King 1980, 1981). The vapor pressure sets the decompression pressure at a pipeline break and then determines if another ductile fracture can occur or not.

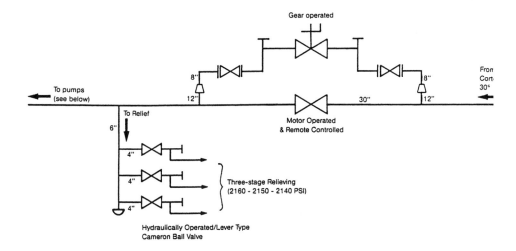

Figure 12-21. Inlet block valve and blowdown piping arrangement

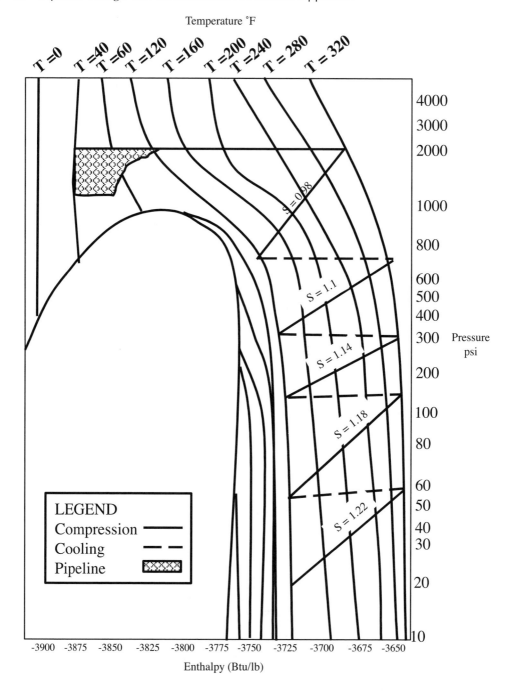

Figure 12-22. Idealized thermodynamic path of compression, cooling, and pipeline operations for carbon dioxide [adapted from GPSA, 1994]

Table 12-14 lists the effects of impurities in CO_2 on pipeline capacity. As much as 20% in capacity reduction of CO_2 transportation can be expected with increases in the level of impurities. Such impurities also affect pump design for CO_2 pipelines. In order to prevent cavitation (see Chapter 5), the minimum pump suction pressure must be set higher than the

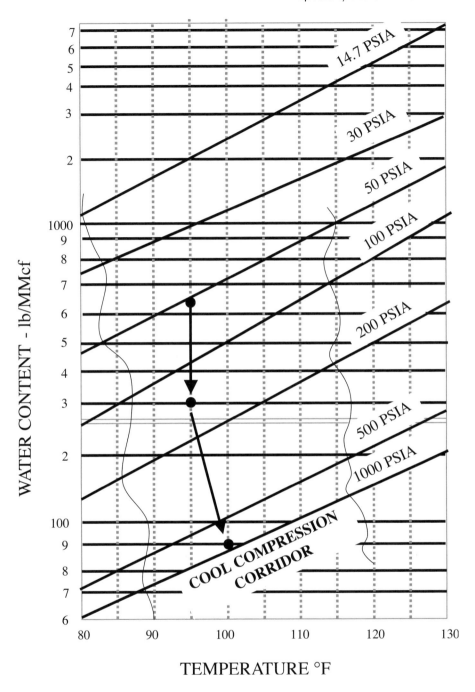

Figure 12-23. Saturated water content of CO_2 [Fesmier 1983]

fluid vapor pressure. A high pump suction pressure requires a corresponding higher maximum operating pressure so that optimum station spacing and flow rate can be attained.

Evaluation of centrifugal pumps for CO_2 pipeline applications is detailed by Eyen (1986). A rigorous hydraulic design analysis for CO_2 pipelining is described in detail by Hein (1986).

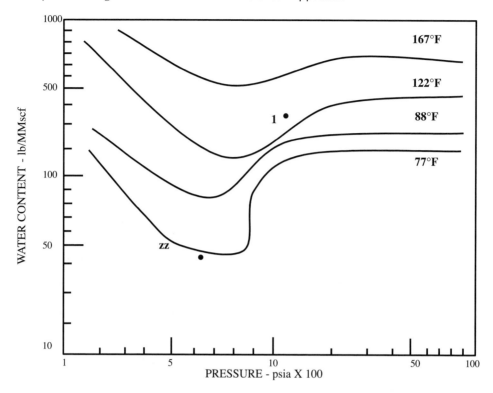

Figure 12-24. Saturated water content of 100 mol. % CO_2 [Renfro 1979]

Material Design

The arrest of propagating ductile fractures is an important criteria that needs to be considered when designing CO_2 pipelines. The fracture arrest criterion stipulated by Battelle (refer to Chapter 8) states that ductile fractures will not propagate if the pipeline is designed such that:

$$3.33 \frac{\sigma_d}{\sigma_f} > \frac{2}{\pi} \cos^{-1} \exp\left(\frac{-\pi E_N}{24}\right) \qquad (12-25)$$

where

$$\sigma_d = \frac{P_d D}{2t} \qquad (12-26)$$

and

$$E_N = \frac{E C_v^{\frac{1}{2}}}{A \sigma_f^2 \left(\frac{Dt}{2}\right)} \qquad (12-27)$$

where A = area beneath Charpy notch, m^2
C_v = material Charpy notch toughness, J
D = pipe outside diameter, m
E = Young's modulus of elasticity, Pa
E_N = normalized toughness parameter
P_d = decompressed pressure, Pa

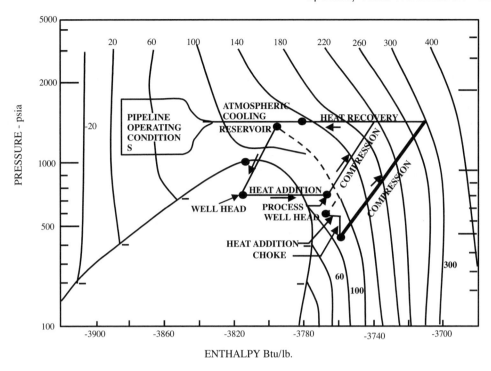

Figure 12-25. Thermodynamic paths from initial reservoir conditions through to processing, compression, and heat recovery to pipeline operating conditions [Renfro 1979]

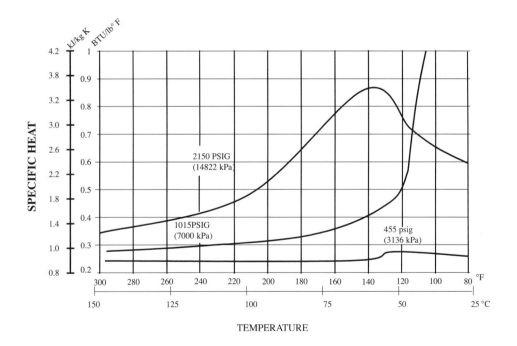

Figure 12-26. Commercial purity CO_2 properties — specific heat

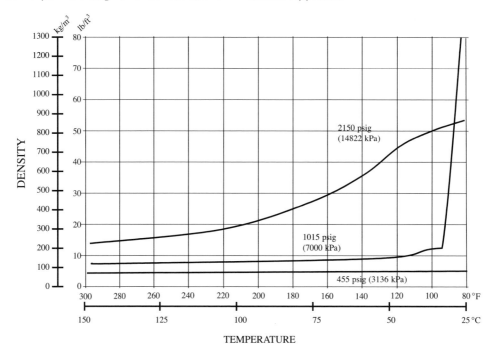

Figure 12-27. Commercial purity CO_2 properties — density

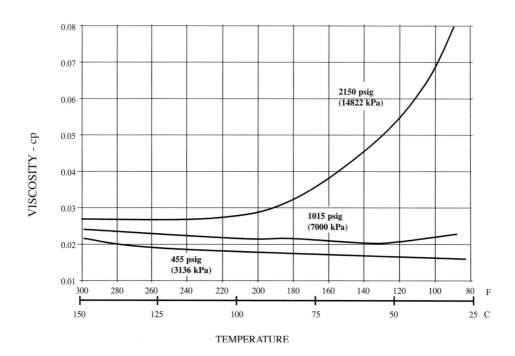

Figure 12-28. Commercial purity CO_2 properties — viscosity

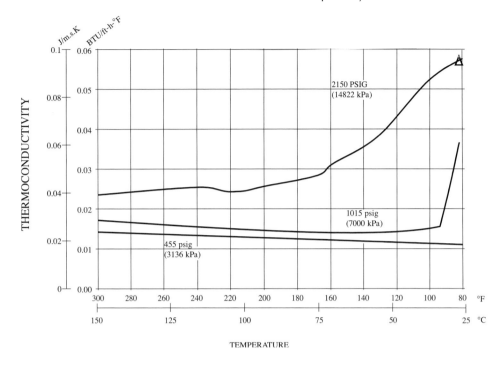

Figure 12-29.   Commercial purity CO_2 properties — thermoconductivity

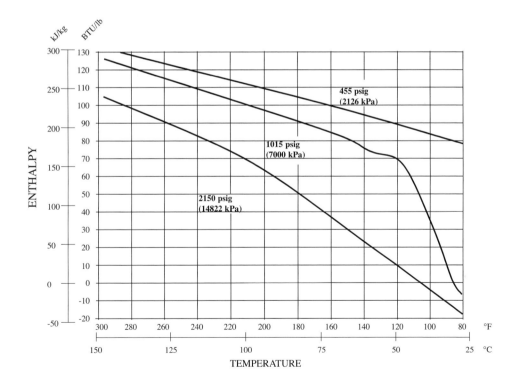

Figure 12-30.   Commercial purity CO_2 properties — enthalpy [King, G.G., 1980–81]

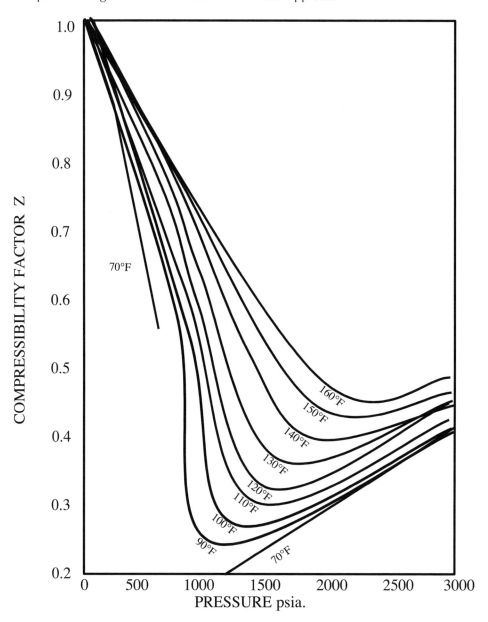

Figure 12-31. Compressibility factors (98% CO_2 – 2% CH_4 mixture) [Recht 1986]

    t = pipe wall thickness, m
    σ_d = decompressed pipe hoop stress, Pa
    σ_f = pipe steel flow stress, $\sigma_f = \sigma_y + 68.95$ MPa
    σ_y = yield stress, MPa

Figure 12-33 is a graphical representation of Equation (12-25) for a pipeline carrying pure CO_2 in a cold region where ground temperatures at pipeline depth do not exceed 15°C.
    Examination of Equation (12-25) indicates that the pipe flow stress σ_f has to be equal to or greater than the decompressed hoop stress σ_d by a factor of 3.33 to ensure avoidance of

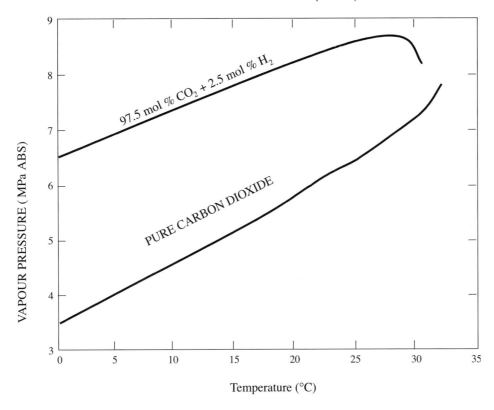

Figure 12-32. Comparison of phase boundaries for pure CO_2 and CO_2/hydrogen mixtures [King, G.G., 1980–81]

ductile fracture propagation in CO_2 pipelines. In this case, for a given Charpy value, either pipe strength or wall thickness has to be increased to satisfy this condition.

For a given pipe diameter and wall thickness, pipe cost does not radically vary with pipe grade to satisfy the Battelle fracture arrest hypotheses. Accordingly, if CO_2 pipeline is to be designed to control propagating fracture, from an economic viewpoint it is best to utilize lower grades of steel.

Therefore, as previously indicated, very high vapor pressure of CO_2 prevents rapid depressurization in the event that the pipeline breaks. On the other hand, high sustained pressures increase the chance of forming propagating ductile fractures. To avoid risks associated with this failure mode, it is best to select lower-strength steels and thicker wall pipe than might otherwise be selected. Nonetheless, high-strength steel (grade API 5LX 80) has been developed by U.S. Steel (Shoemaker et al., 1976) that is suitable for CO_2 transmission.

Hydrogen Pipeline Transmission
Background and Uses
Hydrogen was prepared many years before it was recognized as a distinct substance by Cavendish in 1766. The first recognition of hydrogen is attributed to Paracelsus around

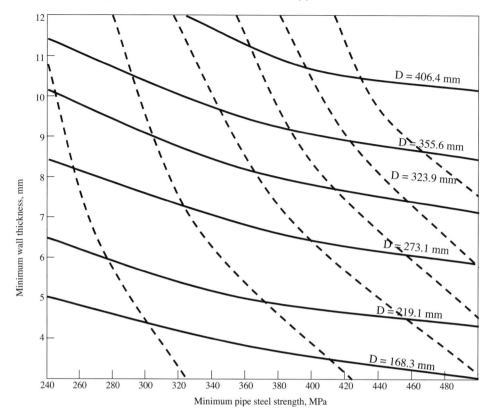

Figure 12-33. Minimum wall thickness of pipe satisfying ductile fracture arrest criterion; material toughness less than 27J (20 ft lb) [King 1981]

1500. It was named by Lavoisier. Hydrogen is the most abundant of all elements in the universe and it is thought that the heavier elements were, and still are, being built from hydrogen and helium. It has been established that hydrogen makes up more than 90% of all atoms or three quarters of the mass of the universe. It is found in the sun and most other stars and plays an important part in the proton-proton reaction and carbon-nitrogen cycle, which account for the energy of stars.

Hydrogen has been produced in metallic form at a pressure of 2.8 Mbar. It is present as a free element in the atmosphere, but only to the extent of less than 1 ppm by volume. It is the lightest of all gases and combines with other elements, sometimes explosively, to form compounds.

Great quantities of hydrogen are required commercially for ammonia production and for the hydrogenation of fat and oils. It is used in large quantities in methanol production, in hydrodealkylation, hydrocracking (for heavy oil production), hydrotreating, and hydro-desulphurization. It is used as rocket fuel, as an alternative to hydrocarbons for aviation and other uses (Reynolds and Slage 1974), for welding, for production of hydrochloric acid, and for the reduction of metallic ores. It is used as an energy source for power generation and transportation. Ammonia and urea production consume more hydrogen than any other applications.

TABLE 12-14. Decrease in pipeline capacity due to impurities in CO_2

Fluid	Decrease in Volumetric Capacity
CO2	1
CO2 + 5% methane	0.906
CO2 + 5% nitrogen	0.874
CO2 + 10% methane	0.837
CO2 + 5% hydrogen	0.817
CO2 + 10% methane and nitrogen	0.808
CO2 + 10% nitrogen	0.782
CO2 + 10% hydrogen	0.734

Note: Based on 82.7 kPa/km pressure drop at 10,341 kPa, NPS 16 pipeline 16°C (King 1981, Farris 1983).

Hydrogen Identification

Hydrogen identification, including its physical and chemical characteristics, is provided in Table 12-15(Environment Canada 1984). Hydrogen is a colorless, odorless gas, which burns in air with an almost-invisible flame. At atmospheric pressure, the ignition temperature of a hydrogen-air mixture can be as low as 500°C. The flammable limits of a hydrogen-air mixture depend upon pressure, temperature, and water vapor content. At atmospheric pressure, the flammable range is approximately 4–75% by volume of hydrogen in air.

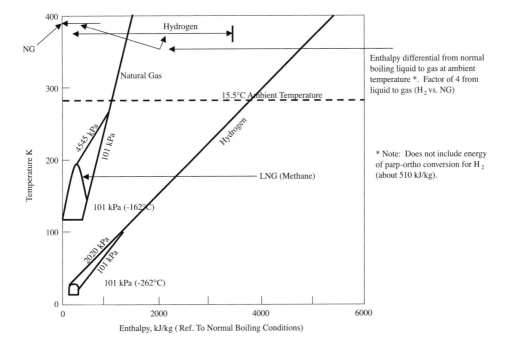

Figure 12-34a. Overall phase diagram and energy content for hydrogen and natural gas

Hydrogen Properties

Hydrogen is a mixture of 75% ortho-hydrogen (o-H_2) and 25% para-hydrogen (p-H_2). At atmospheric conditions (15°C, 101.325 kPa), normal hydrogen is ortho-hydrogen gas. At temperatures below 200K, it exists mainly as para-hydrogen. A comparison between hydrogen gas and natural gas phase diagrams is provided in Figure 12-34. (Mohitpour et al., 1988).

Properties that are important for hydraulics and general design calculations for hydrogen gas pipelines at various temperatures and pressures are: composition, density, specific volume, specific heat, conductivity, viscosity, enthalpy, entropy, Joule-Thompson coefficient, and sound velocity.

While no single equation of state can be recommended to predict some of the foregoing data for hydrogen gas (pure and raw), there are several analytical correlations that can provide computer solutions. To recommend an equation of state for hydrogen gas property prediction, detailed mixture composition and conditions must be known. For pipeline applications, a versatile equation of state is given in Rohleder (1972) A better and more accurate equation of state is given in McCarty (1979). This is a three-term modified Benedict Webb Rubin (MBWR) equation of state. The methods for prediction of various hydrogen properties, including hydrogen gas viscosity, are provided by McCarty (1979) and Matveev et al. (1984).

One of the most important properties useful in thermal analysis of hydrogen pipeline is the Joule-Thompson coefficient. The value of the Joule-Thompson coefficient can be calculated using Table 12-16 or from a paper by Michels et al. (1963).

Additionally, the following provides the equilibrium points for hydrogen:

- Triple point
  - Temperature = –259.203°C
  - Pressure = 7.259 kPa
  - Latent heat of fusion = 13.91 kcal/kg
- Boiling point (1 atm)
  - Temperature = –252.766°C
  - Latent heat = 108.5 kcal/kg
  - Liquid density = 70.973 kg/m3
  - Gas density = 1.312 kg/m3
- Critical point
  - Temperature = –239.91°C
  - Pressure = 1,315.2 kPa
  - Density = 30.08 kg/m3

The compressibility of hydrogen can be derived from Figure 12-35a (McCarty 1979). The temperature-entropy diagram for hydrogen is shown in Figures 12-35b and 12-35c.

The properties of pure hydrogen gas (useful for pipeline design) at the following conditions are shown in Table 12-16:

- Mixture molecular mass = 2.0160
- Base temperature = 0.000°C
- Base pressure = 101.325 kPa-abs
- Density at base conditions = 0.080 kg/m^3

TABLE 12-15. Hydrogen identification and characteristics

Transportation and Storage Information

Shipping state: Gas (compressed) and liquid (compressed gas).
Classification: Flammable gas.

Label(s): Red label — FLAMMABLE GAS; Class 2.1.
Storage temperature: Ambient

Grades or purity: Technical; pure from 99.8% to ultrapure.

Containers and material: As gas in cylinders, tube trailers; steel. As liquid in cargo or portable tanks.

Inert atmosphere: No requirement.

Hose type: Gas — braided-high pressure

Physical and Chemical Characteristics

Physical state (20°C, 1 atm): Gas.

Solubility (water): 0.00015 g/100 mL (20°C)
Molecular weight: 2.0

Vapor pressure: 8.590 mm Hg (−241°C).
Boiling point: −252.8°C

Floatability (water): Floats and boils.
Odor: Odorless flash point: <−50°C
Vapor density: 0.07 (gas) (25°C)
Specific gravity: (Liquid) 0.07 (−250°C)

Color: Colorless.

Explosive limits: 4.0–75% (in air)

Melting point: −259.2°C

Pipeline Quality Hydrogen Mixture

The raw gas from which hydrogen is obtained will have a composition generally consisting of hydrogen, carbon monoxide, carbon dioxide, methane, nitrogen, and traces of other impurities such as oxygen and water (Steinmetz 1982).

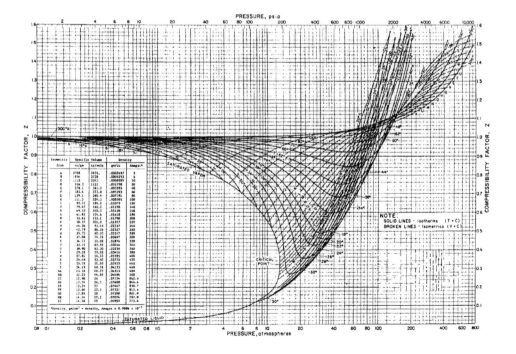

Figure 12-35a   Compressibility factor for normal pure hydrogen gas [Permission CRC Press]

618 ■ Pipeline Design and Construction: A Practical Approach

TABLE 12-16. Properties of pure hydrogen gas (McCarty 1979)

Pressure (kPa)	Temperature (K)	Specific Volume (m³/kg)	Density (kg/m³)	Thermal Conductivity (W/K-m × 10³)	Viscosity (kg/m.s. × 10⁷)	Enthalpy (kJ/kg)	Entropy (kJ/kg -K)	C_v (kJ/kg -K)	C_P (kJ/kg -K)	Velocity of Sound (m/s)
100.0	160	6.60298	0.1514	118.66	63.00	2,297.6	62.014	8.70	12.83	987
	180	7.42873	0.1346	128.16	66.17	2,558.3	63.549	9.09	13.22	1,040
	200	8.25429	0.1211	136.62	69.01	2,825.9	64.959	9.40	13.53	1,090
	220	9.07970	0.1101	144.33	71.66	3,099.2	66.261	9.65	13.78	1,139
	240	9.90502	0.1010	151.40	74.19	3,376.8	67.470	9.85	13.97	1,186
	260	10.73026	0.0932	157.97	76.64	3,657.9	68.595	9.99	14.12	1,232
	280	11.55544	0.0865	164.16	79.04	3,941.5	69.645	10.10	14.23	1,276
	300	12.37968	0.0808	170.01	81.41	4,226.9	70.629	10.18	14.31	1,319
	350	14.44218	0.0692	183.67	87.23	4,945.8	72.846	10.30	14.43	1,423
	400	16.50455	0.0606	196.50	92.97	5,668.5	74.777	10.35	14.48	1,520
1,000	160	0.66386	1.0563	119.03	63.01	2,294.2	52.474	8.71	12.93	996
	180	0.74718	1.3384	128.53	66.22	2,556.6	54.020	9.10	13.30	1,048
	200	0.83031	1.2044	136.99	69.09	2,825.6	55.437	9.41	13.59	1,099
	220	0.91330	1.8949	144.70	71.76	3,099.9	56.744	9.66	13.83	1,147
	240	0.99620	1.0038	151.79	74.30	3,378.5	57.956	9.85	14.02	1,194
	260	1.07902	0.9268	158.37	76.77	3,660.3	59.084	10.00	14.16	1,240
	280	1.16179	0.8607	167.57	79.18	3,944.5	60.137	10.11	14.26	1,284
	300	1.2446	0.8036	170.45	81.56	4,230.7	61.124	10.19	14.33	1,327
	350	1.45108	0.6891	184.16	87.42	4,950.5	63.344	10.30	14.44	1,430
	400	1.65757	0.6033	197.06	93.20	5,673.9	65.276	10.35	14.49	1,527
2,000	160	0.33405	2.9935	119.53	63.09	2,290.8	49.570	8.72	13.04	1,006
	180	0.37609	2.6590	129.01	66.32	2,555.0	51.126	9.11	13.38	1,058
	200	0.41794	2.3927	137.45	69.20	2,825.5	52.550	9.42	13.66	1,108
	220	0.45968	2.1754	145.16	71.88	3,101.0	53.863	9.67	13.88	1,157
	240	0.50131	1.9948	152.24	74.44	3,380.5	55.079	9.86	14.06	1,203
	260	0.54288	1.8420	158.84	76.92	3,663.1	56.211	10.01	14.19	1,249
	280	0.58439	1.7112	165.05	79.35	3,948.0	57.266	10.12	14.29	1,293
	300	0.62587	1.5978	170.94	81.75	4,234.9	58.255	10.20	14.36	1,336
	350	0.72937	1.3718	184.72	87.64	4,955.8	60.478	10.31	14.46	1,438
	400	0.83275	1.2008	197.68	93.45	5,680.0	62.412	10.36	14.50	1,534
3,000	160	0.22421	4.4602	120.12	63.21	2,287.9	47.853	8.74	13.13	1,016
	180	0.25246	3.9610	129.54	66.45	2,553.9	49.419	9.13	13.46	1,068
	200	0.28054	3.5646	137.95	69.35	2,825.7	50.851	9.43	13.72	1,118
	220	0.30851	3.2414	145.64	72.03	3,102.3	52.169	9.68	13.93	1,166
	240	0.33638	2.9728	152.72	74.60	3,382.7	53.390	9.87	14.10	1,213
	260	0.36419	2.7458	159.32	77.09	3,666.1	54.524	10.02	14.23	1,258
	280	0.39195	2.5514	165.54	19.53	3,951.6	55.582	10.12	14.32	1,301
	300	0.41969	2.3827	171.45	81.94	4,239.3	56.573	10.20	14.38	1,344
	350	0.48881	2.0158	185.28	87.86	4,961.2	58.800	10.31	14.48	1,446
	400	0.55781	1.7927	198.30	93.70	5,686.1	60.736	10.636	14.51	1,542
4,000	160	0.163935	5.9050	120.79	63.39	2,285.5	46.624	8.74	13.23	1,027
	180	0.19069	5.2448	130.14	66.63	2,553.1	48.200	9.14	13.53	1,078

(continued on next page)

TABLE 12-16. (Continued)

Pressure (kPa)	Temperature (K)	Specific Volume (m³/kg)	Density (kg/m³)	Thermal Conductivity (W/K-m × 10³)	Viscosity (kg/m.s. × 10⁷)	Enthalpy (kJ/kg)	Entropy (kJ/kg-K)	C_V (kJ/kg-K)	C_P (kJ/kg-K)	Velocity of Sound (m/s)z
	200	0.21188	4.7197	435.51	69.52	2,826.2	46.639	9.44	13.78	1,128
	220	0.23295	4.2927	146.17	72.21	3,103.9	50.962	9.69	13.98	1,176
	240	0.25394	3.9376	153.23	74.78	3,385.1	52.186	9.88	14.14	1,222
	260	0.27486	3.6382	159.82	77.27	2,669.2	53.324	10.02	14.26	1,267
	280	0.29574	3.3814	166.05	79.72	3,955.4	54.384	10.13	14.35	1,310
	300	0.31662	3.1574	171.97	82.13	7,273.9	55.377	10.21	14.41	1,353
	350	0.36854	2.7134	185.84	88.09	4,966.7	57.607	10.32	14.49	1,454
	400	0.42035	2.3790	198.92	93.96	5,692.3	59.545	10.36	14.52	1,550
5,000	160	0.13648	7.3272	1,215.55	63.63	2,283.5	45.664	8.77	13.32	1,038
	180	0.15367	6.5075	130.81	66.84	2,552.6	47.249	9.15	13.60	1,089
	200	0.17071	5.8579	139.10	69.72	2,827.0	48.694	9.46	13.83	1,138
	220	0.18764	5.3293	146.72	72.40	3,105.6	50.022	9.70	14.03	1,185
	240	0.20449	4.8902	153.76	74.97	3,387.7	51.250	9.89	14.18	1,231
	260	0.22128	4.5191	160.5	77.47	3,672.5	52.390	10.03	14.29	1,276
	280	0.23882	4.2013	166.57	79.97	3,959.2	53.452	10.14	14.38	1,319
	300	0.25478	3.9249	172.50	82.34	4,248.5	54.448	10.22	14.43	1,362
	350	0.29639	3.3740	186.41	88.32	4,972.3	56.681	10.32	14.51	1,462
	400	0.33788	2.9596	199.55	94.22	5,698.6	58.620	13.37	14.53	1,557
6,000	160	0.11460	8.7257	122.40	63.92	2,281.9	44.873	8.79	13.40	1,049
	180	0.12901	7.7511	131.53	637.90	2,552.5	46.467	9.16	13.66	1,099
	200	0.14328	6.9792	139.75	69.95	2,828.0	47.919	9.47	13.89	1,148
	220	0.15745	6.3512	147.31	72.62	3,107.6	79.251	9.71	14.07	1,195
	240	0.17154	5.8296	154.32	75.18	3,390.5	50.482	9.90	14.21	1,241
	260	0.18557	5.3888	160.89	77.68	3,675.9	51.625	10.04	14.32	1,285
	280	0.1995	5.0112	167.11	80.13	3,963.2	52689	10.15	14.40	1,328
	300	0.21357	4.6823	173.84	82.56	4,253.2	53.688	10.22	14.45	1,370
	350	0.24829	4.0275	186.98	88.55	4,987.0	55.923	10.33	14.52	1,471
	400	0.2298	3.5348	200.17	97.48	5,704.9	57.864	10.37	14.54	1,565
7,000	160	0.09901	10.1003	123.32	64.26	2,280.7	44.201	8.80	13.48	1,060
	180	0.11143	8.9746	132.31	67.38	2,552.7	45.803	9.18	13.73	1,110
	200	0.12371	8.0834	140.43	70.21	2,829.3	47.260	9.48	13.94	1,158
	220	0.13590	7.3585	147.94	72.83	3,109.8	48.597	9.72	14.11	1,205
	240	0.14801	6.7563	154391	75.41	3,393.4	49.831	9.91	14.25	1,250
	260	0.16007	6.2474	161.45	77.90	3,679.5	50.976	10.05	14.35	1,294
	280	0.17208	5.8112	167.66	80.35	3,967.3	52.043	10.15	14.43	1,337
	300	0.18414	5.4307	173.59	82.78	4,258.31	53.043	10.23	14.48	1,379
	350	0.21394	4.6743	187.56	88.79	4,983.8	55.281	10.33	14.54	1,479
	400	0.24363	4.1046	200.80	94.75	5,711.2	57.225	10.38	14.56	1,573
8,000	160	0.08733	11.4504	124.32	64.65	2,279.8	43.615	8.82	13.56	1,072
	180	0.09825	10.1779	133.15	67.71	2,553.2	45.225	9.19	13.79	1,121
	200	0.10904	9.1706	141.16	70.49	2,830.8	46.688	9.49	13.99	1,169
	220	0.11974	8.3512	148.60	73.12	3,112.2	48.028	9.73	14.15	1,215
	240	0.13037	7.6703	155.52	75.65	3,396.5	49.266	9.91	14.28	1,260

(continued on next page)

TABLE 12-16. (Continued)

Pressure (kPa)	Temperature (K)	Specific Volume (m³/kg)	Density (kg/m³)	Thermal Conductivity (W/K-m × 10³)	Viscosity (kg/m.s. × 10⁷)	Enthalpy (kJ/kg)	Entropy (kJ/kg –K)	C_v (kJ/kg –K)	C_p (kJ/kg –K)	Velocity of Sound (m/s)z
	260	0.14095	7.0948	162.03	78.13	3,683.2	50.413	10.05	14.38	1,303
	280	0.15148	6.6015	168.23	80.58	3,971.6	51.482	10.16	14.45	1,346
	300	0.16207	6.1701	174.15	83.01	4,263.0	52.485	10.24	14.50	1,388
	350	0.18818	5.3141	188.14	89.03	4,989.6	54.725	10.34	14.55	1,487
	400	0.21418	43.6698	201.43	95.02	5,717.7	56.670	10.38	14.57	1,581
10,000	160	0.07104	14.0768	126.53	65.56	2,279.2	42.629	8.85	13.69	1,096
	180	0.07985	12.5242	134.99	68.46	2,554.9	44.253	9.22	13.90	1,144
	200	0.08854	11.2944	142.74	71.14	2,834.5	45.726	9.51	14.07	1,190
	220	0.09715	10.2934	150.00	73.69	3,177.5	47.074	9.75	14.23	1,235
	240	0.10570	9.4610	156.81	76.18	3,403.2	48.317	9.93	14.35	1,279
	260	0.11419	8.7570	163.25	78.63	3,691.0	49.470	10.07	14.44	1,322
	280	0.12265	8.1531	169.41	81.07	3,980.3	50.542	10.17	14.50	1,364
	300	0.13119	7.6226	175.31	83.49	4,273.2	51.549	10.25	14.54	1,406
	350	0.15212	6.5738	189.32	89.53	5,001.4	53.794	10.35	14.58	1,503
	400	0.17295	5.7819	202.69	95.56	5,730.7	55.742	10.39	14.59	1,596
12,000	160	0.06040	16.5567	129.00	66.65	2,281.8	41.827	8.88	13.85	1,126
	180	0.06761	14.7907	137.02	69.34	2,557.4	43.453	9.24	14.03	1,167
	200	0.07490	13.3517	144.47	71.88	2,839.0	44.935	9.53	14.16	1,211
	220	0.08211	12.1790	151.52	74.35	3,123.4	46.290	9.77	14.30	1,255
	240	0.08926	11.2029	158.19	76.77	3,410.3	47.539	9.95	14.40	1,298
	260	0.09637	10.3766	164.54	79.19	3,699.2	48.696	10.09	14.49	1,340
	280	0.10344	9.6672	170.64	81.60	3,989.4	49.771	10.19	14.54	1,382
	300	0.11061	9.0407	176.51	84.01	4,283.8	50.782	10.27	14.57	1,424
	350	0.12809	7.8071	190.52	90.05	5,013.5	53.033	10.36	14.61	1,520
	400	0.14547	6.8741	203.95	96.11	5,743.9	54.983	10.40	14.60	1,611
14,000	160	0.05272	18.9674	131.67	67.87	2,283.8	41.135	8.92	13.96	1,152
	180	0.05909	16.9222	139.22	70.34	2,561.1	42.786	9.28	14.10	1,194
	200	0.06537	15.2983	146.34	72.72	2,847.2	44.278	9.56	14.23	1,275
	220	0.07138	14.0096	153.15	75.07	3,130.0	45.625	9.79	14.36	1,275
	240	0.07753	12.8974	159.65	77.42	3,418.0	46.879	9.97	14.46	1,318
	260	0.08365	11.9548	165.89	79.78	3,707.8	48.039	10.10	14.53	1,359
	280	0.08973	11.1448	171.92	82.16	3,998.9	49.117	10.20	14.58	1,399
	300	0.09592	10.4251	177.75	84.55	4,294.7	50.132	10.28	14.61	1,441
	350	0.11093	9.0149	191.74	90.58	5,025.9	52.388	10.37	14.63	1,536
	400	0.12585	7.9461	205.23	96.66	5,757.3	54.341	10.40	14.62	1,627
16,000	160	0.04700	21.2755	134.51	69.21	2,287.0	40.533	8.95	14.05	1,179
	180	0.05259	19.0141	141.56	71.43	2,569.0	42.195	9.31	14.19	1,219
	200	0.05810	17.2120	148.32	73.63	2,853.7	43.695	9.59	14.30	1,259
	220	0.06353	15.7408	154.87	75.85	3,140.4	45.061	9.82	14.39	1,299
	240	0.03875	14.5460	161.19	78.12	3,426.1	46.305	9.99	14.50	1,336
	260	0.07411	13.4930	167.38	80.41	3,716.8	47.768	10.12	14.58	1,377
	280	0.07945	12.5871	173.24	82.75	4,008.7	48.550	10.22	14.62	1,417
	300	0.08491	11.7769	179.02	85.12	4,305.9	49.569	10.30	14.64	1,459

(continued on next page)

TABLE 12-16. (Continued)

Pressure (kPa)	Temperature (K)	Specific Volume (m^3/kg)	Density (kg/m^3)	Thermal Conductivity (W/K-m $\times 10^3$)	Viscosity (kg/m.s. $\times 10^7$)	Enthalpy (kJ/kg)	Entropy (kJ/kg −K)	C_V (kJ/kg −K)	C_P (kJ/kg −K)	Velocity of Sound (m/s)
	350	0.09806	10.1977	192.98	91.13	5,038.5	51.828	10.38	14.65	1,553
	400	0.11113	8.9985	206.50	97.23	5,770.8	53.784	10.41	14.64	1,642
18,000	160	0.04258	23.4862	137.48	70.64	2,291.2	40.001	8.98	14.12	1,206
	180	0.04756	21.0278	144.02	72.60	2,574.8	41.672	9.34	14.26	1,243
	200	0.05246	19.0621	150.40	74.60	2,860.9	43.179	9.63	14.37	1,281
	220	0.02730	17.4520	156.67	76.69	3,148.9	44.551	9.85	14.45	1,320
	240	0.06208	16.1070	162.79	78.86	3,438.3	45.811	10.02	14.51	1,358
	260	0.06670	14.9925	168.77	81.09	3,726.4	46.965	10.14	14.58	1,394
	280	0.07145	13.9953	174.61	83.38	4,018.7	48.048	10.23	14.66	1,435
	300	0.07635	13.0969	180.33	58.71	4,317.4	49.071	10.31	14.67	1,447
	350	0.08806	11.3563	194.23	91.69	5,051.3	51.334	10.39	14.67	1,569
	400	0.09968	10.0318	207.79	97.80	5,784.6	53.292	10.42	14.65	1,657
20,000	160	0.03906	25.6015	140.53	72.14	2,296.6	39.525	9.01	14.19	1,234
	180	0.04355	22.9647	146.57	73.83	2,581.4	41.203	9.37	14.32	1,268
	200	0.04796	28.8486	152.56	75.64	2,868.7	42.717	9.66	14.43	1,384
	220	0.05233	19.1108	158.55	77.58	3,157.9	44.095	9.88	14.51	1,341
	240	0.05664	17.6519	164.46	79.64	3,448.5	45.359	10.05	14.56	1,378
	260	0.06091	16.4175	170.28	81.79	3,739.8	46.526	10.17	14.59	1,415
	280	0.06514	15.3589	176.02	84.03	4,031.5	47.606	10.26	14.60	1,451
	300	0.06951	14.3862	181.67	86.33	4,329.2	48.624	10.33	14.70	1,495
	350	0.08005	12.4914	195.58	92.27	5,064.3	50.892	10.41	14.69	1,585
	400	0.09053	11.0465	209.07	98.38	5,798.4	52.852	10.43	14.67	1,672

Commercially produced raw hydrogen gas will have composition as depicted in Table 12-17.

High H_2:CO ratio raw gas mixtures are suitable for use in various synthesis operations, such as methanol (methyl alcohol) production (Dotterweich 1978). The preferable H_2:CO ratio for methanol production is two.

For ammonia synthesis, a very high percentage of hydrogen is required. Carbon dioxide impurities must be eliminated to prevent poisoning of the ammonia synthesis process. A trace level of carbon dioxide is, however, beneficial in reducing the formation of hydrogen hard spots, allowing ammonia synthesis to proceed under more favorable conditions.

During ammonia production, oxygen-containing compounds are also poisonous and must be avoided. Typical oxygen compound concentration in catalytic converters usually does not exceed 2 vppm, although 10 vppm compounds are commercially acceptable. Chlorine, sulfur, and phosphorous are severely poisonous to the ammonia synthesis catalyst, so the raw gas required for the production of ammonia must be free of these impurities.

General Hydrogen Production

Hydrogen is prepared in several ways:

- By the action of steam on heated carbon
- By decomposition of certain hydrocarbons with heat
- By the electrolysis of water
- By displacement from acids by certain metals
- By the action of sodium or potassium hydroxide on aluminum

In North America, hydrogen that is used in refinery processes is produced either by catalytic reforming of gasoline fractions or by steam reforming of natural gas (SMR hydrogen). Hydrogen is also produced by delayed or fluid coking in oil sands processing. An additional production process is the partial oxidation of residual oil.

TABLE 12-17. Composition of commercial produced raw gas hydrogen mixture and pure hydrogen

		Mixture Composite (% by Volume)						
Ref.	Process	H_2	CO	CO_2	CH_4	N_2	H_2O	Others
A.	Methanol production							
1.	High H_2:CO ratio	65.0	33.0	0.0	1.4	0.6	0.0	
2.		65.6	32.8	0.0	0.4	1.2	0.0	
3.	Medium H_2:CO ratio	65.2	32.8	0.0	0.2	1.8	0.0	
4.		51.1	48.5	0.0	0.0	0.0	0.0	
5.	Low H_2:CO ratio	47.7	51.2	0.4	0.5	0.2	0.0	
6.		30.5	69.2			0.3	0.0	
7.		17.5	82.2	0.1		0.2	0.0	Trace
8.		7.5	73.4	18.5				
B.	Ammonia production							
9.	Raw gas	82.5	1.8	2.3	12.1	1.1	0.05	0.15
10.	Pure hydrogen 1	99.9				0.1		
11.	Pure hydrogen 2	99.87	0.01	0.01	0.1	0.01		

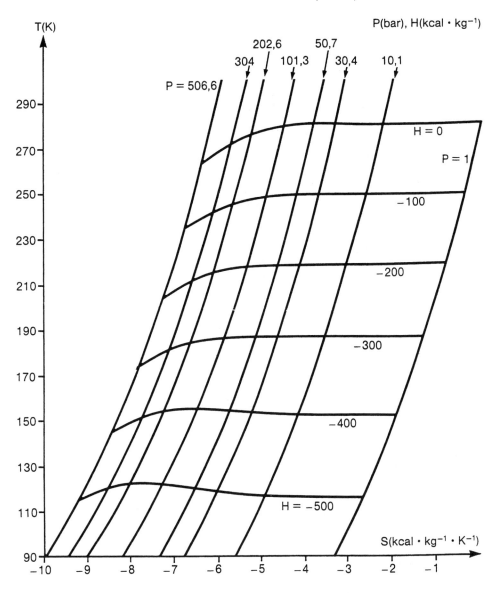

Figure 12-35b. TS Diagram for Normal Hydrogen (-10 < S < -1 kcal.kg^{-1}.K^{-1}) [Michel, et al., 1959, reprinted from PHYSICA, Courtesy of Elsevier Scientific Publishing Company]

The particular method used to produce hydrogen gas depends on a variety of considerations, such as volume, nature, and type of hydrogen usage, plant facilities, and purity of hydrogen desired. Although produced in large volumes as a by-product, very little of such hydrogen is recovered. Compression facilities are not usually available at these sources. The bulk of the hydrogen produced globally is consumed at or within pipeline distance of the production location. The various common hydrogen production methods are described below.

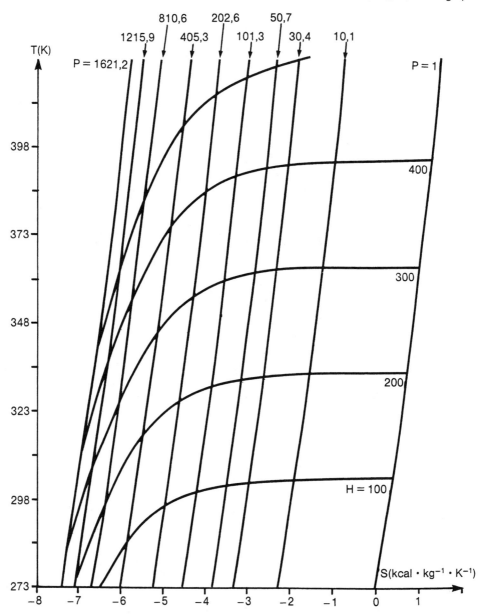

Figure 12-35c. TS Diagram for Normal Hydrogen (-8 < S < -1 kcal.kg⁻¹.K⁻¹) [Courtesy of Elsevier Scientific Publishing Company]

Electrolysis

Electrolysis of water or aqueous acid or alkali solutions is a source of hydrogen. At present there are not many "electrolytic" plants operated solely to produce oxygen and hydrogen from water, and in most of these cases hydrogen is the primary desired product. However, considerable hydrogen is obtained as a by-product in other types of electrolytic operations in the chemical industry where cells are used primarily to produce other commercially valuable end products, such as fertilizers.

Steam Reforming Oxidation from Natural Gas (SMR Hydrogen)

Most hydrogen for ammonia production is manufactured either by partial oxidation of natural gas or by steam reforming of natural gas to a mixture of hydrogen and oxides of carbon. During partial oxidation of natural gas to produce hydrogen and oxides of carbon, additional yield results from the water-gas shift reaction. The water-gas shift reaction converts most of the carbon monoxide into CO_2 and H_2. The CO_2 is removed by alkaline scrubbing, either with an amine solution or a regenerative caustic solution. The hydrogen-rich gas is refrigerated to low temperatures and purified by fractionation. The final liquefaction of hydrogen occurs at temperatures below $-230\,^{\circ}C$. In addition to the usual heat of liquefaction, which has to be removed, there is a high heat of transition involved (resulting from a catalyzed change in the hydrogen molecule).

Hydrocracking

Hydrogen can be produced through the cracking (dissociation) of ammonia. Further, it is obtained frequently as a by-product of cracking operations using petroleum liquids or vapors as feedstock and operated to produce other, more valuable products (Compressed 1974).

Other Methods of Hydrogen Production

Some other means of producing hydrogen (Compressed 1974) include:

- Passing steam over heated iron, which reduce the steam to hydrogen with an accompanying formation of iron oxide. Several variations of this fundamental "steam-iron" process are practiced industrially.
- The reaction of steam with incandescent coke or coal (water-gas reaction) is sometimes used as a source of hydrogen. Carbon monoxide is the other product of this process. There is also a catalytic water-gas process involving the use of excess steam, which breaks down to form more hydrogen while oxidizing the carbon monoxide to carbon dioxide.
- For use in inflating balloons, small quantities of hydrogen are sometimes generated "on the spot" from ferrosilicon or aluminum chips and caustic soda, or from materials such as lithium hydride.

High-Purity Hydrogen Production Plants

A typical plant for the production of high-purity hydrogen from natural gas, propane, or butane will consist of the following five sections:

1. Hydrocarbon handling and purification
2. Hydrogen production
3. Hydrogen purification
4. Amine reactivation
5. Hydrogen compression and storage

An older plant of this type is the Girdler's Hygirtol hydrogen manufacturing plant using propane gas (Read and Eriksen 1948). This type of plant will produce hydrogen sufficiently pure for all applications (for example, hydrogenation of edible oils), except for a few specialized uses such as the heat-treating of vacuum tube electrodes, in which even traces of carbon compounds cannot be tolerated. However, Hygirtol plant hydrogen can be purified further to be suitable for this use. An analysis of high-purity hydrogen gas produced in a Hygirtol plant is listed in Table 12-18.

Pipeline Transportation of Hydrogen
World Hydrogen Pipeline Experience

Pipeline transportation of hydrogen dates back to late 1930's.

Current world experience is of the order of 3000 kilometers (1900 miles) of hydrogen transmission pipelines up to $14''$ diameter, mostly designed to transport hydrogen in-plant for commercial use for feedstock or for pipeline fuel. In USA there is 1300 Kilometers (800 miles) of pipeline infrastructure, predominantly owned by Air Products, Praxair, Air Liquide and El Paso. Materials of construction range from different varieties of stainless steel, high and low grades of carbon steel, ductile cast iron and various alloys of aluminum alloys, copper, nickel and titanium. Polymer/fiber glass reinforced pipes are also used, as in-plant piping at moderate temperatures. These pipelines generally operate at less than 1000 PSI, with a good safety record.

Although produced in large volumes as a by-product, very little of such hydrogen is recovered. Compression facilities are not usually available at these sources. The bulk of the hydrogen produced globally is consumed at or within pipeline distance of the production location. However world experience of hydrogen gas pipeline transmission is increasing because of increasing hydrogen consumption currently 50 million pounds daily in the USA alone. Table 12-19. provide a list of hydrogen transmission pipelines that successfully operate worldwide with the largest networks in USA and Europe, Table 12-19, is an update previously compiled by Mohitpour et. al, 1990.

Hydrogen Pipeline Codes and Standards

There are a number of codes and standards that the industry in general follows for the design, construction and operation of hydrogen systems. Generally, modifications to existing codes related to the design of natural gas pipelines such as the ANSI B31.8 and the U.S. Department of Transportation regulations for Transportation of Natural and Other Gases by Pipelines and Minimum Safety Standards (USDT), approved by regulatory bodies, are used. There are also various codes, regulations and standards which govern or are related to the installation of hydrogen handling systems. Specific codes related to Gaseous Hydrogen Systems and codes for good practice as related to hydrogen handling are listed in Table 12-20. These codes are additional to those listed in Chapter 1, and are necessary when considering the design of hydrogen pipeline and handling facilities systems.

However the codes lack guidance for material selection and design for hydrogen compatible systems particularly high-pressure transmission pipelines.

Following the work of various ASME task groups under the DOE directive (US DOE 2002), the ASME Board on Pressure Technology Codes and Standards (ASME-BPTCS) has initiated the development of an independent consensus standard/code, B31.12 for

TABLE 12-18. High-purity H_2 composition from a typical Hygirtol plant

Component	Mole (%)
CO_2	0.01
CO	0.01
CH_4	0.10
N_2	0.01
H_2	99.87
Total	100.00

TABLE 12-19. Summary of hydrogen pipelines operating worldwide

Location	Pipeline Material	Years of Operation	Diameter (mm)	Length (km)	Pressure (kPa) and Gas Purity (%)	Experience Reported	Status
AGEC, Alberta, Canada	Gr. 290 (5LX X42)	Since 1987	273 x 4.8 WT	3.7	3,790 kPA-99.9	No	Operational
American Air Liquide Texas/Louisa, USA	API 5LX42, X52, X60 and others	?	3" to 14"	390	5100 kPa (740 PSI)	Yes	Operational
Air Products, Texas	Carbon Steel	Since 1967	50-450(2"-18')	357 (222 m)	1825-10030 kPa (Pure H2, 99.999% purity)	No	Operational
Air Products, Louisiana	ASTM 106, Carbon Steel	1983	101.6-304.8	163 (101m)	5514	Yes	Operational
Air Products, Sarnia (Dow to Dome plant)	Carbon Steel	Since 1981	150-300 (2"-6")	9 app.	4238-10030 kPa	No	Operational
Air Products, Cleveland Ohio	Carbon Steel	Since 1979	50& 100 (2" & 4")3.5		1136	Yes	Operational
Air Products, CA	Steel, schedule 40	Since 1982	150-250 (6"-10")	17.4	6651-15407 kPa-Pure H2 (throughput = 50 tons/day)	Yes	Operational
Air products, Netherland				45 Km	to 2,500: raw gas		Operational
Chemische Werke Huis	Seamless equipment to SAE 1016 Steel	Since 1938	168.3-273	215	(throughput = 300 x 106 m^3)	Yes	Operational
AG- Marl., Germany					>30,000.62 to 100% pure H2		
Cominco B.C., Canada	Carbon Steel (ASTM 210 seamless)	Since 1964	5 x 0.8125 WT	06		No	Standby
Gulf Petroleum Cnd, (Petromont- Varnnes)	Carbon Steel, seamless, Sch. 40	-	168.3	16	93.5% H2; 7.5% methane	No	Operational
Hawkeye Chemical, Iowa	ASTM A53 Gr. B	3	152.4	3.2	2.757.6	Yes	Operational
ICI Billingham, UK	Carbon Steel	-	-	15	30,000 kPa, pure	No	-
L'Air Liquide, France, Netherland, Belgium	Carbon Steel, seamless,	Since 1966	sizes up to 12"	879	6,484 - 10,000 kPa; pure and raw	No	Operational
LASL, N.M.	ASME A357-Gr. 5	-	25.4	6.4	13,788	Yes	Abandoned
Los Alamos, N.M.	5 Cr. - Mo (ASME A357 Gr. 5)	>8	30	6	13.790 pure	Yes	Abandoned
Linde, Germany	-	-	-	1.6-3.2	-	-	-
NASA-KSC, Fla	316 SS (austinitic)	>16	50	1.6-2	42,000 kPa	No	Operational
NSA-MSFC, Ala	ASTM A106-B	-	76.2	0.091	34470	Yes	Abandoned
Phillips Petroleum	ASTM A524	4	203.2	20.9	12,133-12,822	Yes	Operational
Praxair, Golf Coast, Tx, Indiana, California, Alabama, Louisiana, Michigan	Carbon Steel			450 Km	Commercial Purity H2 (500 MSCFD)		Operational
Rockwell International S. South Africa	SS-116	>10	250	-	>100,000 kPa; ultra pure	No	-
				80			?

(Dimensions in metric unless otherwise stated).

TABLE 12-20. Design codes for gaseous hydrogen systems and good practice

Code	Description
NFPA Standard 50A	Gaseous hydrogen systems at consumer sites
NFPA Standard #30	Standard for tank vehicles for flammable and combustible liquids
NFPA standard #386	Standard for portable shipping tanks for flammable and combustible liquids
NFPA Standard #68	Guide for explosion venting
Codes for Good Practice	
C-10	Recommendations for changes of service for compressed gas cylinders
C-11	Recommended practices for inspection of compressed gas cylinders at time of manufacture
C-14	Procedures for fire testing of DOT cylinder safety relief device systems
C-15	Procedures for cylinder design proof and service performance tests
G-5	Compressed gas association (CGA) pamphlet #5 "Hydrogen"
G-5.1	CGA "Standard of gaseous hydrogen at consumer sites"
G-5.3	CGA "Commodity specification for hydrogen"
P-1	Safe handling of compressed gas in containers
P-6	Standard density data atmospheric gases in hydrogen
P-12	Safe handling of cryogenic liquids
L'Air Liquide	Guide de securite hydrogen
ICG Code of Practice	Gaseous hydrogen stations

hydrogen piping and pipelines. The B31.12 piping and pipeline standard/code is based on the various current ASME Boiler, and other applicable codes including B31.8. It will contain materials science and engineering research information to provide a necessary confidence level to the users for the design, construction and operation of hydrogen piping and pipeline systems.

Pressure Drop Calculations

Accurate hydrogen pipeline and facilities sizing is possible using techniques similar to those described in Chapter 3 detailing gas pipeline hydraulics formulation and examples. Most computer programs, such as the LIC Energy Pipeline Studio (Houston, TX) have built-in equations of state that predict hydrogen properties important to hydraulic calculations (refer to section on Hydrogen Properties).

A nomograph providing a quick reference to hydrogen pipeline and compression sizing is provided in Figure 12-36.

Transporting Hydrogen as a Fuel Source

In many ways hydrogen is an ideal fuel source. It burns without contamination, uniting with oxygen in the air and producing, as a consequence of combustion, heat and water. However, the heat generated by the combustion of one cubic meter of hydrogen is about 10,000 kJ compared to natural gas, including primary methane, which produces approximately 33,000 kJ per cubic meter [ANSI/NPFA Standard 50A, 1984]. Thus, to transport an equal amount of energy, more than three times the volume of hydrogen must be transported compared to natural gas. In addition, because of the relatively low heat output resulting from the combustion of hydrogen, hydrogen gas must be stored in greater volumes in order to compare favorably with the storage requirements of other fuels.

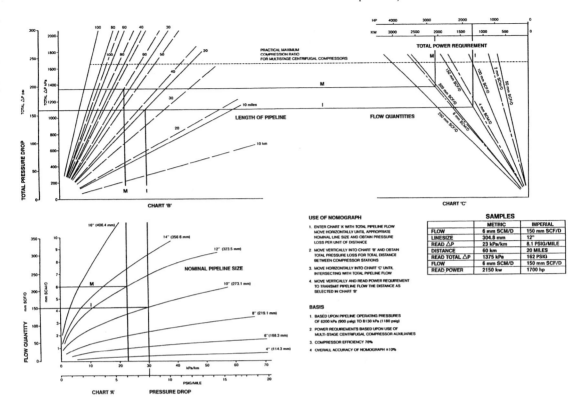

Figure 12-36. Hydrogen pipeline nomogram

Transporting Hydrogen by Pipelines

Due to the difference in physical properties between hydrogen and natural gas, transmission of hydrogen in pipelines will differ from transmission of natural gas. This difference is reflected in the economy of hydrogen pipelines as a whole, whether hydrogen gas is transported as a fuel source or for another use, such as urea production.

If existing gas pipelines are modified to deliver hydrogen as a fuel source, these pipelines would only be capable of delivering approximately 30% of the required energy. As a result, both compression power requirements and capacity have to be augmented to restore the energy capacity of the natural gas pipeline to its original value. Similar discrepancies will be evident if other gases such as oxygen or ammonia are transported.

It has been estimated that for the same energy flow transportation, hydrogen will cost between 30% and 50% more to transport than natural gas. Storage costs for hydrogen over natural gas will be of the same order as transportation cost. However, operation and maintenance costs will be even higher (Thomas 1980; Beghi et al., 1979). Operation and maintenance costs are 50%–70% more for hydrogen service (Konopka and Wurm 1897) due to higher hydrogen fuel cost and other operational limitations.

Economic Comparison

Figure 12-37 presents economic comparisons between hydrogen and other fluids for various flow rates (McAuliffe 1980). The comparisons are based on an idealized and hypothetical pipeline traversing relatively moderate terrain and only offers order of magnitude comparative costs. The "real world" value will depend on actual construction conditions and performance of each pipeline system. The relative energy transmission costs, assuming 100% usage and a 15% fixed charge rate, is provided in Figure 12-38 (McAuliffe 1980). The above figures are included to assist in establishing relative pipeline transportation costs.

Materials

Failure Modes

Hydrogen tends to react with the type of metals that are used to construct pipeline, storage, and other related facilities. These reactions can have a detrimental effect on the toughness, ductility, burst strength, and fatigue life of the pipeline.

The main problem associated with the exposure of pipelines to high-pressure hydrogen gas is the possibility of hydrogen attack or hydrogen embrittlement. Hydrogen embrittlement can be partially inhibited by the presence of trace amounts of impurity gases, such as oxygen or carbon dioxide, as long as the level of these impurities is acceptable to the industrial user of the hydrogen. The presence of carbon monoxide in hydrogen gas inhibits the pipeline material from hydrogen degradation at all stress levels and steel grades, including weld regions and hard spots, so long as the ratio of CO/H_2 is greater than 0.1 (Holbrook and Cialene 1985). Hydrogen-induced enhancement of crack initiation and hydrogen-induced crack growth can thus be avoided. No inhibition effect is possible with additions of N_2 or CH_4.

Factors influencing hydrogen embrittlement and inhibition are summarized in Figure 12-39.

Hydrogen Attack

Hydrogen attack can be distinguished from hydrogen embrittlement in that hydrogen attack occurs at temperatures above 200°C, whereas embrittlement occurs at temperatures below

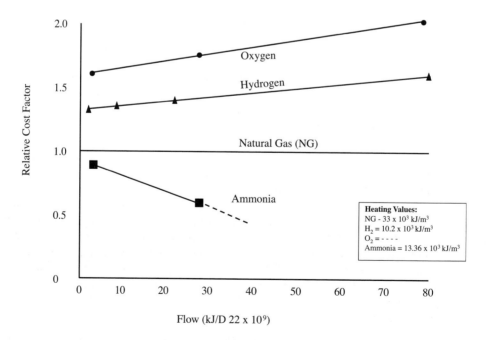

Figure 12-37.   Relative cost of hydrogen transmission versus other fluids [McAuliffe 1980]

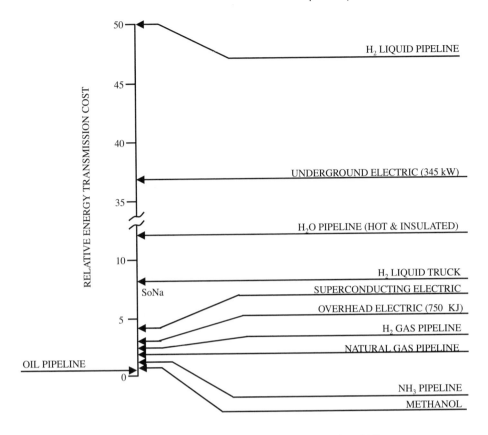

Figure 12-38. Relative energy transmission cost [McAuliffe 1980]

50°C (122°F). The process of hydrogen attack involves the diffusion of hydrogen into the steel and the subsequent nucleation, growth, and coalescence of methane bubbles primarily at the grain boundaries. The methane is produced by the elevated temperature reaction of the hydrogen gas with the carbon in the steel. The methane gas exerts an internal pressure, which can cause fissures or enlarged pores. The overall effect is that the strength of the steel is reduced.

The typical operating temperatures for a hydrogen gas pipeline system are usually in the range of –45°C to +50°C. Below these temperatures, hydrogen attack does not take place.

Hydrogen Embrittlement

Hydrogen embrittlement (HE) occurs at normal pipeline operating temperatures and affects the mechanical properties and behavior of the steel. Due to the complexity of the problem, the actual mechanism by which HE affects the steel is not fully understood. It is believed that hydrogen adsorbs to the surface of the steel and also diffuses into, or is absorbed by, the steel.

When hydrogen is adsorbed onto the steel surface (Figure 12-39) it is thought to reduce the surface energy required to form a surface crack. Diffused hydrogen may coalesce at voids or inclusions, resulting in a hydrogen pressure in the steel that causes an increase in internal stress and thus a lowering of the apparent fracture stress. Absorbed hydrogen affects the lattice of the material, which reduces the cohesive strength. This reduces apparent fracture stress. In all cases, the hydrogen concentration and the strength level of the steel, as well as other factors such as strain rate, chemical composition of steel, etc.,

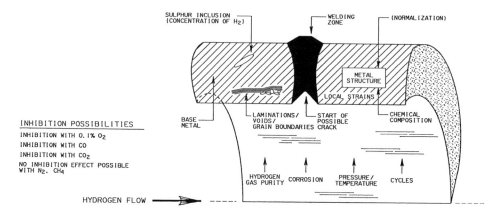

Figure 12-39. Factors affecting hydrogen embrittlement in pipelines

could have an effect. The end result of HE is that the steel can exhibit a loss in ductility, toughness, true stress at fracture, and load-carrying capability.

Hydrogen Induced Cracking

Structural failures resulting from hydrogen induced cracking occur suddenly in a brittle manner. It has been shown (Loginow and Phelps, 1974) that under sustained load, hydrogen degradation can be characterized by threshold stress intensity K_H typical for various type steels. Lower K_H would result in smaller defect size or lower operating pressures. It is not clear, however if this phenomenon varies with the pressure, microstructure and chemistry of steel and increasing pressure. If such a phenomenon can occur in pipelines operating in high pressure in hydrogen service, it is important to be cautious in the choice of materials for which K_H is unknown. In a sustained high-pressure hydrogen environment and with relatively modern high strength steels, any concern is justified because currently no data is available for pipeline steels to judge the severity of hydrogen. See Delayed Failure below.

Crack Growth in Hydrogen Environment

Crack growth process in metals is governed by the mechanical driving force at the crack tip, which is characterized by stress intensity K. The process is stepwise stable growth under sustained loading as well as under cyclical loading. If the environment is hydrogen, the process includes transport of dissociated hydrogen into the lattice structure ahead of the crack tip causing enhanced crack growth in any of the above crack growth processes. Combination of hydrogen enhanced crack growth and embrittlement is the phenomenon that needs utmost attention by the researchers.

Fatigue Crack growth in Hydrogen Environment

The conventional method of fatigue analysis is based on prediction of cyclic life based on applied stress or strain range vs. cycles to failure S-N curves. These curves are developed for each material by testing smooth round bar specimens subjected to full reversal loading. Consequently, the life includes crack initiation and crack propagation. Given the premise that materials do have or generate in service, cracks or crack-like defects, a fracture mechanics approach is warranted (Broek, 1983). Fatigue as a function of crack growth per cycle and stress intensity range at the crack tip is expressed in the form:

$$da/dn = C(\Delta K)^n$$

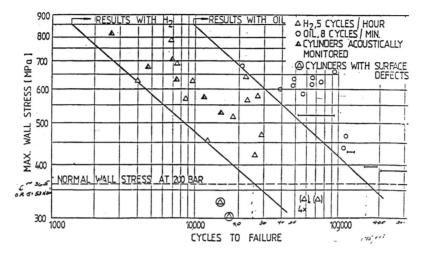

FIGURE 12-39a. Fatigue Data for Pipeline Material (Kisten and Runow 1981)

where C and n are the experimental values that define the fatigue behavior of the material.

Published fatigue data for pipelines that can be used as a base line to determine the effect of hydrogen environment are scarce. There is limited data on high strength quenched and tempered steel cylinders in hydrogen service, (Kisten and Runow 1981). From Figure 12-39a, it can be observed that for the quenched and tempered steels, in a hydrogen environment an order of magnitude loss in cyclic life is possible. There is a need to obtain basic fatigue data in both air and hydrogen environments. The experimental data need to include fatigue threshold, stable fatigue crack growth and low cycle fatigue that can be helpful to develop fatigue design curves for various pipeline steels.

Delayed Failure

One aspect of HE that is of considerable concern is the problem of hydrogen-delayed failure. In this case, steel that contains hydrogen and is subjected to a sustained load may fail at a stress level well below the measured tensile stress of the material. This type of failure is the most common in pipelines employed for hydrogen service and is, in essence, a result of the decreased true stress at fracture. In actual hydrogen service applications, delayed failure has been the most frequently encountered problem. It should be noted that such cases typically involved high-strength materials (SMYS greater than 500 MPa) (Interrante 1982; Tyson 1979).

Material Design

The typical operating temperature conditions of a pipeline system generally preclude hydrogen attack from being a problem. Therefore, hydrogen embrittlement is the major problem to be addressed in material design. Since hydrogen embrittlement affects the mechanical properties of steel, reductions in the material's burst strength, ductility, toughness, and fatigue life may occur. Precautions must, therefore, be taken to ensure the safety and reliability of a hydrogen pipeline system through proper design, material selection, and construction.

Ductility and Strength

The predominant effect of hydrogen on the mechanical properties of carbon steel is decreased ductility and true stress at fracture. This effect is manifested in smooth-bar tensile tests as a decrease in the reduction of area and elongation, and in notched-bar tensile tests as a decrease in the notched tensile strength.

Pipeline systems are designed to prevent plastic strain. However, soil movement or accidental mechanical damage can cause local yielding of the pipe. Should this occur, a reduction in the ductility of the pipe due to effects of hydrogen could lead to failure that would not be expected with "inert" gases. While this ductility loss has been found to be affected by the chemistry, microstructure, and strength level of the steel, the hydrogen content in the soil, the temperature and strain rate effects of the test, and the stress concentration of the specimen, there are no definitive correlations between these variables and the effects observed.

Various studies have been conducted with regard to the above phenomena and it is generally agreed that the effects are more pronounced where high-strength materials are employed. In high-strength steels, comparatively small amounts of hydrogen can lead to large changes in properties. Low-strength steels, however, appear to be affected to a lesser degree. A "rule of thumb" that has been used effectively to minimize the reduction in ductility-caused hydrogen embrittlement is to select steel with a tensile strength less than 700 MPa (Interrante 1982; Tyson 1979; Groeneveld 1974).

Sulfur Content

Another factor that should be considered in the selection of pipeline steel materials is the potential for hydrogen gas to congregate in voids or inclusions. By reducing the number of inclusions or voids in the steel, the possibility of hydrogen embrittlement (HE) occurring may be reduced. In order to achieve this, the sulfur content of the steel can be limited to reduce the number of sulfide inclusions. By limiting the sulfur content to 0.02%, it is believed that the steel will be less susceptible to HE.

Toughness

Loss in toughness is usually associated with the reduced ductility of steel in hydrogen transmission. While limits on strength levels and sulphur content to prevent HE may also prevent toughness loss, it is desirable to ensure that the steel has adequate toughness.

Fracture specimens exposed to hydrogen and tested using the J-integral technique have shown significant losses in toughness. Burst tests performed on preflawed pipes have also shown that the apparent fracture toughness is reduced by hydrogen. Both test methods indicate that maximum toughness losses of approximately 30% are possible even with low-grade steel, such as ASTM A516-70 and A106-B (SMYS of 260 MPa and 241 Mpa, respectively) (Hoover et al., 1981). However, it would be necessary to specify higher toughness values for these grades of pipe than would be required for natural gas service.

Burst Strength

Although the burst strength can be related to the true stress at fracture, burst testing of pipes allows for the direct assessment of the loss in the pressure-carrying ability of a pipe. This information can then be related to a factor of safety. Burst tests conducted on flawed pipes carrying hydrogen gas showed that the burst pressure will generally be 8–16% less than that required for bursting with an inert gas (Hoover et al., 1981). Since the pipes tested were flawed, this reduction in burst pressure is indicative of a loss in toughness. Smooth-bar tensile tests, however, also exhibited approximately 15% loss in fracture stress when exposed to hydrogen. This would indicate that the strength of the steel is also reduced. Thus, for a given pipe size and grade, the operating pressure/design factor of a hydrogen system would also have to be reduced by 8–16% to achieve the same margin of safety as a hydrogen-free pipeline.

Lab tests have demonstrated that the fatigue life of steels in hydrogen service can be reduced by a factor of three to six. These tests on high-strength hydrogen pressure vessels (SMYS greater than 550 MPa) indicate that fatigue failure could become a problem after 30,000 full pressure cycles to a stress level of approximately 40% of SMYS. While this reduction is significant, it is not considered to be a problem in standard gas pipelines that involve primarily static loads. While pipelines are not usually subjected to pressure changes of the same magnitude or number of pressure cycles, it is recommended that the number of pressure reversals involving changes in excess of 3,000 kPa be limited to 100 cycles/year (Kesten and Windgassen 1980; Bartholemy et al., 1980).

Table 12-21 lists specific material design criteria, which can be used to avoid failure problems in hydrogen pipelines. The criteria are similar to those used for conventional pipeline materials operating at normal temperatures, but are based on laboratory tests and experience with operating hydrogen pipelines.

While the normal operating temperature range of a hydrogen pipeline is recommended at below 50°C, certain pipeline facilities may experience higher temperatures and pressures. For example, the operating temperature and pressure in compressor discharge piping might be very high. Also, in plants or refineries or where hydrogen is produced or utilized for process purposes (e.g., for production of synthetic chemicals or fuel), piping materials could be experiencing temperatures of over 200°C. At higher temperatures hydrogen affects the mechanical properties of carbon steel in many ways. This not only applies to high temperature, but also to high hydrogen pressures. An example would be in heater tubes such as catalytic reformers and related piping.

Exposure to hydrogen at elevated temperatures can result in surface or internal decarburization (the latter with blistering). These forms of hydrogen damage can significantly reduce the strength of carbon and low alloy steel. Nelson's diagram (Figure 12-40) serves as a guide for selecting the minimum alloy content required to render piping and fittings suitable for hydrogen service. The carbon steel line in Figure 12-40 should be used for cast carbon steels (i.e., valves, pipe fittings, etc.).

The temperature to be used in Figure 12-40 should be the specified mechanical design metal temperature for the specific piece of equipment. If the temperature and hydrogen partial pressure fall on one of the curves shown in Figure 12-40, the next highest alloy should be used.

The calculation of hydrogen partial pressure should be based on vapor phase only as follows:

$$H_2 \text{ Partial Pressure} = \text{Mole Fraction of } H_2 \text{ in Vapor Phase} \\ \times \text{Total System Pressure(kPa)}$$

Austenitic stainless steels are satisfactory for all temperatures and pressures shown in Figure 12-40. If a corrosive medium is present in addition to hydrogen, the material selected should provide necessary resistance to both the corrosive medium and the hydrogen.

The Nelson diagram, however, does not specify material toughness requirement for high temperature service. This must be addressed separately but similarly to gas pipeline systems (refer to Chapter 8).

Selection

Figure 12-41 summarizes materials that are commonly used for a hydrogen transmission pipeline. It also summarizes the reported operating pressure for the various pipelines.

TABLE 12-21. Pipeline material design criteria for hydrogen gas service

Materials Design Criteria	Material Design Concern Addressed	Basis
Maximum material grade 290 MPa. Actual yield and tensile strengths not to exceed 414 MPa and 700 MPa, respectively.	Ductility and strength Burst strength Delayed failure	Lab Tests: • Steels with tensile strength less than 700 MPa are significantly less susceptible to hydrogen embrittlement (Interrante 1982; Tyson 1979; Groeneveld 1974) • Tests on A106B material with defects: 20%–90% of the material thickness showed no sustained load cracking (i.e., surface defects were stable with constant stress levels of 0.57–0.86 of pipe SMYS). The actual yield and tensile strengths must approximate to 450 MPa and 550 MPa, respectively. These tests results are indicative of materials utilized by the hydrogen industry (Hoover et al., 1981). Operating Pipelines: • Hydrogen pipelines have been in service for years with no reported problems. In all cases, low-strength materials (grade 290 or less) were used (Troiano 1960) • L'Air Liquide in France have hydrogen pipelines and have operated these lines with no reported problems. The yield strength levels used by the industry specified is equivalent to those by L'air Liquide (Interrante 1982).
Specify toughness levels that are at least 30% greater than normally specified for natural gas service.	Toughness	Lab tests: J-integral tests and burst tests on flawed pipes have shown that materials (A516-70, A106-B) subjected to hydrogen can exhibit a 30% loss in toughness as measured by fracture tests and Charpy energy. The toughness level specified by the industry (35J for pipe up to NPS 16) (Girard 1985).
Limit design factor to a maximum of 0.6 for hydrogen gas service.	Burst strength Delayed failure	Lab Tests: • Burst tests conducted with hydrogen on flawed pipes have shown that the burst pressure is reduced by 8–16% when compared to the results obtained for an "inert" gas. This loss in burst pressure is accounted for by a reduced design factor. For a hydrogen pipeline to have a factor of safety equivalent to that of a hydrogen-free pipeline, the design factor must be reduced by at least 16%. The adopted design factor of 0.6 adopted by the industry equates to a 25% reduction in the design factor compared to natural gas service (Hoover et al., 1981). Operating Pipelines: • Hydrogen pipelines operated by L'air Liquide have a design factor of 0.66 (Girard 1985).
Limit on operating design temperature of 40°C.	Delayed Failure	
Limit pressure reversals to 100 cycles/year involving pressure changes in excess of 3,000 kPa.	Fatigue	(Hoover et al., 1981; Bartholemy et al. 1980)

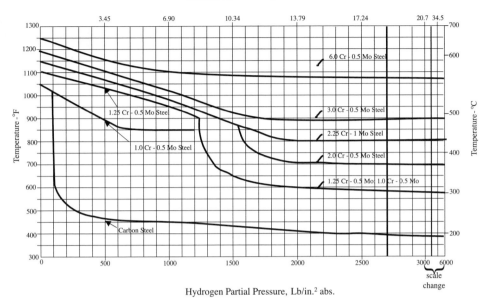

Figure 12-40. Operating limits for steels in high temperature and pressure service [(API 941, 1983) (reprinted courtesy of American Petroleum Institute)]

Most transmission lines operate within the range of 3,000 to 6,000 kPa and use lower grades of steel, typically equal to or lower than API 5L X42 (CSA Grade 290). At these pressures very few or no failures have been reported.

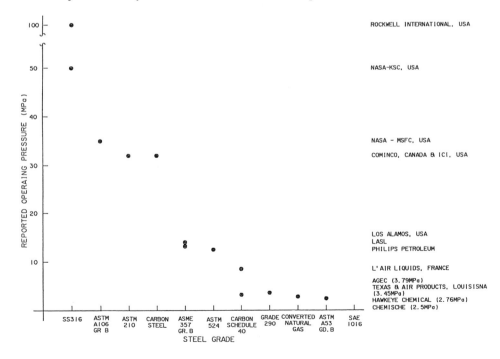

Figure 12-41. Historical pipeline material used for H_2 pipelines

A typical range of pipeline steel and appropriate carbon and manganese compositions well suited to hydrogen services is summarized in Figure 12-42. Steel with a low percentage of carbon and manganese is found to be the least susceptible to hydrogen embrittlement and attack.

Specific design criteria that can be used for hydrogen gas pipelines, and that can be used to prevent failures, are summarized in Table 12-22.

Table 12-22 provides standard notch toughness requirements for hydrogen service line pipe. Other requirements for valves, fittings, and operating limits of materials (temperature and pressure) are provided in Table 12-23.

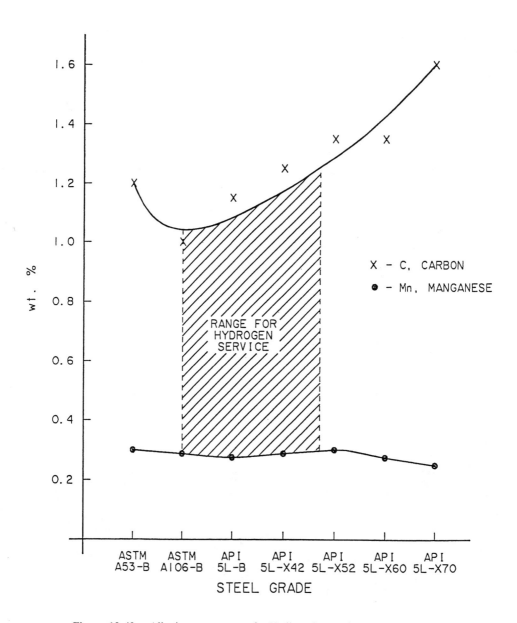

Figure 12-42. Alloying components for H$_2$ line pipe steels

TABLE 12-22. Standard notch toughness requirement for H_2 line pipe service

	Full-Size Charpy V-Notch Absorbed Energy (J)							Pipe Body	
	Pipe Body			Weld [†]				Shear Area (%)	
	Any Heat			Spiral, Circumferential, or Skelp End		Longitudinal			
Pipe size (in.)	Minimum Any Specimen	Average Three Specimens	All Heat Average	Minimum Any Specimen	Average Three Specimens	Minimum Any Specimen	Average Three Specimens	Any Heat Minimum	All Heat Average
NPS 48<	66	88	140	51	66	66	88	78	100
NPS 42–46	53	70	—	40	53	53	70	78	100
NPS 36–40	46	61	—	35	46	46	61	78	100
NPS 30–32	39	52	—	30	39	39	52	78	100
NPS 18–28	36	44	—	27	36	36	44	78	100
NPS 16>	26	35	—	20	26	26	35	78	100

* Measured by the drop weight tear test for wall thickness of 5 mm and larger, and by the Charpy V-notch test for wall thickness smaller than 5 mm.
[†] Applicable to SAW pipe only.

Safety Considerations
Leakage and Fire

Due to its small molecular size, hydrogen is more likely to leak than other gases. It follows that the major hazard associated with pipeline transmission of hydrogen is leakage. The range of flammability limits of hydrogen is unusually large, 4–75% by volume with air at normal temperature and pressure (natural gas limits are 5–15%). The ignition temperature of hydrogen (570°C) is higher than most hydrocarbons (200–370°C), but its ignition energy (the amount of heat required to ignite hydrogen) is an order of magnitude lower (ignition energy: 0.02 mJ at 30% hydrogen in air at standard pressure versus 0.3 mJ for natural gas). As indicated in Figure 12-43, this hydrogen energy decreases with pressure. Consequently, the possibility of sparks affecting hydrogen ignition is far greater at higher pressures.

Another example of safety considerations is that no flashlight has yet been approved for hydrogen-bound facilities. Once ignited, flame speeds are in the order of 10 times higher for hydrogen than for natural gas (0.87–3 m/s versus 0.03–0.3 m/s for natural gas) (ANSI/NFPA Standard 50A 1984).

TABLE 12-23. Standard notch toughness requirements for fittings, valves, and flanges for hydrogen service (NPS 16 and larger)

	Full-Size Charpy V-Notch Absorbed Energy (J)	
	Minimum Any Specimen	Average Three Specimens
Fittings		
NPS 44 and larger	39	52
NPS 42–NPS 38	35	44
NPS 36–NPS 16	26	35
Valves		
NPS 16 and larger	20	26
Flanges		
NPS 44 and larger	39	52
NPS 42–NPS 38	35	44
NPS 36–NPS 16	26	35

Pure hydrogen flames are invisible, and are characterized by a low level of radiation in the surrounding area. A hydrogen fire is less likely than a hydrocarbon fire to ignite other combustible materials by radiation. Since a hydrogen flame is nearly invisible, it is difficult to detect visually, which makes it more of a hazard to personnel than hydrocarbon flames.

Flammability and Combustion Data

Hydrogen flammability details are provided below (Thomas 1980; Drell and Belles 1958; Knowlton 1984; Grumer et al., 1968).

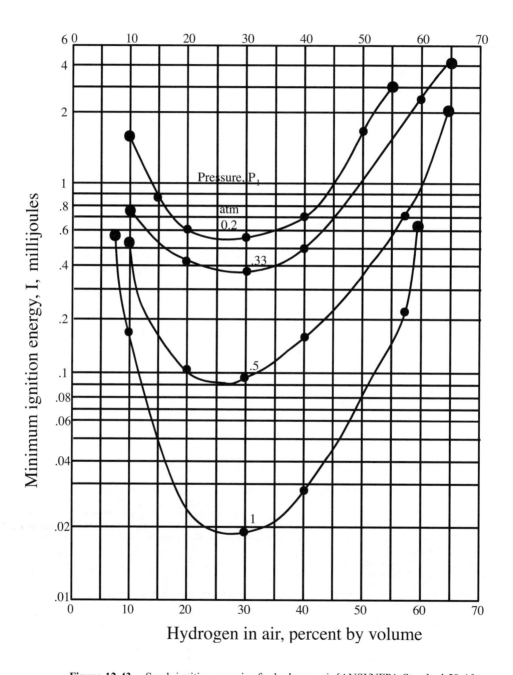

Figure 12-43. Spark ignition energies for hydrogen air [ANSI/NFPA Standard 50 A]

Flammability in Air (Drell and Belles 1958):

- Flammability in air at 20°C and 101.325 kPa: lower at 4%; upper at 74.5%
- Minimum auto-ignition temperature at 101.325 kPa: 520–570°C
- Maximum auto-ignition temperature at 101.325 kPa: 677°C
  Note: The minimum energy of combustion of hydrogen is very low, 0.02 mJ, at least 10 times lower than that of hydrocarbons (0.28 for methane, 0.25 for propane).

- Stoichiometric combustion
  - Flame temperature: 1,430°C to 2,150°C (Knowlton 1984)
  - Flame speed (maximum): 2.6 m/s
  - Flame emissivity 0.1: (Drell and Belles 1958)
  - Maximum burning velocity 3.0 m/s (Figure 12-44)

Flammability in oxygen

- Flammability limits at 20°C, and 101.325 kPa: lower at 4%; upper at 94%
- Minimum auto-ignition temperature at 101.325 kPa, 560°C
- Stoichiometric combustion
  - Flame temperature: 2,830°C
  - Flame speed: 14.46 m/S
  - Heat of combustion: High 3,050 kcal/m^3; low 2,570 kcal/m^3

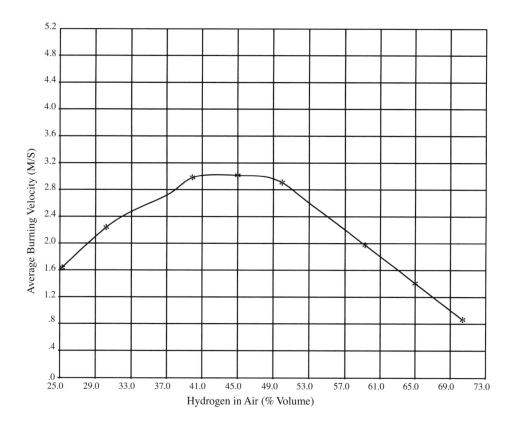

Figure 12-44. Average burning velocities of hydrogen air mixtures

Other properties

- Heat of combustion in air: 242 kJ/g mole
- Air required for combustion: 804 m^3/kcal
- Relative orifice capacity kcal/m^3: H_2 = 0.93; Natural gas = 3.03

Explosion

An explosion of hydrogen-air mixtures is very unlikely to occur as a result of a pipeline leak or rupture. Detonation limits of hydrogen-air mixtures range from 18% to 59% by volume. Since the hydrogen molecule has a very low mass, it rises rapidly in air. Unless the leak occurs in a confined region (i.e., an enclosed space), it is improbable that a concentration of 18% can be reached. Furthermore, the ignition energy is very large compared to that required for combustion.

Release in Confined/Enclosed Areas

The "orifice capacity," or the rate at which gas will escape through a hole of a given size, is three times greater for hydrogen on a volume basis than natural gas but, because of the lower heating capacity, the escape of energy is only 0.93 times as fast (AGA Publication XU0179 1981).

If this hydrogen is escaping into a confined space, it will reach the lower flammable limit of 4% by volume in 0.26 of the time, or 3.78 times faster than natural gas will reach its 5% threshold value. At this point, however, the energy contained within the confined space that would be released in a fire or explosion is only one-quarter of the energy of a 5% natural gas-air mixture occupying the same volume. Thus, relatively small hydrogen explosions can be contained within the walls of laboratory glassware while natural gas explosions cannot.

If no ignition occurs and escape continues into the confined space, the hydrogen will reach its upper flammability limit of 75% in air and become "safe" again in 1.6 times longer than it would take for natural gas to reach its upper "safe" condition of 15% or above. However, in practical cases, some degree of ventilation is likely, and it is very possible that hydrogen will never reach its upper flammable level while natural gas can do so readily.

Release in Unconfined Space

If the limited release of hydrogen occurs in an unconfined space, it is observed that the hydrogen moves away from the release point at a far greater rate than does natural gas for two reasons: (1) the density of hydrogen is only one-fourteenth that of air, compared with about two-third for natural gas (methane), so its tendency to rise is far greater; and (2) its smaller molecular size makes its diffusion rate in air 2.82 times faster than that of methane.

If a flammable mixture is created, the lower ignition energy of hydrogen will cause it to ignite far more readily than natural gas. Since the ignition energy for hydrogen is only 0.02 mJ, an invisible static spark can contain this energy. Thus, it is far more difficult to eliminate ignition sources for hydrogen than for natural gas.

Once a flame occurs, the rate of propagation of the flame is between 6 and 100 times faster for hydrogen than for natural gas, depending on the air ratio. Although extremely rapid combustion of hydrogen can therefore occur, detonation of hydrogen-air mixtures is unlikely in an unconfined space.

Hazard Data/Toxicity and Spill

Hazards related to human health and fire due to hydrogen spill and chemical reactions have been detailed by Environment Canada (1984).

Hydrogen is nontoxic, but can cause anoxia (asphyxiation) when it displaces the normal 21% oxygen in a confined area without adequate ventilation. Because hydrogen is colorless, odorless, and tasteless, its presence cannot be detected by human senses. It is therefore essential that hydrogen detection instrumentation be installed in a confined area.

Emergency measures required to protect humans and the environment in case of a hydrogen spill have also been provided by Environment Canada (1984).

Peculiar Temperature Effects

Hydrogen has a Joule-Thomson inversion temperature of –43°C, far below ambient temperatures. The natural gas inversion temperature is 677°C. At ambient temperatures, unlike natural gas, hydrogen undergoes a temperature increase upon throttled expansion. The extent of the observed temperature rise depends upon the pressure drop, the initial gas temperature, and the rate of heat exchange between the hydrogen and the environment. This Joule-Thomson effect could have an effect in plastic components if used for hydrogen service. Thermoplastic pipe for natural gas service, for instance, does not have to be pressure-tested (at the greater of 345 kPa or 150% of operating pressure) at temperatures above 38°C (Knowlton 1984). The Joule-Thomson effect on temperature rise is demonstrated in Figure 12-45 (Johnston et al., 1946).

Blowdown, Line Purge, and Maintenance

An area of serious safety concern is the blowdown and purging requirements for hydrogen pipelines, both when commissioning new pipelines and when repairing existing systems.

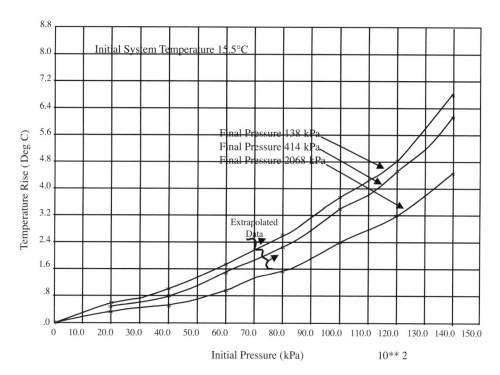

Figure 12-45. Joule-Thomson effect on hydrogen (calculated temperature rise upon expansion) [Jhonston et al., 1946]

The use of an inert gas to sweep out air, or to sweep out natural gas (if a natural gas line is converted for hydrogen service), will be necessary. This is because of the wide flammability limits of hydrogen (4–75%) compared with natural gas (5–15%).

For natural gas pipelines, any welding repair or operation can be and usually is carried out when the gas-air mixture in the pipeline components is well above 15%. But for hydrogen pipelines, the hydrogen-to-air ratio must be safely above 75%. For this purpose, special piping design and purging procedures may be required.

Pipeline/Pipe Design Features
Safety Features

To ensure leak-proof and safely operating hydrogen gas pipelines, certain design features must be considered. These include provisions for a positive shut-off system and a fully welded pipeline facility (where feasible). The positive shut-off purge will ensure prevention of air and moisture entry into the pipeline in case of a blowdown. A design feature incorporating a purge station, with nitrogen providing positive pressure in the vent/blowdown line, is depicted in Figure 12-46.

For safety reasons, prevention measures must be taken when designing pipeline/piping arrangements and components, including:

- Elimination of ignition sources
- Elimination of ingress air or oxygen into the piping system — use welded components as much as possible
- Design against static electricity pickup — hydrogen discharge to atmosphere to be orientated downward
- Elimination of confined areas
- Use of an explosion-proof and pressure-shock-resistant system in confined areas
- Use of an explosion-relief and suppression system inside confined areas

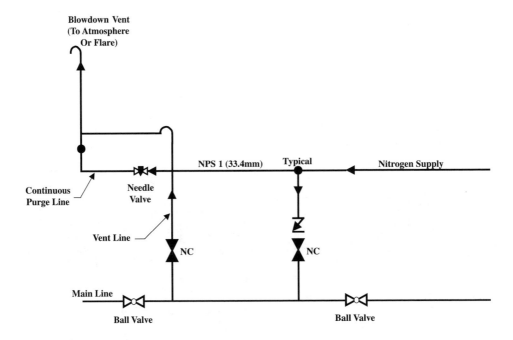

Figure 12-46.   Typical hydrogen pipeline purge piping system

Metering Design

Typical meter station facilities for hydrogen services are schematically depicted in Figure 12-47. A station of meter runs suitable for hydrogen service must be capable of accurately measuring the maximum flow at specified pressure and temperature conditions. Either orifice or turbine meters may be utilized for flow measurement.

In adverse environments, part of one of the meter runs (in multirun stations) might have to be enclosed in an instrumentation building. This will allow for hydrogen samples (for density and moisture analysis) to be obtained at actual flowing conditions. However, a building enclosure limits both the gauge line length and the range of operating temperature.

In most plants, scrubbing and filtering of hydrogen gas is undertaken prior to metering facilities.

If an enclosure or a building is required it must be equipped with hydrogen gas detectors and have continuous ventilation for at least eight air changes per hour (*ASHRAE Handbook* 1984).

Valves, Flanges, and Fittings

Table 12-24 summarizes some types of valves utilized by a few hydrogen pipeline and plant facilities. The industry's experience indicates that "soft-seal" valves normally used for bubble-tight natural gas service are inadequate for high-pressure hydrogen applications. Hard-seal bubble-type values for absolute zero leaking need to be used (Kendrick 1985).

All valves for hydrogen service need to be subjected to a hydrogen gas or a helium-gas seat test. Helium gas is safer and provides a small molecule with which to test the valve seats. Hydrogen molecules are binary molecules, as compared to helium, which is inatomic. Even though the hydrogen atom is the simplest atom known, the hydrogen molecule is larger than the helium molecule.

Where gate valves are used for hydrogen process applications, the valves need to have a stuffing/packing box and valve spindle finish of 32 RMS (roughness measurement system). For hydrogen services, use of welded connection/joints wherever practical is recommended. However, situations arise where flange connections need to be used. In these

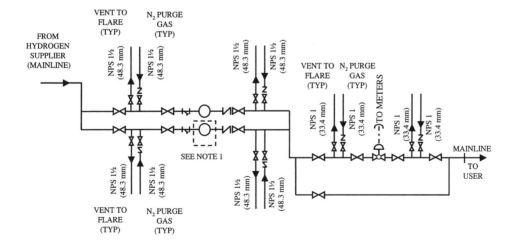

NOTE:

1. BUILDING ENCLOSURE FOR INSTRUMENTATION AND METER RUN ONLY.

Figure 12-47. Typical metering/instrument station for hydrogen pipeline

TABLE 12-24. Types of valves used in hydrogen service

Valve Type	Manufacturer	Condition	Comments
1. Gate/globe/check/ball	Tufflin Jamesbury	—	Do not place valve horizontally in piping
	KYMR	—	
		High pressure/temperature (positive shutoff)	—
2. Ball/gate	Cameron	Moderate pressure/temperature	—
3. Gate	—	Low pressure/temperature	Old styles mainly in use — stem in vertical position
4. Gate	Flexitallic Int. Ltd. (FIVE)	High pressure/temperature positive shutoff	(See Kendrick 1985)

situations, flange finishes need to be specified as 125–200 AARH (average arithmetic roughness height) with a concentric ring on spiral serration design. Gaskets that are generally specified include Graphoil-Filler gaskets or "Flexitallic" asbestos-filled gasket as per API 601 series.

For hydrogen service where blinding of process equipment is anticipated, use of removable spacer rings that are designed and installed to allow for a stress-free situation is recommended.

Weld Joint Design

Numerous welded joints are required in the fabrication of hydrogen transmission systems. Weld joints (weldments) appear to be more susceptible to hydrogen degradation/embrittlement than the parent pipe. Moreover, various welding processes may yield different degrees of susceptibility. As a result, welding processes that minimize fusion- and heat-affected zones (HAZ) and produce high integrity as well as inclusion-free welds should ideally be utilized. Also, it is important to utilize weld electrodes that are compatible with the parent metal, the environment, and which produce welds that have low susceptibility to hydrogen degradation. This requires welds to be of lower toughness than usually obtained for welding of natural gas pipelines.

Weld micro-hardness for hydrogen services is limited to HRC 22. This level of hardness limits hydrogen-induced cracking, as well as laminar tearing, and produces acceptable fracture toughness in the weldments and the adjoining heat-affected zone.

Conventional arc welding processes (with welding directions downhill, uphill, or flat) can be utilized for welding of hydrogen transmission pipeline similar to natural gas. Mostly cellulosic welding electrodes, with downhill welding direction, can be employed for service temperatures down to $-5°C$.

For service temperatures down to $-45°C$ (e.g., above-ground piping and assemblies in very cold environments), the practice is to utilize uphill welding with low hydrogen electrodes. All low temperature service piping requires preheating to $100°C$ prior to welding. The weld direction for shop welds is generally in a flat position.

Protection

Cathodic Protection

In order to avoid hydrogen-induced cracking and to maintain the ductile fracture properties of steel pipe, application off-potential not less than -1.2 volts [with regard to saturated copper-copper sulphate ($Cu/Cu\text{-}SO_4$) reference cell] to pipeline is recommended. Any potential more negative than -1.2 volts could result in excessive hydrogen evolution, and therefore could enhance hydrogen embrittlement/cracking of the steel and would reduce ductility (Hinton and Proctor 1981; Mayer 1982).

Coating

Lower off-potential than suggested in cathodic protection could also cause cathodic disbondment. Insofar as coating selection is concerned, no specific requirement is envisaged for a hydrogen pipeline coating system. Coating materials similar to those used for natural gas transmission can be utilized.

Depth of Cover (DOC) and Clearance

Generally, DOC for hydrogen-pipeline should follow natural gas pipeline codes. However, from a safety point of view, industry practice has been to adapt a minimum of 1.2 meters for depth of cover. Thirty-meter clearance is recommended where a hydrogen pipeline is paralleling another pipeline, except in established pipeline corridors, when 10 meters minimum is accepted by the industry.

FUNDAMENTALS OF LPG PIPELINING

Background

Liquefied Petroleum Gas (LPG) is a mixture of light hydrocarbons, gaseous at normal temperature (15°C) and pressure (101.329 kPa) and maintained in the liquid state by increased pressure or lowered temperature.

The two liquefied petroleum gases in general are referred to as 'commercial butane' and 'commercial propane'. They exist as gases at normal ambient temperatures but can be liquefied under moderate pressure. Propane has a lower 'boiling point' than butane, i.e. it turns into gas at a lower temperature. In order for it to become liquefied, propane is stored under higher pressure, approximately 700 kPa as opposed to about 200 kPa for butane.

Because of their high heating values, their high purity and cleanness of combustion and their easy handling, LPG finds very wide application in a large variety of industrial, commercial, domestic and leisure uses.

Liquid Petroleum Gas (LPG or LP Gas) is normally created as a by-product of petroleum refining and from natural gas production. The main composition of LPG is propane, propylene, butane, and butylene in various mixtures. However, for all fuels typically the normal components of LPG, are propane (C_3H_8, Figure 12-48) and butane (C_4H_{10}, Figure 12-49). Small concentrations of other hydrocarbons may also be present. Depending on the source of the LPG and how it has been produced, components other than hydrocarbons may also be present.

LPG is used as a fuel in heating appliances and vehicles, and increasingly is replacing fluorocarbons as an aerosol propellant and a refrigerant to reduce damage to the ozone layer.

One liter of liquid propane liberates 311 liters of propane gas (15°C - 1 bar) while one liter of liquid butane liberates 239 liters of butane gas (15°C - 1 bar). These physical characteristics provide great advantage concerning storage and transport.

LPG Timeline & Consumptions

In the early 1900's, gasoline used for automobiles was difficult to handle. Due to inadequate refining techniques, the gasoline would quickly evaporate while it was in storage. Under the direction of Dr. Walter Snelling (a Chemist), the U.S. Bureau of Mines began experiments to stabilize gasoline. Through these experiments, Dr. Snelling discovered that the gases which evaporated could be condensed and stored as a liquid at moderate temperatures and pressures. By 1911, Dr. Snelling had isolated and identified these gases as propane and butane, the two major components of LP-gasses. 1912 saw the development of the first propane stove and hence the use of propane gas for cooking food in the home. The first car powered by propane ran in 1913, and by 1915 propane was used in torches to cut through metal. Propane was marketed for flame cutting and cooking applications and LPG was sold commercially by 1920.

In 1913 the first butane lighter "Wonderliter" was produced and in 1914 a patent was grated to the company which later became RONSON (Ronson, 2004). 1928 saw the first automatic lighter patented in America.

The timeline of major events leading to identification, use and the construction of the first LPG pipeline systems is provided in below (Mohitpour, et al, 2006).

900 BC Earliest development of pipeline by Chinese
347 AD Oil wells drilled in China
1264 Mining of seep oil in Persia (Marco Polo)
1594 Oil wells hand dug in Baku, Persia
1735 Mining of oil sands in Alsace, France
1802 First commercial use of natural gas (J Watt steam engine factory)
1806 First gas mains to be laid in a public street, London
1815 First production of oil in the USA
1846 Distillation of kerosene (Dr A Gesner, Canada)
1848 First modern oil well in Asia
1854 World's first oil company (Charles Tripp)
1854 First oil wells in Europe (Bóbrka, Polang: Ignacy Lukasiewicz)
1857 Invention of kerosene lamp (Michael Dietz)
1858 First oil well in North America (Ontario, Canada)
1859 First oil well in US (Titusville, Pennsylvania: Colonel E Drake)
1878 Invention of electric light bulb (Thomas Edison)
1879 First US long distance pipeline (Tidewater Pipeline, 174 km NPS6, pumped crude over the Allegheny)
1886 Gasoline-powered automobiles (Karl Benz and Wilhelm Daimler)
1886 Louis V Aronson forms Art metal Works in New York (RONSON Metal Works)
1911 Identification of Propane & Butane (W Snelling, US Bureau of Mines)
1912 Development of first propane stove
1913 First car powered by propane
1913 First butane lighter "Wonderliter" by RONSON
1914 Patent granted for fuel to sustain a flame
1915 First use of propane in torches for metal cutting
1917 Formation of Phillips Petroleum Company (Oklahoma)
1918 Sale of Propane Patent by Dr Snelling to Frank Phillips
1920 First commercial sale of LPG
1928 First automatic lighter patented in America
1940 First construction of LPG pipelines (Panhandle to Texas regions James Harold Dunn)

In 1927, the total sales of propane in the U.S. were more than one million gallons per year, and after World War II the propane gas annual sales increased to more than 15 billion gallons and in 1994, over 1640 billion gallon of LPG was consumed in the United States alone. In 2002 consumption in the United States exceed 1700 billion gallons, PERC 2004, EIA 2004.

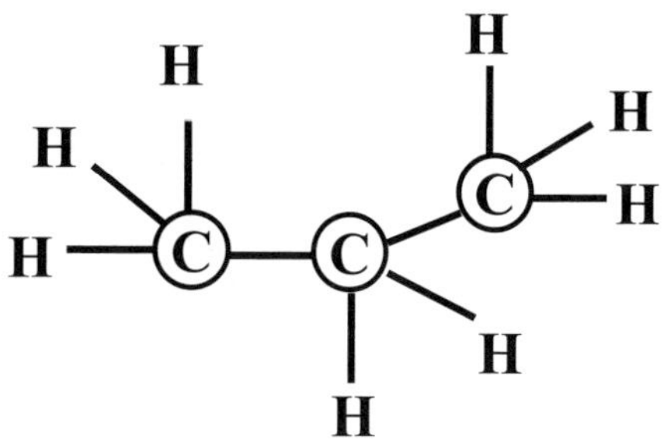

Figure 12-48. Propane Constituents

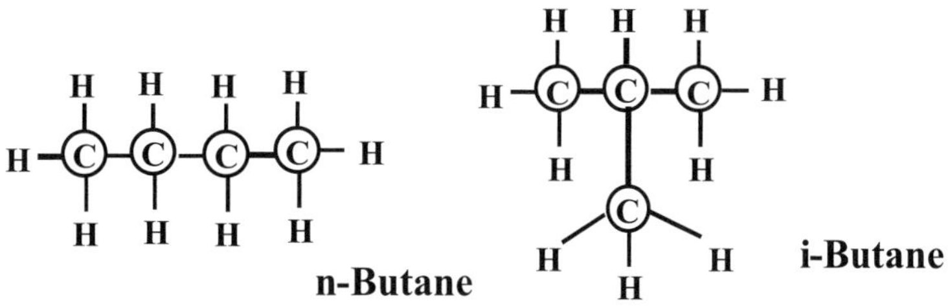

n-Butane   i-Butane

Figure 12-49. Butane Constituents

Figure 12-50 illustrates the total LPG and Propane Consumption in the Unites States. Figure 12-51 provides the breakdown of US industrial propane consumption (excluding feed stocks) for 2002. Comparison of 1997 and 2002 propane market sales indicate that the breakdown for usage had remained relatively constant with residential, chemical/refining and the industrial sectors using the bulk of the propane each respectively 35%, 42% and 11%, (PERC 2004, MCS 2001, NPGA 2004). In 2002, odorized propane sales accounted for 59% of total U.S. propane consumption; the other 41% was consumed as a non-odorized feedstock fuel by the chemical industry and other industrial applications, NPGA, 2004.

By contrast the global consumption of LPG in 1999 was for 192 billion tonnes (>100,000 Billion gallons) and has since risen to over 200 billion tonnes (EIA 2004). The breakdown of world consumption for LPG is shown in Figure 12-52, (MCS 2001). The global consumption by sectors for 2003 is shown in Figure 12-53.

By the 1930s, the Compressed Gas Association (CGA) proposed a set of recommendations to the National Fire Protection Association (NFPA). In 1932, the first pamphlet of standards (No. 58) was adopted for publication, NFPA 2004.

When Dr. Snelling sold his propane patent to Frank Phillips, the founder of Phillips Petroleum Company, his price was $50,000. Today, propane gas is an $8 billion industry in the United States alone and it is still growing.

James Harold Dunn was instrumental in constructing the first LPG pipelines in early 1940 from the Panhandle field to other regions of Texas and the United States in association with the Phillips Petroleum Company, TSHA, 2002.

LPG Properties/Characteristics/Uses

The physical characteristics of typical commercial LPG are provided in Table 12-25 & Figure 12-54, L'Air Liquide, 1976, EIA 1994. A phase diagram for propane is shown in Figure 12-55 (L'Air Liquide, 1976).

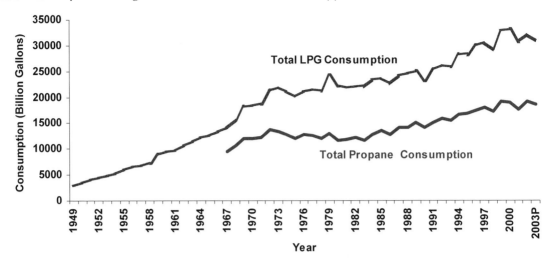

Figure 12-50. Historical Record of Total US LPG and Propane Consumption, EIA, 2004

Because of these characteristics, LPG provides a great advantage concerning storage and transportation. LPG has very high heating value with high purity and cleanness of combustion and easy handling, it finds very wide application in a large variety of industrial, commercial, domestic and leisure uses. Propane will be more used where faster rates of flow are required in particular for industrial applications. Additionally, LPG are safe to be transported by pipelines and cylinders. In case of rupture of the container, LPG will not ignite when combined with air unless the source of ignition reaches at least about 500°C. In contrast, gasoline will ignite when the source of ignition reaches only 220 to 260°C (a very narrow range).

Typical Pipeline Properties of LPG products are provided in Table 12-26, Mohitpour, et al, 2003.

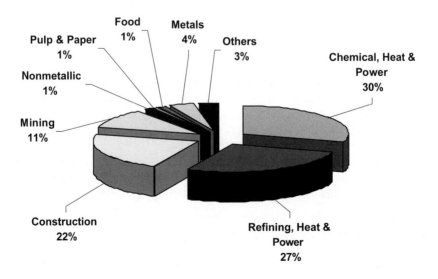

Figure 12-51. 2002 Breakdown of US Propane Consumption By Sector (PERC, 2004)

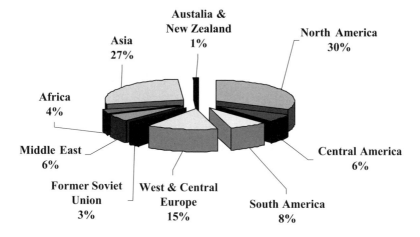

Figure 12-52. Global LPG Consumption - 1999 (After MCS, 2001)

Typically propane is used for residential and commercial heating, automotive propane and industrial fuel gas applications (such as food production, ceramics and metallurgical applications, etc.), while butane is primarily used for seasonal gasoline blending and is stored under pressure at the refineries. A portion is also used as industrial feedstock for other industries such as refrigeration (industrially known as R600a), cigarette lighters and portable stoves. Both products are also used as a raw material for petrochemical applications.

Properties of Commercial LPG

The utilization of LPG as a fuel varies very widely within a country and from one country to another, depending on the cost and availability of the fuel in relation to alternative fuels, notably gasoline and diesel. For example Table 12-27 shows the variation in LPG fuel composition in Europe in 1982, AAA, 2001, Watson and Gowdie 2000.

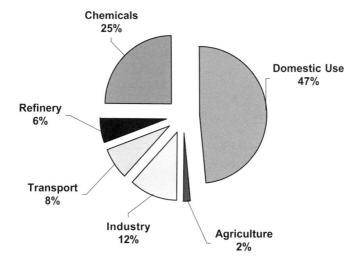

Figure 12-53. Global LPG Consumption by Sector (Feehan, 2003)

TABLE 12-25. Properties of Commercially Available Propane & Butane (L'Air Liquide, 1976, EIA 1994)

Properties	Propane	Butane
Relative density (Specific Gravity) of liquid at 15°C	0.50 to 0.51	0.57 to 0.58
Imperial gallons/ton at 15°C	439 to 448	385 to 393
Litre/tonne at 15°C	1 965 to 2 019	1 726 to 1 760
Relative density of gas @ 15°C and 101.325 kPa, Air=1	1.40 to 1.55	1.90 to 2.10
Volume of gas (litres) per kg of liquid at 15°C and 1 01.35 kPa	537 to 543	406 to 431
Volume of gas (ft^3) per lb of liquid at 60°F and 30 in Hg	8.5 to 8.7	6.5 to 6.9
Boiling point at atmospheric pressure °C (approx.)	-42	-0.5
Vapour pressure (Bar) @ various temperatures	See Figure	
Latent heat of vaporisation (kJ/kg) at 15°C	358.2	372.2
(Btu/lb) at 60°F	154	160
Gas liberation/litre of liquid @ 15°C	311	239
Specific heat of liquid at 15°C (kJ/kg °C)	2,512	2,386
Sulphur content per cent weight	Negligible to 0,02	Negligible to 0,02
Limits of flammability (% by volume of gas in a gas-air mixture to form a combustible mixture)		
Upper	9.5	9
Lower	2.2	1.8
Ignition Temperature Deg C	470-600	420 min
Flame Temperature Dec C	1980	1970
Calorific Values:		
Higher		
(MJ/m^3) dry	93,1	121,8
(Btu/ft^3) dry	2 500	3 270
(MJ/kg)	50,0	49,3
(Btu/lb)	21 500	21 200
Lower		
(MJ/m^3) dry	86,1	112,9
(Btu/ft^3) dry	2 310	3 030
(MJ/kg)	46,3	45,8
(Btu/lb)	19 900	19 700
Air required for combustion (m^3 to burn 1 m^3 of gas)	24	30

In contrast LPG HD5 is used in North America and has specifications to that regulated by the California Air Resources Board, 1997, which is similar to properties defined in Table 12-28.

Pipeline Transmission of LPG

LPG production and distribution is schematically depicted in Figure 12.56. Normally, LPG is stored in liquid form under pressure in a steel container, cylinder or tanks. The pressure inside the container will depend on the type of LPG (commercial butane or commercial propane) and the outside temperature. Propane and butane that are produced from refineries are thus transported by tank trucks and distributed for most domestic use with 24 lb.(10 Kg) cylinders.

Transportation of LPG by pipelines provides an environmentally friendly mode that entails less energy consumption and exhaust emissions than other forms of transportation.

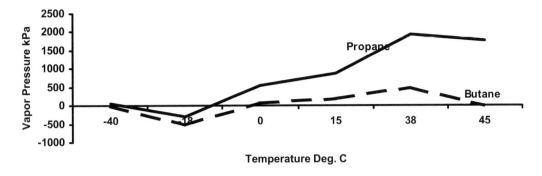

Figure 12-54. Typical Vapor Pressure of Commecially Available LPG

The transportation of LPG through the pipeline system helps reduce air pollutants, including carbon monoxide, suspended particulate matter, unburned hydrocarbons, and sulfur and nitrogen oxides. Noise pollution is also reduced, as the pipeline makes it unnecessary to deploy a fleet of trucks per day, each producing 90 decibels of noise.

Using propane as peak shaving and standby gas is common in the natural gas industry, Figure 12-57. This provides enhanced security and flexibility in a dynamic energy market while reducing the overall cost of energy supply. A typical system produces "propane-air" for direct replacement of natural gas during peak demand periods.

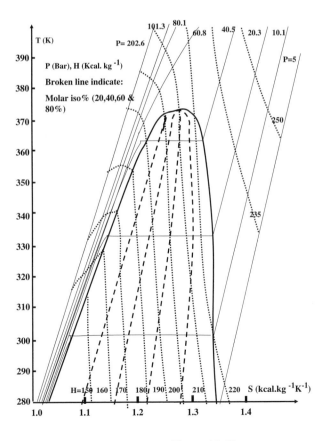

Figure 12-55.

TABLE 12-26. Typical Pipeline Properties of LPG Products

Commodity	Viscosity (CS) mm²/s	Temperature (°C)	Density (Kg/m³)*
Propane (@1000 kPa)	0.218	20	500.6
	0.199	30	483.5
	0.166	52	446.9
Butane (@470-520Kpa)	0.237	25	560.8
	0.235	44	535.12
	0.212	52	529

As depicted in Figure12-57, LPG Transmission facilities consist of the following:

- Fractionation plant/ oil refinery (for production, propane, butane)
- LPG Storage System (above and underground)
- LPG pipeline (including measurement and compression facilities)
- Trucking/ ship transport

LPG Production, Market Influence

The LPG market is comprised of three main components: production (consisting of offshore and onshore facilities, local refineries and imports), plus wholesale and retail.

Generally most of the LPG production is naturally occurring and is sourced directly from underground reserves associated with natural gas. For example in Australia some 77% of LPG production is from underground reserves. The remaining 23% is extracted from crude oil refining at the seven refineries located near major capital cities. Australia in 2005 exported approximately 1.3 million tonnes of LPG but also imported around 300,000 tonnes of LPG (propane) to the large East Coast market, LPG Australia 2005.

The LPG markets are highly competitive and like gasoline and diesel, a range of factors including international prices, domestic demand and product availability and transportation methods affect LPG prices.

LPG prices are influenced by fluctuations in crude oil prices which affect prices of liquid petroleum fuels. Additionally it is influenced by environmental factors. For example the widely reported hurricane activity in the South East USA and Gulf of Mexico region in 2005, reduced domestic supplies of crude oil, natural gas and LPG in the USA and led to

TABLE 12-27. LPG Composition (% by volume) as automotive fuel in Europe in 1982, AAA 2001)

Country	Propane	Butane
Austria	50	50
Belgium	50	50
Denmark	50	50
France	35	65
Greece	20	80
Ireland	100	-
Italy	25	75
Netherlands	50	50
Spain	30	70
Sweden	95	5
UK	100	-
Germany	90	10

TABLE 12-28. Typical UK Propane Specification (Conoco - Phillips Company 2003)

Property	Unit	Limit
Composition	% Mole	
■ Ethylene		1.0 Max
■ Alkynes		0.5 Max
■ C4 & Heavier		10.0 Max
■ C5 & Heavier		2.0 Max
■ Total Dienes		0.5 Max
Vapor Pressure	kPag	1550 Max
Hydrogen Sulphide	Mg/m^3	0.75 Max
Water	Valve Freeze (sec)	60 Min
Total Sulphur(after odorizing)	ppm wt	100 Max
Mercaptan Sulphur	ppm wt	50 Max
Ammonia Content in vapor Phase	Mg/m^3	2.0 Max

increasing demand for imported petroleum products including LPG, sourced from various countries with consequent increases in LPG prices world wide.

Major Existing LPG Pipelines:

Worldwide there are over 220,000 miles (350,000 kilometers) of petroleum, refined products and LPG pipelines. Over 72% of the pipelines are in the United States. Some the refined products pipelines carry LPG in batch form. However there are only about 8000 kilometers of single phase pipelines of varying diameter that transport LPG (Propane or Butane) fluids. Percentage breakdown of such pipelines by country is depicted in Figure 12-58, CIA, 2005. Those percentages appearing as zero in Figure 12-58 are each less than one half percent of the total of 8000 Kilometers. Nonetheless both the batched and dedicated LPG networks are growing all the time as consumers and governments recognize the benefits of the environmental advantages of LPG.

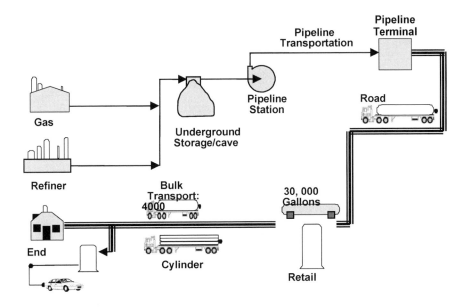

Figure 12-56. Typical LPG Import and Distribution (EIA, 2004)

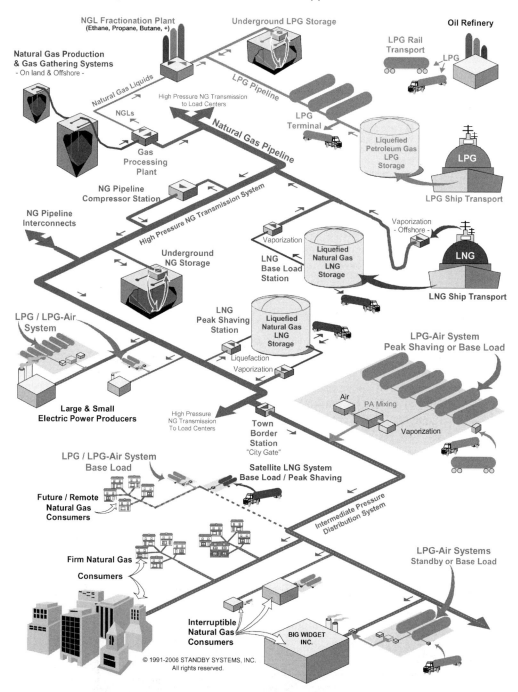

Figure 12-57. Typical LPG Production and Distribution (Courtesy Standby Systems, 2004)

Mexico and Russia each have the next longest length of the total refined products pipelines (about 3%) through which propane or butane is transported in batch form. The structure of LPG imports, exports and production in Mexico is depicted in Figure 11 below.

The United States is a net importer of LPGs, with net imports running about (14600 Billion gallons (100,000 barrels) per day. The majority of this material comes from Canada

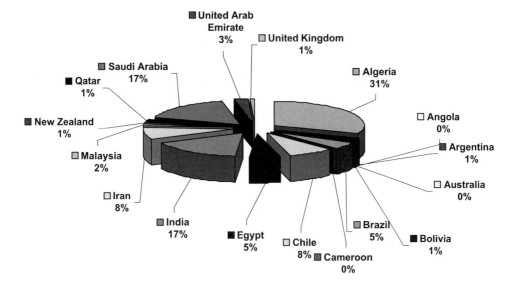

Figure 12-58. Worldwide Distribution of Single Phase Un-batched LPG Pipelines

via batched pipeline, but significant volumes are imported as waterborne cargoes from Algeria and Venezuela. Depending on market conditions in various parts of the world, the United States also imports LPG from Europe (North Sea) and the Middle East (Saudi Arabia, UAE) as well Nigeria.

Of particular interest to Mexico Figure 12-59, is the fact that an annual average of 35 million barrels a day (MBD), 3413 billion gallons/year of LPGs are imported into U.S. Gulf Coast (USCG) region from outside North America, MCS, 2001. About 70

Figure 12-59. Structure of LPG Imports, Exports and Production in Mexico

percent of this material comes from Algeria, the remainder from Venezuela. These imports are landed at Houston, where they can move into storage facilities at Mont Belvieu.

LPG is produced in the oil fields in southern Mexico near Cactus. LPG is imported and exported at Pajaritos by sea. LPG is also exported by sea to Central and South America (Belize, Costa Rica, Guatemala, Ecuador, as well as to other Latin American countries) from the Pacific coast. It is imported on the U.S.-Mexico border by pipeline and truck. Most LPG is consumed in the center of Mexico and this demand is primarily supplied by batched pipelines. The balance of this demand is mostly in the north of Mexico.

Similarly is the production and consumption of LPG in Brazil, Lucon et al., 2004. Brazil can only provide about 65% of its LPG consumption requirement and the rest is imported.

LPG Pipeline Systems Design & Operational Considerations
Codes & Standard Requirement/Legislations

There are a number of codes that industry follows for the design, fabrication/construction and operation of LPG facilities. One specific code that is followed internationally is The National Fire Protection Association "Liquefied Petroleum Gas Code (NFPA 58)", NFPA 2004. However the pipeline industry generally follows pipeline transmission codes such as ASME, ANSI B31.4 & B41.8, Canadian Standard Association (CSA) Z662-03, BSI / European Codes BS-EN (European Committee for Standardization (CEN), 14161) for the design of LPG Transportation and storage facilities. Some of the latter codes are listed below:

- **BS-EN1442** Transportable refillable welded steel cylinders for liquefied petroleum gas (LPG) -Design and construction
- **BS-EN 14161, 2003** Petroleum and natural gas industries - Pipeline transportation systems
- **BS-EN 12252** Equipping of Liquefied Petroleum Gas (LPG) road tankers
- **EN 12493** Welded steel tanks for liquefied petroleum gas (LPG) - Road tankers - Design and manufacture
- **EN 14334 in draft** Inspection and testing of LPG road tankers

Additionally UK, International Institution of Gas Engineers and Managers (IGEM) also provides a professional and policy forum and Recommendations on Transmission and Distribution Practices for engineers, technologists and managers working in the field of fuel gas technology. Some of the practices includes the followings:

- *IGE/TD/13 : Pressure regulating installations for transmission and distribution systems*
- *IGE/SR/18 : Safe working in the vicinity of gas pipelines, mains and associated installations*
- *IGE/SR/22 2nd Impression : Purging operations for fuel gases in transmission, distribution and storage*
- *IGE/SR/23 2nd Impression : Venting of natural gas*
- *IGE/UP/2 : Gas installation pipework, boosters and compressors on industrial and commercial premises*

Regulations/Legislation

Generally the role of a government is the following:

- Establish appropriate fiscal and regulatory framework to correct environmental externalities
- Establish a level playing field for all alternative fuels
- Long-term policy commitment
- Environmental improvement

There is no regulations/legislation that specifically cites the pipeline transportation of LPG. Fundamentally any acts cited are in reference to LPG rates and pricing & safety issues related to LPG usage, storage and release.

Onshore Pipeline Regulations only refer to HVP Pipelines and refer to the applicable codes. For example **National Energy Board (NEB) of Canada** regulations refer to section 10. of **NEB** regulation viz:

1. When an HVP pipeline is to be situated in a Class 1 location and within 500 m of the right-of-way of a railway or paved roadway, a company shall develop a documented risk assessment to determine the need for heavier wall design, taking into account such factors as pipeline diameter and operating pressure, HVP fluid characteristics, topography, and traffic type and density of the traffic on the railway or paved road. Also the regulation will require that:
2. Company submit a documented risk assessment document to the Board when required to do so under section 7 of NEB Act.
3. GUIDANCE NOTES GOAL (section.10): To ensure an acceptable level of safety for HVP pipelines is provided for in the vicinity of railways and roadways.

For assistance in preparing the documented risk assessment, companies are directed to CSA Z662, Appendix B, Guidelines for Risk Assessment of Pipelines. The nature and magnitude of the risk assessment could be commensurate with the nature, magnitude, and potential impact of the activity.

Testing Liquefied Petroleum Gases

The following provides a list of standards for testing of LPG products:

- ASTM D 1142 - Dew Point of Water Vapor in Gaseous Fuels. This method covers the determination of the water vapor content of gaseous fuels by measurement of the dew -point temperature and the calculation of the water vapor content.
- ASTM D 2713 - IP 395 Dryness of Propane: Water vapor content determination by checking the time taken to block an orifice with ice.
- ASTM D 1267 - IP 161 - ISO 4256 - EN 24 256 : Determination of Vapor Pressure of Liquefied Petroleum Gases
- ASTM D 1838 - IP 411 - ISO 6251 - EN 26251 - NF M41-007 Copper Strip Corrosion by Liquefied Petroleum Gases
- ASTM D 1837 - **Volatility of Liquefied Petroleum Gases** - A method to ensure the suitable volatility performance by measuring the relative purity of the various types of liquefied petroleum gases indicating:

- butane in propane-type gas,
- pentane in propane-butane & butane-type fuels
- hydrocarbon compounds by increase in the temperature caused by evaporation

- **IP 59 Method C** - Determination of Density or Relative Density of Gases by Schilling Effusiometer. This standard is no longer in general use, however it exists as a reference document for older systems.

Safety & Environmental Considerations

LPG contains negligible toxic components. However although LPG is non-toxic, its abuse - (like that of solvents) - is highly dangerous. LPG has lower particulate emissions and lower noise levels relative to diesel, making propane attractive for urban areas. Noise levels can be less than 50% of equivalent engines using diesel. Some of the safety & environmental issues related to LPG are highlighted below, NGPA, 2003:

1. LPG emissions are low in greenhouse gases and low in NOx, thus they are low in ozone precursors.
2. Although LPG has a relatively high energy content per unit mass, its energy content per unit volume is low. Therefore LPG tanks take more space and weigh more than diesel fuel tanks of the same energy storage capacity. LPG is typically a colorless and odorless gas to which foul-smelling mercaptan is added so that a leak can be easily detected. LPG is generally a nontoxic, nonpoisonous fuel that doesn't contaminate aquifers or soil.
3. Propane has a high expansion coefficient so that propane tanks can be filled to only 80% of capacity.
4. The pressure of the LPG in steel tanks/container varies with the surrounding temperature. It is also much higher than is needed by the appliances that use it; it needs to be controlled to ensure a steady supply at constant pressure. This is done by a regulator, which limits the pressure to suit the appliance that is being fuelled.
5. LPG is heavier than air, which requires appropriate handling. Therefore any leakage will sink to the ground and accumulate in low lying areas and may be difficult to disperse.
6. Though the lower flammability limit for LPG is actually higher than the lower flammability limit for gasoline, the vapor flammability limits in air are wider than those of gasoline, which makes LPG ignite more easily. LPG will not ignite when combined with air unless the source of ignition reaches at least 420-470 degrees' Celsius. In contrast, gasoline will ignite when the source of ignition reaches only 220 to 260 degrees Celsius.
7. LPG in liquid form can cause cold burns to the skin in case of inappropriate use.
8. LPG will cause natural rubber and some plastics to deteriorate. This is why only hoses and other equipment specifically designed for LPG should be used.

A booklet by American Gas association (AGA 2000) provides further safety guidelines in LPG storage and handling for propane-air plant owners and operators. Also it provides useful tips to make safety professionals familiar with LPG safety in propane-air plants.

Design & Operational Considerations

Because of expansion of LPG at higher temperatures, storage tanks need to be designed to at least allow for about 15% of expansion (at 15°C). A typical of such a design is graphically represented in Figure 12-60.

However design and operational considerations for LPG pipelines are similar to those of liquid pipelines with specific consideration for HVP product transportation. Major issues include:

- hydraulics (setting pressure limits for phase control, or phase issues related to elevation changes, or water hammer effects, column separation, etc.);
- routing criteria such as High Consequence Areas (HCA);
- material fracture control properties (pipe, valves, fittings);
- sealing issues;
- automation and controls requirements (including leak detection, line break controls);
- measurement;
- pumping (maintaining the required vapour pressure);
- control of contaminants in the fluid stream;
- operations procedures such as pressuring or de-pressuring, flushing, isolations & lock out of parallel piping; and
- issues related to batch operation (if applicable)

Vapor pressures of LPG, as they relate to pipeline design and operations, are provided in Table 12-29 below

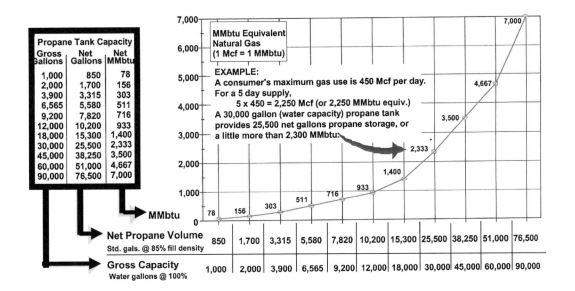

Figure 12-60. Sizing LPG Storage Tanks (Courtesy: Standby Systems, 2004)

Hazards Analysis of LPG Pipeline Transportation

Risk assessment of liquefied petroleum gas (LPG) in pipeline transportation includes accidental releases, evaporation, vapor cloud propagation and dispersion for a number of different accidents involving complete or partial rupture of the main pipeline.

As it is well known, the most dangerous accidents at LPG facilities are accompanied by the formation and evolution of dense vapor clouds. The explosive cloud can travel over a long distance producing large hazardous zone and severe consequences.

The accident scenarios can be: complete or partial rupture of the pipeline without pumping shut-down, stop pumping in a certain period of time (after leakage identification) and isolation of the pipeline damaged section by means of shut-off valves.

The factors that affect the distance to which the flammable boundaries of the vapor cloud travels include:

1. Internal condition of the pipeline at the point of rupture (pressure, flow rate, fluid properties). The internal conditions change with time after rupture
2. Severity of rupture, full break versus partial rupture including orientation (top, side, bottom)
3. Detection time versus pipeline shutdown time (pumps, isolation valves) and isolation conditions (valve spacing and valve closure timing)
4. Surrounding terrain (affecting pooling & evaporation)
5. Upwind environmental conditions (air velocity, terrain condition (rough: buildings in semi urban environment versus cross country: farmland)) which affect the take up rate of vapor into atmosphere
6. Atmospheric conditions: wind velocity, direction and stability

Items 1, 2 & 3 affect the time-dependent mass-venting rate from the rupture. Items 2 & 4 affect the spreading of the liquid/vapor layers. Items 4, 5 & 6 affect the dispersion of the escaping vapor cloud. It may be noted that item 3,(reaction time and isolation valve spacing, are the only controllable factors. Example of such factors in controlling liquid spill to increase oil pipeline safety is described by Platus et al. 1974 & Mohitpour et al. 2003 & 2004.

Since the molecular weight of LPG is greater than that of air, a dense gas dispersion model is required to determine hazard ranges for ambient and low temperature releases. Such predictions by passive dispersion models may be optimistic or pessimistic as the diagram below indicates. A safety report that uses a passive dispersion model (based on Lower Flammability Limit, LFL) will over estimate the down wind extent of a flash fire but under estimate its width. Assessors should expect

TABLE 12-29. Propane and n-Butane Vapor Pressures at Various Temperatures

Temperature °C	Vapor Pressure kPa Propane	Butane
−10	256	−4
0	388	40
10	552	95
20	757	172
30	1004	266
37.8	1218	362

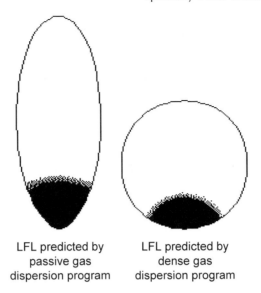

LFL predicted by
passive gas
dispersion program

LFL predicted by
dense gas
dispersion program

Figure 12-61. Passive and Dense Dispersion Modeling (HDI 2001)

to see a sound justification for the use of a passive dispersion model and a discussion of the inability of the model to correctly predict the width of the flammable cloud, HID, 2001, Figure 12-61.

The following summarizes a technique for venting LPG through segment of a pipeline between isolations valves subject to a full rupture (Morrow et.al., 1983), Figure 12-62.

1. Estimate the initial discharge flow rate W_{e1} for P_e-P_i, where W_{e1} = flow rate out of pipe exit under full rupture. This usually equals the pumping flow rate while the pump is still running. It can be assumed that initially it is the pipeline flow rate. Pe = Pressure at rupture, Pi= pressure at the interface (Gas/liquid LPG)
2. Assume that $W_{e2} = 0.95 \, W_{e1}$
3. Calculate the average flow rate $W_{eavg} = (W_{e1} + W_{e2})/2$

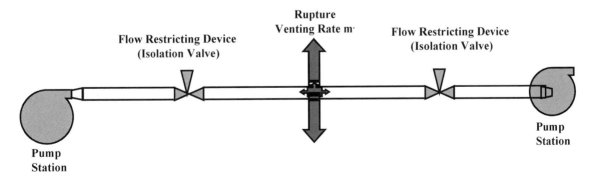

Figure 12-62. Pipeline Configuration and Rupture Schematic

4. Calculate the new exit pressure P_e, corresponding to W_{eavg} using the following equations 12.28 & 12.29
5. Calculate the distance to the interface Z_i from Equation 12.30 below.
6. Calculate the mass removed from the pipeline, M_{yi2} from Equation 12.31. It may be noted that $M_{yi1} = 0$)
7. Calculate the time increment for the mass removed for the time step Δt

$$\Delta t = (M_{yi2} - M_{yi1})/W_{eavg}$$

8. Repeat for each new time step until P_e, atmospheric

Where:

$$\frac{1}{G_{cr}^2} = -\frac{dX}{dP}\left[-2v_f + 2\sqrt{v_f v_g} - 4X\sqrt{v_f v_g} + 2Xv_f + 2Xv_g\right]$$
$$- \frac{dv_g}{dP}\left[X\sqrt{v_f/v_g} - X^2\sqrt{v_f/v_g} + X^2\right] \quad (12.28)$$

$$\frac{dX}{dP} = -\frac{1}{h_{fg}}\left[\frac{dh_f}{dP} + X\frac{dh_{fg}}{dP} + \frac{X}{v_f}\frac{U_f^2}{2}\frac{dv_g}{dP}\right] \quad (12.29)$$

$$Z_i = \frac{3DA^2[F_2(P_i) - F_2(P_e)]}{2fv_f W_e^2\left[1 + \frac{W_i}{W_e} + \left(\frac{W_i}{W_e}\right)^2\right]} \quad (12.30)$$

where

$$F_2(P) = C_1 P + \frac{C_2}{2}P^2 + \frac{C_3}{3}P^3 + \frac{C_4}{4}P^4$$

$$M_{yi} = \frac{3DA^3[F_4(P_i) - F_4(P_e)]}{2fv_f^2 W_e^2\left[1 + \frac{W_i}{W_e} + \left(\frac{W_i}{W_e}\right)^2\right]} \quad (12.31)$$

Notations

A = pipe cross-sectional area
b = width of region around plume centerline that has concentration distribution independent of y
C_A = concentration at vapor layer/air interface on plume centerline
c = local concentration distribution
D = pipe inside diameter
f = fanning friction factor
F_1, F_2 = polynomials in pressure
F_3, F_4 = representing curve fits of thermodynamic functions
G_{cr} = critical mass flow rate per unit area

M_{yi} = quantity of mass removed from pipeline by void formation
$m(t)$ = venting rate
P = pressure
Q_s = atmospheric take-up rate
Q_1 = volume flux of emitted gas
r = radius of heavy gas layer
r_m = position of leading edge
t = time
U_f = velocity of liquid phase
U_{fs} = superficial liquid velocity
U_g = velocity of vapor phase
u_o = velocity at reference height

W_r = flow through rupture (partial break)
X = thermodynamic quality
x = downwind distance
Y = void fraction
y = distance in crosswind direction
Z = distance along pipe from point of rupture
Z_i = distance from point of rupture to point where two-phase flow just starts
z = vertical distance
Z_o = reference height above ground

g = gravitational constant
g' = $g (1 - \rho_o/\rho_e)$
h = height of layer
h_f = specific enthalpy of saturated liquid
h_{fg} = latent heat of vaporization
h_o = specific enthalpy of saturated liquid at pressure of 124.73 lbf/in^2 absolute
K = coefficient in equation (24)
L = length of pipeline segment isolated by valve closure

u_x = wind velocity
V_y = vapor space volume
V_{yi} = vapor space volume from rupture to interface
v_f = specific volume of saturated liquid
v_g = specific volume of saturated vapor
W = mass flow rate
W_d = mass flow rate discharged from pipeline
W_e = flow rate out of pipe exit (complete rupture)
W_i = flow rate at interface (where $Z=Z_i$)

α = velocity profile constant (0.14 for neutral stability)
ρ_e = density of propane vapor
ρ_{gr} = density of gas at ground level
ρ_o = density of air at ground level upwind
σ_y = lateral dispersion coefficient
σ_z = vertical dispersion coefficient
τ_o = shear stress upwind of break site
ϕ_f^2 = Martinelli parameter

Factors that affect LPG modeling are: Wind speed, Ground roughness, Averaging period, Elevation of Fireball, Surface emissive power, Stored Energy in LPG cloud and Substrate

LPG Sources of Information:

The following list organization from which information on LPG/Propane can be obtained

Propane Associations - National & International

- National Propane Gas Association
- Propane Education & Research Council
- Propane Gas Association Group (Canada)
- World LP Gas Association

USA: Propane Associations - State & Regional

- Alabama Propane Gas Association
- Arizona Propane Gas Association
- Illinois Propane Education & Research Council
- Indiana Propane Gas Association
- Kentucky Propane Education & Research Council
- Michigan Propane Gas Association
- Missouri Propage Gas Association
- (New England) Propane Gas Association of New England
- New Jersey Propane Gas Association
- New Mexico Propane Gas Association
- New York Propane Gas Association
- North Carolina Propane Gas Association
- North Dakota Propane Gas Association
- Northwest Propane Gas Association
- Ontario Propane Association
- Pennsylvania Propane Gas Association
- Rocky Mountain Propane Gas Association
- Virginia Propane Gas Assocation
- Western Propane Gas Association

Government Agencies
- U.S. Department of Energy
- U.S. Department of Transportation
- U.S. Energy Information Association
- U.S. Environmental Protection Agency

Other Information Sources

- American Gas Association
- American Petroleum Institute
- CETP Instruction for the Northeast and Mid-Atlantic Regions
- Energy.Com
- Propanegas.Com

Maintenance

Emergency Response

"Emergency " is defined as any unforeseen combination of circumstances or a disruption of normal operating conditions that poses a potential threat to human life, health, environment or property if not contained, controlled or eliminated.

Types of emergencies addressed in LPG transmission situations:

- Vapour/gas Inside or Near a Building
- Fire Near or Involving a Pipeline Facility
- Explosion Near or Involving Pipeline Facility
- Natural Disasters
- Emergencies Involving Other Pipelines
- Arson/Bomb Threats

Protection of people, environment and properties are the top priorities in any pipeline emergency response

Nevertheless, risks to pipeline during construction, operations and maintenance are real. If an incident does occur, both the operator and local emergency responders along the pipeline route must be ready to respond. For example there are three potential risks associated with the failure of the pipeline and the release of LPG into the surrounding environment. This includes the risk of fire, Figure 12-63, a vapor cloud explosion and, as a consequence of small quantities of mercury in the gas/product and the possibility of toxic effects of an un-ignited airborne vapor cloud. Others include oxygen cut off/starvation causing asphyxiation and death.

Training opportunities and materials for first responders and other emergency-oriented organizations are available on liquid and gas transmission and distribution systems through each of the pipeline companies and utilities. They have many common features and each focuses on the type of emergency situation where their product would be involved.

Emergency Response exercises must be conducted on an annual basis to assure the availability of trained responders along the pipeline route during all phases of pipeline work in particular pipeline operation. This may be done whenever possible to do so in conjunction with other pipeline companies and interested parties to assure timely response for securing continued public safety, environmental protection as well as facilities continuing safe functioning and operation.

By combining efforts, pipeline companies & emergency response entities can provide a more complete training program and response to an emergency situation.

Regulations/Standards

There are several regulation and standards/safety rules for the transportation of LPGL and responding to emergency situations. Some of the internationally accepted codes and regulations are:

- ASME ANSI B31.4 " Pipeline Transportation Systems for Liquid Hydrocarbon and Other Liquids"
- API 1160 Integrity Management of Liquid Hydrocarbon Pipelines
- ASMEB31-S
- US Department of Transportation (DOT), Code of Federal Regulations (CFR) - Regulation 195 (Pipeline Safety)
- Occupational Health and Safety Act OHSA (Act 85 of 1993).
- OHSA Process Safety Management

While there are other codes and regulations in place that affect the design and operation/maintenance aspects of LPG transportation by pipeline, the above codes and practices specifically address safety situations and general requirement with respect to emergency response requirements of hydrocarbon including natural gas liquids.

Figure 12-63. An LPG Pipeline Fire, PVFD, 2005 (Image Courtesy of KTAB-TV)

Flaring LPG

Flaring LPG is typically an emergency situation. Flaring and venting is an important safety measure during LPG emergencies, power and equipment failures, or other upsets that might otherwise pose hazards to workers or nearby residents and sites.

Flaring and venting operations may be subject to a variety of conditions, and some countries have adopted operational standards and guidelines. There are however different approaches and results.

Operational standards and guidelines typically include the following aspects:

- Burn technologies and practices
- Timing and duration
- Flare location
- Heat and noise generation
- Smoke and noxious odor

Most pipeline operating profiles can permit the flaring and venting of associated gas under the following circumstances:

- For safety reasons
- For unavoidable technical reasons (such as purge venting/emergency)
- In emergencies (emergency transportation stops, compression, and others)

As most situations in LPG pipeline transmission is related to emergencies, then measurement and monitoring are not prevalent. Flaring and venting under these circumstances normally does not require regulatory approval. Hence only company operating practices are applicable and procedures are written to assure safe release with minimum impact on the public and the environment.

Emergency Levels

Pipeline industry generally respond to emergencies depending on the level associated with a situations. These are defined in Table 12-30.

TABLE 12-30. Emergency Level of Alert

LEVEL 1	An emergency that has the **potential to escalate**, but does not meet any Level Two or Three Alert Criteria. Meets **all** of the following conditions: ✓ No immediate threat to the public ✓ No serious threat to health and safety of workers (**personal protective equipment may be required**) ✓ Minimal environmental impact ✓ Impact confined to company jobsite ✓ Creates little or no media interest ✓ Handled entirely by company personnel
LEVEL 2	An emergency that does not meet any Level Three Alert criteria, but meets **any** of the following conditions: ✓ Definite risk to the public, workers or the environment. ✓ Confirmed incident. Expected to be under control quickly. ✓ Requires involvement of external emergency services, state and federal agencies. ✓ Public safety actions are likely required. ✓ Moderate environmental impact that extends or has the potential to extend beyond company right-of-way. ✓ Creates local/regional media interest.
LEVEL 3	An emergency the meets **any** of the following conditions: ✓ Causes serious threat to the public, workers and /or the environment. ✓ Requires extensive involvement of external emergency services, federal and/or state agencies ✓ Significant and ongoing environmental impact with extend beyond company right-of-way. ✓ Creates national media interest

Guidelines Response to LPG Emergencies

When responding to any LPG pipeline emergency situation, the goal is to **protect people first, then the environment, and finally property**. The response to emergencies include the followings:

LPG Vapour/gas Inside or Near a Building

When approaching any building or confined space that may contain escaped LPG vapor, one must always look for and listen for any signs of escaped vapor. The following are actions to consider when faced with a leaking vapor/gas emergency:

- Do not open any doors until explosive limits have been determined.
- Return to vehicle and reposition upwind, blocking access to the location.
- Evacuate people from adjacent buildings if they are close enough to be injured from an explosion or fire.
- Contact pipeline Control Centre and record the incident. Mobilize personnel with a portable gas detector(s).
- Shut off electrical power to the building and attempt to eliminate other potential ignition sources (vehicles, telephones, and radios) in the area.
- Isolate the building from LPG sources if possible. Close service line valves on buildings receiving domestic service. On measurement buildings, close inlet and outlet block valves.
- After LPG sources are shut off, proceed to the building with a portable gas detector and check door seams for an explosive mixture.
- If safe to enter, attempt to determine the cause of the leak.
- Once the source of the leak has been determined, contact the appropriate personnel for repairs and restoration of service.
- Keep pipeline Control Centre informed of response details at regular intervals.

Fire Near or Involving a Pipeline Facility

Guidelines for responding to this type of emergency are as follows:

- Keep at a safe distance. Account for all personnel.
- Secure the area and restrict access to trained personnel only. Evacuate any adjacent facilities or buildings that may be endangered.
- Contact pipeline Control Centre and record the incident. Wait for assistance before attempting control measures.
- If the fire is to be contained by a fire department and the fire involves escaping LPG in any facilities, inform the fire department immediately.
- If the fire is due to escaping LPG or some similar flammable material, isolate all fuel sources and/or threatened facilities and close doors. **Normally, LPG fires should not be extinguished unless the fuel source can be safely eliminated.** Apply a cooling water spray, if available, to any equipment affected by flames.
- Only trained personnel with backup should operate fire extinguishers:
- Keep pipeline Control Centre informed of response details at regular intervals.

Explosion Near or Involving Pipeline Facilities

General guidelines for responding to this type of emergency are as follows:

- Keep at a safe distance.
- Account for all personnel.

- Secure the area and restrict access to trained personnel only.
- Contact pipeline Control Centre and immediately record the incident. Evacuate any adjacent facilities or building that may be endangered.
- If necessary wait for assistance to arrive before attempting to taking control measures.
- Keep pipeline Control Centre informed of response details at regular intervals

Natural Disasters

General guidelines for responding to natural disasters (for example a tornado) are:

- Monitor weather information. (A tornado watch means atmospheric conditions are favorable for tornadoes and a tornado warning means a tornado has been sighted.)
- If a "tornado warning for the area" is issued, assign an observer to watch storm conditions. If a tornado is sighted, notify all affected personnel. Take the following actions as time permits:

  - Notify pipeline Control Centre and the local field office. Advise them that your facility may be out of radio communication.
  - Extinguish all unnecessary fires and lights.
  - Switch over to auxiliary power.
  - Do not trip the Emergency Shut Down (ESD) system. It will usually function automatically if a fault occurs. Leave facilities in operation and seek shelter.

- When a tornado approaches personnel in vehicles they should:
  - If possible, drive at right angles, away from the tornado.
  - If the tornado cannot be avoided, seek shelter in a ditch or other low-lying area. Avoid locations under electrical power lines

Emergencies Involving Other Pipelines

In emergencies involving other pipeline operators, response guidelines are as follows:

- Verify that no LPG/Gas pipeline facilities/operations are involved or threatened. If potential for involvement/damage exists, begin the internal notification process.
- Exchange resources and information with other pipeline operators as necessary.
- Equipment and services of pipeline contractors located in the vicinity of the pipeline system could be shared as deemed necessary.
- Upon request, provide appropriate assistance, as fast as practical, to other pipeline companies, which have an emergency on their pipeline systems.
- Continue to monitor emergency until assured there is no potential impact on operations.
- Keep pipeline Control Centre updated of all actions taken

Arson/Bomb Threats/Terrorist Attack

General guidelines for dealing with such threats are as follows:

- Notify appropriate agencies, i.e., local law enforcement, etc.
- Alert LPG pipeline facilities of a possible search, shutdown, and/or evacuation.

- All searches should be coordinated with law enforcement agencies. If it is determined that searches are to be performed under pipeline company authority, the company will usually authorize the search.
- All search teams should be two member teams with priority assigned areas.
- Personnel who have daily contact with specific areas are to be assigned those areas to search. Be suspicious of ordinary looking objects such as lunch pails, boxes, unopened packages and objects that appear out of the ordinary such as a newly placed fire extinguisher.
- If a suspected device is located, report it immediately. **Do not touch the object.**
- If an object is located, emergency shutdown, evacuation, or other appropriate measures may be initiated by the company.

LPG Leak Response Guidelines

The following provides an overview of LPG leak identification and response.

I Identifying LPG Leaks

Pressure drops related to LPG leaks are rarely identified at a pipeline Control Center, except in the event of a major rupture, due to LPG's high vapor pressure and high compressibility (e.g., propane can be compressed by as much as 5% if pressure is raised to 1,000 psi).

Leaks are generally reported by someone in the vicinity of the leak site. Some indications of an LPG leak are listed below:

- A cloud of steam or mist, caused by condensation and freezing moisture., See Figure 12-64.
- Ice build-up on exposed pipe and frozen ground around an underground pipe.
- Brown vegetation, which is an indication of soil saturation.
- Yellow-stained snow, which may be an indication of LPG accumulation under the snow.
- The odor of the condensate fraction of LPG.

Figure 12-64. LPG Cloud (inset) and Ignited LPG Vapor

II LPG Safety Precautions
General Precautions
The following safety precautions apply to all operations involving LPG leaks:

- Notify pipeline Control Centre immediately concerning any potential leaks or gas alarms.
- Prior to proceeding to a potential leak site, verifiers should ensure that they are wearing the appropriate fire retardant clothing and that the required personal protective equipment (flare gun, gas detector, etc.) is readily available.
- Route vehicles arriving on-site around any vapor clouds (monitoring wind conditions and considering elevations).
- Park vehicles in vapor-free areas (Cold Zone, see below for definition) and on high ground, if possible.
- Shut down vehicles to eliminate the possibility of vapor ignition.
- Use intrinsically safe equipment (e.g., flashlights, radios, continuous gas detectors with audible alarms).
- Eliminate or shut off all potential sources of ignition in the immediate area.

Exploring for Leaks
Standard Precautions: The following standard precautions apply to leak exploration both indoors and outdoors:

- Narrow down the leak area as closely as possible, using all available sources of information.
- Explore on foot, using the buddy system if possible.
- Wear appropriate protective clothing (fire retardant clothes, splash resistant gloves, etc.).
- Do not carry any ignition sources.
- Use gas detectors to monitor leak sites and identify areas containing vapors.
- Monitor regularly during an emergency.
- Record all readings in the Incident Log.

Exploring Indoors: In addition to the standard precautions, the following precautions apply to indoor leak exploration:

- Use gas level detectors to test for possible vapor concentrations before entering any building, room or confined space.
- If hazardous levels are present, vent the room/confined space and re-test before entry.
- Enter the room confined space only after the atmosphere is at a non-explosive, safe breathing level.
- If entering an unsafe atmosphere, wear a self-contained breathing apparatus and have a second employee on standby with similar breathing apparatus.

CAUTION: Exercise caution, as LPG may accumulate in low-lying areas, enclosed spaces or areas with poor circulation.

Exploring Outdoors: In addition to the standard precautions, the following precautions apply to outdoor leak exploration:

- Use a gas detector to determine the presence of a vapor cloud.
- Initiate monitoring upwind of the suspected leak site to ensure personnel are not overcome by vapors.

- Monitor regularly to establish the distance and direction of vapor progression.
- Use a wind sock to monitor wind direction and avoid being trapped in the vapor cloud.
- Determine the area and perimeter of any vapor clouds which may be present.
- Use a self-contained breathing apparatus in all areas where gas detector readings are between 11 and 20% of the lower flammability/explosive limit (LF/EL).
- Restrict entry of personnel into areas where gas detector readings exceed 20% of the LF/EL.

WARNING: Never enter a vapor cloud.

CAUTION: Check low-lying areas. Probe under any covered area in the vicinity of a suspected leak, as LPG may accumulate in these areas.

Night Exploration: Night exploration for LPG leaks can be extremely difficult as the hazards associated with LPG are compounded by lack of visibility. In addition to the standard exploration procedures for leaks, extra caution should be exercised for the following procedures:

- Try to narrow down the suspected leak site as closely as possible before arriving on-site, using all available sources of information.
- Determine the meteorological conditions present in affected areas.
- Approach the suspected leak site slowly on foot from the upwind side, using a gas detector to continuously monitor gas levels.
- Use all appropriate safety precautions including protective clothing, intrinsically safe equipment and lighting sources. Do not carry any ignition sources.
- Avoid being trapped in the vapor cloud by possible wind shifts.

III Emergency Response Procedure

Protecting the Public

Due to the highly flammable nature of LPG and the possibility of vapor cloud formation, evacuation or other emergency procedures may be required to protect the public and property.

Pipeline Operations

Pipe Operating Procedures usually provide specific procedures and instructions for line operation and shutdown in LPG leak situations.

Guidelines for Small Leaks

The preferred procedure for small LPG leaks is to continue pumping until the Segment of the pipeline can be safely isolated to make the appropriate repairs using standard procedures. The prime consideration in deciding whether to continue pumping represents a hazard to the public and the surrounding property. Where available, water fog can be used to break up and disperse small vapor clouds.

Guidelines for Large Leaks

If the LPG leak is large or the LPG batch cannot be pumped past the leak location, ignition of the leaking LPG is recommended. The prime consideration in deciding whether to ignite an LPG cloud is the hazard to the public and emergency response personnel posed by a growing and drifting explosive cloud.

Igniting an LPG Cloud

Before igniting an LPG cloud, obtain clearance or authorization if possible. Isolation or shutting down the line may be considered. However, if clearance cannot be obtained quickly, any personnel trained in ignition procedures can ignite leaking LPG.

Each pipeline maintenance vehicle, Electrical or personnel's vehicle, or manned station should be equipped with a flare launcher and signal flares. These flares can ignite natural gas vapors from approximately 100 - 200 yards (90 - 180 m), depending on the type of launcher and flare used.

When igniting an LPG cloud, the following procedure are generally used:

1. Select an upwind site and test for vapors.
2. Determine the perimeter and extent of the vapor cloud.
3. Evacuate all response personnel to a safe, upwind location (usually called Cold zone see below) before ignition to ensure no personnel are caught in the flash fire.
4. Aim the flare at the perimeter of the vapor cloud, where air to fuel mixtures will allow ignition.
5. Perform initial ignition attempts approximately 200 - 250 yards (180 - 225 m) from the outer edge of the vapor cloud. If ignition is not achieved, move closer to the cloud and attempt ignition again.
6. Maintain a minimum of 100 yards (90 m) from the perimeter vapor cloud.

Depending on conditions (extended period of leaking, accumulations in low-lying areas, etc.), take extra precautions to ensure the safety of personnel during the initial flash.

Repair Procedures

If the LPG cloud cannot be ignited and repair procedures must begin, locate the work trailer and equipment upwind in a safe area, a minimum distance of 0.5 miles (0.8 km) from the leak site. Monitor the perimeter of the vapor area continuously to detect any shift in the vapor cloud. Air movers are also an effective method of providing air circulation in confined areas or in buildings.

IV LPG control and clean up procedures

Isolating the Pipeline Section

When LPG is escaping uncontrolled, isolate the affected section by closing the appropriate mainline sectionalizing valves. Whether the leak is burning or not, the amount of escaping LPG can then be reduced using blow-down valves that are generally installed on either side of the sectionalizing valves.

Relieving Pressure

To relieve pressure at a leaking pipeline section, one of the following procedures is generally adopted by the industry:

1. Install a pipe discharge line.

If LPG is present at the blow-down valve, install a steel discharge line with a double valve to flare the LPG at least 100 feet (30 m) from the pipeline. Construct an earthen berm over this discharge pipe to stabilize the pipe and provide a heat shield.

The double valves provide a shutoff system. If the downstream valve becomes inoperative due to ice build-up, caused by vaporizing LPG, then the upstream valve can be used to control flow or isolate the line

When burning is in progress, closely monitor the prevailing wind direction.

Control the outflow to the fire to prevent excessive heat build-up at the blow-down valve. Do not allow the blow-down valve to become inaccessible to manual operation.

2. Transfer the material to a tanker truck.

If LPG is present at the blow-down valve, transfer the fluid to tanker trucks equipped to handle LPG to reduce the intensity of the LPG leak. A transfer pump connected to the blow-down valve may be required to fill the tanker truck if elevation does not provide a standing head in the isolated section.

3. Transfer at downstream sectionalizing valve.

At the downstream sectionalizing valve, pressure may also be relieved by installing a pump connected to each blow-down valve on either side of the mainline sectionalizing valve. To accomplish this safely, a check valve is required in the discharge line of the pump to prevent an accidental back-blow if the pump fails. Line pressure and relief of volumes being transferred can be controlled at a downstream tank (as available).

Digging Out a Leak Site

Repair operations involving LPG are difficult, slow and hazardous. Exercise caution, as pockets of gas may be trapped in the ground. If the LPG leak has existed for some time, the condensate portion may have saturated the soil for a considerable distance around the site.

CAUTION: All active LPG leaks should be ignited or left burning before beginning excavation or repair preparation.

The following precautions should be observed when digging out a leak site:

- Determine the perimeter of LPG or condensate saturation by digging test holes. Using a combustible gas meter, determine the limits of underground migration by progressive testing of holes.
- Ensure fire extinguishing equipment is readily available and manned during excavation work at the leak site.
- Always dig from the upwind side, starting from the outside edge of any ground frozen due to the LPG leak and working in and down.
- If no wind is present, use air movers to keep air moving across the worksite and away from workers.
- Continuously monitor air using a gas detector.
- Continuously monitor wind direction.

Zones Established for LPG Release

When an LPG release incident occurs and whether it causes fire or LPG is ignited, generally the industry divides the areas surrounding the release into hot, warm and cold zones, Figure 12-65. This is for the purpose of responding to the release, placing emergency response personnel and equipment and performing the necessary work. These zones are described below:

Hot Zone: This the zone which is highly explosive and extremely dangerous and should only be approached and entered in by trained personnel equipped with the appropriate facilities including Personal Protection Equipment (PPE) and self contained breathing apparatus (SCBA). This zone has LPG concentration level exceeding the lower flammability/Explosive limit (LFL/LEL) and up to Upper Flammability/ Explosive Limit (UFL/UEL) of LPG vapor. *No equipment is installed in this zone.*

Warm Zone: This is the radial area next to the Hot Zone, which should still be considered dangerous, but workers can approach through this zone without self contained breathing apparatus (SCBA). Liquefied Petroleum Gas level monitoring should be conducted at all times. Typically the zone has concentrations LFL to LFL/2. This means that the zone is generally not explosive, but could change to explosive depending on the leak circumstances. Because there is still a small risk of explosion, no

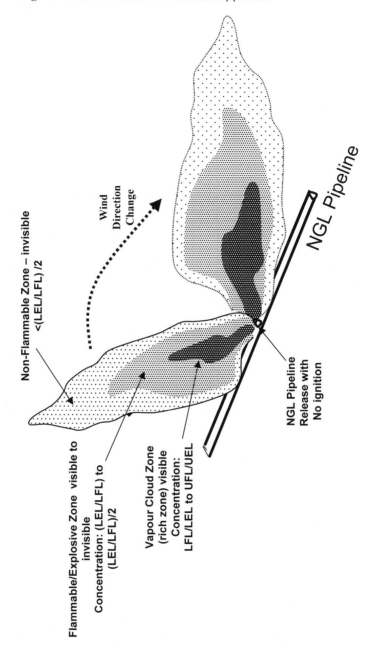

Wind Direction Change

Non-Flammable Zone – invisible <(LEL/LFL)/2

Flammable/Explosive Zone visible to invisible Concentration: (LEL/LFL) to (LEL/LFL)/2

Vapour Cloud Zone (rich zone) visible Concentration: LFL/LEL to UFL/UEL

NGL Pipeline Release with No ignition

NGL Pipeline

Figure 12-65. Vapor Cloud From a Pipeline Rupture & Zones In LPG Release

equipment is installed in this zone. The warm zone is also an area where workers can be affected by the force of an explosion in the hot zone (that is, the blast effects can knock people down).

Cold Zone: This is the radial area next to the warm zone which should be safe for all work and workers. Typically this zone will have Liquefied Petroleum Gas concentrations less than LFL/2 and is the zone where equipment and trailers can be set up without concern for explosion and fire (or blast effects from explosions). Usually this is where the incident command post for responding to emergencies is stationed and rescue equipment is available.

Equipment Needs

Equipments to respond to LPG emergency situation include **Safety Equipment including**, firefighting and first aid equipment to be placed at strategic locations and in vehicles. Regular inspection of safety equipment must be performed and this includes those listed in Table 12-31 below:

TABLE 12-31. Inspection of Emergency Equipment

Equipment	Type of Inspection
containment boom	inspect for tears, wear or defects that may affect performance
portable tanks	
skimmers	service and operate to ensure functioning properly
gas/air operated pumps and engines	
compressors	
generators	
boats and motors	service and operate to ensure functioning properly
	ensure batteries are charged
	ensure on-board equipment is operable
communications	operate radios to ensure proper function

LPG Conversion Table

The following Table 12-32, (**Lange Gas, 2004**) provides approximate conversion factors for LPG volume and calorific values.

TABLE 12-32. LPG Conversion Factors

LPG quantities	Kilogram	Liter (at 8 deg C)	Tons	US Gallons
1 Kilogram	*	1,908	0,001	0,521
1 Liter (at 8 deg C)	0,524	*	0,000524	0,273
1 ton	1000	1908	*	521
1 US Gallon	1,983	3,785	0,001983	*
1 Cubic meter - m^3	2,05	3,93	1970	1,0265

EXAMPLE 1. Liter is 0.524 kg at 8 degrees Celsius 1 ton equals 521 US Gallons

Calorific Equivalents	Kilojoule	kcal	kWh	BTU
1 Kilojoule	*	0,2388	0,0003	0,9478
1 kcal	4,1868	*	0,001163	3,968
1 kWh	3600	860	*	3411
1 BTU	1,0551	0,252	0,000293	*

REFERENCES

AAA (Australian Automobile Association), 2001, "Standards for Liquefied Petroleum Gas (Autogas)" http://www.aaa.asn.au/submis/2001/EA_LPG_Dec01.pdf and http://www.greenhouse.gov.au/transport/comparison/pubs/2ch10.

AGA (American Gas Association), 2000, " Introduction to LPG Safety for Propane Air Plant Operators" http://www.aga.org/Template.cfm?Section=Browse_by_Topic&template=/Ecommerce/ProductDisplay.cfm&ProductID=188.

AGA Publication XU019, 1981, Underground Storage of Gaseous Fuels: A Bibliography, 1926–1980.

Ahluwalla, M. S., and Gupta, G. D., 1985, "Composite Reinforced Pipelines," Disc 6th Int. Conf; Internal & Ext Protection of Pipes, Nice, France, Nov. 5–7.

Ahmed K., and Baker R. C., 1986, "Computation and Experimental Results of Wear in a Slurry Pump Impeller," *Proc. Inst. Mech. Engineers*, Vol. 200, # C6, pp. 439–445.

Air Resolution Board, 1997, 'Title 13, California Code of Regulations, Section 2292.6" http://www.arb.ca.gov/regact/lpgpro/res9715.wpd.

Alexander, L. W., and Lee, F. T., 1986, "CO_2 Pipelining in the U.S.A.," Int. Pipeline Symp. BQL98610601, Vancouver, British Columbia, Canada, May 7, Report.

ANSI/NFPA STD 50A, 1984, Gaseous Hydrogen System.

API 941, 1983, Steels for Hydrogen Service at Elevated Temperatures and Pressures in Petroleum Refineries and Petrochemical Plants.

ASHRAE Handbook, 1984.

Austin, J. E., and Palfrey, J. R., 1989, "Mixing of Miscible but Dissimilar Liquids in Serial Flow in a Pipeline."

Bain, A. G., and Bannington, A. G., 1970, "The Hydraulic Transport of Solids by Pipeline," *Program Press.*

Bartholemy, H., Bryselbout, J., Galea, G., and Barbe, C., 1980, "Effect of Hydrogen on the Behaviour of European Steels Used for Hydrogen Pressure Vessels," Proc. 3rd Int. Conf. on Effects of Hydrogen on Behaviour of Material, Ed. Bernstein and Thompson, Moran, WY, Ref. (9), pp. 1027–1035.

Beghi, G., et al., 1979, *Economics of Pipeline Transportation for Hydrogen and Oxygen*, Entropie No. 59, 1974 and Hydrogen Energy, Part A., Plenum Press, New York, NY.

Broek, D., 1983, "Elementary Engineering Fracture Mechanics" Martinus Nijhoff, The Hague.

Canjar L. N., Pollack E. K., Cadman T. W., Lee W. E., and Maning F. S., 1966, Hydrocarbon Processing Petroleum Refiners, January.

Carleton, A. J., and Cheng, D. C. H., 1974, "Design Velocities for Hydraulic Conveying of Settling Suspension," *Proc. Third Int. Conf.*, Hyd. Transport of Solids, BHRA, U.K.

Cathro, D. L., Good, C. K., and Mohitpour M., 1986, "CO_2 Facilities Report," NOVA and Alberta Corporation Internal Report, Vol. 1 & Appendices A & B.

CIA, The World Factbook, 2005, *"Pipelines"*, http://www.odci.gov/cia/publications/factbook/fields/2117.html.

Compressed Gas Association, 1974, "Hydrogen," CGA Pamphlet, No. 5.

Cox, K. E., and Willams Jr., K. D., 1977, "Hydrogen: Its Technology and Implications," Vol. II Transmission and Storage, CRC Press, Cleveland, OH.

Crane Supply Handbook.

Davis, D., 1999, "Sand and Gravel Pipeline Connects Texas and Oklahoma," Pipeline and Gas Journal, Sept 1999, pp. 40–42.

Dotterweich, F. H., 1978, *Production of Synthesis Gases and Hydrogen*, Chemicals from Natural Gas, Chapter 2, pp. 13/6–13/8.

EIA (Energy Information Administration), 1994, "Alternatives to Traditional Transportation Fuels :An Overview" Report DOE/EIA-0585/O - Distribution Category UC-98, June, http://tonto.eia.doe.gov/FTPROOT/alternativefuels/0585o.pdf.

EIA (Energy Information Administration), 2004, "Energy Information Sheet, Propane" http://www.eia.doe.gov/neic/infosheets/propane.htm.

Drell, I. L., and Belles, F. E., 1958, *Survey of Hydrogen Combustion Properties*, NACA Report 1383 — Washington, D.C., National Advisory Committee for Aeronautics.

Environment Canada, 1984, *Manuals of Spills of Hazardous Materials*, Technical Services Branch, Environment Services Protection, Ottawa, Ontario, Canada.

Eyen, J. J., 1986, "How to Evaluate Centrifugal Pumps for CO_2 Pipeline & Injection Use," Pipeline Industry, pp. 29–31, June.

Farris, C. B., 1983, "Unusual Design Factors for Supercritical CO_2 Pipeline," Energy Prog. Vol. 3, #3, pp. 150–158, September.

Feehan, B., 2003 "Global LPG Motor Fuel Development", US DOT Clean Cities, LPG Workshop, Mexico City, Nov 6. http://www.eere.energy.gov/cleancities/international/pdfs/brian_freehan.pdf.

Fesmire, C. J., 1983, "531 BCF of CO_2 Through the CRC System," Eng. Prog., Vol. 3, No. 4, pp. 203–206, December.

Gas Processors Suppliers Association (GPSA), 1994, *Engineering Data Book*, Tulsa, OK, USA.

Girard, J., 1985, *L' Air Liquide*, Meeting Communications.

Groeneveld, T. P., 1974, *Hydrogen Stress Cracking*, 5th Symposium on Line Pipe Research, Houston, TX.

Grumer, J., Harris, M. E., and Rowe, V. E., 1968, *Fundamental Flashback Blowoff and Yellow Tip Limits of Fuel Gas Air Mixtures*, U.S. Bureau of Mines Report 5225, Washington, D.C., U.S. Department of Interior.

Guran, B., 1985, "Slurry Pipeline System," *Internal Report*, NOVA Corporation, Calgary, Alberta, Canada.

Hanna, B. W., 1978, "Cochin Pipeline Projects," *Energy Processing/Canada*, January–February.

Hanna, B. W., 1979, "Cochin Line System Will Move HVP Liquids," *Oil and Gas Journal*, September 24.

Hanna, B. W., and Bourbonnie, T. V., 1978, "Cochin Pipeline Project," *Energy Processing/Canada*, January – February.

Hein, M. A., 1986, "Rigorous and Approximate Method for CO_2 Pipeline Analysis," Sim Sci Inc, Houston, TX.

HID (Hazardous Installation Directorate, 2001 "Safety Report Assessment Guide: LPG - Criteria", http://www.hse.gov.uk/comah/index.htm

Hinton, B. R. W., and Proctor, R. P. M., 1981, "The Effect of Cathodic Protection and Over Protection on the Tensile Ductility of Corrosion Fatigue Behaviour of X-65 Pipeline Steel," Hyd-Eff. Met. Proc. 3rd Int. Conf., pp. 1005–1015.

Holbrok, J. H., and Cialene, H. J., 1985, "Effects of SNC/Hydrogen Gas Mixture or High Pressure Pipeline," Report to Line Pressure Research, Supervisory Committee, AGA, BaHalle, Columbus Labs, OH, NG 18-February.

Hoover, W. R., Robinson, S. L., Stoltz, R. L., and Springarn, J. R., 1981, Hydrogen Compatability of Structural Materials for Energy Storage and Transmission, SAND81-8006.

Interrante, C. G., 1982, *Basic Aspects of the Problems of Hydrogen in Steels*, Proc. 1st Int. Conf. on Current Solutions to Hydrogen Problems in Steels, Eds., Interrante & Pressouyre, Washington, D.C., pp. 3–17.

Jasionowski, W. J., Pangborn, J. B., and Johnson, D. G., Distribution of Gaseous Hydrogen — Technology Evaluation, IGT Report, pp. 485–508.

Johnston, H. G., Bezinan, I. I., and Hood, C. B., 1946, *Joule Thompson Effects in Hydrogen at Liquid Air and at Room Temperatures*, J. Am. Chem. Soc., 68, pp. 2367–77.

Kisten, M., and Runow, P., 1981, "On the Kinetics of Environmentally Assisted Fatigue Cracking of Pressure Vessel Steels in the Presence of Hydrogen Gas". Proc. 8th Int'l Congress of Metallic Corrosion, Making, W. Germany.

Kisten, M., and Windgassen, K.F., 1980, *Hydrogen Assisted Fatigue of Periodically Pressurized Steel Cylinders*, Proc. 3rd Int. Conf. on Effect of Hydrogen on Behaviour of Materials, Eds., Bernstein & Thompson, Moran, WY, pp. 1017–1025.

Kendrick, A., 1985, "Bubble — Tight Valve, Stop High Pressure Hydrogen Leakage," Process Industry, Oct–Nov, pp. 9–13.

King, G. G., 1980, "Design of Carbon Dioxide Pipeline," Proc. ASME, ETC&E Conference, January 28–22, Houston, TX.

King, G. G., 1981, "Design Consideration for Carbon Dioxide Pipe Line," Pipe Line Industry, Nov., pp. 125–132.

Knowlton, R. E., 1984, An Investigation of the Safety Aspects in the Use of Hydrogen as a Ground Transport Fuel, Hydrogen Energy Progress V, pp. 1881–1886.

Konopka, A., and Wurm, J., 1978, *Transmission of Gaseous Hydrogen*, pp. 405–412.

Kumano, A., 1980, "An Investigation of Wave Propagation in Fluid Containing Particles."

Kung, P. and Mohitpour, M., 1986, "Non-Newtonian Liquid Pipeline Hydraulic Design and Simulation Using Microcomputer," *Proc. ASME ETCE Conference*, Pipeline Engineering Symposium, pp. 73–78.

Lange Gas, 2004, http://www.langegas.com/umrele.htm.

L'air L., 1976, *Gas Encyclopedia*, Elsevier Scientific Publishing Company, Amsterdam, The Netherlands.

Lee J. I., and Sigmond P. M., 1978, "Phase Behaviour and Displacement Studies of Carbon Dioxide — HydroCarbon Systems," Research Report RR-37 Pet Rec Inst.

Loginow, A.W., and Phelps, E.H., 1974, " Steels for seamless Hydrogen Pressure Vessels", J. of Engineering for Industry, Trans.ASME, October.

LPG Australia, 2005 http://www.lpgaustralia.com.au/displaycommon.cfm?an=1&subarticlenbr=10.

Lucon, O., Coelho S.T., and Goldemberg, J., 2004, " LPG in Brazil: lessons and challenges", energy for Sustainable Development, Vol. VIII, No 3, Sept. pp 82-90.

Luk, B., 1987, "High Vapour Pressure HVVP Liquid Pipeline Design Manual," *NOVA Internal Report*, Pipeline Engineering Department.

Kisten, M., and Runow, P., 1981, "On the kinetics of Environmentally assisted Fatigue Cracking of Pressure Vessel Steels in the Presence of Hydrogen Gas" 8th International Congress of Metallic Corrosion, Making, W. Germany.

Matveev, V. V., Mayer, S., and McCarty R. D., 1984, "Adiabatic Equation of state for Hydrogen up to 150kBar," Inst. Atomic Energy, U.S.S.R, PISMA.V.ZH. EKSP and Teor Riz. 29, 5, pp. 219–221.

Mayer, S., 1982, *Cathodic Protection Design*, A Tutorial Course, University of Wisconsin, July 12–16.

McAuliffe, C. A., 1980, *Hydrogen and Energy*, The MacMillan Press Ltd., London.

McCarty, R. D., 1979, *Hydrogen: It's Technology and Implications*, Hydrogen Properties, Vol. III, CRC Press Inc., 2nd Edition.

MCS (Management Consulting Services Inc), 2001, " Propane Business Primer", October, Washington DC.

Michel, et al., 1959, "Compressibility Isotherms of Hydrogen T-S Diagrams," *PHYSIAC*, Vol. 25, p. 25.

Michels, A., DeGraff, W., and Wolkers, G. J., 1963, *Applied Science Section A*, 12 (1), p. 9.

Mohitpour, M., 1987, "A Guideline Manual for Design of Hydrogen Pipeline System," Internal Report, NOVA Corporation, Calgary, Alberta, Canada.

Mohitpour, M., Jenkins, A., and Babuk T., 2006, " Pipelining Liquefied Petroleum Gas (LPG)", Proceeding 5th ASME International Pipeline Conference. Hayatt Regency, Calgary, AB, CANADA, Sept. 25-29.

Mohitpour, M., Pierce C.L., and Graham P. 1990, "Design basis developed for H2 Pipeline", Oil & Gas Journal, May 14 pp. 83-94.

Mohitpour, M., Pierce, C. L., and Hopper, R., 1988, "The Design and Engineering of Cross Country Hydrogen Pipeline" Jr. Energy Resource Technology, Vol. 110, pp. 203–107, Dec.

Mohitpour, M., Szabo J., and Van Hardeveld, T., 2003 " Pipeline Operation & Maintenance - A Practical Approach", ASME Press New York.

Mohitpour, M., Trefanenko, W., Tolmasquim, S. T., and Kossatz, H., 2003, "Oil Pipeline valve Automation for Spill Reduction", presented at Rio Pipeline Conference, Hotel Continental, Oct., 21-24.

Mohitpour, M., Trefanenko, W., Tolmasquim, S.T., and Kossatz, H., 2004, " Valve Automation to Increase Oil Pipeline Safety" AMSE 5th International Pipeline Conference, Hyatt Regency Calgary, AB, CND, Oct 4-8.

Morrow, T. B., Bass, R. L., & Lock, J. A, 1983, " An LPG Pipeline Break Flow Model " ASME Transaction, Jr. Energy Resources Tech. Vol. 105 pp 379-387 Sept.

NPGA (National Propane Gas Association), 2003, *"Propane Education & Research Council".*

NPGA (National Propane Gas Association), 2004, General U.S. Industry Statistics and Characteristics of Propane" http://www.npga.org/i4a/pages/index.cfm?pageid=633.

NFPA (National Fire Protection Association), 2004, " Liquefied Petroleum Gas Code (NFPA 58)".

PERC (Propane Education & Research Council), 2004 " Propane Industry Issues & Trends" June.

Petroleum Frontiers, Feb 1984, "Carbon Dioxide and its applications to Enhanced Oil Recovery," Pet. Info Crop. Vol 2#1.

Pipeline Gas Journal 2002, Code of Federal Regulations Subpart 1 Requirement 192.629, Part 192, Title 49, 52.

Platus, D. L., Mackenzie, D. W., and Morse, C. P., 1974, " Rapid Shutdown of Failed Pipeline Systems and Limiting Pressure to Prevent Pipeline failure Due to Over Pressure" Part 1, Report MRI-2628- TRI, Oct.

PVFD (Potosi Volunteer Fire Department), 2005, "Pictures of 9/7/2000 LPG Pipeline Fire" http://www.angelfire.com/tx/pvfd/pipeline.html.

Quarles, W. R., 1983, "Wilbros near completion Cortez CO_2 Trunkline," Pipe Line Industry, Aug., pp. 31–34.

Read, R. M., and Eriksen, A., 1948, *Hydrogen and Synthesis Gas Production*, Gas 24, 53–56.

Recht, D. L., 1984, "Carbon Dioxide Pipeline Design Considerations," Proc. Symp Valve and Wellheads, API, 1984, Stand Conf. June 11–15 Miami, FL.

Recht, D. L., 1986, "CO_2 Line Design Needs," Pipeline and Gas Journal, Jan., 41–44.

Renfro, J. J., 1979, "Sheep Mountain CO_2 Production Facilities — A Conceptual Design," SPE Pub 7796.

Rohleder, G. V., 1972, "How High Vapour Pressure Products are Transported Safely by Pipeline," *Pipeline and Gas Journal*.

Ronson, 2004, "History" http://www.ronson.com/about/index.html.

Sayegh, S., and Najman, J. G., 1984, "CO_2 – SO2 – Brine Phase Behaviour Studies," Pet. Rec. Inst Report, May.

Shoemaker, A. K., Dobkowski, D. S., Kenkos, P. J., and Zgonc, J. H., 1976, "Development of 80/90 KSI Yield Strength ERW Line Pipe for CO_2 Pipeline Construction."

Standby Systems Inc., 2004 "Propane Standby Systems . . ..an overview" , 1991, revised 10-29-2004 http://standby.com.

Steinmetz, G. F., 1982, Review Presentation for Transmission, Distribution of Bulk Storage of Hydrogen Relative to Natural Gas Supplementation, Hydrogen Energy Process V, pp. 1187–1200.

Streeter, V. C., and Wylie, E. B., 1967, "Hydraulic Transients," McGraw Hill, New York, NY.

Swink, M. N., 1982, "Design Consideration for Sheep Mountain CO_2 Line," Pipe Line Ind. June, pp. 41–92.

Tampkins, L. I., and Johnson, L., 1986, "Chevron Lays Co., PO4, Lines," Pipeline & Gas Journal, Jan., pp. 29–34.

Thomas, 1980, *Method of Transporting Hydrogen*, U.S. Patent #4, 183, 369.

Thompson, A. W., 1977, *Materials for Hydrogen Services, Hydrogen*: Its Technology and Implications, Chapter 4.

Troiano, A. R., 1960, The Role of Hydrogen and Other Interstitials on the Mechanical Behaviour of Metals, Trans ASME 52, pp. 54–80.

TSHA (The Texas State Historical Association), 2002, "The Handbook of Texas online" http://standby.com.

Tyson, W., 1979, *Hydrogen in Metals*, Canadian Metallurgical Quarterly, Vol. 18, pp. 1–11.

Warren, R. N., 1983, "Sheep Mountain CO_2 Pipeline Facilities and Initial Operation," Pipe Line Industry, June, pp. 21–22.

Wasp, E. J., Kenny, J. P., and Gandhi, R. L., 1977, "Solid-Liquid Flow Slurry Pipeline Transportation," *TransTech Publication*, Series or Bulk Material Handling, Vol. 1, No. 4.

Watkins, M. J., 1983, "Deep Well Conditions Pose Stiff Challenges for Elastometers," Pet. Eng. Int., April, pp. 51–56.

Watson, H. C., Gowdie, R. R., 2000, "The Systematic Evaluation of Twelve LP Gas Fuels for Emissions and Fuel Consumption", SAE Technical Paper, 2000-01-1867.

Webb, W. P., and Gupta, S. C., 1984, *Metals for Hydrogen Service*, Chemical Engineering, pp. 113–116.

Wiebe, R., 1941, *Chem. Rev.*

Yedigarov, A., 1996, "Hazards Analysis of LPG Pipeline Transportation" Presented at The 1996 Annual Meeting of the Society for Risk Analysis-Europe, Institute of Natural Gases & Gas Technology 142717, Russia, Moskovskaya obl, Leninsky raion, P.Razvilka, VNIIGAS, Fax: 7-095-399-16-77 http://www.riskworld.com/Abstract/1996/sraeurop/ab6ad061.htm.

GLOSSARY OF TERMS

Alignment sheet/ drawing	A drawing that illustrates the pipeline route, pipeline specifications, land ownership, topography, and installation information. In most cases an aerial photo-mosaic is used.
Axial stress	Stress in the longitudinal direction of the pipe.
Bead weld	First weld joining two pipe segments.
Berms	Mounds of earth or rock used to divert surface water flow/provide slope stability.
Bevels	The tapered end preparation on pipe joints to facilitate welding.
Blind flange	A temporary or permanent end closure on a pipe opening.
Block valves	Major valves used to isolate sections of pipeline in the event of a line break or need for major maintenance.
Blowdown	A valve used in the process of venting a pressurized gas line in a controlled manner.
Bulk modulus	A three-dimensional property of materials and fluids, which defines the relationship between applied load and volumetric strain.
Buoyancy	Buoyancy is the ability of a liquid (such as water) to support or hold up other liquids or objects. The buoyancy of a liquid is closely related to its density because the denser a liquid is, the more it can support. For example, salt water is denser than fresh water so things are supported more in salt water and float higher or more easily in it. Buoyancy can be measured as the mass of material that can be supported by one milliliter (1 ml) of liquid. For example, in fresh water, 1 gram of material can be supported by 1 ml of water; therefore the buoyancy of fresh water is 1 gram per ml (1 g/ml) or 1000 kg/m^3.
Caliper pig	An in-line tool used to measure the ovality of a pipe in service.
Cavitation	A fluid phenomenon, which occurs when gas or vapor comes out of solution at low partial pressure. The subsequent collapse of the gas bubbles can cause pitting on adjacent material surfaces.

Chainages	Contour plus chainages, which are measured along the contour of the pipe or along the ground surface above the pipeline.
	Horizontal plus chainages measured horizontally between points either along the pipeline or along some other line.
	Imperial plus chainages, which are measured in feet and represented with the hundreds to the left of the plus sign (e.g., 2 + 34.78 = 234.78).
	Metric plus chainages, which are measured in meters and represented with the thousands to the left of the plus (e.g., 7 + 109.45 = 7,109.45 m).
	Equations corrections or adjustments, which are applied to chainages to compensate for changes to the continuous length from a selected point or origin. At the point of correction, the back chainage is equated to the ahead chainage.
Charpy notch	A notch machined in a test specimen to initiate cracking under load.
Combination bends	A side bend in the pipe that is combined with a sag or overbend.
Crossing angle	The angle at which the pipeline, utilities, roads, or rail lines cross.
Cut-out	A portion of pipeline that is cut out of the pipeline to affect a repair.
Dead weight	A measurement devices used in hydrostatic testing.
Drag section	The name for a section of pipeline that is installed between open ends of mainline pipeline. Typically this is done at crossings of roads, railways, pipelines, rivers, etc.
Diametral strain	Caused by pressurizing the pipe. It is the ratio of the increase in diameter to the original pipe diameter.
Ditch line	The centerline of the ditch where the new pipeline is to be located.
Double-jointing	Two segments of pipe welded together to save making a field weld.
Ductile fracture	A fracture, which initiates and propagates in the pipeline material and has significant plastic deformation at the fracture surface.
Downstream (D/S)	In the direction of the flow of the fluid in the pipe.
Elastic limit	Maximum stress a material is capable of sustaining without causing any permanent (plastic) deformation upon release of the load (also known as proportional limit).
Electronic distance measured (EDM)	An instrument with the capabilities of measuring distances electronically from a reference point.

Electronic pipe and cable locator	An electronic device with the capability of locating buried pipeline and cables by inducing or transmitting a signal to the pipeline or cable.
Emergency shutdown	A process used to isolate compressor station yard piping and equipment from the consequences of an unwanted event.
Engineering survey	Surveys performed on pipeline right-of-way for engineering design/construction purposes.
Final tie-in	The final weld joining two sections of pipeline. Final tie-in welds are those that are not subjected to a hydrostatic test.
Flow regime	Describes the flow within a pipe that is characterized as being either smooth (laminar) or turbulent.
Flow direction	The direction in which the gas flows in the pipeline.
Flume	A method for ensuring construction does not introduce sediments in critical spawning creek/rivers.
Foam pads	Foam that is used as padding to protect the pipeline in areas where rock, gravel, or frozen material is used as backfill.
Foreign pipelines	Pipelines that belong to other operators.
Gabions	Wire mesh containment structures holding stones and rocks used to stabilize the toes of steep slopes and to divert water.
Gaps	Distance between sections of mainline pipeline.
Geodetic elevation	The vertical distance of a point above mean sea level.
Girth weld	A circumferential weld.
Grade of pipe	The classification of the material used to manufacture the line pipe.
Head	An alternative means of describing pressure in terms of the height of column of a fluid.
Heat number	The number assigned to a joint of pipe by the pipe mill to indicate the batch of molten steel from which that joint of pipe is produced.
Holiday	A breach or void in the protective coating of a pipe.
Hoop stress	The stress in the circumferential direction of a pipe or pressure vessel.

Hot bend	Heat is used on a thick-walled pipe, or where a tight radius of curvature is required, in order to facilitate bending.
Hot line	An existing pipeline, which is loaded with fluid in a pressurized state.
Hot pass	The welding pass following the bead weld where the highest heat input is required.
Hot tap	Control drilling for an entry into pipeline under operating pressure.
Hydrostatic test	A test to which all new pipe is subjected. It is performed by filling the pipe with water and increasing the pressure a set amount above the normal operating pressure.
Iron bar (IB)	A monument indicating a legally surveyed line or boundary.
J-integral	A fracture mechanics parameter relating crack size, geometry, and stress acting on a crack. This parameter accounts for plasticity effects in crack growth.
Jeeping (of pipe)	Testing of the electrical insulation of the pipe coating.
Joint	An individual length of pipe usually 12.2 meters (40 feet) long.
Land agent	A company representative who is responsible for all dealings with landowners for the purpose of acquisition or lease.
Landowners	A person (or persons) who owns the lands on which the pipeline facilities are located.
Leave open end	The downstream end of a section of mainline pipeline (unless the pipeline is laid in reverse flow direction).
Loop line	A new pipeline, which is parallel and connected to an existing pipeline.
Monumented line	The baseline containing the survey monuments from which the pipeline easement limits are established.
Muskeg	Organic material typically located in wet boreal regions.
NDE	Nondestructive examination of pipeline welds by X-rays, gamma rays, etc.
NEB	National Energy Board, the Canadian federal authority that regulates the construction and operation of interprovincial pipelines.

PI	The point of intersection of two straight survey lines.
Profile	The determination of the horizontal and vertical relationships of points along a line.
Pup	A portion of a joint of pipe.
Pipeline construction specification	Specifications prepared to govern the construction of a high-pressure gas transmission pipeline.
Pipe-laying contractor	The contractor that is awarded the contract for the construction of the pipeline.
Pipe number	The number assigned consecutively to each pipe joint by the pipe mill.
Plastic deformation	Material straining that occurs beyond the elastic limit which results in permanent set, i.e., it is nonrecoverable.
Poisson effect	The phenomenon in solids whereby straining in one direction causes strains in the mutually perpendicular directions.
Primer	Coating applied to pipe before wrapping.
Priming	The start-up phase for a pump in which the impeller needs sufficient working fluid on which to act.
Probing	The use of a steel probe to determine the amount of cover over a pipeline.
Project description	An outline of the work to be performed on the project.
Right-of-way	The strip of land running the length of pipeline, for which an easement or permission has been granted within which to construct and operate the pipeline.
Roach	The remnant backfill material over a pipeline centerline.
Rock shield sheets	Used in lieu of sand padding in areas where it is warranted by the condition of the backfill materials.
Sag bend	A vertical (upward) bend in the pipe.
Saddle weights	Concrete weights placed on top of the pipeline at locations where it is susceptible to floating.
Sand padding	Layer of sand placed under/over the pipeline to protect it from damage from stones or rocks.

SCADA	Acronym for supervisor control and data acquisition. A process that enables the critical operating parameters of the system to be recorded and transmitted for control purposes.
Shear rate	A measure of the ratio of the shearing force and the velocity at a given point in viscous fluids.
Shear stress	A tensile stress, which attempts to cause surfaces to slide over one another.
Shock loading	Loads that are imparted to a structure through sudden or violent action, such as an impact or earthquake.
Shrink sleeve	A composite material shrink wrapped over the joint between two pipe sections so as to provide corrosion protection.
Side bends	A horizontal bend in the pipe.
Side boom	A specialized construction pipe layer.
Slip bore	The process of installing pipe under roads, railways, etc., by boring.
Slabs	Concrete slabs placed above the pipeline for protection at ditch locations.
Slug catcher	A device fitted at intervals along the length of gas pipelines that is used to collect and remove slugs of liquid, that have accumulated in the line over time.
Sour gas	Natural gas that contains elements of hydrogen sulphide (H_2S).
Staging areas	Locations along the right-of-way where pipe material is off-loaded and stacked prior to stringing.
Standard iron bar (SIB)	A monument that is 25-mm square and 1.2 m in length, usually used to indicate a legally surveyed line or boundary.
Swabbing	Removing foreign material from inside a pipe.
Tarpaulins	Covers used on either side of a pipe joint to protect the coated pipe from weld spatter.
Test head	A device attached to the end of a length of pipeline for hydrostatic testing.
Test lead	A wire attached to the pipeline that is used primarily to measure the electrical current to and from the pipeline to monitor corrosion rate.
Toe erosion	The gradual removal of material by natural means from the base of slopes.

Topographic survey	A survey of all significant land features along or near a survey line.
Trigonometric leveling	The determination of the difference in elevation between two points by measuring the vertical angle and the slope distance between the points.
Trunk line	The principal component of the pipeline transmission system into which gathering lines flow or from which distribution lines emanate.
Underfiling	Removal of material from the impeller in order to alter performance.
Upstream (U/S)	Direction against the fluid flow.
Vertical angle	The angular dimension of the sight line above or below the horizontal plane at the survey instrument.
Volute chipping	Removal of small amounts of the impeller casing in order to change the flow and thus alter the performance characteristics of a pump.
Water hammer	A transient phenomenon produced by a sudden valve closure in which a traveling wave moves along the length of the pipe, reversing direction when it encounters a reflective boundary.
Work room	An area of land required usually for a temporary period for construction activities.
Wrap	Inner tape applied over the pipe after primer is applied; outer tape applied over the inner wrap.
Yield plot	Created during a hydrostatic test, it is a plot of increasing pressure versus added volume of water. It also has the form of a stress versus strain plot and is curtailed when a condition of 0.2% strain is reached.

Appendix *A*

Refer to Chapter 2. Reprint from ASME Proceedings of the International Pipeline Conference Book No. G1075A-1998

Route Selection for Project Success: Addressing "Feeling/Perception" Issues

M. Mohitpour, G. Von Bassenheim and Ardean Braun

NOVA Gas International Ltd. Calgary, Alberta, Canada

ABSTRACT

Selecting a route for a pipeline right-of-way (ROW) generally consists of engineering (technical and economic), socioeconomic and biophysical components. To effectively select a route, simultaneous consideration must be given to all the components from the initiation of a project to the integration of all aspects of each throughout the route selection process. To successfully select a route that creates a win-win situation for all the stakeholders of a pipeline project, political/governmental issues, community and landowner views, public perceptions, and other similar controlling factors [such as Safety, Health, Environment and Risk (SHER)] must be carefully analyzed and integrated into the process. It is the consideration of all these issues that will lead to a ROW that will provide a technically acceptable solution, which is at the same time the least expensive, economically viable, and acceptable to the community it traverses.

This paper will provide an overview of route selection techniques (including new technologies) used and the process generally practiced by pipeline designers, highlighting controlling issues and optimization methods that need to be utilized in order to achieve a cost-effective route selection. It provides details on significant "Feeling/Perception" issues that can either thwart or, by careful consideration of these issues, lead to a successful pipeline project. An example of such a route selection process will be provided on a project located in rough and mountainous terrain, which has significant regulatory/governmental, land, environmental, indigenous and geological issues.

BACKGROUND

Pipeline route selection is as old as pipelines themselves. In 1991, United Nations Gas Facts indicated that over 3.5 million kilometers of gas lines alone were installed around the world. Route selection has played a significant role in these pipeline installations.

Originally, route selection for most pipelines were generally based on the location of supply and delivery points and on the identification of controlling factors such as

691

population densities, lakes and rivers, mountain passes, energy corridors, proximity to other facilities, etc. Consideration of socioeconomic and biophysical components is only a recent phenomenon. These considerations have been necessitated by regulatory and environmental issues and generally addressed in earlier years through Environmental Impact Assessment Techniques (White A.D. 1979; Bosenbury 1983).

In the early days of pipeline route selection, the process was based on major linear development projects, such as roads and railways. Major considerations were length, terrain, climate, strategic nodes and consideration for safety, reliability, protection/ maintenance and operation. Biophysical setting such as land use and capabilities, hydrological systems, environmental hazards and climatic conditions as well as socioeconomic issues played a role in pipeline route selection in later years. Socioeconomic components, such as employment, population growth, service, community, land and regional infrastructures, along with consideration of social adjustments by communities that a pipeline route traversed is a subsequent phenomenon.

This three-dimensional approach has been utilized by the pipeline industry for project developments since the late 1970s and through the early 1980s (Passey and Wooley 1980).

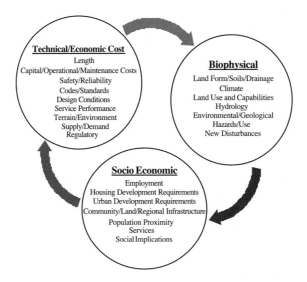

Figure 1. Route selection process in the 1980s

With the increase in urban sophistication and ever-increasing awareness of the impacts of pipelines on people's daily lives, a fourth dimension, the "Feeling/Perception" issue, needs to be added to this route selection process. This is particularly important if the pipeline is to transport natural gas through a community or region of the world where no pipeline construction/operation has been previously experienced, but where the awareness of safety related incidents/issues occurring elsewhere is routinely headlined and monitored closely by populace/community.

For the route selection process to succeed therefore, consideration must be given from the onset of a project to not only address and integrate significant aspects of each of the components of engineering/cost, biophysical, and socioeconomic, but to concurrently include a consultation process addressing community feelings and perceptions. Significant problems and cost/schedule impacts may be encountered if the issues around the affected communities are not properly considered during the route selection process. Such public/community consultation is distinct from those that address affected regulatory and landowner concerns, in that it addresses the feelings and perceptions of communities whose lives would be affected and disturbed by the pipeline.

CURRENT/EVOLVING TECHNOLOGIES

The current pipeline industry requires a route to be cost-efficient, ensure pipeline integrity, minimize adverse environmental impacts and risk to public, and recognize land use, and constraints, and proximity restrictions to existing facilities. A single technique or a combination of methodologies (Exner 1983), such as the committee approach, check lists, overlays, matrixes, or a computer assisted information processing, integration, and analysis. A step-by-step method of route selection using the combination of approaches is described elsewhere (Mohitpour 1985–1996).

Recently detailed imagery, aerial photography and GIS (Geographic Information System) have been used for route planning assessment and regulatory reporting, etc. (Gale 1997; Feldman et al. 1994).

Data for GIS is obtained by various sources, including:

- Optical such as:
  - Land Sat 1, 2, 3, 4 and 5
  - Spot 1, 2, 3
  - IRS 1A, 1B and 1C
- Radar such as:
  - JERS 1
  - ERS 1, 2
  - Radar Set 1
- Satellite photos such as:
  - Corrona spy satellite archieves
  - Russian photographic data
- Digital ortho imagery (Airborne)
  - Aerial photomosasic

The selection of Image Data Source depends on many factors, like:

- Suitability for application such as
  - Information content (topographic, geological/geophysical, land information, etc.)
  - Cost
  - Space
  - Scale and accuracy, resolution
- Data of acquisition and availability
- Cloud cover

A number of applications have been developed that support data gathering generation, analysis, and decision making. An example of the third one is the "Pipeline Manager™"

used for route planning/assessment and selection decision making (Maybury 1997). This type of tool is utilized to effect a route solution by managing maps, databases, and pictorial information in conjunction with a portable or fixed satellite gathered information.

GIS combines the following information to produce the most cost-effective route:

- Accurate and up-to-date map base
- Current geological engineering information (interpreted from imagery)
- Accurate hydrography and geomorphology
- Quick assessment of construction and operational risks and hazards, including existing environmental hazards/impacts
- Assessment of land use, land cover, and significant habitation
- Assessment of existing infrastructure, roads, routes, energy corridors, airports, etc.
- Facility siting.

An example of the usefulness of GIS is shown in Figure 2, where GIS images and aerial photography coupled with a route from topographic maps are superimposed to provide key information on constructibility features of pipelines for a particular route (Gale 1997).

"FEELING/PERCEPTION" ISSUE IN ROUTE SELECTION

All approaches/methodologies and tools outlined above, whether current or evolving, have their application and can be, and are, valuable depending on the size of the project and the degree of environmental/ecological sensitivities. These provide information of substance for ranking of routes based on engineering/cost-efficiency as well as all biophysical and some socioeconomic advantages.

What these methodologies lack or the tools cannot track, analyze, or detect is communities' "feelings and perceptions" when an intended pipeline corridor is perceived to affect their lives and their normal day-to-day activities.

Negative feeling/perceptions usually do not have any bounds. They are initiated by many factors. These include, among others, the following:

- Fear of unknown/danger
- Media depiction of danger and safety issues
- Peer talks/pressure
- Lack of experience/information
- Lack of understanding/trust
- Lack of open and honest communication
- Fear of disturbance
- Fear of devaluation of property
- Fear of deterioration of social conditions/quality of life
- Prior deception
- Prior poor experience
- Agitation by interested groups/parties
- Customs/traditions
- Money and greed
- Income level
- Method and timing of approach and consultation by the pipeline company

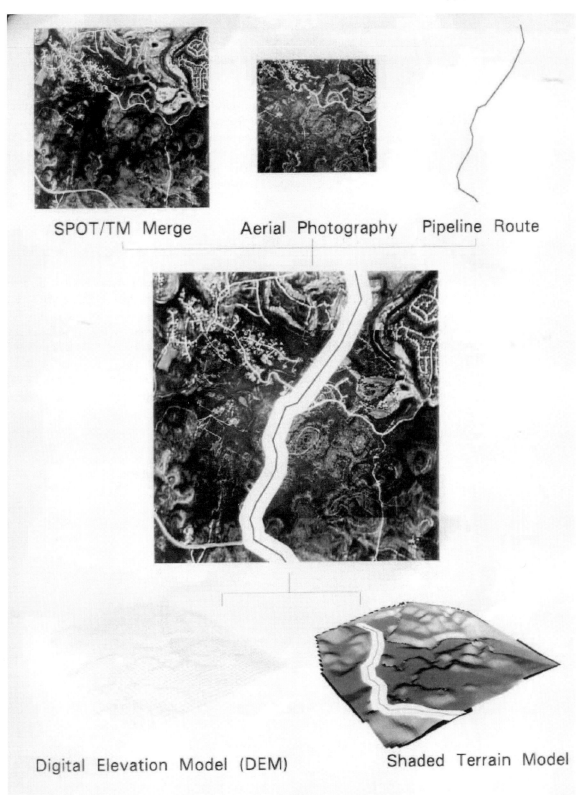

SPOT/TM Merge   Aerial Photography   Pipeline Route

Digital Elevation Model (DEM)   Shaded Terrain Model

Figure 2. TM digital image/route integration for pipeline constructibility assessment (Courtesy Logicon Geodyanmics Service, Englewood, Co., USA)

- Safety, health and environmental issues
- Availability of route alternatives
- Socioeconomic characteristics of the communities/land owners
- Others

While alleviating an individual's negative feelings/perceptions is a simple task to deal with and remedy, altering perceptions and feelings of a community from a negative to a positive stance is much more complicated and requires careful planning as well as continuous monitoring and rebuttal. This is even more complex and difficult to achieve in a foreign environment, particularly in developing countries (such as Chile) where experience with pipelines is scarce and knowledge of pipelines limited.

ADDRESSING FEELING/PERCEPTION ISSUES

It is no coincidence that an extensive pipeline network goes hand-in-hand with high standards of living and technological progress. Pipelines are the most efficient and safest means of transporting energy such as oil and gas. However, gas pipelines and explosions may be perceived to go hand-in-hand. In order to address and overcome negative feelings a fourth dimension, the "Public Information/Consultation" process needs to be incorporated and integrated within the route selection process, specifically the part related to ROW acquisition and permitting process. The integration of the distinct dimension as seen by the authors is depicted in Figure 3.

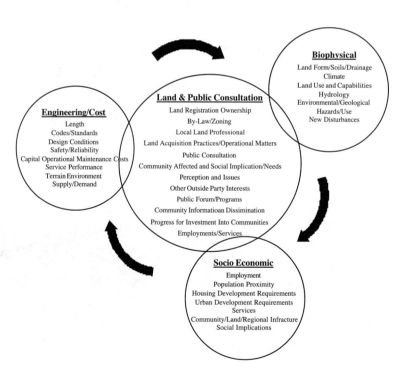

Figure 3. Route selection process with a public consultation dimension

It is this consultation process that can effectively highlight the advantages of pipelines not only providing employment opportunities and benefits to the affected communities, but also the long-term contribution to the economic and technological progress and success of the region it traverses. The process by which the consultation can be effective is its incorporation in the land usage and acquisition component of the route selection, as described below.

Before selecting a route, a thorough understanding of the procedures for obtaining easement and land access and acquiring land needs to be assessed.

The process begins by having an understanding of the procedures that are normally followed for linear development projects. In addition, meetings need to take place with the applicable regulatory boards, land service companies, and state and federal law offices as well as community agents to obtain a detailed understanding of processes and issues. The following provide some of the key points related to the assessment of land and regulatory and public/community requirements and issues:

- Determine which are the applicable regulatory boards and their jurisdiction
- Investigate subsurface access process
- Review application processes
- Review landowners' rights
- Determine landowner concerns
- Investigate landowner compensation process
- Review possible community concerns
- Review process for obtaining access to state-owned land
- Determine survey requirements (for access to land)
- Establish availability of local land service companies (if applicable)
- Determine a community and public consultation programs
- Determine community perceptions
- Determine the operational impact on the landowners and the community
- Determine applicable laws and regulations
- Be aware of pipeline company rights
- Determine the implications related to the encroachment of third-party facilities
- Investigate land usage process and procedures
- Review land use (current and proposed)
- Review requirements associated with ROW/land sharing
- Review requirements associated with crossings
- Identify aboriginal communities

The outcome of such an assessment is a detailed report, highlighting the following:

- Extent of land requirement
- Process of acquiring land
- Compensation and damage settlement
- Land application/permit requirements
- Documentation needs
- Land agents and analysis
- Constraints/opportunities
- Public and government participation/consultation program (with schedule as applicable)
  - Initial community contact
  - Issues identified and development plans
  - Method of resolving identified issues

Registration/Ownership
• Land registry system
• Land registry system applicable to regions/state versus federal
• Land transfer system
• Rights conveyed and secured against property
• Registry records and force of law
• Land registration/records location (federal, regional, state or municipal)
• Method of accessing land registration
Bylaw/Zoning
• Zoning bylaws/regulations in effect
• Level of legislative bylaws (federal, regional, state or municipal
Land Professional/Companies
• Requirement for registry or licensing land professionals
Land Acquisition Practices/Operational Matters
• Who is responsible for land usage document preparations (land firms, legal firms or others)
• Current land usage practices for pipelines or other similar types of utilities or facilities
• Operational impacts on land and how to handle them
• Land negotiating basis (land productivity, crop type, market land value, area price, best deal, government land)
• Rate of encroachment of development in areas of concern
• Landowners view/acceptance of construction/operation issues and strategy to deal with them
• Method for selling and buying land
• Permitting process
• Property/land taxes
• Operational taxes and levies
Public Consultation Matters
• Determination of public communities affected and social implication
• Aboriginal people lands and communities affected and their rights
• Aboriginal people structure of governance
• Community perception of the pipeline as a whole, issues and strategy to deal with
• Affected private and public sectors and their perception, interest, constraints and opportunities
• Determine government agencies having jurisdiction over ROW acquisition, permitting
• Determine government agencies position and interest in the project
• Interact with government or community issues
• Public forum program
• Community information dissemination seminar , location and schedule
• Compensation or programs for community involvement
• Programs for investment in the community (such as building roads, schools, etc.)

Figure 4. Land acquisition/easement information requirement

- public forums
- Documentation of all public feedback
- Outline and scope of agreement with landowners
- Extent of construction attendance
- Physical asset operational strategies (such as blow down location, approaches

The information that needs to be gathered in order to undertake a meaningful assessment is outlined in Figure 4.

APPLICATION TO A SOUTH AMERICAN PROJECT

Supply and transportation of natural gas to Chile is a recent occurrence (Mohitpour et al. 1996). The first gas pipeline, GasAndes (Figure 5) to serve the Santiago Metropolitan area was inaugurated in August 7, 1997. During the GasAndes route selection process a great number of lessons were learned and insight gained in selecting a route that not only had to overcome many physical boundaries—such as a 3,700-meter-high elevation of the Andes Mountain, significant geological, geophysical (volcanoes, seismic conditions), and environmental constraints—but, above all, had to overcome the perceptions and feelings of the people whose lives were impacted.

Figure 5. GasAndes pipeline

One single issue affecting the project was the negative community perception.

The final and timely project completion owed its success to an extensive public consultation process that was integrated in the land acquisition process and complemented by rigorous environmental planning and impact assessment studies.

The lessons learned was incorporated into selecting a route corridor for a new natural gas pipeline (Figure 6) that will stretch between the Province of Neuquen (supply point in Argentina) and the Concepcion area markets in Chile.

For the project, the objective of corridor selection and analysis was to decide on the best possible strip of land in terms of economics and other factors that would affect

Figure 6. Neuguen-conception pipeline general route corridor (Courtesy EISYS, Austin, Texas, USA)

construction of the proposed pipeline. In the context of this selection, this strip was to be approximately 2 km wide. This corridor selection should not be confused with micro routing, which is done at a later time during fine-tuning and finalizing the route. Micro routing involves picking the pipeline centerline within the recommended corridor and will involve a series of additional steps.

For the determination of the proposed corridor, the following steps were taken:

1. Identify gas source, customers, and locations.
2. Identify the corridor study area. Acquire high level mapping information for study area (topographical maps, environmental information, aerial photographis, roadmaps, landowner information, etc.).
3. Identify all Andes Mountain passes located within study area (a north-south approach as opposed to east-west approach was followed).
4. Determine a preliminary corridor and prepare high level estimate of project cost based on assumed facility need and size.
5. Determine cost to ascertain economically optimum facilities solution.
6. Acquire additional engineering, geological, construction, environmental, regulatory, community, pipeline, and facilities solutions; and a more detailed level of land information for the study area.
7. Identify specific technical, environmental, geological, aboriginal community, and landowner and social issues that will require additional meetings with regulators, meetings with people living and communities affected in study area, and on-site evaluation of community and construction conditions.
8. Perform high level but rigorous government, regulatory, and community consultation and information dissemination sessions.
9. Perform a high level reconnaissance to review potential corridors.
10. From the available information, derive potential corridors and display corridors on 1:50,000 scale maps. Corridor selection was based on:
    - Minimizing the number of river and stream crossings
    - Minimizing pipeline length

- Optimum location of suitable mountain passes
- Minimizing project costs
- Maximizing system reliability and availability
- Proximity to additional potential customers
- Avoidance of known geologically/geotechnically, socially, and environmentally sensitive areas
- Avoidance of vineyards and orchards where possible
- Avoidance of developed areas and high population densities
- Following existing corridors such as train tracks, roads, power lines where reasonably possible
- Choosing the most constructible locations for river crossings (easy approaches, narrow, no rock)
- Avoidance of marshy areas, sand, and salt flats

11. Perform a low level helicopter reconnaissance of the terrain along each corridor. Take photographs and identify preferred Andes pass and corridors.
12. Evaluate the preferred corridors based on corridor reconnaissance, project economics, social and environmental concerns, and future expansion. Propose an alternative.
13. Acquire existing aerial photographs of proposed corridor and landownership and community information.
14. Acquire panoramic photos of proposed corridor and indicate the location of the centerline for the proposed corridor (see example Figure 7).

Figure 7. Panoramic photo sample—Andes

The new pipeline is now being implemented. Design and construction activities are proceeding. However, the effect of such a public communication and consultation can only, and truly, be realized at a future date as construction proceeds and the line becomes operational. It is then that current perception can be gauged against what was perceived.

CONCLUSION

Economic and technological developments are very much related to energy transportation structures such as pipelines. Where there is a large network of pipelines and facilities, systematic progress is more evident. However, to some people pipelines, especially those intended to carry natural gas, explosions, environmental risks, and safety issues are perceived to go hand-in-hand.

Over the years of pipeline development, route selection has progressed from just considering a straight line between supply and demand points to consideration of natural and physical controlling factors, such as population densities, high mountains, rivers and lakes, etc., to the present day use of sophisticated data acquisition technologies such as GIS. The tools, however, are tools only and do not by themselves consider the real interaction of engineering/cost, biophysical, and socioeconomic, as well as the public consultation process necessary in modern day route selection.

With an increase in urban sophistication and ever-increasing awareness of the impact of pipelines on people and the environment, a consultative and information dissemination process must be truly followed right from initiation of the project, in order to combat any negative issues and feelings on the impact of the pipeline, to alleviate concerns of stakeholders, and to ensure a winning project.

ACKNOWLEDGMENT

The authors wish to express their appreciation to the management of NOVA for permission granted to publish this paper.

REFERENCES

Bonsenbury, T. R., 1983, "Current Topics and Approaches to Environmental Planning for Linear Development," PJTS Seminar on Environmental Planning for Linear Development, April 18, Calgary, Alberta, Canada.

Exner, K., 1983, "Route Selection Methodologies for the Linear Development Planning Process," PITS Seminar, Environmental Plan for Linear Development, April 18, Calgary, Alberta, Canada.

Feldman, S. C., Pelletier, R. E., Walsar, W. E., Smooth, J. C., 1994, "Integration of Remotely Sensor Data and Geographic Information System Analysis for Routing of the Caspian Pipeline," 10th Environmental Res. Inst. Mich. Geol., Remote Sensing Thematic Conference, San Antonio, Proc. V2, PP. 206–213.

Gale T., 1997, "EISYS" presentation, NGI, Nov.

Maybury, P., 1997, "Private Communications," "The Pipeline Manager™" RSK Enviro. Ltd., Chester, England.

Mohitpour, M., 1985–1996, "Pipeline Design and Construction," Seminar, Faculty of Continuing Education, University of Calgary, Alberta, Canada, October.

Mohitpour, M., Thompson, W., and Asante, B., 1996, "The Importance of Dynamic Simulation on the Design and Optimization of Pipeline Transmission Systems," Proc. ASME 1st International Pipeline Conference (IPC), Vol. 2., 1183, Calgary, Alberta, Canada.

Passey, G. H., and Wooley, D. R., 1980, "The Route Selection Process, A Biophysical Perspective," and Alberta Energy and Natural Resources, Edmonton, Alberta, Canada.

White, A. D., 1979, "Overview of Corridor Selection and Routing Methodologies," 2nd Symposium Environmental Concerns in ROW Management, Ann Arbor, MI, October 16–18.

Appendix *B*

Refer to Chapter 3

IMPACT OF DIFFERENT GAS AND PIPELINE PARAMETERS ON FLOW EFFICIENCY OF THE PIPELINE

This study uses the hydraulic analysis and resource tool (HART), a simulation model incorporated by the Pipeline System Design Department at TransCanada Pipelines to design its pipeline network. The objective of the study is to compare the impact of changing some of the main gas and pipeline parameters on gas flow efficiency. As described in Chapter 3, simulations was done for the three major gas flow equations (AGA Fully Turbulent, Panhandle B and Colebrook-White), and the results are presented both in tables and in graphs. Data used in the simulations are given at the bottom of the figures; a roughness of 700 micro-inches is used in all simulations when roughness is fined, except for the case where pipeline inlet gas temperature is variable (temperature profile), in which a roughness factor of 400 micro-inches is being used.

AGA fully turbulent equation
K_e, Roughness factors (micro-inches)

NPS	250	400	700	1,000	2,000	2,953
10	42.9	41.2	39.2	37.9	35.4	34
12	70.3	67.5	64.3	62.2	58.1	55.9
16	151.8	146.1	139.2	134.8	126.3	121.5
20	274.6	264.3	252.1	244.3	229.2	220.6
24	444.3	427.9	408.4	396	371.9	358.3
30	798.3	769.4	735	713.1	670.5	646.6
36	1,286.3	1,240.4	1,185.8	1,150.9	1,083.2	1,045.2
42	1,923.2	1,855.3	1,774.6	1,723.1	1,623.1	1,566.9
48	2,722.7	2,627.7	2,514.5	2,442.4	2,302.2	2,223.4

CBW equation
K_e, Roughness factors (micro-inches)

NPS	250	400	700	1,000	2,000	2,953
10	40.7	39.7	37.7	37.2	35.1	33.8
12	67	65.3	62.9	61.2	57.7	55.6
16	145.8	142	136.8	133.1	125.5	121
20	264.8	257.8	248.3	241.7	228	220
24	429.8	418.4	402.9	392.2	370.2	357.4
30	775.3	754.3	726.4	707.3	668.1	645.3
36	1,252.6	1,218.5	1,173.4	1,142.7	1,080	1,043.6
42	1,876.7	1,825.4	1,757.9	1,712.1	1,619	1,565
48	2,661.5	2,588.5	2,492.9	2,428.2	2,297.1	2,221.3

Panhandle B equation
Pipeline efficiencies

NPS	1	0.95	0.9	0.85	0.8	0.75	0.7
10	46.6	44.5	42.1	39.8	37.4	35.1	32.8
12	75.9	72.1	68.3	64.5	60.7	56.7	53.1
16	160.9	152.9	144.8	136.8	128.7	120.7	113
20	287.7	273.3	258.9	244.5	230.1	215.7	202
24	461.2	438.1	415.1	392	369	345.9	323.9
30	822.5	781.4	740.2	699.1	658	616.9	575.7
36	1,313.9	1,248.2	1,182.5	1,116.8	1,051.1	985.4	919.7
42	1,950.4	1,852.9	1,755.4	1,657.9	1,560.4	1,462.8	1,365.3
48	2,744.8	2,607.5	2,470.3	2,333.1	2,195.8	2,058.6	1,921.3

All volume gas flow rates calculated and presented in the tables are in MMSCFD; 2,953 micro-inches is the largest roughness factor that HART simulation software will accept. ($L = 40$ mi; $P_1 = 900$ Psi; $P_2 = 700$ Psi; $WT = 0.25$ inch; $G = 0.58$; $T_{ave} = 60$ °F; $\Delta H = 0$).

AGA fully turbulent equation
Isothermal gas temperature in (°F)

NPS	30	40	50	60	70	80	90
10	43.9	43.2	42.5	41.9	41.3	40.7	40.2
12	71.9	70.7	69.7	68.6	67.7	66.8	65.9
16	155.5	153	150.7	148.5	146.4	144.4	142.6
20	281.5	276.9	272.7	268.7	264.9	261.4	258
24	455.7	448.3	441.4	434.9	428.9	423.1	417.7
30	819.4	806.2	793.8	782.1	771.2	760.9	751.1
36	1,321	1,299.6	1,279.6	1,260.9	1,243.3	1,226.6	1,210.8
42	1,975.8	1,943.9	1,914	1,886	1,859.6	1,834.6	1,811
48	2,798.3	2,753.1	2,710.8	2,671	2,633.7	2,598.4	2,564.9

CBW equation
Isothermal gas temperature (°F)

NPS	30	40	50	60	70	80	90
10	42.2	41.5	40.8	40.2	39.6	39	38.5
12	69.4	68.2	67.1	66.1	65.1	64.2	63.3
16	150.8	148.3	145.9	143.7	141.6	139.6	137.7
20	273.8	269.2	264.9	260.8	257	253.4	250
24	444.3	436.8	429.9	423.3	417.2	411.3	405.8
30	800.9	787.5	775	763.2	752.2	741.7	731.8
36	1,293.5	1,271.9	1,251.8	1,232.8	1,215	1,198.2	1,182.2
42	1,937.4	1,905.2	1,875.1	1,846.8	1,820.2	1,795	1,771.1
48	2,747.1	2,701.5	2,658.9	2,618.8	2,581.1	2,545.5	2,511.7

Panhandle B equation
Isothermal gas temperature (°F)

NPS	30	40	50	60	70	80	90
10	42.2	41.5	40.8	40.2	39.7	39.1	38.6
12	68.4	67.3	66.2	65.2	64.3	63.4	62.6
16	145.6	143.2	140.9	138.8	136.8	135	133.2
20	260.2	255.9	251.9	248.1	244.6	241.3	238.1
24	417.2	410.3	403.9	397.8	392.2	386.8	381.7
30	741.6	729.4	717.9	707.2	697.1	687.6	678.6
36	1,184.7	1,165.1	1,146.9	1,129.7	1,113.6	1,098.4	1,084
42	1,758.7	1,729.7	1,702.5	1,677.1	1,653.2	1,630.6	1,609.2
48	2,474.9	2,434.1	2,395.9	2,360.1	2,326.5	2,294.6	2,264.5

All volume gas flow rates calculated and presented in the tables are in MMSCFD (L=40 mi; P_1 = 900 Psi; P_2 = 700 Psi; WT = 0.25 inch; G = 0.58; K_e = 700 micro-inches; η_p = 0.86; ΔH = 0).

AGA fully turbulent equation
Value of gas gravity used in the hydraulic equation

NPS	0.573	0.58	0.63	0.653
10	41.6	41.2	40.2	39.8
12	68.2	67.5	65.9	65.2
16	147.5	146.1	142.6	141.1
20	266.8	264.3	258.1	255.4
24	432	427.9	417.8	413.5
30	776.8	769.4	751.3	743.5
36	1,252.4	1,240.4	1,211.2	1,198.6
42	1,873.2	1,855.3	1,811.7	1,792.8
48	2,653	2,627.7	2,565.8	2,539.1

CBW equation
Value of gas gravity used in the hydraulic equation

NPS	0.573	0.58	0.63	0.653
10	40	39.7	38.8	38.4
12	65.8	65.3	63.8	63.1
16	143	142	138.6	137.2
20	259.7	257.8	251.6	249.1
24	421.5	418.4	408.2	404.2
30	760	754.3	736	728.7
36	1,227.7	1,218.5	1,188.8	1,176.9
42	1,839.2	1,825.4	1,780.8	1,762.9
48	2,608.1	2,588.5	2,525	2,499.7

Panhandle B equation
Value of gas gravity used in the hydraulic equation

NPS	0.573	0.58	0.63	0.653
10	39.9	39.8	38.7	38.3
12	64.8	64.5	62.7	62.1
16	137.8	136.8	133.5	132.1
20	246.4	244.5	238.6	236.2
24	395	392	382.5	378.7
30	702.2	699.1	680	673.2
36	1,121.7	1,116.8	1,086.2	1,075.4
42	1,665.1	1,657.9	1,612.5	1,596.4
48	2,343.3	2,333.1	2,269.1	2,246.6

All volume gas flow rates calculated and presented in the tables are in MMSCFD ($L=40$ mi; $P_1 = 900$ Psi; $P_2 = 700$ Psi; $WT = 0.25$ inch; $K_e = 700$ micro-inches; $T_{ave} = 60\ °F$; $\eta_p = 0.86$; $\Delta H = 0$).

AGA fully turbulent equation
Elevation change of node 2 expressed in feet

NPS	0	500	1,000	1,500
10	41.2	40.4	39.4	38.4
12	67.5	66.3	64.6	63
16	146.1	143.4	139.8	136.2
20	264.3	259.4	253	246.5
24	427.9	420	409.6	399.1
30	769.4	755.2	736.6	717.6
36	1,240.4	1,217.4	1,187.5	1,156.9
42	1,855.3	1,821	1,776.2	1,730.4
48	2,627.7	2,579	2,515.6	2,450.8

CBW equation
Elevation change of node 2 expressed in feet

NPS	0	500	1,000	1,500
10	39.7	38.7	37.8	36.8
12	65.3	63.7	62.1	60.5
16	142	138.6	135.1	131.5
20	257.8	251.6	245.3	238.8
24	418.4	408.3	398.1	387.6
30	754.3	736.3	717.9	699
36	1,218.5	1,189.4	1,159.7	1,129.2
42	1,825.4	1,781.9	1,737.4	1,691.8
48	2,588.5	2,526.8	2,463.8	2,399.2

Panhandle B equation
Elevation change of node 2 expressed in feet

NPS	0	500	1,000	1,500
10	39.8	38.8	37.8	36.8
12	64.5	62.9	61.4	59.8
16	136.8	133.9	130.6	127.1
20	244.5	239.4	233.4	227.3
24	392	383.9	374.2	364.4
30	699.1	682.4	665.2	647.7
36	1,116.8	1,090	1,062.7	1,034.7
42	1,657.9	1,618.1	1,577.6	1,536
48	2,333.1	2,277.1	2,220	2,161.6

All volume gas flow rates calculated and presented in the tables are in MMSCFD (L=40 mi; P_1 = 900 Psi; P_2 = 700 Psi; WT = 0.25 inch; K_e = 700 micro-inches; T_{ave} = 60 °F; η_p = 0.86; G = 0.58).

Appendix *C*

Reprint from ASME, PD-VOL 34,
Pipeline Engineering Book No. G00587-1991

Temperature Computations in Fluid Transmission Pipelines

M. Mohitpour

Novacorp International Consulting, Incorporated Calgary, Alberta, Canada

ABSTRACT

Temperature has considerable influence on the design of pipelines and related facilities, including the establishment of facilities sizing and optimization, economic and technical evaluation, etc. Computing steady-state flow temperature in a fluid pipeline may provide adequate information for some design purposes; however, it cannot adequately address soil/pipe/environment interaction information when time- and temperature-dependent parameters are required.

In this paper a comprehensive formulation for computing a steady-state temperature profile along a buried or exposed pipeline is provided. Additionally, an alternative finite-element formulation technique for a time-dependent three-dimensional heat transfer condition prevalent in pipeline situations with non-homogenous soil properties along and through the soil depth surrounding the pipe is demonstrated.

The steady-state approach provides satisfactory temperature solutions for most design purposes. The alternative method will however provide a total picture of time-dependent heat transfer solutions for fluid, pipe, and surroundings in a three-dimensional situation. The paper presents the application of the alternative approach.

INTRODUCTION

Temperature and pressure influence all fluid properties. In fluid transmission pipelines, both pressure and temperature vary along the pipeline length. In long-distance-transmission pipelines traversing varied terrain, from perma-frost regions to moderate climate conditions, pipelines experience significant temperature changes. Temperature change affects viscosity, density, and specific heat in liquid lines, particularly in crude oil pipelines. With some fluids there will actually be a temperature change during transportation independent of heat transfer from soil to pipe.

While a temperature rise is generally beneficial in liquid pipelines as it lowers viscosity and density, thereby lowering the pressure drop, such a temperature rise lowers transmissibility in gas pipelines due to an increase in pressure drop. This results in a net increase in compressor power requirements for a given flow rate. The value of absolute (dynamic) viscosity ($v = \mu/\rho$) for gas increases with an increase in temperature and to some lesser degree with pressure. Such an increase will result in an increase in frictional loss along the length of the pipeline.

Pipeline temperature also has an impact on the environment. Heated liquid lines that are not insulated can cause crop damage in farmland during summer seasons (Toogood 1979). In winter, the cold soil temperature can affect the pipe and the fluid it is transporting. Cooling of uninsulated liquid pipelines by frozen ground increases liquid viscosity and density, thereby requiring a greater pumping power. This cooling effect is more dramatic in non-Newtonian fluids (Kung and Mohitpour 1986). Such a cooling can also affect the notch toughness properties of steel material used in pipelines (Slimmon 1989). In an Arctic environment, lower pipeline temperature can significantly enhance the formation of frost heave and, as a consequence, may cause undue stresses in the pipeline. In discontinuous perma-frost areas, pipeline temperature changes and freezing and thawing of ground can induce cyclical strain on the pipe, affecting it's structural integrity. Accurate temperature computations are therefore essential both for steady-state situations and under transient conditions. These influence pipeline structural integrity and operation in the long term and pipeline design, construction, and commissioning in the short term.

TEMPERATURE COMPUTATION

In any pipeline segment, the overall temperature change (ΔT) is computed by considering the temperature changes due to conduction and convection (ΔT_c), isentropic expansion caused by elevation change (ΔT_e) and due to isenthalpic expansion caused by friction (ΔT_f), i.e.,

$$\Delta T = \Delta T_c + \Delta T_e + \Delta T_f \tag{1}$$

The following expressions summarize the computation of temperature components referred to above (Mohitpour 1987).

For a pipeline (Figure 1) buried at a finite depth (h_o) with insulation, the following expression for computing fluid flow temperature To is applicable (Holman 1981). For nomenclature refer to Figure 1.

$$T_o = \frac{m \cdot Cp_g \, T_i + \frac{kgS}{1+\propto} Tg}{\frac{kg}{1+\propto} + m \cdot Cp_g} \tag{2}$$

Where:

$$\propto = \frac{kg}{k_i} \, \ln\left(\frac{Di}{D_p}\right) \tag{3}$$

and

$$S = \text{shapefactor} = \frac{2\Pi\Delta L}{\ln\left[\frac{2h_o}{D_p} + \sqrt{\left(\frac{2h_o}{D_p}\right)^2 - 1}\right]} \tag{4}$$

For uninsulated pipeline $\propto = 1$

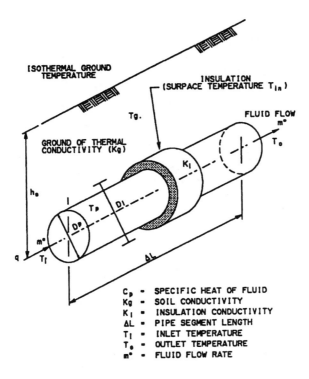

C_p - SPECIFIC HEAT OF FLUID
Kg - SOIL CONDUCTIVITY
K_I - INSULATION CONDUCTIVITY
ΔL - PIPE SEGMENT LENGTH
T_I - INLET TEMPERATURE
T_o - OUTLET TEMPERATURE
m° - FLUID FLOW RATE

Figure 1. Heat transfer from a buried pipeline

For an above-ground or offshore pipeline (Figure 2), the corresponding fluid flow temperature is:

$$T_o = T_a + (T_i - T_a)\, \text{Exp}\left(\frac{-\Delta L}{m \cdot Cp_g U}\right) \tag{5}$$

Where U = overall heat transfer coefficient and is given by:

$$U = \frac{1}{\Pi}\left[\frac{1}{h_f d} + \frac{1}{2k_p}\ln\frac{D}{d} + \frac{1}{2k_i}\ln\frac{D_i}{d} + \frac{1}{h_a\, D_i}\right] \tag{6}$$

Other notations are defined in Figure 2.

In Equation (6) radiative heat losses are ignored as they are small at most normal pipeline operating temperatures.

When no insulation exists, the third term in Equation (6) is reduced to zero and D_i in the fourth term is set equal to D (i.e., outside diameter of the pipe).

For above-ground pipeline, the film coefficient (h_a) for air can be calculated from the following equation recommended by Dittus and Boelter (Holman 1981).

$$N_u = 0.023(\text{Re})^{0.8}(P_r)^n \tag{7}$$

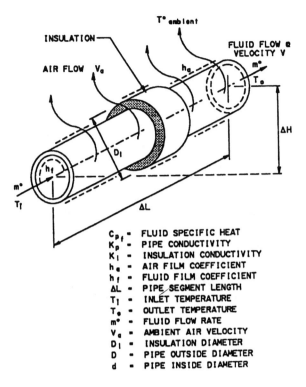

C_{p_f} = FLUID SPECIFIC HEAT
K_p = PIPE CONDUCTIVITY
K_I = INSULATION CONDUCTIVITY
h_a = AIR FILM COEFFICIENT
h_f = FLUID FILM COEFFICIENT
ΔL = PIPE SEGMENT LENGTH
T_I = INLET TEMPERATURE
T_o = OUTLET TEMPERATURE
m° = FLUID FLOW RATE
V_a = AMBIENT AIR VELOCITY
D_I = INSULATION DIAMETER
D = PIPE OUTSIDE DIAMETER
d = PIPE INSIDE DIAMETER

Figure 2. Heat transfer from an above-ground pipeline

Where

$$N_u = \text{Nussett number} = \left(\frac{h_a D}{k_a}\right)$$

$$\text{Re} = \text{Reynolds number} = \left(\frac{\rho_a V_a D}{\mu_a}\right)$$

$$P_r = \text{Prandtle number} = \left(-\frac{\mu_a Cp_a}{k_a}\right)$$

Fluid film coefficient (h_f) for fluid flowing at velocity V_f through the pipe segment is given by:

$$h_f = 0.023 \, k_f^{0.6} \left(\frac{Cp_f}{\mu_f}\right)^{0.4} \frac{\left(\rho_f V_f\right)^{0.8}}{d^{0.2}} \tag{8}$$

In the above equations, Cp, μ, ρ, and k, respectively, refer to specific heat (at constant pressure), viscosity, density, and conductivity of the flowing medium in the either denoted as suffix (a) for air (f) for fluid.

For wind blowing over a pipe segment at a velocity of (V_a), the film coefficient (h_a) can be calculated from the following equation:

$$h_a = C \frac{k_a}{D_i} \left(\frac{\rho_a V_a D_i}{\mu_a} \right)^n \tag{9}$$

Properties of air are provided elsewhere (Holman 1981). Values of constant C and exponent n are dependent on the Reynolds number and are also given elsewhere (Holman 1981).

For an offshore pipeline, h_a can be calculated from Equation (9) with appropriate values of C, ρ, k, and μ for sea water, and knowing the current velocity.

The following expressions summarize the computation of ΔT_e and ΔT_f (Mohitpour 1987).

$$\Delta T_e = -E_s P_h = \left(\frac{\partial T}{\partial P} \right)_s$$
$$= \frac{T}{C_p} \left(\frac{\partial V}{\partial T} \right)_p P_h \tag{10}$$

and

$$\Delta T_f = -JP_f = -\left(\frac{\partial H}{\partial p} \right)_T \bigg/ \left(\frac{\partial H}{\partial T} \right)_p P_f$$
$$= \frac{-1}{C_p} \left(\frac{\partial H}{\partial p} \right)_T P_f \tag{11}$$

Where E_s and J are elevation sensitivity and Joule Thompson coefficients, respectively, P_h is pressure loss due to elevation change, and P_f is pressure loss in overcoming friction.

E_s can be computed from graphs of pressure (P) and temperature (T) at constant entropy (S), and P_J can be calculated from graphs of enthalply (H) versus pressure (P) at constant temperature (T).

The sign of Joule Thompson coefficient J indicates whether fluid expansion or compression will cause an increase or decrease in the temperature. As an example, in an expanding gas if J is positive, the gas will cool. A negative J in an expanding gas indicates temperature rise, and is observed in expansion of some special gases, e.g., hydrogen (Mohitpour 1988). Methods for calculating E_s and J are given elsewhere (Edminster 1974).

APPLICATIONS AND LIMITATIONS

The procedure outlined above provides an accurate prediction of fluid flow temperature under steady-state condition for buried and exposed pipelines. Sample plots of temperature profiles for a liquid pipeline (carrying bitumen/condensate mixture) and for a gas pipeline (sweet gas) are provided in Figures 3 and 4, respectively.

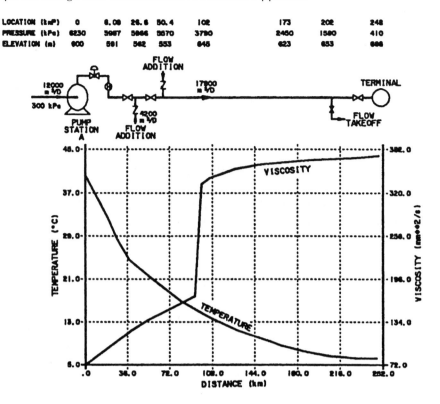

Figure 3. Temperature profile in a buried non-Newtonian liquid pipeline

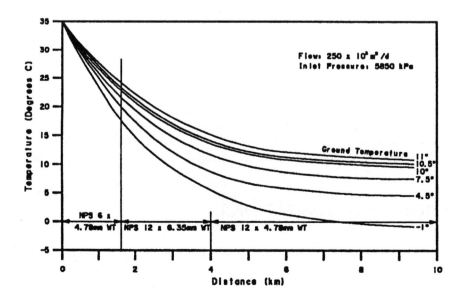

Figure 4. Temperature profiles in a buried gas pipeline

Figure 3 also shows the viscosity profile. The discontinuity is due to fluid changing its behaviour from Newtonian to non-Newtonian (Kung and Mohitpour 1986).

The procedure is a useful design tool from which one can establish and optimize pipeline design features such as hydraulics, size, capacity, power and insulation, requirement, heating equipment sizing, etc. (Haynes 1975) under worst operating scenarios. However, this approach cannot adequately provide an analytical under-standing of soil/pipe or environment/pipe interactions due to temperature effects. Complete temperaturedistribution/pattern for various segments (fluid/pipe/soil/environ-ment) along pipeline lengths will be required to determine adequacy in design and operation due to extreme environmental conditions. For example, monthly temperature variations can affect pipeline operation and determination of accurate line capacity in a gas pipeline system.

ALTERNATIVE PROCEDURE

To predict temperature pattern and histories in a pipeline segment, a finite-element technique where pipe/soil is replaced by a continuum of finite elements (e.g., triangulations), with finer elements placed near discontinuities/boundaries, is proposed as an alternative for both steady-state and transient conditions. The application of the technique to solve such time-dependent heat transfer problems in a planar pipe segment has been previously reported and is widely applied (Norton 1982). In this paper the application will be extended to time-dependent, three-dimensional pipeline situations where the temperature pattern of each section is dependent on adjacent sections and that nonhomogenous and temperature-dependent soil properties (conductivity, specific heat, and density) exist along the length of the pipeline and through the depth of the surrounding soil.

FINITE-ELEMENT METHODOLOGY

The methodology is a solution of time-dependent heat flux in an axisymmetric body using a numerical recursive procedure and is described elsewhere (Mohitpour 1975). Briefly, since the initial value of temperature T (i.e., T at time t) is known, the temperature field at section/segment can be established at subsequent time $t + \Delta t$ in the following manner:

$$[K]\{T\}_{t+\Delta t} + [H]\{T\}_{t+\Delta t} + \frac{2}{\Delta t}[C]\{T\}_{t+\Delta t} + \{Q\}$$

$$= \{q\} + \frac{2}{\Delta t}[C]\{T\}_t + [C]\frac{\partial}{\partial t}\{T\}_t \quad (12)$$

Initially and at the beginning of each time step, the temperature field is known. The temperature gradient is then computed and used for subsequent computation of temperature $\{T\}$ at time $t+\Delta t$.

The time increment Δt is determined by knowing pipeline segment ΔL and computing the fluid flow velocity through the segment (Marks 1978). Assuming that this fluid velocity V_f remains constant through each pipe segment, Δt is then computed by dividing the pipe length ΔL to the velocity V_f. Once the initial temperature of the segment is known, subsequent temperature is computed for time $t+\Delta t$, and the procedure is repeated from one segment to next and the preceding segment(s) until the total length of the pipeline is covered for the total duration anticipated or a steady-state condition is achieved. The higher the number of segments, the more accurate will be the segment/pipeline temperature pattern and history. However, computation time can be enormous if a large number of elements and many pipe segments are considered.

METHOD AND RESULTS

The computer program TEMPSYS, devised by the author (Mohitpour 1975) utilizing the above finite-element heat transfer technique, was used to assess the viability of the procedure to pipeline applications. To check the accuracy of the program, a prediction was made for soil temperature at depths at which pipelines are normally buried by comparison with previously known recorded mean monthly and corresponding soil temperatures (Coskey 1990; Toogood 1979). The result of such a test is shown in Figure 5.

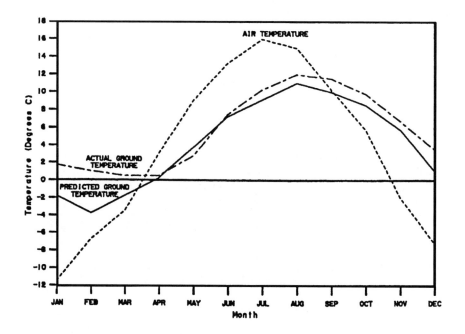

Figure 5. Actual and predicted soil temperatures at pipeline depth (Coskey 1990)

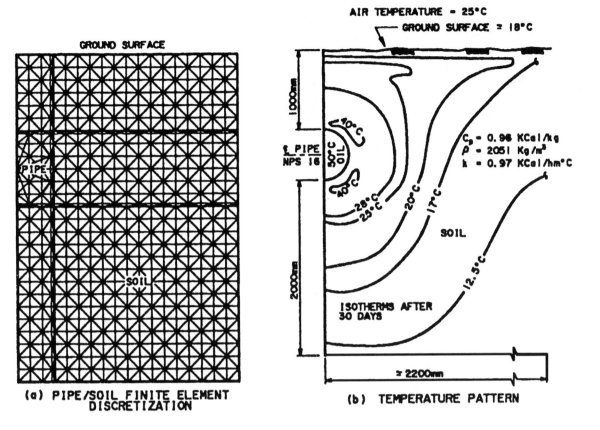

GROUND SURFACE

(a) PIPE/SOIL FINITE ELEMENT DISCRETIZATION

AIR TEMPERATURE = 25°C

GROUND SURFACE ≅ 18°C

£ PIPE
NPS 16

C_p = 0.96 KCal/kg
ρ = 2051 Kg/m³
k = 0.97 KCal/hm°C

SOIL

ISOTHERMS AFTER 30 DAYS

≅ 2200mm

(b) TEMPERATURE PATTERN

Figure 6. Temperature isotherms around buried pipeline summer season - after 30 days

Actual and predicted soil temperatures compare well. The discrepancy that is evident can be attributed to the temperature-independent and homogenous soil properties assumed. The soil properties (ρ_g, Cp_g, and k_g) are temperature-dependent (Lunardini 1981) and usually nonhomogenous through the depth and along the length of pipeline. For this exercise, summer condition values for these properties were used and assumed as temperature-independent and homogenous. Therefore, it is not surprising that differences between actual and computed temperatures are greater for the winter season when compared with those for warmer seasons. Such a ground temperature prediction model was utilized to assess ice formation in a gas pipeline system due to water moisture dropout at low points during winter, and subsequent remelting during warmer summer months (Pierce and Mohitpour 1990).

Figure 6 shows the temperature contours in modeling a very hot, bitumen pipeline maintaining its temperature for a 30-day period during the summer months. In this case initial ground temperature was assumed at 12°C. The study was to determine the effect of maintaining high pipeline and air temperature on crop survival within the pipeline right-of-way crossing farm lands.

Figure 7 shows temperature history for a segment during a time interval of filling 25 km of pipeline with warm water for hydrostate testing purposes in winter. This illustrates a further application of the technique elaborated above.

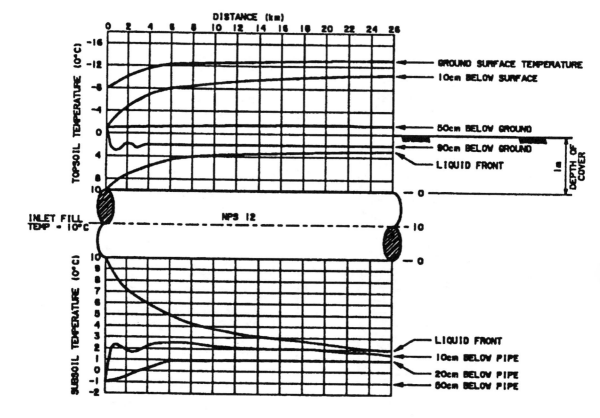

Figure 7. Temperature pattern along a buried pipeline

CONCLUSION

Conventional heat-transfer solution techniques provide an accurate assessment of fluid temperature profiles in a buried or exposed pipelines. An alternative finite-element heat transfer solution technique described in this paper for a time-dependent three-dimensionalheat transfer condition prevalent in a pipeline situation with temperature-dependent and nonhomgenous soil properties provides temperature patterns and histories along the pipeline length and through the depth of soil.

Initial examples have been provided to illustrate the results and application of the alternative technique. These examples confirm that large computing time and power will be required if temperature histories of many pipeline segments are considered and that temperature-dependent and nonhomgenous soil properties must be utilized if accurate assessment of pipe/soil interaction due to temperature effects are needed.

ACKNOWLEDGMENT

The author gratefully acknowledges the permission granted by management of Novacorp International Consulting Inc. for publication of the paper.

REFERENCES

Coskey, T., 1990, "Private Communication with University of Calgary".

Edminster, W.C., 1972, "Applied Hydrocarbon Thermodynamics," Vol. 2, Gulf Publishing Co., Houston, TX.

Haynes, C.D., 1975, "Calculating Temperature Profile Along a Buried Gas Pipeline," Oil and Gas Pipeline Handbook, 3rd Ed., Pet. Eng. Pub. Co., Dallas, TX.

Holman, J.P., 1981, "Heat Transfer," 5th Edition, McGraw Hill Book Co.

Kung, P. and Mohitpour, M., 1986, "Non-Newtonian Liquid Pipeline, Hydraulics Design and Simulation using Microcomputer," Proc. Pipeline Engineering Symposium, 9th ETCE Conference, Vol. 3, pp. 73–78.

Lunardini, V.J., 1981, "Heat Transfer in Cold Climates," Van Nostrand Reinhold Company, New York.

Marks, A., 1978, "Handbook of Pipeline Engineering Computation," Petroleum Publishing Co., Tulsa, OK.

Mohitpour, M., 1975, "Application of Heat Transfer, Finite Element Analysis to Process Design," Proc. 2nd Chemical Engineering Cong., Tehran, Iran, May.

Mohitpour, M., 1975, Pierce, C.L. and Hooper, R., 1988, "The Design and Engineering of Cross-Country Hydrogen Pipelines," Jr. of Eng. Res. Tech., Vol. 110, pp. 203–207.

Mohitpour, M., 1987, "A Guideline Manual for Design of Hydrogen Pipeline Systems," NOVA Corporation of Alberta, Rev. 1, June, 1981.

Norton, W.R., 1982, "Explanatory and Technical Information on the Computer Program Geodyn—A Finite Element Model for Simulation of Temperatures in Freezing and Thawing Media," Resources Management Associates, Lafayette, CA.

Pierce, C.L. and Mohitpour, M., 1990, "Flat Lake Sales Temperature Effect," Private Communication.

Slimmon, T.C., 1989, "Material Selection and Quality Management," Seminar on Practical Pipeline System Design and Innovation, University of Calgary—Continuing Education Department.

Toogood, J.A., 1979, "Comparison of Soil Temperatures under Different Vegetative Covers at Edmonton," Can. J. Soil, Aug., Vol. 59, pp. 329–335.

Appendix *D*

Refer to Chapter 9

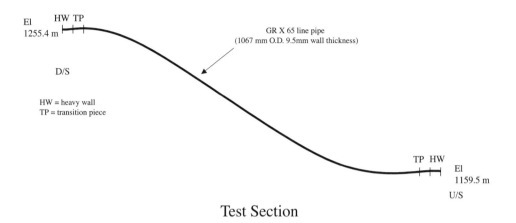

Test Section

Figure D-1. Example of a hydrostatic test section

Elevation Data:

Test point:	1,159.5 m
Low point:	1,159.5 m
High point:	1,255.4 m

Pipe Data:

Outside diameter (OD):	1,067 mm
Overall length:	5,176.5 m
$SMYS_1$ (s_1):	448,000 kPa
Wall thickness (t_1):	9.5 mm

Transition Piece:

$SMYS_2$ (s_2):	448,000 kPa
Wall thickness (t_2):	9.5 mm
MAOP:	6,205 kPa

Sample Calculations for $t_1=9.5$ mm:

Inside diameter (ID)$=$OD-2 $t_1=1,067$ mm $-$ $(2\times9.5$ mm$)=1,048$ mm

Design pressure $(P)=2ts_1/D=(2\times9.5\times448,000/1,067)=7,980$ kPa

Difference in elevation $(H)=1,255.4$ m $-1,159.5$ m$=95.9$ m

Static pressure $=\rho gh=9.795\times95.9=940$ kPa

Test pressure $=1.1\times P=1.1\times7,980=8,778$ kPa

Leak test pressure$=0.8 \times P=0.8 \times 7{,}980=6{,}384$ kPa

8-hour minimum test pressure$=$static head$+(1.25 \times$MAOP$)$

$$=940+(1.25 \times 6{,}205)$$

$$=8{,}696 \text{ kPa}$$

Maximum test pressure$=$Lesser of 8-hour minimum and test pressure

$$=8{,}696 \text{ kPa or } 8{,}778 \text{ kPa}$$

Maximum test pressure$=8{,}696$ kPa

Test section volume$=\pi D^2 L/4=0.785 \times 1.048^2 \times 5{,}176.5=4{,}463 \text{ m}^3$

0.2% offset volume$=$ total volume $\times 0.0002 \text{ m}^3=4{,}463 \times 0.0002=8.926 \text{ m}^3$

Temperature Effects:
To account for possible changes in pressure due to an increase in temperature over the 8-hour test period:
Coefficient of expansion of water at temperature T:

$$B \times 10^6 = -64.26 + 17{,}105T - 0.203T^2 + 0.0016048T^3$$

So, for example, if $T=10°$C, then $B \times 10^6=87.0728$

The corresponding change in water pressure is given by

$$\Delta P = \frac{B - 2\alpha}{\dfrac{D}{Et}(1-\nu^2)+c}$$

where $\alpha=1.17 \times 10-5/°$C
$c=4.86 \times 10-7/$kPa

$$\Delta P = \frac{87.0728 \times 10^{-6} - 2 \times 1.17 \times 10^{-5}}{\dfrac{1{,}067}{200 \times 10^6 \times 9.5}(1 - 0.3^2) + 4.86 \times 10^{-7}} = 63.86 \text{ kPa/°C}$$

or 638.6 kPa/10°C.

Appendix *E*

Cathodic Protection Problem Solution

SOLUTION FOR CASE A (SINGLE 12″ LINE) ONLY

a) Coverage

Steel pipe resistivity $\rho_s = 18 \, \mu$ ohm/cm $= 18 \times 10^{-6}$ ohm/cm

For one mile $\rho_s = 1 \times 5280 \times 12 \times 2.54$ ohm

Using Equation (10-6)

$$R_s = \rho_s \frac{L}{A} = \frac{\rho_s L}{\pi D t}$$

$$= 18 \times 10^{-6} \times \frac{1 \times 5280 \times 12 \times 2.54}{3.14 \times 12 \times 2.5 \times 0.25 \times 2.54}$$

$$= 0.0461 \text{ ohm/mile}$$

Using Equation (10-7)

$$R_L = \frac{1}{gAL} \text{ ohm/km}$$

But $\frac{1}{g} = 500,000$ ohm/ft^2 (Coating Conductivity)

$$\therefore R_L = 500000 \times \frac{1 \times 12}{\pi \times 12 \times 5280}$$

$$= 30.16 \text{ ohm/mile}$$

Using Equation (10-9)

$$\alpha = \text{Attenuation factor}$$

$$= \sqrt{\frac{\text{Pipe resistance}}{\text{Coating resistance}}} = \sqrt{\frac{R_s}{R_L}}$$

$$= \sqrt{\frac{0.0461}{30.16}} = \sqrt{0.0015}$$

Using Equation (10-8)

$$R_o = \sqrt{R_s \cdot R_L}$$

$$\text{or } R_o^2 = R_s \cdot R_L$$

$$\text{or } \alpha = \sqrt{\frac{R_s}{R_L}} = \sqrt{\frac{R_o^2}{R_L^2}} = \frac{R_o}{R_L} = \frac{R_s I}{2E_A I}$$

I = current at drain point (amps)

Using Equation (10-5) Voltage Potential $E_x = E_o e^{-\alpha x}$
where x = distance to be covered

Assuming voltage potential E_o at drain point = 2 volts, then;

$$\frac{E_x}{E_o} = \frac{0.2}{2} = e^{-\alpha x} = 0.1$$

or $\alpha x = e^{0.1} = 2.3$

or $x = \frac{2.3}{\alpha} = \frac{2.3}{\sqrt{0.0015}}$

= 60 miles in one direction
or 120 miles in two directions

b) Rectifier Current Requirement

$$E_o = IR_o = I\sqrt{R_s R_L}$$

or $2 = I\sqrt{0.00461 \times 30.16}$

∴ I = 1.5 amps in one direction
or I = 3 amps in two directions

Assuming interference of about 2 to 3 amps then;
Total rectifier amperage requirement = 3 + 3 = 6 amps
   Select amplifier for 8 amps; i.e. I = 8 amps to allow for valves, interference abatements etc.

Assuming zinc anode
      W_m = Anode weight (kg)
      I_a = Anode current output
then using equation (10-11a) with anode life of 20 years;
   Anode life = $20 \times 71.4 \times \frac{20}{I_a}$
   or I_a = 71.4 milliamperes

c) Number of anode required
   Using equation (10-1);

$$n = \frac{I \times Y \times C}{\text{weight of anode} \times \text{safety factor} \,(0.6)}$$

$$= \frac{8 \times 20 \times 0.5}{20 \times 0.6}$$

$\simeq$ 7 anodes

INDEX